STUDENT'S SOLUTIONS MANUAL

TIM BRITT
Jackson State Community College

AF217980

COLLEGE ALGEBRA ENHANCED WITH GRAPHING UTILITIES
SEVENTH EDITION

Michael Sullivan
Chicago State University

Michael Sullivan III
Joliet Junior College

PEARSON

Boston Columbus Indianapolis New York San Francisco
Amsterdam Cape Town Dubai London Madrid Milan Munich Paris Montreal Toronto
Delhi Mexico City São Paulo Sydney Hong Kong Seoul Singapore Taipei Tokyo

Table of Contents

Chapter 4 Linear and Quadratic Functions

Chapter 5 Polynomial and Rational Functions

Chapter 6 Exponential and Logarithmic Functions

Chapter 7 Analytic Geometry

Chapter 8 Systems of Equations and Inequalities

Chapter 9 Sequences; Induction; the Binomial Theorem

Chapter 10 Counting and Probability

Chapter R
Review

1. rational

3. Distributive

5. a

7. True

9. False; 6 is the Greatest Common Factor of 12 and 18. The Least Common Multiple is the smallest value that both numbers will divide evenly. The LCM for 12 and 18 is 36.

11. $A \cup B = \{1, 3, 4, 5, 9\} \cup \{2, 4, 6, 7, 8\}$
    $= \{1, 2, 3, 4, 5, 6, 7, 8, 9\}$

13. $A \cap B = \{1, 3, 4, 5, 9\} \cap \{2, 4, 6, 7, 8\} = \{4\}$

15. $(A \cup B) \cap C$
    $= (\{1, 3, 4, 5, 9\} \cup \{2, 4, 6, 7, 8\}) \cap \{1, 3, 4, 6\}$
    $= \{1, 2, 3, 4, 5, 6, 7, 8, 9\} \cap \{1, 3, 4, 6\}$
    $= \{1, 3, 4, 6\}$

17. $\overline{A} = \{0, 2, 6, 7, 8\}$

19. $\overline{A \cap B} = \overline{\{1, 3, 4, 5, 9\} \cap \{2, 4, 6, 7, 8\}}$
    $= \overline{\{4\}} = \{0, 1, 2, 3, 5, 6, 7, 8, 9\}$

21. $\overline{A} \cup \overline{B} = \{0, 2, 6, 7, 8\} \cup \{0, 1, 3, 5, 9\}$
    $= \{0, 1, 2, 3, 5, 6, 7, 8, 9\}$

23. a. $\{2, 5\}$

    b. $\{-6, 2, 5\}$

    c. $\left\{-6, \frac{1}{2}, -1.333... = -1.\overline{3}, 2, 5\right\}$

    d. $\{\pi\}$

    e. $\left\{-6, \frac{1}{2}, -1.333... = -1.\overline{3}, \pi, 2, 5\right\}$

25. a. $\{1\}$

    b. $\{0, 1\}$

    c. $\left\{0, 1, \frac{1}{2}, \frac{1}{3}, \frac{1}{4}\right\}$

    d. None

    e. $\left\{0, 1, \frac{1}{2}, \frac{1}{3}, \frac{1}{4}\right\}$

27. a. None

    b. None

    c. None

    d. $\left\{\sqrt{2}, \pi, \sqrt{2}+1, \pi+\frac{1}{2}\right\}$

    e. $\left\{\sqrt{2}, \pi, \sqrt{2}+1, \pi+\frac{1}{2}\right\}$

29. a. 18.953   b. 18.952

31. a. 28.653   b. 28.653

33. a. 0.063   b. 0.062

35. a. 9.999   b. 9.998

37. a. 0.429   b. 0.428

39. a. 34.733   b. 34.733

41. $3 + 2 = 5$

43. $x + 2 = 3 \cdot 4$

45. $3y = 1 + 2$

47. $x - 2 = 6$

49. $\frac{x}{2} = 6$

51. $9 - 4 + 2 = 5 + 2 = 7$

53. $-6 + 4 \cdot 3 = -6 + 12 = 6$

55. $4 + 5 - 8 = 9 - 8 = 1$

1

57. $4 + \dfrac{1}{3} = \dfrac{12+1}{3} = \dfrac{13}{3}$

59. $6 - \left[3 \cdot 5 + 2 \cdot (3-2)\right] = 6 - \left[15 + 2 \cdot (1)\right]$
$$= 6 - 17$$
$$= -11$$

61. $2 \cdot (3-5) + 8 \cdot 2 - 1 = 2 \cdot (-2) + 16 - 1$
$$= -4 + 16 - 1$$
$$= 12 - 1$$
$$= 11$$

63. $10 - \left[6 - 2 \cdot 2 + (8-3)\right] \cdot 2 = 10 - \left[6 - 4 + 5\right] \cdot 2$
$$= 10 - \left[2 + 5\right] \cdot 2$$
$$= 10 - \left[7\right] \cdot 2$$
$$= 10 - 14$$
$$= -4$$

65. $(5-3)\dfrac{1}{2} = (2)\dfrac{1}{2} = 1$

67. $\dfrac{4+8}{5-3} = \dfrac{12}{2} = 6$

69. $\dfrac{3}{5} \cdot \dfrac{10}{21} = \dfrac{3 \cdot 2 \cdot 5}{5 \cdot 3 \cdot 7} = \dfrac{\cancel{3} \cdot 2 \cdot \cancel{5}}{\cancel{5} \cdot \cancel{3} \cdot 7} = \dfrac{2}{7}$

71. $\dfrac{6}{25} \cdot \dfrac{10}{27} = \dfrac{2 \cdot 3 \cdot 5 \cdot 2}{5 \cdot 5 \cdot 3 \cdot 9} = \dfrac{2 \cdot \cancel{3} \cdot \cancel{5} \cdot 2}{\cancel{5} \cdot 5 \cdot \cancel{3} \cdot 9} = \dfrac{4}{45}$

73. $\dfrac{3}{4} + \dfrac{2}{5} = \dfrac{15+8}{20} = \dfrac{23}{20}$

75. $\dfrac{5}{6} + \dfrac{9}{5} = \dfrac{25+54}{30} = \dfrac{79}{30}$

77. $\dfrac{5}{18} + \dfrac{1}{12} = \dfrac{10+3}{36} = \dfrac{13}{36}$

79. $\dfrac{1}{30} - \dfrac{7}{18} = \dfrac{3-35}{90} = -\dfrac{32}{90} = -\dfrac{16}{45}$

81. $\dfrac{3}{20} - \dfrac{2}{15} = \dfrac{9-8}{60} = \dfrac{1}{60}$

83. $\dfrac{\left(\dfrac{5}{18}\right)}{\left(\dfrac{11}{27}\right)} = \dfrac{5}{18} \cdot \dfrac{27}{11} = \dfrac{5 \cdot 9 \cdot 3}{9 \cdot 2 \cdot 11} = \dfrac{5 \cdot \cancel{9} \cdot 3}{\cancel{9} \cdot 2 \cdot 11} = \dfrac{15}{22}$

85. $\dfrac{1}{2} \cdot \dfrac{3}{5} + \dfrac{7}{10} = \dfrac{3}{10} + \dfrac{7}{10} = \dfrac{3+7}{10} = \dfrac{10}{10} = 1$

87. $2 \cdot \dfrac{3}{4} + \dfrac{3}{8} = \dfrac{2}{1} \cdot \dfrac{3}{4} + \dfrac{3}{8} = \dfrac{6}{4} + \dfrac{3}{8} = \dfrac{6}{4} \cdot \dfrac{2}{2} + \dfrac{3}{8}$
$$= \dfrac{12}{8} + \dfrac{3}{8} = \dfrac{12+3}{8} = \dfrac{15}{8}$$

89. $6(x+4) = 6x + 24$

91. $x(x-4) = x^2 - 4x$

93. $2\left(\dfrac{3}{4}x - \dfrac{1}{2}\right) = 2 \cdot \dfrac{3}{4}x - 2 \cdot \dfrac{1}{2} = \dfrac{2 \cdot 3x}{2 \cdot 2} - \dfrac{2}{2}$
$$= \dfrac{\cancel{2} \cdot 3x}{\cancel{2} \cdot 2} - \dfrac{2}{2} = \dfrac{3}{2}x - 1$$

95. $(x+2)(x+4) = x^2 + 4x + 2x + 8$
$$= x^2 + 6x + 8$$

97. $(x-2)(x+1) = x^2 + x - 2x - 2$
$$= x^2 - x - 2$$

99. $(x-8)(x-2) = x^2 - 2x - 8x + 16$
$$= x^2 - 10x + 16$$

101. $2x + 3x = 2 \cdot x + 3 \cdot x$
$$= (2+3) \cdot x$$
$$= (5) \cdot x$$
$$= 5x$$

103. $2(3 \cdot 4) = 2(12) = 24$
$(2 \cdot 3) \cdot (2 \cdot 4) = (6)(8) = 48$

105. Subtraction is not commutative; for example: $2 - 3 = -1 \neq 1 = 3 - 2$.

107. Division is not commutative; for example: $\dfrac{2}{3} \neq \dfrac{3}{2}$.

109. The Symmetric Property implies that if $2 = x$, then $x = 2$.

111. There are no real numbers that are both rational and irrational, since an irrational number, by definition, is a number that cannot be expressed as the ratio of two integers; that is, not a rational number

Every real number is either a rational number or an irrational number, since the decimal form of a real number either involves an infinitely repeating pattern of digits or an infinite, non-repeating string of digits.

113. Answers will vary.

115. Answers will vary.

Section R.2

1. variable

3. strict

5. 1.2345678×10^3

7. b

9. True

11. False; a number in scientific notation is expressed as the product of a number, x, $1 \le x < 10$ or $-10 < x \le -1$, and a power of 10.

13.

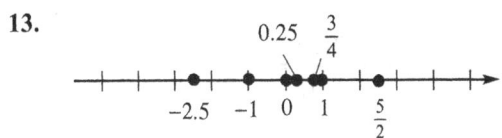

15. $\frac{1}{2} > 0$

17. $-1 > -2$

19. $\pi > 3.14$

21. $\frac{1}{2} = 0.5$

23. $\frac{2}{3} < 0.67$

25. $x > 0$

27. $x < 2$

29. $x \le 1$

31. Graph on the number line: $x \ge -2$

33. Graph on the number line: $x > -1$

35. $d(C,D) = d(0,1) = |1-0| = |1| = 1$

37. $d(D,E) = d(1,3) = |3-1| = |2| = 2$

39. $d(A,E) = d(-3,3) = |3-(-3)| = |6| = 6$

41. $x + 2y = -2 + 2 \cdot 3 = -2 + 6 = 4$

43. $5xy + 2 = 5(-2)(3) + 2 = -30 + 2 = -28$

45. $\dfrac{2x}{x-y} = \dfrac{2(-2)}{-2-3} = \dfrac{-4}{-5} = \dfrac{4}{5}$

47. $\dfrac{3x+2y}{2+y} = \dfrac{3(-2)+2(3)}{2+3} = \dfrac{-6+6}{5} = \dfrac{0}{5} = 0$

49. $|x+y| = |3+(-2)| = |1| = 1$

51. $|x| + |y| = |3| + |-2| = 3 + 2 = 5$

53. $\dfrac{|x|}{x} = \dfrac{|3|}{3} = \dfrac{3}{3} = 1$

55. $|4x - 5y| = |4(3) - 5(-2)|$
$= |12 + 10|$
$= |22|$
$= 22$

3

57. $\left| \, |4x| - |5y| \, \right| = \left| \, |4(3)| - |5(-2)| \, \right|$
$= \left| \, |12| - |-10| \, \right|$
$= |12 - 10|$
$= |2|$
$= 2$

59. $\dfrac{x^2 - 1}{x}$

Part (c) must be excluded. The value $x = 0$ must be excluded from the domain because it causes division by 0.

61. $\dfrac{x}{x^2 - 9} = \dfrac{x}{(x-3)(x+3)}$

Part (a), $x = 3$, must be excluded because it causes the denominator to be 0.

63. $\dfrac{x^2}{x^2 + 1}$

None of the given values are excluded. The domain is all real numbers.

65. $\dfrac{x^2 + 5x - 10}{x^3 - x} = \dfrac{x^2 + 5x - 10}{x(x-1)(x+1)}$

Parts (b), (c), and (d) must be excluded. The values $x = 0$, $x = 1$, and $x = -1$ must be excluded from the domain because they cause division by 0.

67. $\dfrac{4}{x - 5}$

$x = 5$ must be exluded because it makes the denominator equal 0.
Domain $= \{x \mid x \neq 5\}$

69. $\dfrac{x}{x + 4}$

$x = -4$ must be excluded sine it makes the denominator equal 0.
Domain $= \{x \mid x \neq -4\}$

71. $C = \dfrac{5}{9}(F - 32) = \dfrac{5}{9}(32 - 32) = \dfrac{5}{9}(0) = 0°C$

73. $C = \dfrac{5}{9}(F - 32) = \dfrac{5}{9}(77 - 32) = \dfrac{5}{9}(45) = 25°C$

75. $(-4)^2 = (-4)(-4) = 16$

77. $4^{-2} = \dfrac{1}{4^2} = \dfrac{1}{16}$

79. $3^{-6} \cdot 3^4 = 3^{-6+4} = 3^{-2} = \dfrac{1}{3^2} = \dfrac{1}{9}$

81. $\left(3^{-2}\right)^{-1} = 3^{(-2)(-1)} = 3^2 = 9$

83. $\sqrt{25} = \sqrt{5^2} = 5$

85. $\sqrt{(-4)^2} = |-4| = 4$

87. $\left(8x^3\right)^2 = 8^2 \left(x^3\right)^2 = 64x^6$

89. $\left(x^2 y^{-1}\right)^2 = \left(x^2\right)^2 \cdot \left(y^{-1}\right)^2 = x^4 y^{-2} = \dfrac{x^4}{y^2}$

91. $\dfrac{x^2 y^3}{xy^4} = x^{2-1} y^{3-4} = x^1 y^{-1} = \dfrac{x}{y}$

93. $\dfrac{(-2)^3 x^4 (yz)^2}{3^2 x y^3 z} = \dfrac{-8x^4 y^2 z^2}{9x y^3 z}$
$= \dfrac{-8}{9} x^{4-1} y^{2-3} z^{2-1}$
$= \dfrac{-8}{9} x^3 y^{-1} z^1$
$= -\dfrac{8x^3 z}{9y}$

95. $\left(\dfrac{3x^{-1}}{4y^{-1}}\right)^{-2} = \left(\dfrac{3y}{4x}\right)^{-2} = \left(\dfrac{4x}{3y}\right)^2 = \dfrac{4^2 x^2}{3^2 y^2} = \dfrac{16x^2}{9y^2}$

97. $2xy^{-1} = \dfrac{2x}{y} = \dfrac{2(2)}{(-1)} = -4$

99. $x^2 + y^2 = (2)^2 + (-1)^2 = 4 + 1 = 5$

101. $(xy)^2 = (2 \cdot (-1))^2 = (-2)^2 = 4$

103. $\sqrt{x^2} = |x| = |2| = 2$

105. $\sqrt{x^2 + y^2} = \sqrt{(2)^2 + (-1)^2} = \sqrt{4 + 1} = \sqrt{5}$

107. $x^y = 2^{-1} = \dfrac{1}{2}$

4

109. If $x = 2$,
$$2x^3 - 3x^2 + 5x - 4 = 2 \cdot 2^3 - 3 \cdot 2^2 + 5 \cdot 2 - 4$$
$$= 16 - 12 + 10 - 4$$
$$= 10$$

If $x = 1$,
$$2x^3 - 3x^2 + 5x - 4 = 2 \cdot 1^3 - 3 \cdot 1^2 + 5 \cdot 1 - 4$$
$$= 2 - 3 + 5 - 4$$
$$= 0$$

111. $\dfrac{(666)^4}{(222)^4} = \left(\dfrac{666}{222}\right)^4 = 3^4 = 81$

112. $(0.1)^3 (20)^3 = \left(\dfrac{1}{10}\right)^3 \cdot (2 \cdot 10)^3$
$$= \dfrac{1}{10^3} \cdot 2^3 \cdot 10^3$$
$$= 2^3 = 8$$

113. $(8.2)^6 \approx 304{,}006.671$

115. $(6.1)^{-3} \approx 0.004$

117. $(-2.8)^6 \approx 481.890$

119. $(-8.11)^{-4} \approx 0.000$

121. $454.2 = 4.542 \times 10^2$

123. $0.013 = 1.3 \times 10^{-2}$

125. $32{,}155 = 3.2155 \times 10^4$

127. $0.000423 = 4.23 \times 10^{-4}$

129. $6.15 \times 10^4 = 61{,}500$

131. $1.214 \times 10^{-3} = 0.001214$

133. $1.1 \times 10^8 = 110{,}000{,}000$

135. $8.1 \times 10^{-2} = 0.081$

137. $A = lw$

139. $C = \pi d$

141. $A = \dfrac{\sqrt{3}}{4} x^2$

143. $V = \dfrac{4}{3} \pi r^3$

145. $V = x^3$

147. **a.** If $x = 1000$,
$$C = 4000 + 2x$$
$$= 4000 + 2(1000)$$
$$= 4000 + 2000$$
$$= \$6000$$
The cost of producing 1000 watches is $6000.

b. If $x = 2000$,
$$C = 4000 + 2x$$
$$= 4000 + 2(2000)$$
$$= 4000 + 4000$$
$$= \$8000$$
The cost of producing 2000 watches is $8000.

149. We want the difference between x and 4 to be at least 6 units. Since we don't care whether the value for x is larger or smaller than 4, we take the absolute value of the difference. We want the inequality to be non-strict since we are dealing with an 'at least' situation. Thus, we have
$$|x - 4| \geq 6$$

151. **a.** $|x - 110| = |108 - 110| = |-2| = 2 \leq 5$
108 volts is acceptable.

b. $|x - 110| = |104 - 110| = |-6| = 6 > 5$
104 volts is *not* acceptable.

153. **a.** $|x - 3| = |2.999 - 3|$
$$= |-0.001|$$
$$= 0.001 \leq 0.01$$
A radius of 2.999 centimeters is acceptable.

b. $|x-3| = |2.89-3|$

$\qquad = |-0.11|$

$\qquad = 0.11 \not\leq 0.01$

A radius of 2.89 centimeters is *not* acceptable.

155. The distance from Earth to the Moon is about $4 \times 10^8 = 400,000,000$ meters.

157. The wavelength of visible light is about $5 \times 10^{-7} = 0.0000005$ meters.

159. The smallest commercial copper wire has a diameter of about $0.0005 = 5 \times 10^{-4}$ inches.

161. $186,000 \cdot 60 \cdot 60 \cdot 24 \cdot 365$

$= \left(1.86 \times 10^5\right)\left(6 \times 10^1\right)^2 \left(2.4 \times 10^1\right)\left(3.65 \times 10^2\right)$

$= 586.5696 \times 10^{10} = 5.865696 \times 10^{12}$

There are about 5.9×10^{12} miles in one light-year.

163. $\dfrac{1}{3} = 0.333333\ldots > 0.333$

$\dfrac{1}{3}$ is larger by approximately $0.0003333\ldots$

165. No. For any positive number a, the value $\dfrac{a}{2}$ is smaller and therefore closer to 0.

167. Answers will vary.

Section R.3

1. right; hypotenuse

3. $C = 2\pi r$

5. c

7. True.

9. False; the surface area of a sphere of radius r is given by $V = 4\pi r^2$.

11. True. Two corresponding angles are equal.

13. $a = 5,\ b = 12,$

$c^2 = a^2 + b^2$

$\qquad = 5^2 + 12^2$

$\qquad = 25 + 144$

$\qquad = 169 \Rightarrow c = 13$

15. $a = 10,\ b = 24,$

$c^2 = a^2 + b^2$

$\qquad = 10^2 + 24^2$

$\qquad = 100 + 576$

$\qquad = 676 \Rightarrow c = 26$

17. $a = 7,\ b = 24,$

$c^2 = a^2 + b^2$

$\qquad = 7^2 + 24^2$

$\qquad = 49 + 576$

$\qquad = 625 \Rightarrow c = 25$

19. $5^2 = 3^2 + 4^2$

$25 = 9 + 16$

$25 = 25$

The given triangle is a right triangle. The hypotenuse is 5.

21. $6^2 = 4^2 + 5^2$

$36 = 16 + 25$

$36 = 41$ false

The given triangle is not a right triangle.

23. $25^2 = 7^2 + 24^2$

$625 = 49 + 576$

$625 = 625$

The given triangle is a right triangle. The hypotenuse is 25.

25. $6^2 = 3^2 + 4^2$

$36 = 9 + 16$

$36 = 25$ false

The given triangle is not a right triangle.

27. $A = l \cdot w = 4 \cdot 2 = 8 \text{ in}^2$

29. $A = \dfrac{1}{2} b \cdot h = \dfrac{1}{2}(2)(4) = 4 \text{ in}^2$

31. $A = \pi r^2 = \pi(5)^2 = 25\pi$ m^2
$C = 2\pi r = 2\pi(5) = 10\pi$ m

33. $V = lwh = 8 \cdot 4 \cdot 7 = 224$ ft^3
$S = 2lw + 2lh + 2wh$
$\quad = 2(8)(4) + 2(8)(7) + 2(4)(7)$
$\quad = 64 + 112 + 56$
$\quad = 232$ ft^2

35. $V = \frac{4}{3}\pi r^3 = \frac{4}{3}\pi \cdot 4^3 = \frac{256}{3}\pi$ cm^3
$S = 4\pi r^2 = 4\pi \cdot 4^2 = 64\pi$ cm^2

37. $V = \pi r^2 h = \pi(9)^2(8) = 648\pi$ in^3
$S = 2\pi r^2 + 2\pi rh$
$\quad = 2\pi(9)^2 + 2\pi(9)(8)$
$\quad = 162\pi + 144\pi$
$\quad = 306\pi$ in^2

39. The diameter of the circle is 2, so its radius is 1.
$A = \pi r^2 = \pi(1)^2 = \pi$ square units

41. The diameter of the circle is the length of the diagonal of the square.
$$d^2 = 2^2 + 2^2$$
$$= 4 + 4$$
$$= 8$$
$$d = \sqrt{8} = 2\sqrt{2}$$
$$r = \frac{d}{2} = \frac{2\sqrt{2}}{2} = \sqrt{2}$$
The area of the circle is:

43. Since the triangles are similar, the lengths of corresponding sides are proportional. Therefore, we get
$$\frac{8}{4} = \frac{x}{2}$$
$$\frac{8 \cdot 2}{4} = x$$
$$4 = x$$
In addition, corresponding angles must have the same angle measure. Therefore, we have $A = 90°$, $B = 60°$, and $C = 30°$.

45. Since the triangles are similar, the lengths of corresponding sides are proportional. Therefore, we get
$$\frac{30}{20} = \frac{x}{45}$$
$$\frac{30 \cdot 45}{20} = x$$
$$\frac{135}{2} = x \text{ or } x = 67.5$$
In addition, corresponding angles must have the same angle measure. Therefore, we have $A = 60°$, $B = 95°$, and $C = 25°$.

47. The total distance traveled is 4 times the circumference of the wheel.
Total Distance $= 4C = 4(\pi d) = 4\pi \cdot 16$
$\quad\quad = 64\pi \approx 201.1$ inches ≈ 16.8 feet

49. Area of the border = area of EFGH − area of ABCD $= 10^2 - 6^2 = 100 - 36 = 64$ ft^2

51. Area of the window = area of the rectangle + area of the semicircle.
$$A = (6)(4) + \frac{1}{2} \cdot \pi \cdot 2^2 = 24 + 2\pi \approx 30.28 \text{ ft}^2$$
Perimeter of the window = 2 heights + width + one-half the circumference.
$$P = 2(6) + 4 + \frac{1}{2} \cdot \pi(4) = 12 + 4 + 2\pi$$
$$= 16 + 2\pi \approx 22.28 \text{ feet}$$

53. We can form similar triangles using the Great Pyramid's height/shadow and Thales' height/shadow:

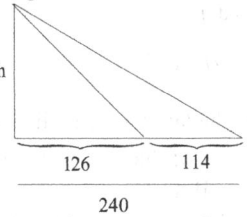
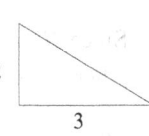

This allows us to write
$$\frac{h}{240} = \frac{2}{3}$$
$$h = \frac{2 \cdot 240}{3} = 160$$
The height of the Great Pyramid is 160 paces.

55. Convert 20 feet to miles, and solve the Pythagorean Theorem to find the distance:

$$20 \text{ feet } = 20 \text{ feet } \cdot \frac{1 \text{ mile}}{5280 \text{ feet}} = 0.003788 \text{ miles}$$

$$d^2 = (3960 + 0.003788)^2 - 3960^2 = 30 \text{ sq. miles}$$

$$d \approx 5.477 \text{ miles}$$

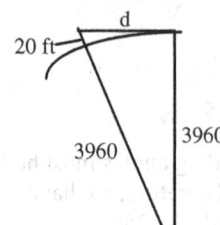

57. Convert 100 feet to miles, and solve the Pythagorean Theorem to find the distance:

$$100 \text{ feet } = 100 \text{ feet } \cdot \frac{1 \text{ mile}}{5280 \text{ feet}} = 0.018939 \text{ miles}$$

$$d^2 = (3960 + 0.018939)^2 - 3960^2 \approx 150 \text{ sq. miles}$$

$$d \approx 12.2 \text{ miles}$$

Convert 150 feet to miles, and solve the Pythagorean Theorem to find the distance:

$$150 \text{ feet } = 150 \text{ feet } \cdot \frac{1 \text{ mile}}{5280 \text{ feet}} = 0.028409 \text{ miles}$$

$$d^2 = (3960 + 0.028409)^2 - 3960^2 \approx 225 \text{ sq. miles}$$

$$d \approx 15.0 \text{ miles}$$

59. Let l = length of the rectangle and w = width of the rectangle. Notice that

$$(l+w)^2 - (l-w)^2$$
$$= [(l+w) + (l-w)][(l+w) - (l-w)]$$
$$= (2l)(2w) = 4lw = 4A$$

So $A = \frac{1}{4}[(l+w)^2 - (l-w)^2]$

Since $(l-w)^2 \geq 0$, the largest area will occur when $l - w = 0$ or $l = w$; that is, when the rectangle is a square. But

$$1000 = 2l + 2w = 2(l+w)$$
$$500 = l + w = 2l$$
$$250 = l = w$$

The largest possible area is $250^2 = 62500$ sq ft. A circular pool with circumference = 1000 feet yields the equation: $2\pi r = 1000 \Rightarrow r = \dfrac{500}{\pi}$

The area enclosed by the circular pool is:

$$A = \pi r^2 = \pi \left(\frac{500}{\pi}\right)^2 = \frac{500^2}{\pi} \approx 79577.47 \text{ ft}^2$$

Thus, a circular pool will enclose the most area.

Section R.4

1. 4; 3

3. $x^3 - 8$

5. c

7. True

9. $2x^3$ Monomial; Variable: x; Coefficient: 2; Degree: 3

11. $\dfrac{8}{x} = 8x^{-1}$ Not a monomial; when written in the form ax^k, the variable has a negative exponent.

13. $-2xy^2$ Monomial; Variable: x, y; Coefficient: −2; Degree: 3

15. $\dfrac{8x}{y} = 8xy^{-1}$ Not a monomial; when written in the form $ax^n y^m$, the exponent on the variable y is negative.

17. $x^2 + y^2$ Not a monomial; the expression contains more than one term. This expression is a binomial.

19. $3x^2 - 5$ Polynomial; Degree: 2

21. 5   Polynomial; Degree: 0

23. $3x^2 - \dfrac{5}{x}$ Not a polynomial; the variable in the denominator results in an exponent that is not a nonnegative integer.

25. $2y^3 - \sqrt{2}$ Polynomial; Degree: 3

27. $\dfrac{x^2+5}{x^3-1}$ Not a polynomial; the polynomial in the denominator has a degree greater than 0.

29. $(x^2+4x+5)+(3x-3)$
$= x^2+(4x+3x)+(5-3)$
$= x^2+7x+2$

31. $(x^3-2x^2+5x+10)-(2x^2-4x+3)$
$= x^3-2x^2+5x+10-2x^2+4x-3$
$= x^3+(-2x^2-2x^2)+(5x+4x)+(10-3)$
$= x^3-4x^2+9x+7$

33. $\left(6x^5+x^3+x\right)+\left(5x^4-x^3+3x^2\right)$
$= 6x^5+5x^4+3x^2+x$

35. $(x^2-3x+1)+2(3x^2+x-4)$
$= x^2-3x+1+6x^2+2x-8$
$= 7x^2-x-7$

37. $6(x^3+x^2-3)-4(2x^3-3x^2)$
$= 6x^3+6x^2-18-8x^3+12x^2$
$= -2x^3+18x^2-18$

39. $\left(x^2-x+2\right)+\left(2x^2-3x+5\right)-\left(x^2+1\right)$
$= x^2-x+2+2x^2-3x+5-x^2-1$
$= 2x^2-4x+6$

41. $9\left(y^2-3y+4\right)-6\left(1-y^2\right)$
$= 9y^2-27y+36-6+6y^2$
$= 15y^2-27y+30$

43. $x(x^2+x-4)=x^3+x^2-4x$

45. $-2x^2(4x^3+5)=-8x^5-10x^2$

47. $(x+1)(x^2+2x-4)$
$= x(x^2+2x-4)+1(x^2+2x-4)$
$= x^3+2x^2-4x+x^2+2x-4$
$= x^3+3x^2-2x-4$

49. $(x+2)(x+4)=x^2+4x+2x+8$
$\qquad = x^2+6x+8$

51. $(2x+5)(x+2)=2x^2+4x+5x+10$
$\qquad = 2x^2+9x+10$

53. $(x-4)(x+2)=x^2+2x-4x-8$
$\qquad = x^2-2x-8$

55. $(x-3)(x-2)=x^2-2x-3x+6$
$\qquad = x^2-5x+6$

57. $(2x+3)(x-2)=2x^2-4x+3x-6$
$\qquad = 2x^2-x-6$

59. $(-2x+3)(x-4)=-2x^2+8x+3x-12$
$\qquad = -2x^2+11x-12$

61. $(-x-2)(-2x-4)=2x^2+4x+4x+8$
$\qquad = 2x^2+8x+8$

63. $(x-2y)(x+y)=x^2+xy-2xy-2y^2$
$\qquad = x^2-xy-2y^2$

65. $(-2x-3y)(3x+2y)=-6x^2-4xy-9xy-6y^2$
$\qquad = -6x^2-13xy-6y^2$

67. $(x-7)(x+7)=x^2-7^2=x^2-49$

69. $(2x+3)(2x-3)=(2x)^2-3^2=4x^2-9$

71. $(x+4)^2=x^2+2\cdot x\cdot 4+4^2=x^2+8x+16$

73. $(x-4)^2=x^2-2\cdot x\cdot 4+4^2=x^2-8x+16$

75. $(3x+4)(3x-4)=(3x)^2-4^2=9x^2-16$

77. $(2x-3)^2=(2x)^2-2(2x)(3)+3^2$
$\qquad = 4x^2-12x+9$

79. $(x+y)(x-y)=(x)^2-\left(y\right)^2=x^2-y^2$

81. $(3x+y)(3x-y)=(3x)^2-\left(y\right)^2=9x^2-y^2$

83. $(x+y)^2=x^2+2xy+y^2$

85. $(x-2y)^2=x^2+2\left(x\cdot(-2y)\right)+(2y)^2$
$\qquad = x^2-4xy+4y^2$

87. $(x-2)^3 = x^3 - 3 \cdot x^2 \cdot 2 + 3 \cdot x \cdot 2^2 - 2^3$
$= x^3 - 6x^2 + 12x - 8$

89. $(2x+1)^3 = (2x)^3 + 3(2x)^2(1) + 3(2x) \cdot 1^2 + 1^3$
$= 8x^3 + 12x^2 + 6x + 1$

91.
$$
\begin{array}{r}
4x^2 - 11x + 23 \\
x+2\overline{)4x^3 - 3x^2 + x + 1} \\
\underline{4x^3 + 8x^2} \\
-11x^2 + x \\
\underline{-11x^2 - 22x} \\
23x + 1 \\
\underline{23x + 46} \\
-45
\end{array}
$$

Check:

$(x+2)(4x^2 - 11x + 23) + (-45)$

$= 4x^3 - 11x^2 + 23x + 8x^2 - 22x + 46 - 45$

$= 4x^3 - 3x^2 + x + 1$

The quotient is $4x^2 - 11x + 23$; the remainder is -45.

93.
$$
\begin{array}{r}
4x - 3 \\
x^2\overline{)4x^3 - 3x^2 + x + 1} \\
\underline{4x^3 } \\
-3x^2 + x + 1 \\
\underline{-3x^2 } \\
x + 1
\end{array}
$$

Check:

$(x^2)(4x-3) + (x+1) = 4x^3 - 3x^2 + x + 1$

The quotient is $4x - 3$; the remainder is $x + 1$.

95.
$$
\begin{array}{r}
5x^2 - 13 \\
x^2+2\overline{)5x^4 + 0x^3 - 3x^2 + x + 1} \\
\underline{5x^4 + 10x^2} \\
-13x^2 + x + 1 \\
\underline{-13x^2 - 26} \\
x + 27
\end{array}
$$

Check:

$\left(x^2 + 2\right)\left(5x^2 - 13\right) + (x + 27)$

$= 5x^4 + 10x^2 - 13x^2 - 26 + x + 27$

$= 5x^4 - 3x^2 + x + 1$

The quotient is $5x^2 - 13$; the remainder is $x + 27$.

97.
$$
\begin{array}{r}
2x^2 \\
2x^3-1\overline{)4x^5 + 0x^4 + 0x^3 - 3x^2 + x + 1} \\
\underline{4x^5 - 2x^2} \\
-x^2 + x + 1
\end{array}
$$

Check:

$\left(2x^3 - 1\right)\left(2x^2\right) + \left(-x^2 + x + 1\right)$

$= 4x^5 - 2x^2 - x^2 + x + 1 = 4x^5 - 3x^2 + x + 1$

The quotient is $2x^2$; the remainder is $-x^2 + x + 1$.

99.
$$
\begin{array}{r}
x^2 - 2x + \frac{1}{2} \\
2x^2+x+1\overline{)2x^4 - 3x^3 + 0x^2 + x + 1} \\
\underline{2x^4 + x^3 + x^2} \\
-4x^3 - x^2 + x \\
\underline{-4x^3 - 2x^2 - 2x} \\
x^2 + 3x + 1 \\
\underline{x^2 + \frac{1}{2}x + \frac{1}{2}} \\
\frac{5}{2}x + \frac{1}{2}
\end{array}
$$

Check:

$\left(2x^2 + x + 1\right)\left(x^2 - 2x + \frac{1}{2}\right) + \frac{5}{2}x + \frac{1}{2}$

$= 2x^4 - 4x^3 + x^2 + x^3 - 2x^2 + \frac{1}{2}x$

$ + x^2 - 2x + \frac{1}{2} + \frac{5}{2}x + \frac{1}{2}$

$= 2x^4 - 3x^3 + x + 1$

The quotient is $x^2 - 2x + \frac{1}{2}$; the remainder is $\frac{5}{2}x + \frac{1}{2}$.

101.
$$x-1\overline{)\begin{array}{r} -4x^2-3x-3 \\ -4x^3+x^2+0x-4 \end{array}}$$
$$\underline{-4x^3+4x^2}$$
$$-3x^2$$
$$\underline{-3x^2+3x}$$
$$-3x-4$$
$$\underline{-3x+3}$$
$$-7$$

Check:

$(x-1)(-4x^2-3x-3)+(-7)$

$= -4x^3-3x^2-3x+4x^2+3x+3-7$

$= -4x^3+x^2-4$

The quotient is $-4x^2-3x-3$; the remainder is -7.

103.
$$x^2+x+1\overline{)\begin{array}{r} x^2-x-1 \\ x^4+0x^3-x^2+0x+1 \end{array}}$$
$$\underline{x^4+x^3+x^2}$$
$$-x^3-2x^2$$
$$\underline{-x^3-x^2-x}$$
$$-x^2+x+1$$
$$\underline{-x^2-x-1}$$
$$2x+2$$

Check:

$(x^2+x+1)(x^2-x-1)+2x+2$

$= x^4+x^3+x^2-x^3-x^2-x-x^2-x$

$\quad -1+2x+2$

$= x^4-x^2+1$

The quotient is x^2-x-1; the remainder is $2x+2$.

105.
$$x-a\overline{)\begin{array}{r} x^2+ax+a^2 \\ x^3+0x^2+0x-a^3 \end{array}}$$
$$\underline{x^3-ax^2}$$
$$ax^2$$
$$\underline{ax^2-a^2x}$$
$$a^2x-a^3$$
$$\underline{a^2x-a^3}$$
$$0$$

Check:

$(x-a)(x^2+ax+a^2)+0$

$= x^3+ax^2+a^2x-ax^2-a^2x-a^3$

$= x^3-a^3$

The quotient is x^2+ax+a^2; the remainder is 0.

107. When we multiply polynomials $p_1(x)$ and $p_2(x)$, each term of $p_1(x)$ will be multiplied by each term of $p_2(x)$. So when the highest-powered term of $p_1(x)$ multiplies by the highest powered term of $p_2(x)$, the exponents on the variables in those terms will add according to the basic rules of exponents. Therefore, the highest powered term of the product polynomial will have degree equal to the sum of the degrees of $p_1(x)$ and $p_2(x)$.

109. When we add two polynomials $p_1(x)$ and $p_2(x)$, where the degree of $p_1(x) =$ the degree of $p_2(x)$, the new polynomial will have degree $\leq$ the degree of $p_1(x)$ and $p_2(x)$.

111. Answers will vary.

Section R.5

1. $3x(x-2)(x+2)$

3. c

5. d

7. True; x^2+4 is prime over the set of real numbers.

9. $3x+6=3(x+2)$

11. $ax^2+a=a(x^2+1)$

13. $x^3+x^2+x=x(x^2+x+1)$

15. $2x^2-2x=2x(x-1)$

17. $3x^2y-6xy^2+12xy=3xy(x-2y+4)$

19. $x^2-1=x^2-1^2=(x-1)(x+1)$

21. $4x^2-1=(2x)^2-1^2=(2x-1)(2x+1)$

23. $x^2-16=x^2-4^2=(x-4)(x+4)$

25. $25x^2-4=(5x-2)(5x+2)$

27. $x^2+2x+1=(x+1)^2$

29. $x^2+4x+4=(x+2)^2$

31. $x^2-10x+25=(x-5)^2$

33. $4x^2+4x+1=(2x+1)^2$

35. $16x^2+8x+1=(4x+1)^2$

37. $x^3-27=x^3-3^3=(x-3)(x^2+3x+9)$

39. $x^3+27=x^3+3^3=(x+3)(x^2-3x+9)$

41. $8x^3+27=(2x)^3+3^3$
$\qquad =(2x+3)(4x^2-6x+9)$

43. $x^2+5x+6=(x+2)(x+3)$

45. $x^2+7x+6=(x+6)(x+1)$

47. $x^2+7x+10=(x+2)(x+5)$

49. $x^2-10x+16=(x-2)(x-8)$

51. $x^2-7x-8=(x+1)(x-8)$

53. $x^2+7x-8=(x+8)(x-1)$

55. $2x^2+4x+3x+6=2x(x+2)+3(x+2)$
$\qquad =(x+2)(2x+3)$

57. $2x^2-4x+x-2=2x(x-2)+1(x-2)$
$\qquad =(x-2)(2x+1)$

59. $18x^2+27x+12x+18=3(6x^2+9x+4x+6)$
$\qquad =3[3x(2x+3)+2(2x+3)]$
$\qquad =3(2x+3)(3x+2)$

61. $3x^2+4x+1=(3x+1)(x+1)$

63. $2z^2+5z+3=(2z+3)(z+1)$

65. $3x^2+2x-8=(3x-4)(x+2)$

67. $3x^2-2x-8=(3x+4)(x-2)$

69. $12x^4+56x^3+32x^2=4x^2(3x^2+14x+8)$
$\qquad =4x^2(3x+2)(x+4)$

71. $3x^2+10x-8=(3x-2)(x+4)$

73. Since B is 10 then we need half of 10 squared to be the last term in our trinomial. Thus
$\frac{1}{2}(10)=5;\ (5)^2=25$
$x^2+10x+25=(x+5)^2$

75. Since B is -6 then we need half of -6 squared to be the last term in our trinomial. Thus
$\frac{1}{2}(-6)=-3;\ (-3)^2=9$
$y^2-6y+9=(y-3)^2$

77. Since B is $-\frac{1}{2}$ then we need half of $-\frac{1}{2}$ squared to be the last term in our trinomial. Thus
$$\frac{1}{2}\left(-\frac{1}{2}\right) = -\frac{1}{4}; \quad \left(-\frac{1}{4}\right)^2 = \frac{1}{16}$$
$$x^2 - \frac{1}{2}x + \frac{1}{16} = \left(x - \frac{1}{4}\right)^2$$

79. $x^2 - 36 = (x-6)(x+6)$

81. $2 - 8x^2 = 2(1 - 4x^2) = 2(1 - 2x)(1 + 2x)$

83. $8x^2 + 88x + 80 = 8(x^2 + 11x + 10)$
$$= 8(x+1)(x+10)$$

85. $x^2 - 10x + 21 = (x-7)(x-3)$

87. $4x^2 - 8x + 32 = 4(x^2 - 2x + 8)$

89. $x^2 + 4x + 16$ is prime over the reals because there are no factors of 16 whose sum is 4.

91. $15 + 2x - x^2 = -(x^2 - 2x - 15) = -(x-5)(x+3)$

93. $3x^2 - 12x - 36 = 3(x^2 - 4x - 12)$
$$= 3(x-6)(x+2)$$

95. $y^4 + 11y^3 + 30y^2 = y^2(y^2 + 11y + 30)$
$$= y^2(y+5)(y+6)$$

97. $8x^2 + 24x + 18 = 2(4x^2 + 12x + 9)$
$$= 2(2x+3)^2$$

99. $6x^2 + 8x + 2 = 2(3x^2 + 4x + 1)$
$$= 2(3x+1)(x+1)$$

101. $x^4 - 81 = \left(x^2\right)^2 - 9^2 = (x^2 - 9)(x^2 + 9)$
$$= (x-3)(x+3)(x^2 + 9)$$

103. $x^6 - 2x^3 + 1 = (x^3 - 1)^2$
$$= \left[(x-1)(x^2 + x + 1)\right]^2$$
$$= (x-1)^2(x^2 + x + 1)^2$$

105. $x^7 - x^5 = x^5(x^2 - 1) = x^5(x-1)(x+1)$

107. $16x^2 + 24x + 9 = (4x+3)^2$

109. $5 + 16x - 16x^2 = -(16x^2 - 16x - 5)$
$$= -(4x-5)(4x+1)$$

111. $4y^2 - 16y + 15 = (2y-5)(2y-3)$

113. $1 - 8x^2 - 9x^4 = -(9x^4 + 8x^2 - 1)$
$$= -(9x^2 - 1)(x^2 + 1)$$
$$= -(3x-1)(3x+1)(x^2 + 1)$$

115. $x(x+3) - 6(x+3) = (x+3)(x-6)$

117. $(x+2)^2 - 5(x+2) = (x+2)\left[(x+2) - 5\right]$
$$= (x+2)(x-3)$$

119. $(3x-2)^3 - 27$
$$= (3x-2)^3 - 3^3$$
$$= \left[(3x-2) - 3\right]\left[(3x-2)^2 + 3(3x-2) + 9\right]$$
$$= (3x-5)\left(9x^2 - 12x + 4 + 9x - 6 + 9\right)$$
$$= (3x-5)\left(9x^2 - 3x + 7\right)$$

121. $3(x^2 + 10x + 25) - 4(x+5)$
$$= 3(x+5)^2 - 4(x+5)$$
$$= (x+5)\left[3(x+5) - 4\right]$$
$$= (x+5)(3x+15-4)$$
$$= (x+5)(3x+11)$$

123. $x^3 + 2x^2 - x - 2 = x^2(x+2) - 1(x+2)$
$$= (x+2)(x^2 - 1)$$
$$= (x+2)(x-1)(x+1)$$

125. $x^4 - x^3 + x - 1 = x^3(x-1) + 1(x-1)$
$$= (x-1)(x^3 + 1)$$
$$= (x-1)(x+1)(x^2 - x + 1)$$

127. $2(3x+4)^2 + (2x+3) \cdot 2(3x+4) \cdot 3$
$$= 2(3x+4)\left((3x+4) + (2x+3) \cdot 3\right)$$
$$= 2(3x+4)(3x+4+6x+9)$$
$$= 2(3x+4)(9x+13)$$

13

129. $2x(2x+5)+x^2 \cdot 2 = 2x((2x+5)+x)$

$\qquad = 2x(2x+5+x)$

$\qquad = 2x(3x+5)$

131. $2(x+3)(x-2)^3+(x+3)^2 \cdot 3(x-2)^2$

$= (x+3)(x-2)^2(2(x-2)+(x+3)\cdot 3)$

$= (x+3)(x-2)^2(2x-4+3x+9)$

$= (x+3)(x-2)^2(5x+5)$

$= 5(x+3)(x-2)^2(x+1)$

133. $(4x-3)^2+x\cdot 2(4x-3)\cdot 4$

$= (4x-3)((4x-3)+8x)$

$= (4x-3)(4x-3+8x)$

$= (4x-3)(12x-3)$

$= 3(4x-3)(4x-1)$

135. $2(3x-5)\cdot 3(2x+1)^3+(3x-5)^2 \cdot 3(2x+1)^2 \cdot 2$

$= 6(3x-5)(2x+1)^2((2x+1)+(3x-5))$

$= 6(3x-5)(2x+1)^2(2x+1+3x-5)$

$= 6(3x-5)(2x+1)^2(5x-4)$

137. Factors of 4: $\quad 1,4 \quad 2,2 \quad -1,-4 \quad -2,-2$

$\qquad$ Sum: $\qquad 5 \qquad 4 \qquad -5 \qquad -4$

$\qquad$ None of the sums of the factors is 0, so

x^2+4 is prime.

Alternatively, the possibilities are

$(x\pm 1)(x\pm 4)=x^2\pm 5x+4$ or

$(x\pm 2)(x\pm 2)=x^2\pm 4x+4$, none of which

equals x^2+4.

139. Answers will vary.

Section R.6

1. quotient; divisor; remainder

3. d

5. True

7. $2\overline{)1 \quad -1 \quad 2 \quad 4}$

$\underline{\qquad 2 \quad 2 \quad 8}$

$\quad 1 \quad 1 \quad 4 \quad 12$

Quotient: x^2+x+4

Remainder: 12

9. $3\overline{)3 \quad 2 \quad -1 \quad 3}$

$\underline{\qquad 9 \quad 33 \quad 96}$

$\quad 3 \quad 11 \quad 32 \quad 99$

Quotient: $3x^2+11x+32$

Remainder: 99

11. $-3\overline{)1 \quad 0 \quad -4 \quad 0 \quad 1 \quad 0}$

$\underline{\qquad -3 \quad 9 \quad -15 \quad 45 \quad -138}$

$\quad 1 \quad -3 \quad 5 \quad -15 \quad 46 \quad -138$

Quotient: $x^4-3x^3+5x^2-15x+46$

Remainder: -138

13. $1\overline{)4 \quad 0 \quad -3 \quad 0 \quad 1 \quad 0 \quad 5}$

$\underline{\qquad 4 \quad 4 \quad 1 \quad 1 \quad 2 \quad 2}$

$\quad 4 \quad 4 \quad 1 \quad 1 \quad 2 \quad 2 \quad 7$

Quotient: $4x^5+4x^4+x^3+x^2+2x+2$

Remainder: 7

15. $-1.1\overline{)0.1 \quad 0 \quad 0.2 \quad 0}$

$\underline{\qquad -0.11 \quad 0.121 \quad -0.3531}$

$\quad 0.1 \quad -0.11 \quad 0.321 \quad -0.3531$

Quotient: $0.1x^2-0.11x+0.321$

Remainder: -0.3531

17. $1\overline{)1 \quad 0 \quad 0 \quad 0 \quad 0 \quad -1}$

$\underline{\qquad 1 \quad 1 \quad 1 \quad 1 \quad 1}$

$\quad 1 \quad 1 \quad 1 \quad 1 \quad 1 \quad 0$

Quotient: $x^4+x^3+x^2+x+1$

Remainder: 0

19. $2\overline{)4 \quad -3 \quad -8 \quad 4}$

$\underline{\qquad 8 \quad 10 \quad 4}$

$\quad 4 \quad 5 \quad 2 \quad 8$

Remainder $= 8 \neq 0$. Therefore, $x-2$ is not a

factor of $4x^3-3x^2-8x+4$.

21. $2\overline{)3 \quad -6 \quad 0 \quad -5 \quad 10}$
$\phantom{2\overline{)3}} \quad \underline{6 \quad 0 \quad 0 \quad -10}$
$\phantom{2\overline{)}} 3 \quad 0 \quad 0 \quad -5 \quad 0$

Remainder $= 0$. Therefore, $x - 2$ is a factor of $3x^4 - 6x^3 - 5x + 10$.

23. $-3\overline{)3 \quad 0 \quad 0 \quad 82 \quad 0 \quad 0 \quad 27}$
$\phantom{-3\overline{)3}} \quad \underline{-9 \quad 27 \quad -81 \quad -3 \quad 9 \quad -27}$
$\phantom{-3\overline{)}} 3 \quad -9 \quad 27 \quad 1 \quad -3 \quad 9 \quad 0$

Remainder $= 0$. Therefore, $x + 3$ is a factor of $3x^6 + 82x^3 + 27$.

25. $-4\overline{)4 \quad 0 \quad -64 \quad 0 \quad 1 \quad 0 \quad -15}$
$\phantom{-4\overline{)4}} \quad \underline{-16 \quad 64 \quad 0 \quad 0 \quad -4 \quad 16}$
$\phantom{-4\overline{)}} 4 \quad -16 \quad 0 \quad 0 \quad 1 \quad -4 \quad 1$

Remainder $= 1 \neq 0$. Therefore, $x + 3$ is not a factor of $4x^6 - 64x^4 + x^2 - 15$.

27. $\frac{1}{2}\overline{)2 \quad -1 \quad 0 \quad 2 \quad -1}$
$\phantom{\frac{1}{2}\overline{)2}} \quad \underline{1 \quad 0 \quad 0 \quad 1}$
$\phantom{\frac{1}{2}\overline{)}} 2 \quad 0 \quad 0 \quad 2 \quad 0$

Remainder $= 0$; therefore $x - \frac{1}{2}$ is a factor of $2x^4 - x^3 + 2x - 1$.

29. $-2\overline{)1 \quad -2 \quad 3 \quad 5}$
$\phantom{-2\overline{)1}} \quad \underline{-2 \quad 8 \quad -22}$
$\phantom{-2\overline{)}} 1 \quad -4 \quad 11 \quad -17$

$\dfrac{x^3 - 2x^2 + 3x + 5}{x + 2} = x^2 - 4x + 11 + \dfrac{-17}{x+2}$
$a + b + c + d = 1 - 4 + 11 - 17 = -9$

Section R.7

1. lowest terms

3. d

5. True; $\dfrac{2x^3 - 4x}{x - 2} = \dfrac{2x(x^2 - 2)}{x - 2}$

7. $\dfrac{3x + 9}{x^2 - 9} = \dfrac{3(x + 3)}{(x - 3)(x + 3)} = \dfrac{3}{x - 3}$

9. $\dfrac{x^2 - 2x}{3x - 6} = \dfrac{x(x - 2)}{3(x - 2)} = \dfrac{x}{3}$

11. $\dfrac{24x^2}{12x^2 - 6x} = \dfrac{24x^2}{6x(2x - 1)} = \dfrac{4x}{2x - 1}$

13. $\dfrac{y^2 - 25}{2y^2 - 8y - 10} = \dfrac{(y + 5)(y - 5)}{2(y^2 - 4y - 5)}$
$= \dfrac{(y + 5)(y - 5)}{2(y - 5)(y + 1)}$
$= \dfrac{y + 5}{2(y + 1)}$

15. $\dfrac{x^2 + 4x - 5}{x^2 - 2x + 1} = \dfrac{(x + 5)(x - 1)}{(x - 1)(x - 1)} = \dfrac{x + 5}{x - 1}$

17. $\dfrac{x^2 + 5x - 14}{2 - x} = \dfrac{(x + 7)(x - 2)}{2 - x}$
$= \dfrac{(x + 7)(x - 2)}{-1(x - 2)}$
$= -(x + 7)$
$= -x - 7$

19. $\dfrac{3x + 6}{5x^2} \cdot \dfrac{x}{x^2 - 4} = \dfrac{3(x + 2)}{5x^2} \cdot \dfrac{x}{(x - 2)(x + 2)}$
$= \dfrac{3}{5x(x - 2)}$

21. $\dfrac{4x^2}{x^2 - 16} \cdot \dfrac{x^3 - 64}{2x}$
$= \dfrac{4x^2}{(x - 4)(x + 4)} \cdot \dfrac{(x - 4)(x^2 + 4x + 16)}{2x}$
$= \dfrac{2x \cdot 2x(x - 4)(x^2 + 4x + 16)}{2x(x - 4)(x + 4)}$
$= \dfrac{2x(x^2 + 4x + 16)}{x + 4}$

23. $\dfrac{4x-8}{-3x} \cdot \dfrac{12}{12-6x} = \dfrac{4(x-2)}{-3x} \cdot \dfrac{12}{6(2-x)}$

$\qquad = \dfrac{4(x-2)}{-3x} \cdot \dfrac{2}{(-1)(x-2)}$

$\qquad = \dfrac{8}{3x}$

25. $\dfrac{x^2-3x-10}{x^2+2x-35} \cdot \dfrac{x^2+4x-21}{x^2+9x+14}$

$= \dfrac{(x-5)(x+2)}{(x+7)(x-5)} \cdot \dfrac{(x+7)(x-3)}{(x+7)(x+2)}$

$= \dfrac{x-3}{x+7}$

27. $\dfrac{\dfrac{6x}{x^2-4}}{\dfrac{3x-9}{2x+4}} = \dfrac{6x}{x^2-4} \cdot \dfrac{2x+4}{3x-9}$

$\qquad = \dfrac{6x}{(x-2)(x+2)} \cdot \dfrac{2(x+2)}{3(x-3)}$

$\qquad = \dfrac{4x}{(x-2)(x-3)}$

29. $\dfrac{\dfrac{8x}{x^2-1}}{\dfrac{10x}{x+1}} = \dfrac{8x}{x^2-1} \cdot \dfrac{x+1}{10x}$

$\qquad = \dfrac{8x}{(x-1)(x+1)} \cdot \dfrac{x+1}{10x}$

$\qquad = \dfrac{4}{5(x-1)}$

31. $\dfrac{\dfrac{4-x}{4+x}}{\dfrac{4x}{x^2-16}} = \dfrac{4-x}{4+x} \cdot \dfrac{x^2-16}{4x}$

$\qquad = \dfrac{4-x}{4+x} \cdot \dfrac{(x+4)(x-4)}{4x}$

$\qquad = \dfrac{(4-x)(x-4)}{4x}$

$\qquad = -\dfrac{(x-4)^2}{4x}$

33. $\dfrac{\dfrac{x^2+7x+12}{x^2-7x+12}}{\dfrac{x^2+x-12}{x^2-x-12}} = \dfrac{x^2+7x+12}{x^2-7x+12} \cdot \dfrac{x^2-x-12}{x^2+x-12}$

$\qquad = \dfrac{(x+3)(x+4)}{(x-3)(x-4)} \cdot \dfrac{(x-4)(x+3)}{(x+4)(x-3)}$

$\qquad = \dfrac{(x+3)^2}{(x-3)^2}$

35. $\dfrac{\dfrac{2x^2-x-28}{3x^2-x-2}}{\dfrac{4x^2+16x+7}{3x^2+11x+6}} = \dfrac{2x^2-x-28}{3x^2-x-2} \cdot \dfrac{3x^2+11x+6}{4x^2+16x+7}$

$\qquad = \dfrac{(2x+7)(x-4)}{(3x+2)(x-1)} \cdot \dfrac{(3x+2)(x+3)}{(2x+7)(2x+1)}$

$\qquad = \dfrac{(x-4)(x+3)}{(x-1)(2x+1)}$

37. $\dfrac{x}{2} + \dfrac{5}{2} = \dfrac{x+5}{2}$

39. $\dfrac{x^2}{2x-3} - \dfrac{4}{2x-3} = \dfrac{x^2-4}{2x-3} = \dfrac{(x+2)(x-2)}{2x-3}$

41. $\dfrac{x+1}{x-3} + \dfrac{2x-3}{x-3} = \dfrac{x+1+2x-3}{x-3} = \dfrac{3x-2}{x-3}$

43. $\dfrac{3x+5}{2x-1} - \dfrac{2x-4}{2x-1} = \dfrac{(3x+5)-(2x-4)}{2x-1}$

$\qquad = \dfrac{3x+5-2x+4}{2x-1}$

$\qquad = \dfrac{x+9}{2x-1}$

45. $\dfrac{4}{x-2} + \dfrac{x}{2-x} = \dfrac{4}{x-2} - \dfrac{x}{x-2} = \dfrac{4-x}{x-2}$

47. $\dfrac{4}{x-1} - \dfrac{2}{x+2} = \dfrac{4(x+2)}{(x-1)(x+2)} - \dfrac{2(x-1)}{(x+2)(x-1)}$

$\qquad = \dfrac{4x+8-2x+2}{(x+2)(x-1)}$

$\qquad = \dfrac{2x+10}{(x+2)(x-1)}$

$\qquad = \dfrac{2(x+5)}{(x+2)(x-1)}$

49. $\dfrac{x}{x+1} + \dfrac{2x-3}{x-1} = \dfrac{x(x-1)}{(x+1)(x-1)} + \dfrac{(2x-3)(x+1)}{(x-1)(x+1)}$

$\qquad = \dfrac{x^2 - x + 2x^2 - x - 3}{(x-1)(x+1)}$

$\qquad = \dfrac{3x^2 - 2x - 3}{(x-1)(x+1)}$

51. $\dfrac{x-3}{x+2} - \dfrac{x+4}{x-2} = \dfrac{(x-3)(x-2)}{(x+2)(x-2)} - \dfrac{(x+4)(x+2)}{(x-2)(x+2)}$

$\qquad = \dfrac{x^2 - 5x + 6 - (x^2 + 6x + 8)}{(x+2)(x-2)}$

$\qquad = \dfrac{x^2 - 5x + 6 - x^2 - 6x - 8}{(x+2)(x-2)}$

$\qquad = \dfrac{-11x - 2}{(x+2)(x-2)} \;\; \text{or} \;\; \dfrac{-(11x+2)}{(x+2)(x-2)}$

53. $\dfrac{x}{x^2-4} + \dfrac{1}{x} = \dfrac{x^2 + x^2 - 4}{x(x^2 - 4)}$

$\qquad = \dfrac{2x^2 - 4}{x(x^2 - 4)}$

$\qquad = \dfrac{2(x^2 - 2)}{x(x-2)(x+2)}$

55. $x^2 - 4 = (x+2)(x-2)$

$x^2 - x - 2 = (x+1)(x-2)$

Therefore, $\text{LCM} = (x+2)(x-2)(x+1)$.

57. $x^3 - x = x(x^2 - 1) = x(x+1)(x-1)$

$x^2 - x = x(x-1)$

Therefore, $\text{LCM} = x(x+1)(x-1)$.

59. $4x^3 - 4x^2 + x = x(4x^2 - 4x + 1)$

$\qquad\qquad\qquad = x(2x-1)(2x-1)$

$2x^3 - x^2 = x^2(2x-1)$

x^3

Therefore, $\text{LCM} = x^3(2x-1)^2$.

61. $x^3 - x = x(x^2 - 1) = x(x+1)(x-1)$

$x^3 - 2x^2 + x = x(x^2 - 2x + 1) = x(x-1)^2$

$x^3 - 1 = (x-1)(x^2 + x + 1)$

Therefore, $\text{LCM} = x(x+1)(x-1)^2(x^2 + x + 1)$.

63. $\dfrac{x}{x^2 - 7x + 6} - \dfrac{x}{x^2 - 2x - 24}$

$= \dfrac{x}{(x-6)(x-1)} - \dfrac{x}{(x-6)(x+4)}$

$= \dfrac{x(x+4)}{(x-6)(x-1)(x+4)} - \dfrac{x(x-1)}{(x-6)(x+4)(x-1)}$

$= \dfrac{x^2 + 4x - x^2 + x}{(x-6)(x+4)(x-1)} = \dfrac{5x}{(x-6)(x+4)(x-1)}$

65. $\dfrac{4x}{x^2 - 4} - \dfrac{2}{x^2 + x - 6}$

$= \dfrac{4x}{(x-2)(x+2)} - \dfrac{2}{(x+3)(x-2)}$

$= \dfrac{4x(x+3)}{(x-2)(x+2)(x+3)} - \dfrac{2(x+2)}{(x+3)(x-2)(x+2)}$

$= \dfrac{4x^2 + 12x - 2x - 4}{(x-2)(x+2)(x+3)}$

$= \dfrac{4x^2 + 10x - 4}{(x-2)(x+2)(x+3)}$

$= \dfrac{2(2x^2 + 5x - 2)}{(x-2)(x+2)(x+3)}$

67. $\dfrac{3}{(x-1)^2(x+1)} + \dfrac{2}{(x-1)(x+1)^2}$

$= \dfrac{3(x+1) + 2(x-1)}{(x-1)^2(x+1)^2}$

$= \dfrac{3x + 3 + 2x - 2}{(x-1)^2(x+1)^2}$

$= \dfrac{5x + 1}{(x-1)^2(x+1)^2}$

69. $\dfrac{x+4}{x^2-x-2} - \dfrac{2x+3}{x^2+2x-8}$

$= \dfrac{x+4}{(x-2)(x+1)} - \dfrac{2x+3}{(x+4)(x-2)}$

$= \dfrac{(x+4)(x+4)}{(x-2)(x+1)(x+4)} - \dfrac{(2x+3)(x+1)}{(x+4)(x-2)(x+1)}$

$= \dfrac{x^2+8x+16-(2x^2+5x+3)}{(x-2)(x+1)(x+4)}$

$= \dfrac{-x^2+3x+13}{(x-2)(x+1)(x+4)}$

71. $\dfrac{1}{x} - \dfrac{2}{x^2+x} + \dfrac{3}{x^3-x^2}$

$= \dfrac{1}{x} - \dfrac{2}{x(x+1)} + \dfrac{3}{x^2(x-1)}$

$= \dfrac{x(x+1)(x-1)-2x(x-1)+3(x+1)}{x^2(x+1)(x-1)}$

$= \dfrac{x(x^2-1)-2x^2+2x+3x+3}{x^2(x+1)(x-1)}$

$= \dfrac{x^3-x-2x^2+5x+3}{x^2(x+1)(x-1)}$

$= \dfrac{x^3-2x^2+4x+3}{x^2(x+1)(x-1)}$

73. $\dfrac{1}{h}\left(\dfrac{1}{x+h} - \dfrac{1}{x}\right) = \dfrac{1}{h}\left(\dfrac{1 \cdot x}{(x+h)x} - \dfrac{1(x+h)}{x(x+h)}\right)$

$= \dfrac{1}{h}\left(\dfrac{x-x-h}{x(x+h)}\right)$

$= \dfrac{-h}{hx(x+h)}$

$= \dfrac{-1}{x(x+h)}$

75. $\dfrac{1+\dfrac{1}{x}}{1-\dfrac{1}{x}} = \dfrac{\left(\dfrac{x}{x}+\dfrac{1}{x}\right)}{\left(\dfrac{x}{x}-\dfrac{1}{x}\right)} = \dfrac{\left(\dfrac{x+1}{x}\right)}{\left(\dfrac{x-1}{x}\right)} = \dfrac{x+1}{x} \cdot \dfrac{x}{x-1} = \dfrac{x+1}{x-1}$

77. $\dfrac{2-\dfrac{x+1}{x}}{3+\dfrac{x-1}{x+1}} = \dfrac{\dfrac{2x}{x}-\dfrac{x+1}{x}}{\dfrac{3(x+1)}{x+1}+\dfrac{x-1}{x+1}} = \dfrac{\dfrac{2x-x-1}{x}}{\dfrac{3x+3+x-1}{x+1}}$

$= \dfrac{\dfrac{x-1}{x}}{\dfrac{4x+2}{x+1}} = \dfrac{x-1}{x} \cdot \dfrac{x+1}{2(2x+1)}$

$= \dfrac{(x-1)(x+1)}{2x(2x+1)}$

79. $\dfrac{\dfrac{x+4}{x-2} - \dfrac{x-3}{x+1}}{x+1}$

$= \dfrac{\left(\dfrac{(x+4)(x+1)}{(x-2)(x+1)} - \dfrac{(x-3)(x-2)}{(x+1)(x-2)}\right)}{x+1}$

$= \dfrac{\left(\dfrac{x^2+5x+4-(x^2-5x+6)}{(x-2)(x+1)}\right)}{x+1}$

$= \dfrac{10x-2}{(x-2)(x+1)} \cdot \dfrac{1}{x+1}$

$= \dfrac{2(5x-1)}{(x-2)(x+1)^2}$

81. $\dfrac{\dfrac{x-2}{x+2}+\dfrac{x-1}{x+1}}{\dfrac{x}{x+1}-\dfrac{2x-3}{x}}$

$$=\dfrac{\left(\dfrac{(x-2)(x+1)}{(x+2)(x+1)}+\dfrac{(x-1)(x+2)}{(x+1)(x+2)}\right)}{\left(\dfrac{x^2}{(x+1)(x)}-\dfrac{(2x-3)(x+1)}{x(x+1)}\right)}$$

$$=\dfrac{\left(\dfrac{x^2-x-2+x^2+x-2}{(x+2)(x+1)}\right)}{\left(\dfrac{x^2-(2x^2-x-3)}{x(x+1)}\right)}$$

$$=\dfrac{\left(\dfrac{2x^2-4}{(x+2)(x+1)}\right)}{\left(\dfrac{-x^2+x+3}{x(x+1)}\right)}$$

$$=\dfrac{2(x^2-2)}{(x+2)(x+1)}\cdot\dfrac{x(x+1)}{-(x^2-x-3)}$$

$$=\dfrac{2x(x^2-2)}{-(x+2)(x^2-x-3)}$$

$$=\dfrac{-2x(x^2-2)}{(x+2)(x^2-x-3)}$$

83. $1-\dfrac{1}{1-\dfrac{1}{x}}=1-\dfrac{1}{\dfrac{x-1}{x}}$

$$=1-\dfrac{x}{x-1}$$

$$=\dfrac{x-1-x}{x-1}$$

$$=\dfrac{-1}{x-1}$$

85. $\dfrac{2(x-1)^{-1}+3}{3(x-1)^{-1}+2}=\dfrac{\dfrac{2}{x-1}+3}{\dfrac{3}{x-1}+2}=\dfrac{\dfrac{2}{x-1}+\dfrac{3(x-1)}{x-1}}{\dfrac{3}{x-1}+\dfrac{2(x-1)}{x-1}}$

$$=\dfrac{\dfrac{2+3(x-1)}{x-1}}{\dfrac{3+2(x-1)}{x-1}}$$

$$=\dfrac{2+3(x-1)}{x-1}\cdot\dfrac{x-1}{3+2(x-1)}$$

$$=\dfrac{2+3(x-1)}{3+2(x-1)}=\dfrac{2+3x-3}{3+2x-2}$$

$$=\dfrac{3x-1}{2x+1}$$

87. $\dfrac{(2x+3)\cdot 3-(3x-5)\cdot 2}{(3x-5)^2}=\dfrac{6x+9-6x+10}{(3x-5)^2}$

$$=\dfrac{19}{(3x-5)^2}$$

89. $\dfrac{x\cdot 2x-(x^2+1)\cdot 1}{(x^2+1)^2}=\dfrac{2x^2-x^2-1}{(x^2+1)^2}$

$$=\dfrac{x^2-1}{(x^2+1)^2}$$

$$=\dfrac{(x-1)(x+1)}{(x^2+1)^2}$$

91. $\dfrac{(3x+1)\cdot 2x-x^2\cdot 3}{(3x+1)^2}=\dfrac{6x^2+2x-3x^2}{(3x+1)^2}$

$$=\dfrac{3x^2+2x}{(3x+1)^2}$$

$$=\dfrac{x(3x+2)}{(3x+1)^2}$$

93. $\dfrac{(x^2+1)\cdot 3 - (3x+4)\cdot 2x}{(x^2+1)^2} = \dfrac{3x^2 + 3 - 6x^2 - 8x}{(x^2+1)^2}$

$\qquad = \dfrac{-3x^2 - 8x + 3}{(x^2+1)^2}$

$\qquad = \dfrac{-(3x^2 + 8x - 3)}{(x^2+1)^2}$

$\qquad = -\dfrac{(3x-1)(x+3)}{(x^2+1)^2}$

95. $\dfrac{1}{f} = (n-1)\left(\dfrac{1}{R_1} + \dfrac{1}{R_2}\right)$

$\qquad \dfrac{1}{f} = (n-1)\left(\dfrac{R_2 + R_1}{R_1 \cdot R_2}\right)$

$\qquad \dfrac{R_1 \cdot R_2}{f} = (n-1)(R_2 + R_1)$

$\qquad \dfrac{f}{R_1 \cdot R_2} = \dfrac{1}{(n-1)(R_2 + R_1)}$

$\qquad f = \dfrac{R_1 \cdot R_2}{(n-1)(R_2 + R_1)}$

$\qquad f = \dfrac{0.1(0.2)}{(1.5-1)(0.2+0.1)}$

$\qquad = \dfrac{0.02}{0.5(0.3)} = \dfrac{0.02}{0.15} = \dfrac{2}{15}$ meters

97. $1 + \dfrac{1}{x} = \dfrac{x+1}{x} \Rightarrow a = 1, b = 1, c = 0$

$\qquad 1 + \dfrac{1}{1 + \dfrac{1}{x}} = 1 + \dfrac{1}{\left(\dfrac{x+1}{x}\right)} = 1 + \dfrac{x}{x+1}$

$\qquad\qquad = \dfrac{x+1+x}{x+1} = \dfrac{2x+1}{x+1}$

$\qquad\qquad \Rightarrow a = 2, b = 1, c = 1$

$\qquad 1 + \dfrac{1}{1 + \dfrac{1}{1 + \dfrac{1}{x}}} = 1 + \dfrac{1}{\left(\dfrac{2x+1}{x+1}\right)} = 1 + \dfrac{x+1}{2x+1}$

$\qquad\qquad = \dfrac{2x+1+x+1}{2x+1} = \dfrac{3x+2}{2x+1}$

$\qquad\qquad \Rightarrow a = 3, b = 2, c = 1$

$1 + \dfrac{1}{1 + \dfrac{1}{1 + \dfrac{1}{1 + \dfrac{1}{x}}}} = 1 + \dfrac{1}{\left(\dfrac{3x+2}{2x+1}\right)} = 1 + \dfrac{2x+1}{3x+2}$

$\qquad = \dfrac{3x+2+2x+1}{3x+2} = \dfrac{5x+3}{3x+2}$

$\qquad \Rightarrow a = 5, b = 3, c = 2$

If we continue this process, the values of a, b and c produce the following sequences:

$a: 1, 2, 3, 5, 8, 13, 21, ...$

$b: 1, 1, 2, 3, 5, 8, 13, 21, ...$

$c: 0, 1, 1, 2, 3, 5, 8, 13, 21, ...$

In each case we have a *Fibonacci Sequence*, where the next value in the list is obtained from the sum of the previous 2 values in the list.

99. Answers will vary.

Section R.8

1. 9; −9

3. index

5. b

7. c

9. true

11. $\sqrt[3]{27} = \sqrt[3]{3^3} = 3$

13. $\sqrt[3]{-8} = \sqrt[3]{(-2)^3} = -2$

15. $\sqrt{8} = \sqrt{4 \cdot 2} = 2\sqrt{2}$

17. $\sqrt{700} = \sqrt{100 \cdot 7} = 10\sqrt{7}$

19. $\sqrt[3]{32} = \sqrt[3]{8 \cdot 4} = 2\sqrt[3]{4}$

21. $\sqrt[3]{-8x^4} = \sqrt[3]{-8x^3 \cdot x} = -2x\sqrt[3]{x}$

23. $\sqrt[4]{243} = \sqrt[4]{81 \cdot 3} = 3\sqrt[4]{3}$

25. $\sqrt[4]{x^{12}y^8} = \sqrt[4]{(x^3)^4 (y^2)^4} = x^3 y^2$

27. $\sqrt[4]{\dfrac{x^9 y^7}{xy^3}} = \sqrt[4]{x^8 y^4} = x^2 y$

29. $\sqrt{36x} = 6\sqrt{x}$

31. $\sqrt[4]{162 x^9 y^{12}} = \sqrt[4]{2(3)^4 x \left(x^2\right)^4 \left(y^3\right)^4}$
$= 3x^2 y^3 \sqrt[4]{2x}$

33. $\sqrt{3x^2}\sqrt{12x} = \sqrt{36x^2 \cdot x} = 6x\sqrt{x}$

35. $\left(\sqrt{5}\sqrt[3]{9}\right)^2 = \left(\sqrt{5}\right)^2 \left(\sqrt[3]{9}\right)^2$
$= 5 \cdot \sqrt[3]{9^2} = 5\sqrt[3]{81} = 5 \cdot 3\sqrt[3]{3} = 15\sqrt[3]{3}$

37. $\left(3\sqrt{6}\right)\left(2\sqrt{2}\right) = 6\sqrt{12} = 6\sqrt{4 \cdot 3} = 12\sqrt{3}$

39. $3\sqrt{2} + 4\sqrt{2} = (3+4)\sqrt{2} = 7\sqrt{2}$

41. $-\sqrt{18} + 2\sqrt{8} = -\sqrt{9\cdot 2} + 2\sqrt{4\cdot 2}$
$= -3\sqrt{2} + 4\sqrt{2}$
$= (-3+4)\sqrt{2}$
$= \sqrt{2}$

43. $\left(\sqrt{3}+3\right)\left(\sqrt{3}-1\right) = \left(\sqrt{3}\right)^2 + 3\sqrt{3} - \sqrt{3} - 3$
$= 3 + 2\sqrt{3} - 3$
$= 2\sqrt{3}$

45. $5\sqrt[3]{2} - 2\sqrt[3]{54} = 5\sqrt[3]{2} - 2\cdot 3\sqrt[3]{2}$
$= 5\sqrt[3]{2} - 6\sqrt[3]{2}$
$= (5-6)\sqrt[3]{2}$
$= -\sqrt[3]{2}$

47. $\left(\sqrt{x}-1\right)^2 = \left(\sqrt{x}\right)^2 - 2\sqrt{x} + 1$
$= x - 2\sqrt{x} + 1$

49. $\sqrt[3]{16x^4} - \sqrt[3]{2x} = \sqrt[3]{8x^3 \cdot 2x} - \sqrt[3]{2x}$
$= 2x\sqrt[3]{2x} - \sqrt[3]{2x}$
$= (2x-1)\sqrt[3]{2x}$

51. $\sqrt{8x^3} - 3\sqrt{50x} = \sqrt{4x^2 \cdot 2x} - 3\sqrt{25\cdot 2x}$
$= 2x\sqrt{2x} - 15\sqrt{2x}$
$= (2x-15)\sqrt{2x}$

53. $\sqrt[3]{16x^4 y} - 3x\sqrt[3]{2xy} + 5\sqrt[3]{-2xy^4}$
$= \sqrt[3]{8x^3 \cdot 2xy} - 3x\sqrt[3]{2xy} + 5\sqrt[3]{-y^3 \cdot 2xy}$
$= 2x\sqrt[3]{2xy} - 3x\sqrt[3]{2xy} - 5y\sqrt[3]{2xy}$
$= (2x - 3x - 5y)\sqrt[3]{2xy}$
$= (-x-5y)\sqrt[3]{2xy}$ or $-(x+5y)\sqrt[3]{2xy}$

55. $\dfrac{1}{\sqrt{2}} = \dfrac{1}{\sqrt{2}} \cdot \dfrac{\sqrt{2}}{\sqrt{2}} = \dfrac{\sqrt{2}}{2}$

57. $\dfrac{-\sqrt{3}}{\sqrt{5}} = \dfrac{-\sqrt{3}}{\sqrt{5}} \cdot \dfrac{\sqrt{5}}{\sqrt{5}} = \dfrac{-\sqrt{15}}{5}$

59. $\dfrac{\sqrt{3}}{5-\sqrt{2}} = \dfrac{\sqrt{3}}{5-\sqrt{2}} \cdot \dfrac{5+\sqrt{2}}{5+\sqrt{2}}$
$= \dfrac{\sqrt{3}\left(5+\sqrt{2}\right)}{25-2}$
$= \dfrac{\sqrt{3}\left(5+\sqrt{2}\right)}{23}$ or $\dfrac{5\sqrt{3}+\sqrt{6}}{23}$

61. $\dfrac{2-\sqrt{5}}{2+3\sqrt{5}} = \dfrac{2-\sqrt{5}}{2+3\sqrt{5}} \cdot \dfrac{2-3\sqrt{5}}{2-3\sqrt{5}}$
$= \dfrac{4 - 2\sqrt{5} - 6\sqrt{5} + 15}{4-45}$
$= \dfrac{19 - 8\sqrt{5}}{-41} = \dfrac{8\sqrt{5}-19}{41}$

63. $\dfrac{5}{\sqrt{2}-1} = \dfrac{5}{\sqrt{2}-1} \cdot \dfrac{\sqrt{2}+1}{\sqrt{2}+1}$
$= \dfrac{5\sqrt{2}+5}{2-1} = 5\sqrt{2} + 5$

65. $\dfrac{5}{\sqrt[3]{2}} = \dfrac{5}{\sqrt[3]{2}} \cdot \dfrac{\sqrt[3]{4}}{\sqrt[3]{4}} = \dfrac{5\sqrt[3]{4}}{2}$

67. $\dfrac{\sqrt{x+h}-\sqrt{x}}{\sqrt{x+h}+\sqrt{x}} = \dfrac{\sqrt{x+h}-\sqrt{x}}{\sqrt{x+h}+\sqrt{x}} \cdot \dfrac{\sqrt{x+h}-\sqrt{x}}{\sqrt{x+h}-\sqrt{x}}$

$\qquad = \dfrac{(x+h)-2\sqrt{x(x+h)}+x}{(x+h)-x}$

$\qquad = \dfrac{x+h-2\sqrt{x^2+xh}+x}{x+h-x}$

$\qquad = \dfrac{2x+h-2\sqrt{x^2+xh}}{h}$

69. $8^{2/3} = \left(\sqrt[3]{8}\right)^2 = 2^2 = 4$

71. $(-27)^{1/3} = \sqrt[3]{-27} = -3$

73. $16^{3/2} = \left(\sqrt{16}\right)^3 = 4^3 = 64$

75. $9^{-3/2} = \dfrac{1}{9^{3/2}} = \dfrac{1}{\left(\sqrt{9}\right)^3} = \dfrac{1}{3^3} = \dfrac{1}{27}$

77. $\left(\dfrac{9}{8}\right)^{3/2} = \sqrt{\dfrac{9}{8}}^{\,3} = \left(\dfrac{3}{2\sqrt{2}}\right)^3 = \dfrac{3^3}{2^3\left(\sqrt{2}\right)^3}$

$\qquad = \dfrac{27}{8 \cdot 2\sqrt{2}} = \dfrac{27}{16\sqrt{2}} = \dfrac{27}{16\sqrt{2}} \cdot \dfrac{\sqrt{2}}{\sqrt{2}} = \dfrac{27\sqrt{2}}{32}$

79. $\left(\dfrac{8}{9}\right)^{-3/2} = \left(\dfrac{9}{8}\right)^{3/2} = \left(\sqrt{\dfrac{9}{8}}\right)^3 = \left(\dfrac{3}{2\sqrt{2}}\right)^3$

$\qquad = \dfrac{3^3}{2^3\left(\sqrt{2}\right)^3} = \dfrac{27}{8 \cdot 2\sqrt{2}} = \dfrac{27}{16\sqrt{2}}$

$\qquad = \dfrac{27}{16\sqrt{2}} \cdot \dfrac{\sqrt{2}}{\sqrt{2}} = \dfrac{27\sqrt{2}}{32}$

81. $(-1000)^{-1/3} = \left(-\dfrac{1}{1000}\right)^{1/3} = \sqrt[3]{-\dfrac{1}{1000}} = -\dfrac{1}{10}$

83. $\left(-\dfrac{64}{125}\right)^{-2/3} = \left(-\dfrac{125}{64}\right)^{2/3} = \sqrt[3]{-\dfrac{125}{64}}^{\,2}$

$\qquad = \left(-\dfrac{5}{4}\right)^2 = \dfrac{25}{16}$

85. $x^{3/4}x^{1/3}x^{-1/2} = x^{3/4+1/3-1/2} = x^{7/12}$

87. $\left(x^3 y^6\right)^{1/3} = \left(x^3\right)^{1/3}\left(y^6\right)^{1/3} = xy^2$

89. $\dfrac{\left(x^2 y\right)^{1/3}\left(xy^2\right)^{2/3}}{x^{2/3}y^{2/3}} = \dfrac{\left(x^2\right)^{1/3}(y)^{1/3}(x)^{2/3}\left(y^2\right)^{2/3}}{x^{2/3}y^{2/3}}$

$\qquad = \dfrac{x^{2/3}y^{1/3}x^{2/3}y^{4/3}}{x^{2/3}y^{2/3}}$

$\qquad = x^{2/3+2/3-2/3}y^{1/3+4/3-2/3}$

$\qquad = x^{2/3}y^1 = x^{2/3}y$

91. $\dfrac{\left(16x^2 y^{-1/3}\right)^{3/4}}{\left(xy^2\right)^{1/4}} = \dfrac{16^{3/4}\left(x^2\right)^{3/4}\left(y^{-1/3}\right)^{3/4}}{x^{1/4}\left(y^2\right)^{1/4}}$

$\qquad = \dfrac{\left(\sqrt[4]{16}\right)^3 x^{3/2}y^{-1/4}}{x^{1/4}y^{1/2}}$

$\qquad = 2^3 x^{3/2-1/4}y^{-1/4-1/2}$

$\qquad = 8x^{5/4}y^{-3/4} = \dfrac{8x^{5/4}}{y^{3/4}}$

93. $\dfrac{x}{(1+x)^{1/2}} + 2(1+x)^{1/2} = \dfrac{x+2(1+x)^{1/2}(1+x)^{1/2}}{(1+x)^{1/2}}$

$\qquad = \dfrac{x+2(1+x)}{(1+x)^{1/2}}$

$\qquad = \dfrac{x+2+2x}{(1+x)^{1/2}} = \dfrac{3x+2}{(1+x)^{1/2}}$

95. $2x\left(x^2+1\right)^{1/2} + x^2 \cdot \dfrac{1}{2}\left(x^2+1\right)^{-1/2} \cdot 2x$

$\qquad = 2x\left(x^2+1\right)^{1/2} + \dfrac{x^3}{\left(x^2+1\right)^{1/2}}$

$\qquad = \dfrac{2x\left(x^2+1\right)^{1/2} \cdot \left(x^2+1\right)^{1/2} + x^3}{\left(x^2+1\right)^{1/2}}$

$\qquad = \dfrac{2x\left(x^2+1\right)^{1/2+1/2} + x^3}{\left(x^2+1\right)^{1/2}} = \dfrac{2x\left(x^2+1\right)^1 + x^3}{\left(x^2+1\right)^{1/2}}$

$\qquad = \dfrac{2x^3+2x+x^3}{\left(x^2+1\right)^{1/2}} = \dfrac{3x^3+2x}{\left(x^2+1\right)^{1/2}}$

$\qquad = \dfrac{x\left(3x^2+2\right)}{\left(x^2+1\right)^{1/2}}$

97. $\sqrt{4x+3}\cdot\dfrac{1}{2\sqrt{x-5}}+\sqrt{x-5}\cdot\dfrac{1}{5\sqrt{4x+3}},\,x>5$

$=\dfrac{\sqrt{4x+3}}{2\sqrt{x-5}}+\dfrac{\sqrt{x-5}}{5\sqrt{4x+3}}$

$=\dfrac{\sqrt{4x+3}\cdot5\cdot\sqrt{4x+3}+\sqrt{x-5}\cdot2\cdot\sqrt{x-5}}{10\sqrt{x-5}\sqrt{4x+3}}$

$=\dfrac{5(4x+3)+2(x-5)}{10\sqrt{(x-5)(4x+3)}}$

$=\dfrac{20x+15+2x-10}{10\sqrt{(x-5)(4x+3)}}$

$=\dfrac{22x+5}{10\sqrt{(x-5)(4x+3)}}$

99. $\dfrac{\left(\sqrt{1+x}-x\cdot\dfrac{1}{2\sqrt{1+x}}\right)}{1+x}=\dfrac{\left(\sqrt{1+x}-\dfrac{x}{2\sqrt{1+x}}\right)}{1+x}$

$=\dfrac{\left(\dfrac{2\sqrt{1+x}\sqrt{1+x}-x}{2\sqrt{1+x}}\right)}{1+x}$

$=\dfrac{2(1+x)-x}{2(1+x)^{1/2}}\cdot\dfrac{1}{1+x}$

$=\dfrac{2+x}{2(1+x)^{3/2}}$

101. $\dfrac{(x+4)^{1/2}-2x(x+4)^{-1/2}}{x+4}$

$=\dfrac{\left((x+4)^{1/2}-\dfrac{2x}{(x+4)^{1/2}}\right)}{x+4}$

$=\dfrac{\left((x+4)^{1/2}\cdot\dfrac{(x+4)^{1/2}}{(x+4)^{1/2}}-\dfrac{2x}{(x+4)^{1/2}}\right)}{x+4}$

$=\dfrac{\left(\dfrac{x+4-2x}{(x+4)^{1/2}}\right)}{x+4}$

$=\dfrac{-x+4}{(x+4)^{1/2}}\cdot\dfrac{1}{x+4}$

$=\dfrac{-x+4}{(x+4)^{3/2}}$

$=\dfrac{4-x}{(x+4)^{3/2}}$

103. $\dfrac{\dfrac{x^2}{\left(x^2-1\right)^{1/2}}-\left(x^2-1\right)^{1/2}}{x^2},\,x<-1\ \text{or}\ x>1$

$=\dfrac{\left(\dfrac{x^2-\left(x^2-1\right)^{1/2}\cdot\left(x^2-1\right)^{1/2}}{\left(x^2-1\right)^{1/2}}\right)}{x^2}$

$=\dfrac{x^2-\left(x^2-1\right)^{1/2}\cdot\left(x^2-1\right)^{1/2}}{\left(x^2-1\right)^{1/2}}\cdot\dfrac{1}{x^2}$

$=\dfrac{x^2-\left(x^2-1\right)}{\left(x^2-1\right)^{1/2}}\cdot\dfrac{1}{x^2}$

$=\dfrac{x^2-x^2+1}{\left(x^2-1\right)^{1/2}}\cdot\dfrac{1}{x^2}$

$=\dfrac{1}{x^2\left(x^2-1\right)^{1/2}}$

105. $\dfrac{\dfrac{1+x^2}{2\sqrt{x}}-2x\sqrt{x}}{\left(1+x^2\right)^2},\,x>0$

$=\dfrac{\left(\dfrac{1+x^2-\left(2\sqrt{x}\right)\left(2x\sqrt{x}\right)}{2\sqrt{x}}\right)}{\left(1+x^2\right)^2}$

$=\dfrac{1+x^2-\left(2\sqrt{x}\right)\left(2x\sqrt{x}\right)}{2\sqrt{x}}\cdot\dfrac{1}{\left(1+x^2\right)^2}$

$=\dfrac{1+x^2-4x^2}{2\sqrt{x}}\cdot\dfrac{1}{\left(1+x^2\right)^2}=\dfrac{1-3x^2}{2\sqrt{x}\left(1+x^2\right)^2}$

107. $(x+1)^{3/2}+x\cdot\dfrac{3}{2}(x+1)^{1/2}$

$=(x+1)^{1/2}\left[x+1+\dfrac{3}{2}x\right]$

$=(x+1)^{1/2}\left[\dfrac{5}{2}x+1\right]=\dfrac{1}{2}(x+1)^{1/2}(5x+2)$

109. $6x^{1/2}\left(x^2+x\right)-8x^{3/2}-8x^{1/2}$

$=2x^{1/2}\left(3(x^2+x)-4x-4\right)$

$=2x^{1/2}\left(3x^2-x-4\right)$

$=2x^{1/2}(3x-4)(x+1)$

111. $3\left(x^2+4\right)^{4/3}+x\cdot4\left(x^2+4\right)^{1/3}\cdot2x$

$=\left(x^2+4\right)^{1/3}\left[3\left(x^2+4\right)+8x^2\right]$

$=\left(x^2+4\right)^{1/3}\left[3x^2+12+8x^2\right]$

$=\left(x^2+4\right)^{1/3}\left(11x^2+12\right)$

113. $4(3x+5)^{1/3}(2x+3)^{3/2}+3(3x+5)^{4/3}(2x+3)^{1/2}$

$=(3x+5)^{1/3}(2x+3)^{1/2}\left[4(2x+3)+3(3x+5)\right]$

$=(3x+5)^{1/3}(2x+3)^{1/2}(8x+12+9x+15)$

$=(3x+5)^{1/3}(2x+3)^{1/2}(17x+27)$

where $x\ge-\dfrac{3}{2}$.

115. $3x^{-1/2}+\dfrac{3}{2}x^{1/2},x>0$

$=\dfrac{3}{x^{1/2}}+\dfrac{3}{2}x^{1/2}=\dfrac{3\cdot2+3x^{1/2}\cdot x^{1/2}}{2x^{1/2}}$

$=\dfrac{6+3x}{2x^{1/2}}=\dfrac{3(x+2)}{2x^{1/2}}$

117. $\sqrt{2}\approx1.41$

```
√(2)
        1.414213562
```

119. $\sqrt[3]{4}\approx1.59$

```
³√(4)
        1.587401052
```

121. $\dfrac{2+\sqrt{3}}{3-\sqrt{5}}\approx4.89$

```
(2+√(3))/(3-√(5))
        4.885317931
```

123. $\dfrac{3\sqrt[3]{5}-\sqrt{2}}{\sqrt{3}}\approx2.15$

```
(3*³√(5)-√(2))/√
(3)
        2.145268638
```

125. a. $V=40(12)^2\sqrt{\dfrac{96}{12}-0.608}$

$\approx15,660.4$ gallons

b. $V=40(1)^2\sqrt{\dfrac{96}{1}-0.608}\approx390.7$ gallons

127. $T=2\pi\sqrt{\dfrac{64}{32}}=2\pi\sqrt{2}\approx8.89$ seconds

129. Answers may vary. One possibility follows: If

$a=-5$, then $\sqrt{a^2}=\sqrt{(-5)^2}=\sqrt{25}=5\ne a$.

Since we use the principal square root, which is always non-negative,

$\sqrt{a^2}=\begin{cases}a & \text{if } a\ge0\\ -a & \text{if } a<0\end{cases}$

which is the definition of $|a|$, so $\sqrt{a^2}=|a|$.

Chapter 1
Graphs, Equations, and Inequalities

Section 1.1

1. 0

3. $\sqrt{3^2 + 4^2} = \sqrt{25} = 5$

5. $\dfrac{1}{2}bh$

7. x-coordinate; y-coordinate

9. midpoint

11. False; points that lie in Quadrant IV will have a positive x-coordinate and a negative y-coordinate. The point $(-1, 4)$ lies in Quadrant II.

13. d

15. (a) Quadrant II
    (b) x-axis
    (c) Quadrant III
    (d) Quadrant I
    (e) y-axis
    (f) Quadrant IV

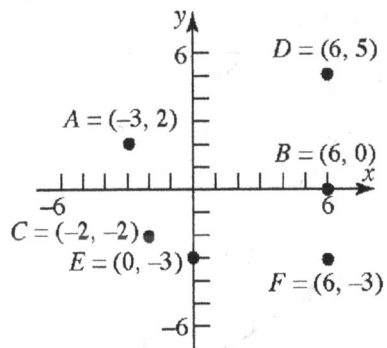

17. The points will be on a vertical line that is two units to the right of the y-axis.

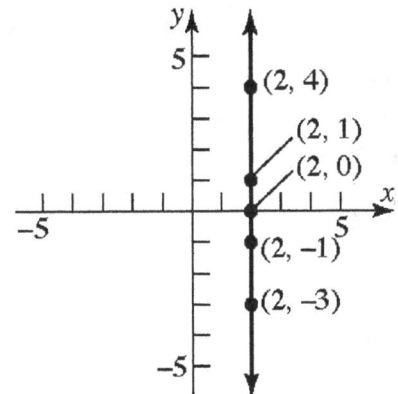

19. $(-1, 4)$; Quadrant II

21. (3, 1); Quadrant I

23. $X\min = -11$
    $X\max = 5$
    $X\,\text{scl} = 1$
    $Y\min = -3$
    $Y\max = 6$
    $Y\,\text{scl} = 1$

25. $X\min = -30$
    $X\max = 50$
    $X\,\text{scl} = 10$
    $Y\min = -90$
    $Y\max = 50$
    $Y\,\text{scl} = 10$

27. $X\min = -10$
    $X\max = 110$
    $X\,\text{scl} = 10$
    $Y\min = -10$
    $Y\max = 160$
    $Y\,\text{scl} = 10$

29. X min $= -6$
X max $= 6$
X scl $= 2$
Y min $= -4$
Y max $= 4$
Y scl $= 2$

31. X min $= -6$
X max $= 6$
X scl $= 2$
Y min $= -1$
Y max $= 3$
Y scl $= 1$

33. X min $= 3$
X max $= 9$
X scl $= 1$
Y min $= 2$
Y max $= 10$
Y scl $= 2$

35. $d(P_1, P_2) = \sqrt{(2-0)^2 + (1-0)^2} = \sqrt{4+1} = \sqrt{5}$

37. $d(P_1, P_2) = \sqrt{(-2-1)^2 + (2-1)^2} = \sqrt{9+1} = \sqrt{10}$

39. $d(P_1, P_2) = \sqrt{(5-3)^2 + (4-(-4))^2} = \sqrt{2^2 + (8)^2}$
$= \sqrt{4+64} = \sqrt{68} = 2\sqrt{17}$

41. $d(P_1, P_2) = \sqrt{(11-(-5))^2 + (9-(-3))^2}$
$= \sqrt{16^2 + 12^2} = \sqrt{256+144}$
$= \sqrt{400} = 20$

43. $d(P_1, P_2) = \sqrt{(6-4)^2 + (4-(-3))^2} = \sqrt{2^2 + 7^2}$
$= \sqrt{4+49} = \sqrt{53}$

45. $d(P_1, P_2) = \sqrt{(0-a)^2 + (0-b)^2} = \sqrt{a^2 + b^2}$

47. $P_1 = (1,3); P_2 = (5,15)$
$d(P_1, P_2) = \sqrt{(5-1)^2 + (15-3)^2}$
$= \sqrt{(4)^2 + (12)^2}$
$= \sqrt{16+144}$
$= \sqrt{160} = 4\sqrt{10}$

49. $P_1 = (-4,6); P_2 = (4,-8)$
$d(P_1, P_2) = \sqrt{(4-(-4))^2 + (-8-6)^2}$
$= \sqrt{(8)^2 + (-14)^2}$
$= \sqrt{64+196}$
$= \sqrt{260} = 2\sqrt{65}$

51. $A = (-2,5), \quad B = (1,3), \quad C = (-1,0)$
$d(A,B) = \sqrt{(1-(-2))^2 + (3-5)^2}$
$= \sqrt{3^2 + (-2)^2} = \sqrt{9+4}$
$= \sqrt{13}$
$d(B,C) = \sqrt{(-1-1)^2 + (0-3)^2}$
$= \sqrt{(-2)^2 + (-3)^2} = \sqrt{4+9}$
$= \sqrt{13}$
$d(A,C) = \sqrt{(-1-(-2))^2 + (0-5)^2}$
$= \sqrt{1^2 + (-5)^2} = \sqrt{1+25}$
$= \sqrt{26}$

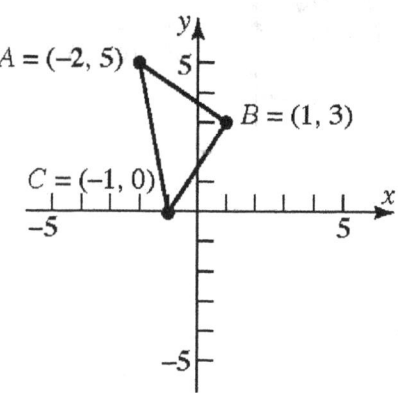

Verifying that $\triangle$ ABC is a right triangle by the Pythagorean Theorem:

$$[d(A,B)]^2 + [d(B,C)]^2 = [d(A,C)]^2$$
$$\left(\sqrt{13}\right)^2 + \left(\sqrt{13}\right)^2 = \left(\sqrt{26}\right)^2$$
$$13 + 13 = 26$$
$$26 = 26$$

The area of a triangle is $A = \frac{1}{2} \cdot bh$. In this problem,

$$A = \frac{1}{2} \cdot [d(A,B)] \cdot [d(B,C)]$$
$$= \frac{1}{2} \cdot \sqrt{13} \cdot \sqrt{13}$$
$$= \frac{1}{2} \cdot 13$$
$$= \frac{13}{2} \text{ square units}$$

53. $A = (-5, 3), \ B = (6, 0), \ C = (5, 5)$

$$d(A,B) = \sqrt{(6-(-5))^2 + (0-3)^2}$$
$$= \sqrt{11^2 + (-3)^2} = \sqrt{121+9}$$
$$= \sqrt{130}$$

$$d(B,C) = \sqrt{(5-6)^2 + (5-0)^2}$$
$$= \sqrt{(-1)^2 + 5^2} = \sqrt{1+25}$$
$$= \sqrt{26}$$

$$d(A,C) = \sqrt{(5-(-5))^2 + (5-3)^2}$$
$$= \sqrt{10^2 + 2^2} = \sqrt{100+4}$$
$$= \sqrt{104}$$
$$= 2\sqrt{26}$$

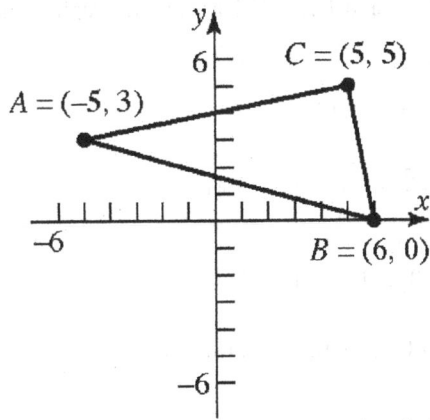

Verifying that $\triangle ABC$ is a right triangle by the Pythagorean Theorem:

$$[d(A,C)]^2 + [d(B,C)]^2 = [d(A,B)]^2$$
$$\left(\sqrt{104}\right)^2 + \left(\sqrt{26}\right)^2 = \left(\sqrt{130}\right)^2$$
$$104 + 26 = 130$$
$$130 = 130$$

The area of a triangle is $A = \frac{1}{2} bh$. In this problem,

$$A = \frac{1}{2} \cdot [d(A,C)] \cdot [d(B,C)]$$
$$= \frac{1}{2} \cdot \sqrt{104} \cdot \sqrt{26}$$
$$= \frac{1}{2} \cdot 2\sqrt{26} \cdot \sqrt{26}$$
$$= \frac{1}{2} \cdot 2 \cdot 26$$
$$= 26 \text{ square units}$$

55. $A = (4, -3), \ B = (0, -3), \ C = (4, 2)$

$$d(A,B) = \sqrt{(0-4)^2 + (-3-(-3))^2}$$
$$= \sqrt{(-4)^2 + 0^2} = \sqrt{16+0}$$
$$= \sqrt{16}$$
$$= 4$$

$$d(B,C) = \sqrt{(4-0)^2 + (2-(-3))^2}$$
$$= \sqrt{4^2 + 5^2} = \sqrt{16+25}$$
$$= \sqrt{41}$$

$$d(A,C) = \sqrt{(4-4)^2 + (2-(-3))^2}$$
$$= \sqrt{0^2 + 5^2} = \sqrt{0+25}$$
$$= \sqrt{25}$$
$$= 5$$

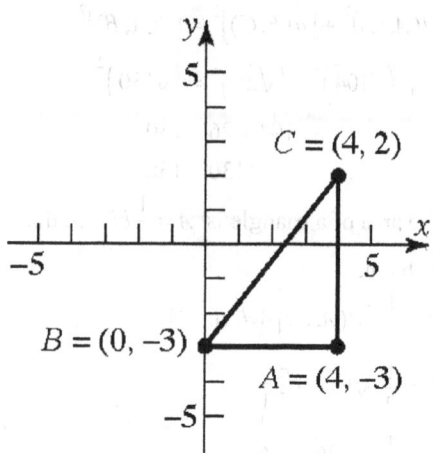

Verifying that $\triangle ABC$ is a right triangle by the Pythagorean Theorem:

$$\left[d(A,B)\right]^2 + \left[d(A,C)\right]^2 = \left[d(B,C)\right]^2$$
$$4^2 + 5^2 = \left(\sqrt{41}\right)^2$$
$$16 + 25 = 41$$
$$41 = 41$$

The area of a triangle is $A = \frac{1}{2}bh$. In this problem,

$$A = \frac{1}{2} \cdot \left[d(A,B)\right] \cdot \left[d(A,C)\right]$$
$$= \frac{1}{2} \cdot 4 \cdot 5$$
$$= 10 \text{ square units}$$

57. The coordinates of the midpoint are:

$$(x, y) = \left(\frac{x_1 + x_2}{2}, \frac{y_1 + y_2}{2}\right)$$
$$= \left(\frac{3+5}{2}, \frac{-4+4}{2}\right)$$
$$= \left(\frac{8}{2}, \frac{0}{2}\right)$$
$$= (4, 0)$$

59. The coordinates of the midpoint are:

$$(x, y) = \left(\frac{x_1 + x_2}{2}, \frac{y_1 + y_2}{2}\right)$$
$$= \left(\frac{-5+11}{2}, \frac{-3+9}{2}\right)$$
$$= \left(\frac{6}{2}, \frac{6}{2}\right)$$
$$= (3, 3)$$

61. The coordinates of the midpoint

$$\text{are } (x, y) = \left(\frac{x_1 + x_2}{2}, \frac{y_1 + y_2}{2}\right)$$
$$= \left(\frac{4+6}{2}, \frac{-3+1}{2}\right)$$
$$= \left(\frac{10}{2}, \frac{-2}{2}\right)$$
$$= (5, -1)$$

63. The coordinates of the midpoint are:

$$(x, y) = \left(\frac{x_1 + x_2}{2}, \frac{y_1 + y_2}{2}\right)$$
$$= \left(\frac{a+0}{2}, \frac{b+0}{2}\right)$$
$$= \left(\frac{a}{2}, \frac{b}{2}\right)$$

65. $y = x^4 - \sqrt{x}$

$$\begin{array}{lll} 0 = 0^4 - \sqrt{0} & 1 = 1^4 - \sqrt{1} & 0 = (-1)^4 - \sqrt{-1} \\ 0 = 0 & 1 \neq 0 & 0 \neq 1 - \sqrt{-1} \end{array}$$

$(0, 0)$ is on the graph of the equation.

67. $y^2 = x^2 + 9$

$$\begin{array}{lll} 3^2 = 0^2 + 9 & 0^2 = 3^2 + 9 & 0^2 = (-3)^2 + 9 \\ 9 = 9 & 0 \neq 18 & 0 \neq 18 \end{array}$$

$(0, 3)$ is on the graph of the equation.

69. $x^2 + y^2 = 4$

$$\begin{array}{lll} 0^2 + 2^2 = 4 & (-2)^2 + 2^2 = 4 & \sqrt{2}^2 + \sqrt{2}^2 = 4 \\ 4 = 4 & 8 \neq 4 & 4 = 4 \end{array}$$

$(0, 2)$ and $\left(\sqrt{2}, \sqrt{2}\right)$ are on the graph of the equation.

71. $(-1, 0), (1, 0)$

73. $\left(-\frac{\pi}{2}, 0\right), \left(\frac{\pi}{2}, 0\right), (0, 1)$

75. $(1, 0), (0, 2), (0, -2)$

77. $(-4, 0), (-1, 0), (4, 0), (0, -3)$

79. $y = x + 2$

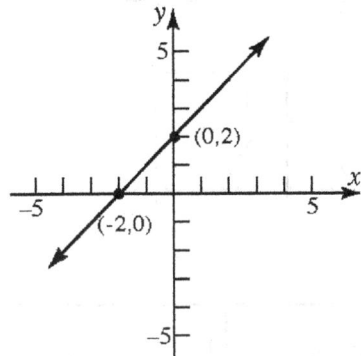

81. $y = 2x + 8$

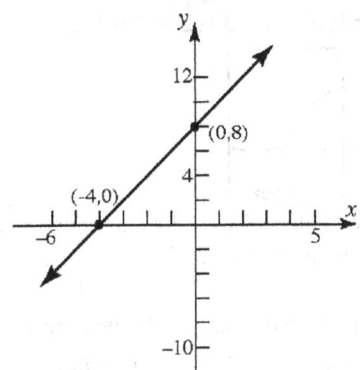

83. $y = x^2 - 1$

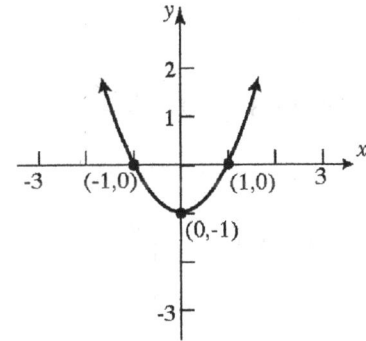

85. $y = -x^2 + 4$

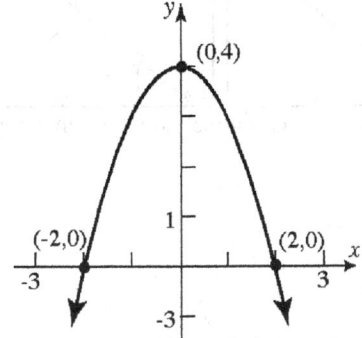

87. $2x + 3y = 6$

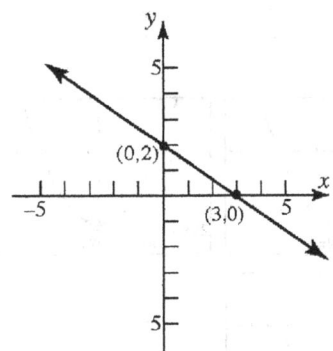

89. $9x^2 + 4y = 36$

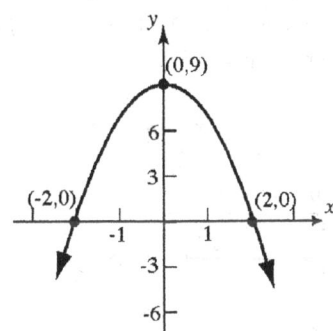

91. $y = 2x - 13$

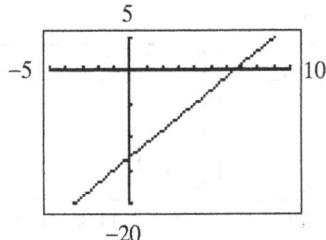

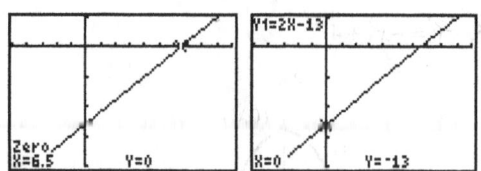

The x-intercept is $x = 6.5$ and the y-intercept is $y = -13$.

93. $y = 2x^2 - 15$

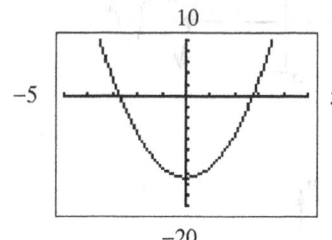

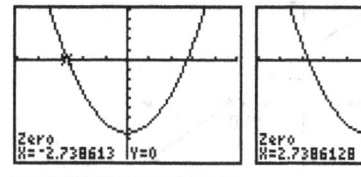

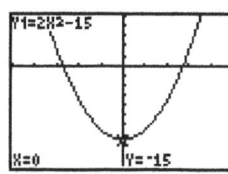

The x-intercepts are $x = -2.74$ and $x = 2.74$.
The y-intercept is $y = -15$.

95. $3x - 2y = 43$ or $y = \dfrac{3x - 43}{2}$

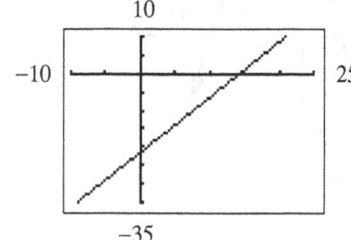

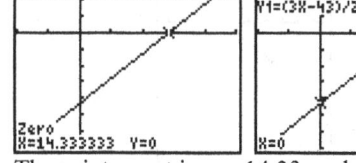

The x-intercept is $x = 14.33$ and the y-intercept is $y = -21.5$.

97. $5x^2 + 3y = 37$ or $y = \dfrac{-5x^2 + 37}{3}$

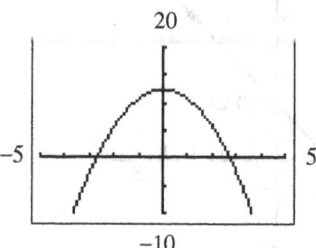

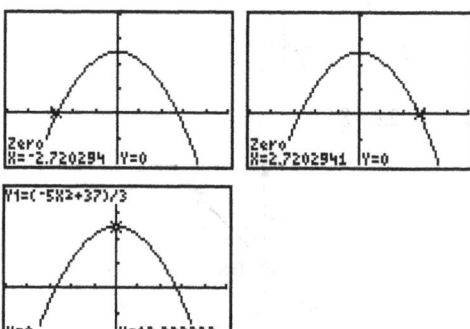

The x-intercepts are $x = -2.72$ and $x = 2.72$.
The y-intercept is $y = 12.33$.

99. If $(2, 5)$ is shifted 3 units right then the x coordinate would be $2 + 3$. If it is shifted 2 units down then the y-coordinate would be $5 + (-2)$. Thus the new point would be
$(2 + 3, 5 + (-2)) = (5, 3)$.

101. The midpoint of AB is: $D = \left(\dfrac{0+6}{2}, \dfrac{0+0}{2} \right)$
$$= (3, 0)$$

The midpoint of AC is: $E = \left(\dfrac{0+4}{2}, \dfrac{0+4}{2} \right)$
$$= (2, 2)$$

The midpoint of BC is: $F = \left(\dfrac{6+4}{2}, \dfrac{0+4}{2} \right)$
$$= (5, 2)$$

$$d(C,D) = \sqrt{(0-4)^2 + (3-4)^2}$$
$$= \sqrt{(-4)^2 + (-1)^2} = \sqrt{16+1}$$
$$= \sqrt{17}$$

$$d(B,E) = \sqrt{(2-6)^2 + (2-0)^2}$$
$$= \sqrt{(-4)^2 + 2^2} = \sqrt{16+4}$$
$$= \sqrt{20} = 2\sqrt{5}$$

$$d(A,F) = \sqrt{(2-0)^2 + (5-0)^2}$$
$$= \sqrt{2^2 + 5^2}$$
$$= \sqrt{4+25}$$
$$= \sqrt{29}$$

103. $d(P_1,P_2) = \sqrt{(-4-2)^2 + (1-1)^2}$
$$= \sqrt{(-6)^2 + 0^2}$$
$$= \sqrt{36}$$
$$= 6$$

$$d(P_2,P_3) = \sqrt{(-4-(-4))^2 + (-3-1)^2}$$
$$= \sqrt{0^2 + (-4)^2}$$
$$= \sqrt{16}$$
$$= 4$$

$$d(P_1,P_3) = \sqrt{(-4-2)^2 + (-3-1)^2}$$
$$= \sqrt{(-6)^2 + (-4)^2}$$
$$= \sqrt{36+16}$$
$$= \sqrt{52}$$
$$= 2\sqrt{13}$$

Since $[d(P_1,P_2)]^2 + [d(P_2,P_3)]^2 = [d(P_1,P_3)]^2$,
the triangle is a right triangle.

105. $d(P_1,P_2) = \sqrt{(0-(-2))^2 + (7-(-1))^2}$
$$= \sqrt{2^2 + 8^2} = \sqrt{4+64} = \sqrt{68}$$
$$= 2\sqrt{17}$$

$$d(P_2,P_3) = \sqrt{(3-0)^2 + (2-7)^2}$$
$$= \sqrt{3^2 + (-5)^2} = \sqrt{9+25}$$
$$= \sqrt{34}$$

$$d(P_1,P_3) = \sqrt{(3-(-2))^2 + (2-(-1))^2}$$
$$= \sqrt{5^2 + 3^2} = \sqrt{25+9}$$
$$= \sqrt{34}$$

Since $d(P_2,P_3) = d(P_1,P_3)$, the triangle is isosceles.

Since $[d(P_1,P_3)]^2 + [d(P_2,P_3)]^2 = [d(P_1,P_2)]^2$, the triangle is also a right triangle.

Therefore, the triangle is an isosceles right triangle.

107. Let the coordinates of point B be (x,y). Using the midpoint formula, we can write

$$(2,3) = \left(\frac{-1+x}{2}, \frac{8+y}{2}\right).$$

This leads to two equations we can solve.

$$\frac{-1+x}{2} = 2 \qquad \frac{8+y}{2} = 3$$
$$-1+x = 4 \qquad 8+y = 6$$
$$x = 5 \qquad y = -2$$

Point B has coordinates $(5,-2)$.

109. Using the Pythagorean Theorem:

$$90^2 + 90^2 = d^2$$
$$8100 + 8100 = d^2$$
$$16200 = d^2$$
$$d = \sqrt{16200} = 90\sqrt{2} \approx 127.28 \text{ feet}$$

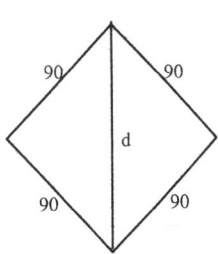

111. a. First: (90, 0), Second: (90, 90)
Third: (0, 90)

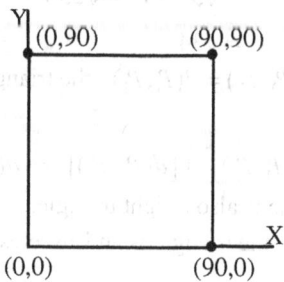

b. Using the distance formula:
$$d = \sqrt{(310-90)^2 + (15-90)^2}$$
$$= \sqrt{220^2 + (-75)^2} = \sqrt{54025}$$
$$= 5\sqrt{2161} \approx 232.43 \text{ feet}$$

c. Using the distance formula:
$$d = \sqrt{(300-0)^2 + (300-90)^2}$$
$$= \sqrt{300^2 + 210^2} = \sqrt{134100}$$
$$= 30\sqrt{149} \approx 366.20 \text{ feet}$$

113. The car heading east moves a distance $30t$ after t hours. The truck heading south moves a distance $40t$ after t hours. Their distance apart after t hours is:

$$d = \sqrt{(30t)^2 + (40t)^2}$$
$$= \sqrt{900t^2 + 1600t^2}$$
$$= \sqrt{2500t^2}$$
$$= 50t$$

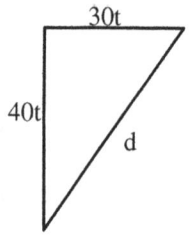

115. a. The shortest side is between $P_1 = (2.6, 1.5)$ and $P_2 = (2.7, 1.7)$. The estimate for the desired intersection point is:
$$\left(\frac{x_1+x_2}{2}, \frac{y_1+y_2}{2}\right) = \left(\frac{2.6+2.7}{2}, \frac{1.5+1.7}{2}\right)$$
$$= \left(\frac{5.3}{2}, \frac{3.2}{2}\right)$$
$$= (2.65, 1.6)$$

b. Using the distance formula:
$$d = \sqrt{(2.65-1.4)^2 + (1.6-1.3)^2}$$
$$= \sqrt{(1.25)^2 + (0.3)^2}$$
$$= \sqrt{1.5625 + 0.09}$$
$$= \sqrt{1.6525}$$
$$\approx 1.285 \text{ units}$$

117. For 2004 we have the ordered pair $(2004, 19157)$ and for 2014 we have the ordered pair $(2014, 24008)$. The midpoint is
$$(\text{year, }\$) = \left(\frac{2004+2014}{2}, \frac{19157+24008}{2}\right)$$
$$= \left(\frac{4018}{2}, \frac{43465}{2}\right)$$
$$= (2009, 21582.5)$$

Using the midpoint, we estimate the poverty level in 2009 to be $21,582.50. This is slightly lower than the actual value.

119. Answers will vary

121. Answers will vary. A complete graph presents enough of the graph to the viewer so they can "see" the rest of the graph as an obvious continuation of what is shown.

Section 1.2

1. Using the distributive property:
   $-3(x-5) = -3x + 15$

3. No, $x = 4$ is not in the domain of $\dfrac{3}{x-4}$ because it makes the denominator equal 0.

5. b

7. linear; first-degree

9. True; equations that are contradictions have no solution.

11. Divide both sides by 7. The solution set is $\{3\}$.

13. Subtract 15 from both sides, then divide both sides by 3. The solution set is $\{-5\}$.

15. Add 3 to both sides, then divide both sides by 2. The solution set is $\left\{\dfrac{3}{2}\right\}$.

17. Multiply both sides by 3. The solution set is $\left\{\dfrac{5}{4}\right\}$.

19. $x^3 - 4x + 2 = 0$; Use ZERO (or ROOT) on the graph of $y_1 = x^3 - 4x + 2$.

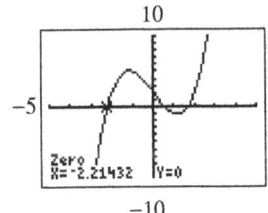 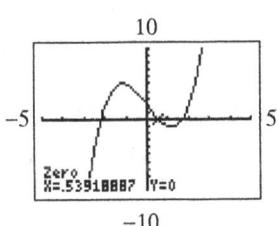

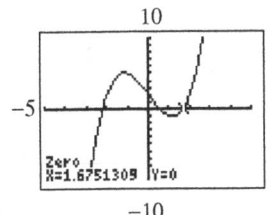

The solution set is $\{-2.21, 0.54, 1.68\}$.

21. $-2x^4 + 5 = 3x - 2$; Use INTERSECT on the graphs of $y_1 = -2x^4 + 5$ and $y_2 = 3x - 2$.

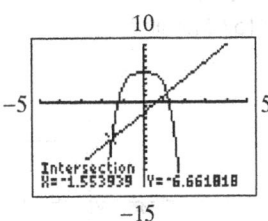 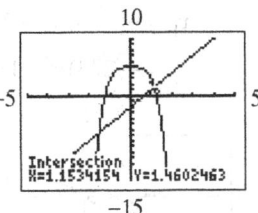

The solution set is $\{-1.55, 1.15\}$.

23. $x^4 - 2x^3 + 3x - 1 = 0$; Use ZERO (or ROOT) on the graph of $y_1 = x^4 - 2x^3 + 3x - 1$.

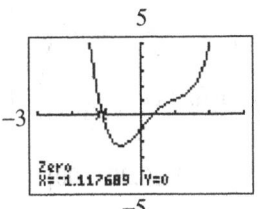 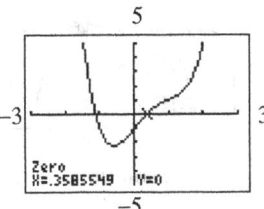

The solution set is $\{-1.12, 0.36\}$.

25. $-x^3 - \dfrac{5}{3}x^2 + \dfrac{7}{2}x + 2 = 0$;

    Use ZERO (or ROOT) on the graph of
    $y_1 = -x^3 - (5/3)x^2 + (7/2)x + 2$.

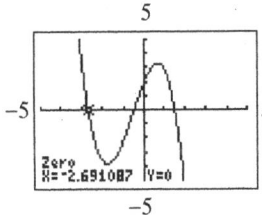 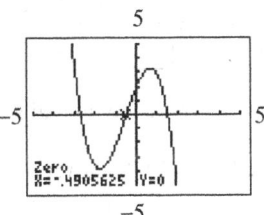

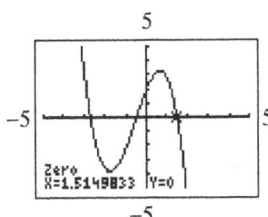

The solution set is $\{-2.69, -0.49, 1.51\}$.

27. $-\dfrac{2}{3}x^4 - 2x^3 + \dfrac{5}{2}x = -\dfrac{2}{3}x^2 + \dfrac{1}{2}$

Use INTERSECT on the graphs of

$y_1 = -(2/3)x^4 - 2x^3 + (5/2)x$ and

$y_2 = -(2/3)x^2 + 1/2$.

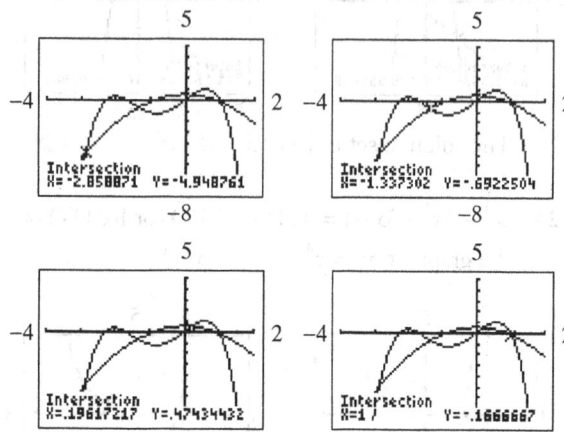

The solution set is $\{-2.86, -1.34, 0.20, 1.00\}$.

29. $x^4 - 5x^2 + 2x + 11 = 0$

Use ZERO (or ROOT) on the graph of

$y_1 = x^4 - 5x^2 + 2x + 11$.

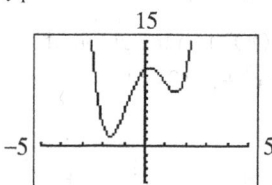

There are no real solutions.

31.
$$3x + 4 = x$$
$$3x + 4 - 4 = x - 4$$
$$3x = x - 4$$
$$3x - x = x - 4 - x$$
$$2x = -4$$
$$\frac{2x}{2} = \frac{-4}{2}$$
$$x = -2$$

The solution set is $\{-2\}$.

33.
$$2t - 6 = 3 - t$$
$$2t - 6 + 6 = 3 - t + 6$$
$$2t = 9 - t$$
$$2t + t = 9 - t + t$$
$$3t = 9$$
$$\frac{3t}{3} = \frac{9}{3}$$
$$t = 3$$

The solution set is $\{3\}$.

35.
$$6 - x = 2x + 9$$
$$6 - x - 6 = 2x + 9 - 6$$
$$-x = 2x + 3$$
$$-x - 2x = 2x + 3 - 2x$$
$$-3x = 3$$
$$\frac{-3x}{-3} = \frac{3}{-3}$$
$$x = -1$$

The solution set is $\{-1\}$.

37.
$$2(3 + 2x) = 3(x - 4)$$
$$6 + 4x = 3x - 12$$
$$6 + 4x - 6 = 3x - 12 - 6$$
$$4x = 3x - 18$$
$$4x - 3x = 3x - 18 - 3x$$
$$x = -18$$

The solution set is $\{-18\}$.

39.
$$8x - (3x + 2) = 3x - 10$$
$$8x - 3x - 2 = 3x - 10$$
$$5x - 2 = 3x - 10$$
$$5x - 2 + 2 = 3x - 10 + 2$$
$$5x = 3x - 8$$
$$5x - 3x = 3x - 8 - 3x$$
$$2x = -8$$
$$\frac{2x}{2} = \frac{-8}{2}$$
$$x = -4$$

The solution set is $\{-4\}$.

41. $2(3x-5)+6(x-3)=-3(4-5x)+5x-6$

$6x-10+6x-18=-12+15x+5x-6$

$12x-28=20x-18$

$-8x-28=-18$

$-8x=10$

$x=-\dfrac{5}{4}$

The solution set is $\left\{-\dfrac{5}{4}\right\}$.

43. $\dfrac{3}{2}x+2=\dfrac{1}{2}-\dfrac{1}{2}x$

$2\left(\dfrac{3}{2}x+2\right)=2\left(\dfrac{1}{2}-\dfrac{1}{2}x\right)$

$3x+4=1-x$

$3x+4-4=1-x-4$

$3x=-x-3$

$3x+x=-x-3+x$

$4x=-3$

$\dfrac{4x}{4}=\dfrac{-3}{4}$

$x=-\dfrac{3}{4}$

The solution set is $\left\{-\dfrac{3}{4}\right\}$.

45. $\dfrac{2}{3}p=\dfrac{1}{2}p-\dfrac{1}{3}$

$6\left(\dfrac{2}{3}p\right)=6\left(\dfrac{1}{2}p-\dfrac{1}{3}\right)$

$4p=3p-2$

$4p-3p=3p-2-3p$

$p=-2$

The solution set is $\{-2\}$.

47. $0.9t=0.4+0.1t$

$0.9t-0.1t=0.4+0.1t-0.1t$

$0.8t=0.4$

$\dfrac{0.8t}{0.8}=\dfrac{0.4}{0.8}$

$t=0.5$

The solution set is $\{0.5\}$.

49. $\dfrac{x+1}{3}+\dfrac{x+2}{7}=2$

$21\left(\dfrac{x+1}{3}\right)+21\left(\dfrac{x+2}{7}\right)=21(2)$

$7x+7+3x+6=42$

$10x+13=42$

$10x+13-13=42-13$

$10x=29$

$\dfrac{10x}{10}=\dfrac{29}{10}$

$x=\dfrac{29}{10}$

The solution set is $\left\{\dfrac{29}{10}\right\}$.

51. $\dfrac{2}{y}+\dfrac{4}{y}=3$

$\dfrac{6}{y}=3$

$\left(\dfrac{6}{y}\right)^{-1}=3^{-1}$

$\dfrac{y}{6}=\dfrac{1}{3}$

$6\cdot\dfrac{y}{6}=6\cdot\dfrac{1}{3}$

$y=2$

The solution set is $\{2\}$.

53. $\dfrac{1}{2}+\dfrac{2}{x}=\dfrac{3}{4}$

$4x\left(\dfrac{1}{2}+\dfrac{2}{x}\right)=4x\left(\dfrac{3}{4}\right)$

$2x+8=3x$

$2x+8-2x=3x-2x$

$8=x$ or $x=8$

The solution set is $\{8\}$.

35

55. $(x+7)(x-1)=(x+1)^2$

$x^2-x+7x-7=x^2+2x+1$

$x^2+6x-7=x^2+2x+1$

$x^2+6x-7-x^2=x^2+2x+1-x^2$

$6x-7=2x+1$

$6x-7-2x=2x+1-2x$

$4x-7=1$

$4x-7+7=1+7$

$4x=8$

$\dfrac{4x}{4}=\dfrac{8}{4}\Rightarrow x=2$

The solution set is $\{2\}$.

57. $x(2x-3)=(2x+1)(x-4)$

$2x^2-3x=2x^2-7x-4$

$2x^2-3x-2x^2=2x^2-7x-4-2x^2$

$-3x=-7x-4$

$-3x+7x=-7x-4+7x$

$4x=-4$

$\dfrac{4x}{4}=\dfrac{-4}{4}\Rightarrow x=-1$

The solution set is $\{-1\}$.

59. $z\left(z^2+1\right)=3+z^3$

$z^3+z=3+z^3$

$z^3+z-z^3=3+z^3-z^3$

$z=3$

The solution set is $\{3\}$.

61. $\dfrac{x}{x-2}+3=\dfrac{2}{x-2}$

$(x-2)\left(\dfrac{x}{x-2}+3\right)=(x-2)\left(\dfrac{2}{x-2}\right)$

$x+3(x-2)=2$

$x+3x-6=2$

$4x-6=2$

$4x=8$

$x=2$

This solution is not in the domain of the variable so it must be discarded. The equation has no solution. The solution set is $\{\ \}$ or $\varnothing$.

63. $\dfrac{2x}{x^2-4}=\dfrac{4}{x^2-4}-\dfrac{3}{x+2}$

$\left(x^2-4\right)\left(\dfrac{2x}{x^2-4}\right)=\left(x^2-4\right)\left(\dfrac{4}{x^2-4}-\dfrac{3}{x+2}\right)$

$2x=4-3(x-2)$

$2x=4-3x+6$

$2x=10-3x$

$5x=10$

$x=2$

This solution is not in the domain of the variable so it must be discarded. The equation has no solution. The solution set is $\{\ \}$ or $\varnothing$.

65. $\dfrac{x}{x+2}=\dfrac{3}{2}$

$2x=3(x+2)$

$2x=3x+6$

$-x=6$

$x=-6$

The solution set is $\{-6\}$.

67. $\dfrac{5}{2x-3}=\dfrac{3}{x+5}$

$5(x+5)=3(2x-3)$

$5x+25=6x-9$

$5x=6x-34$

$-x=-34$

$x=34$

The solution set is $\{34\}$.

69. $\dfrac{6t+7}{4t-1}=\dfrac{3t+8}{2t-4}$

$(6t+7)(2t-4)=(3t+8)(4t-1)$

$12t^2+14t-24t-28=12t^2+32t-3t-8$

$12t^2-10t-28=12t^2+29t-8$

$-10t-28=29t-8$

$-10t=29t+20$

$-39t=20$

$t=\dfrac{20}{-39}$

$t=-\dfrac{20}{39}$

The solution set is $\left\{-\dfrac{20}{39}\right\}$.

71. $\dfrac{4}{x-2} = \dfrac{-3}{x+5} + \dfrac{7}{(x+5)(x-2)}$

LCD $= (x-2)(x+5)$

$\dfrac{4(x+5)}{(x-2)(x+5)} = \dfrac{-3(x-2)}{(x-2)(x+5)} + \dfrac{7}{(x+5)(x-2)}$

$4(x+5) = -3(x-2)+7$

$4x+20 = -3x+6+7$

$4x+20 = -3x+13$

$4x = -3x-7$

$7x = -7$

$x = -1$

The solution set is $\{-1\}$.

73. $\dfrac{2}{y+3} + \dfrac{3}{y-4} = \dfrac{5}{y+6}$

LCD $= (y+3)(y-4)(y+6)$

$(y+3)(y-4)(y+6)\left(\dfrac{2}{y+3} + \dfrac{3}{y-4} = \dfrac{5}{y+6}\right)$

$2(y-4)(y+6) + 3(y+3)(y+6) = 5(y+3)(y-4)$

$2(y^2+2y-24) + 3(y^2+9y+18) = 5(y^2-y-12)$

$2y^2+4y-48+3y^2+27y+54 = 5y^2-5y-60$

$5y^2+31y+6 = 5y^2-5y-60$

$31y+6 = -5y-60$

$31y = -5y-66$

$36y = -66$

$y = -\dfrac{66}{36}$

$y = -\dfrac{11}{6}$

The solution set is $\left\{-\dfrac{11}{6}\right\}$.

75. $\dfrac{x}{x^2-1} - \dfrac{x+3}{x^2-x} = \dfrac{-3}{x^2+x}$

LCD $= x(x+1)(x-1)$

$x(x+1)(x-1)\left(\dfrac{x}{x^2-1} - \dfrac{x+3}{x^2-x} = \dfrac{-3}{x^2+x}\right)$

$x^2 - (x+1)(x+3) = -3(x-1)$

$x^2 - (x^2+4x+3) = -3x+3$

$x^2 - x^2 - 4x - 3 = -3x+3$

$-4x-3 = -3x+3$

$-4x = -3x+6$

$-x = 6$

$x = -6$

The solution set is $\{-6\}$.

77. $y = 5x+4$

$2 = 5a+4$

$-2 = 5a$

$a = -\dfrac{2}{5}$

79. $2x+3y = 6$

$2a+3b = 6$

$3b = -2a+6$

$b = -\dfrac{2}{3}a+2$

81. $ax - b = c, \quad a \neq 0$

$ax - b + b = c + b$

$ax = c+b$

$\dfrac{ax}{a} = \dfrac{c+b}{a}$

$x = \dfrac{b+c}{a}$

83. $\dfrac{x}{a} + \dfrac{x}{b} = c, \quad a \neq 0, b \neq 0, a \neq -b$

$ab\left(\dfrac{x}{a} + \dfrac{x}{b}\right) = ab \cdot c$

$bx + ax = abc$

$x(a+b) = abc$

$\dfrac{x(a+b)}{a+b} = \dfrac{abc}{a+b}$

$x = \dfrac{abc}{a+b}$

37

85. $\dfrac{1}{x-a}+\dfrac{1}{x+a}=\dfrac{2}{x-1}$

Multiply both sides by the LCD,
$(x-a)(x+a)(x-1)$, to get.

$$(x+a)(x-1)+(x-a)(x-1)=2(x-a)(x+a)$$

$$x^2+ax-x-a+x^2-ax-x+a=2x^2-2a^2$$

$$2x^2-2x=2x^2-2a^2$$

$$2x^2-2x-2x^2=2x^2-2a^2-2x^2$$

$$-2x=-2a^2$$

$$\dfrac{-2x}{-2}=\dfrac{-2a^2}{-2}$$

$$x=a^2$$

where $a\neq -1,0,1$.

87. $x+2a=16+ax-6a$

when $x=4$:

$$4+2a=16+a(4)-6a$$

$$4+2a=16+4a-6a$$

$$4+2a=16-2a$$

$$4a=12$$

$$a=3$$

89. Solving for R:

$$\dfrac{1}{R}=\dfrac{1}{R_1}+\dfrac{1}{R_2}$$

$$RR_1R_2\left(\dfrac{1}{R}\right)=RR_1R_2\left(\dfrac{1}{R_1}+\dfrac{1}{R_2}\right)$$

$$R_1R_2=RR_2+RR_1$$

$$R_1R_2=R(R_2+R_1)$$

$$\dfrac{R_1R_2}{R_2+R_1}=\dfrac{R(R_2+R_1)}{R_2+R_1}$$

$$\dfrac{R_1R_2}{R_1+R_2}=R$$

91. Solving for R:

$$F=\dfrac{mv^2}{R}$$

$$RF=R\left(\dfrac{mv^2}{R}\right)$$

$$RF=mv^2$$

$$\dfrac{RF}{F}=\dfrac{mv^2}{F}\Rightarrow R=\dfrac{mv^2}{F}$$

93. Solving for r:

$$S=\dfrac{a}{1-r}$$

$$(1-r)\cdot S=(1-r)\cdot\dfrac{a}{1-r}$$

$$(1-r)S=a$$

$$\dfrac{(1-r)S}{S}=\dfrac{a}{S}$$

$$1-r=\dfrac{a}{S}$$

$$1-r-1=\dfrac{a}{S}-1$$

$$-r=\dfrac{a}{S}-1$$

$$r=1-\dfrac{a}{S}\quad\text{or}\quad r=\dfrac{S-a}{S}$$

95. Let x = amount invested in bonds.

Invested in CDs: $x-3000$

$$x+(x-3,000)=20,000$$

$$2x-3,000=20,000$$

$$2x=23,000$$

$$x=11,500$$

$11,500 will be invested in bonds and $8,500 will be invested in CD's.

97. Let x = amount that Scott gets.

Amount for Alice: $\dfrac{3}{4}x$

Amount for Tricia: $\dfrac{1}{2}x$

$$x+\left(\dfrac{3}{4}x\right)+\left(\dfrac{1}{2}x\right)=900,000$$

$$\dfrac{9}{4}x=900,000$$

$$x=\dfrac{4}{9}(900,000)$$

$$x=400,000$$

Scott receives $400,000; Alice receives $300,000; Tricia receives $200,000.

99. Let x Sandra's regular hourly wage.
$$40x + (1.5x)(8) = 598$$
$$40x + 12x = 598$$
$$52x = 598$$
$$x = \frac{598}{52} = 11.50$$
Sandra's regular hourly wage is $11.50.

101. Let x = final exam score.
Compute the final average and set equal to 80.
$$\left(\frac{1}{7}\right)(80 + 83 + 71 + 61 + 95 + x + x) = 80$$

Now solve for x:
$$\left(\frac{1}{7}\right)(390 + 2x) = 80$$
$$390 + 2x = 560$$
$$2x = 170$$
$$x = 85$$
Brooke needs to score an 85 on the final exam to get an average of 80 in the course.

103. Let x represent the original price of the house. Then $0.15x$ represents the reduction in the price of the house.
original price – reduction = new price
$$x - 0.15x = 170,000$$
$$0.85x = 170,000$$
$$x = 200,000$$
The original price of the house was $200,000. The amount of the savings is $0.15(\$200,000) = \$30,000$.

105. Let x represent the price the theater pays for the candy.
Then $2.75x$ represents the markup on the candy. The selling price of the candy is $3.00.
suppier price + markup = selling price
$$x + 2.75x = 3.00$$
$$3.75x = 3.00$$
$$x = 0.80$$
The theater paid $0.80 for the candy.

107. Let a = number of adult patrons.
Child patrons: $5200 - a$
$$8.50a + 6.00(5200 - a) = 32,200$$
$$8.50a + 31,200 - 6.00a = 32,200$$
$$2.50a + 31,200 = 32,200$$
$$2.50a = 1,000$$
$$\frac{2.50a}{2.50} = \frac{1,000}{2.50}$$
$$a = 400$$
The theater had 400 adult patrons.

109. Let w = width.
Length: $l = w + 8$
Perimeter: $P = 2l + 2w$
$$2l + 2w = 60$$
$$2(w + 8) + 2w = 60$$
$$2w + 16 + 2w = 60$$
$$4w + 16 = 60$$
$$4w = 44$$
$$w = 11$$
The rectangle has a width of 11 feet and a length of 19 feet.

111. Let x = the total number of people who owned a smart phone in January, 2015. Then $0.532x$ represents the number of people who owned a smart phone that ran the Google Android OS.
$$0.532x = 97.89$$
$$x = \frac{97.89}{0.532} \approx 184$$
There were about 184 million people who owned a smart phone in January, 2015.

113. Step 7 is only allowed if $x \neq 2$, otherwise we are dividing by 0. But step 1 states that $x = 2$, so we have a contradiction.

115. Answers will vary. One example is $3x + 1 = 3x + 6$.

Section 1.3

1. $x^2 - 5x - 6 = (x - 6)(x + 1)$

3. $\left\{-\frac{5}{3}, 3\right\}$

39

5. $\dfrac{1}{2} \cdot 5 = \dfrac{5}{2}; \left(\dfrac{5}{2}\right)^2 = \dfrac{25}{4}; x^2 + 5x + \dfrac{25}{4}$

$x^2 + 5x + \dfrac{25}{4} = \left(x + \dfrac{5}{2}\right)^2$

7. discriminant; negative

9. b

11. $x^2 - 9x = 0$

$x(x - 9) = 0$

$x = 0$ or $x - 9 = 0$

$x = 0$ or $\quad x = 9$

The solution set is $\{0, 9\}$.

13. $x^2 - 25 = 0$

$(x + 5)(x - 5) = 0$

$x + 5 = 0 \quad$ or $\quad x - 5 = 0$

$x = -5$ or $\quad x = 5$

The solution set is $\{-5, 5\}$.

15. $z^2 + z - 6 = 0$

$(z + 3)(z - 2) = 0$

$z + 3 = 0 \quad$ or $\quad z - 2 = 0$

$z = -3$ or $\quad z = 2$

The solution set is $\{-3, 2\}$.

17. $2x^2 - 5x - 3 = 0$

$(2x + 1)(x - 3) = 0$

$2x + 1 = 0 \quad$ or $x - 3 = 0$

$x = -\dfrac{1}{2}$ or $\quad x = 3$

The solution set is $\left\{-\dfrac{1}{2}, 3\right\}$

19. $3t^2 - 48 = 0$

$3(t^2 - 16) = 0$

$3(t + 4)(t - 4) = 0$

$t + 4 = 0 \quad$ or $\quad t - 4 = 0$

$t = -4$ or $\quad t = 4$

The solution set is $\{-4, 4\}$.

21. $x(x - 8) + 12 = 0$

$x^2 - 8x + 12 = 0$

$(x - 6)(x - 2) = 0$

$x - 6 = 0 \quad$ or $\quad x - 2 = 0$

$x = 6 \quad$ or $\quad x = 2$

The solution set is $\{2, 6\}$.

23. $4x^2 + 9 = 12x$

$4x^2 - 12x + 9 = 0$

$(2x - 3)^2 = 0$

$2x - 3 = 0$

$x = \dfrac{3}{2}$

The solution set is $\left\{\dfrac{3}{2}\right\}$.

25. $6(p^2 - 1) = 5p$

$6p^2 - 6 = 5p$

$6p^2 - 5p - 6 = 0$

$(3p + 2)(2p - 3) = 0$

$3p + 2 = 0 \quad$ or $\quad 2p - 3 = 0$

$p = -\dfrac{2}{3}$ or $\quad p = \dfrac{3}{2}$

The solution set is $\left\{-\dfrac{2}{3}, \dfrac{3}{2}\right\}$.

27. $6x - 5 = \dfrac{6}{x}$

$(6x - 5)x = \left(\dfrac{6}{x}\right)x$

$6x^2 - 5x = 6$

$6x^2 - 5x - 6 = 0$

$(3x + 2)(2x - 3) = 0$

$3x + 2 = 0 \quad$ or $\quad 2x - 3 = 0$

$x = -\dfrac{2}{3}$ or $\quad x = \dfrac{3}{2}$

Neither of these values causes a denominator to equal zero, so the solution set is $\left\{-\dfrac{2}{3}, \dfrac{3}{2}\right\}$.

29.
$$\frac{4(x-2)}{x-3}+\frac{3}{x}=\frac{-3}{x(x-3)}$$
$$\left(\frac{4(x-2)}{x-3}+\frac{3}{x}\right)x(x-3)=\left(\frac{-3}{x(x-3)}\right)x(x-3)$$
$$4x(x-2)+3(x-3)=-3$$
$$4x^2-8x+3x-9=-3$$
$$4x^2-5x-6=0$$
$$(4x+3)(x-2)=0$$
$$4x+3=0 \quad \text{or} \quad x-2=0$$
$$x=-\frac{3}{4} \quad \text{or} \quad x=2$$

Neither of these values causes a denominator to equal zero, so the solution set is $\left\{-\frac{3}{4},2\right\}$.

31.
$$x^2=25$$
$$x=\pm\sqrt{25}$$
$$x=\pm5$$
The solution set is $\{-5,5\}$.

33. $(x-1)^2=4$
$$x-1=\pm\sqrt{4}$$
$$x-1=\pm2$$
$$x-1=2 \quad \text{or} \quad x-1=-2$$
$$x=3 \quad \text{or} \quad x=-1$$
The solution set is $\{-1,3\}$.

35. $(2x+3)^2=9$
$$2x+3=\pm\sqrt{9}$$
$$2x+3=\pm3$$
$$2x+3=3 \quad \text{or} \quad 2x+3=-3$$
$$2x=0 \quad \text{or} \quad 2x=-6$$
$$x=0 \quad \text{or} \quad x=-3$$
The solution set is $\{-3,0\}$.

37.
$$x^2+4x=21$$
$$x^2+4x+4=21+4$$
$$(x+2)^2=25$$
$$x+2=\pm\sqrt{25}$$
$$x+2=\pm5$$
$$x=-2\pm5$$
$$x=3 \quad \text{or} \quad x=-7$$
The solution set is $\{-7,3\}$.

39. $x^2-\frac{1}{2}x-\frac{3}{16}=0$
$$x^2-\frac{1}{2}x=\frac{3}{16}$$
$$x^2-\frac{1}{2}x+\frac{1}{16}=\frac{3}{16}+\frac{1}{16}$$
$$\left(x-\frac{1}{4}\right)^2=\frac{1}{4}$$
$$x-\frac{1}{4}=\pm\sqrt{\frac{1}{4}}=\pm\frac{1}{2}$$
$$x=\frac{1}{4}\pm\frac{1}{2}$$
$$x=\frac{3}{4} \quad \text{or} \quad x=-\frac{1}{4}$$
The solution set is $\left\{-\frac{1}{4},\frac{3}{4}\right\}$.

41. $3x^2+x-\frac{1}{2}=0$
$$x^2+\frac{1}{3}x-\frac{1}{6}=0$$
$$x^2+\frac{1}{3}x=\frac{1}{6}$$
$$x^2+\frac{1}{3}x+\frac{1}{36}=\frac{1}{6}+\frac{1}{36}$$
$$\left(x+\frac{1}{6}\right)^2=\frac{7}{36}$$
$$x+\frac{1}{6}=\pm\sqrt{\frac{7}{36}}$$
$$x+\frac{1}{6}=\pm\frac{\sqrt{7}}{6}$$
$$x=\frac{-1\pm\sqrt{7}}{6}$$
The solution set is $\left\{\frac{-1-\sqrt{7}}{6},\frac{-1+\sqrt{7}}{6}\right\}$.

43. $x^2 - 4x + 2 = 0$

$a = 1, \quad b = -4, \quad c = 2$

$x = \dfrac{-(-4) \pm \sqrt{(-4)^2 - 4(1)(2)}}{2(1)} = \dfrac{4 \pm \sqrt{16-8}}{2}$

$= \dfrac{4 \pm \sqrt{8}}{2} = \dfrac{4 \pm 2\sqrt{2}}{2} = 2 \pm \sqrt{2}$

The solution set is $\left\{2 - \sqrt{2},\, 2 + \sqrt{2}\right\}$.

45. $x^2 - 4x - 1 = 0$

$a = 1, \quad b = -4, \quad c = -1$

$x = \dfrac{-(-4) \pm \sqrt{(-4)^2 - 4(1)(-1)}}{2(1)} = \dfrac{4 \pm \sqrt{16+4}}{2}$

$= \dfrac{4 \pm \sqrt{20}}{2} = \dfrac{4 \pm 2\sqrt{5}}{2} = 2 \pm \sqrt{5}$

The solution set is $\left\{2 - \sqrt{5},\, 2 + \sqrt{5}\right\}$.

47. $2x^2 - 5x + 3 = 0$

$a = 2, \quad b = -5, \quad c = 3$

$x = \dfrac{-(-5) \pm \sqrt{(-5)^2 - 4(2)(3)}}{2(2)}$

$= \dfrac{5 \pm \sqrt{25-24}}{4} = \dfrac{5 \pm \sqrt{1}}{4} = \dfrac{5 \pm 1}{4}$

$x = \dfrac{5+1}{4} \quad \text{or} \quad x = \dfrac{5-1}{4}$

$x = \dfrac{6}{4} \quad \text{or} \quad x = \dfrac{4}{4}$

$x = \dfrac{3}{2} \quad \text{or} \quad x = 1$

The solution set is $\left\{1,\, \dfrac{3}{2}\right\}$.

49. $4y^2 - y + 2 = 0$

$a = 4, \quad b = -1, \quad c = 2$

$y = \dfrac{-(-1) \pm \sqrt{(-1)^2 - 4(4)(2)}}{2(4)}$

$= \dfrac{1 \pm \sqrt{1-32}}{8} = \dfrac{1 \pm \sqrt{-31}}{8}$

No real solution.

51. $4x^2 = 1 - 2x$

$4x^2 + 2x - 1 = 0$

$a = 4, \quad b = 2, \quad c = -1$

$x = \dfrac{-2 \pm \sqrt{2^2 - 4(4)(-1)}}{2(4)}$

$= \dfrac{-2 \pm \sqrt{4+16}}{8} = \dfrac{-2 \pm \sqrt{20}}{8}$

$= \dfrac{-2 \pm 2\sqrt{5}}{8} = \dfrac{-1 \pm \sqrt{5}}{4}$

The solution set is $\left\{\dfrac{-1-\sqrt{5}}{4},\, \dfrac{-1+\sqrt{5}}{4}\right\}$.

53. $4x^2 = 9x$

$4x^2 - 9x = 0$

$x(4x - 9) = 0$

$x = 0 \quad \text{or} \quad 4x - 9 = 0$

$x = 0 \quad \text{or} \qquad x = \dfrac{9}{4}$

The solution set is $\left\{0,\, \dfrac{9}{4}\right\}$.

55. $9t^2 - 6t + 1 = 0$

$a = 9, \quad b = -6, \quad c = 1$

$t = \dfrac{-(-6) \pm \sqrt{(-6)^2 - 4(9)(1)}}{2(9)}$

$= \dfrac{6 \pm \sqrt{36-36}}{18} = \dfrac{6 \pm 0}{18} = \dfrac{1}{3}$

The solution set is $\left\{\dfrac{1}{3}\right\}$.

57. $\dfrac{3}{4}x^2 - \dfrac{1}{4}x - \dfrac{1}{2} = 0$

$4\left(\dfrac{3}{4}x^2 - \dfrac{1}{4}x - \dfrac{1}{2}\right) = 4(0)$

$3x^2 - x - 2 = 0$

$a = 3, \quad b = -1, \quad c = -2$

$$x = \frac{-(-1)\pm\sqrt{(-1)^2-4(3)(-2)}}{2(3)}$$

$$= \frac{1\pm\sqrt{1+24}}{6} = \frac{1\pm\sqrt{25}}{6} = \frac{1\pm5}{6}$$

$$x = \frac{1+5}{6} \quad \text{or} \quad x = \frac{1-5}{6}$$

$$x = \frac{6}{6} \quad \text{or} \quad x = \frac{-4}{6}$$

$$x = 1 \quad \text{or} \quad x = -\frac{2}{3}$$

The solution set is $\left\{-\frac{2}{3}, 1\right\}$.

59. $\frac{5}{3}x^2 - x = \frac{1}{3}$

$$3\left(\frac{5}{3}x^2 - x\right) = 3\left(\frac{1}{3}\right)$$

$$5x^2 - 3x = 1$$

$$5x^2 - 3x - 1 = 0$$

$$a = 5, \quad b = -3, \quad c = -1$$

$$x = \frac{-(-3)\pm\sqrt{(-3)^2-4(5)(-1)}}{2(5)}$$

$$= \frac{3\pm\sqrt{9+20}}{10} = \frac{3\pm\sqrt{29}}{10}$$

The solution set is $\left\{\frac{3-\sqrt{29}}{10}, \frac{3+\sqrt{29}}{10}\right\}$.

61. $2x(x+2) = 3$

$$2x^2 + 4x - 3 = 0$$

$$a = 2, \quad b = 4, \quad c = -3$$

$$x = \frac{-4\pm\sqrt{4^2-4(2)(-3)}}{2(2)} = \frac{-4\pm\sqrt{16+24}}{4}$$

$$= \frac{-4\pm\sqrt{40}}{4} = \frac{-4\pm2\sqrt{10}}{4} = \frac{-2\pm\sqrt{10}}{2}$$

The solution set is $\left\{\frac{-2-\sqrt{10}}{2}, \frac{-2+\sqrt{10}}{2}\right\}$.

63. $4 - \frac{1}{x} - \frac{2}{x^2} = 0$

$$x^2\left(4 - \frac{1}{x} - \frac{2}{x^2}\right) = x^2(0)$$

$$4x^2 - x - 2 = 0$$

$$a = 4, \quad b = -1, \quad c = -2$$

$$x = \frac{-(-1)\pm\sqrt{(-1)^2-4(4)(-2)}}{2(4)}$$

$$= \frac{1\pm\sqrt{1+32}}{8} = \frac{1\pm\sqrt{33}}{8}$$

Neither of these values causes a denominator to equal zero, so the solution set is

$$\left\{\frac{1-\sqrt{33}}{8}, \frac{1+\sqrt{33}}{8}\right\}.$$

65. $\frac{3x}{x-2} + \frac{1}{x} = 4$

$$\left(\frac{3x}{x-2} + \frac{1}{x}\right)x(x-2) = 4x(x-2)$$

$$3x(x) + (x-2) = 4x^2 - 8x$$

$$3x^2 + x - 2 = 4x^2 - 8x$$

$$0 = x^2 - 9x + 2$$

$$a = 1, \quad b = -9, \quad c = 2$$

$$x = \frac{-(-9)\pm\sqrt{(-9)^2-4(1)(2)}}{2(1)}$$

$$= \frac{9\pm\sqrt{81-8}}{2} = \frac{9\pm\sqrt{73}}{2}$$

Neither of these values causes a denominator to equal zero, so the solution set is

$$\left\{\frac{9-\sqrt{73}}{2}, \frac{9+\sqrt{73}}{2}\right\}.$$

67. $2x^2 - 6x + 7 = 0$

$$a = 2, \quad b = -6, \quad c = 7$$

$$b^2 - 4ac = (-6)^2 - 4(2)(7) = 36 - 56 = -20$$

Since the $b^2 - 4ac < 0$, the equation has no real solution.

69. $9x^2 - 30x + 25 = 0$

$$a = 9, \quad b = -30, \quad c = 25$$

$$b^2 - 4ac = (-30)^2 - 4(9)(25) = 900 - 900 = 0$$

Since $b^2 - 4ac = 0$, the equation has one repeated real solution.

71. $3x^2 + 5x - 8 = 0$

$a = 3, \quad b = 5, \quad c = -8$

$b^2 - 4ac = (5)^2 - 4(3)(-8) = 25 + 96 = 121$

Since $b^2 - 4ac > 0$, the equation has two unequal real solutions.

73. $x^2 - 5 = 0$

$x^2 = 5$

$x = \pm\sqrt{5}$

The solution set is $\left\{-\sqrt{5}, \sqrt{5}\right\}$.

75. $16x^2 - 8x + 1 = 0$

$(4x - 1)(4x - 1) = 0$

$4x - 1 = 0$

$x = \dfrac{1}{4}$

The solution set is $\left\{\dfrac{1}{4}\right\}$.

77. $10x^2 - 19x - 15 = 0$

$(5x + 3)(2x - 5) = 0$

$5x + 3 = 0 \quad$ or $\quad 2x - 5 = 0$

$x = -\dfrac{3}{5} \quad$ or $\qquad x = \dfrac{5}{2}$

The solution set is $\left\{-\dfrac{3}{5}, \dfrac{5}{2}\right\}$.

79. $2 + z = 6z^2$

$0 = 6z^2 - z - 2$

$0 = (3z - 2)(2z + 1)$

$3z - 2 = 0 \quad$ or $\quad 2z + 1 = 0$

$z = \dfrac{2}{3} \quad$ or $\qquad z = -\dfrac{1}{2}$

The solution set is $\left\{-\dfrac{1}{2}, \dfrac{2}{3}\right\}$.

81. $x^2 + \sqrt{2}x = \dfrac{1}{2}$

$x^2 + \sqrt{2}x - \dfrac{1}{2} = 0$

$a = 1, b = \sqrt{2}, c = -\dfrac{1}{2}$

$x = \dfrac{-(\sqrt{2}) \pm \sqrt{(\sqrt{2})^2 - 4(1)\left(-\dfrac{1}{2}\right)}}{2(1)}$

$= \dfrac{-\sqrt{2} \pm \sqrt{2 + 2}}{2} = \dfrac{-\sqrt{2} \pm 2}{2}$

The solution set is $\left\{\dfrac{-\sqrt{2} - 2}{2}, \dfrac{-\sqrt{2} + 2}{2}\right\}$.

83. $x^2 + x = 4$

$x^2 + x - 4 = 0$

$a = 1, \quad b = 1, \quad c = -4$

$x = \dfrac{-(1) \pm \sqrt{(1)^2 - 4(1)(-4)}}{2(1)}$

$= \dfrac{-1 \pm \sqrt{1 + 16}}{2} = \dfrac{-1 \pm \sqrt{17}}{2}$

The solution set is $\left\{\dfrac{-1 - \sqrt{17}}{2}, \dfrac{-1 + \sqrt{17}}{2}\right\}$.

85. $5x(x - 1) = -7x^2 + 2$

$5x^2 - 5x = -7x^2 + 2$

$12x^2 - 5x - 2 = 0$

$(4x + 1)(3x - 2) = 0$

$4x + 1 = 0 \quad$ or $\quad 3x - 2 = 0$

$4x = -1 \qquad\qquad 3x = 2$

$x = -\dfrac{1}{4} \qquad\qquad x = \dfrac{2}{3}$

The solution set is $\left\{-\dfrac{1}{4}, \dfrac{2}{3}\right\}$.

87.
$$\frac{x}{x-2}+\frac{2}{x+1}=\frac{7x+1}{x^2-x-2}$$
$$\frac{x}{x-2}+\frac{2}{x+1}=\frac{7x+1}{(x-2)(x+1)}$$
$$\left(\frac{x}{x-2}+\frac{2}{x+1}\right)(x-2)(x+1)=\left(\frac{7x+1}{(x-2)(x+1)}\right)(x-2)(x+1)$$
$$x(x+1)+2(x-2)=7x+1$$
$$x^2+x+2x-4=7x+1$$
$$x^2+3x-4=7x+1$$
$$x^2-4x-5=0$$
$$(x+1)(x-5)=0$$
$$x+1=0 \quad\text{or}\quad x-5=0$$
$$x=-1 \quad\text{or}\quad x=5$$

The value $x=-1$ causes a denominator to equal zero, so we disregard it. Thus, the solution set is $\{5\}$.

89. Since this is a right triangle then we can use the Pythagorean Theorem. So
$$(2x+3)^2=(2x-5)^2+(x+7)^2$$
$$4x^2+12x+9=4x^2-20x+25+x^2+14x+49$$
$$12x+9=x^2-6x+74$$
$$0=x^2-18x+65$$
$$0=(x-5)(x-13)$$
$$x-5=0 \quad\text{or}\quad x-13=0$$
$$x=5 \quad\text{or}\quad x=13$$

This means there are 2 possible that meet these requirements. Substituting x into the given sides gives:
When x = 5: 5m, 12m, 13m
When x = 13: 20m, 21m, 29m
Thus there are 2 solutions.

91. Let w represent the width of window.
Then $l=w+2$ represents the length of the window.
Since the area is 143 square feet, we have:
$$w(w+2)=143$$
$$w^2+2w-143=0$$
$$(w+13)(w-11)=0$$
$$w=-13 \quad\text{or}\quad w=11$$

Discard the negative solution since width cannot be negative. The width of the rectangular window is 11 feet and the length is 13 feet.

93. Let l represent the length of the rectangle.
Let w represent the width of the rectangle.
The perimeter is 26 meters and the area is 40 square meters.
$$2l+2w=26$$
$$l+w=13 \quad\text{so}\quad w=13-l$$
$$lw=40$$
$$l(13-l)=40$$
$$13l-l^2=40$$
$$l^2-13l+40=0$$
$$(l-8)(l-5)=0$$
$$l=8 \quad\text{or}\quad l=5$$
$$w=5 \qquad w=8$$

The dimensions are 5 meters by 8 meters.

95. Let x = length of side of original sheet in feet.
Length of box: $x-2$ feet
Width of box: $x-2$ feet
Height of box: 1 foot
$$V=l\cdot w\cdot h$$
$$4=(x-2)(x-2)(1)$$
$$4=x^2-4x+4$$
$$0=x^2-4x$$
$$0=x(x-4)$$
$$x=0 \quad\text{or}\quad x=4$$

Discard $x=0$ since that is not a feasible length for the original sheet. Therefore, the original sheet should measure 4 feet on each side.

97. a. When the ball strikes the ground, the distance from the ground will be 0. Therefore, we solve

$$96 + 80t - 16t^2 = 0$$

$$-16t^2 + 80t + 96 = 0$$

$$t^2 - 5t - 6 = 0$$

$$(t-6)(t+1) = 0$$

$$t = 6 \quad \text{or} \quad t = -1$$

Discard the negative solution since the time of flight must be positive. The ball will strike the ground after 6 seconds.

b. When the ball passes the top of the building, it will be 96 feet from the ground. Therefore, we solve

$$96 + 80t - 16t^2 = 96$$

$$-16t^2 + 80t = 0$$

$$t^2 - 5t = 0$$

$$t(t-5) = 0$$

$$t = 0 \quad \text{or} \quad t = 5$$

The ball is at the top of the building at time $t = 0$ when it is thrown. It will pass the top of the building on the way down after 5 seconds.

99. Let x represent the number of centimeters the length and width should be reduced.
$12 - x$ = the new length, $7 - x$ = the new width. The new volume is 90% of the old volume.

$$(12-x)(7-x)(3) = 0.9(12)(7)(3)$$

$$3x^2 - 57x + 252 = 226.8$$

$$3x^2 - 57x + 25.2 = 0$$

$$x^2 - 19x + 8.4 = 0$$

$$x = \frac{-(-19) \pm \sqrt{(-19)^2 - 4(1)(8.4)}}{2(1)} = \frac{19 \pm \sqrt{327.4}}{2}$$

$$x \approx 0.45 \quad \text{or} \quad x \approx 18.55$$

Since 18.55 exceeds the dimensions, it is discarded. The dimensions of the new chocolate bar are: 11.55 cm by 6.55 cm by 3 cm.

101. Let x represent the width of the border measured in feet. The radius of the pool is 5 feet. Then $x + 5$ represents the radius of the circle, including both the pool and the border. The total area of the pool and border is

$$A_T = \pi(x+5)^2.$$

The area of the pool is $A_P = \pi(5)^2 = 25\pi$.

The area of the border is

$$A_B = A_T - A_P = \pi(x+5)^2 - 25\pi.$$

Since the concrete is 3 inches or 0.25 feet thick, the volume of the concrete in the border is

$$0.25 A_B = 0.25\left(\pi(x+5)^2 - 25\pi\right)$$

Solving the volume equation:

$$0.25\left(\pi(x+5)^2 - 25\pi\right) = 27$$

$$\pi\left(x^2 + 10x + 25 - 25\right) = 108$$

$$\pi x^2 + 10\pi x - 108 = 0$$

$$x = \frac{-10\pi \pm \sqrt{(10\pi)^2 - 4(\pi)(-108)}}{2(\pi)}$$

$$= \frac{-31.42 \pm \sqrt{100\pi^2 + 432\pi}}{6.28}$$

$$x \approx 2.71 \quad \text{or} \quad x \approx -12.71$$

Discard the negative solution. The width of the border is roughly 2.71 feet.

103. Let x represent the width of the border measured in feet.
The total area is $A_T = (6+2x)(10+2x)$.
The area of the garden is $A_G = 6 \cdot 10 = 60$.
The area of the border is

$$A_B = A_T - A_G = (6+2x)(10+2x) - 60.$$

Since the concrete is 3 inches or 0.25 feet thick, the volume of the concrete in the border is

$$0.25 A_B = 0.25\left((6+2x)(10+2x) - 60\right)$$

Solving the volume equation:

$$0.25\left((6+2x)(10+2x) - 60\right) = 27$$

$$60 + 32x + 4x^2 - 60 = 108$$

$$4x^2 + 32x - 108 = 0$$

$$x^2 + 8x - 27 = 0$$

$$x = \frac{-8 \pm \sqrt{8^2 - 4(1)(-27)}}{2(1)} = \frac{-8 \pm \sqrt{172}}{2}$$

$$x \approx 2.56 \quad \text{or} \quad x \approx -10.56$$

Discard the negative solution. The width of the border is approximately 2.56 feet.

105. Let x = the length of a 9.7-inch iPad Air™ in a 4:3 format.

Then $\frac{3}{4}x$ = the width of the iPad. The diagonal of the 9.7-inch iPad is 9.7 inches, so by the Pythagorean Theorem we have:

$$x^2 + \left(\frac{3}{4}x\right)^2 = 9.7^2$$

$$x^2 + \frac{9}{16}x^2 = 94.09$$

$$16\left(x^2 + \frac{9}{16}x^2\right) = 16(94.09)$$

$$16x^2 + 9x^2 = 1505.44$$

$$25x^2 = 1505.44$$

$$x^2 = 60.2176$$

$$x = \pm\sqrt{60.2176} = \pm 7.76$$

Since the length cannot be negative, the length of the iPad is 7.76 inches and the width is

$\frac{3}{4}(7.76) = 5.82$ inches. Thus, the area of the

iPad is $(7.76)(5.85) = 45.1632$ square inches.

Let $y =$ the length of a 10-inch 16:10 format Google Nexus 10™.

Then $\frac{10}{16}y =$ the width of the Nexus.

The diagonal of a 10-inch Nexus is 10 inches, so by the Pythagorean Theorem we have:

$$y^2 + \left(\frac{10}{16}y\right)^2 = 10^2$$

$$y^2 + \frac{100}{256}y^2 = 100$$

$$256\left(y^2 + \frac{100}{256}y^2\right) = 256(100)$$

$$256y^2 + 100y^2 = 25600$$

$$356y^2 = 25600$$

$$y^2 = \frac{25600}{356}$$

$$y = \pm\sqrt{\frac{25600}{356}} \approx \pm 8.48$$

Since the length cannot be negative, the length of

the Nexus is $\sqrt{\frac{25600}{356}} \approx 8.48$ inches and the width

is $\frac{10}{16}\sqrt{\frac{25600}{356}} \approx 5.30$ inches. Thus, the area of the

10-inch 16:10 format Nexus is

$$\left(\sqrt{\frac{25600}{356}}\right)\left(\frac{10}{16}\sqrt{\frac{25600}{356}}\right)$$

≈ 44.94 square inches.

The iPad Air™ 4:3 format has the larger screen since its area is larger.

107. Let x be 35. Then

$$h = -0.00025x^2 + 0.04x$$

$$= -0.00025(35)^2 + 0.04(35)$$

$$\approx 1.1 \text{ ft}$$

109. We will set g = 2.97 and solve for h using the Quadratic Formula.

$$g = -0.0006x^2 + 0.015x + 3.04$$

$$2.97 = -0.0006x^2 + 0.015x + 3.04$$

$$0 = -0.0006x^2 + 0.015x + 0.07$$

$$x = \frac{-0.015 \pm \sqrt{(0.015)^2 - 4(-0.0006)(0.07)}}{2(-0.0006)}$$

$$= \frac{-0.015 \pm \sqrt{0.000393}}{-0.0012}$$

$$x \approx 29 \text{ or } x \approx -4.02$$

So the estimated numbers of hours worked by a student with a GPA of 2.97 is 29 hours. The value -4.02 has no meaning since it is negative.

111.
$$\frac{1}{2}n(n+1) = 666$$

$$n(n+1) = 1332$$

$$n^2 + n - 1332 = 0$$

$$(n-36)(n+37) = 0$$

$$n = 36 \quad \text{or} \quad n = -37$$

Since the number of consecutive integers cannot be negative, we discard the negative value. We must add 36 consecutive integers, beginning at 1, in order to get a sum of 666.

113. The roots of a quadratic equation are

$$x_1 = \frac{-b - \sqrt{b^2 - 4ac}}{2a} \text{ and } x_2 = \frac{-b + \sqrt{b^2 - 4ac}}{2a}$$

$$x_1 + x_2 = \frac{-b - \sqrt{b^2 - 4ac}}{2a} + \frac{-b + \sqrt{b^2 - 4ac}}{2a}$$

$$= \frac{-b - \sqrt{b^2 - 4ac} - b + \sqrt{b^2 - 4ac}}{2a}$$

$$= \frac{-2b}{2a}$$

$$= -\frac{b}{a}$$

47

115. In order to have one repeated solution, we need the discriminant to be 0.

$$b^2 - 4ac = 0$$
$$1^2 - 4(k)(k) = 0$$
$$1 - 4k^2 = 0$$
$$4k^2 = 1$$
$$k^2 = \frac{1}{4}$$
$$k = \pm\sqrt{\frac{1}{4}}$$
$$k = \frac{1}{2} \quad \text{or} \quad k = -\frac{1}{2}$$

117. For $ax^2 + bx + c = 0$:

$$x_1 = \frac{-b - \sqrt{b^2 - 4ac}}{2a} \quad \text{and} \quad x_2 = \frac{-b + \sqrt{b^2 - 4ac}}{2a}$$

For $ax^2 - bx + c = 0$:

$$x_1^* = \frac{-(-b) - \sqrt{(-b)^2 - 4ac}}{2a}$$
$$= \frac{b - \sqrt{b^2 - 4ac}}{2a}$$
$$= -\left(\frac{-b + \sqrt{b^2 - 4ac}}{2a}\right)$$
$$= -x_2$$

and

$$x_2^* = \frac{-(-b) + \sqrt{(-b)^2 - 4ac}}{2a}$$
$$= \frac{b + \sqrt{b^2 - 4ac}}{2a}$$
$$= -\left(\frac{-b - \sqrt{b^2 - 4ac}}{2a}\right)$$
$$= -x_1$$

119. a. $x^2 = 9$ and $x = 3$ are not equivalent because they do not have the same solution set. In the first equation we can also have $x = -3$.

b. $x = \sqrt{9}$ and $x = 3$ are equivalent because $\sqrt{9} = 3$.

c. $(x-1)(x-2) = (x-1)^2$ and $x - 2 = x - 1$ are not equivalent because they do not have the same solution set.

The first equation has the solution set $\{1\}$ while the second equation has no solutions.

121. Answers will vary. Knowing the discriminant allows us to know how many real solutions the equation will have.

123. Answers will vary.

Section 1.4

1. Integers: $\{-3, 0\}$

Rationals: $\left\{-3, 0, \frac{6}{5}\right\}$

3. $\dfrac{3}{2+\sqrt{3}} = \dfrac{3}{2+\sqrt{3}} \cdot \dfrac{2-\sqrt{3}}{2-\sqrt{3}}$

$$= \frac{3(2-\sqrt{3})}{2^2 - (\sqrt{3})^2}$$
$$= \frac{3(2-\sqrt{3})}{4-3}$$
$$= 3(2-\sqrt{3})$$

5. False; the conjugate of $2 + 5i$ is $2 - 5i$.

7. False; if $2 - 3i$ is a solution of a quadratic equation with real coefficients, then its conjugate, $2 + 3i$, is also a solution.

9. a

11. $(2 - 3i) + (6 + 8i) = (2 + 6) + (-3 + 8)i = 8 + 5i$

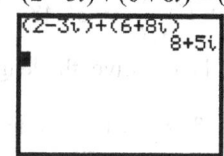

13. $(-3 + 2i) - (4 - 4i) = (-3 - 4) + (2 - (-4))i$

$$= -7 + 6i$$

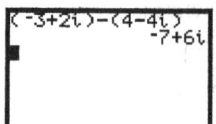

48

15. $(2 - 5i) - (8 + 6i) = (2 - 8) + (-5 - 6)i$
$$= -6 - 11i$$

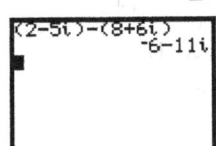

17. $3(2 - 6i) = 6 - 18i$

19. $2i(2 - 3i) = 4i - 6i^2 = 4i - 6(-1) = 6 + 4i$

21. $(3 - 4i)(2 + i) = 6 + 3i - 8i - 4i^2$
$$= 6 - 5i - 4(-1)$$
$$= 10 - 5i$$

23. $(-6 + i)(-6 - i) = 36 + 6i - 6i - i^2$
$$= 36 - (-1)$$
$$= 37$$

25. $\dfrac{10}{3 - 4i} = \dfrac{10}{3 - 4i} \cdot \dfrac{3 + 4i}{3 + 4i} = \dfrac{30 + 40i}{9 + 12i - 12i - 16i^2}$
$$= \dfrac{30 + 40i}{9 - 16(-1)} = \dfrac{30 + 40i}{25}$$
$$= \dfrac{30}{25} + \dfrac{40}{25}i$$
$$= \dfrac{6}{5} + \dfrac{8}{5}i$$

27. $\dfrac{2 + i}{i} = \dfrac{2 + i}{i} \cdot \dfrac{-i}{-i} = \dfrac{-2i - i^2}{-i^2}$
$$= \dfrac{-2i - (-1)}{-(-1)} = \dfrac{1 - 2i}{1}$$
$$= 1 - 2i$$

29. $\dfrac{6 - i}{1 + i} = \dfrac{6 - i}{1 + i} \cdot \dfrac{1 - i}{1 - i} = \dfrac{6 - 6i - i + i^2}{1 - i + i - i^2}$
$$= \dfrac{6 - 7i + (-1)}{1 - (-1)} = \dfrac{5 - 7i}{2}$$
$$= \dfrac{5}{2} - \dfrac{7}{2}i$$

31. $\left(\dfrac{1}{2} + \dfrac{\sqrt{3}}{2}i\right)^2 = \dfrac{1}{4} + 2\left(\dfrac{1}{2}\right)\left(\dfrac{\sqrt{3}}{2}i\right) + \dfrac{3}{4}i^2$
$$= \dfrac{1}{4} + \dfrac{\sqrt{3}}{2}i + \dfrac{3}{4}(-1)$$
$$= -\dfrac{1}{2} + \dfrac{\sqrt{3}}{2}i$$

49

33. $(1+i)^2 = 1+2i+i^2 = 1+2i+(-1) = 2i$

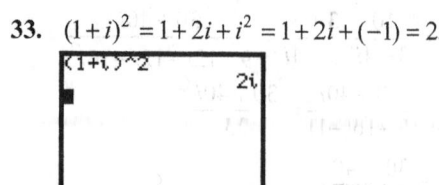

35. $i^{23} = i^{22+1} = i^{22} \cdot i = \left(i^2\right)^{11} \cdot i = (-1)^{11} i = -i$

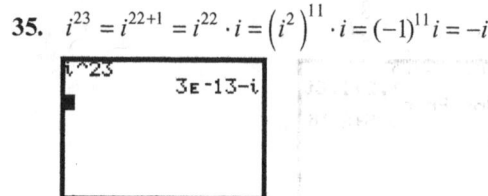

37. $i^{-15} = \dfrac{1}{i^{15}} = \dfrac{1}{i^{14+1}} = \dfrac{1}{i^{14} \cdot i} = \dfrac{1}{\left(i^2\right)^7 \cdot i}$

$$= \dfrac{1}{(-1)^7 i} = \dfrac{1}{-i} = \dfrac{1}{-i} \cdot \dfrac{i}{i} = \dfrac{i}{-i^2} = \dfrac{i}{-(-1)}$$

$$= i$$

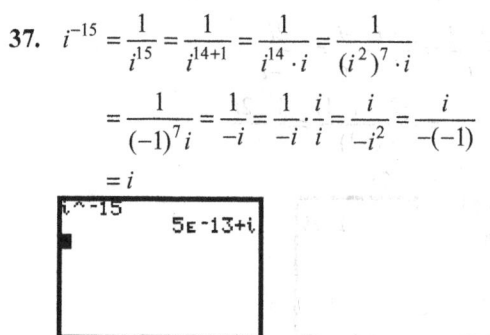

39. $i^6 - 5 = \left(i^2\right)^3 - 5 = (-1)^3 - 5 = -1 - 5 = -6$

41. $6i^3 - 4i^5 = i^3\left(6 - 4i^2\right)$

$$= i^2 \cdot i\left(6 - 4(-1)\right)$$

$$= -1 \cdot i(10)$$

$$= -10i$$

43. $(1+i)^3 = (1+i)(1+i)(1+i) = \left(1+2i+i^2\right)(1+i)$

$$= (1+2i-1)(1+i) = 2i(1+i)$$

$$= 2i + 2i^2 = 2i + 2(-1)$$

$$= -2 + 2i$$

45. $i^7\left(1+i^2\right) = i^7\left(1+(-1)\right) = i^7(0) = 0$

47. $i^6 + i^4 + i^2 + 1 = \left(i^2\right)^3 + \left(i^2\right)^2 + i^2 + 1$

$$= (-1)^3 + (-1)^2 + (-1) + 1$$

$$= -1 + 1 - 1 + 1$$

$$= 0$$

49. $\sqrt{-4} = 2i$

51. $\sqrt{-25} = 5i$

53. $\sqrt{-12} = i\sqrt{4 \cdot 3} = 2i\sqrt{3}$

55. $\sqrt{-200} = i\sqrt{100 \cdot 2} = 10i\sqrt{2}$

57. $\sqrt{(3+4i)(4i-3)} = \sqrt{12i - 9 + 16i^2 - 12i}$

$$= \sqrt{-9 + 16(-1)}$$

$$= \sqrt{-25}$$

$$= 5i$$

59. $x^2 + 4 = 0$

$x^2 = -4$

$x = \pm\sqrt{-4}$

$x = \pm 2i$

The solution set is $\{\pm 2i\}$.

```
(-2i)^2+4
              0
(2i)^2+4
              0
```

61. $x^2 - 16 = 0$

$(x+4)(x-4) = 0 \Rightarrow x = -4, x = 4$

The solution set is $\{\pm 4\}$.

63. $x^2 - 6x + 13 = 0$

$a = 1, b = -6, c = 13,$

$b^2 - 4ac = (-6)^2 - 4(1)(13) = 36 - 52 = -16$

$x = \dfrac{-(-6)\pm\sqrt{-16}}{2(1)} = \dfrac{6\pm 4i}{2} = 3\pm 2i$

The solution set is $\{3 - 2i, 3 + 2i\}$.

```
(3-2i)^2-6(3-2i)
+13
              0
(3+2i)^2-6(3+2i)
+13
              0
```

65. $x^2 - 6x + 10 = 0$

$a = 1, b = -6, c = 10$

$b^2 - 4ac = (-6)^2 - 4(1)(10) = 36 - 40 = -4$

$x = \dfrac{-(-6)\pm\sqrt{-4}}{2(1)} = \dfrac{6\pm 2i}{2} = 3\pm i$

The solution set is $\{3 - i, 3 + i\}$.

67. $8x^2 - 4x + 1 = 0$

$a = 8, b = -4, c = 1$

$b^2 - 4ac = (-4)^2 - 4(8)(1) = 16 - 32 = -16$

$x = \dfrac{-(-4)\pm\sqrt{-16}}{2(8)} = \dfrac{4\pm 4i}{16} = \dfrac{1}{4}\pm\dfrac{1}{4}i$

The solution set is $\left\{\dfrac{1}{4}-\dfrac{1}{4}i,\ \dfrac{1}{4}+\dfrac{1}{4}i\right\}$.

69. $5x^2 + 1 = 2x$

$5x^2 - 2x + 1 = 0$

$a = 5, b = -2, c = 1$

$b^2 - 4ac = (-2)^2 - 4(5)(1) = 4 - 20 = -16$

$x = \dfrac{-(-2)\pm\sqrt{-16}}{2(5)} = \dfrac{2\pm 4i}{10} = \dfrac{1}{5}\pm\dfrac{2}{5}i$

The solution set is $\left\{\dfrac{1}{5}-\dfrac{2}{5}i,\ \dfrac{1}{5}+\dfrac{2}{5}i\right\}$.

71. $x^2 + x + 1 = 0$

$a = 1, b = 1, c = 1,$

$b^2 - 4ac = 1^2 - 4(1)(1) = 1 - 4 = -3$

$x = \dfrac{-1\pm\sqrt{-3}}{2(1)} = \dfrac{-1\pm\sqrt{3}\,i}{2} = -\dfrac{1}{2}\pm\dfrac{\sqrt{3}}{2}i$

The solution set is $\left\{-\dfrac{1}{2}-\dfrac{\sqrt{3}}{2}i,\ -\dfrac{1}{2}+\dfrac{\sqrt{3}}{2}i\right\}$.

73. $x^3 - 8 = 0$

$(x-2)(x^2 + 2x + 4) = 0$

$x - 2 = 0 \Rightarrow x = 2$

$x^2 + 2x + 4 = 0$

$a = 1, b = 2, c = 4$

$b^2 - 4ac = 2^2 - 4(1)(4) = 4 - 16 = -12$

$x = \dfrac{-2\pm\sqrt{-12}}{2(1)} = \dfrac{-2\pm 2\sqrt{3}\,i}{2} = -1\pm\sqrt{3}\,i$

The solution set is $\left\{2,\ -1-\sqrt{3}\,i,\ -1+\sqrt{3}\,i\right\}$.

75. $x^4 = 16$

$x^4 - 16 = 0$

$(x^2 - 4)(x^2 + 4) = 0$

$(x-2)(x+2)(x^2 + 4) = 0$

$x - 2 = 0$ or $x + 2 = 0$ or $x^2 + 4 = 0$

$x = 2$ or $x = -2$ or $x^2 = -4$

$x = 2$ or $x = -2$ or $x = \pm\sqrt{-4} = \pm 2i$

The solution set is $\{-2, 2, -2i, 2i\}$.

77. $x^4 + 13x^2 + 36 = 0$

$(x^2 + 9)(x^2 + 4) = 0$

$x^2 + 9 = 0 \Rightarrow x = \pm 3i$

$x^2 + 4 = 0 \Rightarrow x = \pm 2i$

The solution set is $\{-3i, \ 3i, \ -2i, \ 2i\}$.

79. $3x^2 - 3x + 4 = 0$

$a = 3, b = -3, c = 4$

$b^2 - 4ac = (-3)^2 - 4(3)(4) = 9 - 48 = -39$

The equation has two complex conjugate solutions.

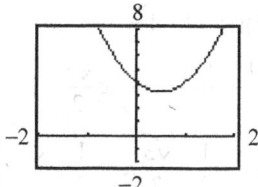

81. $2x^2 + 3x = 4$

$2x^2 + 3x - 4 = 0$

$a = 2, b = 3, c = -4$

$b^2 - 4ac = 3^2 - 4(2)(-4) = 9 + 32 = 41$

The equation has two unequal real solutions.

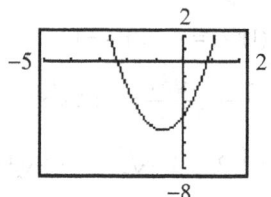

83. $9x^2 - 12x + 4 = 0$

$a = 9, b = -12, c = 4$

$b^2 - 4ac = (-12)^2 - 4(9)(4) = 144 - 144 = 0$

The equation has a repeated real solution.

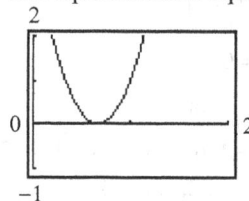

85. The other solution is the conjugate of $2 + 3i$, or $2 - 3i$.

87. $z + \bar{z} = 3 - 4i + \overline{3 - 4i} = 3 - 4i + 3 + 4i = 6$

89. $z \cdot \bar{z} = (3 - 4i)(\overline{3 - 4i}) = (3 - 4i)(3 + 4i)$

$= 9 + 12i - 12i - 16i^2 = 9 - 16(-1)$

$- 25$

91. $Z = \dfrac{V}{I} = \dfrac{18 + i}{3 - 4i} = \dfrac{18 + i}{3 - 4i} \cdot \dfrac{3 + 4i}{3 + 4i}$

$= \dfrac{54 + 72i + 3i + 4i^2}{9 + 12i - 12i - 16i^2} = \dfrac{54 + 75i - 4}{9 + 16} = \dfrac{50 + 75i}{25}$

$= 2 + 3i$

The impedance is $2 + 3i$ ohms.

93. $z + \bar{z} = a + bi + \overline{a + bi}$

$= a + bi + a - bi$

$= 2a$

$z - \bar{z} = a + bi - (\overline{a + bi})$

$= a + bi - (a - bi)$

$= a + bi - a + bi$

$= 2bi$

95. $\overline{z + w} = \overline{(a + bi) + (c + di)}$

$= \overline{(a + c) + (b + d)i}$

$= (a + c) - (b + d)i$

$= (a - bi) + (c - di)$

$= \overline{a + bi} + \overline{c + di}$

$= \bar{z} + \bar{w}$

97. Answers will vary.

99. Answers will vary.

101. Although the set of real numbers is a subset of the set of complex numbers, not all rules that work in the real number system can be used in the larger complex number system. The rule that allows us to write the product of two square roots as the square root of the product only works in the real number system. That is, $\sqrt{a} \cdot \sqrt{b} = \sqrt{ab}$ only when $\sqrt{a}$ and $\sqrt{b}$ are real numbers. In the complex number system we must first convert the radicals to complex form. In this case this means we need to write $\sqrt{-9}$ as $\sqrt{-1 \cdot 9} = \sqrt{9} \cdot \sqrt{-1} = 3i$. Then we can multiply to get $\sqrt{-9} \cdot \sqrt{-9} = 3i \cdot 3i = 9i^2 = 9(-1) = -9$.

Section 1.5

1. True

3. $2x^2 - 7x - 4 = (x-4)(2x+1)$

5. The distance from the origin to -4 on a real number line is 4 units.

7. quadratic in form

9. a

11. $\sqrt{y+3} = 5$

$\left(\sqrt{y+3}\right)^2 = 5^2$

$y+3 = 25 \rightarrow y = 22$

Check: $\sqrt{22+3} = \sqrt{25} = 5$

The solution set is $\{22\}$.

13. $\sqrt{2t-1} = 1$

$\left(\sqrt{2t-1}\right)^2 = 1^2$

$2t-1 = 1 \rightarrow 2t = 2 \rightarrow t = 1$

Check: $\sqrt{2(1)-1} = \sqrt{1} = 1$

The solution set is $\{1\}$.

15. $\sqrt{3t+4} = -6$

Since the principal square root is never negative, the equation has no real solution.

17. $\sqrt[3]{1-2x} - 3 = 0$

$\sqrt[3]{1-2x} = 3$

$\left(\sqrt[3]{1-2x}\right)^3 = 3^3$

$1-2x = 27 \rightarrow -2x = 26 \rightarrow x = -13$

Check: $\sqrt[3]{1-2(-13)} - 3 = \sqrt[3]{27} - 3 = 0$

The solution set is $\{-13\}$.

19. $\sqrt[4]{5x-4} = 2$

$\left(\sqrt[4]{5x-4}\right)^4 = 2^4$

$5x-4 = 16$

$5x = 20$

$x = 4$

Check:

$\sqrt[4]{5(4)-4} = \sqrt[4]{16} = 2\,\mathsf{T}$

The solution set is $\{4\}$.

21. $\sqrt[5]{x^2+2x} = -1$

$\left(\sqrt[5]{x^2+2x}\right)^5 = (-1)^5$

$x^2 + 2x = -1$

$x^2 + 2x + 1 = 0$

$(x+1)^2 = 0$

$x+1 = 0$

$x = -1$

Check:

$\sqrt[5]{(-1)^2 + 2(-1)} = \sqrt[5]{1-2} = \sqrt[5]{-1} = -1\,\mathsf{T}$

The solution set is $\{-1\}$.

23. $x = 8\sqrt{x}$

$(x)^2 = \left(8\sqrt{x}\right)^2$

$x^2 = 64x$

$x^2 - 64x = 0$

$x(x-64) = 0$

$x = 0$ or $x = 64$

Check:

$0 = 8\sqrt{0} \qquad 64 = 8\sqrt{64}$

$0 = 0\,\mathsf{T} \qquad 64 = 8\cdot 8$

$\qquad\qquad\quad 64 = 64\,\mathsf{T}$

The solution set is $\{0, 64\}$.

25. $\sqrt{15-2x} = x$

$\left(\sqrt{15-2x}\right)^2 = x^2$

$15-2x = x^2 \rightarrow x^2 + 2x - 15 = 0$

$(x+5)(x-3) = 0 \rightarrow x = -5$ or $x = 3$

Check -5: $\sqrt{15-2(-5)} = \sqrt{25}$

$\qquad\qquad\qquad = 5 \neq -5$

Check 3: $\sqrt{15-2(3)} = \sqrt{9} = 3 = 3$

The solution set is $\{3\}$.

27. $\sqrt{3(x+10)} - 4 = x$

$\sqrt{3(x+10)} = x+4$

$\left(\sqrt{3(x+10)}\right)^2 = (x+4)^2$

$3x+30 = x^2+8x+16$

$0 = x^2+5x-14$

$0 = (x+7)(x-2)$

$x = -7 \text{ or } x = 2$

Check -7: $\sqrt{3(-7+10)} - 4 = \sqrt{9} - 4 = -1 \neq -7$

Check 2: $\sqrt{3(2+10)} - 4 = \sqrt{36} - 4 = 2 = 2$

Discard $x = -7$ as extraneous.

The solution set is $\{2\}$.

29. $\sqrt{x^2-x-4} = x+2$

$\left(\sqrt{x^2-x-4}\right)^2 = (x+2)^2$

$x^2-x-4 = x^2+4x+4$

$-8 = 5x \rightarrow -\dfrac{8}{5} = x$

Check:

$\sqrt{\left(-\dfrac{8}{5}\right)^2 - \left(-\dfrac{8}{5}\right) - 4} = \left(-\dfrac{8}{5}\right) + 2$

$\sqrt{\dfrac{64}{25} + \dfrac{8}{5} - 4} = \dfrac{2}{5}$

$\sqrt{\dfrac{4}{25}} = \dfrac{2}{5}$

$\dfrac{2}{5} = \dfrac{2}{5}$ T

The solution set is $\left\{-\dfrac{8}{5}\right\}$.

31. $3 + \sqrt{3x+1} = x$

$\sqrt{3x+1} = x-3$

$\left(\sqrt{3x+1}\right)^2 = (x-3)^2$

$3x+1 = x^2-6x+9$

$0 = x^2-9x+8$

$0 = (x-1)(x-8)$

$x = 1 \text{ or } x = 8$

Check 1: $3 + \sqrt{3(1)+1} = 3 + \sqrt{4} = 5 \neq 1$

Check 8: $3 + \sqrt{3(8)+1} = 3 + \sqrt{25} = 8 = 8$

Discard $x = 1$ as extraneous.

The solution set is $\{8\}$.

33. $\sqrt{2x+3} - \sqrt{x+1} = 1$

$\sqrt{2x+3} = 1 + \sqrt{x+1}$

$\left(\sqrt{2x+3}\right)^2 = \left(1 + \sqrt{x+1}\right)^2$

$2x+3 = 1 + 2\sqrt{x+1} + x+1$

$x+1 = 2\sqrt{x+1}$

$(x+1)^2 = \left(2\sqrt{x+1}\right)^2$

$x^2+2x+1 = 4(x+1)$

$x^2+2x+1 = 4x+4$

$x^2-2x-3 = 0$

$(x+1)(x-3) = 0 \rightarrow x = -1 \text{ or } x = 3$

Check -1: $\sqrt{2(-1)+3} - \sqrt{-1+1} = 1 - 0 = 1$

Check 3: $\sqrt{2(3)+3} - \sqrt{3+1} = 3 - 2 = 1$

The solution set is $\{-1, 3\}$.

35. $\sqrt{3x+1} - \sqrt{x-1} = 2$

$\sqrt{3x+1} = 2 + \sqrt{x-1}$

$\left(\sqrt{3x+1}\right)^2 = \left(2 + \sqrt{x-1}\right)^2$

$3x+1 = 4 + 4\sqrt{x-1} + x-1$

$2x-2 = 4\sqrt{x-1}$

$(2x-2)^2 = \left(4\sqrt{x-1}\right)^2$

$4x^2-8x+4 = 16(x-1)$

$x^2-2x+1 = 4x-4$

$x^2-6x+5 = 0$

$(x-1)(x-5) = 0 \rightarrow x = 1 \text{ or } x = 5$

Check 1: $\sqrt{3(1)+1} - \sqrt{1-1}$

$= \sqrt{4} - \sqrt{0} = 2 - 0 = 2 = 2$

Check 5: $\sqrt{3(5)+1} - \sqrt{5-1}$

$= \sqrt{16} - \sqrt{4} = 4 - 2 = 2 = 2$

The solution set is $\{1, 5\}$.

37. $\sqrt{3-2\sqrt{x}} = \sqrt{x}$

$\left(\sqrt{3-2\sqrt{x}}\right)^2 = \left(\sqrt{x}\right)^2$

$3-2\sqrt{x} = x$

$-2\sqrt{x} = x-3$

$\left(-2\sqrt{x}\right)^2 = (x-3)^2$

$4x = x^2 - 6x + 9$

$0 = x^2 - 10x + 9$

$0 = (x-9)(x-1)$

$x = 1$ or $x = 9$

Check:

$\sqrt{3-2\sqrt{1}} = \sqrt{1}$ $\sqrt{3-2\sqrt{9}} = \sqrt{9}$

$\sqrt{3-2} = 1$ $\sqrt{3-2\cdot 3} = 3$

$\sqrt{1} = 1$ $\sqrt{-3} \neq 3$

$1 = 1$ T

Discard $x = 9$ as extraneous. The solution set is $\{1\}$.

39. $(3x+1)^{1/2} = 4$

$\left((3x+1)^{1/2}\right)^2 = (4)^2$

$3x+1 = 16 \rightarrow 3x = 15 \rightarrow x = 5$

Check:

$(3(5)+1)^{1/2} = 4$

$16^{1/2} = 4$

$4 = 4$ T

The solution set is $\{5\}$.

41. $(5x-2)^{1/3} = 2$

$\left((5x-2)^{1/3}\right)^3 = (2)^3$

$5x-2 = 8 \rightarrow 5x = 10 \rightarrow x = 2$

Check:

$(5(2)-2)^{1/3} = 2$

$8^{1/3} = 2$

$2 = 2$

The solution set is $\{2\}$.

43. $(x^2+9)^{1/2} = 5$

$\left((x^2+9)^{1/2}\right)^2 = (5)^2$

$x^2 + 9 = 25 \rightarrow x^2 = 16$

$x = -4$ or $x = 4$

Check:

$((-4)^2+9)^{1/2} = 5$ $((4)^2+9)^{1/2} = 5$

$25^{1/2} = 5$ $25^{1/2} = 5$

$5 = 5$ $5 = 5$

The solution set is $\{-4,4\}$.

45. $t^4 - 16 = 0$

$(t^2+4)(t^2-4) = 0$

$(t^2+4)(t+2)(t-2) = 0$

$t^2 + 4 = 0$ has no real solution, so we only need to consider

$t+2 = 0$ or $t-2 = 0$

$t = -2$ $t = 2$

The solution set is $\{-2,2\}$

47. $x^4 - 5x^2 + 4 = 0$

$(x^2-4)(x^2-1) = 0$

$x^2-4 = 0$ or $x^2-1 = 0$

$x = \pm 2$ or $x = \pm 1$

The solution set is $\{-2,-1,1,2\}$.

49. $3x^4 - 2x^2 - 1 = 0$

$\left(3x^2 + 1\right)\left(x^2 - 1\right) = 0$

$3x^2 + 1 = 0$ or $x^2 - 1 = 0$

$3x^2 = -1$,which is impossible

or $x = \pm 1$

The solution set is $\{-1, 1\}$.

51. $x^6 + 7x^3 - 8 = 0$

$\left(x^3 + 8\right)\left(x^3 - 1\right) = 0$

$x^3 + 8 = 0$ or $x^3 - 1 = 0$

$x^3 = -8 \rightarrow x = -2$

or $x^3 = 1 \rightarrow x = 1$

The solution set is $\{-2, 1\}$.

53. $(x+2)^2 + 7(x+2) + 12 = 0$

let $p = x + 2 \rightarrow p^2 = (x+2)^2$

$p^2 + 7p + 12 = 0$

$(p+3)(p+4) = 0$

$p + 3 = 0$ or $p + 4 = 0$

$p = -3 \rightarrow x + 2 = -3 \rightarrow x = -5$

or $p = -4 \rightarrow x + 2 = -4 \rightarrow x = -6$

The solution set is $\{-6, -5\}$.

55. $2(s+1)^2 - 5(s+1) = 3$

let $p = s + 1 \rightarrow p^2 = (s+1)^2$

$2p^2 - 5p = 3$

$2p^2 - 5p - 3 = 0$

$(2p+1)(p-3) = 0$

$2p + 1 = 0$ or $p - 3 = 0$

$p = -\dfrac{1}{2} \rightarrow s + 1 = -\dfrac{1}{2} \rightarrow s = -\dfrac{3}{2}$

or $p = 3 \rightarrow s + 1 = 3 \rightarrow s = 2$

The solution set is $\left\{-\dfrac{3}{2}, 2\right\}$.

57. $x - 4\sqrt{x} = 0$

$x = 4\sqrt{x}$

$(x)^2 = \left(4\sqrt{x}\right)^2$

$x^2 = 16x \rightarrow 0 = 16x - x^2$

$0 = x(16 - x) \rightarrow x = 0$

or $16 - x = 0 \rightarrow x = 16$

Check:

$x = 0: \quad 0 - 4\sqrt{0} = 0$

$\qquad\qquad 0 = 0$

$x = 16: \quad (16) - 4\sqrt{16} = 0$

$\qquad\qquad 16 - 16 = 0$

$\qquad\qquad 0 = 0$

The solution set is $\{0, 16\}$.

59. $x + \sqrt{x} = 20$

let $p = \sqrt{x} \rightarrow p^2 = x$

$p^2 + p = 20$

$p^2 + p - 20 = 0$

$(p+5)(p-4) = 0$

$p + 5 = 0$ or $p - 4 = 0$

$p = -5 \rightarrow \sqrt{x} = -5$ non-real

or $p = 4 \rightarrow \sqrt{x} = 4 \rightarrow x = 16$

Check:

$x = 16: \quad 16 + \sqrt{16} = 20$

$\qquad\qquad 16 + 4 = 20$

The solution set is $\{16\}$.

61. $t^{1/2} - 2t^{1/4} + 1 = 0$

let $p = t^{1/4} \rightarrow p^2 = t^{1/2}$

$p^2 - 2p + 1 = 0$

$(p-1)(p-1) = 0$

$p - 1 = 0$

$p = 1 \rightarrow t^{1/4} = 1 \rightarrow t = 1$

Check:

$t = 1: \quad 1^{1/2} - 2(1)^{1/4} + 1 = 0$

$\qquad\qquad 1 - 2 + 1 = 0 \rightarrow 0 = 0$

The solution set is $\{1\}$.

63. $4x^{1/2} - 9x^{1/4} + 4 = 0$

let $p = x^{1/4} \rightarrow p^2 = x^{1/2}$

$4p^2 - 9p + 4 = 0$

$p = \dfrac{9 \pm \sqrt{81-64}}{8} = \dfrac{9 \pm \sqrt{17}}{8}$

$x^{1/4} = \dfrac{9 \pm \sqrt{17}}{8} \rightarrow x = \left(\dfrac{9 \pm \sqrt{17}}{8}\right)^4$

Check:

$4\left(\left(\dfrac{9+\sqrt{17}}{8}\right)^4\right)^{1/2} - 9\left(\left(\dfrac{9+\sqrt{17}}{8}\right)^4\right)^{1/4} + 4 = 0$

$4\left(\dfrac{9+\sqrt{17}}{8}\right)^2 - 9\left(\dfrac{9+\sqrt{17}}{8}\right) + 4 = 0$

$4\dfrac{\left(9+\sqrt{17}\right)^2}{64} - 9\left(\dfrac{9+\sqrt{17}}{8}\right) + 4 = 0$

$64\left(4\dfrac{\left(9+\sqrt{17}\right)^2}{64} - 9\left(\dfrac{9+\sqrt{17}}{8}\right) + 4\right) = (0)(64)$

$4\left(9+\sqrt{17}\right)^2 - 72\left(9+\sqrt{17}\right) + 256 = 0$

$4\left(81+18\sqrt{17}+17\right) - 72\left(9+\sqrt{17}\right) + 256 = 0$

$324 + 72\sqrt{17} + 68 - 648 - 72\sqrt{17} + 256 = 0$

$0 = 0$

$4\left(\left(\dfrac{9-\sqrt{17}}{8}\right)^4\right)^{1/2} - 9\left(\left(\dfrac{9-\sqrt{17}}{8}\right)^4\right)^{1/4} + 4 = 0$

$4\left(\dfrac{9-\sqrt{17}}{8}\right)^2 - 9\left(\dfrac{9-\sqrt{17}}{8}\right) + 4 = 0$

$4\left(81-18\sqrt{17}+17\right) - 72\left(9-\sqrt{17}\right) + 256 = 0$

$324 - 72\sqrt{17} + 68 - 648 + 72\sqrt{17} + 256 = 0$

$0 = 0$

The solution set is $\left\{\left(\dfrac{9-\sqrt{17}}{8}\right)^4, \left(\dfrac{9+\sqrt{17}}{8}\right)^4\right\}$.

65. $\dfrac{1}{(x+1)^2} = \dfrac{1}{x+1} + 2$

let $p = \dfrac{1}{x+1} \rightarrow p^2 = \left(\dfrac{1}{x+1}\right)^2$

$p^2 = p + 2 \rightarrow p^2 - p - 2 = 0$

$(p+1)(p-2) = 0 \rightarrow p = -1 \text{ or } p = 2$

$p = -1 \rightarrow \dfrac{1}{x+1} = -1 \rightarrow 1 = -x - 1 \rightarrow x = -2$

or

$p = 2 \rightarrow \dfrac{1}{x+1} = 2 \rightarrow 1 = 2x + 2 \rightarrow x = -\dfrac{1}{2}$

Check:

$x = -2: \dfrac{1}{(-2+1)^2} = \dfrac{1}{-2+1} + 2$

$1 = -1 + 2$

$1 = 1$

$x = -\dfrac{1}{2}: \dfrac{1}{\left(-\frac{1}{2}+1\right)^2} = \dfrac{1}{\left(-\frac{1}{2}+1\right)} + 2$

$4 = 2 + 2$

$4 = 4$

The solution set is $\left\{-2, -\dfrac{1}{2}\right\}$.

67. $3x^{-2} - 7x^{-1} - 6 = 0$

let $p = x^{-1} \rightarrow p^2 = x^{-2}$

$3p^2 - 7p - 6 = 0$

$(3p+2)(p-3) = 0$

$p = -\dfrac{2}{3} \text{ or } p = 3$

$p = -\dfrac{2}{3} \rightarrow x^{-1} = -\dfrac{2}{3} \rightarrow \left(x^{-1}\right)^{-1} = \left(-\dfrac{2}{3}\right)^{-1}$

$\rightarrow x = -\dfrac{3}{2}$

$p = 3 \rightarrow x^{-1} = 3 \rightarrow \left(x^{-1}\right)^{-1} = (3)^{-1} \rightarrow x = \dfrac{1}{3}$

Check:

$$x = -\frac{3}{2}: \ 3\left(-\frac{3}{2}\right)^{-2} - 7\left(-\frac{3}{2}\right)^{-1} - 6 = 0$$

$$3\left(\frac{4}{9}\right) - 7\left(-\frac{2}{3}\right) - 6 = 0$$

$$\frac{4}{3} + \frac{14}{3} - 6 = 0$$

$$0 = 0$$

$$x = \frac{1}{3}: \ 3\left(\frac{1}{3}\right)^{-2} - 7\left(\frac{1}{3}\right)^{-1} - 6 = 0$$

$$3(9) - 7(3) - 6 = 0$$

$$27 - 21 - 6 = 0$$

$$0 = 0$$

The solution set is $\left\{-\frac{3}{2}, \frac{1}{3}\right\}$.

69. $2x^{2/3} - 5x^{1/3} - 3 = 0$

let $p = x^{1/3} \to p^2 = x^{2/3}$

$2p^2 - 5p - 3 = 0 \to (2p + 1)(p - 3) = 0$

$p = -\frac{1}{2}$ or $p = 3$

$p = -\frac{1}{2} \to x^{1/3} = -\frac{1}{2}$

$$\to \left(x^{1/3}\right)^3 = \left(-\frac{1}{2}\right)^3 \to x = -\frac{1}{8}$$

or

$p = 3 \to x^{1/3} = 3 \to \left(x^{1/3}\right)^3 = (3)^3$

$$\to x = 27$$

Check:

$$x = -\frac{1}{8}: \ 2\left(-\frac{1}{8}\right)^{2/3} - 5\left(-\frac{1}{8}\right)^{1/3} - 3 = 0$$

$$2\left(\frac{1}{4}\right) - 5\left(-\frac{1}{2}\right) - 3 = 0$$

$$\frac{1}{2} + \frac{5}{2} - 3 = 0$$

$$3 - 3 = 0 \to 0 = 0$$

$$x = 27: \ 2(27)^{2/3} - 5(27)^{1/3} - 3 = 0$$

$$2(9) - 5(3) - 3 = 0$$

$$18 - 15 - 3 = 0$$

$$3 - 3 = 0 \to 0 = 0$$

The solution set is $\left\{-\frac{1}{8}, 27\right\}$.

71. $|2x + 3| = 5$

$2x + 3 = 5$ or $2x + 3 = -5$

$2x = 2$ or $\quad 2x = -8$

$x = 1$ or $\quad\quad x = -4$

The solution set is $\{-4, 1\}$.

73. $|1 - 4t| + 8 = 13 \Rightarrow |1 - 4t| = 5$

$1 - 4t = 5$ or $1 - 4t = -5$

$-4t = 4$ or $\quad -4t = -6$

$t = -1$ or $\quad\quad t = \frac{3}{2}$

The solution set is $\left\{-1, \frac{3}{2}\right\}$.

75. $|-2x| = 8$

$-2x = 8 \quad$ or $\quad -2x = -8$

$x = -4$ or $\quad\quad x = 4$

The solution set is $\{-4, 4\}$.

77. $4 - |2x| = 3 \Rightarrow |2x| = 1$

$2x = 1$ or $2x = -1$

$x = \frac{1}{2}$ or $x = -\frac{1}{2}$

The solution set is $\left\{-\frac{1}{2}, \frac{1}{2}\right\}$.

79. $\frac{2}{3}|x| = 9$

$|x| = \frac{27}{2} \to x = \frac{27}{2}$ or $x = -\frac{27}{2}$

The solution set is $\left\{-\frac{27}{2}, \frac{27}{2}\right\}$.

81. $\left|\frac{x}{3} + \frac{2}{5}\right| = 2$

$\frac{x}{3} + \frac{2}{5} = 2 \quad$ or $\quad \frac{x}{3} + \frac{2}{5} = -2$

$5x + 6 = 30$ or $5x + 6 = -30$

$5x = 24$ or $\quad 5x = -36$

$x = \frac{24}{5}$ or $\quad\quad x = -\frac{36}{5}$

The solution set is $\left\{-\frac{36}{5}, \frac{24}{5}\right\}$.

83. $|u-2| = -\dfrac{1}{2}$

impossible, since absolute value always yields a non-negative number.

85. $|x^2 - 9| = 0$

$x^2 - 9 = 0$

$x^2 = 9$

$x = \pm 3$

The solution set is $\{-3, 3\}$.

87. $|x^2 - 2x| = 3$

$x^2 - 2x = 3$ or $x^2 - 2x = -3$

$x^2 - 2x - 3 = 0$ or $x^2 - 2x + 3 = 0$

$(x-3)(x+1) = 0$ or $x = \dfrac{2 \pm \sqrt{4-12}}{2}$

$= \dfrac{2 \pm \sqrt{-8}}{2}$ no real sol.

$x = 3$ or $x = -1$

The solution set is $\{-1, 3\}$.

89. $x^3 - 9x = 0$

$x(x^2 - 9) = 0$

$x(x-3)(x+3) = 0$

$x = 0$ or $x - 3 = 0$ $x + 3 = 0$

$x = 3$ $x = -3$

The solution set is $\{-3, 0, 3\}$.

91. $x^3 + x^2 - 20x = 0$

$x(x^2 + x - 20) = 0$

$x(x+5)(x-4) = 0$

$x = 0$ or $x + 5 = 0$ or $x - 4 = 0$

$x = -5$ $x = 4$

The solution set is $\{-5, 0, 4\}$.

93. $x^3 + x^2 - x - 1 = 0$

$x^2(x+1) - 1(x+1) = 0$

$(x+1)(x^2 - 1) = 0$

$(x+1)(x-1)(x+1) = 0$

$x + 1 = 0$ or $x - 1 = 0$

$x = -1$ $x = 1$

The solution set is $\{-1, 1\}$.

95. $x^3 - 3x^2 - 4x + 12 = 0$

$x^2(x-3) - 4(x-3) = 0$

$(x-3)(x^2 - 4) = 0$

$(x-3)(x-2)(x+2) = 0$

$x - 3 = 0$ or $x - 2 = 0$ or $x + 2 = 0$

$x = 3$ $x = 2$ $x = -2$

The solution set is $\{-2, 2, 3\}$.

97. $2x^3 + 4 = x^2 + 8x$

$2x^3 - x^2 - 8x + 4 = 0$

$x^2(2x-1) - 4(2x-1) = 0$

$(2x-1)(x^2 - 4) = 0$

$(2x-1)(x-2)(x+2) = 0$

$2x - 1 = 0$ or $x - 2 = 0$ or $x + 2 = 0$

$2x = 1$ $x = 2$ $x = -2$

$x = \dfrac{1}{2}$

The solution set is $\left\{-2, \dfrac{1}{2}, 2\right\}$.

99. $x - 4x^{1/2} + 2 = 0$; Use ZERO (or ROOT) on the graph of $y_1 = x - 4\sqrt{x} + 2$.

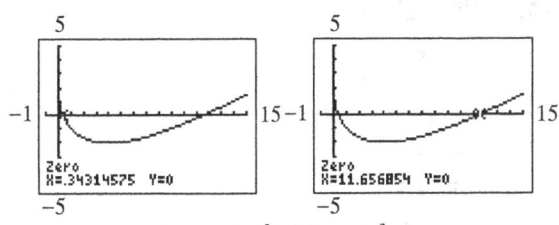

The solution set is $\{0.34, 11.66\}$.

101. $x^4 + \sqrt{3}x^2 - 3 = 0$; Use ZERO (or ROOT) on the graph of $y_1 = x^4 + \sqrt{3}x^2 - 3$.

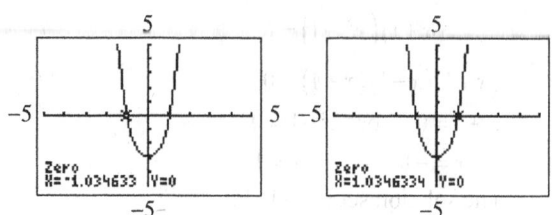

The solution set is $\{-1.03, 1.03\}$.

103. $3x^2 + 7x - 20 = 0$

$(3x - 5)(x + 4) = 0$

$3x - 5 = 0$ or $x + 4 = 0$

$3x = 5 \qquad x = -4$

$x = \dfrac{5}{3}$

The solution set is $\left\{-4, \dfrac{5}{3}\right\}$.

105. $5a^3 - 45a = -2a^2 + 18$

$5a^3 + 2a^2 - 45a - 18 = 0$

$a^2(5a + 2) - 9(5a + 2) = 0$

$(a^2 - 9)(5a + 2) = 0$

$(a - 3)(a + 3)(5a + 2) = 0$

$a - 3 = 0$ or $a + 3 = 0$ or $5a + 2 = 0$

$a = 3 \qquad a = -3 \qquad 5a = -2$

$a = -\dfrac{2}{5}$

The solution set is $\left\{-3, -\dfrac{2}{5}, 3\right\}$.

107. $-3|5x - 2| + 9 = 0$

$-3|5x - 2| = -9$

$|5x - 2| = 3$

$5x - 2 = 3$ or $5x - 2 = -3$

$5x = 5 \qquad 5x = -1$

$x = 1 \qquad x = -\dfrac{1}{5}$

The solution set is $\left\{-\dfrac{1}{5}, 1\right\}$.

109. $4(w - 3) = w + 3$

$4w - 12 = w + 3$

$3w = 15$

$w = 5$

The solution set is $\{5\}$.

111. $\left(\dfrac{v}{v + 1}\right)^2 + \dfrac{2v}{v + 1} = 8$

Let $u = \dfrac{v}{v + 1}$. Rewrite the equation:

$u^2 + 2u = 8$

$u^2 + 2u - 8 = 0$

$(u - 2)(u + 4) = 0$

$u = 2$ or $u = -4$

Go back in terms of v and solve:

$\dfrac{v}{v + 1} = 2 \qquad$ or $\qquad \dfrac{v}{v + 1} = -4$

$v = 2v + 2 \qquad\qquad v = -4v - 4$

$-v = 2 \qquad\qquad 5v = -4$

$v = -2 \qquad\qquad v = -\dfrac{4}{5}$

The solution set is $\left\{-2, -\dfrac{4}{5}\right\}$.

113. $|-3x + 2| = x + 10$

$-3x + 2 = -(x + 10)$ or $-3x + 2 = x + 10$

$-3x + 2 = -x - 10 \qquad\qquad -4x = 8$

$-2x = -12 \qquad\qquad\qquad x = -2$

$x = 6$

The solution set is $\{-2, 6\}$.

115. $\sqrt{2x + 5} - x = 1$

$\sqrt{2x + 5} = x + 1$

$\left(\sqrt{2x + 5}\right)^2 = (x + 1)^2$

$2x + 5 = x^2 + 2x + 1$

$x^2 - 4 = 0$

$(x - 2)(x + 2) = 0$

$x = 2$ or $x = -2$

Check:

$\sqrt{2(-2)+5}-(-2)=1 \quad \sqrt{2(2)+5}-(2)=1$

$\sqrt{1}+2=1 \qquad \sqrt{9}-2=1$

$3 \neq 1 \qquad\qquad 1=1 \text{ T}$

The solution set is $\{2\}$.

117. $3m^2+6m=-1$

$3m^2+6m+1=0$

$a=3,\ b=6,\ c=1$

$m=\dfrac{-6\pm\sqrt{6^2-4(3)(1)}}{2(3)}=\dfrac{-6\pm\sqrt{24}}{6}$

$=\dfrac{-6\pm2\sqrt{6}}{6}=\dfrac{-3\pm\sqrt{6}}{3}$

The solution set is $\left\{\dfrac{-3-\sqrt{6}}{3},\dfrac{-3+\sqrt{6}}{3}\right\}$.

119. $\left|x^2+x-1\right|=1$

$x^2+x-1=1 \qquad$ or $\quad x^2+x-1=-1$

$x^2+x-2=0 \qquad\qquad x^2+x=0$

$(x+2)(x-1)=0 \qquad x(x+1)=0$

$x=-2$ or $x=1 \qquad x=0$ or $x=-1$

The solution set is $\{-2,-1,0,1\}$.

121. $\sqrt[4]{5x^2-6}=x$

$\left(\sqrt[4]{5x^2-6}\right)^4=x^4$

$5x^2-6=x^4$

$x^4-5x^2+6=0$

$(x^2-3)(x^2-2)=0$

$x^2-3=0 \qquad$ or $\quad x^2-2=0$

$x^2=3 \qquad\qquad x^2=2$

$x=\pm\sqrt{3} \qquad\qquad x=\pm\sqrt{2}$

Since an even root cannot be negative, we can exclude $-\sqrt{3}$ and $-\sqrt{2}$ as solutions.

Check:

$\sqrt[4]{5\left(\sqrt{2}\right)^2-6}=\sqrt{2} \qquad \sqrt[4]{5\left(\sqrt{3}\right)^2-6}=\sqrt{3}$

$\sqrt[4]{4}=\sqrt{2} \qquad\qquad \sqrt[4]{9}=\sqrt{3}$

$\sqrt[4]{2^2}=\sqrt{2} \qquad\qquad \sqrt[4]{3^2}=\sqrt{3}$

$\sqrt{2}=\sqrt{2} \text{ T} \qquad\qquad \sqrt{3}=\sqrt{3} \text{ T}$

The solution set is $\left\{\sqrt{2},\sqrt{3}\right\}$.

123. $t^4-16=0$

$(t^2-4)(t^2+4)=0$

$t^2-4=0 \quad$ or $\quad t^2+4=0$

$t^2=4 \qquad\qquad t^2=-4$

$t=\pm2 \qquad\qquad t=\pm2i$

The solution set is $\{-2i,2i,-2,2\}$.

125. $x^6-9x^3+8=0$

$(x^3-8)(x^3-1)=0$

$(x-2)(x^2+2x+4)(x-1)(x^2+x+1)=0$

$x-2=0 \quad$ or $\quad x-1=0$

$x=2 \qquad\qquad x=1$

or $\quad x^2+2x+4=0$

$a=1,b=2,c=4$

$x=\dfrac{-2\pm\sqrt{2^2-4(1)(4)}}{2(1)}=\dfrac{-2\pm\sqrt{-12}}{2}$

$=\dfrac{-2\pm2i\sqrt{3}}{2}=-1\pm i\sqrt{3}$

or

$x^2+x+1=0$

$a=1,b=1,c=1$

$x=\dfrac{-1\pm\sqrt{1^2-4(1)(1)}}{2(1)}=\dfrac{-1\pm\sqrt{-3}}{2}=\dfrac{-1\pm i\sqrt{3}}{2}$

The solution set is $\left\{-\dfrac{1}{2}-\dfrac{\sqrt{3}}{2}i,-1-i\sqrt{3},\right.$

$\left.-\dfrac{1}{2}+\dfrac{\sqrt{3}}{2}i,-1+i\sqrt{3},\ 1,\ 2\right\}$.

127. $k^2 - k = 12 \Rightarrow k^2 - k - 12 = 0$

$(k-4)(k+3) = 0$

$k = 4$ or $k = -3$

$\dfrac{x+3}{x-3} = 4$ $\dfrac{x+3}{x-3} = -3$

$x + 3 = 4x - 12$ $x + 3 = -3x + 9$

$3x = 15$ $x = \dfrac{3}{2}$

$x = 5$

And since neither of these x values causes a denominator to equal zero, the solution set is $\left\{ \dfrac{3}{2}, 5 \right\}$.

129. All points having an x-coordinate of 2 are of the form $(2, y)$. Those which are 5 units from $(-2, -1)$ are:

$\sqrt{(2-(-2))^2 + (y-(-1))^2} = 5$

$\sqrt{4^2 + (y+1)^2} = 5$

Squaring both sides:

$4^2 + (y+1)^2 = 25$

$16 + (y+1)^2 = 25$

$(y+1)^2 = 9$

$y + 1 = \pm 3$

$y = -1 \pm 3$

$y = -4$ or $y = 2$

Therefore, the points are $(2, 4)$ and $(2, 2)$.

131. All points on the x-axis are of the form $(x, 0)$. Those which are 5 units from $(4, -3)$ are:

$\sqrt{(x-4)^2 + (0-(-3))^2} = 5$

$\sqrt{(x-4)^2 + 3^2} = 5$

Squaring both sides:

$(x-4)^2 + 9 = 25$

$x^2 - 8x + 16 + 9 = 25$

$x^2 - 8x = 0$

$x(x-8) = 0$

$x = 0$ or $x = 8$

Therefore, the points are $(0, 0)$ and $(8, 0)$.

133. $|8 - 3x| = |2x - 7|$

$8 - 3x = 2x - 7$ or $8 - 3x = -(2x - 7)$

$-3x - 2x = -7 - 8$ $8 - 3x = -2x + 7$

$-5x = -15$ $-3x + 2x = 7 - 8$

$x = -3$ $-x = -1$

$x = 1$

The solution set is $\{-3, 1\}$.

135. Graph the equations $y_1 = \sqrt{x}/4 + x/1100$ and $y_2 = 4$; then use INTERSECT to find the x-coordinate of the points of intersection:

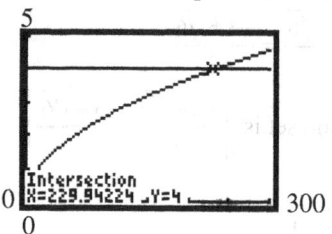

The distance to the water's surface is approximately 229.94 feet.

137. $T = 2\pi \sqrt{\dfrac{l}{32}}$

Let $T = 16.5$ and solve for l.

$16.5 = 2\pi \sqrt{\dfrac{l}{32}}$

$\dfrac{16.5}{2\pi} = \sqrt{\dfrac{l}{32}}$

$\left(\dfrac{16.5}{2\pi}\right)^2 = \left(\sqrt{\dfrac{l}{32}}\right)^2$

$\left(\dfrac{16.5}{2\pi}\right)^2 = \dfrac{l}{32}$

$l = 32\left(\dfrac{16.5}{2\pi}\right)^2 \approx 220.7$

The length was approximately 221 feet.

139. Answers will vary, one example is

$x - \sqrt{x} - 2 = 0$

141. $|x|$ represents the distance from x to the origin. Since distance is never negative, $|x| = -2$ has no real solution.

Section 1.6

1. mathematical modeling

3. uniform motion

5. True; this is the uniform motion formula.

7. b

9. Let A represent the area of the circle and r the radius. The area of a circle is the product of π times the square of the radius: $A = \pi r^2$

11. Let A represent the area of the square and s the length of a side. The area of the square is the square of the length of a side: $A = s^2$

13. Let F represent the force, m the mass, and a the acceleration. Force equals the product of the mass times the acceleration: $F = ma$

15. Let W represent the work, F the force, and d the distance. Work equals force times distance: $W = Fd$

17. C = total variable cost in dollars, x = number of dishwashers manufactured: $C = 150x$

19. Let x represent the amount of money invested in bonds. Then $50,000 - x$ represents the amount of money invested in CD's. Since the total interest is to be $6,000, we have:
$$0.15x + 0.07(50,000 - x) = 6,000$$
$$(100)(0.15x + 0.07(50,000 - x)) = (6,000)(100)$$
$$15x + 7(50,000 - x) = 600,000$$
$$15x + 350,000 - 7x = 600,000$$
$$8x + 350,000 = 600,000$$
$$8x = 250,000$$
$$x = 31,250$$
$31,250 should be invested in bonds at 15% and $18,750 should be invested in CD's at 7%.

21. Let x represent the amount of money loaned at 8%. Then $12,000 - x$ represents the amount of money loaned at 18%. Since the total interest is to be $1,000, we have:

$$0.08x + 0.18(12,000 - x) = 1,000$$
$$(100)(0.08x + 0.18(12,000 - x)) = (1,000)(100)$$
$$8x + 18(12,000 - x) = 100,000$$
$$8x + 216,000 - 18x = 100,000$$
$$-10x + 216,000 = 100,000$$
$$-10x = -116,000$$
$$x = 11,600$$
$11,600 is loaned at 8% and $400 is at 18%.

23. Let x represent the number of pounds of Earl Gray tea. Then $100 - x$ represents the number of pounds of Orange Pekoe tea.
$$5x + 3(100 - x) = 4.50(100)$$
$$5x + 300 - 3x = 450$$
$$2x + 300 = 450$$
$$2x = 150$$
$$x = 75$$
75 pounds of Earl Gray tea must be blended with 25 pounds of Orange Pekoe.

25. Let x represent the number of pounds of cashews. Then $x + 60$ represents the number of pounds in the mixture.
$$9x + 3.50(60) = 7.50(x + 60)$$
$$9x + 210 = 7.50x + 450$$
$$1.5x = 240$$
$$x = 160$$
160 pounds of cashews must be added to the 60 pounds of almonds.

27. Let r represent the speed of the current.

	Rate	Time	Distance
Upstream	$16 - r$	$\frac{20}{60} = \frac{1}{3}$	$\frac{16-r}{3}$
Downstream	$16 + r$	$\frac{15}{60} = \frac{1}{4}$	$\frac{16+r}{4}$

Since the distance is the same in each direction:
$$\frac{16-r}{3} = \frac{16+r}{4}$$
$$4(16 - r) = 3(16 + r)$$
$$64 - 4r = 48 + 3r$$
$$16 = 7r$$
$$r = \frac{16}{7} \approx 2.286$$
The speed of the current is approximately 2.286 miles per hour.

29. Let r represent the speed of the current.

	Rate	Time	Distance
Upstream	$15-r$	$\dfrac{10}{15-r}$	10
Downstream	$15+r$	$\dfrac{10}{15+r}$	10

Since the total time is 1.5 hours, we have:

$$\frac{10}{15-r}+\frac{10}{15+r}=1.5$$
$$10(15+r)+10(15-r)=1.5(15-r)(15+r)$$
$$150+10r+150-10r=1.5(225-r^2)$$
$$300=1.5(225-r^2)$$
$$200=225-r^2$$
$$r^2-25=0$$
$$(r-5)(r+5)=0$$
$$r=5 \text{ or } r=-5$$

Speed must be positive, so disregard $r=-5$.
The speed of the current is 5 miles per hour.

31. Let r represent Karen's normal walking speed.

	Rate	Time	Distance
With walkway	$r+2.5$	$\dfrac{50}{r+2.5}$	50
Against walkway	$r-2.5$	$\dfrac{50}{r-2.5}$	50

Since the total time is 40 seconds:

$$\frac{50}{r+2.5}+\frac{50}{r-2.5}=40$$
$$50(r-2.5)+50(r+2.5)=40(r-2.5)(r+2.5)$$
$$50r-125+50r+125=40(r^2-6.25)$$
$$100r=40r^2-250$$
$$0=40r^2-100r-250$$
$$0=4r^2-10r-25$$
$$r=\frac{-(-10)\pm\sqrt{(-10)^2-4(4)(-25)}}{2(4)}$$
$$=\frac{10\pm\sqrt{500}}{8}=\frac{10\pm10\sqrt{5}}{8}=\frac{5\pm5\sqrt{5}}{4}$$
$$r\approx4.05 \text{ or } r\approx-1.55$$

Speed must be positive, so disregard $r\approx-1.55$.
Karen' normal walking speed is approximately 4.05 feet per second.

33. Let w represent the width of a regulation doubles tennis court. Then $2w+6$ represents the length. The area is 2808 square feet:

$$w(2w+6)=2808$$
$$2w^2+6w=2808$$
$$2w^2+6w-2808=0$$
$$w^2+3w-1404=0$$
$$(w+39)(w-36)=0$$
$$w+39=0 \quad \text{or } w-36=0$$
$$w=-39 \text{ or } \qquad w=36$$

The width must be positive, so disregard $w=-39$. The width of a regulation doubles tennis court is 36 feet and the length is $2(36)+6=78$ feet.

35. Let t represent the time it takes to do the job together.

	Time to do job	Part of job done in one minute
Trent	30	$\dfrac{1}{30}$
Lois	20	$\dfrac{1}{20}$
Together	t	$\dfrac{1}{t}$

$$\frac{1}{30}+\frac{1}{20}=\frac{1}{t}$$
$$2t+3t=60$$
$$5t=60$$
$$t=12$$

Working together, the job can be done in 12 minutes.

37. $l=$ length of the garden
$w=$ width of the garden

a. The length of the garden is to be twice its width. Thus, $l=2w$.
The dimensions of the fence are $l+4$ and $w+4$.
The perimeter is 46 feet, so:

$$2(l+4)+2(w+4)=46$$
$$2(2w+4)+2(w+4)=46$$
$$4w+8+2w+8=46$$
$$6w+16=46$$
$$6w=30$$
$$w=5$$

The dimensions of the garden are 5 feet by 10 feet.

b. Area $= l \cdot w = 5 \cdot 10 = 50$ square feet

c. If the dimensions of the garden are the same, then the length and width of the fence are also the same $(l + 4)$. The perimeter is 46 feet, so:

$$2(l+4) + 2(l+4) = 46$$
$$2l + 8 + 2l + 8 = 46$$
$$4l + 16 = 46$$
$$4l = 30$$
$$l = 7.5$$

The dimensions of the garden are 7.5 feet by 7.5 feet.

d. Area $= l \cdot w = 7.5(7.5) = 56.25$ square feet.

39. Let t represent the time it takes for the defensive back to catch the tight end.

	Time to run 100 yards	Time	Rate	Distance
Tight End	12 sec	t	$\frac{100}{12} = \frac{25}{3}$	$\frac{25}{3}t$
Def. Back	10 sec	t	$\frac{100}{10} = 10$	$10t$

Since the defensive back has to run 5 yards farther, we have:

$$\frac{25}{3}t + 5 = 10t$$
$$25t + 15 = 30t$$
$$15 = 5t$$
$$t = 3 \quad \rightarrow \quad 10t = 30$$

The defensive back will catch the tight end at the 45 yard line (15 + 30 = 45).

41. Let x represent the number of gallons of pure water. Then $x + 1$ represents the number of gallons in the 60% solution.

$$(\%)(\text{gallons}) + (\%)(\text{gallons}) = (\%)(\text{gallons})$$
$$0(x) + 1(1) = 0.60(x+1)$$
$$1 = 0.6x + 0.6$$
$$0.4 = 0.6x$$
$$x = \frac{4}{6} = \frac{2}{3}$$

$\frac{2}{3}$ gallon of pure water should be added.

43. Let x represent the number of ounces of water to be evaporated; the amount of salt remains the same. Therefore, we get

$$0.04(32) = 0.06(32 - x)$$
$$1.28 = 1.92 - 0.06x$$
$$0.06x = 0.64$$
$$x = \frac{0.64}{0.06} = \frac{64}{6} = \frac{32}{3} = 10\frac{2}{3}$$

$10\frac{2}{3} \approx 10.67$ ounces of water need to be evaporated.

45. Let x represent the number of grams of pure gold. Then $60 - x$ represents the number of grams of 12 karat gold to be used.

$$x + \frac{1}{2}(60 - x) = \frac{2}{3}(60)$$
$$x + 30 - 0.5x = 40$$
$$0.5x = 10$$
$$x = 20$$

20 grams of pure gold should be mixed with 40 grams of 12 karat gold.

47. Let t represent the time it takes for Mike to catch up with Dan. Since the distances are the same, we have:

$$\frac{1}{6}t = \frac{1}{9}(t+1)$$
$$3t = 2t + 2$$
$$t = 2$$

Mike will pass Dan after 2 minutes, which is a distance of $\frac{1}{3}$ mile.

49. Let t represent the time the auxiliary pump needs to run. Since the two pumps are emptying one tanker, we have:

$$\frac{3}{4} + \frac{t}{9} = 1$$
$$27 + 4t = 36$$
$$4t = 9$$
$$t = \frac{9}{4} = 2.25$$

The auxiliary pump must run for 2.25 hours. It must be started at 9:45 a.m.

51. Let t represent the time for the tub to fill with the faucets on and the stopper removed. Since one tub is being filled, we have:

$$\frac{t}{15} + \left(-\frac{t}{20}\right) = 1$$

$$4t - 3t = 60$$

$$t = 60$$

60 minutes is required to fill the tub.

53. Let t represent the time spent running. Then $5 - t$ represents the time spent biking.

	Rate	Time	Distance
Run	6	t	$6t$
Bike	25	$5-t$	$25(5-t)$

The total distance is 87 miles:

$$6t + 25(5-t) = 87$$

$$6t + 125 - 25t = 87$$

$$-19t + 125 = 87$$

$$-19t = -38$$

$$t = 2$$

The time spent running is 2 hours, so the distance of the run is $6(2) = 12$ miles. The distance of the bicycle race is $25(5-2) = 75$ miles.

55. Burke's rate is $\frac{100}{12}$ meters/sec. In 9.69 seconds, Burke will run $\frac{100}{12}(9.69) = 80.75$ meters. Bolt would win by 19.25 meters.

57. Let x be the original selling price of the shirt.
    Profit = Revenue − Cost

$$4 = x - 0.40x - 20 \rightarrow 24 = 0.60x \rightarrow x = 40$$

The original price should be \$40 to ensure a profit of \$4 after the sale.

If the sale is 50% off, the profit is:
$$40 - 0.50(40) - 20 = 40 - 20 - 20 = 0$$
At 50% off there will be no profit.

59. It is impossible to mix two solutions with a lower concentration and end up with a new solution with a higher concentration.

Algebraic Solution:
Let $x =$ the number of liters of 25% solution.
$$(\%)(\text{liters}) + (\%)(\text{liters}) = (\%)(\text{liters})$$
$$0.25x + 0.48(20) = 0.58(20 + x)$$
$$0.25x + 9.6 = 10.6 + 0.58x$$
$$-0.33x = 1$$
$$x \approx -3.03 \text{ liters}$$
$$(\text{not possible})$$

61. The time traveled with the tail wind was:
$$t = \frac{919}{550} \approx 1.67091 \text{ hours}.$$

Since they were 20 minutes $\left(\frac{1}{3} \text{ hour}\right)$ early, the time in still air would have been:
$$1.67091 \text{ hrs} + 20 \text{ min} = (1.67091 + 0.33333) \text{ hrs}$$
$$\approx 2.00424 \text{ hrs}$$
Thus, with no wind, the ground speed is
$$\frac{919}{2.00424} \approx 458.53.$$ Therefore, the tail wind is
$$550 - 458.53 = 91.47 \text{ knots}.$$

Section 1.7

1. $x \geq -2$

3. $A \cup B = \{a, b, c, d, e, i, o, u\}$

5. closed interval

7. $(-\infty, a]$

9. True; this follows from the multiplication property for inequalities.

11. a

13. Interval: $[0, 2]$
    Inequality: $0 \leq x \leq 2$

15. Interval: $[2, \infty)$
    Inequality: $x \geq 2$

66

17. Interval: $[0,3)$

Inequality: $0 \leq x < 3$

19. a. $\quad 3 < 5$

$\quad 3+3 < 5+3$

$\quad 6 < 8$

b. $\quad 3 < 5$

$\quad 3-5 < 5-5$

$\quad -2 < 0$

c. $\quad 3 < 5$

$\quad 3(3) < 3(5)$

$\quad 9 < 15$

d. $\quad 3 < 5$

$\quad -2(3) > -2(5)$

$\quad -6 > -10$

21. a. $\quad 2x+1 < 2$

$\quad 2x+1+3 < 2+3$

$\quad 2x+4 < 5$

b. $\quad 2x+1 < 2$

$\quad 2x+1-5 < 2-5$

$\quad 2x-4 < -3$

c. $\quad 2x+1 < 2$

$\quad 3(2x+1) < 3(2)$

$\quad 6x+3 < 6$

d. $\quad 2x+1 < 2$

$\quad -2(2x+1) > -2(2)$

$\quad -4x-2 > -4$

23. $[0, 4]$

25. $[4, 6)$

27. $[4, \infty)$

29. $(-\infty, -4)$

31. $2 \leq x \leq 5$

33. $-3 < x < -2$

35. $x \geq 4$

37. $x < -3$

39. If $x < 5$, then $x - 5 < 0$.

41. If $x > -4$, then $x + 4 > 0$.

43. If $x \geq -4$, then $3x \geq -12$.

45. If $x > 6$, then $-2x < -12$.

47. If $2x < 6$, then $x < 3$.

49. If $-\dfrac{1}{2}x \leq 3$, then $x \geq -6$.

51. $3x - 7 > 2$

$\quad 3x > 9$

$\quad x > 3$

$\quad \{x \mid x > 3\}$ or $(3, \infty)$

53. $1 - 2x \leq 3$

$\quad -2x \leq 2$

$\quad x \geq -1$

$\quad \{x \mid x \geq -1\}$ or $[-1, \infty)$

67

55. $3x - 1 \geq 3 + x$

$2x \geq 4$

$x \geq 2$

$\{x \mid x \geq 2\}$ or $[2, \infty)$

57. $-2(x+3) < 8$

$-2x - 6 < 8$

$-2x < 14$

$x > -7$

$\{x \mid x > -7\}$ or $(-7, \infty)$

59. $4 - 3(1-x) \leq 3$

$4 - 3 + 3x \leq 3$

$3x + 1 \leq 3$

$3x \leq 2$

$x \leq \dfrac{2}{3}$

$\left\{x \mid x \leq \dfrac{2}{3}\right\}$ or $\left(-\infty, \dfrac{2}{3}\right]$

61. $\dfrac{1}{2}(x-4) > x + 8$

$\dfrac{1}{2}x - 2 > x + 8$

$-\dfrac{1}{2}x > 10$

$x < -20$

$\{x \mid x < -20\}$ or $(-\infty, -20)$

63. $0 \leq 2x - 6 \leq 4$

$6 \leq 2x \leq 10$

$3 \leq x \leq 5$

$\{x \mid 3 \leq x \leq 5\}$ or $[3, 5]$

65. $-5 \leq 4 - 3x \leq 2$

$-9 \leq -3x \leq -2$

$3 \geq x \geq \dfrac{2}{3}$

$\left\{x \mid \dfrac{2}{3} \leq x \leq 3\right\}$ or $\left[\dfrac{2}{3}, 3\right]$

67. $-3 < \dfrac{2x-1}{4} < 0$

$-12 < 2x - 1 < 0$

$-11 < 2x < 1$

$-\dfrac{11}{2} < x < \dfrac{1}{2}$

$\left\{x \mid -\dfrac{11}{2} < x < \dfrac{1}{2}\right\}$ or $\left(-\dfrac{11}{2}, \dfrac{1}{2}\right)$

69. $1 < 1 - \dfrac{1}{2}x < 4$

$0 < -\dfrac{1}{2}x < 3$

$0 > x > -6$ or $-6 < x < 0$

$\{x \mid -6 < x < 0\}$ or $(-6, 0)$

71. $(x+2)(x-3) > (x-1)(x+1)$

$x^2 - x - 6 > x^2 - 1$

$-x - 6 > -1$

$-x > 5$

$x < -5$

$\{x \mid x < -5\}$ or $(-\infty, -5)$

73. $|2x| < 8$

$-8 < 2x < 8$

$-4 < x < 4$

$\{x \mid -4 < x < 4\}$ or $(-4, 4)$

68

75. $|3x| > 12$

$3x < -12$ or $3x > 12$

$x < -4$ or $x > 4$

$\{x \mid x < -4 \text{ or } x > 4\}$ or $(-\infty, -4) \cup (4, \infty)$

77. $|3t - 2| \le 4$

$-4 \le 3t - 2 \le 4$

$-2 \le 3t \le 6$

$-\dfrac{2}{3} \le t \le 2$

$\left\{t \mid -\dfrac{2}{3} \le t \le 2\right\}$ or $\left[-\dfrac{2}{3}, 2\right]$

79. $|x - 2| + 2 < 3$

$|x - 2| < 1$

$-1 < x - 2 < 1$

$1 < x < 3$

$\{x \mid 1 < x < 3\}$ or $(1, 3)$

81. $|x - 3| \ge 2$

$x - 3 \le -2$ or $x - 3 \ge 2$

$x \le 1$ or $x \ge 5$

$\{x \mid x \le 1 \text{ or } x \ge 5\}$ or $(-\infty, 1] \cup [5, \infty)$

83. $|1 - 2x| > |-3|$

$|1 - 2x| > 3$

$1 - 2x < -3$ or $1 - 2x > 3$

$-2x < -4$ or $-2x > 2$

$x > 2$ or $x < -1$

$\{x \mid x < -1 \text{ or } x > 2\}$ or $(-\infty, -1) \cup (2, \infty)$

85. $|1 - 4x| - 7 < -2$

$|1 - 4x| < 5$

$-5 < 1 - 4x < 5$

$-6 < -4x < 4$

$\dfrac{-6}{-4} > x > \dfrac{4}{-4}$

$\dfrac{3}{2} > x > -1$ or $-1 < x < \dfrac{3}{2}$

$\left\{x \mid -1 < x < \dfrac{3}{2}\right\}$ or $\left(-1, \dfrac{3}{2}\right)$

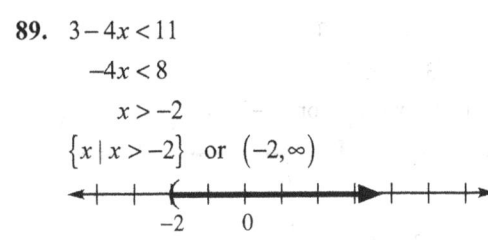

87. $|2x + 1| < -1$

No solution since absolute value is always non-negative.

89. $3 - 4x < 11$

$-4x < 8$

$x > -2$

$\{x \mid x > -2\}$ or $(-2, \infty)$

91. $|2x + 1| - 5 \ge -1$

$|2x + 1| \ge 4$

$2x + 1 \le -4$ or $2x + 1 \ge 4$

$2x \le -5$ $2x \ge 3$

$x \le -\dfrac{5}{2}$ $x \ge \dfrac{3}{2}$

$\left\{x \mid x \le -\dfrac{5}{2} \text{ or } x \ge \dfrac{3}{2}\right\}$ or $\left(-\infty, -\dfrac{5}{2}\right] \cup \left[\dfrac{3}{2}, \infty\right)$

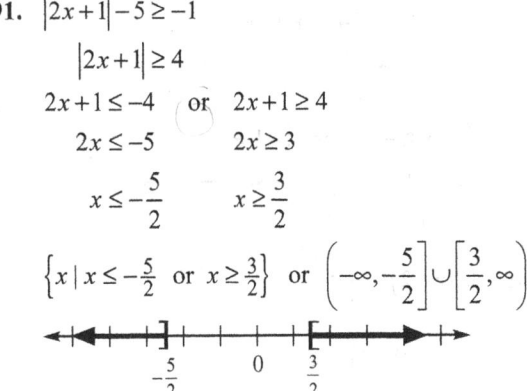

93. $\dfrac{x}{2} \geq 1 - \dfrac{x}{4}$

$4 \cdot \dfrac{x}{2} \geq 4 \cdot \left(1 - \dfrac{x}{4}\right)$

$2x \geq 4 - x$

$3x \geq 4$

$x \geq \dfrac{4}{3}$

$\left\{x \mid x \geq \dfrac{4}{3}\right\}$ or $\left[\dfrac{4}{3}, \infty\right)$

95. $-\dfrac{1}{3} \leq \dfrac{x+1}{6} < \dfrac{4}{3}$

$6\left(-\dfrac{1}{3}\right) \leq 6 \cdot \dfrac{x+1}{6} < 6 \cdot \dfrac{4}{3}$

$-2 \leq x+1 < 8$

$-3 \leq x < 7$

$\{x \mid -3 \leq x < 7\}$ or $[-3, 7)$

97. $x(4x+3) \leq (2x+1)^2$

$4x^2 + 3x \leq 4x^2 + 4x + 1$

$-x \leq 1$

$x \geq -1$

$\{x \mid x \geq -1\}$ or $[-1, \infty)$

99. $\left|(3x-2)-7\right| < \dfrac{1}{2}$

$|3x-9| < \dfrac{1}{2}$

$-\dfrac{1}{2} < 3x-9 < \dfrac{1}{2}$

$\dfrac{17}{2} < 3x < \dfrac{19}{2}$

$\dfrac{17}{6} < x < \dfrac{19}{6}$

$\left\{x \mid \dfrac{17}{6} < x < \dfrac{19}{6}\right\}$ or $\left(\dfrac{17}{6}, \dfrac{19}{6}\right)$

101. $-3 < 5 - 2x \leq 11$

$-8 < -2x \leq 6$

$4 > x \geq -3$

$-3 \leq x < 4$

$\{x \mid -3 \leq x < 4\}$ or $[-3, 4)$

103. $7 - |x-1| > 4$

$-|x-1| > -3$

$|x-1| < 3$

$-3 < x-1 < 3$

$-2 < x < 4$

$\{x \mid -2 < x < 4\}$ or $(-2, 4)$

105. $-3 < x+5 < 2x$

$-3 < x+5$ and $x+5 < 2x$

$-8 < x$ $\qquad$ $5 < x$

$x > -8$ $\qquad$ $x > 5$

We need both $x > -8$ and $x > 5$. Therefore, $x > 5$ is sufficient to satisfy both inequalities.

$\{x \mid x > 5\}$ or $(5, \infty)$

107. $x+2 < 2x-1 < 5x$

$x+2 < 2x-1$ and $2x-1 < 5x$

$2 < x-1$ $\qquad$ $-3x-1 < 0$

$3 < x$ $\qquad$ $-3x < 1$

$x > 3$ $\qquad\qquad$ $x > -\dfrac{1}{3}$

We need both $x > -\dfrac{1}{3}$ and $x > 3$. Therefore, $x > 3$ is sufficient to satisfy both inequalities.

$\{x \mid x > 3\}$ or $(3, \infty)$.

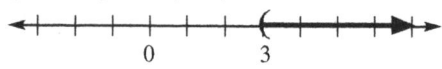

109. $|x-2| < 0.5$

$-0.5 < x-2 < 0.5$

$-0.5+2 < x < 0.5+2$

$1.5 < x < 2.5$

Solution set: $\{x|\ 1.5 < x < 2.5\}$

111. $|x-(-3)| > 2$

$x-(-3) < -2 \quad$ or $\quad x-(-3) > 2$

$x+3 < -2 \quad$ or $\quad x+3 > 2$

$x < -5 \quad\quad$ or $\quad x > -1$

Solution set: $\{x|\ x < -5 \text{ or } x > -1\}$

113. $21 <$ young adult's age < 30

115. A temperature x that differs from $98.6°$ F by at least $1.5°F$.

$|x-98.6°| \geq 1.5°$

$x-98.6° \leq -1.5°$ or $x-98.6° \geq 1.5°$

$\quad x \leq 97.1°$ or $\quad\quad x \geq 100.1°$

The temperatures that are considered unhealthy are those that are less than 97.1°F or greater than 100.1°F, inclusive.

117. a. Let $x =$ age at death.

$x-30 \geq 51.9$

$\quad x \geq 81.9$

Therefore, the average life expectancy for a 30-year-old male in 2014 will be greater than or equal to 81.9 years.

b. Let $x =$ age at death.

$x-30 \geq 55.6$

$\quad x \geq 85.6$

Therefore, the average life expectancy for a 30-year-old female in 2014 will be greater than or equal to 85.6 years.

c. By the given information, a female can expect to live $85.6 - 81.9 = 3.7$ years longer.

119. Let P represent the selling price and C represent the commission.

Calculating the commission:

$C = 45,000 + 0.25(P - 900,000)$

$\quad = 45,000 + 0.25P - 225,000$

$\quad = 0.25P - 180,000$

Calculate the commission range, given the price range:

$900,000 \leq P \leq 1,100,000$

$0.25(900,000) \leq 0.25P \leq 0.25(1,100,000)$

$225,000 \leq 0.25P \leq 275,000$

$225,000 - 180,000 \leq 0.25P - 180,000 \leq 275,000 - 180,000$

$45,000 \leq C \leq 95,000$

The agent's commission ranges from \$45,000 to \$95,000, inclusive.

$\dfrac{45,000}{900,000} = 0.05 = 5\%$ to

$\dfrac{95,000}{1,100,000} = 0.086 = 8.6\%$, inclusive.

As a percent of selling price, the commission ranges from 5% to 8.6%, inclusive.

121. Let $W =$ weekly wages and $T =$ tax withheld. Calculating the withholding tax range, given the range of weekly wages:

$900 \leq W \leq 1100$

$900 - 753 \leq W - 753 \leq 1100 - 753$

$147 \leq W - 753 \leq 347$

$0.25(147) \leq 0.25(W - 753) \leq 0.25(347)$

$36.75 \leq 0.25(W - 753) \leq 86.75$

$36.75 + 97.75 \leq 0.25(W - 753) + 97.75 \leq 86.75 + 97.75$

$134.50 \leq T \leq 184.50$

The amount of withholding tax ranges from \$134.50 to \$184.50, inclusive.

123. Let K represent the monthly usage in kilowatt-hours and let C represent the monthly customer bill.

Calculating the bill:

$C = 0.0821K + 15.37$

Calculating the range of kilowatt-hours, given the range of bills:

$72.84 \leq C \leq 237.04$

$72.84 \leq 0.0821K + 15.37 \leq 237.04$

$57.47 \leq 0.0821K \leq 221.67$

$700 \leq K \leq 2700$

The range of usage in kilowatt-hours varied from 700 to 2700, inclusive.

125. Let C represent the dealer's cost and M represent the markup over dealer's cost. If the price is \$8800, then

$8800 = C + MC = C(1+M)$

Solving for C yields: $C = \dfrac{8800}{1+M}$

71

Calculating the range of dealer costs, given the range of markups:

$$0.12 \le M \le 0.18$$

$$1.12 \le 1 + M \le 1.18$$

$$\frac{1}{1.12} \ge \frac{1}{1+M} \ge \frac{1}{1.18}$$

$$\frac{8800}{1.12} \ge \frac{8800}{1+M} \ge \frac{8800}{1.18}$$

$$7857.14 \ge C \ge 7457.63$$

The dealer's cost ranged from \$7457.63 to \$7857.14, inclusive.

127. a. Let T represent the score on the last test and G represent the course grade. Calculating the course grade and solving for the last test:

$$G = \frac{68 + 82 + 87 + 89 + T}{5}$$

$$= \frac{326 + T}{5}$$

$$5G = 326 + T$$

$$T = 5G - 326$$

Calculating the range of scores on the last test, given the grade range:

$$80 \le G < 90$$

$$400 \le 5G < 450$$

$$74 \le 5G - 326 < 124$$

$$74 \le T < 124$$

The fifth test must be greater than or equal to 74.

b. Let T represent the score on the last test and G represent the course grade. Calculating the course grade and solving for the last test:

$$G = \frac{68 + 82 + 87 + 89 + 2T}{6} = \frac{326 + 2T}{6} = \frac{163 + T}{3}$$

$$T = 3G - 163$$

Calculating the range of scores on the last test, given the grade range:

$$80 \le G < 90$$

$$240 \le 3G < 270$$

$$77 \le 3G - 163 < 107$$

$$77 \le T < 107$$

The fifth test must be greater than or equal to 77 to get a B.

129. The true average number of books read x should differ from 13.6 by less than 1.8 books.

$$|x - 13.6| < 1.8$$

$$-1.8 < x - 13.6 < 1.8$$

$$11.8 < x < 15.4$$

Gallup is 99% confident that the actual average number of books read per year is between 11.8 and 15.4 books.

131. Since $a < b$

$$\frac{a}{2} < \frac{b}{2} \qquad\qquad \frac{a}{2} < \frac{b}{2}$$

$$\frac{a}{2} + \frac{a}{2} < \frac{a}{2} + \frac{b}{2} \qquad \frac{a}{2} + \frac{b}{2} < \frac{b}{2} + \frac{b}{2}$$

$$a < \frac{a+b}{2} \qquad\qquad \frac{a+b}{2} < b$$

Thus, $a < \dfrac{a+b}{2} < b$.

133. If $0 < a < b$, then

$$ab > a^2 > 0 \qquad\qquad b^2 > ab > 0$$

$$\left(\sqrt{ab}\right)^2 > a^2 \qquad\qquad b^2 > \left(\sqrt{ab}\right)^2$$

$$\sqrt{ab} > a \qquad\qquad b > \sqrt{ab}$$

Thus, $a < \sqrt{ab} < b$

135. For $0 < a < b$, $\dfrac{1}{h} = \dfrac{1}{2}\left(\dfrac{1}{a} + \dfrac{1}{b}\right)$

$$h \cdot \frac{1}{h} = \frac{1}{2}\left(\frac{b+a}{ab}\right) \cdot h$$

$$1 = \frac{1}{2}\left(\frac{b+a}{ab}\right) \cdot h$$

$$\frac{2ab}{a+b} = h$$

$$h - a = \frac{2ab}{a+b} - a = \frac{2ab - a(a+b)}{a+b}$$

$$= \frac{2ab - a^2 - ab}{a+b} = \frac{ab - a^2}{a+b}$$

$$= \frac{a(b-a)}{a+b} > 0$$

Therefore, $h > a$.

$$b - h = b - \frac{2ab}{a+b} = \frac{b(a+b) - 2ab}{a+b}$$
$$= \frac{ab + b^2 - 2ab}{a+b} = \frac{b^2 - ab}{a+b}$$
$$= \frac{b(b-a)}{a+b} > 0$$

Therefore, $h < b$ and we get $a < h < b$.

137. Answers will vary. One possibility:

No solution: $4x + 6 \le 2(x-5) + 2x$

One solution: $3x + 5 \le 2(x+3) + 1 \le 3(x+2) - 1$

139. Since $x^2 \ge 0$, we have

$$x^2 + 1 \ge 0 + 1$$
$$x^2 + 1 \ge 1$$

Therefore, the expression $x^2 + 1$ can never be less than -5.

Chapter 1 Review Exercises

1. $2 - \frac{x}{3} = 8$

$$6 - x = 24$$
$$x = -18$$

The solution set is $\{-18\}$.

2. $-2(5 - 3x) + 8 = 4 + 5x$

$$-10 + 6x + 8 = 4 + 5x$$
$$6x - 2 = 4 + 5x$$
$$x = 6$$

The solution set is $\{6\}$.

3. $\frac{3x}{4} - \frac{x}{3} = \frac{1}{12}$

$$9x - 4x = 1$$
$$5x = 1$$
$$x = \frac{1}{5}$$

The solution set is $\left\{\frac{1}{5}\right\}$

4. $\frac{x}{x-1} = \frac{6}{5}$

$$5x = 6x - 6$$
$$6 = x$$

Since $x = 6$ does not cause a denominator to equal zero, the solution set is $\{6\}$.

5. $x(1-x) = 6$

$$x - x^2 = 6$$
$$0 = x^2 - x + 6$$
$$b^2 - 4ac = (-1)^2 - 4(1)(6)$$
$$= 1 - 24 = -23$$

Therefore, there are no real solutions.

6. $\frac{1-3x}{4} = \frac{x+6}{3} + \frac{1}{2}$

$$(12)\left(\frac{1-3x}{4}\right) = \left(\frac{x+6}{3} + \frac{1}{2}\right)(12)$$
$$3(1-3x) = 4(x+6) + 6$$
$$3 - 9x = 4x + 24 + 6$$
$$-13x = 27$$
$$x = -\frac{27}{13}$$

The solution set is $\left\{-\frac{27}{13}\right\}$.

7. $(x-1)(2x+3) = 3$

$$2x^2 + x - 3 = 3$$
$$2x^2 + x - 6 = 0$$
$$(2x-3)(x+2) = 0 \Rightarrow x = \frac{3}{2} \text{ or } x = -2$$

The solution set is $\left\{-2, \frac{3}{2}\right\}$.

8. $2x + 3 = 4x^2$

$$0 = 4x^2 - 2x - 3$$
$$x = \frac{2 \pm \sqrt{4+48}}{8} = \frac{2 \pm \sqrt{52}}{8}$$
$$= \frac{2 \pm 2\sqrt{13}}{8} = \frac{1 \pm \sqrt{13}}{4}$$

The solution set is $\left\{\frac{1-\sqrt{13}}{4}, \frac{1+\sqrt{13}}{4}\right\}$.

9. $\sqrt[3]{x^2-1}=2$

$\left(\sqrt[3]{x^2-1}\right)^3=(2)^3$

$x^2-1=8$

$x^2=9 \Rightarrow x=\pm 3$

Check: $x=-3$ Check: $x=3$

$\sqrt[3]{(-3)^2-1}=2$ $\sqrt[3]{(3)^2-1}=2$

$\sqrt[3]{9-1}=2$ $\sqrt[3]{9-1}=2$

$\sqrt[3]{8}=2$ $\sqrt[3]{8}=2$

$2=2$ $2=2$

The solution set is $\{-3,3\}$.

10. $x(x+1)+2=0$

$x^2+x+2=0$

$x=\dfrac{-1\pm\sqrt{1-8}}{2}=\dfrac{-1\pm\sqrt{-7}}{2}$

No real solutions.

11. $3x^2-x+1=0$

$x=\dfrac{1\pm\sqrt{1-12}}{6}=\dfrac{1\pm\sqrt{-11}}{6}$

No real solutions.

12. $x^4-5x^2+4=0$

$\left(x^2-4\right)\left(x^2-1\right)=0$

$x^2-4=0$ or $x^2-1=0$

$x=\pm 2$ or $x=\pm 1$

The solution set is $\{-2,-1,1,2\}$.

13. $\sqrt{2x-3}+x=3$

$\sqrt{2x-3}=3-x$

$2x-3=9-6x+x^2$

$x^2-8x+12=0 \Rightarrow (x-2)(x-6)=0$

$x=2$ or $x=6$

Check $x=2$:

$\sqrt{2(2)-3}+2=\sqrt{1}+2=3$

Check $x=6$:

$\sqrt{2(6)-3}+6=\sqrt{9}+6=9\neq 3$

The solution set is $\{2\}$.

14. $\sqrt[4]{2x+3}=2$

$\left(\sqrt[4]{2x+3}\right)^4=2^4$

$2x+3=16$

$2x=13$

$x=\dfrac{13}{2}$

Check $x=\dfrac{13}{2}$:

$\sqrt[4]{2\left(\dfrac{13}{2}\right)+3}=\sqrt[4]{13+3}=\sqrt[4]{16}=2$

The solution set is $\left\{\dfrac{13}{2}\right\}$.

15. $\sqrt{x+1}+\sqrt{x-1}=\sqrt{2x+1}$

$\left(\sqrt{x+1}+\sqrt{x-1}\right)^2=\left(\sqrt{2x+1}\right)^2$

$x+1+2\sqrt{x+1}\sqrt{x-1}+x-1=2x+1$

$2x+2\sqrt{x+1}\sqrt{x-1}=2x+1$

$2\sqrt{x+1}\sqrt{x-1}=1$

$\left(2\sqrt{x+1}\sqrt{x-1}\right)^2=(1)^2$

$4(x+1)(x-1)=1$

$4x^2-4=1$

$4x^2=5$

$x^2=\dfrac{5}{4}$

$x=\pm\dfrac{\sqrt{5}}{2}$

Check $x=\dfrac{\sqrt{5}}{2}$:

$\sqrt{\dfrac{\sqrt{5}}{2}+1}+\sqrt{\dfrac{\sqrt{5}}{2}-1}=\sqrt{2\left(\dfrac{\sqrt{5}}{2}\right)+1}$

$1.79890743995=1.79890743995$

Check $x=-\dfrac{\sqrt{5}}{2}$:

$\sqrt{-\dfrac{\sqrt{5}}{2}+1}+\sqrt{-\dfrac{\sqrt{5}}{2}-1}=\sqrt{2\left(-\dfrac{\sqrt{5}}{2}\right)+1}$,

The second solution is not possible because it makes the radicand negative.

The solution set is $\left\{\dfrac{\sqrt{5}}{2}\right\}$.

16. $2x^{1/2} - 3 = 0$

$$2x^{1/2} = 3$$

$$\left(2x^{1/2}\right)^2 = 3^2$$

$$4x = 9$$

$$x = \frac{9}{4}$$

Check $x = \frac{9}{4}$:

$$2\left(\frac{9}{4}\right)^{1/2} - 3 = 2\left(\frac{3}{2}\right) - 3 = 3 - 3 = 0$$

The solution set is $\left\{\frac{9}{4}\right\}$.

17. $x^{-6} - 7x^{-3} - 8 = 0$

let $p = x^{-3} \Rightarrow p^2 = x^{-6}$

$$p^2 - 7p - 8 = 0$$

$$(p-8)(p+1) = 0$$

$$p = 8 \quad \text{or} \quad p = -1$$

$$p = 8 \Rightarrow x^{-3} = 8$$

$$\left(x^{-3}\right)^{-1/3} = (8)^{-1/3} \Rightarrow x = \frac{1}{2}$$

$$p = -1 \Rightarrow x^{-3} = -1$$

$$\left(x^{-3}\right)^{-1/3} = (-1)^{-1/3} \Rightarrow x = -1$$

Check $x = \frac{1}{2}$:

$$\left(\frac{1}{2}\right)^{-6} - 7\left(\frac{1}{2}\right)^{-3} - 8 = 64 - 56 - 8 = 0$$

Check $x = -1$:

$$(-1)^{-6} - 7(-1)^{-3} - 8 = 1 + 7 - 8 = 0$$

The solution set is $\left\{-1, \frac{1}{2}\right\}$.

18.

$$x^2 + m^2 = 2mx + (nx)^2$$

$$x^2 + m^2 = 2mx + n^2x^2$$

$$x^2 - n^2x^2 - 2mx + m^2 = 0$$

$$\left(1 - n^2\right)x^2 - 2mx + m^2 = 0$$

$$x = \frac{2m \pm \sqrt{4m^2 - 4m^2\left(1 - n^2\right)}}{2\left(1 - n^2\right)}$$

$$= \frac{2m \pm \sqrt{4m^2\left(1 - \left(1 - n^2\right)\right)}}{2\left(1 - n^2\right)}$$

$$= \frac{2m \pm 2m\sqrt{1 - \left(1 - n^2\right)}}{2\left(1 - n^2\right)}$$

$$= \frac{m \pm m\sqrt{n^2}}{1 - n^2} = \frac{m \pm mn}{1 - n^2} = \frac{m(1 \pm n)}{1 - n^2}$$

$$x = \frac{m(1 + n)}{1 - n^2} = \frac{m(1 + n)}{(1 + n)(1 - n)} = \frac{m}{1 - n}$$

or

$$x = \frac{m(1 - n)}{1 - n^2} = \frac{m(1 - n)}{(1 + n)(1 - n)} = \frac{m}{1 + n}$$

The solution set is $\left\{\frac{m}{1-n}, \frac{m}{1+n}\right\}$, $n \neq 1$, $n \neq -1$.

19. $10a^2x^2 - 2abx - 36b^2 = 0$

$$5a^2x^2 - abx - 18b^2 = 0$$

$$(5ax + 9b)(ax - 2b) = 0$$

$$5ax + 9b = 0 \quad \text{or} \quad ax - 2b = 0$$

$$5ax = -9b \qquad ax = 2b$$

$$x = -\frac{9b}{5a} \qquad x = \frac{2b}{a}$$

The solution set is $\left\{-\frac{9b}{5a}, \frac{2b}{a}\right\}$, $a \neq 0$.

20. $\sqrt{x^2+3x+7}-\sqrt{x^2-3x+9}+2=0$

$\sqrt{x^2+3x+7}=\sqrt{x^2-3x+9}-2$

$\left(\sqrt{x^2+3x+7}\right)^2=\left(\sqrt{x^2-3x+9}-2\right)^2$

$x^2+3x+7=x^2-3x+9-4\sqrt{x^2-3x+9}+4$

$6x-6=-4\sqrt{x^2-3x+9}$

$\left(6(x-1)\right)^2=\left(-4\sqrt{x^2-3x+9}\right)^2$

$36\left(x^2-2x+1\right)=16\left(x^2-3x+9\right)$

$36x^2-72x+36=16x^2-48x+144$

$20x^2-24x-108=0$

$5x^2-6x-27=0$

$(5x+9)(x-3)=0 \Rightarrow x=-\dfrac{9}{5}$ or $x=3$

Check $x=-\dfrac{9}{5}$:

$\sqrt{\left(-\dfrac{9}{5}\right)^2+3\left(-\dfrac{9}{5}\right)+7}-\sqrt{\left(-\dfrac{9}{5}\right)^2-3\left(-\dfrac{9}{5}\right)+9}+2$

$=\sqrt{\dfrac{81}{25}-\dfrac{27}{5}+7}-\sqrt{\dfrac{81}{25}+\dfrac{27}{5}+9}+2$

$=\sqrt{\dfrac{81-135+175}{25}}-\sqrt{\dfrac{81+135+225}{25}}+2$

$=\sqrt{\dfrac{121}{25}}-\sqrt{\dfrac{441}{25}}+2=\dfrac{11}{5}-\dfrac{21}{5}+2=0$

Check $x=3$:

$\sqrt{(3)^2+3(3)+7}-\sqrt{(3)^2-3(3)+9}+2$

$=\sqrt{9+9+7}-\sqrt{9-9+9}+2$

$=\sqrt{25}-\sqrt{9}+2=2+2$

$=4\neq0$

The solution set is $\left\{-\dfrac{9}{5}\right\}$.

21. $|2x+3|=7$

$2x+3=7$ or $2x+3=-7$

$2x=4$ or $2x=-10$

$x=2$ or $x=-5$

The solution set is $\{-5, 2\}$.

22. $|2-3x|+2=9$

$|2-3x|=7$

$2-3x=7$ or $2-3x=-7$

$-3x=5$ $\qquad -3x=-9$

$x=-\dfrac{5}{3}$ $\qquad x=3$

The solution set is $\left\{-\dfrac{5}{3},3\right\}$

23. $2x^3=3x^2$

$2x^3-3x^2=0$

$x^2(2x-3)=0$

$x^2=0 \Rightarrow x=0$

$2x-3=0 \Rightarrow x=\dfrac{3}{2}$

The solution set is $\left\{0,\dfrac{3}{2}\right\}$.

24. $2x^3+5x^2-8x-20=0$

$x^2(2x+5)-4(2x+5)=0$

$(2x+5)(x^2-4)=0$

$2x+5=0$ or $x^2-4=0$

$x=-\dfrac{5}{2}$ or $x=\pm2$

The solution set is $\left\{-\dfrac{5}{2},-2,2\right\}$.

25. $\dfrac{1}{x-1}+\dfrac{3}{x+2}=\dfrac{11}{x^2+x-2}$

$\dfrac{(x+2)+3(x-1)}{(x-1)(x+2)}=\dfrac{11}{(x-1)(x+2)}$

$(x+2)+3(x-1)=11$

$x+2+3x-3=11$

$4x-1=11$

$4x=12$

$x=3$

Since 3 does not make any denominator equal to 0, the solution set is $\{3\}$.

26. $(x-2)^2 = 9$

$$x-2 = \pm\sqrt{9}$$
$$x-2 = \pm 3$$
$$x = 2 \pm 3$$
$$x = 5 \quad \text{or} \quad x = -1$$

The solution set is $\{-1, 5\}$.

27. $x^3 - 5x + 3 = 0$

Use the Zero option from the CALC menu.

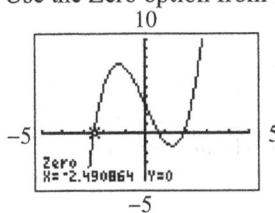

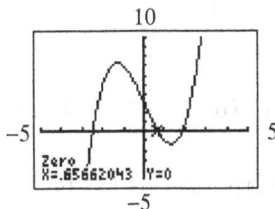

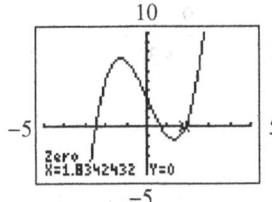

The solution set is $\{-2.49, 0.66, 1.83\}$.

28. $x^4 - 3 = 2x + 1$

Use the Intersect option on the CALC menu.

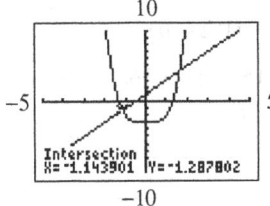

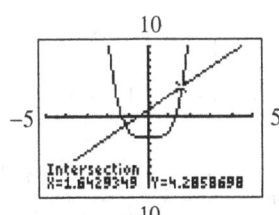

The solution set is $\{-1.14, 1.64\}$.

29.
$$\frac{2x-3}{5} + 2 \le \frac{x}{2}$$
$$2(2x-3) + 10(2) \le 5x$$
$$4x - 6 + 20 \le 5x$$
$$14 \le x$$
$$x \ge 14$$
$$\{x \mid x \ge 14\} \text{ or } [14, \infty)$$

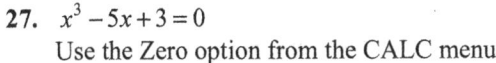

30.
$$-9 \le \frac{2x+3}{-4} \le 7$$
$$36 \ge 2x + 3 \ge -28$$
$$33 \ge 2x \ge -31$$
$$\frac{33}{2} \ge x \ge -\frac{31}{2}$$
$$-\frac{31}{2} \le x \le \frac{33}{2}$$
$$\left\{x \mid -\frac{31}{2} \le x \le \frac{33}{2}\right\} \text{ or } \left[-\frac{31}{2}, \frac{33}{2}\right]$$

31.
$$2 < \frac{3-3x}{12} < 6$$
$$24 < 3 - 3x < 72$$
$$21 < -3x < 69$$
$$-7 > x > -23$$
$$\{x \mid -23 < x < -7\} \text{ or } (-23, -7)$$

32. $\left|3x+4\right| < \dfrac{1}{2}$

$$-\dfrac{1}{2} < 3x+4 < \dfrac{1}{2}$$

$$-\dfrac{9}{2} < \quad 3x \quad < -\dfrac{7}{2}$$

$$-\dfrac{3}{2} < \quad x \quad < -\dfrac{7}{6}$$

$$\left\{x \mid -\dfrac{3}{2} < x < -\dfrac{7}{6}\right\} \text{ or } \left(-\dfrac{3}{2}, -\dfrac{7}{6}\right)$$

33. $\left|2x-5\right| \geq 9$

$$2x-5 \leq -9 \text{ or } 2x-5 \geq 9$$

$$2x \leq -4 \text{ or } \quad 2x \geq 14$$

$$x \leq -2 \text{ or } \quad x \geq 7$$

$$\left\{x \mid x \leq -2 \text{ or } x \geq 7\right\} \text{ or}$$

$$(-\infty, -2] \text{ or } [7, \infty)$$

34. $2+\left|2-3x\right| \leq 4$

$$\left|2-3x\right| \leq 2$$

$$-2 \leq 2-3x \leq 2$$

$$-4 \leq -3x \leq 0$$

$$\dfrac{4}{3} \geq x \geq 0$$

$$0 \leq x \leq \dfrac{4}{3}$$

$$\left\{x \mid 0 \leq x \leq \dfrac{4}{3}\right\} \text{ or } \left[0, \dfrac{4}{3}\right]$$

35. $1-\left|2-3x\right| < -4$

$$-\left|2-3x\right| < -5$$

$$\left|2-3x\right| > 5$$

$$2-3x < -5 \text{ or } 2-3x > 5$$

$$7 < 3x \text{ or } \quad -3 > 3x$$

$$\dfrac{7}{3} < x \text{ or } \quad -1 > x$$

$$x < -1 \text{ or } \quad x > \dfrac{7}{3}$$

$$\left\{x \mid x < -1 \text{ or } x > \dfrac{7}{3}\right\} \text{ or}$$

$$(-\infty, -1) \text{ or } \left(\dfrac{7}{3}, \infty\right)$$

36. $(6+3i)-(2-4i) = (6-2)+(3-(-4))i = 4+7i$

37. $4(3-i)+3(-5+2i) = 12-4i-15+6i = -3+2i$

38. $\dfrac{3}{3+i} = \dfrac{3}{3+i} \cdot \dfrac{3-i}{3-i} = \dfrac{9-3i}{9-3i+3i-i^2} = \dfrac{9-3i}{10}$

$$= \dfrac{9}{10} - \dfrac{3}{10}i$$

39. $i^{50} = i^{48} \cdot i^2 = \left(i^4\right)^{12} \cdot i^2 = 1^{12}(-1) = -1$

40. $(2+3i)^3 = (2+3i)^2(2+3i)$

$$= \left(4+12i+9i^2\right)(2+3i)$$

$$= (-5+12i)(2+3i)$$

$$= -10-15i+24i+36i^2$$

$$= -46+9i$$

41. $x^2+x+1 = 0$

$$a=1, b=1, c=1,$$

$$b^2-4ac = 1^2-4(1)(1) = 1-4 = -3$$

$$x = \dfrac{-1 \pm \sqrt{-3}}{2(1)} = \dfrac{-1 \pm \sqrt{3}\,i}{2} = -\dfrac{1}{2} \pm \dfrac{\sqrt{3}}{2}i$$

The solution set is $\left\{-\dfrac{1}{2} - \dfrac{\sqrt{3}}{2}i, \ -\dfrac{1}{2} + \dfrac{\sqrt{3}}{2}i\right\}$.

42. $2x^2 + x - 2 = 0$

$a = 2, b = 1, c = -2,$

$b^2 - 4ac = 1^2 - 4(2)(-2) = 1 + 16 = 17$

$x = \dfrac{-1 \pm \sqrt{17}}{2(2)} = \dfrac{-1 \pm \sqrt{17}}{4}$

The solution set is $\left\{ \dfrac{-1 - \sqrt{17}}{4}, \dfrac{-1 + \sqrt{17}}{4} \right\}$.

43. $x^2 + 3 = x$

$x^2 - x + 3 = 0$

$a = 1, b = -1, c = 3,$

$b^2 - 4ac = (-1)^2 - 4(1)(3) = 1 - 12 = -11$

$x = \dfrac{-(-1) \pm \sqrt{-11}}{2(1)} = \dfrac{1 \pm \sqrt{11}\, i}{2} = \dfrac{1}{2} \pm \dfrac{\sqrt{11}}{2} i$

The solution set is $\left\{ \dfrac{1}{2} - \dfrac{\sqrt{11}}{2} i, \dfrac{1}{2} + \dfrac{\sqrt{11}}{2} i \right\}$.

44. $x(1 + x) = 2$

$x^2 + x - 2 = 0$

$(x + 2)(x - 1) = 0 \Rightarrow x = -2$ or $x = 1$

The solution set is $\{-2, 1\}$.

45. $(0,0), (4,2)$

a. distance $= \sqrt{(4-0)^2 + (2-0)^2}$

$= \sqrt{16 + 4} = \sqrt{20}$

$= 2\sqrt{5}$

b. midpoint $= \left(\dfrac{0+4}{2}, \dfrac{0+2}{2} \right) = \left(\dfrac{4}{2}, \dfrac{2}{2} \right) = (2,1)$

46. $(1,-1), (-2,3)$

a. distance $= \sqrt{(-2-1)^2 + (3-(-1))^2}$

$= \sqrt{9 + 16} = \sqrt{25} = 5$

b. midpoint $= \left(\dfrac{1+(-2)}{2}, \dfrac{-1+3}{2} \right)$

$= \left(\dfrac{-1}{2}, \dfrac{2}{2} \right) = \left(-\dfrac{1}{2}, 1 \right)$

47. $(4,-4), (4,8)$

a. distance $= \sqrt{(4-4)^2 + (8-(-4))^2}$

$= \sqrt{0 + 144} = \sqrt{144} = 12$

b. midpoint $= \left(\dfrac{4+4}{2}, \dfrac{-4+8}{2} \right)$

$= \left(\dfrac{8}{2}, \dfrac{4}{2} \right) = (4,2)$

48. $y = -x^2 + 15$

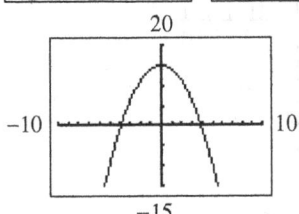

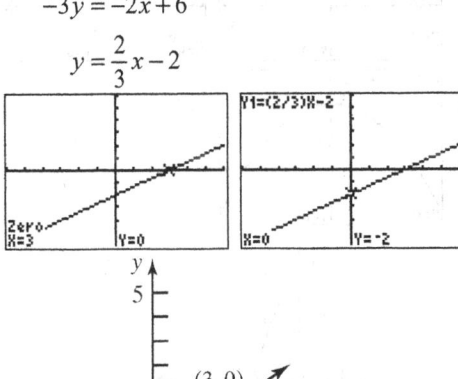

49. $(-4,0), (0,0), (2,0), (0,-2), (0,0), (0,2)$

x-intercepts: $-4, 0, 2$

y-intercepts: $-2, 0, 2$

50. $2x - 3y = 6$

$-3y = -2x + 6$

$y = \dfrac{2}{3}x - 2$

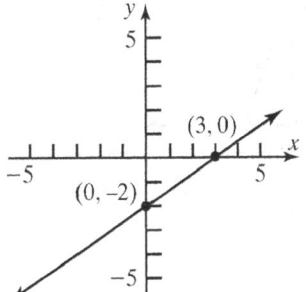

79

51. $y = x^2 - 9$

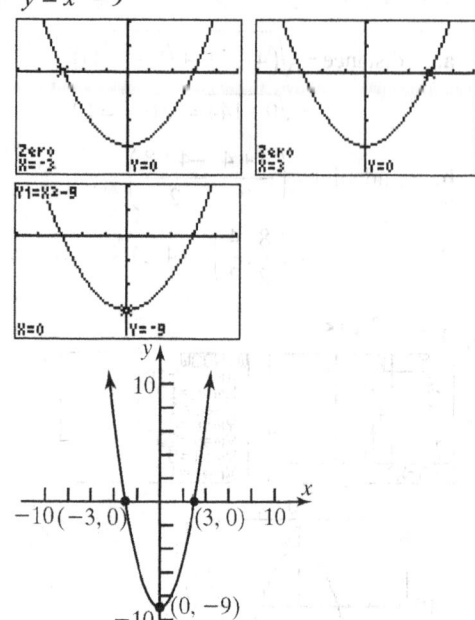

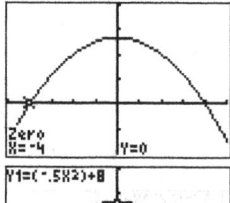

52. $x^2 + 2y = 16$

$$2y = -x^2 + 16$$

$$y = -\frac{1}{2}x^2 + 8$$

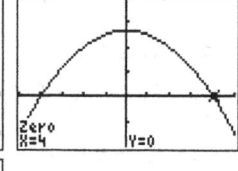

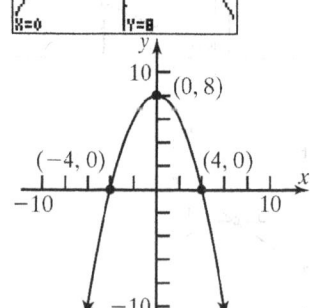

53. Find the distance between each pair of points.

$$d(A,B) = \sqrt{(1-3)^2 + (1-4)^2}$$
$$= \sqrt{4+9} = \sqrt{13}$$
$$d(B,C) = \sqrt{(-2-1)^2 + (3-1)^2}$$
$$= \sqrt{9+4} = \sqrt{13}$$
$$d(A,C) = \sqrt{(-2-3)^2 + (3-4)^2}$$
$$= \sqrt{25+1} = \sqrt{26}$$

Since $AB = BC$, triangle ABC is isosceles.

54. Using the distance formula on the points $(-3, 2)$ and $(5, y)$ yields

$$d = \sqrt{(5-(-3))^2 + (y-2)^2} = \sqrt{64 + (y-2)^2}$$

Now set $d = 10$ and solve for y.

$$10 = \sqrt{64 + (y-2)^2}$$
$$10^2 = \left(\sqrt{64 + (y-2)^2}\right)^2$$
$$100 = 64 + (y-2)^2$$
$$36 = (y-2)^2$$
$$\pm 6 = y - 2$$

so $y - 2 = 6 \rightarrow y = 8$

and $y - 2 = -6 \rightarrow y = -4$

55. $p = 2l + 2w$

56. $I = P \cdot r \cdot t \Rightarrow I = (9000)(0.07)(1) = \$630$

57. Let x represent the amount of money invested in bonds. Then $70,000 - x$ represents the amount of money invested in CD's.

Since the total interest is to be $5000, we have:

$$0.08x + 0.05(70,000 - x) = 5000$$
$$(100)(0.08x + 0.05(70,000 - x)) = (5000)(100)$$
$$8x + 350,000 - 5x = 500,000$$
$$3x + 350,000 = 500,000$$
$$3x = 150,000$$
$$x = 50,000$$

$50,000 should be invested in bonds at 8% and $20,000 should be invested in CD's at 5%.

58. Using $s = vt$, we have $t = 3$ and $v = 1100$.

Finding the distance s in feet:

$s = 1100(3) = 3300$

The storm is 3300 feet away.

59. $1600 \le I \le 3600$

$$1600 \le \frac{900}{x^2} \le 3600$$

$$\frac{1}{1600} \ge \frac{x^2}{900} \ge \frac{1}{3600}$$

$$\frac{9}{16} \ge x^2 \ge \frac{1}{4}$$

$$\frac{3}{4} \ge x \ge \frac{1}{2}$$

The range of distances is from 0.5 meters to 0.75 meters, inclusive.

60. Let s represent the distance the plane can travel.

	With wind	Against wind
Rate	$250+30=280$	$250-30=220$
Time	$\dfrac{(s/2)}{280}$	$\dfrac{(s/2)}{220}$
Dist.	$\dfrac{s}{2}$	$\dfrac{s}{2}$

Since the total time is at most 5 hours, we have:

$$\frac{(s/2)}{280}+\frac{(s/2)}{220} \le 5$$

$$\frac{s}{560}+\frac{s}{440} \le 5$$

$$11s+14s \le 5(6160)$$

$$25s \le 30,800$$

$$s \le 1232$$

The plane can travel at most 1232 miles or 616 miles one way and return 616 miles.

61. Let t represent the time it takes the helicopter to reach the raft.

	Raft	Helicopter
Rate	5	90
Time	t	t
Dist.	$5t$	$90t$

Since the total distance is 150 miles, we have:

$$5t+90t=150$$

$$95t=150$$

$$t \approx 1.58 \text{ hours} \approx 1 \text{ hour and } 35 \text{ minutes}$$

The helicopter will reach the raft in about 1 hour and 35 minutes.

62. Let d represent the distance flown by the bee traveling at 3 meters per second.

$$\frac{d}{3}=\frac{150-d}{5} \quad \text{(Times needed to meet are equal.)}$$

$$5d=450-3d$$

$$8d=450$$

$$d=56.25 \text{ meters} \Rightarrow t=\frac{56.25}{3}=18.75 \text{ seconds}$$

The bees meet for the first time after 18.75 seconds.

The bees will meet a second time on the second lap. The first bee will have traveled $150+x$ meters and the second bee will have traveled $150+(150-x)$ meters.
Solving for time, we have:

$$\frac{150+x}{3}=\frac{150+(150-x)}{5}$$

$$\frac{150+x}{3}=\frac{300-x}{5}$$

$$750+5x=900-3x$$

$$8x=150$$

$$x=18.75 \text{ meters into the second lap}$$

$$t=\frac{168.75}{3}=56.25 \text{ seconds}$$

The bees meet the second time after 56.25 seconds (37.5 seconds later).

63. Given that $s=1280-32t-16t^2$,

a. The object hits the ground when $s=0$.

$$0=1280-32t-16t^2$$

$$t^2+2t-80=0$$

$$(t+10)(t-8)=0 \Rightarrow t=-10, t=8$$

The object hits the ground after 8 seconds.

b. After 4 seconds, the object's height is

$$s=1280-32(4)-16(4)^2=896 \text{ feet.}$$

64. Let t represent the time it takes Clarissa to complete the job by herself.

	Clarissa	Shawna
Time to do job alone	t	$t+5$
Part of job done in 1 day	$\dfrac{1}{t}$	$\dfrac{1}{t+5}$
Time on job (days)	6	6
Part of job done by each person	$\dfrac{6}{t}$	$\dfrac{6}{t+5}$

Since the two people paint one house, we have:

$$\frac{6}{t} + \frac{6}{t+5} = 1$$

$$6(t+5) + 6t = t(t+5)$$

$$6t + 30 + 6t = t^2 + 5t$$

$$t^2 - 7t - 30 = 0$$

$$(t-10)(t+3) = 0 \Rightarrow t = 10 \text{ or } t = -3$$

It takes Clarissa 10 days to paint the house when working by herself.

65. Let x represent the amount of the $8-per-pound coffee.

Amt. of coffee (pounds)	Price ($)	Total $
20	4	$(20)(4)$
x	8	$(8)(x)$
$20+x$	5	$(5)(20+x)$

$$80 + 8x = (5)(20+x)$$

$$80 + 8x = 100 + 5x$$

$$3x = 20$$

$$x = \frac{20}{3} = 6\frac{2}{3}$$

Add $6\frac{2}{3}$ pounds of $8/lb coffee to get $26\frac{2}{3}$ pounds of $5/lb coffee.

66. Let x represent the amount of water evaporated.

% salt	Tot. amt.	amt. of salt
2%	64	$(0.02)(64)$
0%	x	$(0.00)(x)$
10%	$64-x$	$(0.10)(64-x)$

$$(0.02)(64) - (0.00)(x) = (0.10)(64-x)$$

$$1.28 = 6.4 - 0.10x$$

$$0.10x = 5.12$$

$$x = 51.2$$

51.2 ounces of water must be evaporated.

67. Let the length of leg 1 = x.
Then the length of leg 2 = $17 - x$.
By the Pythagorean Theorem we have

$$x^2 + (17-x)^2 = (13)^2$$

$$x^2 + x^2 - 34x + 289 = 169$$

$$2x^2 - 34x + 120 = 0$$

$$x^2 - 17x + 60 = 0$$

$$(x-12)(x-5) = 0 \Rightarrow x = 12 \text{ or } x = 5$$

the legs are 5 cm and 12 cm long.

68. The effective speed of the train (i.e., relative to the man) is $30 - 4 = 26$ miles per hour. The time is $5 \sec = \dfrac{5}{60} \min = \dfrac{5}{3600} \text{ hr} = \dfrac{1}{720} \text{ hr}$.

$$s = vt$$

$$= 26\left(\frac{1}{720}\right)$$

$$= \frac{26}{720} \text{ miles}$$

$$= \frac{26}{720} \cdot 5280 \approx 190.67 \text{ feet}$$

The freight train is about 190.67 feet long.

69. a. Consider the following diagram:

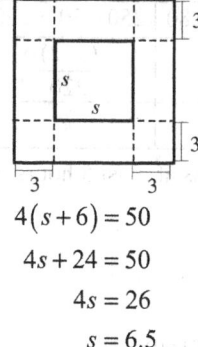

$$4(s+6) = 50$$

$$4s + 24 = 50$$

$$4s = 26$$

$$s = 6.5$$

The painting is 6.5 inches by 6.5 inches.
$s + 6 = 12.5$, so the frame is 12.5 inches by 12.5 inches.

b. Consider the following diagram:

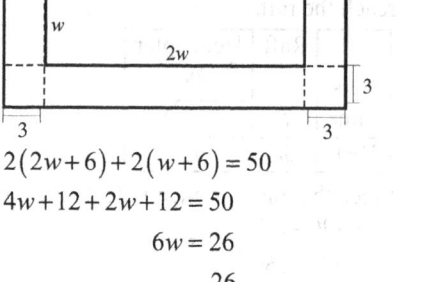

$$2(2w+6) + 2(w+6) = 50$$

$$4w + 12 + 2w + 12 = 50$$

$$6w = 26$$

$$w = \frac{26}{6} = 4\frac{1}{3} \Rightarrow l = 2w = 8\frac{2}{3}$$

The painting is $8\frac{2}{3}$ inches by $4\frac{1}{3}$ inches.

The frame is $14\frac{2}{3}$ inches by $10\frac{1}{3}$ inches.

70. Let t represent the time it takes the smaller pump to finish filling the tank.

	3hp Pump	8hp Pump
Time to do job alone	12	8
Part of job done in 1 hr	$\dfrac{1}{12}$	$\dfrac{1}{8}$
Time on job (hrs)	$t+4$	4
Part of job done by each pump	$\dfrac{t+4}{12}$	$\dfrac{4}{8}$

Since the two pumps fill one tank, we have:

$$\frac{t+4}{12}+\frac{4}{8}=1$$

$$\frac{t+4}{12}=\frac{1}{2}$$

$$t+4=6$$

$$t=2$$

It takes the small pump a total of 2 more hours to fill the tank.

71. Let $w=4$. Solve for the length:

$$l^2 = 4(l+4)$$

$$l^2 = 4l+16$$

$$l^2 - 4l - 16 = 0$$

$$l = \frac{-(-4)+\sqrt{(-4)^2 - 4(1)(-16)}}{2(1)}$$

$$= \frac{4+\sqrt{80}}{2} = 2+2\sqrt{5} \approx 6.47$$

The length of the plasterboard should be cut to a length of approximately 6.47 feet.

72. Let x represent the number of passengers over 20. Then $20+x$ represents the total number of passengers, and $15-0.1x$ represents the fare for each passenger. Solving the equation for total cost, \$482.40, we have:

$$(20+x)(15-0.1x) = 482.40$$

$$300+13x-0.1x^2 = 482.40$$

$$-0.1x^2 + 13x - 182.40 = 0$$

$$x^2 - 130x + 1824 = 0$$

$$(x-114)(x-16) = 0 \Rightarrow x = 114 \text{ or } x = 16$$

Since the capacity of the bus is 44, we discard

the 114. Therefore, $20+16=36$ people went on the trip; each person paid $15-0.1(16) = \$13.40$.

73. Let r_S represent Scott's rate and let r_T represent Todd's rate. The time for Scott to run 95 meters is the same as for Todd to run 100 meters.

$$\frac{95}{r_S} = \frac{100}{r_T}.$$

$$r_S = 0.95r_T$$

$$d_S = t \cdot r_s = t(0.95r_T) = 0.95d_T$$

If Todd starts from 5 meters behind the start:

$$d_T = 105$$

$$d_S = 0.95d_T = 0.95(105) = 99.75$$

a. The race does not end in a tie.

b. Todd wins the race.

c. Todd wins by 0.25 meters.

d. To end in a tie:
$$100 = 0.95(100 + x)$$
$$100 = 95 + 0.95x$$
$$5 = 0.95x$$
$$x = 5.263 \text{ meters}$$

e. $95 = 0.95(100)$ Therefore, the race ends in a tie.

f,g,h. Race is a tie.

74. In each problem, we need to use the Least Common Multiple of the expressions $x-2$ and x^2-4, namely x^2-4.

a. We use x^2-4 as the LCD in order to combine the given expressions.

b. We multiply each side of the equation by x^2-4 in order to clear out the denominators before solving the equation.

c. We use x^2-4 as the LCD in order to combine the terms on the left hand side before solving the inequality.

Chapter 1 Test

1. a. $d = \sqrt{(x_2 - x_1)^2 + (y_2 - y_1)^2}$

$= \sqrt{(4 - (-2))^2 + (5 - (-3))^2}$

$= \sqrt{6^2 + 8^2}$

$= \sqrt{36 + 64}$

$= \sqrt{100}$

$= 10$

b. $M = \left(\dfrac{x_1 + x_2}{2}, \dfrac{y_1 + y_2}{2} \right)$

$= \left(\dfrac{4 + (-2)}{2}, \dfrac{5 + (-3)}{2} \right)$

$= \left(\dfrac{2}{2}, \dfrac{2}{2} \right)$

$= (1,1)$

2. $2x^2 + 6x = x - 3$

$2x^2 + 5x + 3 = 0$

$(2x+3)(x+1) = 0$

$2x + 3 = 0 \quad$ or $\quad x + 1 = 0$

$2x = -3 \qquad\qquad x = -1$

$x = -\dfrac{3}{2}$

The solution set is $\left\{ -\frac{3}{2}, -1 \right\}$.

3. $x + 1 = \sqrt{x + 7}$

$(x+1)^2 = \left(\sqrt{x+7} \right)^2$

$x^2 + 2x + 1 = x + 7$

$x^2 + x - 6 = 0$

$(x+3)(x-2) = 0$

$x + 3 = 0 \quad$ or $\quad x - 2 = 0$

$x = -3 \qquad\qquad x = 2$

Check:

$-3 + 1 \overset{?}{=} \sqrt{-3 + 7} \qquad 2 + 1 \overset{?}{=} \sqrt{2 + 7}$

$-2 \overset{?}{=} \sqrt{4} \qquad\qquad 3 \overset{?}{=} \sqrt{9}$

$-2 \neq 2 \qquad\qquad\qquad 3 \overset{?}{=} 3 \ \text{T}$

The solution set is $\{2\}$.

4. $2 - \dfrac{3}{m} = \dfrac{2}{m+2}$

LCD: $m(m+2)$

Restricted values: $m = -2, m = 0$

$\dfrac{2m(m+2)}{m(m+2)} - \dfrac{3(m+2)}{m(m+2)} = \dfrac{2m}{m(m+2)}$

$2m(m+2) - 3(m+2) = 2m$

$2m^2 + 4m - 3m - 6 = 2m$

$2m^2 - m - 6 = 0$

$(2m+3)(m-2) = 0$

$2m + 3 = 0 \quad$ or $\quad m - 2 = 0$

$m = -\dfrac{3}{2} \qquad\qquad m = 2$

Since neither solution is a restricted value, the solution set is $\left\{ -\frac{3}{2}, 2 \right\}$.

5. $5x - 8 = -4(x-1) + 6$

$5x - 8 = -4x + 4 + 6$

$5x - 8 = -4x + 10$

$9x = 18$

$x = 2$

The solution set is $\{2\}$.

6. $5|3 - 2b| - 7 = 8$

$5|3 - 2b| = 15$

$|3 - 2b| = 3$

$3 - 2b = 3 \quad$ or $\quad 3 - 2b = -3$

$-2b = 0 \qquad\qquad -2b = -6$

$b = 0 \qquad\qquad\quad b = 3$

The solution set is $\{0, 3\}$.

7. $x^4 + x^2 = 3x^2 + 8$

$x^4 - 2x^2 - 8 = 0$

Let $u = x^2$. Then $u^2 = \left(x^2 \right)^2 = x^4$, and we have

$u^2 - 2u - 8 = 0$

$(u-4)(u+2) = 0$

$u = 4 \ $ or $\ u = -2$

Since we are solving for x, we get

$x^2 = 4 \quad$ or $\quad x^2 = -2$

$x = \pm 2 \qquad\qquad x = \pm\sqrt{-2}$

$\qquad\qquad\qquad\qquad$ no real solution

The solution set is $\{-2, 2\}$.

8. $x^2 - 4x + 2 = 0$

$a = 1, b = -4, c = 2$

$x = \dfrac{-b \pm \sqrt{b^2 - 4ac}}{2a}$

$\quad = \dfrac{-(-4) \pm \sqrt{(-4)^2 - 4(1)(2)}}{2(1)}$

$\quad = \dfrac{4 \pm \sqrt{16 - 8}}{2}$

$\quad = \dfrac{4 \pm \sqrt{8}}{2}$

$\quad = \dfrac{4 \pm 2\sqrt{2}}{2}$

$\quad = 2 \pm \sqrt{2}$

The solution set is $\left\{ 2 - \sqrt{2}, 2 + \sqrt{2} \right\}$.

9. $2x^2 + x - 1 = x(x + 7) + 2$

$2x^2 + x - 1 = x^2 + 7x + 2$

$x^2 - 6x - 3 = 0$

$a = 1, b = -6, c = -3$

$x = \dfrac{-b \pm \sqrt{b^2 - 4ac}}{2a}$

$\quad = \dfrac{-(-6) \pm \sqrt{(-6)^2 - 4(1)(-3)}}{2(1)}$

$\quad = \dfrac{6 \pm \sqrt{36 + 12}}{2}$

$\quad = \dfrac{6 \pm \sqrt{48}}{2}$

$\quad = \dfrac{6 \pm 4\sqrt{3}}{2}$

$\quad = 3 \pm 2\sqrt{3}$

The solution set is $\left\{ 3 - 2\sqrt{3}, 3 + 2\sqrt{3} \right\}$.

10. $2x - 7y = 21$

$\quad -7y = -2x + 21$

$\quad\quad y = \dfrac{2}{7}x - 3$

x	$y = \frac{2}{7}x - 3$	(x, y)
-7	$y = \frac{2}{7}(-7) - 3 = -5$	$(-7, -5)$
0	$y = \frac{2}{7}(0) - 3 = -3$	$(0, -3)$
7	$y = \frac{2}{7}(7) - 3 = -1$	$(7, -1)$

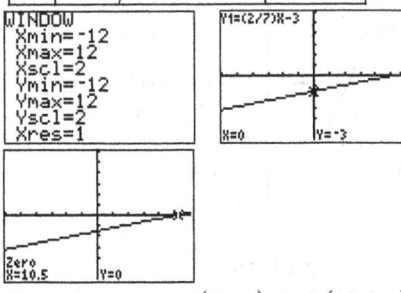

The intercepts are $(0, -3)$ and $(10.5, 0)$.

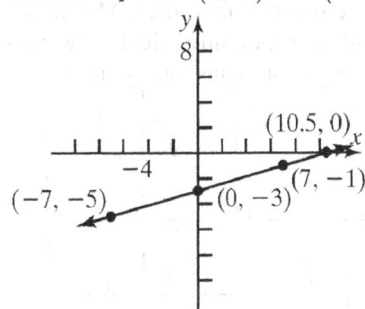

11. $y = x^2 - 5$

x	$y = x^2 - 5$	(x, y)
-3	$y = (-3)^2 - 5 = 4$	$(-3, 4)$
-1	$y = (-1)^2 - 5 = -4$	$(-1, -4)$
0	$y = (0)^2 - 5 = -5$	$(0, -5)$
1	$y = (1)^2 - 5 = -4$	$(1, -4)$
3	$y = (3)^2 - 5 = 4$	$(3, 4)$

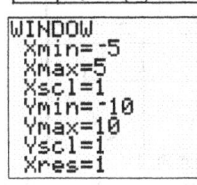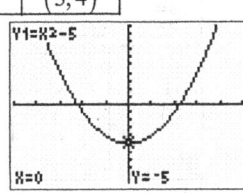

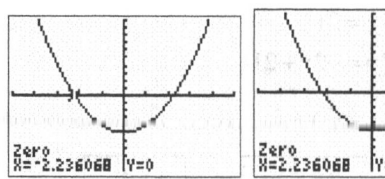

The intercepts are $(0,-5)$, $\approx(-2.24,0)$, and $\approx(2.24,0)$.

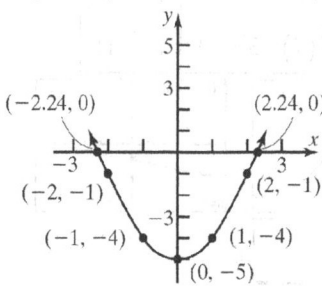

12. $2x^3 - x^2 - 2x + 1 = 0$

Since this equation has 0 on one side, we will use the Zero option from the CALC menu. It will be important to carefully select the window settings so as not to miss any solutions.

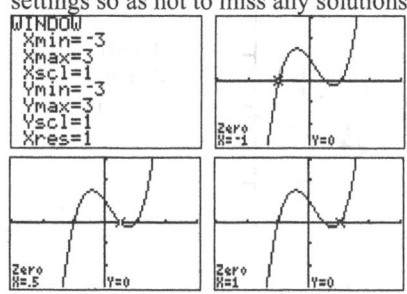

The solutions to the equation are -1, 0.5, and 1.

13. $x^4 - 5x^2 - 8 = 0$

Since this equation has 0 on one side, we will use the Zero option from the CALC menu.

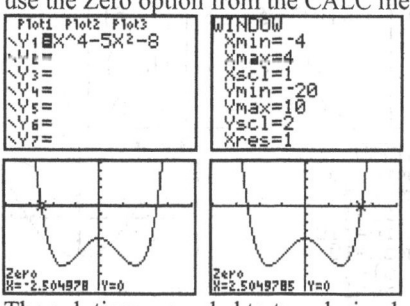

The solutions, rounded to two decimal places, are -2.50 and 2.50.

14. $-x^3 + 7x - 2 = x^2 + 3x - 3$

Since there are nonzero expressions on both sides of the equation, we will use the Intersect

option from the CALC menu. Enter the left side of the equation in Y1 and the right side in Y2.

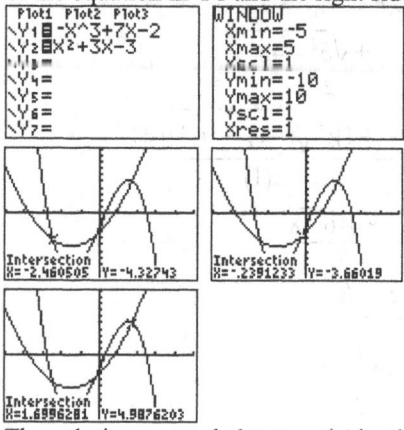

The solutions, rounded to two decimal places, are -2.46, -0.24, and 1.70.

15. $\dfrac{2x+3}{4} < -2$

$$4 \cdot \dfrac{2x+3}{4} < 4 \cdot (-2)$$

$$2x + 3 < -8$$

$$2x = -11$$

$$x < -\dfrac{11}{2}$$

Solution set: $\left\{ x \mid x < -\dfrac{11}{2} \right\}$

Interval: $\left(-\infty, -\dfrac{11}{2} \right)$

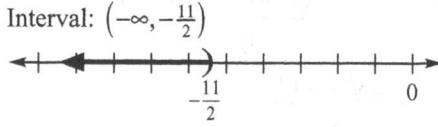

16. $|2x+3| - 4 \geq 3$

$$|2x+3| - 4 + 4 \geq 3 + 4$$

$$|2x+3| \geq 7$$

$$2x + 3 \leq -7 \quad \text{or} \quad 2x + 3 \geq 7$$

$$2x \leq -10 \qquad\qquad 2x \geq 4$$

$$x \leq 5 \qquad\qquad x \geq 2$$

Solution set: $\{ x \mid x \leq -5 \ \text{or} \ x \geq 2 \}$

Interval: $(-\infty, -5] \cup [2, \infty)$

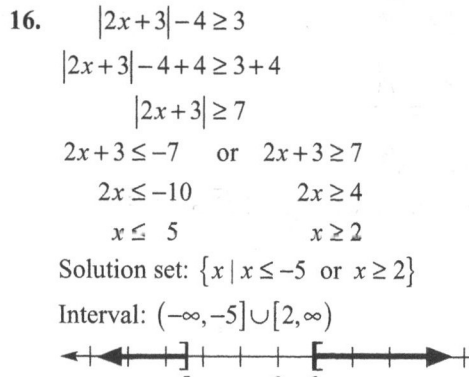

86

17. $\quad -7 < 3 - 5x \le 8$

$\quad -7 - 3 < 3 - 5x - 3 \le 8 - 3$

$\quad\quad -10 < -5x \le 5$

$\quad\quad \dfrac{-10}{-5} > \dfrac{-5x}{-5} \ge \dfrac{5}{-5}$

$\quad\quad\quad 2 > x \ge -1$

$\quad\quad\quad -1 \le x < 2$

Solution set: $\{x \mid -1 \le x < 2\}$

Interval: $[-1, 2)$

18. $\quad |3x + 4| < 8$

$\quad\quad -8 < 3x + 4 < 8$

$\quad\quad -12 < 3x < 4$

$\quad\quad -4 < x < \dfrac{4}{3}$

$\quad \left\{ x \middle| -4 < x < \dfrac{4}{3} \right\}$ or $\left(-4, \dfrac{4}{3} \right)$

19. $2(3 - 7i) - (4 + 11i) = 6 - 14i - 4 - 11i$

$\quad\quad\quad\quad\quad\quad\quad = 6 - 4 - 14i - 11i$

$\quad\quad\quad\quad\quad\quad\quad = 2 - 25i$

20. $(3 + 10i)(8 + i) = 3(8 + i) + 10i(8 + i)$

$\quad\quad\quad\quad\quad\quad = 24 + 3i + 80i + 10i^2$

$\quad\quad\quad\quad\quad\quad = 24 + 83i + 10(-1)$

$\quad\quad\quad\quad\quad\quad = 14 + 83i$

21. $\dfrac{2+i}{5-3i} = \dfrac{2+i}{5-3i} \cdot \dfrac{5+3i}{5+3i}$

$\quad\quad = \dfrac{10 + 6i + 5i + 3i^2}{25 - 9i^2}$

$\quad\quad = \dfrac{10 + 11i + 3(-1)}{25 - 9(-1)}$

$\quad\quad = \dfrac{7 + 11i}{34}$

$\quad\quad = \dfrac{7}{34} + \dfrac{11}{34}i$

22. $4x^2 - 4x + 5 = 0$

$\quad a = 4, b = -4, c = 5$

$\quad x = \dfrac{-(-4) \pm \sqrt{(-4)^2 - 4(4)(5)}}{2(4)} = \dfrac{4 \pm \sqrt{16 - 80}}{8}$

$\quad\quad = \dfrac{4 \pm \sqrt{-64}}{8} = \dfrac{4 \pm 8i}{8} = \dfrac{1}{2} \pm i$

The solution set is $\left\{ \dfrac{1}{2} - i, \dfrac{1}{2} + i \right\}$.

23. We can use a table to summarize the given information.

	minutes per customer	# of cust. in 1 minute
Jamie	5	$\dfrac{1}{5}$
Scott	8.5	$\dfrac{1}{8.5}$
Together	t	$\dfrac{1}{t}$

This leads to the following:

$(\text{Jamie's rate}) + (\text{Scott's rate}) = (\text{rate together})$

$\quad\quad \dfrac{1}{5} + \dfrac{1}{8.5} = \dfrac{1}{t}$

$\quad\quad \dfrac{8.5 + 5}{42.5} = \dfrac{1}{t}$

$\quad\quad \dfrac{13.5}{42.5} = \dfrac{1}{t}$

$\quad\quad 13.5t = 42.5$

$\quad\quad t = \dfrac{42.5}{13.5} = \dfrac{85}{27}$

Since $65 \cdot \dfrac{85}{27} \approx 204.63$, it will take Jamie and Scott about 204.63 minutes (3.41 hours) to check out 65 customers working together.

24. Let $x = $ pounds of banana chips. Then the total pounds of the mix will be $x + 40$.

Since there is to be no loss in revenue, we get

$\text{revenue}_{\text{new mix}} = \text{revenue}_{\text{banana}} + \text{revenue}_{\text{cherry mix}}$

$(\text{price})(\text{lbs})_{\text{new}} = (\text{price})(\text{lbs})_{\text{ban.}} + (\text{price})(\text{lbs})_{\text{cher.}}$

$(10.25)(x + 40) = (2.25)(x) + (15.00)(40)$

$\quad 10.25x + 410 = 2.25x + 600$

$\quad\quad\quad 8x = 190$

$\quad\quad\quad x = 23.75$

The retailer needs to mix 23.75 pounds of banana chips with the 40 pounds of the original mix.

25. Let x = sale price. We know that the discount is $0.42(275.00) = 115.50$. Therefore,

$$\text{sale price} = \text{original price} - \text{discount}$$
$$x = 275 - 115.50$$
$$x = 159.50$$

The sale price is $159.50.

26. Here we need the simple interest formula, $I = P \cdot r \cdot t$. In this case we have $r = 0.04$,

$$t = \frac{3 \text{ months}}{12 \text{ months/yr}} = \frac{1}{4} \text{ yr , and } P = 10,000.$$

$$I = P \cdot r \cdot t$$
$$= (10,000)(0.04)\left(\frac{1}{4}\right)$$
$$= 100$$

Glenn will earn $100.00 in interest after 3 months.

Chapter 1 Projects

Project I

Internet-based Project

Project II

1. $T = \dfrac{n}{Cnp + L + M}$, $n = 3$, $L = 5$, $M = 1$, $C = 0.2$

$$T = \frac{3}{0.2(3)p + 5 + 1} = \frac{3}{0.6p + 6} = \frac{1}{0.2p + 2}$$

2. All of the times given in problem 1 were in seconds, so $T = 0.1$ board per second needs to used as the value for T in the equation found in problem 1.

$$0.1 = \frac{1}{0.2p + 2}$$
$$(0.2p + 2)(0.1) = 1$$
$$0.02p + 0.2 = 1$$
$$0.02p = 0.8$$
$$p = 40 \text{ parts per board}$$

3. $T = 0.15$ board per second

$$0.15 = \frac{1}{0.2p + 2}$$
$$(0.2p + 2)(0.15) = 1$$
$$0.03p + 0.3 = 1$$
$$0.03p = 0.7$$
$$p \approx 23.3 \text{ parts per board}$$

Thus, only 23 parts per board will work.

For problems $4 - 6$, C is requested, so solve for C first:

$$T = \frac{n}{Cnp + L + M}$$
$$(Cnp + L + M)T = n$$
$$CnpT + LT + MT = n$$
$$CnpT = n - LT - MT$$
$$C = \frac{n - LT - MT}{npT}$$

4. $T = 0.06$, $n = 3$, $p = 100$, $M = 1$, $L = 5$
$$C = \frac{3 - 5(0.06) - 1(0.06)}{3(100)(0.06)} \approx 0.147 \text{ sec}$$

5. $T = 0.06$, $n = 3$, $p = 150$, $M = 1$, $L = 5$
$$C = \frac{3 - 5(0.06) - 1(0.06)}{3(150)(0.06)} \approx 0.098 \text{ sec}$$

6. $T = 0.06$, $n = 3$, $p = 200$, $M = 1$, $L = 5$
$$C = \frac{3 - 5(0.06) - 1(0.06)}{3(200)(0.06)} \approx 0.073 \text{ sec}$$

7. As the number of parts per board increases, the tact time decreases, if all the other factors remain constant.

Chapter 2
Graphs

Section 2.1

1. $y = 2x - 4$

x	$y = 2x - 4$	(x, y)
-2	$y = 2(-2) - 4 = -8$	$(-2, -8)$
0	$y = 2(0) - 4 = -4$	$(0, -4)$
2	$y = 2(2) - 4 = 0$	$(2, 0)$

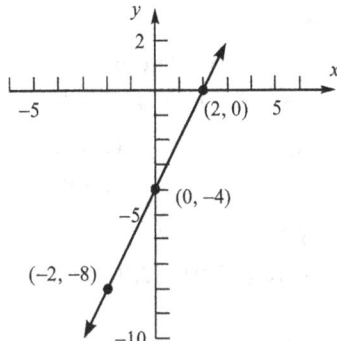

Based on the graph, the intercepts are $(2, 0)$ and $(0, -4)$.

3. intercepts

5. 4

7. True

9. a

11. $y = x + 2$

   x-intercept: y-intercept:
     $0 = x + 2$ $y = 0 + 2$
    $-2 = x$ $y = 2$

The intercepts are $(-2, 0)$ and $(0, 2)$.

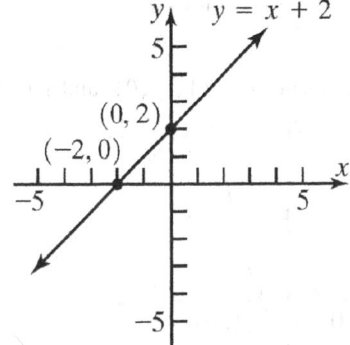

13. $y = 2x + 8$

   x-intercept: y-intercept:
     $0 = 2x + 8$ $y = 2(0) + 8$
    $2x = -8$ $y = 8$
     $x = -4$

The intercepts are $(-4, 0)$ and $(0, 8)$.

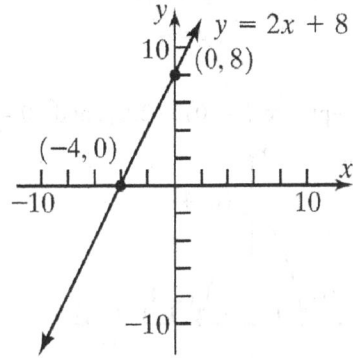

15. $y = x^2 - 1$

x-intercepts: y-intercept:

$0 = x^2 - 1$ $y = 0^2 - 1$

$x^2 = 1$ $y = -1$

$x = \pm 1$

The intercepts are $(-1,0)$, $(1,0)$, and $(0,-1)$.

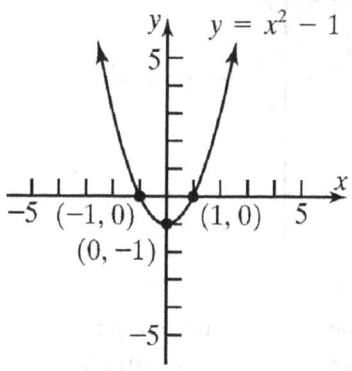

17. $y = -x^2 + 4$

x-intercepts: y-intercepts:

$0 = -x^2 + 4$ $y = -(0)^2 + 4$

$x^2 = 4$ $y = 4$

$x = \pm 2$

The intercepts are $(-2,0)$, $(2,0)$, and $(0,4)$.

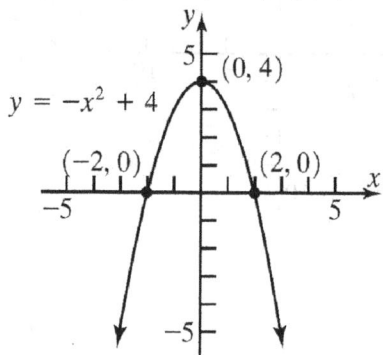

19. $2x + 3y = 6$

x-intercepts: y-intercept:

$2x + 3(0) = 6$ $2(0) + 3y = 6$

$2x = 6$ $3y = 6$

$x = 3$ $y = 2$

The intercepts are $(3,0)$ and $(0,2)$.

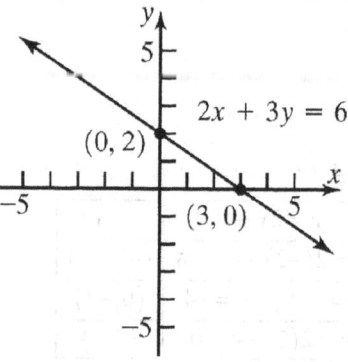

21. $9x^2 + 4y = 36$

x-intercepts: y-intercept:

$9x^2 + 4(0) = 36$ $9(0)^2 + 4y = 36$

$9x^2 = 36$ $4y = 36$

$x^2 = 4$ $y = 9$

$x = \pm 2$

The intercepts are $(-2,0)$, $(2,0)$, and $(0,9)$.

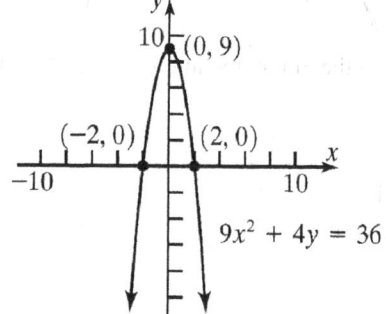

23.

25.

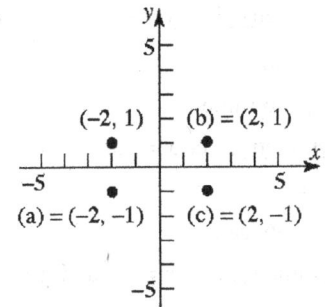

27.

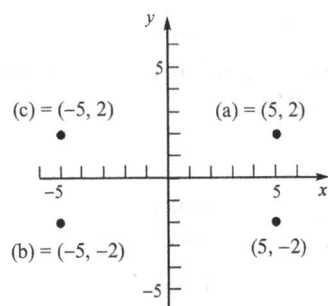

29.

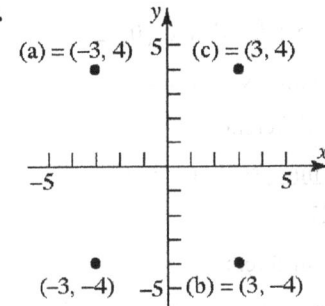

31.

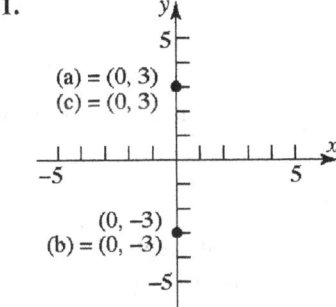

33. a. Intercepts: $(-1,0)$ and $(1,0)$

    b. Symmetric with respect to the x-axis, y-axis, and the origin.

35. a. Intercepts: $\left(-\frac{\pi}{2},0\right)$, $(0,1)$, and $\left(\frac{\pi}{2},0\right)$

37. a. Intercepts: $(0,0)$

    b. Symmetric with respect to the x-axis.

39. a. Intercepts: $(-2,0)$, $(0,0)$, and $(2,0)$

    b. Symmetric with respect to the origin.

41. a. x-intercept: $[-2,1]$, y-intercept 0

    b. Not symmetric to x-axis, y-axis, or origin.

43. a. Intercepts: none

    b. Symmetric with respect to the origin.

45.

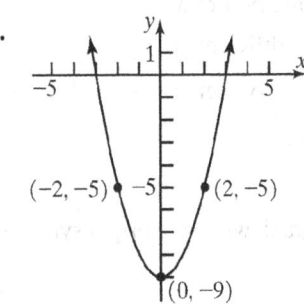

47.

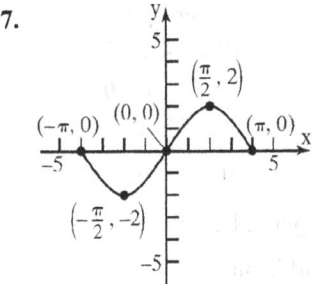

49. $y^2 = x + 4$

    x-intercepts: y-intercepts:

    $0^2 = x + 4$ $y^2 = 0 + 4$

    $-4 = x$ $y^2 = 4$

                           $y = \pm 2$

The intercepts are $(-4,0)$, $(0,-2)$ and $(0,2)$.

Test x-axis symmetry: Let $y = -y$

$$(-y)^2 = x + 4$$
$$y^2 = x + 4 \text{ same}$$

Test y-axis symmetry: Let $x = -x$

$$y^2 = -x + 4 \text{ different}$$

Test origin symmetry: Let $x = -x$ and $y = -y$.

$$(-y)^2 = -x + 4$$
$$y^2 = -x + 4 \text{ different}$$

Therefore, the graph will have x-axis symmetry.

51. $y = \sqrt[3]{x}$

x-intercepts: y-intercepts:

$0 = \sqrt[3]{x}$ $y = \sqrt[3]{0} = 0$

$0 = x$

The only intercept is $(0,0)$.

Test x-axis symmetry: Let $y = -y$

$-y = \sqrt[3]{x}$ different

Test y-axis symmetry: Let $x = -x$

$y = \sqrt[3]{-x} = -\sqrt[3]{x}$ different

Test origin symmetry: Let $x = -x$ and $y = -y$

$-y = \sqrt[3]{-x} = -\sqrt[3]{x}$

$y = \sqrt[3]{x}$ same

Therefore, the graph will have origin symmetry.

53. $x^2 + y - 9 = 0$

x-intercepts: y-intercepts:

$x^2 - 9 = 0$ $0^2 + y - 9 = 0$

$x^2 = 9$ $y = 9$

$x = \pm 3$

The intercepts are $(-3,0)$, $(3,0)$, and $(0,9)$.

Test x-axis symmetry: Let $y = -y$

$x^2 - y - 9 = 0$ different

Test y-axis symmetry: Let $x = -x$

$(-x)^2 + y - 9 = 0$

$x^2 + y - 9 = 0$ same

Test origin symmetry: Let $x = -x$ and $y = -y$

$(-x)^2 - y - 9 = 0$

$x^2 - y - 9 = 0$ different

Therefore, the graph will have y-axis symmetry.

55. $9x^2 + 4y^2 = 36$

x-intercepts: y-intercepts:

$9x^2 + 4(0)^2 = 36$ $9(0)^2 + 4y^2 = 36$

$9x^2 = 36$ $4y^2 = 36$

$x^2 = 4$ $y^2 = 9$

$x = \pm 2$ $y = \pm 3$

The intercepts are $(-2,0)$, $(2,0)$, $(0,-3)$, and $(0,3)$.

Test x-axis symmetry: Let $y = -y$

$9x^2 + 4(-y)^2 = 36$

$9x^2 + 4y^2 = 36$ same

Test y-axis symmetry: Let $x = -x$

$9(-x)^2 + 4y^2 = 36$

$9x^2 + 4y^2 = 36$ same

Test origin symmetry: Let $x = -x$ and $y = -y$

$9(-x)^2 + 4(-y)^2 = 36$

$9x^2 + 4y^2 = 36$ same

Therefore, the graph will have x-axis, y-axis, and origin symmetry.

57. $y = x^3 - 27$

x-intercepts: y-intercepts:

$0 = x^3 - 27$ $y = 0^3 - 27$

$x^3 = 27$ $y = -27$

$x = 3$

The intercepts are $(3,0)$ and $(0,-27)$.

Test x-axis symmetry: Let $y = -y$

$-y = x^3 - 27$ different

Test y-axis symmetry: Let $x = -x$

$y = (-x)^3 - 27$

$y = -x^3 - 27$ different

Test origin symmetry: Let $x = -x$ and $y = -y$

$-y = (-x)^3 - 27$

$y = x^3 + 27$ different

Therefore, the graph has none of the indicated symmetries.

59. $y = x^2 - 3x - 4$

x-intercepts: y-intercepts:

$0 = x^2 - 3x - 4$ $y = 0^2 - 3(0) - 4$

$0 = (x-4)(x+1)$ $y = -4$

$x = 4$ or $x = -1$

The intercepts are $(4,0)$, $(-1,0)$, and $(0,-4)$.

Test x-axis symmetry: Let $y = -y$

$-y = x^2 - 3x - 4$ different

Test y-axis symmetry: Let $x = -x$

$y = (-x)^2 - 3(-x) - 4$

$y = x^2 + 3x - 4$ different

Test origin symmetry: Let $x = -x$ and $y = -y$

$-y = (-x)^2 - 3(-x) - 4$

$-y = x^2 + 3x - 4$ different

Therefore, the graph has none of the indicated symmetries.

61. $y = \dfrac{3x}{x^2 + 9}$

x-intercepts:

$0 = \dfrac{3x}{x^2 + 9}$

$3x = 0$

$x = 0$

y-intercepts:

$y = \dfrac{3(0)}{0^2 + 9} = \dfrac{0}{9} = 0$

The only intercept is $(0, 0)$.

Test x-axis symmetry: Let $y = -y$

$-y = \dfrac{3x}{x^2 + 9}$ different

Test y-axis symmetry: Let $x = -x$

$y = \dfrac{3(-x)}{(-x)^2 + 9}$

$y = -\dfrac{3x}{x^2 + 9}$ different

Test origin symmetry: Let $x = -x$ and $y = -y$

$-y = \dfrac{3(-x)}{(-x)^2 + 9}$

$-y = -\dfrac{3x}{x^2 + 9}$

$y = \dfrac{3x}{x^2 + 9}$ same

Therefore, the graph has origin symmetry.

63. $y = \dfrac{-x^3}{x^2 - 9}$

x-intercepts:

$0 = \dfrac{-x^3}{x^2 - 9}$

$-x^3 = 0$

$x = 0$

y-intercepts:

$y = \dfrac{-0^3}{0^2 - 9} = \dfrac{0}{-9} = 0$

The only intercept is $(0, 0)$.

Test x-axis symmetry: Let $y = -y$

$-y = \dfrac{-x^3}{x^2 - 9}$

$y = \dfrac{x^3}{x^2 - 9}$ different

Test y-axis symmetry: Let $x = -x$

$y = \dfrac{-(-x)^3}{(-x)^2 - 9}$

$y = \dfrac{x^3}{x^2 - 9}$ different

Test origin symmetry: Let $x = -x$ and $y = -y$

$-y = \dfrac{-(-x)^3}{(-x)^2 - 9}$

$-y = \dfrac{x^3}{x^2 - 9}$

$y = \dfrac{-x^3}{x^2 - 9}$ same

Therefore, the graph has origin symmetry.

65. $y = x^3$

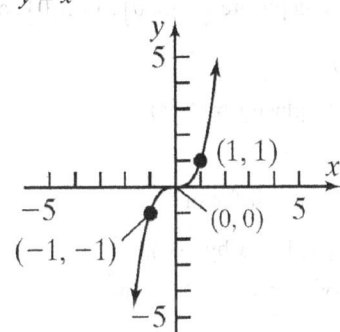

67. $y = \sqrt{x}$

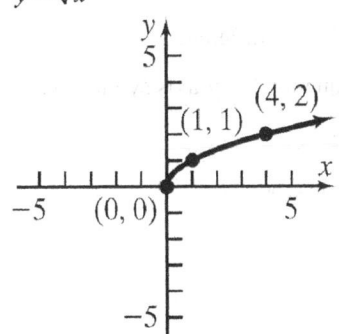

69. If the point $(3,b)$ is on the graph of $y=4x+1$,

then we have $b=4(3)+1=12+1=13$

Thus, $b=13$.

71. If the point $(a,4)$ is on the graph of

$y=x^2+3x$, then we have

$4=a^2+3a$

$0=a^2+3a-4$

$0=(a+4)(a-1)$

$a+4=0$ or $a-1=0$

$a=-4$ $a=1$

Thus, $a=-4$ or $a=1$.

73. a. $0=x^2-5$

$x^2=5$

$x=\pm\sqrt{5}$

The x-intercepts are $x=-\sqrt{5}$ and $x=\sqrt{5}$.

$y=(0)^2-5=-5$

The y-intercept is $y=-5$.

The intercepts are $\left(-\sqrt{5},0\right)$, $\left(\sqrt{5},0\right)$, and $(0,-5)$.

b. x-axis (replace y by $-y$):

$-y=x^2-5$

$y=5-x^2$ different

y-axis (replace x by $-x$):

$y=(-x)^2-5=x^2-5$ same

origin (replace x by $-x$ and y by $-y$):

$-y=(-x)^2-5$

$y=5-x^2$ different

The equation has y-axis symmetry.

c. $y=x^2-5$

Additional points:

x	$y=x^2-5$	(x,y)
1	$y=1^2-5=-4$	$(1,-4)$
-1	from symmetry	$(-1,-4)$
2	$y=2^2-5=-1$	$(2,-1)$
-2	from symmetry	$(-2,-1)$

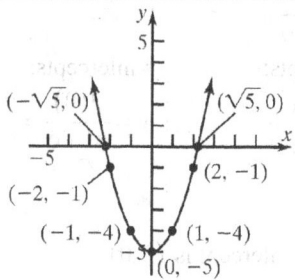

75. a. $x-(0)^2=-9$

$x=-9$

The x-intercept is $x=-9$.

$(0)-y^2=-9$

$-y^2=-9$

$y^2=9 \;\rightarrow\; y=\pm3$

The y-intercepts are $y=-3$ and $y=3$.

The intercepts are $(-9,0)$, $(0,-3)$, and $(0,3)$.

b. x-axis (replace y by $-y$):

$x-(-y)^2=-9$

$x-y^2=-9$ same

y-axis (replace x by $-x$):

$-x-y^2=-9$

$x+y^2=9$ different

origin (replace x by $-x$ and y by $-y$):

$-x-(-y)^2=-9$

$-x-y^2=-9$

$x+y^2=9$ different

The equation has x-axis symmetry.

c. $x - y^2 = -9$ or $x = y^2 - 9$

Additional points:

y	$x = y^2 - 9$	(x, y)
2	$x = 2^2 - 9 = -5$	$(-5, 2)$
-2	from symmetry	$(-5, -2)$

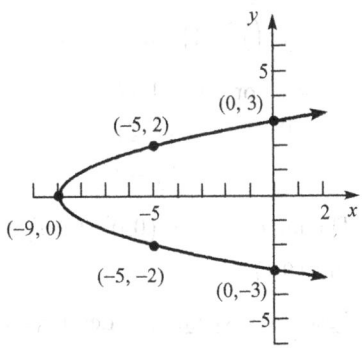

77. a. $x^2 + (0)^2 = 9$

$$x^2 = 9$$
$$x = \pm 3$$

The x-intercepts are $x = -3$ and $x = 3$.

$$(0)^2 + y^2 = 9$$
$$y^2 = 9$$
$$y = \pm 3$$

The y-intercepts are $y = -3$ and $y = 3$.

The intercepts are $(-3, 0)$, $(3, 0)$, $(0, -3)$, and $(0, 3)$.

b. x-axis (replace y by $-y$):

$$x^2 + (-y)^2 = 9$$
$$x^2 + y^2 = 9 \quad \text{same}$$

y-axis (replace x by $-x$):

$$(-x)^2 + y^2 = 9$$
$$x^2 + y^2 = 9 \quad \text{same}$$

origin (replace x by $-x$ and y by $-y$):

$$(-x)^2 + (-y)^2 = 9$$
$$x^2 + y^2 = 9 \quad \text{same}$$

The equation has x-axis symmetry, y-axis symmetry, and origin symmetry.

c. $x^2 + y^2 = 9$

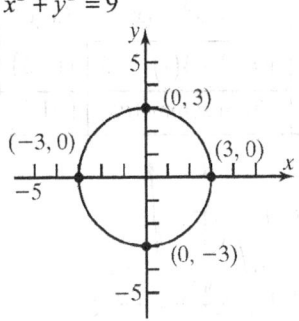

79. a. $0 = x^3 - 4x$

$$0 = x(x^2 - 4)$$
$$x = 0 \quad \text{or} \quad x^2 - 4 = 0$$
$$x^2 = 4$$
$$x = \pm 2$$

The x-intercepts are $x = 0$, $x = -2$, and $x = 2$.

$$y = 0^3 - 4(0) = 0$$

The y-intercept is $y = 0$.

The intercepts are $(0, 0)$, $(-2, 0)$, and $(2, 0)$.

b. x-axis (replace y by $-y$):

$$-y = x^3 - 4x$$
$$y = 4x - x^3 \quad \text{different}$$

y-axis (replace x by $-x$):

$$y = (-x)^3 - 4(-x)$$
$$y = -x^3 + 4x \quad \text{different}$$

origin (replace x by $-x$ and y by $-y$):

$$-y = (-x)^3 - 4(-x)$$
$$-y = -x^3 + 4x$$
$$y = x^3 - 4x \quad \text{same}$$

The equation has origin symmetry.

c. $y = x^3 - 4x$

Additional points:

95

Copyright © 2017 Pearson Education, Inc.

x	$y = x^3 - 4x$	(x, y)
1	$y = 1^3 - 4(1) = -3$	$(1, -3)$
-1	from symmetry	$(-1, 3)$

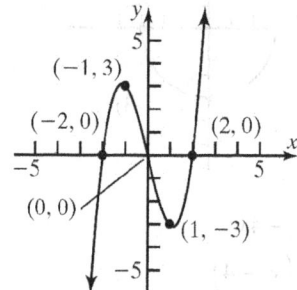

81. For a graph with origin symmetry, if the point (a, b) is on the graph, then so is the point $(-a, -b)$. Since the point $(1, 2)$ is on the graph of an equation with origin symmetry, the point $(-1, -2)$ must also be on the graph.

83. For a graph with origin symmetry, if the point (a, b) is on the graph, then so is the point $(-a, -b)$. Since -4 is an x-intercept in this case, the point $(-4, 0)$ is on the graph of the equation. Due to the origin symmetry, the point $(4, 0)$ must also be on the graph. Therefore, 4 is another x-intercept.

85. **a.** $\left(x^2 + y^2 - x\right)^2 = x^2 + y^2$

 x-intercepts:

 $$\left(x^2 + (0)^2 - x\right)^2 = x^2 + (0)^2$$
 $$\left(x^2 - x\right)^2 = x^2$$
 $$x^4 - 2x^3 + x^2 = x^2$$
 $$x^4 - 2x^3 = 0$$
 $$x^3(x - 2) = 0$$
 $$x^3 = 0 \quad \text{or} \quad x - 2 = 0$$
 $$x = 0 \qquad\qquad x = 2$$

y-intercepts:

$$\left((0)^2 + y^2 - 0\right)^2 = (0)^2 + y^2$$
$$\left(y^2\right)^2 = y^2$$
$$y^4 = y^2$$
$$y^4 - y^2 = 0$$
$$y^2\left(y^2 - 1\right) = 0$$
$$y^2 = 0 \quad \text{or} \quad y^2 - 1 = 0$$
$$y = 0 \qquad\qquad y^2 = 1$$
$$y = \pm 1$$

The intercepts are $(0, 0)$, $(2, 0)$, $(0, -1)$, and $(0, 1)$.

b. Test x-axis symmetry: Let $y = -y$

$$\left(x^2 + (-y)^2 - x\right)^2 = x^2 + (-y)^2$$
$$\left(x^2 + y^2 - x\right)^2 = x^2 + y^2 \quad \text{same}$$

Test y-axis symmetry: Let $x = -x$

$$\left((-x)^2 + y^2 - (-x)\right)^2 = (-x)^2 + y^2$$
$$\left(x^2 + y^2 + x\right)^2 = x^2 + y^2 \quad \text{different}$$

Test origin symmetry: Let $x = -x$ and $y = -y$

$$\left((-x)^2 + (-y)^2 - (-x)\right)^2 = (-x)^2 + (-y)^2$$
$$\left(x^2 + y^2 + x\right)^2 = x^2 + y^2 \quad \text{different}$$

Thus, the graph will have x-axis symmetry.

87. **a.**

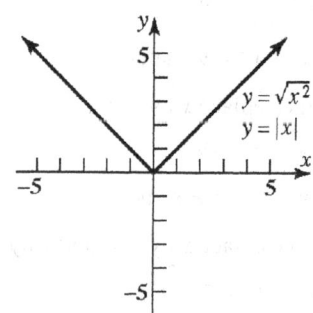

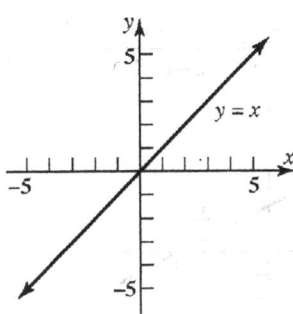

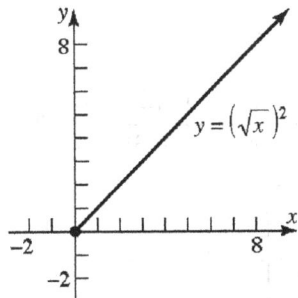

b. Since $\sqrt{x^2} = |x|$ for all x, the graphs of $y = \sqrt{x^2}$ and $y = |x|$ are the same.

c. For $y = \left(\sqrt{x}\right)^2$, the domain of the variable x is $x \geq 0$; for $y = x$, the domain of the variable x is all real numbers. Thus, $\left(\sqrt{x}\right)^2 = x$ only for $x \geq 0$.

d. For $y = \sqrt{x^2}$, the range of the variable y is $y \geq 0$; for $y = x$, the range of the variable y is all real numbers. Also, $\sqrt{x^2} = x$ only if $x \geq 0$. Otherwise, $\sqrt{x^2} = -x$.

89. Answers will vary. One example:

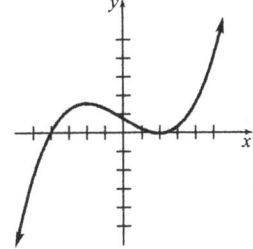

91. Answers will vary

93. Answers may vary. The graph must contain the points $(-2,5)$, $(-1,3)$, and $(0,2)$. For the graph to be symmetric about the y-axis, the graph must also contain the points $(2,5)$ and $(1,3)$ (note that $(0, 2)$ is on the y-axis).

For the graph to also be symmetric with respect to the x-axis, the graph must also contain the points $(-2,-5)$, $(-1,-3)$, $(0,-2)$, $(2,-5)$, and $(1,-3)$. Recall that a graph with two of the symmetries (x-axis, y-axis, origin) will necessarily have the third. Therefore, if the original graph with y-axis symmetry also has x-axis symmetry, then it will also have origin symmetry.

95. $3x^2 - 30x + 75 =$
$3(x^2 - 10x + 25) =$
$3(x-5)(x-5) = 3(x-5)^2$

97. $x^2 - 8x + 4 = 0$
$x^2 - 8x = -4$
$x^2 - 8x + 16 = -4 + 16$
$(x-4)^2 = 12$
$x - 4 = \pm\sqrt{12}$
$x = 4 \pm \sqrt{12}$
$= 4 \pm 2\sqrt{3}$

97

Section 2.2

1. undefined; 0

3. True

5. True; $2(1)+(2)\overset{?}{=}4$
$$2+2\overset{?}{=}4$$
$$4=4 \text{ True}$$

7. 2

9. False; perpendicular lines have slopes that are opposite-reciprocals of each other.

11. c

13. a. $\text{Slope}=\dfrac{1-0}{2-0}=\dfrac{1}{2}$

    b. If x increases by 2 units, y will increase by 1 unit.

15. a. $\text{Slope}=\dfrac{1-2}{1-(-2)}=-\dfrac{1}{3}$

    b. If x increases by 3 units, y will decrease by 1 unit.

17. $\text{Slope}=\dfrac{y_2-y_1}{x_2-x_1}=\dfrac{0-3}{4-2}=-\dfrac{3}{2}$

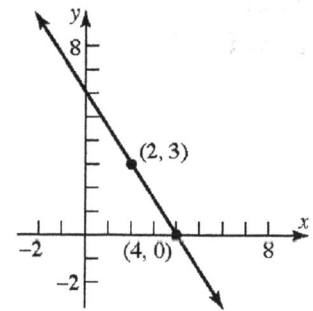

19. $\text{Slope}=\dfrac{y_2-y_1}{x_2-x_1}=\dfrac{1-3}{2-(-2)}=\dfrac{-2}{4}=-\dfrac{1}{2}$

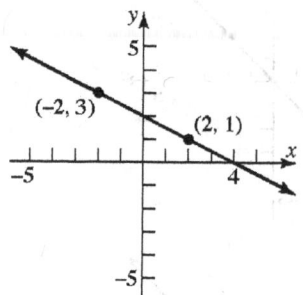

21. $\text{Slope}=\dfrac{y_2-y_1}{x_2-x_1}=\dfrac{-1-(-1)}{2-(-3)}=\dfrac{0}{5}=0$

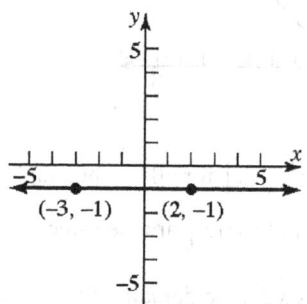

23. $\text{Slope}=\dfrac{y_2-y_1}{x_2-x_1}=\dfrac{-2-2}{-1-(-1)}=\dfrac{-4}{0}$ undefined.

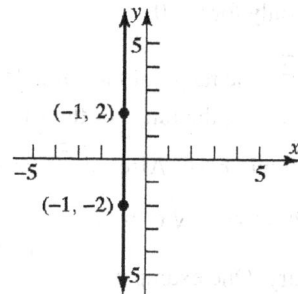

25. $P=(1,2); m=3$

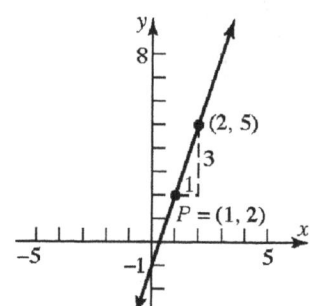

27. $P = (2,4); m = -\dfrac{3}{4}$

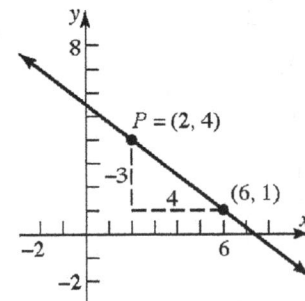

29. $P = (-1,3); m = 0$

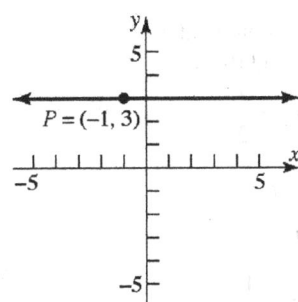

31. $P = (0,3)$; slope undefined

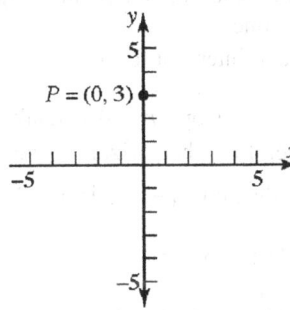

(note: the line is the y-axis)

33. Slope $= 4 = \dfrac{4}{1}$; point: $(1,2)$

If x increases by 1 unit, then y increases by 4 units.

Answers will vary. Three possible points are:
$x = 1 + 1 = 2$ and $y = 2 + 4 = 6$
$(2,6)$
$x = 2 + 1 = 3$ and $y = 6 + 4 = 10$
$(3,10)$
$x = 3 + 1 = 4$ and $y = 10 + 4 = 14$
$(4,14)$

35. Slope $= -\dfrac{3}{2} = \dfrac{-3}{2}$; point: $(2,-4)$

If x increases by 2 units, then y decreases by 3 units.

Answers will vary. Three possible points are:
$x = 2 + 2 = 4$ and $y = -4 - 3 = -7$
$(4,-7)$
$x = 4 + 2 = 6$ and $y = -7 - 3 = -10$
$(6,-10)$
$x = 6 + 2 = 8$ and $y = -10 - 3 = -13$
$(8,-13)$

37. Slope $= -2 = \dfrac{-2}{1}$; point: $(-2,-3)$

If x increases by 1 unit, then y decreases by 2 units.

Answers will vary. Three possible points are:
$x = -2 + 1 = -1$ and $y = -3 - 2 = -5$
$(-1,-5)$
$x = -1 + 1 = 0$ and $y = -5 - 2 = -7$
$(0,-7)$
$x = 0 + 1 = 1$ and $y = -7 - 2 = -9$
$(1,-9)$

39. (0, 0) and (2, 1) are points on the line.

Slope $= \dfrac{1-0}{2-0} = \dfrac{1}{2}$

y-intercept is 0; using $y = mx + b$:

$$y = \frac{1}{2}x + 0$$
$$2y = x$$
$$0 = x - 2y$$
$$x - 2y = 0 \text{ or } y = \frac{1}{2}x$$

41. $(-1, 3)$ and $(1, 1)$ are points on the line.

Slope $= \dfrac{1-3}{1-(-1)} = \dfrac{-2}{2} = -1$

Using $y - y_1 = m(x - x_1)$
$$y - 1 = -1(x - 1)$$
$$y - 1 = -x + 1$$
$$y = -x + 2$$
$$x + y = 2 \text{ or } y = -x + 2$$

99

43. $y - y_1 = m(x - x_1), \ m = 2$

$\quad y - 3 = 2(x - 3)$

$\quad y - 3 = 2x - 6$

$\quad y = 2x - 3$

$\quad 2x - y = 3 \ \text{or} \ y = 2x - 3$

45. $y - y_1 = m(x - x_1), \ m = -\dfrac{1}{2}$

$\quad y - 2 = -\dfrac{1}{2}(x - 1)$

$\quad y - 2 = -\dfrac{1}{2}x + \dfrac{1}{2}$

$\quad y = -\dfrac{1}{2}x + \dfrac{5}{2}$

$\quad x + 2y = 5 \ \text{or} \ y = -\dfrac{1}{2}x + \dfrac{5}{2}$

47. Slope = 3; containing $(-2, 3)$

$\quad y - y_1 = m(x - x_1)$

$\quad y - 3 = 3(x - (-2))$

$\quad y - 3 = 3x + 6$

$\quad y = 3x + 9$

$\quad 3x - y = -9 \ \text{or} \ y = 3x + 9$

49. Slope $= -\dfrac{2}{3}$; containing $(1, -1)$

$\quad y - y_1 = m(x - x_1)$

$\quad y - (-1) = -\dfrac{2}{3}(x - 1)$

$\quad y + 1 = -\dfrac{2}{3}x + \dfrac{2}{3}$

$\quad y = -\dfrac{2}{3}x - \dfrac{1}{3}$

$\quad 2x + 3y = -1 \ \text{or} \ y = -\dfrac{2}{3}x - \dfrac{1}{3}$

51. Slope = -3; y-intercept $= 3$

$\quad y = mx + b$

$\quad y = -3x + 3$

$\quad 3x + y = 3 \ \text{or} \ y = -3x + 3$

53. Containing $(1, 3)$ and $(-1, 2)$

$\quad m = \dfrac{2 - 3}{-1 - 1} = \dfrac{-1}{-2} = \dfrac{1}{2}$

$\quad y - y_1 = m(x - x_1)$

$\quad y - 3 = \dfrac{1}{2}(x - 1)$

$\quad y - 3 = \dfrac{1}{2}x - \dfrac{1}{2}$

$\quad y = \dfrac{1}{2}x + \dfrac{5}{2}$

$\quad x - 2y = -5 \ \text{or} \ y = \dfrac{1}{2}x + \dfrac{5}{2}$

55. x-intercept = 2; y-intercept = -1
Points are $(2,0)$ and $(0,-1)$

$\quad m = \dfrac{-1 - 0}{0 - 2} = \dfrac{-1}{-2} = \dfrac{1}{2}$

$\quad y = mx + b$

$\quad y = \dfrac{1}{2}x - 1$

$\quad x - 2y = 2 \ \text{or} \ y = \dfrac{1}{2}x - 1$

57. Slope undefined; containing the point $(2, 4)$
This is a vertical line.
$\quad x = 2 \qquad$ No slope-intercept form.

59. Horizontal lines have slope $m = 0$ and take the form $y = b$. Therefore, the horizontal line passing through the point $(-3, 2)$ is $y = 2$.

61. Parallel to $y = 4x$; Slope = 4
Containing $(-1, 2)$

$\quad y - y_1 = m(x - x_1)$

$\quad y - 2 = 4(x - (-1))$

$\quad y - 2 = x + 4 \rightarrow y = 4x + 6$

$\quad 4x - y = -6 \ \text{or} \ y = 4x + 6$

63. Parallel to $5x - y = -2$; Slope = 5
Containing the point (0, 0)
$$y - y_1 = m(x - x_1)$$
$$y - 0 = 5(x - 0)$$
$$y = 5x$$
$$5x - y = 0 \text{ or } y = 5x$$

65. Parallel to $x = 5$; Containing (4,2)
This is a vertical line.
$x = 4$ No slope-intercept form.

67. Perpendicular to $y = \frac{1}{6}x + 4$; Containing (1, –2)
Slope of perpendicular = –6
$$y - y_1 = m(x - x_1)$$
$$y - (-2) = -6(x - 1)$$
$$y + 2 = -6x + 2 \rightarrow y = -6x$$
$$6x + y = 0 \text{ or } y = -6x$$

69. Perpendicular to $2x + 5y = 2$; Containing the point (–3, –6)
Slope of perpendicular $= \frac{5}{2}$
$$y - y_1 = m(x - x_1)$$
$$y - (-6) = \frac{5}{2}(x - (-3)) \rightarrow y + 6 = \frac{5}{2}x + \frac{15}{2}$$
$$y = \frac{5}{2}x + \frac{3}{2}$$
$$5x - 2y = -3 \text{ or } y = \frac{5}{2}x + \frac{3}{2}$$

71. Perpendicular to $x = 8$; Containing (3, 4)
Slope of perpendicular = 0 (horizontal line)
$y = 4$

73. $y = 2x + 3$; Slope = 2; y-intercept = 3

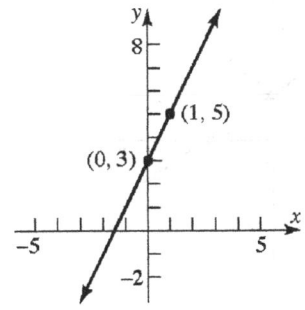

75. $\frac{1}{4}y = x - 1$; $y = 4x - 4$
Slope = 4; y-intercept = –4

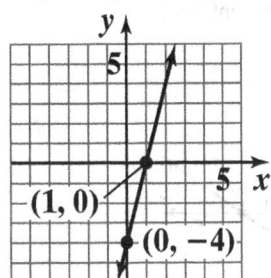

77. $y = \frac{1}{2}x + 2$; Slope $= \frac{1}{2}$; y-intercept = 2

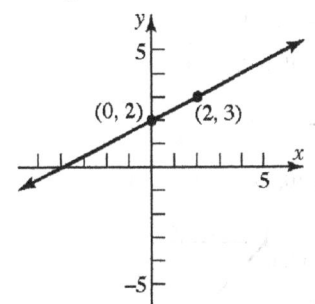

79. $x + 4y = 4$; $4y = -x + 4 \rightarrow y = -\frac{1}{4}x + 1$
Slope $= -\frac{1}{4}$; y-intercept = 1

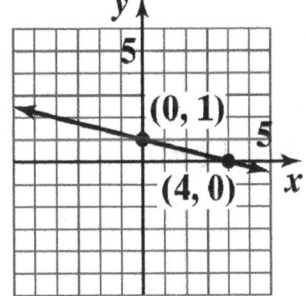

81. $2x - 3y = 6$; $-3y = -2x + 6 \rightarrow y = \dfrac{2}{3}x - 2$

Slope $= \dfrac{2}{3}$, y-intercept $= -2$

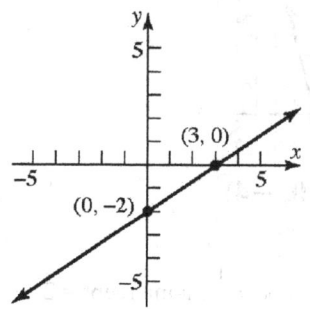

83. $x + y = 1$; $y = -x + 1$
Slope $= -1$; y-intercept $= 1$

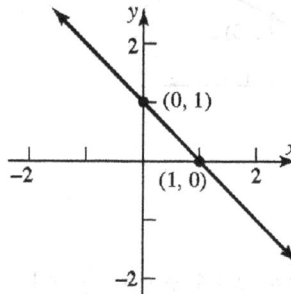

85. $x = -4$; Slope is undefined
y-intercept - none

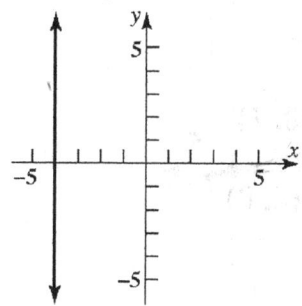

87. $y = 5$; Slope $= 0$; y-intercept $= 5$

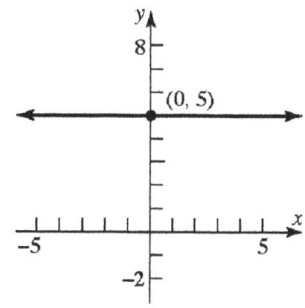

89. $y - x = 0$; $y = x$
Slope $= 1$; y-intercept $= 0$

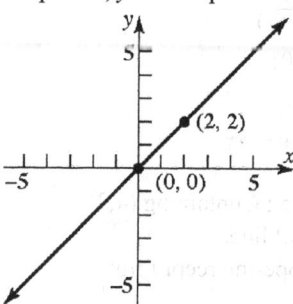

91. $2y - 3x = 0$; $2y = 3x \rightarrow y = \dfrac{3}{2}x$

Slope $= \dfrac{3}{2}$; y-intercept $= 0$

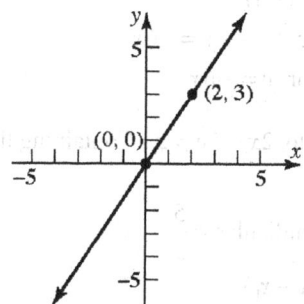

93. a. x-intercept: $2x + 3(0) = 6$

$$2x = 6$$
$$x = 3$$

The point $(3, 0)$ is on the graph.

y-intercept: $2(0) + 3y = 6$

$$3y = 6$$
$$y = 2$$

The point $(0, 2)$ is on the graph.

b.

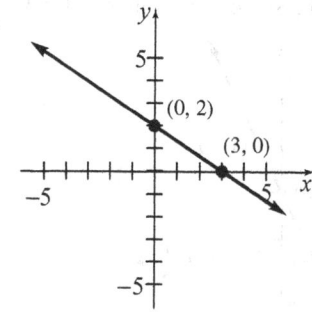

102

95. a. x-intercept: $-4x+5(0)=40$

$$-4x=40$$
$$x=-10$$

The point $(-10,0)$ is on the graph.

y-intercept: $-4(0)+5y=40$

$$5y=40$$
$$y=8$$

The point $(0,8)$ is on the graph.

b.

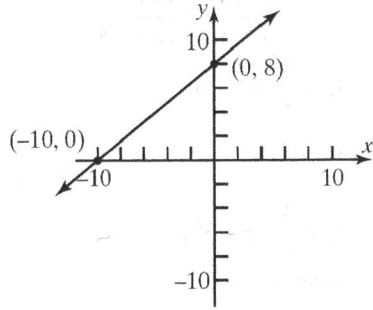

97. a. x-intercept: $7x+2(0)=21$

$$7x=21$$
$$x=3$$

The point $(3,0)$ is on the graph.

y-intercept: $7(0)+2y=21$

$$2y=21$$
$$y=\frac{21}{2}$$

The point $\left(0,\frac{21}{2}\right)$ is on the graph.

b.

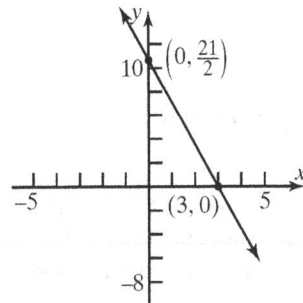

99. a. x-intercept: $\frac{1}{2}x+\frac{1}{3}(0)=1$

$$\frac{1}{2}x=1$$
$$x=2$$

The point $(2,0)$ is on the graph.

y-intercept: $\frac{1}{2}(0)+\frac{1}{3}y=1$

$$\frac{1}{3}y=1$$
$$y=3$$

The point $(0,3)$ is on the graph.

b.

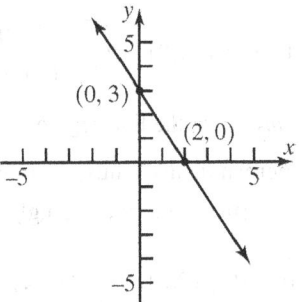

101. a. x-intercept: $0.2x-0.5(0)=1$

$$0.2x=1$$
$$x=5$$

The point $(5,0)$ is on the graph.

y-intercept: $0.2(0)-0.5y=1$

$$-0.5y=1$$
$$y=-2$$

The point $(0,-2)$ is on the graph.

b.

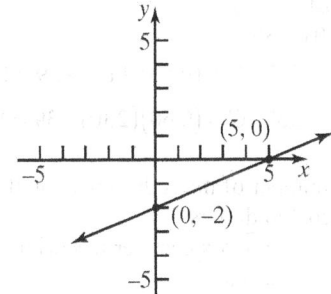

103. The equation of the x-axis is $y=0$. (The slope is 0 and the y-intercept is 0.)

105. The slopes are the same but the y-intercepts are different. Therefore, the two lines are parallel.

107. The slopes are different and their product does not equal -1. Therefore, the lines are neither parallel nor perpendicular.

109. Intercepts: $(0,2)$ and $(-2,0)$. Thus, slope $=1$.

$$y=x+2 \text{ or } x-y=-2$$

103

111. Intercepts: $(3,0)$ and $(0,1)$. Thus, slope $= -\frac{1}{3}$.

$$y = -\frac{1}{3}x+1 \text{ or } x+3y-3$$

113. $P_1 = (-2,5)$, $P_2 = (1,3)$: $m_1 = \frac{5-3}{-2-1} = \frac{2}{-3} = -\frac{2}{3}$

$P_2 = (1,3)$, $P_3 = (-1,0)$: $m_2 = \frac{3-0}{1-(-1)} = \frac{3}{2}$

Since $m_1 \cdot m_2 = -1$, the line segments $\overline{P_1 P_2}$ and $\overline{P_2 P_3}$ are perpendicular. Thus, the points P_1, P_2, and P_3 are vertices of a right triangle.

115. $P_1 = (-1,0)$, $P_2 = (2,3)$, $P_3 = (1,-2)$, $P_4 = (4,1)$

$m_{12} = \frac{3-0}{2-(-1)} = \frac{3}{3} = 1$; $m_{24} = \frac{1-3}{4-2} = -1$;

$m_{34} = \frac{1-(-2)}{4-1} = \frac{3}{3} = 1$; $m_{13} = \frac{-2-0}{1-(-1)} = -1$

Opposite sides are parallel (same slope) and adjacent sides are perpendicular (product of slopes is -1). Therefore, the vertices are for a rectangle.

117. Let x = number of miles driven, and let C = cost in dollars.
Total cost = (cost per mile)(number of miles) + fixed cost
$C = 0.60x+39$
When $x = 110$, $C = (0.60)(110)+39 = \$105.00$.
When $x = 230$, $C = (0.60)(230)+39 = \$177.00$.

119. Let x = number of miles driven annually, and let C = cost in dollars.
Total cost = (approx cost per mile)(number of miles) + fixed cost
$C = 0.16x+1461$

121. a. $C = 0.0757x+15.14$; $0 \le x \le 800$

b.

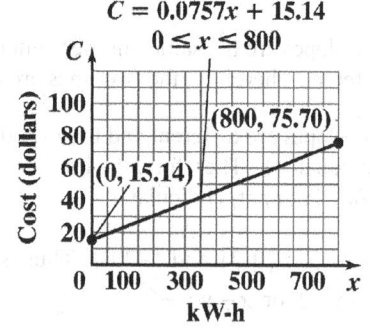

$$C = 0.0757x + 15.14$$
$$0 \le x \le 800$$

c. For 200 kWh,
$C = 0.0757(200)+15.14 = \$30.28$

d. For 500 kWh,
$C = 0.0757(500)+15.14 = \$52.99$

e. For each usage increase of 1 kWh, the monthly charge increases by $0.0757 (that is, 7.57 cents).

123. $(^{\circ}C, ^{\circ}F) = (0, 32)$; $(^{\circ}C, ^{\circ}F) = (100, 212)$

slope $= \frac{212-32}{100-0} = \frac{180}{100} = \frac{9}{5}$

$^{\circ}F - 32 = \frac{9}{5}(^{\circ}C - 0)$

$^{\circ}F - 32 = \frac{9}{5}(^{\circ}C)$

$^{\circ}C = \frac{5}{9}(^{\circ}F - 32)$

If $^{\circ}F = 70$, then

$^{\circ}C = \frac{5}{9}(70-32) = \frac{5}{9}(38)$

$^{\circ}C \approx 21.1^{\circ}$

125. a. The y-intercept is $(0, 30)$, so $b = 30$. Since the ramp drops 2 inches for every 25 inches of run, the slope is $m = \frac{-2}{25} = -\frac{2}{25}$. Thus, the equation is $y = -\frac{2}{25}x+30$.

b. Let $y = 0$.

$$0 = -\frac{2}{25}x+30$$

$$\frac{2}{25}x = 30$$

$$\frac{25}{2}\left(\frac{2}{25}x\right) = \frac{25}{2}(30)$$

$$x = 375$$

The x-intercept is $(375, 0)$. This means that the ramp meets the floor 375 inches (or 31.25 feet) from the base of the platform.

c. No. From part (b), the run is 31.25 feet which exceeds the required maximum of 30 feet.

d. First, design requirements state that the maximum slope is a drop of 1 inch for each 12 inches of run. This means $|m| \le \dfrac{1}{12}$.

Second, the run is restricted to be no more than 30 feet = 360 inches. For a rise of 30 inches, this means the minimum slope is $\dfrac{30}{360} = \dfrac{1}{12}$. That is, $|m| \ge \dfrac{1}{12}$. Thus, the only possible slope is $|m| = \dfrac{1}{12}$. The diagram indicates that the slope is negative. Therefore, the only slope that can be used to obtain the 30-inch rise and still meet design requirements is $m = -\dfrac{1}{12}$. In words, for every 12 inches of run, the ramp must drop *exactly* 1 inch.

127. a. Let x = number of boxes to be sold, and A = money, in dollars, spent on advertising. We have the points
$(x_1, A_1) = (100,000, 40,000);$
$(x_2, A_2) = (200,000, 60,000)$

$$\text{slope} = \frac{60,000 - 40,000}{200,000 - 100,000}$$

$$= \frac{20,000}{100,000} = \frac{1}{5}$$

$$A - 40,000 = \frac{1}{5}(x - 100,000)$$

$$A - 40,000 = \frac{1}{5}x - 20,000$$

$$A = \frac{1}{5}x + 20,000$$

b. If $x = 300,000$, then
$$A = \frac{1}{5}(300,000) + 20,000 = \$80,000$$

c. Each additional box sold requires an additional $0.20 in advertising.

129. $2x - y = C$
Graph the lines:
$2x - y = -4$
$2x - y = 0$
$2x - y = 2$
All the lines have the same slope, 2. The lines are parallel.

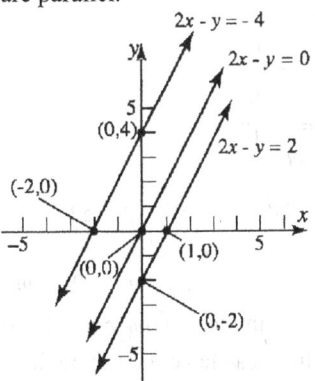

131. (b), (c), (e) and (g)
The line has positive slope and positive y-intercept.

133. (c)
The equation $x - y = -2$ has slope 1 and y-intercept $(0, 2)$. The equation $x - y = 1$ has slope 1 and y-intercept $(0, -1)$. Thus, the lines are parallel with positive slopes. One line has a positive y-intercept and the other with a negative y-intercept.

135 – 137. Answers will vary.

139. No, a line does not need to have both an x-intercept and a y-intercept. Vertical and horizontal lines have only one intercept (unless they are a coordinate axis). Every line must have at least one intercept.

141. Two lines that have the same x-intercept and y-intercept (assuming the x-intercept is not 0) are the same line since a line is uniquely defined by two distinct points.

143. Yes. Two distinct lines with the same y-intercept, but different slopes, can have the same x-intercept if the x-intercept is $x = 0$.
Assume Line 1 has equation $y = m_1 x + b$ and Line 2 has equation $y = m_2 x + b$,
Line 1 has x-intercept $-\dfrac{b}{m_1}$ and y-intercept b.

Line 2 has x-intercept $-\dfrac{b}{m_2}$ and y-intercept b.

Assume also that Line 1 and Line 2 have unequal slopes, that is $m_1 \neq m_2$.

If the lines have the same x-intercept, then

$$-\frac{b}{m_1} = -\frac{b}{m_2}.$$

$$-\frac{b}{m_1} = -\frac{b}{m_2}$$

$$-m_2 b = -m_1 b$$

$$-m_2 b + m_1 b = 0$$

But $-m_2 b + m_1 b = 0 \Rightarrow b(m_1 - m_2) = 0$

$$\Rightarrow b = 0$$

$$\text{or } m_1 - m_2 = 0 \Rightarrow m_1 = m_2$$

Since we are assuming that $m_1 \neq m_2$, the only way that the two lines can have the same x-intercept is if $b = 0$.

145. $m = \dfrac{y_2 - y_1}{x_2 - x_1} = \dfrac{-4-2}{1-(-3)} = \dfrac{-6}{4} = -\dfrac{3}{2}$

It appears that the student incorrectly found the slope by switching the direction of one of the subtractions.

147. $h^2 = a^2 + b^2$

$$= 8^2 + 15^2$$

$$= 16 + 225$$

$$= 289$$

$$h = \sqrt{289} = 17$$

149. $|2x - 5| + 7 < 10$

$$|2x - 5| < 3$$

$$-3 < 2x - 5 < 3$$

$$2 < 2x < 8$$

$$1 < x < 4$$

The solution set is: $\{x \mid 1 < x < 4\}$.

Interval notation: $(1, 4)$

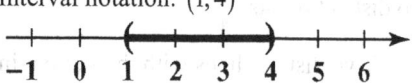

Section 2.3

1. add; $\left(\frac{1}{2} \cdot 10\right)^2 = 25$

3. False. For example, $x^2 + y^2 + 2x + 2y + 8 = 0$ is not a circle. It has no real solutions.

5. True; $r^2 = 9 \rightarrow r = 3$

7. d

9. Center $= (2, 1)$
   Radius $=$ distance from $(0,1)$ to $(2,1)$
   $$= \sqrt{(2-0)^2 + (1-1)^2} = \sqrt{4} = 2$$
   Equation: $(x-2)^2 + (y-1)^2 = 4$

11. Center $=$ midpoint of $(1, 2)$ and $(4, 2)$
    $$= \left(\frac{1+4}{2}, \frac{2+2}{2}\right) = \left(\frac{5}{2}, 2\right)$$
    Radius $=$ distance from $\left(\frac{5}{2}, 2\right)$ to $(4,2)$
    $$= \sqrt{\left(4 - \frac{5}{2}\right)^2 + (2-2)^2} = \sqrt{\frac{9}{4}} = \frac{3}{2}$$
    Equation: $\left(x - \frac{5}{2}\right)^2 + (y-2)^2 = \frac{9}{4}$

13. $(x-h)^2 + (y-k)^2 = r^2$
    $$(x-0)^2 + (y-0)^2 = 2^2$$
    $$x^2 + y^2 = 4$$
    General form: $x^2 + y^2 - 4 = 0$

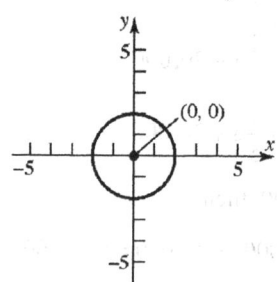

15. $(x-h)^2 + (y-k)^2 = r^2$
    $$(x-0)^2 + (y-2)^2 = 2^2$$
    $$x^2 + (y-2)^2 = 4$$

General form: $x^2 + y^2 - 4y + 4 = 4$

$$x^2 + y^2 - 4y = 0$$

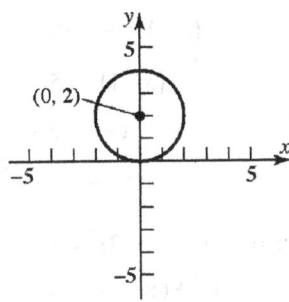

17. $(x-h)^2 + (y-k)^2 = r^2$

$(x-4)^2 + (y-(-3))^2 = 5^2$

$(x-4)^2 + (y+3)^2 = 25$

General form:

$x^2 - 8x + 16 + y^2 + 6y + 9 = 25$

$$x^2 + y^2 - 8x + 6y = 0$$

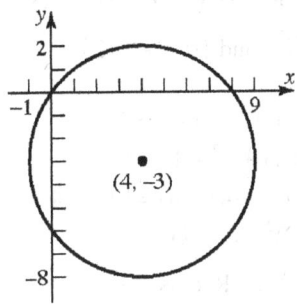

19. $(x-h)^2 + (y-k)^2 = r^2$

$(x-(-2))^2 + (y-1)^2 = 4^2$

$(x+2)^2 + (y-1)^2 = 16$

General form: $x^2 + 4x + 4 + y^2 - 2y + 1 = 16$

$$x^2 + y^2 + 4x - 2y - 11 = 0$$

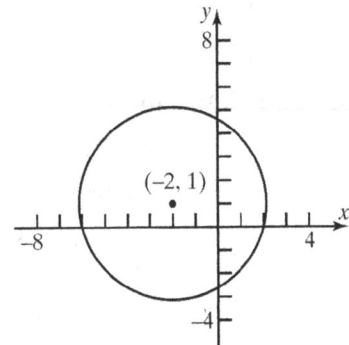

21. $(x-h)^2 + (y-k)^2 = r^2$

$$\left(x - \frac{1}{2}\right)^2 + (y-0)^2 = \left(\frac{1}{2}\right)^2$$

$$\left(x - \frac{1}{2}\right)^2 + y^2 = \frac{1}{4}$$

General form: $x^2 - x + \frac{1}{4} + y^2 = \frac{1}{4}$

$$x^2 + y^2 - x = 0$$

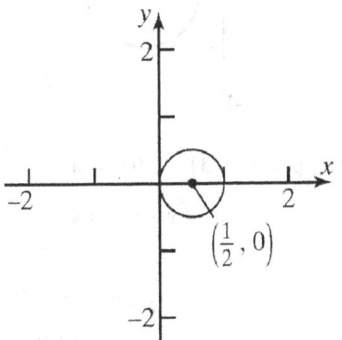

23. $x^2 + y^2 = 4$

$x^2 + y^2 = 2^2$

a. Center: $(0,0)$; Radius $= 2$

b.

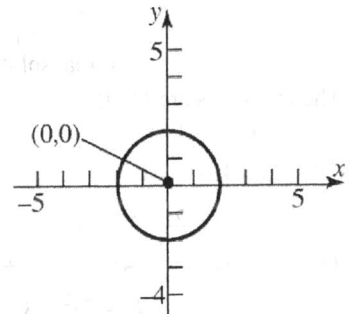

c. x-intercepts: $x^2 + (0)^2 = 4$

$$x^2 = 4$$

$$x = \pm\sqrt{4} = \pm 2$$

y-intercepts: $(0)^2 + y^2 = 4$

$$y^2 = 4$$

$$y = \pm\sqrt{4} = \pm 2$$

The intercepts are $(-2, 0), (2, 0), (0, -2)$, and $(0, 2)$.

25. $2(x-3)^2 + 2y^2 = 8$

$(x-3)^2 + y^2 = 4$

a. Center: $(3, 0)$; Radius $= 2$

b.

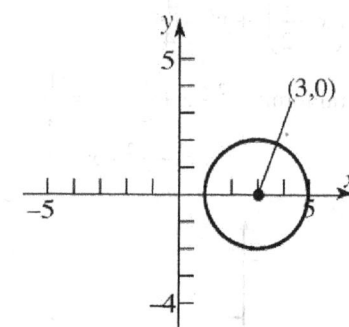

c. x-intercepts: $(x-3)^2 + (0)^2 = 4$

$(x-3)^2 = 4$

$x - 3 = \pm\sqrt{4}$

$x - 3 = \pm 2$

$x = 3 \pm 2$

$x = 5 \quad \text{or} \quad x = 1$

y-intercepts: $(0-3)^2 + y^2 = 4$

$(-3)^2 + y^2 = 4$

$9 + y^2 = 4$

$y^2 = -5$

No real solution.

The intercepts are $(1, 0)$ and $(5, 0)$.

27. $x^2 + y^2 - 2x - 4y - 4 = 0$

$x^2 - 2x + y^2 - 4y = 4$

$(x^2 - 2x + 1) + (y^2 - 4y + 4) = 4 + 1 + 4$

$(x-1)^2 + (y-2)^2 = 3^2$

a. Center: $(1, 2)$; Radius $= 3$

b.

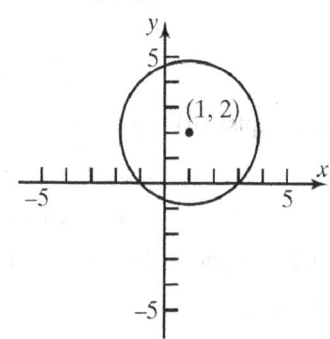

c. x-intercepts: $(x-1)^2 + (0-2)^2 = 3^2$

$(x-1)^2 + (-2)^2 = 3^2$

$(x-1)^2 + 4 = 9$

$(x-1)^2 = 5$

$x - 1 = \pm\sqrt{5}$

$x = 1 \pm \sqrt{5}$

y-intercepts: $(0-1)^2 + (y-2)^2 = 3^2$

$(-1)^2 + (y-2)^2 = 3^2$

$1 + (y-2)^2 = 9$

$(y-2)^2 = 8$

$y - 2 = \pm\sqrt{8}$

$y - 2 = \pm 2\sqrt{2}$

$y = 2 \pm 2\sqrt{2}$

The intercepts are $\left(1-\sqrt{5}, 0\right), \left(1+\sqrt{5}, 0\right),$ $\left(0, 2-2\sqrt{2}\right)$, and $\left(0, 2+2\sqrt{2}\right)$.

29. $x^2 + y^2 + 4x - 4y - 1 = 0$

$x^2 + 4x + y^2 - 4y = 1$

$(x^2 + 4x + 4) + (y^2 - 4y + 4) = 1 + 4 + 4$

$(x+2)^2 + (y-2)^2 = 3^2$

a. Center: $(-2, 2)$; Radius $= 3$

b.

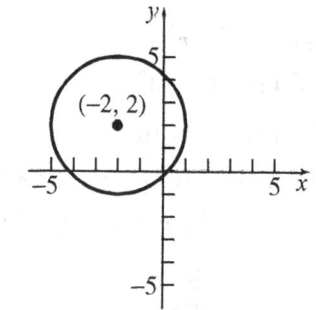

c. x-intercepts: $(x+2)^2 + (0-2)^2 = 3^2$

$(x+2)^2 + 4 = 9$

$(x+2)^2 = 5$

$x + 2 = \pm\sqrt{5}$

$x = -2 \pm \sqrt{5}$

y-intercepts: $(0+2)^2 + (y-2)^2 = 3^2$

$$4 + (y-2)^2 = 9$$
$$(y-2)^2 = 5$$
$$y - 2 = \pm\sqrt{5}$$
$$y = 2 \pm \sqrt{5}$$

The intercepts are $\left(-2-\sqrt{5}, 0\right)$,

$\left(-2+\sqrt{5}, 0\right)$, $\left(0, 2-\sqrt{5}\right)$, and $\left(0, 2+\sqrt{5}\right)$.

31. $\quad x^2 + y^2 - x + 2y + 1 = 0$

$$x^2 - x + y^2 + 2y = -1$$
$$\left(x^2 - x + \frac{1}{4}\right) + (y^2 + 2y + 1) = -1 + \frac{1}{4} + 1$$
$$\left(x - \frac{1}{2}\right)^2 + (y+1)^2 = \left(\frac{1}{2}\right)^2$$

a. Center: $\left(\frac{1}{2}, -1\right)$; Radius $= \frac{1}{2}$

b.

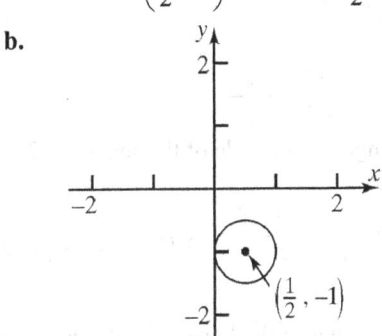

c. x-intercepts: $\left(x - \frac{1}{2}\right)^2 + (0+1)^2 = \left(\frac{1}{2}\right)^2$

$$\left(x - \frac{1}{2}\right)^2 + 1 = \frac{1}{4}$$
$$\left(x - \frac{1}{2}\right)^2 = -\frac{3}{4}$$

No real solutions

y-intercepts: $\left(0 - \frac{1}{2}\right)^2 + (y+1)^2 = \left(\frac{1}{2}\right)^2$

$$\frac{1}{4} + (y+1)^2 = \frac{1}{4}$$
$$(y+1)^2 = 0$$
$$y + 1 = 0$$
$$y = -1$$

The only intercept is $(0, -1)$.

33. $\quad 2x^2 + 2y^2 - 12x + 8y - 24 = 0$

$$x^2 + y^2 - 6x + 4y = 12$$
$$x^2 - 6x + y^2 + 4y = 12$$
$$(x^2 - 6x + 9) + (y^2 + 4y + 4) = 12 + 9 + 4$$
$$(x-3)^2 + (y+2)^2 = 5^2$$

a. Center: $(3, -2)$; Radius $= 5$

b.

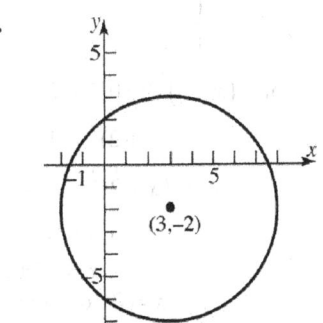

c. x-intercepts: $(x-3)^2 + (0+2)^2 = 5^2$

$$(x-3)^2 + 4 = 25$$
$$(x-3)^2 = 21$$
$$x - 3 = \pm\sqrt{21}$$
$$x = 3 \pm \sqrt{21}$$

y-intercepts: $(0-3)^2 + (y+2)^2 = 5^2$

$$9 + (y+2)^2 = 25$$
$$(y+2)^2 = 16$$
$$y + 2 = \pm 4$$
$$y = -2 \pm 4$$
$$y = 2 \quad \text{or} \quad y = -6$$

The intercepts are $\left(3 - \sqrt{21}, 0\right)$, $\left(3 + \sqrt{21}, 0\right)$,

$(0, -6)$, and $(0, 2)$.

35. $\quad 2x^2 + 8x + 2y^2 = 0$

$$x^2 + 4x + y^2 = 0$$
$$x^2 + 4x + 4 + y^2 = 0 + 4$$
$$(x+2)^2 + y^2 = 2^2$$

a. Center: $(-2, 0)$; Radius: $r = 2$

b.

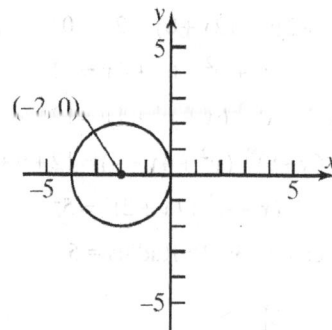

c. x-intercepts: $(x+2)^2 + (0)^2 = 2^2$

$$(x+2)^2 = 4$$
$$(x+2)^2 = \pm\sqrt{4}$$
$$x+2 = \pm 2$$
$$x = -2 \pm 2$$
$$x = 0 \quad \text{or} \quad x = -4$$

y-intercepts: $(0+2)^2 + y^2 = 2^2$

$$4 + y^2 = 4$$
$$y^2 = 0$$
$$y = 0$$

The intercepts are $(-4, 0)$ and $(0, 0)$.

37. Center at (0, 0); containing point (–2, 3).

$$r = \sqrt{(-2-0)^2 + (3-0)^2} = \sqrt{4+9} = \sqrt{13}$$

Equation: $(x-0)^2 + (y-0)^2 = \left(\sqrt{13}\right)^2$

$$x^2 + y^2 = 13$$

39. Center at (2, 3); tangent to the x-axis.

$r = 3$

Equation: $(x-2)^2 + (y-3)^2 = 3^2$

$$(x-2)^2 + (y-3)^2 = 9$$

41. Endpoints of a diameter are (1, 4) and (–3, 2). The center is at the midpoint of that diameter:

Center: $\left(\dfrac{1+(-3)}{2}, \dfrac{4+2}{2}\right) = (-1, 3)$

Radius: $r = \sqrt{(1-(-1))^2 + (4-3)^2} = \sqrt{4+1} = \sqrt{5}$

Equation: $(x-(-1))^2 + (y-3)^2 = \left(\sqrt{5}\right)^2$

$$(x+1)^2 + (y-3)^2 = 5$$

43. Center at (–1, 3); tangent to the line $y = 2$. This means that the circle contains the point (–1, 2), so the radius is $r = 1$.

Equation: $(x+1)^2 + (y-3)^2 = (1)^2$

$$(x+1)^2 + (y-3)^2 = 1$$

45. (c); Center: $(1, -2)$; Radius = 2

47. (b); Center: $(-1, 2)$; Radius = 2

49. Let the upper-right corner of the square be the point (x, y). The circle and the square are both centered about the origin. Because of symmetry, we have that $x = y$ at the upper-right corner of the square. Therefore, we get

$$x^2 + y^2 = 9$$
$$x^2 + x^2 = 9$$
$$2x^2 = 9$$
$$x^2 = \frac{9}{2}$$
$$x = \sqrt{\frac{9}{2}} = \frac{3\sqrt{2}}{2}$$

The length of one side of the square is $2x$. Thus, the area is

$$A = s^2 = \left(2 \cdot \frac{3\sqrt{2}}{2}\right)^2 = \left(3\sqrt{2}\right)^2 = 18 \text{ square units.}$$

51. The diameter of the Ferris wheel was 250 feet, so the radius was 125 feet. The maximum height was 264 feet, so the center was at a height of $264 - 125 = 139$ feet above the ground. Since the center of the wheel is on the y-axis, it is the point (0, 139). Thus, an equation for the wheel is:

$$(x-0)^2 + (y-139)^2 = 125^2$$
$$x^2 + (y-139)^2 = 15,625$$

53. $x^2 + y^2 + 2x + 4y - 4091 = 0$

$$x^2 + 2x + y^2 + 4y - 4091 = 0$$
$$x^2 + 2x + 1 + y^2 + 4y + 4 = 4091 + 5$$
$$(x+1)^2 + (y+2)^2 = 4096$$

The circle representing Earth has center $(-1, -2)$ and radius $= \sqrt{4096} = 64$.

So the radius of the satellite's orbit is $64 + 0.6 = 64.6$ units.

The equation of the orbit is
$$(x+1)^2 + (y+2)^2 = (64.6)^2$$
$$x^2 + y^2 + 2x + 4y - 4168.16 = 0$$

55. $x^2 + y^2 = 9$

Center: $(0, 0)$

Slope from center to $\left(1, 2\sqrt{2}\right)$ is
$$\frac{2\sqrt{2}-0}{1-0} = \frac{2\sqrt{2}}{1} = 2\sqrt{2} \ .$$

Slope of the tangent line is $\dfrac{-1}{2\sqrt{2}} = -\dfrac{\sqrt{2}}{4}$.

Equation of the tangent line is:
$$y - 2\sqrt{2} = -\frac{\sqrt{2}}{4}(x-1)$$
$$y - 2\sqrt{2} = -\frac{\sqrt{2}}{4}x + \frac{\sqrt{2}}{4}$$
$$4y - 8\sqrt{2} = -\sqrt{2}x + \sqrt{2}$$
$$\sqrt{2}x + 4y = 9\sqrt{2}$$
$$\sqrt{2}x + 4y - 9\sqrt{2} = 0$$

57. Let (h, k) be the center of the circle.
$$x - 2y + 4 = 0$$
$$2y = x + 4$$
$$y = \frac{1}{2}x + 2$$

The slope of the tangent line is $\dfrac{1}{2}$. The slope from (h, k) to $(0, 2)$ is -2.
$$\frac{2-k}{0-h} = -2$$
$$2 - k = 2h$$

The other tangent line is $y = 2x - 7$, and it has slope 2.

The slope from (h, k) to $(3, -1)$ is $-\dfrac{1}{2}$.
$$\frac{-1-k}{3-h} = -\frac{1}{2}$$
$$2 + 2k = 3 - h$$
$$2k = 1 - h$$
$$h = 1 - 2k$$

Solve the two equations in h and k :

$$2 - k = 2(1 - 2k)$$
$$2 - k = 2 - 4k$$
$$3k = 0$$
$$k = 0$$
$$h = 1 - 2(0) = 1$$

The center of the circle is $(1, 0)$.

59. Consider the following diagram:

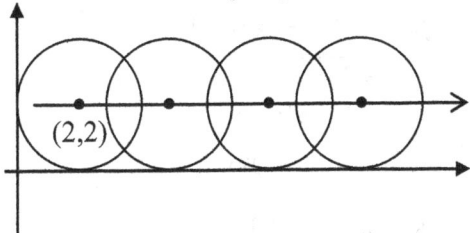

Therefore, the path of the center of the circle has the equation $y = 2$.

61. (b), (c), (e) and (g)

We need $h, k > 0$ and $(0,0)$ on the graph.

63. Answers will vary.

65. $A = \pi r^2$
$$= \pi(13)^2$$
$$= 169\pi \ \text{cm}^2$$
$$C = 2\pi r$$
$$= 2\pi(13)$$
$$= 26\pi \ \text{cm}$$

67. $\sqrt{2x^2 + 3x - 1} = x + 1$
$$2x^2 + 3x - 1 = (x+1)^2$$
$$2x^2 + 3x - 1 = x^2 + 2x + 1$$
$$x^2 + x - 2 = 0$$
$$(x+2)(x-1) = 0$$
$$x = -2 \ \text{or} \ x = 1$$

We need to check each possible solution:

Check $x = -2$

$$\sqrt{2(-2)^2 + 3(-2) - 1} = (-2) + 1$$

$$\sqrt{2(4) - 6 - 1} = -1$$

no

Check $x = 1$

$$\sqrt{2(1)^2 + 3(1) - 1} = (1) + 1$$

$$\sqrt{2 + 3 - 1} = 2$$

$$\sqrt{4} = 2$$

yes

The solution is $\{1\}$

Section 2.4

1. $y = kx$

3. b

5. $y = kx$

$$2 = 10k$$

$$k = \frac{2}{10} = \frac{1}{5}$$

$$y = \frac{1}{5}x$$

7. $A = kx^2$

$$4\pi = k(2)^2$$

$$4\pi = 4k$$

$$\pi - k$$

$$A = \pi x^2$$

9. $F = \dfrac{k}{d^2}$

$$10 = \frac{k}{5^2}$$

$$10 = \frac{k}{25}$$

$$k = 250$$

$$F = \frac{250}{d^2}$$

11. $z = k\left(x^2 + y^2\right)$

$$5 = k\left(3^2 + 4^2\right)$$

$$5 = k(25)$$

$$k = \frac{5}{25} = \frac{1}{5}$$

$$z = \frac{1}{5}\left(x^2 + y^2\right)$$

13. $M = \dfrac{kd^2}{\sqrt{x}}$

$$24 = \frac{k\left(4^2\right)}{\sqrt{9}}$$

$$24 = \frac{16k}{3}$$

$$k = 24\left(\frac{3}{16}\right) = \frac{9}{2}$$

$$M = \frac{9d^2}{2\sqrt{x}}$$

15. $T^2 = \dfrac{ka^3}{d^2}$

$$2^2 = \frac{k\left(2^3\right)}{4^2}$$

$$4 = \frac{k(8)}{16}$$

$$4 = \frac{k}{2}$$

$$k = 8$$

$$T^2 = \frac{8a^3}{d^2}$$

17. $V = \dfrac{4\pi}{3}r^3$

19. $A = \dfrac{1}{2}bh$

21. $F = \left(6.67 \times 10^{-11}\right)\left(\dfrac{mM}{d^2}\right)$

23.
$$p = kB$$
$$6.49 = k(1000)$$
$$0.00649 = k$$
Therefore we have the linear equation
$p = 0.00649B$.
If $B = 145000$, then
$p = 0.00649(145000) = \$941.05$.

25. $s = kt^2$
$$16 = k(1)^2$$
$$k = 16$$
Therefore, we have equation $s = 16t^2$.

If $t = 3$ seconds, then $s = 16(3)^2 = 144$ feet.

If $s = 64$ feet, then
$$64 = 16t^2$$
$$t^2 = 4$$
$$t = \pm 2$$
Time must be positive, so we disregard $t = -2$.
It takes 2 seconds to fall 64 feet.

27. $E = kW$
$$3 = k(20)$$
$$k = \frac{3}{20}$$

Therefore, we have the linear equation $E = \frac{3}{20}W$.

If $W = 15$, then $E = \frac{3}{20}(15) = 2.25$.

29. $R = kg$
$$47.40 = k(12)$$
$$3.95 = k$$
Therefore, we have the linear equation $R = 3.95g$.
If $g = 10.5$, then $R = (3.95)(10.5) \approx \$41.48$.

31. $D = \dfrac{k}{p}$

    a. $D = 156$, $p = 2.75$;

$$156 = \frac{k}{2.75}$$
$$k = 429$$

So, $D = \dfrac{429}{p}$.

    b. $D = \dfrac{429}{3} = 143$ bags of candy

33. $V = \dfrac{k}{P}$

$V = 600$, $P = 150$;

$$600 = \frac{k}{150}$$
$$k = 90,000$$

So, we have the equation $V = \dfrac{90,000}{P}$

If $P = 200$, then $V = \dfrac{90,000}{200} = 450$ cm^3.

35. $W = \dfrac{k}{d^2}$

If $W = 125$, $d = 3960$ then

$$125 = \frac{k}{3960^2} \text{ and } k = 1,960,200,000$$

So, we have the equation $W = \dfrac{1,960,200,000}{d}$.

At the top of Mt. McKinley, we have
$d = 3960 + 3.8 = 3963.8$, so
$$W = \frac{1,960,200,000}{(3963.8)^2} \approx 124.76 \text{ pounds.}$$

37. $V = \pi r^2 h$

39. $I = \dfrac{k}{d^2}$

If $I = 0.075$, $d = 2$, then

$$0.075 = \frac{k}{2^2} \text{ and } k = 0.3.$$

So, we have the equation $I = \dfrac{0.3}{d^2}$.

If $d = 5$, then $I = \dfrac{0.3}{5^2} = 0.012$ foot-candles.

41. $h = ksd^3$
$$36 = k(75)(2)^3$$
$$36 = 600k$$
$$0.06 = k$$
So, we have the equation $h = 0.06sd^3$.

If $h = 45$ and $s = 125$, then

$45 = (0.06)(125)d^3$

$45 = 7.5d^3$

$6 = d^3$

$d = \sqrt[3]{6} \approx 1.82$ inches

43. $K = kmv^2$

$1250 = k(25)(10)^2$

$1250 = 2500k$

$k = 0.5$

So, we have the equation $K = 0.5mv^2$.

If $m = 25$ and $v = 15$, then

$K = 0.5(25)(15)^2 = 2812.5$ Joules

45. $S = \dfrac{kpd}{t}$

$100 = \dfrac{k(25)(5)}{0.75}$

$75 = 125k$

$0.6 = k$

So, we have the equation $S = \dfrac{0.6pd}{t}$.

If $p = 40$, $d = 8$, and $t = 0.50$, then

$S = \dfrac{0.6(40)(8)}{0.50} = 384$ psi.

47 – 49. Answers will vary.

51. $3x^3 + 25x^2 - 12x - 100$

$= (3x^3 + 25x^2) - (12x + 100)$

$= x^2(3x + 25) - 4(3x + 25)$

$= (x^2 - 4)(3x + 25)$

$= (x - 2)(x + 2)(3x + 25)$

53. $\left(\dfrac{4}{25}\right)^{\frac{3}{2}} = \left(\left(\dfrac{4}{25}\right)^{\frac{1}{2}}\right)^3$

$= \left(\dfrac{2}{5}\right)^3 = \dfrac{8}{125}$

Chapter 2 Review Exercises

1. $P_1 = (0,0)$ and $P_2 = (4,2)$

a. slope $= \dfrac{\Delta y}{\Delta x} = \dfrac{2-0}{4-0} = \dfrac{2}{4} = \dfrac{1}{2}$

b. For each run of 2, there is a rise of 1.

2. $P_1 = (1,-1)$ and $P_2 = (-2,3)$

a. slope $= \dfrac{\Delta y}{\Delta x} = \dfrac{3-(-1)}{-2-1} = \dfrac{4}{-3} = -\dfrac{4}{3}$

b. For each run of 3, there is a rise of –4.

3. $P_1 = (4,-4)$ and $P_2 = (4,8)$

a. slope $= \dfrac{\Delta y}{\Delta x} = \dfrac{8-(-4)}{4-4} = \dfrac{12}{0}$, undefined

b. An undefined slope means the points lie on a vertical line. There is no change in x.

4. $P_1 = (-2,-1)$ and $P_2 = (3,-1)$

a. slope $= \dfrac{\Delta y}{\Delta x} = \dfrac{-1-(-1)}{3-(-2)} = \dfrac{0}{5} = 0$

b. A slope of 0 means the points lie on a horizontal line. There is no change in y.

5. $2x = 3y^2$

x-intercepts: y-intercepts:

$2x = 3(0)^2$ $2(0) = 3y^2$

$2x = 0$ $0 = y^2$

$x = 0$ $y = 0$

The only intercept is $(0, 0)$.

Test x-axis symmetry: Let $y = -y$

$2x = 3(-y)^2$

$2x = 3y^2$ same

Test y-axis symmetry: Let $x = -x$

$2(-x) = 3y^2$

$-2x = 3y^2$ different

Test origin symmetry: Let $x = -x$ and $y = -y$.

$2(-x) = 3(-y)^2$

$-2x = 3y^2$ different

Therefore, the graph will have x-axis symmetry.

6. $x^2 + 4y^2 = 16$

x-intercepts:
$$x^2 + 4(0)^2 = 16$$
$$x^2 = 16$$
$$x = \pm 4$$

y-intercepts:
$$(0)^2 + 4y^2 = 16$$
$$4y^2 = 16$$
$$y^2 = 4$$
$$y = \pm 2$$

The intercepts are $(-4, 0)$, $(4, 0)$, $(0, -2)$, and $(0, 2)$.

Test x-axis symmetry: Let $y = -y$
$$x^2 + 4(-y)^2 = 16$$
$$x^2 + 4y^2 = 16 \quad \text{same}$$

Test y-axis symmetry: Let $x = -x$
$$(-x)^2 + 4y^2 = 16$$
$$x^2 + 4y^2 = 16 \quad \text{same}$$

Test origin symmetry: Let $x = -x$ and $y = -y$.
$$(-x)^2 + 4(-y)^2 = 16$$
$$x^2 + 4y^2 = 16 \quad \text{same}$$

Therefore, the graph will have x-axis, y-axis, and origin symmetry.

7. $y = x^4 - 3x^2 - 4$

x-intercepts:
$$0 = x^4 - 3x^2 - 4$$
$$0 = (x^2 - 4)(x^2 + 1)$$
$$x^2 - 4 = 0$$
$$x^2 = 4$$
$$x = \pm 2$$

y-intercepts:
$$y = (0)^4 - 3(0)^2 - 4$$
$$= -4$$

The intercepts are $(-2, 0)$, $(2, 0)$, $(0, -4)$, and $(0, 2)$.

Test x-axis symmetry: Let $y = -y$
$$-y = x^4 - 3x^2 - 4$$
$$y = -x^4 + 3x^2 + 4 \quad \text{different}$$

Test y-axis symmetry: Let $x = -x$
$$y = (-x)^4 - 3(-x)^2 - 4$$
$$y = x^4 - 3x^2 - 4 \quad \text{same}$$

Test origin symmetry: Let $x = -x$ and $y = -y$.
$$-y = (-x)^4 - 3(-x)^2 - 4$$
$$-y = x^4 - 3x^2 - 4$$
$$y = -x^4 + 3x^2 + 4 \quad \text{different}$$

Therefore, the graph will have y-axis symmetry.

8. $y = x^3 - x$

x-intercepts:
$$0 = x^3 - x$$
$$0 = x(x^2 - 1)$$
$$0 = x(x+1)(x-1)$$
$$x = 0, \ x = -1, \ x = 1$$

y-intercepts:
$$y = (0)^3 - 0$$
$$= 0$$

The intercepts are $(-1, 0)$, $(0, 0)$, and $(1, 0)$.

Test x-axis symmetry: Let $y = -y$
$$-y = x^3 - x$$
$$y = -x^3 + x \quad \text{different}$$

Test y-axis symmetry: Let $x = -x$
$$y = (-x)^3 - (-x)$$
$$y = -x^3 + x \quad \text{different}$$

Test origin symmetry: Let $x = -x$ and $y = -y$.
$$-y = (-x)^3 - (-x)$$
$$-y = -x^3 + x$$
$$y = x^3 - x \quad \text{same}$$

Therefore, the graph will have origin symmetry.

9. $x^2 + x + y^2 + 2y = 0$

x-intercepts: $x^2 + x + (0)^2 + 2(0) = 0$
$$x^2 + x = 0$$
$$x(x+1) = 0$$
$$x = 0, \ x = -1$$

y-intercepts: $(0)^2 + 0 + y^2 + 2y = 0$
$$y^2 + 2y = 0$$
$$y(y+2) = 0$$
$$y = 0, \ y = -2$$

The intercepts are $(-1, 0)$, $(0, 0)$, and $(0, -2)$.

Test x-axis symmetry: Let $y = -y$
$$x^2 + x + (-y)^2 + 2(-y) = 0$$
$$x^2 + x + y^2 - 2y = 0 \quad \text{different}$$

Test y-axis symmetry: Let $x = -x$
$$(-x)^2 + (-x) + y^2 + 2y = 0$$
$$x^2 - x + y^2 + 2y = 0 \quad \text{different}$$

Test origin symmetry: Let $x = -x$ and $y = -y$.
$$(-x)^2 + (-x) + (-y)^2 + 2(-y) = 0$$
$$x^2 - x + y^2 - 2y = 0 \quad \text{different}$$

The graph has none of the indicated symmetries.

10. $(x-h)^2 + (y-k)^2 = r^2$

$(x-(-2))^2 + (y-3)^2 = 4^2$

$(x+2)^2 + (y-3)^2 = 16$

11. $(x-h)^2 + (y-k)^2 = r^2$

$(x-(-1))^2 + (y-(-2))^2 = 1^2$

$(x+1)^2 + (y+2)^2 = 1$

12. $x^2 + (y-1)^2 = 4$

$x^2 + (y-1)^2 = 2^2$

Center: $(0,1)$; Radius $= 2$

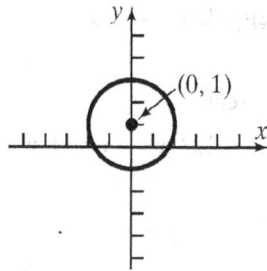

x-intercepts: $x^2 + (0-1)^2 = 4$

$x^2 + 1 = 4$

$x^2 = 3$

$x = \pm\sqrt{3}$

y-intercepts: $0^2 + (y-1)^2 = 4$

$(y-1)^2 = 4$

$y - 1 = \pm 2$

$y = 1 \pm 2$

$y = 3$ or $y = -1$

The intercepts are $\left(-\sqrt{3}, 0\right)$, $\left(\sqrt{3}, 0\right)$, $(0,-1)$,

and $(0,3)$.

13. $x^2 + y^2 - 2x + 4y - 4 = 0$

$x^2 - 2x + y^2 + 4y = 4$

$\left(x^2 - 2x + 1\right) + \left(y^2 + 4y + 4\right) = 4 + 1 + 4$

$(x-1)^2 + (y+2)^2 = 3^2$

Center: $(1,-2)$ Radius $= 3$

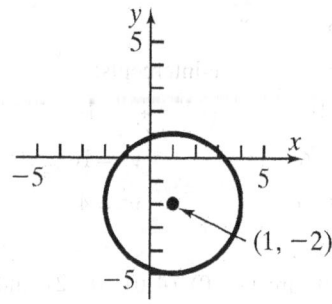

x-intercepts: $(x-1)^2 + (0+2)^2 = 3^2$

$(x-1)^2 + 4 = 9$

$(x-1)^2 = 5$

$x - 1 = \pm\sqrt{5}$

$x = 1 \pm \sqrt{5}$

y-intercepts: $(0-1)^2 + (y+2)^2 = 3^2$

$1 + (y+2)^2 = 9$

$(y+2)^2 = 8$

$y + 2 = \pm\sqrt{8}$

$y + 2 = \pm 2\sqrt{2}$

$y = -2 \pm 2\sqrt{2}$

The intercepts are $\left(1-\sqrt{5}, 0\right)$, $\left(1+\sqrt{5}, 0\right)$,

$\left(0, -2 - 2\sqrt{2}\right)$, and $\left(0, -2 + 2\sqrt{2}\right)$.

14. $3x^2 + 3y^2 - 6x + 12y = 0$

$x^2 + y^2 - 2x + 4y = 0$

$x^2 - 2x + y^2 + 4y = 0$

$\left(x^2 - 2x + 1\right) + \left(y^2 + 4y + 4\right) = 1 + 4$

$(x-1)^2 + (y+2)^2 = \left(\sqrt{5}\right)^2$

Center: $(1,-2)$ Radius $= \sqrt{5}$

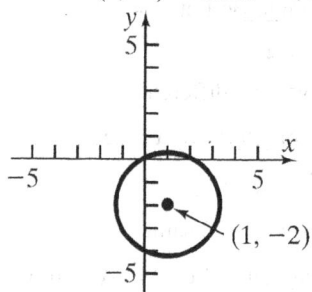

x-intercepts: $(x-1)^2 + (0+2)^2 = \left(\sqrt{5}\right)^2$

$$(x-1)^2 + 4 = 5$$
$$(x-1)^2 = 1$$
$$x-1 = \pm 1$$
$$x = 1 \pm 1$$
$$x = 2 \quad \text{or} \quad x = 0$$

y-intercepts: $(0-1)^2 + (y+2)^2 = \left(\sqrt{5}\right)^2$

$$1 + (y+2)^2 = 5$$
$$(y+2)^2 = 4$$
$$y+2 = \pm 2$$
$$y = -2 \pm 2$$
$$y = 0 \quad \text{or} \quad y = -4$$

The intercepts are $(0,0)$, $(2,0)$, and $(0,-4)$.

15. Slope $= -2$; containing $(3,-1)$
$$y - y_1 = m(x - x_1)$$
$$y - (-1) = -2(x-3)$$
$$y + 1 = -2x + 6$$
$$y = -2x + 5 \quad \text{or} \quad 2x + y = 5$$

16. Slope $= 0$; containing the point $(-5, 4)$
$$y - y_1 = m(x - x_1)$$
$$y - 4 = 0(x - (-5))$$
$$y - 4 = 0$$
$$y = 4$$

17. vertical; containing $(-3,4)$
Vertical lines have equations of the form $x = a$, where a is the x-intercept. Now, a vertical line containing the point $(-3, 4)$ must have an x-intercept of -3, so the equation of the line is $x = -3$. The equation does not have a slope-intercept form.

18. x-intercept $= 2$; containing the point $(4, -5)$
Points are $(2, 0)$ and $(4, -5)$.
$$m = \frac{-5-0}{4-2} = -\frac{5}{2}$$
$$y - y_1 = m(x - x_1)$$
$$y - 0 = -\frac{5}{2}(x-2)$$
$$y = -\frac{5}{2}x + 5 \quad \text{or} \quad 5x + 2y = 10$$

19. y-intercept $= -2$; containing $(5,-3)$
Points are $(5,-3)$ and $(0,-2)$
$$m = \frac{-2-(-3)}{0-5} = \frac{1}{-5} = -\frac{1}{5}$$
$$y = mx + b$$
$$y = -\frac{1}{5}x - 2 \quad \text{or} \quad x + 5y = -10$$

20. Containing the points $(3,-4)$ and $(2, 1)$
$$m = \frac{1-(-4)}{2-3} = \frac{5}{-1} = -5$$
$$y - y_1 = m(x - x_1)$$
$$y - (-4) = -5(x - 3)$$
$$y + 4 = -5x + 15$$
$$y = -5x + 11 \quad \text{or} \quad 5x + y = 11$$

21. Parallel to $2x - 3y = -4$
$$2x - 3y = -4$$
$$-3y = -2x - 4$$
$$\frac{-3y}{-3} = \frac{-2x-4}{-3}$$
$$y = \frac{2}{3}x + \frac{4}{3}$$

Slope $= \frac{2}{3}$; containing $(-5,3)$

$$y - y_1 = m(x - x_1)$$
$$y - 3 = \frac{2}{3}(x - (-5))$$
$$y - 3 = \frac{2}{3}(x + 5)$$
$$y - 3 = \frac{2}{3}x + \frac{10}{3}$$
$$y = \frac{2}{3}x + \frac{19}{3} \quad \text{or} \quad 2x - 3y = -19$$

22. Perpendicular to $3x - y = -4$
$$3x - y = -4$$
$$y = 3x + 4$$
The slope of this line is 3, so the slope of a line perpendicular to it is $-\frac{1}{3}$.

Slope $= -\frac{1}{3}$; containing $(-2, 4)$

$$y - y_1 = m(x - x_1)$$

$$y - 4 = -\frac{1}{3}(x - (-2))$$

$$y - 4 = -\frac{1}{3}x - \frac{2}{3}$$

$$y = -\frac{1}{3}x + \frac{10}{3} \quad \text{or} \quad x + 3y = 10$$

23. $4x - 5y = -20$

$$-5y = -4x - 20$$

$$y = \frac{4}{5}x + 4$$

slope $= \dfrac{4}{5}$; y-intercept $= 4$

x-intercept: Let $y = 0$.

$$4x - 5(0) = -20$$

$$4x = -20$$

$$x = -5$$

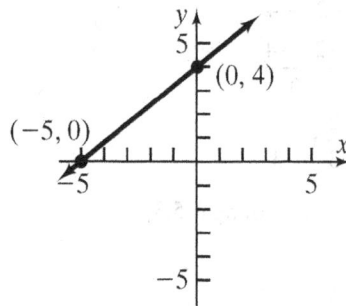

24. $\dfrac{1}{2}x - \dfrac{1}{3}y = -\dfrac{1}{6}$

$$-\frac{1}{3}y = -\frac{1}{2}x - \frac{1}{6}$$

$$y = \frac{3}{2}x + \frac{1}{2}$$

slope $= \dfrac{3}{2}$; y-intercept $= \dfrac{1}{2}$

x-intercept: Let $y = 0$.

$$\frac{1}{2}x - \frac{1}{3}(0) = -\frac{1}{6}$$

$$\frac{1}{2}x = -\frac{1}{6}$$

$$x = -\frac{1}{3}$$

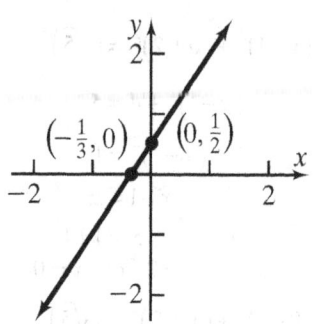

25. $2x - 3y = 12$

x-intercept:	y-intercept:
$2x - 3(0) = 12$	$2(0) - 3y = 12$
$2x = 12$	$-3y = 12$
$x = 6$	$y = -4$

The intercepts are $(6, 0)$ and $(0, -4)$.

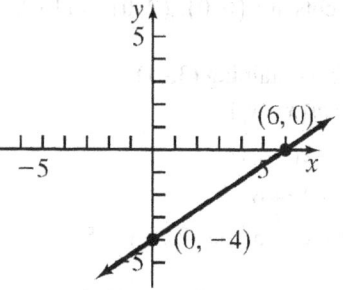

26. $\dfrac{1}{2}x + \dfrac{1}{3}y = 2$

x-intercept:	y-intercept:
$\dfrac{1}{2}x + \dfrac{1}{3}(0) = 2$	$\dfrac{1}{2}(0) + \dfrac{1}{3}y = 2$
$\dfrac{1}{2}x = 2$	$\dfrac{1}{3}y = 2$
$x = 4$	$y = 6$

The intercepts are $(4, 0)$ and $(0, 6)$.

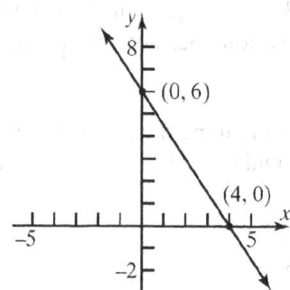

27. $y = x^3$

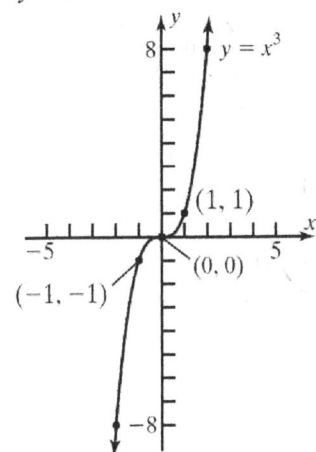

28. slope $= \dfrac{2}{3}$, containing the point $(1,2)$

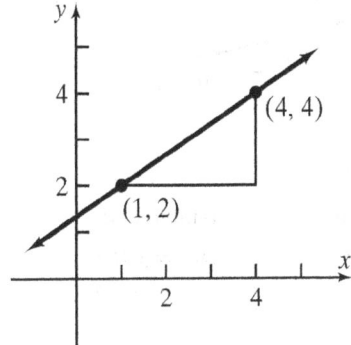

29. Given the points $A = (-2, 0)$, $B = (-4, 4)$, and $C = (8, 5)$.

a. Find the distance between each pair of points.

$$d(A,B) = \sqrt{(-4-(-2))^2 + (4-0)^2}$$
$$= \sqrt{4+16}$$
$$= \sqrt{20} = 2\sqrt{5}$$

$$d(B,C) = \sqrt{(8-(-4))^2 + (5-4)^2}$$
$$= \sqrt{144+1}$$
$$= \sqrt{145}$$

$$d(A,C) = \sqrt{(8-(-2))^2 + (5-0)^2}$$
$$= \sqrt{100+25}$$
$$= \sqrt{125} = 5\sqrt{5}$$

$$\left[d(A,B)\right]^2 + \left[d(A,C)\right]^2 = \left[d(B,C)\right]^2$$
$$\left(\sqrt{20}\right)^2 + \left(\sqrt{125}\right)^2 = \left(\sqrt{145}\right)^2$$
$$20+125 = 145$$
$$145 = 145$$

The Pythagorean Theorem is satisfied, so this is a right triangle.

b. Find the slopes:

$$m_{AB} = \frac{4-0}{-4-(-2)} = \frac{4}{-2} = -2$$

$$m_{BC} = \frac{5-4}{8-(-4)} = \frac{1}{12}$$

$$m_{AC} = \frac{5-0}{8-(-2)} = \frac{5}{10} = \frac{1}{2}$$

Since $m_{AB} \cdot m_{AC} = -2 \cdot \dfrac{1}{2} = -1$, the sides AB and AC are perpendicular and the triangle is a right triangle.

30. Endpoints of the diameter are $(-3, 2)$ and $(5, -6)$. The center is at the midpoint of the diameter:

Center: $\left(\dfrac{-3+5}{2}, \dfrac{2+(-6)}{2}\right) = (1, -2)$

Radius: $r = \sqrt{(1-(-3))^2 + (-2-2)^2}$
$$= \sqrt{16+16}$$
$$= \sqrt{32} = 4\sqrt{2}$$

Equation: $(x-1)^2 + (y+2)^2 = \left(4\sqrt{2}\right)^2$
$$(x-1)^2 + (y+2)^2 = 32$$

31. slope of $\overline{AB} = \dfrac{1-5}{6-2} = -1$

slope of $\overline{AC} = \dfrac{-1-5}{8-2} = -1$

slope of $\overline{BC} = \dfrac{-1-1}{8-6} = -1$

Therefore, the points lie on a line.

32. $p = kB$

$854 = k(130,000)$

$k = \dfrac{854}{130,000} = \dfrac{427}{65,000}$

Therefore, we have the equation $p = \dfrac{427}{65,000}B$.

If $B = 165,000$, then

$p = \dfrac{427}{65,000}(165,000) = \$1083.92$.

33. $w = \dfrac{k}{d^2}$

$200 = \dfrac{k}{3960^2}$

$k = (200)(3960^2) = 3,136,320,000$

Therefore, we have the equation

$w = \dfrac{3,136,320,000}{d^2}$.

If $d = 3960 + 1 = 3961$ miles, then

$w = \dfrac{3,136,320,000}{3961^2} \approx 199.9$ pounds.

34. $H = ksd$

$135 = k(7.5)(40)$

$135 = 300k$

$k = 0.45$

So, we have the equation $H = 0.45sd$.

If $s = 12$ and $d = 35$, then

$H = 0.45(12)(35) = 189$ BTU

Chapter 2 Test

1. a. $m = \dfrac{y_2 - y_1}{x_2 - x_1} = \dfrac{-1-3}{5-(-1)} = \dfrac{-4}{6} = -\dfrac{2}{3}$

b. If x increases by 3 units, y will decrease by 2 units.

2. $y = x^2 - 9$

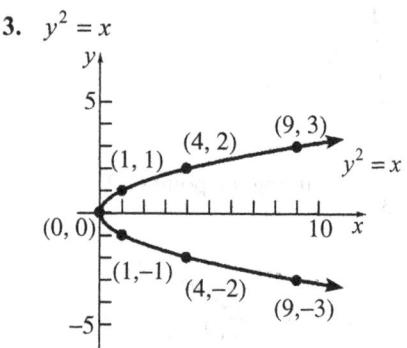

3. $y^2 = x$

4. $x^2 + y = 9$

x-intercepts: y-intercept:

$x^2 + 0 = 9$ $(0)^2 + y = 9$

$x^2 = 9$ $y = 9$

$x = \pm 3$

The intercepts are $(-3,0)$, $(3,0)$, and $(0,9)$.

<u>Test x-axis symmetry:</u> Let $y = -y$

$x^2 + (-y) = 9$

$x^2 - y = 9$ different

<u>Test y-axis symmetry:</u> Let $x = -x$

$(-x)^2 + y = 9$

$x^2 + y = 9$ same

<u>Test origin symmetry:</u> Let $x = -x$ and $y = -y$

$(-x)^2 + (-y) = 9$

$x^2 - y = 9$ different

Therefore, the graph will have y-axis symmetry.

5. Slope $= -2$; containing $(3, -4)$

$y - y_1 = m(x - x_1)$

$y - (-4) = -2(x - 3)$

$y + 4 = -2x + 6$

$y = -2x + 2$

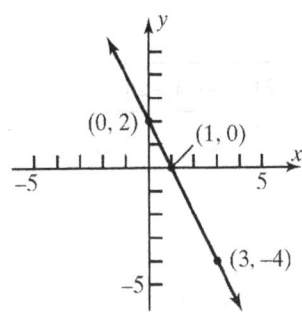

6. $2x + 3y = 9$

$3y = -2x + 9$

$y = -\dfrac{2}{3}x + 3$

slope $= -\dfrac{2}{3}$; y-intercept $= 3$

x-intercept: Let $y = 0$.

$2x + 3(0) = 9$

$3x = 9$

$x = 3$

7. $3x - 4y = 24$

x-intercepts:	y-intercept:
$3x - 4(0) = 24$	$3(0) - 4y = 24$
$3x = 24$	$y = -6$
$x = 8$	

The intercepts are $(8, 0)$ and $(0, -6)$.

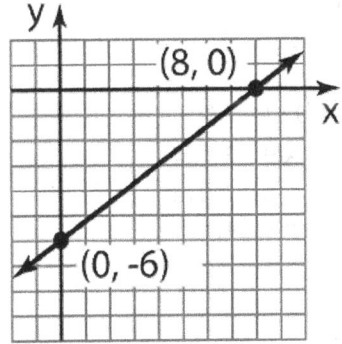

8. $(x-h)^2 + (y-k)^2 = r^2$

$(x-4)^2 + (y-(-3))^2 = 5^2$

$(x-4)^2 + (y+3)^2 = 25$

General form: $(x-4)^2 + (y+3)^2 = 25$

$x^2 - 8x + 16 + y^2 + 6y + 9 = 25$

$x^2 + y^2 - 8x + 6y = 0$

9. $x^2 + y^2 + 4x - 2y - 4 = 0$

$x^2 + 4x + y^2 - 2y = 4$

$(x^2 + 4x + 4) + (y^2 - 2y + 1) = 4 + 4 + 1$

$(x+2)^2 + (y-1)^2 = 3^2$

Center: $(-2, 1)$; Radius $= 3$

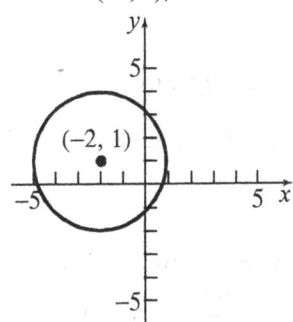

10. $2x + 3y = 6$

$3y = -2x + 6$

$y = -\dfrac{2}{3}x + 2$

Parallel line

Any line parallel to $2x + 3y = 6$ has slope

$m = -\dfrac{2}{3}$. The line contains $(1, -1)$:

$y - y_1 = m(x - x_1)$

$y - (-1) = -\dfrac{2}{3}(x - 1)$

$y + 1 = -\dfrac{2}{3}x + \dfrac{2}{3}$

$y = -\dfrac{2}{3}x - \dfrac{1}{3}$

Perpendicular line

Any line perpendicular to $2x + 3y = 6$ has slope

$m = \dfrac{3}{2}$. The line contains $(0, 3)$:

$y - y_1 = m(x - x_1)$

$y - 3 = \dfrac{3}{2}(x - 0)$

$y - 3 = \dfrac{3}{2}x$

$y = \dfrac{3}{2}x + 3$

11. Let R = the resistance, l = length, and r = radius.

Then $R = k \cdot \dfrac{l}{r^2}$. Now, $R = 10$ ohms, when

$l = 50$ feet and $r = 6 \times 10^{-3}$ inch, so

$$10 = k \cdot \frac{50}{\left(6 \times 10^{-3}\right)^2}$$

$$k = 10 \cdot \frac{\left(6 \times 10^{-3}\right)^2}{50} = 7.2 \times 10^{-6}$$

Therefore, we have the equation

$$R = \left(7.2 \times 10^{-6}\right) \frac{l}{r^2}.$$

If $l = 100$ feet and $r = 7 \times 10^{-3}$ inch, then

$$R = \left(7.2 \times 10^{-6}\right) \frac{100}{\left(7 \times 10^{-3}\right)^2} \approx 14.69 \text{ ohms.}$$

Chapter 2 Cumulative Review

1. $3x - 5 = 0$

$3x = 5$

$x = \dfrac{5}{3}$

The solution set is $\left\{ \dfrac{5}{3} \right\}$.

2. $x^2 - x - 12 = 0$

$(x - 4)(x + 3) = 0$

$x = 4$ or $x = -3$

The solution set is $\{-3, 4\}$.

3. $2x^2 - 5x - 3 = 0$

$(2x + 1)(x - 3) = 0$

$x = -\dfrac{1}{2}$ or $x = 3$

The solution set is $\left\{ -\dfrac{1}{2}, 3 \right\}$.

4. $x^2 - 2x - 2 = 0$

$$x = \frac{-(-2) \pm \sqrt{(-2)^2 - 4(1)(-2)}}{2(1)}$$

$$= \frac{2 \pm \sqrt{4 + 8}}{2}$$

$$= \frac{2 \pm \sqrt{12}}{2}$$

$$= \frac{2 \pm 2\sqrt{3}}{2}$$

$$= 1 \pm \sqrt{3}$$

The solution set is $\left\{ 1 - \sqrt{3}, 1 + \sqrt{3} \right\}$.

5. $x^2 + 2x + 5 = 0$

$$x = \frac{-2 \pm \sqrt{2^2 - 4(1)(5)}}{2(1)}$$

$$= \frac{-2 \pm \sqrt{4 - 20}}{2}$$

$$= \frac{-2 \pm \sqrt{-16}}{2}$$

No real solutions

6. $\sqrt{2x + 1} = 3$

$\left(\sqrt{2x + 1}\right)^2 = 3^2$

$2x + 1 = 9$

$2x = 8$

$x = 4$

Check: $\sqrt{2(4) + 1} = 3$?

$\sqrt{9} = 3$?

$3 = 3$ True

The solution set is $\{4\}$.

7. $|x - 2| = 1$

$x - 2 = 1$ or $x - 2 = -1$

$x = 3$ $\qquad x = 1$

The solution set is $\{1, 3\}$.

8. $\sqrt{x^2+4x}=2$

$\left(\sqrt{x^2+4x}\right)^2=2^2$

$x^2+4x=4$

$x^2+4x-4=0$

$x=\dfrac{-4\pm\sqrt{4^2-4(1)(-4)}}{2(1)}=\dfrac{-4\pm\sqrt{16+16}}{2}$

$=\dfrac{-4\pm\sqrt{32}}{2}=\dfrac{-4\pm4\sqrt{2}}{2}=-2\pm2\sqrt{2}$

Check $x=-2+2\sqrt{2}$:

$\sqrt{\left(-2+2\sqrt{2}\right)^2+4\left(-2+2\sqrt{2}\right)}=2?$

$\sqrt{4-8\sqrt{2}+8-8+8\sqrt{2}}=2?$

$\sqrt{4}=2$ True

Check $x=-2-2\sqrt{2}$:

$\sqrt{\left(-2-2\sqrt{2}\right)^2+4\left(-2-2\sqrt{2}\right)}=2?$

$\sqrt{4+8\sqrt{2}+8-8-8\sqrt{2}}=2?$

$\sqrt{4}=2$ True

The solution set is $\left\{-2-2\sqrt{2},-2+2\sqrt{2}\right\}$.

9. $x^2=-9$

$x=\pm\sqrt{-9}$

$x=\pm3i$

The solution set is $\{-3i,3i\}$.

10. $x^2-2x+5=0$

$x=\dfrac{-(-2)\pm\sqrt{(-2)^2-4(1)(5)}}{2(1)}=\dfrac{2\pm\sqrt{4-20}}{2}$

$=\dfrac{2\pm\sqrt{-16}}{2}=\dfrac{2\pm4i}{2}=1\pm2i$

The solution set is $\{1-2i,1+2i\}$.

11. $2x-3\le7$

$2x\le10$

$x\le5$

$\{x\,|\,x\le5\}$ or $(-\infty,5]$

12. $-1<x+4<5$

$-5<x<1$

$\{x\,|\,-5<x<1\}$ or $(-5,1)$

13. $|x-2|\le1$

$-1\le x-2\le1$

$1\le x\le3$

$\{x\,|\,1\le x\le3\}$ or $[1,3]$

14. $|2+x|>3$

$2+x<-3$ or $2+x>3$

$x<-5$ or $x>1$

$\{x\,|\,x<-5\text{ or }x>1\}$ or $(-\infty,-5)\cup(1,\infty)$

15. $d(P,Q)=\sqrt{(-1-4)^2+(3-(-2))^2}$

$=\sqrt{(-5)^2+(5)^2}$

$=\sqrt{25+25}$

$=\sqrt{50}=5\sqrt{2}$

Midpoint $=\left(\dfrac{-1+4}{2},\dfrac{3+(-2)}{2}\right)=\left(\dfrac{3}{2},\dfrac{1}{2}\right)$

16. $y=x^3-3x+1$

 a. $(-2,-1)$:

 $(-2)^3-(3)(-2)+1=-8+6+1=-1$

 $(-2,-1)$ is on the graph.

 b. $(2,3)$:

 $(2)^3-(3)(2)+1=8-6+1=3$

 $(2,3)$ is on the graph.

 c. $(3,1)$:

 $(3)^3-(3)(3)+1=27-9+1=19\ne1$

 $(3,1)$ is not on the graph.

17. $y = x^3$

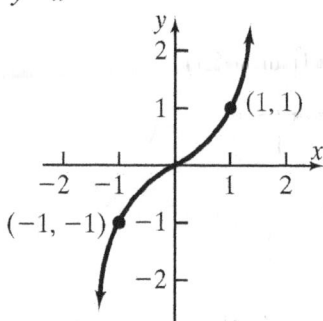

18. The points $(-1,4)$ and $(2,-2)$ are on the line.

$$\text{Slope} = \frac{-2-4}{2-(-1)} = \frac{-6}{3} = -2$$

$$y - y_1 = m(x - x_1)$$

$$y - 4 = -2(x - (-1))$$

$$y - 4 = -2(x + 1)$$

$$y = -2x - 2 + 4$$

$$y = -2x + 2$$

19. Perpendicular to $y = 2x + 1$; Contains $(3,5)$

$$\text{Slope of perpendicular} = -\frac{1}{2}$$

$$y - y_1 = m(x - x_1)$$

$$y - 5 = -\frac{1}{2}(x - 3)$$

$$y - 5 = -\frac{1}{2}x + \frac{3}{2}$$

$$y = -\frac{1}{2}x + \frac{13}{2}$$

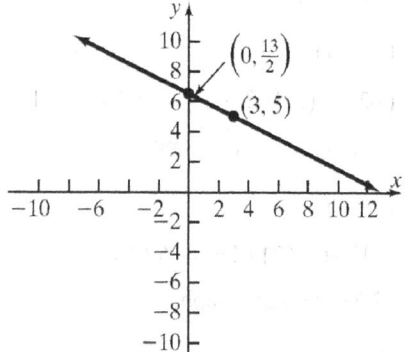

20.
$$x^2 + y^2 - 4x + 8y - 5 = 0$$
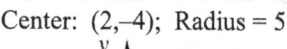
$$(x^2 - 4x + 4) + (y^2 + 8y + 16) = 5 + 4 + 16$$
$$(x - 2)^2 + (y + 4)^2 = 25$$
$$(x - 2)^2 + (y + 4)^2 = 5^2$$

Center: $(2,-4)$; Radius $= 5$

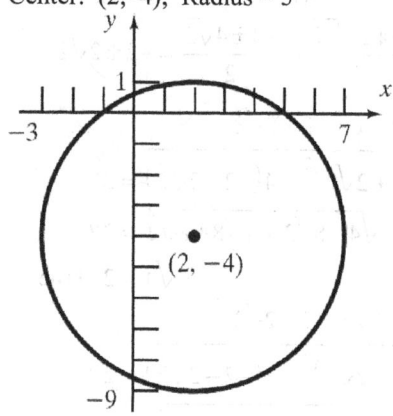

Chapter 2 Project

Internet-based Project

124

Chapter 3

Functions and Their Graphs

Section 3.1

1. $(-1, 3)$

3. We must not allow the denominator to be 0.
$x + 4 \neq 0 \Rightarrow x \neq -4$; Domain: $\{x \mid x \neq -4\}$.

5. $\sqrt{5} + 2$

7. independent; dependent

9. $\neq$; f; g

11. False; every function is a relation, but not every relation is a function. For example, the relation $x^2 + y^2 = 1$ is not a function.

13. False; if the domain is not specified, we assume it is the largest set of real numbers for which the value of f is a real number.

15. a

17. d

19. Function
Domain: {Elvis, Colleen, Kaleigh, Marissa}
Range: {Jan. 8, Mar. 15, Sept. 17}

21. Not a function

23. Not a function

25. Function
Domain: {1, 2, 3, 4}
Range: {3}

27. Not a function

29. Function
Domain: {−2, −1, 0, 1}
Range: {0, 1, 4}

31. Graph $y = 2x^2 - 3x + 4$. The graph passes the vertical line test. Thus, the equation represents a function.

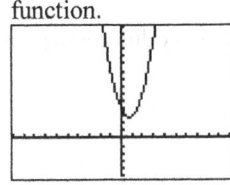

33. Graph $y = \dfrac{1}{x}$. The graph passes the vertical line test. Thus, the equation represents a function.

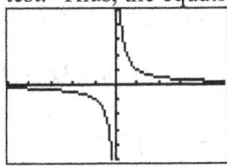

35. $y^2 = 4 - x^2$
Solve for y: $y = \pm\sqrt{4 - x^2}$
For $x = 0$, $y = \pm 2$. Thus, (0, 2) and (0, −2) are on the graph. This is not a function, since a distinct x-value corresponds to two different y-values.

37. $x = y^2$
Solve for y: $y = \pm\sqrt{x}$
For $x = 1$, $y = \pm 1$. Thus, (1, 1) and (1, −1) are on the graph. This is not a function, since a distinct x-value corresponds to two different y-values.

39. Graph $y = x^2$. The graph passes the vertical line test. Thus, the equation represents a function.

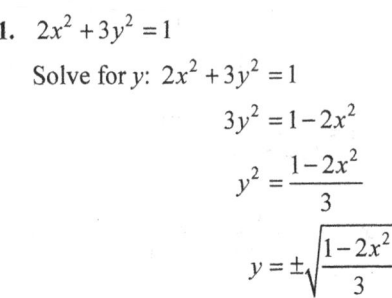

41. $2x^2 + 3y^2 = 1$
Solve for y: $2x^2 + 3y^2 = 1$
$$3y^2 = 1 - 2x^2$$
$$y^2 = \frac{1 - 2x^2}{3}$$
$$y = \pm\sqrt{\frac{1 - 2x^2}{3}}$$

For $x = 0$, $y = \pm\sqrt{\dfrac{1}{3}}$. Thus, $\left(0, \sqrt{\dfrac{1}{3}}\right)$ and

$\left(0, -\sqrt{\dfrac{1}{3}}\right)$ are on the graph. This is not a

function, since a distinct x-value corresponds to two different y-values.

43. $f(x) = 3x^2 + 2x - 4$

a. $f(0) = 3(0)^2 + 2(0) - 4 = -4$

b. $f(1) = 3(1)^2 + 2(1) - 4 = 3 + 2 - 4 = 1$

c. $f(-1) = 3(-1)^2 + 2(-1) - 4 = 3 - 2 - 4 = -3$

d. $f(-x) = 3(-x)^2 + 2(-x) - 4 = 3x^2 - 2x - 4$

e. $-f(x) = -(3x^2 + 2x - 4) = -3x^2 - 2x + 4$

f. $f(x+1) = 3(x+1)^2 + 2(x+1) - 4$

$= 3(x^2 + 2x + 1) + 2x + 2 - 4$

$= 3x^2 + 6x + 3 + 2x + 2 - 4$

$= 3x^2 + 8x + 1$

g. $f(2x) = 3(2x)^2 + 2(2x) - 4 = 12x^2 + 4x - 4$

h. $f(x+h) = 3(x+h)^2 + 2(x+h) - 4$

$= 3(x^2 + 2xh + h^2) + 2x + 2h - 4$

$= 3x^2 + 6xh + 3h^2 + 2x + 2h - 4$

45. $f(x) = \dfrac{x}{x^2 + 1}$

a. $f(0) = \dfrac{0}{0^2 + 1} = \dfrac{0}{1} = 0$

b. $f(1) = \dfrac{1}{1^2 + 1} = \dfrac{1}{2}$

c. $f(-1) = \dfrac{-1}{(-1)^2 + 1} = \dfrac{-1}{1+1} = -\dfrac{1}{2}$

d. $f(-x) = \dfrac{-x}{(-x)^2 + 1} = \dfrac{-x}{x^2 + 1}$

e. $-f(x) = -\left(\dfrac{x}{x^2 + 1}\right) = \dfrac{-x}{x^2 + 1}$

f. $f(x+1) = \dfrac{x+1}{(x+1)^2 + 1}$

$= \dfrac{x+1}{x^2 + 2x + 1 + 1}$

$= \dfrac{x+1}{x^2 + 2x + 2}$

g. $f(2x) = \dfrac{2x}{(2x)^2 + 1} = \dfrac{2x}{4x^2 + 1}$

h. $f(x+h) = \dfrac{x+h}{(x+h)^2 + 1} = \dfrac{x+h}{x^2 + 2xh + h^2 + 1}$

47. $f(x) = |x| + 4$

a. $f(0) = |0| + 4 = 0 + 4 = 4$

b. $f(1) = |1| + 4 = 1 + 4 = 5$

c. $f(-1) = |-1| + 4 = 1 + 4 = 5$

d. $f(-x) = |-x| + 4 = |x| + 4$

e. $-f(x) = -(|x| + 4) = -|x| - 4$

f. $f(x+1) = |x+1| + 4$

g. $f(2x) = |2x| + 4 = 2|x| + 4$

h. $f(x+h) = |x+h| + 4$

49. $f(x) = \dfrac{2x+1}{3x-5}$

a. $f(0) = \dfrac{2(0)+1}{3(0)-5} = \dfrac{0+1}{0-5} = -\dfrac{1}{5}$

b. $f(1) = \dfrac{2(1)+1}{3(1)-5} = \dfrac{2+1}{3-5} = \dfrac{3}{-2} = -\dfrac{3}{2}$

c. $f(-1) = \dfrac{2(-1)+1}{3(-1)-5} = \dfrac{-2+1}{-3-5} = \dfrac{-1}{-8} = \dfrac{1}{8}$

d. $f(-x) = \dfrac{2(-x)+1}{3(-x)-5} = \dfrac{-2x+1}{-3x-5} = \dfrac{2x-1}{3x+5}$

e. $-f(x) = -\left(\dfrac{2x+1}{3x-5}\right) = \dfrac{-2x-1}{3x-5}$

f. $f(x+1) = \dfrac{2(x+1)+1}{3(x+1)-5} = \dfrac{2x+2+1}{3x+3-5} = \dfrac{2x+3}{3x-2}$

g. $f(2x) = \dfrac{2(2x)+1}{3(2x)-5} = \dfrac{4x+1}{6x-5}$

h. $f(x+h) = \dfrac{2(x+h)+1}{3(x+h)-5} = \dfrac{2x+2h+1}{3x+3h-5}$

51. $f(x) = -5x + 4$

Domain: $\{x \mid x \text{ is any real number}\}$

53. $f(x) = \dfrac{x}{x^2+1}$

Domain: $\{x \mid x \text{ is any real number}\}$

55. $g(x) = \dfrac{x}{x^2-16}$

$x^2 - 16 \neq 0$

$x^2 \neq 16 \Rightarrow x \neq \pm 4$

Domain: $\{x \mid x \neq -4,\ x \neq 4\}$

57. $F(x) = \dfrac{x-2}{x^3+x}$

$x^3 + x \neq 0$

$x(x^2+1) \neq 0$

$x \neq 0, \quad x^2 \neq -1$

Domain: $\{x \mid x \neq 0\}$

59. $h(x) = \sqrt{3x-12}$

$3x - 12 \geq 0$

$3x \geq 12$

$x \geq 4$

Domain: $\{x \mid x \geq 4\}$

61. $p(x) = \sqrt{\dfrac{2}{x-1}} = \dfrac{\sqrt{2}}{\sqrt{x-1}}$

$x - 1 > 0$

$x > 1$

Domain: $\{x \mid x > 1\}$

63. $f(x) = \dfrac{x}{\sqrt{x-4}}$

$x - 4 > 0$

$x > 4$

Domain: $\{x \mid x > 4\}$

65. $P(t) = \dfrac{\sqrt{t-4}}{3t-21}$

$t - 4 \geq 0$

$t \geq 4$

Also $3t - 21 \neq 0$

$3t - 21 \neq 0$

$3t \neq 21$

$t \neq 7$

Domain: $\{t \mid t \geq 4, t \neq 7\}$

67. $f(x) = 3x + 4 \qquad g(x) = 2x - 3$

a. $(f+g)(x) = 3x+4+2x-3 = 5x+1$

Domain: $\{x \mid x \text{ is any real number}\}$.

b. $(f-g)(x) = (3x+4)-(2x-3)$

$= 3x+4-2x+3$

$= x+7$

Domain: $\{x \mid x \text{ is any real number}\}$.

c. $(f \cdot g)(x) = (3x+4)(2x-3)$

$= 6x^2 - 9x + 8x - 12$

$= 6x^2 - x - 12$

Domain: $\{x \mid x \text{ is any real number}\}$.

d. $\left(\dfrac{f}{g}\right)(x) = \dfrac{3x+4}{2x-3}$

$2x - 3 \neq 0 \Rightarrow 2x \neq 3 \Rightarrow x \neq \dfrac{3}{2}$

Domain: $\left\{x \mid x \neq \dfrac{3}{2}\right\}$.

e. $(f+g)(3) = 5(3)+1 = 15+1 = 16$

f. $(f-g)(4) = 4+7 = 11$

g. $(f \cdot g)(2) = 6(2)^2 - 2 - 12 = 24 - 2 - 12 = 10$

h. $\left(\dfrac{f}{g}\right)(1) = \dfrac{3(1)+4}{2(1)-3} = \dfrac{3+4}{2-3} = \dfrac{7}{-1} = -7$

69. $f(x) = x - 1 \qquad g(x) = 2x^2$

   a. $(f+g)(x) = x - 1 + 2x^2 = 2x^2 + x - 1$

      Domain: $\{x \mid x \text{ is any real number}\}$.

   b. $(f-g)(x) = (x-1) - (2x^2)$
$$= x - 1 - 2x^2$$
$$= -2x^2 + x - 1$$

      Domain: $\{x \mid x \text{ is any real number}\}$.

   c. $(f \cdot g)(x) = (x-1)(2x^2) = 2x^3 - 2x^2$

      Domain: $\{x \mid x \text{ is any real number}\}$.

   d. $\left(\dfrac{f}{g}\right)(x) = \dfrac{x-1}{2x^2}$

      Domain: $\{x \mid x \neq 0\}$.

   e. $(f+g)(3) = 2(3)^2 + 3 - 1$
$$= 2(9) + 3 - 1$$
$$= 18 + 3 - 1 = 20$$

   f. $(f-g)(4) = -2(4)^2 + 4 - 1$
$$= -2(16) + 4 - 1$$
$$= -32 + 4 - 1 = -29$$

   g. $(f \cdot g)(2) = 2(2)^3 - 2(2)^2$
$$= 2(8) - 2(4)$$
$$= 16 - 8 = 8$$

   h. $\left(\dfrac{f}{g}\right)(1) = \dfrac{1-1}{2(1)^2} = \dfrac{0}{2(1)} = \dfrac{0}{2} = 0$

71. $f(x) = \sqrt{x} \qquad g(x) = 3x - 5$

   a. $(f+g)(x) = \sqrt{x} + 3x - 5$

      Domain: $\{x \mid x \geq 0\}$.

   b. $(f-g)(x) = \sqrt{x} - (3x-5) = \sqrt{x} - 3x + 5$

      Domain: $\{x \mid x \geq 0\}$.

   c. $(f \cdot g)(x) = \sqrt{x}(3x - 5) = 3x\sqrt{x} - 5\sqrt{x}$

      Domain: $\{x \mid x \geq 0\}$.

   d. $\left(\dfrac{f}{g}\right)(x) = \dfrac{\sqrt{x}}{3x-5}$

      $x \geq 0$ and $3x - 5 \neq 0$

$$3x \neq 5 \Rightarrow x \neq \frac{5}{3}$$

      Domain: $\left\{x \mid x \geq 0 \text{ and } x \neq \dfrac{5}{3}\right\}$.

   e. $(f+g)(3) = \sqrt{3} + 3(3) - 5$
$$= \sqrt{3} + 9 - 5 = \sqrt{3} + 4$$

   f. $(f-g)(4) = \sqrt{4} - 3(4) + 5$
$$= 2 - 12 + 5 = -5$$

   g. $(f \cdot g)(2) = 3(2)\sqrt{2} - 5\sqrt{2}$
$$= 6\sqrt{2} - 5\sqrt{2} = \sqrt{2}$$

   h. $\left(\dfrac{f}{g}\right)(1) = \dfrac{\sqrt{1}}{3(1)-5} = \dfrac{1}{3-5} = \dfrac{1}{-2} = -\dfrac{1}{2}$

73. $f(x) = 1 + \dfrac{1}{x} \qquad g(x) = \dfrac{1}{x}$

   a. $(f+g)(x) = 1 + \dfrac{1}{x} + \dfrac{1}{x} = 1 + \dfrac{2}{x}$

      Domain: $\{x \mid x \neq 0\}$.

   b. $(f-g)(x) = 1 + \dfrac{1}{x} - \dfrac{1}{x} = 1$

      Domain: $\{x \mid x \neq 0\}$.

   c. $(f \cdot g)(x) = \left(1 + \dfrac{1}{x}\right)\dfrac{1}{x} = \dfrac{1}{x} + \dfrac{1}{x^2}$

      Domain: $\{x \mid x \neq 0\}$.

   d. $\left(\dfrac{f}{g}\right)(x) = \dfrac{1 + \dfrac{1}{x}}{\dfrac{1}{x}} = \dfrac{\dfrac{x+1}{x}}{\dfrac{1}{x}} = \dfrac{x+1}{x} \cdot \dfrac{x}{1} = x + 1$

      Domain: $\{x \mid x \neq 0\}$.

   e. $(f+g)(3) = 1 + \dfrac{2}{3} = \dfrac{5}{3}$

   f. $(f-g)(4) = 1$

   g. $(f \cdot g)(2) = \dfrac{1}{2} + \dfrac{1}{(2)^2} = \dfrac{1}{2} + \dfrac{1}{4} = \dfrac{3}{4}$

h. $\left(\dfrac{f}{g}\right)(1) = 1 + 1 = 2$

75. $f(x) = \dfrac{2x+3}{3x-2} \qquad g(x) = \dfrac{4x}{3x-2}$

a. $(f+g)(x) = \dfrac{2x+3}{3x-2} + \dfrac{4x}{3x-2}$

$\qquad = \dfrac{2x+3+4x}{3x-2} = \dfrac{6x+3}{3x-2}$

$3x - 2 \neq 0$

$\qquad 3x \neq 2 \Rightarrow x \neq \dfrac{2}{3}$

Domain: $\left\{x \middle| x \neq \dfrac{2}{3}\right\}$.

b. $(f-g)(x) = \dfrac{2x+3}{3x-2} - \dfrac{4x}{3x-2}$

$\qquad = \dfrac{2x+3-4x}{3x-2} = \dfrac{-2x+3}{3x-2}$

$3x - 2 \neq 0$

$\qquad 3x \neq 2 \Rightarrow x \neq \dfrac{2}{3}$

Domain: $\left\{x \middle| x \neq \dfrac{2}{3}\right\}$.

c. $(f \cdot g)(x) = \left(\dfrac{2x+3}{3x-2}\right)\left(\dfrac{4x}{3x-2}\right) = \dfrac{8x^2+12x}{(3x-2)^2}$

$3x - 2 \neq 0$

$\qquad 3x \neq 2 \Rightarrow x \neq \dfrac{2}{3}$

Domain: $\left\{x \middle| x \neq \dfrac{2}{3}\right\}$.

d. $\left(\dfrac{f}{g}\right)(x) = \dfrac{\dfrac{2x+3}{3x-2}}{\dfrac{4x}{3x-2}} = \dfrac{2x+3}{3x-2} \cdot \dfrac{3x-2}{4x} = \dfrac{2x+3}{4x}$

$3x - 2 \neq 0 \quad$ and $\quad x \neq 0$

$\qquad 3x \neq 2$

$\qquad x \neq \dfrac{2}{3}$

Domain: $\left\{x \middle| x \neq \dfrac{2}{3} \text{ and } x \neq 0\right\}$.

e. $(f+g)(3) = \dfrac{6(3)+3}{3(3)-2} = \dfrac{18+3}{9-2} = \dfrac{21}{7} = 3$

f. $(f-g)(4) = \dfrac{-2(4)+3}{3(4)-2} = \dfrac{-8+3}{12-2} = \dfrac{-5}{10} = -\dfrac{1}{2}$

g. $(f \cdot g)(2) = \dfrac{8(2)^2+12(2)}{(3(2)-2)^2}$

$\qquad = \dfrac{8(4)+24}{(6-2)^2} = \dfrac{32+24}{(4)^2} = \dfrac{56}{16} = \dfrac{7}{2}$

h. $\left(\dfrac{f}{g}\right)(1) = \dfrac{2(1)+3}{4(1)} = \dfrac{2+3}{4} = \dfrac{5}{4}$

77. $f(x) = 3x+1 \qquad (f+g)(x) = 6 - \dfrac{1}{2}x$

$6 - \dfrac{1}{2}x = 3x + 1 + g(x)$

$5 - \dfrac{7}{2}x = g(x)$

$\qquad g(x) = 5 - \dfrac{7}{2}x$

79. $f(x) = 4x+3$

$\dfrac{f(x+h)-f(x)}{h} = \dfrac{4(x+h)+3-(4x+3)}{h}$

$\qquad = \dfrac{4x+4h+3-4x-3}{h}$

$\qquad = \dfrac{4h}{h} = 4$

81. $f(x) = x^2 - 4$

$\dfrac{f(x+h)-f(x)}{h}$

$= \dfrac{(x+h)^2-4-(x^2-4)}{h}$

$= \dfrac{x^2+2xh+h^2-4-x^2+4}{h}$

$= \dfrac{2xh+h^2}{h}$

$= 2x + h$

83. $f(x) = x^2 - x + 4$

$$\frac{f(x+h) - f(x)}{h}$$

$$= \frac{(x+h)^2 - (x+h) + 4 - (x^2 - x + 4)}{h}$$

$$= \frac{x^2 + 2xh + h^2 - x - h + 4 - x^2 + x - 4}{h}$$

$$= \frac{2xh + h^2 - h}{h}$$

$$= 2x + h - 1$$

85. $f(x) = \dfrac{1}{x^2}$

$$\frac{f(x+h) - f(x)}{h} = \frac{\dfrac{1}{(x+h)^2} - \dfrac{1}{x^2}}{h}$$

$$= \frac{\dfrac{x^2 - (x+h)^2}{x^2(x+h)^2}}{h}$$

$$= \frac{\dfrac{x - (x^2 + 2xh + h^2)}{x^2(x+h)^2}}{h}$$

$$= \left(\frac{1}{h}\right) \frac{-2xh - h^2}{x^2(x+h)^2}$$

$$= \left(\frac{1}{h}\right) \frac{h(-2x - h)}{x^2(x+h)^2}$$

$$= \frac{-2x - h}{x^2(x+h)^2} = \frac{-(2x + h)}{x^2(x+h)^2}$$

87. $f(x) = \dfrac{2x}{x+3}$

$$\frac{f(x+h) - f(x)}{h} = \frac{\dfrac{2(x+h)}{x+h+3} - \dfrac{2x}{x+3}}{h}$$

$$= \frac{\dfrac{2(x+h)(x+3) - 2x(x+3+h)}{(x+h+3)(x+3)}}{h}$$

$$= \frac{\dfrac{2x^2 + 6x + 2hx + 6h - 2x^2 - 6x - 2xh}{(x+h+3)(x+3)}}{h}$$

$$= \left(\frac{6h}{(x+h+3)(x+3)}\right)\left(\frac{1}{h}\right)$$

$$= \frac{6}{(x+h+3)(x+3)}$$

89. $f(x) = \sqrt{x-2}$

$$\frac{f(x+h) - f(x)}{h}$$

$$= \frac{\sqrt{x+h-2} - \sqrt{x-2}}{h}$$

$$= \frac{\sqrt{x+h-2} - \sqrt{x-2}}{h} \cdot \frac{\sqrt{x+h-2} + \sqrt{x-2}}{\sqrt{x+h-2} + \sqrt{x-2}}$$

$$= \frac{x+h-2-x+2}{h\left(\sqrt{x+h-2} + \sqrt{x-2}\right)}$$

$$= \frac{h}{h\left(\sqrt{x+h-2} + \sqrt{x-2}\right)}$$

$$= \frac{1}{\sqrt{x+h-2} + \sqrt{x-2}}$$

91.
$$11 = x^2 - 2x + 3$$
$$0 = x^2 - 2x - 8$$
$$0 = (x-4)(x+2)$$
$$x - 4 = 0 \quad \text{or} \quad x + 2 = 0$$
$$x = 4 \qquad \text{or} \quad x = -2$$

The solution set is: $\{-2, 4\}$

93. $f(x) = 2x^3 + Ax^2 + 4x - 5$ and $f(2) = 5$

$f(2) = 2(2)^3 + A(2)^2 + 4(2) - 5$

$5 = 16 + 4A + 8 - 5$

$5 = 4A + 19$

$-14 = 4A$

$A = \dfrac{-14}{4} = -\dfrac{7}{2}$

95. $f(x) = \dfrac{3x+8}{2x-A}$ and $f(0) = 2$

$f(0) = \dfrac{3(0)+8}{2(0)-A}$

$2 = \dfrac{8}{-A}$

$-2A = 8$

$A = -4$

97. Let x represent the length of the rectangle.

Then, $\dfrac{x}{2}$ represents the width of the rectangle

since the length is twice the width. The function

for the area is: $A(x) = x \cdot \dfrac{x}{2} = \dfrac{x^2}{2} = \dfrac{1}{2}x^2$

99. Let x represent the number of hours worked.
The function for the gross salary is:
$G(x) = 14x$

101. a. P is the dependent variable; a is the
independent variable

b. $P(20) = 0.014(20)^2 - 5.073(20) + 327.287$

$= 5.6 - 101.46 + 327.287$

$= 231.427$

In 2012 there are 231.427 million people
who are 20 years of age or older.

c. $P(0) = 0.014(0)^2 - 5.073(0) + 327.287$

$= 327.287$

In 2012 there are 327.237 million people.

103. a. $H(1) = 20 - 4.9(1)^2$

$= 20 - 4.9 = 15.1 \text{ meters}$

$H(1.1) = 20 - 4.9(1.1)^2$

$= 20 - 4.9(1.21)$

$= 20 - 5.929 = 14.071 \text{ meters}$

$H(1.2) = 20 - 4.9(1.2)^2$

$= 20 - 4.9(1.44)$

$= 20 - 7.056 = 12.944 \text{ meters}$

$H(1.3) = 20 - 4.9(1.3)^2$

$= 20 - 4.9(1.69)$

$= 20 - 8.281 = 11.719 \text{ meters}$

b. $H(x) = 15:$

$15 = 20 - 4.9x^2$

$-5 = -4.9x^2$

$x^2 \approx 1.0204$

$x \approx 1.01 \text{ seconds}$

$H(x) = 10:$

$10 = 20 - 4.9x^2$

$-10 = -4.9x^2$

$x^2 \approx 2.0408$

$x \approx 1.43 \text{ seconds}$

$H(x) = 5:$

$5 = 20 - 4.9x^2$

$-15 = -4.9x^2$

$x^2 \approx 3.0612$

$x \approx 1.75 \text{ seconds}$

c. $H(x) = 0$

$0 = 20 - 4.9x^2$

$-20 = -4.9x^2$

$x^2 \approx 4.0816$

$x \approx 2.02 \text{ seconds}$

105. $C(x) = 100 + \dfrac{x}{10} + \dfrac{36,000}{x}$

a. $C(500) = 100 + \dfrac{500}{10} + \dfrac{36,000}{500}$

$= 100 + 50 + 72$

$= \$222$

b. $C(450) = 100 + \dfrac{450}{10} + \dfrac{36,000}{450}$

$= 100 + 45 + 80$

$= \$225$

c. $C(600) = 100 + \dfrac{600}{10} + \dfrac{36,000}{600}$

$= 100 + 60 + 60$

$= \$220$

d. $C(400) = 100 + \dfrac{400}{10} + \dfrac{36,000}{400}$

$= 100 + 40 + 90$

$= \$230$

107. $R(x) = \left(\dfrac{L}{P}\right)(x) = \dfrac{L(x)}{P(x)}$

109. $H(x) = (P \cdot I)(x) = P(x) \cdot I(x)$

111. a. $P(x) = R(x) - C(x)$

$= \left(-1.2x^2 + 220x\right) - \left(0.05x^3 - 2x^2 + 65x + 500\right)$

$= -1.2x^2 + 220x - 0.05x^3 + 2x^2 - 65x - 500$

$= -0.05x^3 + 0.8x^2 + 155x - 500$

b. $P(15) = -0.05(15)^3 + 0.8(15)^2 + 155(15) - 500$

$= -168.75 + 180 + 2325 - 500$

$= \$1836.25$

c. When 15 hundred cell phones are sold, the profit is $1836.25.

113. a. $R(v) = 2.2v;\ B(v) = 0.05v^2 + 0.4v - 15$

$D(v) = R(v) + B(v)$

$= 2.2v + 0.05v^2 + 0.4v - 15$

$= 0.05v^2 + 2.6v - 15$

b. $D(60) = 0.05(60)^2 + 2.6(60) - 15$

$= 180 + 156 - 15$

$= 321$

c. The car will need 321 feet to stop once the impediment is observed.

115. No. The domain of f is $\{x \mid x$ is any real number$\}$,

but the domain of g is $\{x \mid x \neq -1\}$.

117. $\dfrac{3x - x^3}{(your\ age)}$

119. $y = 3x^2 - 8\sqrt{x}$

$y = 3(-1)^2 - 8\sqrt{-1}$

There is no solution so (-1,-5) is NOT a solution.

$y = 3x^2 - 8\sqrt{x}$

$y = 3(4)^2 - 8\sqrt{4}$

$= 48 - 16 = 32$

So (4,32) is a solution.

$y = 3x^2 - 8\sqrt{x}$

$y = 3(9)^2 - 8\sqrt{9}$

$= 243 - 24 = 219 \neq 171$

So (9,171) is NOT a solution.

121. $x^3 - 9x = 2x^2 - 18$

$x^3 - 2x^2 - 9x + 18 = 0$

$(x^3 - 2x^2) - (9x - 18) = 0$

$x^2(x - 2) - 9(x - 2) = 0$

$(x^2 - 9)(x - 2) = 0$

$(x - 3)(x + 3)(x - 2) = 0$

$(x - 3) = 0$ or $(x + 3) = 0$ or $(x - 2) = 0$

$x = 3, x = -3, x = 2$

The solution set is: $\{3, -3, 2\}$

Section 3.2

1. $x^2 + 4y^2 = 16$

x-intercepts:

$x^2 + 4(0)^2 = 16$

$x^2 = 16$

$x = \pm 4 \Rightarrow (-4,0), (4,0)$

y-intercepts:

$(0)^2 + 4y^2 = 16$

$4y^2 = 16$

$y^2 = 4$

$y = \pm 2 \Rightarrow (0,-2), (0,2)$

3. vertical

5. $f(x) = ax^2 + 4$

   $a(-1)^2 + 4 = 2 \Rightarrow a = -2$

7. False; e.g. $y = \dfrac{1}{x}$.

9. c

11. a. $f(0) = 3$ since $(0, 3)$ is on the graph.

    $f(-6) = -3$ since $(-6, -3)$ is on the graph.

    b. $f(6) = 0$ since $(6, 0)$ is on the graph.

    $f(11) = 1$ since $(11, 1)$ is on the graph.

    c. $f(3)$ is positive since $f(3) \approx 3.7$.

    d. $f(-4)$ is negative since $f(-4) \approx -1$.

    e. $f(x) = 0$ when $x = -3$, $x = 6$, and $x = 10$.

    f. $f(x) > 0$ when $-3 < x < 6$, and $10 < x \le 11$.

    g. The domain of f is $\{x | -6 \le x \le 11\}$ or $[-6, 11]$.

    h. The range of f is $\{y | -3 \le y \le 4\}$ or $[-3, 4]$.

    i. The x-intercepts are -3, 6, and 10.

    j. The y-intercept is 3.

    k. The line $y = \dfrac{1}{2}$ intersects the graph 3 times.

    l. The line $x = 5$ intersects the graph 1 time.

    m. $f(x) = 3$ when $x = 0$ and $x = 4$.

    n. $f(x) = -2$ when $x = -5$ and $x = 8$.

13. Not a function since vertical lines will intersect the graph in more than one point.

15. Function

    a. Domain: $\{x | -\pi \le x \le \pi\}$;

    Range: $\{y | -1 \le y \le 1\}$

    b. Intercepts: $\left(-\dfrac{\pi}{2}, 0\right)$, $\left(\dfrac{\pi}{2}, 0\right)$, $(0, 1)$

    c. Symmetry about y-axis.

17. Not a function since vertical lines will intersect the graph in more than one point.

19. Function

    a. Domain: $\{x | 0 < x < 3\}$;

    Range: $\{y | y < 2\}$

    b. Intercepts: $(1, 0)$

    c. None

21. Function

    a. Domain: $\{x | x \text{ is any real number}\}$;

    Range: $\{y | y \le 2\}$

    b. Intercepts: $(-3, 0)$, $(3, 0)$, $(0, 2)$

    c. Symmetry about y-axis.

23. Function

    a. Domain: $\{x | x \text{ is any real number}\}$;

    Range: $\{y | y \ge -3\}$

    b. Intercepts: $(1, 0)$, $(3, 0)$, $(0, 9)$

    c. None

25. $f(x) = 2x^2 - x - 1$

    a. $f(-1) = 2(-1)^2 - (-1) - 1 = 2$

    The point $(-1, 2)$ is on the graph of f.

    b. $f(-2) = 2(-2)^2 - (-2) - 1 = 9$

    The point $(-2, 9)$ is on the graph of f.

    c. Solve for x:

    $-1 = 2x^2 - x - 1$

    $0 = 2x^2 - x$

    $0 = x(2x - 1) \Rightarrow x = 0, x = \dfrac{1}{2}$

    $(0, -1)$ and $\left(\dfrac{1}{2}, -1\right)$ are on the graph of f.

    d. The domain of f is $\{x | x \text{ is any real number}\}$.

    e. x-intercepts:

    $f(x) = 0 \Rightarrow 2x^2 - x - 1 = 0$

    $(2x + 1)(x - 1) = 0 \Rightarrow x = -\dfrac{1}{2}, x = 1$

    $\left(-\dfrac{1}{2}, 0\right)$ and $(1, 0)$

    f. y-intercept:

    $f(0) = 2(0)^2 - 0 - 1 = -1 \Rightarrow (0, -1)$

27. $f(x) = \dfrac{x+2}{x-6}$

   a. $f(3) = \dfrac{3+2}{3-6} = -\dfrac{5}{3} \neq 14$

   The point $(3,14)$ is not on the graph of f.

   b. $f(4) = \dfrac{4+2}{4-6} = \dfrac{6}{-2} = -3$

   The point $(4,-3)$ is on the graph of f.

   c. Solve for x:

   $2 = \dfrac{x+2}{x-6}$

   $2x - 12 = x + 2$

   $x = 14$

   $(14, 2)$ is a point on the graph of f.

   d. The domain of f is $\{x \mid x \neq 6\}$.

   e. x-intercepts:

   $f(x) = 0 \Rightarrow \dfrac{x+2}{x-6} = 0$

   $x + 2 = 0 \Rightarrow x = -2 \Rightarrow (-2,0)$

   f. y-intercept: $f(0) = \dfrac{0+2}{0-6} = -\dfrac{1}{3} \Rightarrow \left(0, -\dfrac{1}{3}\right)$

29. $f(x) = \dfrac{2x^2}{x^4+1}$

   a. $f(-1) = \dfrac{2(-1)^2}{(-1)^4+1} = \dfrac{2}{2} = 1$

   The point $(-1,1)$ is on the graph of f.

   b. $f(2) = \dfrac{2(2)^2}{(2)^4+1} = \dfrac{8}{17}$

   The point $\left(2, \dfrac{8}{17}\right)$ is on the graph of f.

   c. Solve for x:

   $1 = \dfrac{2x^2}{x^4+1}$

   $x^4 + 1 = 2x^2$

   $x^4 - 2x^2 + 1 = 0$

   $(x^2 - 1)^2 = 0$

   $x^2 - 1 = 0 \Rightarrow x = \pm 1$

   $(1,1)$ and $(-1,1)$ are on the graph of f.

   d. The domain of f is $\{x \mid x \text{ is any real number}\}$.

   e. x-intercept:

   $f(x) = 0 \Rightarrow \dfrac{2x^2}{x^4+1} = 0$

   $2x^2 = 0 \Rightarrow x = 0 \Rightarrow (0,0)$

   f. y-intercept:

   $f(0) = \dfrac{2(0)^2}{0^4+1} = \dfrac{0}{0+1} = 0 \Rightarrow (0,0)$

31. a. $(f+g)(2) = f(2) + g(2) = 2+1 = 3$

   b. $(f+g)(4) = f(4) + g(4) = 1 + (-3) = -2$

   c. $(f-g)(6) = f(6) - g(6) = 0 - 1 = -1$

   d. $(g-f)(6) = g(6) - f(6) = 1 - 0 = 1$

   e. $(f \cdot g)(2) = f(2) \cdot g(2) = 2(1) = 2$

   f. $\left(\dfrac{f}{g}\right)(4) = \dfrac{f(4)}{g(4)} = \dfrac{1}{-3} = -\dfrac{1}{3}$

33. $h(x) = -\dfrac{44x^2}{v^2} + x + 6$

   a. $h(8) = -\dfrac{44(8)^2}{28^2} + (8) + 6$

   $= -\dfrac{2816}{784} + 14$

   $\approx 10.4 \text{ feet}$

   b. $h(12) = -\dfrac{44(12)^2}{28^2} + (12) + 6$

   $= -\dfrac{6336}{784} + 18$

   $\approx 9.9 \text{ feet}$

   c. From part (a) we know the point $(8,10.4)$ is on the graph and from part (b) we know the point $(12,9.9)$ is on the graph. We could evaluate the function at several more values of x (e.g. $x = 0$, $x = 15$, and $x = 20$) to obtain additional points.

134

$h(0) = -\dfrac{44(0)^2}{28^2} + (0) + 6 = 6$

$h(15) = -\dfrac{44(15)^2}{28^2} + (15) + 6 \approx 8.4$

$h(20) = -\dfrac{44(20)^2}{28^2} + (20) + 6 \approx 3.6$

Some additional points are $(0,6)$, $(15,8.4)$ and $(20,3.6)$. The complete graph is given below.

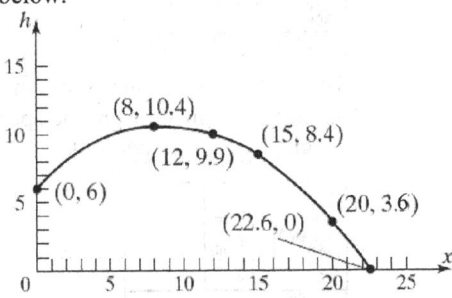

d. $h(15) = -\dfrac{44(15)^2}{28^2} + (15) + 6 \approx 8.4$ feet

No; when the ball is 15 feet in front of the foul line, it will be below the hoop. Therefore it cannot go through the hoop.

In order for the ball to pass through the hoop, we need to have $h(15) = 10$.

$10 = -\dfrac{44(15)^2}{v^2} + (15) + 6$

$-11 = -\dfrac{44(15)^2}{v^2}$

$v^2 = 4(225)$

$v^2 = 900$

$v = 30$ ft/sec

The ball must be shot with an initial velocity of 30 feet per second in order to go through the hoop.

35. $h(x) = \dfrac{-32x^2}{130^2} + x$

a. $h(100) = \dfrac{-32(100)^2}{130^2} + 100$

$= \dfrac{-320,000}{16,900} + 100 \approx 81.07$ feet

b. $h(300) = \dfrac{-32(300)^2}{130^2} + 300$

$= \dfrac{-2,880,000}{16,900} + 300 \approx 129.59$ feet

c. $h(500) = \dfrac{-32(500)^2}{130^2} + 500$

$= \dfrac{-8,000,000}{16,900} + 500 \approx 26.63$ feet

d. Solving $h(x) = \dfrac{-32x^2}{130^2} + x = 0$

$\dfrac{-32x^2}{130^2} + x = 0$

$x\left(\dfrac{-32x}{130^2} + 1\right) = 0$

$x = 0$ or $\dfrac{-32x}{130^2} + 1 = 0$

$1 = \dfrac{32x}{130^2}$

$130^2 = 32x$

$x = \dfrac{130^2}{32} = 528.13$ feet

Therefore, the golf ball travels 528.13 feet.

e. $y_1 = \dfrac{-32x^2}{130^2} + x$

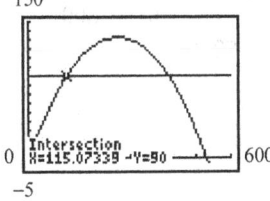

f. Use INTERSECT on the graphs of $y_1 = \dfrac{-32x^2}{130^2} + x$ and $y_2 = 90$.

135

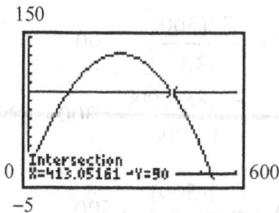

The ball reaches a height of 90 feet twice. The first time is when the ball has traveled approximately 115.07 feet, and the second time is when the ball has traveled about 413.05 feet.

g. The ball travels approximately 275 feet before it reaches its maximum height of approximately 131.8 feet.

h. The ball travels approximately 264 feet before it reaches its maximum height of approximately 132.03 feet.

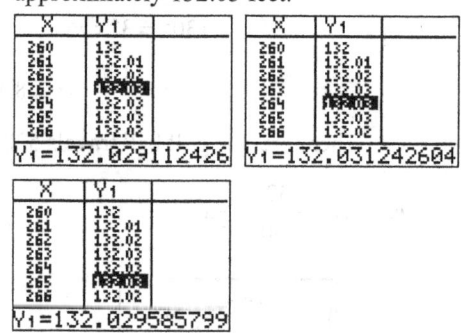

37. $C(x) = 100 + \dfrac{x}{10} + \dfrac{36000}{x}$

a. $C(480) = 100 + \dfrac{480}{10} + \dfrac{36000}{480}$

$= \$223$

$C(600) = 100 + \dfrac{600}{10} + \dfrac{36000}{600}$

$= \$220$

b. $\{x \mid x > 0\}$

c. Graphing:

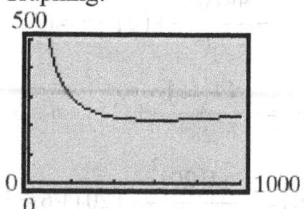

d. TblStart = 0; ΔTbl = 50

e. The cost per passenger is minimized to about \$220 when the ground speed is roughly 600 miles per hour.

39. **a.** $C(0) = \$30$

It costs \$30 if you use 0 gigabytes.

b. $C(5) = \$30$

It costs \$30 if you use 5 gigabytes.

c. $C(15) = \$90$

It costs \$90 if you use 15 gigabytes.

d. The domain is $\{g \mid 0 \le g \le 60\}$. This indicates that there are at most 60 gigabytes in a month.

e. The graph is flat at first and then rises in a straight line.

41. The graph of a function can have any number of x-intercepts. The graph of a function can have at most one y-intercept (otherwise the graph would fail the vertical line test).

43. (a) III; (b) IV; (c) I; (d) V; (e) II

45.

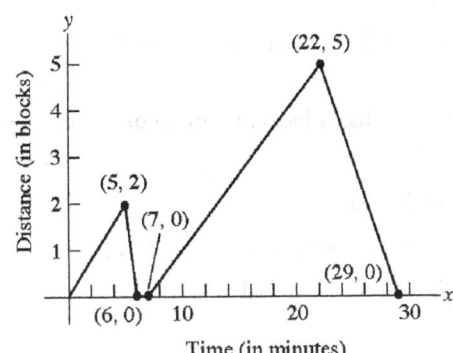

47. a. 2 hours elapsed; Kevin was between 0 and 3 miles from home.

b. 0.5 hours elapsed; Kevin was 3 miles from home.

c. 0.3 hours elapsed; Kevin was between 0 and 3 miles from home.

d. 0.2 hours elapsed; Kevin was at home.

e. 0.9 hours elapsed; Kevin was between 0 and 2.8 miles from home.

f. 0.3 hours elapsed; Kevin was 2.8 miles from home.

g. 1.1 hours elapsed; Kevin was between 0 and 2.8 miles from home.

h. The farthest distance Kevin is from home is 3 miles.

i. Kevin returned home 2 times.

49. Answers (graphs) will vary. Points of the form $(5, y)$ and of the form $(x, 0)$ cannot be on the graph of the function.

51. Answers may vary.

53. $d = \sqrt{(1-3)^2 + (0-(-6))^2}$

$= \sqrt{(-2)^2 + (-6)^2}$

$= \sqrt{4+36} = \sqrt{40} = 2\sqrt{10}$

55. $(4x^3 - 5x^2 + 2) - (3x^2 + 5x - 2)$

$= 4x^3 - 5x^2 + 2 - 3x^2 - 5x + 2$

$= 4x^3 - 8x^2 - 5x + 4$

Section 3.3

1. $2 < x < 5$

3. x-axis: $y \rightarrow -y$

$(-y) = 5x^2 - 1$

$-y = 5x^2 - 1$

$y = -5x^2 + 1$ different

y-axis: $x \rightarrow -x$

$y = 5(-x)^2 - 1$

$y = 5x^2 - 1$ same

origin: $x \rightarrow -x$ and $y \rightarrow -y$

$(-y) = 5(-x)^2 - 1$

$-y = 5x^2 - 1$

$y = -5x^2 + 1$ different

The equation has symmetry with respect to the y-axis only.

5. $y = x^2 - 9$

x-intercepts:

$0 = x^2 - 9$

$x^2 = 9 \rightarrow x = \pm 3$

y-intercept:

$y = (0)^2 - 9 = -9$

The intercepts are $(-3, 0)$, $(3, 0)$, and $(0, -9)$.

7. even; odd

9. True

11. c

13. Yes

15. No

17. f is increasing on the intervals $[-8, -2]$, $[0, 2]$, $[5, 7]$.

19. Yes. The local maximum at $x = 2$ is 10.

21. f has local maxima at $x = -2$ and $x = 2$. The local maxima are 6 and 10, respectively.

23. f has absolute minimum of -4 at $x = -8$.

137

25. a. Intercepts: $(-2, 0)$, $(2, 0)$, and $(0, 3)$.

 b. Domain: $\{x \mid -4 \leq x \leq 4\}$ or $[-4, 4]$;

 Range: $\{y \mid 0 \leq y \leq 3\}$ or $[0, 3]$.

 c. Increasing: $[-2, 0]$ and $[2, 4]$;
 Decreasing: $[-4, -2]$ and $[0, 2]$.

 d. Since the graph is symmetric with respect to the y-axis, the function is <u>even</u>.

27. a. Intercepts: $(0, 1)$.

 b. Domain: $\{x \mid x \text{ is any real number}\}$;

 Range: $\{y \mid y > 0\}$ or $(0, \infty)$.

 c. Increasing: $(-\infty, \infty)$; Decreasing: never.

 d. Since the graph is not symmetric with respect to the y-axis or the origin, the function is <u>neither</u> even nor odd.

29. a. Intercepts: $(-\pi, 0)$, $(\pi, 0)$, and $(0, 0)$.

 b. Domain: $\{x \mid -\pi \leq x \leq \pi\}$ or $[-\pi, \pi]$;

 Range: $\{y \mid -1 \leq y \leq 1\}$ or $[-1, 1]$.

 c. Increasing: $\left[-\dfrac{\pi}{2}, \dfrac{\pi}{2}\right]$;

 Decreasing: $\left[-\pi, -\dfrac{\pi}{2}\right]$ and $\left[\dfrac{\pi}{2}, \pi\right]$.

 d. Since the graph is symmetric with respect to the origin, the function is <u>odd</u>.

31. a. Intercepts: $\left(\dfrac{1}{3}, 0\right)$, $\left(\dfrac{5}{2}, 0\right)$, and $\left(0, \dfrac{1}{2}\right)$.

 b. Domain: $\{x \mid -3 \leq x \leq 3\}$ or $[-3, 3]$;

 Range: $\{y \mid -1 \leq y \leq 2\}$ or $[-1, 2]$.

 c. Increasing: $[2, 3]$; Decreasing: $[-1, 1]$;

 Constant: $[-3, -1]$ and $[1, 2]$

 d. Since the graph is not symmetric with respect to the y-axis or the origin, the

33. a. f has a local maximum of 3 at $x = 0$.

 b. f has a local minimum of 0 at both

35. a. f has a local maximum of 1 at $x = \dfrac{\pi}{2}$.

 b. f has a local minimum of -1 at $x = -\dfrac{\pi}{2}$

37. $f(x) = 4x^3$

$$f(-x) = 4(-x)^3 = -4x^3 = -f(x)$$

Therefore, f is odd.

39. $g(x) = -3x^2 - 5$

$$g(-x) = -3(-x)^2 - 5 = -3x^2 - 5 = g(x)$$

Therefore, g is even.

41. $F(x) = \sqrt[3]{x}$

$$F(-x) = \sqrt[3]{-x} = -\sqrt[3]{x} = -F(x)$$

Therefore, F is odd.

43. $f(x) = x + |x|$

$$f(-x) = -x + |-x| = -x + |x|$$

f is neither even nor odd.

45. $g(x) = \dfrac{1}{x^2}$

$$g(-x) = \dfrac{1}{(-x)^2} = \dfrac{1}{x^2} = g(x)$$

Therefore, g is even.

47. $h(x) = \dfrac{-x^3}{3x^2 - 9}$

$$h(-x) = \dfrac{-(-x)^3}{3(-x)^2 - 9} = \dfrac{x^3}{3x^2 - 9} = -h(x)$$

Therefore, h is odd.

49. f has an absolute maximum of 4 at $x = 1$
f has an absolute minimum of 1 at $x = 5$.
f has an local maximum of 3 at $x = 3$.
f has an local minimum of 2 at $x = 2$.

51. f has an absolute minimum of 1 at $x = 1$.
f has an absolute maximum of 4 at $x = 3$.
f has an local minimum of 1 at $x = 1$.
f has an local maximum of 4 at $x = 3$.

138

53. f has an absolute minimum of 0 at $x = 0$.

f has no absolute maximum.

f has an local minimum of 0 at $x = 0$.

f has an local minimum of 2 at $x = 3$.

f has an local maximum of 3 at $x = 2$.

55. f has no absolute maximum or minimum.

f has no local maximum or minimum.

57. $f(x) = x^3 - 3x + 2$ on the interval $(-2, 2)$

Use MAXIMUM and MINIMUM on the graph of $y_1 = x^3 - 3x + 2$.

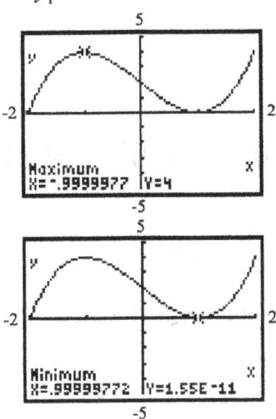

local maximum: $f(-1) = 4$

local minimum: $f(1) = 0$

f is increasing on: $[-2, -1]$ and $[1, 2]$;

59. $f(x) = x^5 - x^3$ on the interval $(-2, 2)$

Use MAXIMUM and MINIMUM on the graph of $y_1 = x^5 - x^3$.

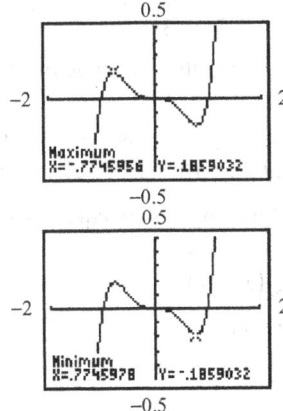

local maximum: $f(-0.77) = 0.19$

local minimum: $f(0.77) = -0.19$

f is increasing on: $[-2, -0.77]$ and $[0.77, 2]$;

f is decreasing on: $[-0.77, 0.77]$

61. $f(x) = -0.2x^3 - 0.6x^2 + 4x - 6$ on the interval $(-6, 4)$

Use MAXIMUM and MINIMUM on the graph of $y_1 = -0.2x^3 - 0.6x^2 + 4x - 6$.

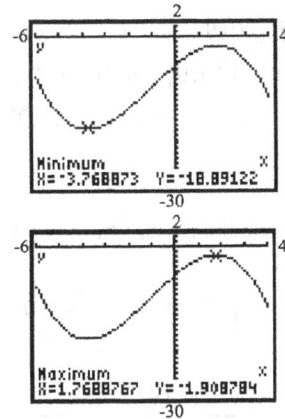

local maximum: $f(1.77) = -1.91$

local minimum: $f(-3.77) = -18.89$

f is increasing on: $[-3.77, 1.77]$;

f is decreasing on: $[-6, -3.77]$ and $[1.77, 4]$

63. $f(x) = 0.25x^4 + 0.3x^3 - 0.9x^2 + 3$ on the interval $(-3, 2)$

Use MAXIMUM and MINIMUM on the graph of $y_1 = 0.25x^4 + 0.3x^3 - 0.9x^2 + 3$.

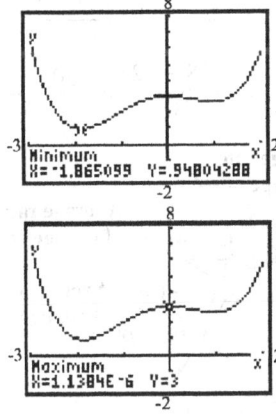

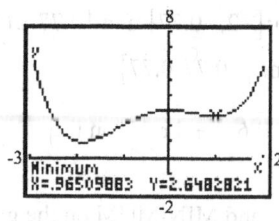

local maximum: $f(0) = 3$

local minimum:

$f(-1.87) = 0.95$, $f(0.97) = 2.65$

f is increasing on: $[-1.87, 0]$ and $[0.97, 2]$;

f is decreasing on: $[-3, -1.87]$ and $[0, 0.97]$

65. $f(x) = -2x^2 + 4$

a. Average rate of change of f from $x = 0$ to $x = 2$

$$\frac{f(2) - f(0)}{2 - 0} = \frac{\left(-2(2)^2 + 4\right) - \left(-2(0)^2 + 4\right)}{2}$$

$$= \frac{(-4) - (4)}{2} = \frac{-8}{2} = -4$$

b. Average rate of change of f from $x = 1$ to $x = 3$:

$$\frac{f(3) - f(1)}{3 - 1} = \frac{\left(-2(3)^2 + 4\right) - \left(-2(1)^2 + 4\right)}{2}$$

$$= \frac{(-14) - (2)}{2} = \frac{-16}{2} = -8$$

c. Average rate of change of f from $x = 1$ to $x = 4$:

$$\frac{f(4) - f(1)}{4 - 1} = \frac{\left(-2(4)^2 + 4\right) - \left(-2(1)^2 + 4\right)}{3}$$

$$= \frac{(-28) - (2)}{3} = \frac{-30}{3} = -10$$

d.

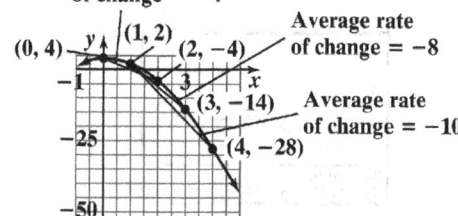

67. $g(x) = x^3 - 2x + 1$

a. Average rate of change of g from $x = -3$ to $x = -2$:

$$\frac{g(-2) - g(-3)}{-2 - (-3)}$$

$$= \frac{\left[(-2)^3 - 2(-2) + 1\right] - \left[(-3)^3 - 2(-3) + 1\right]}{1}$$

$$= \frac{(-3) - (-20)}{1} = \frac{17}{1} = 17$$

b. Average rate of change of g from $x = -1$ to $x = 1$:

$$\frac{g(1) - g(-1)}{1 - (-1)}$$

$$= \frac{\left[(1)^3 - 2(1) + 1\right] - \left[(-1)^3 - 2(-1) + 1\right]}{2}$$

$$= \frac{(0) - (2)}{2} = \frac{-2}{2} = -1$$

c. Average rate of change of g from $x = 1$ to $x = 3$:

$$\frac{g(3) - g(1)}{3 - 1}$$

$$= \frac{\left[(3)^3 - 2(3) + 1\right] - \left[(1)^3 - 2(1) + 1\right]}{2}$$

$$= \frac{(22) - (0)}{2} = \frac{22}{2} = 11$$

69. $f(x) = 5x - 2$

a. Average rate of change of f from 1 to 3:

$$\frac{\Delta y}{\Delta x} = \frac{f(3) - f(1)}{3 - 1} = \frac{13 - 3}{3 - 1} = \frac{10}{2} = 5$$

Thus, the average rate of change of f from 1 to 3 is 5.

b. From (a), the slope of the secant line joining $(1, f(1))$ and $(3, f(3))$ is 5. We use the point-slope form to find the equation of the secant line:

$$y - y_1 = m_{sec}(x - x_1)$$

$$y - 3 = 5(x - 1)$$

$$y - 3 = 5x - 5$$

$$y = 5x - 2$$

140

71. $g(x) = x^2 - 2$

   a. Average rate of change of g from -2 to 1:

$$\frac{\Delta y}{\Delta x} = \frac{g(1) - g(-2)}{1-(-2)} = \frac{-1-2}{1-(-2)} = \frac{-3}{3} = -1$$

Therefore, the average rate of change of g from -2 to 1 is -1.

   b. From (a), the slope of the secant line joining $(-2, g(-2))$ and $(1, g(1))$ is -1. We use the point-slope form to find the equation of the secant line:

$$y - y_1 = m_{sec}(x - x_1)$$
$$y - 2 = -1(x - (-2))$$
$$y - 2 = -x - 2$$
$$y = -x$$

   c.

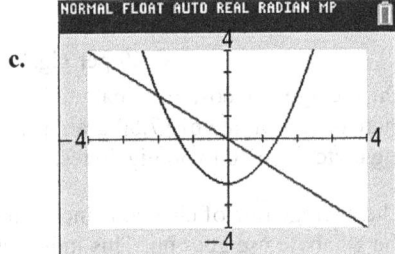

73. $h(x) = x^2 - 2x$

   a. Average rate of change of h from 2 to 4:

$$\frac{\Delta y}{\Delta x} = \frac{h(4) - h(2)}{4-2} = \frac{8-0}{4-2} = \frac{8}{2} = 4$$

Therefore, the average rate of change of h from 2 to 4 is 4.

   b. From (a), the slope of the secant line joining $(2, h(2))$ and $(4, h(4))$ is 4. We use the point-slope form to find the equation of the secant line:

$$y - y_1 = m_{sec}(x - x_1)$$
$$y - 0 = 4(x - 2)$$
$$y = 4x - 8$$

   c.

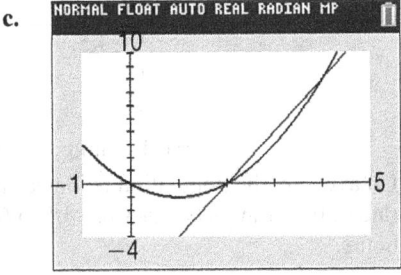

75. a. $g(x) = x^3 - 27x$

$$g(-x) = (-x)^3 - 27(-x)$$
$$= -x^3 + 27x$$
$$= -(x^3 - 27x)$$
$$= -g(x)$$

Since $g(-x) = -g(x)$, the function is odd.

   b. Since $g(x)$ is odd then it is symmetric about the origin so there exist a local maximum at $x = -3$.

$$g(-3) = (-3)^3 - 27(-3) = -27 + 81 = 54$$

So there is a local maximum of 54 at $x = -3$.

77. $F(x) = -x^4 + 8x^2 + 8$

   a. $F(-x) = -(-x)^4 + 8(-x)^2 + 8$
$$= -x^4 + 8x + 8$$
$$= F(x)$$

Since $F(-x) = F(x)$, the function is even.

   b. Since the function is even, its graph has y-axis symmetry. The second local maximum is in Quadrant II and is 24 and occurs at $x = -2$.

   c. Because the graph has y-axis symmetry, the area under the graph between $x = 0$ and $x = 3$ bounded below by the x-axis is the same as the area under the graph between $x = -3$ and $x = 0$ bounded below the x-axis. Thus, the area is 47.4 square units.

79. $\overline{C}(x) = 0.3x^2 + 21x - 251 + \dfrac{2500}{x}$

   a. $y_1 = 0.3x^2 + 21x - 251 + \dfrac{2500}{x}$

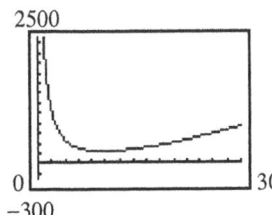

b. Use MINIMUM. Rounding to the nearest whole number, the average cost is minimized when approximately 10 lawnmowers are produced per hour.

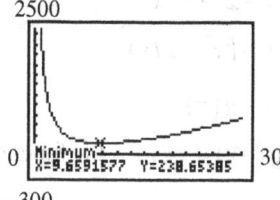

c. The minimum average cost is approximately $239 per mower.

81. a. and b.

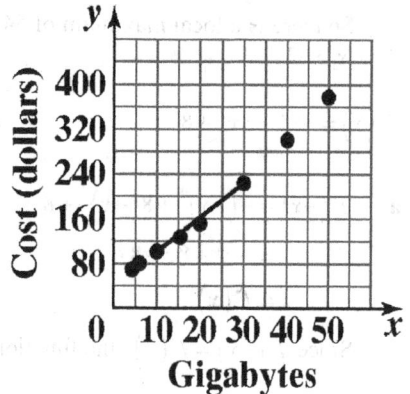

Gigabytes

The slope represents the average rate of change of the cost of the plan from 10 to 30 gigabytes.

c. avg. rate of change $= \dfrac{C(10)-C(4)}{10-4}$

$= \dfrac{100-70}{6}$

$= \dfrac{30}{6}$

$= \$5$ per gigabyte

On average, the cost per gigabyte is increasing at a rate of $5 gram per gigabyte from 4 to 10 gigabytes.

d. avg. rate of change $= \dfrac{C(30)-C(10)}{30-10}$

$= \dfrac{225-100}{20}$

$= \dfrac{125}{20}$

$= \$6.25$ per gigabyte

On average, the cost per gigabyte is increasing at a rate of $6.25 gram per gigabyte from 10 to 30 gigabytes.

e. avg. rate of change $= \dfrac{C(50)-C(30)}{50-30}$

$= \dfrac{375-225}{20}$

$= \dfrac{150}{20}$

$= \$7.50$ per gigabyte

On average, the cost per gigabyte is increasing at a rate of $7.50 gram per gigabyte from 30 to 50 gigabytes.

f. The average rate of change is increasing as the gigabyte use goes up. This indicates that the cost is increasing at an increasing rate.

83. a. avg. rate of change $= \dfrac{P(2.5)-P(0)}{2.5-0}$

$= \dfrac{0.18-0.09}{2.5-0}$

$= \dfrac{0.09}{2.5}$

$= 0.036$ gram per hour

On average, the population is increasing at a rate of 0.036 gram per hour from 0 to 2.5 hours.

b. avg. rate of change $= \dfrac{P(6)-P(4.5)}{6-4.5}$

$= \dfrac{0.50-0.35}{6-4.5}$

$= \dfrac{0.15}{1.5}$

$= 0.1$ gram per hour

On average, the population is increasing at a rate of 0.1 gram per hour from 4.5 to 6 hours.

c. The average rate of change is increasing as time passes. This indicates that the population is increasing at an increasing rate.

85. $f(x) = x^2$

a. Average rate of change of f from $x = 0$ to $x = 1$:
$$\frac{f(1) - f(0)}{1 - 0} = \frac{1^2 - 0^2}{1} = \frac{1}{1} = 1$$

b. Average rate of change of f from $x = 0$ to $x = 0.5$:
$$\frac{f(0.5) - f(0)}{0.5 - 0} = \frac{(0.5)^2 - 0^2}{0.5} = \frac{0.25}{0.5} = 0.5$$

c. Average rate of change of f from $x = 0$ to $x = 0.1$:
$$\frac{f(0.1) - f(0)}{0.1 - 0} = \frac{(0.1)^2 - 0^2}{0.1} = \frac{0.01}{0.1} = 0.1$$

d. Average rate of change of f from $x = 0$ to $x = 0.01$:
$$\frac{f(0.01) - f(0)}{0.01 - 0} = \frac{(0.01)^2 - 0^2}{0.01}$$
$$= \frac{0.0001}{0.01} = 0.01$$

e. Average rate of change of f from $x = 0$ to $x = 0.001$:
$$\frac{f(0.001) - f(0)}{0.001 - 0} = \frac{(0.001)^2 - 0^2}{0.001}$$
$$= \frac{0.000001}{0.001} = 0.001$$

f. Graphing the secant lines:

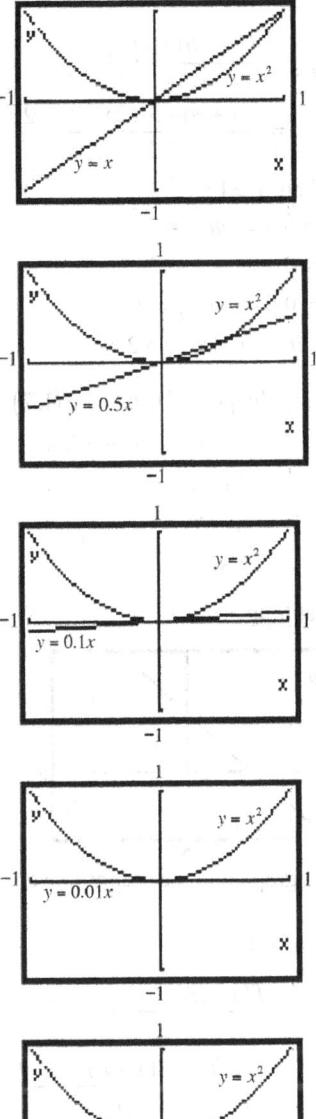

g. The secant lines are beginning to look more and more like the tangent line to the graph of f at the point where $x = 0$.

h. The slopes of the secant lines are getting smaller and smaller. They seem to be approaching the number zero.

143

87. $f(x) = 2x + 5$

   a. $m_{sec} = \dfrac{f(x+h) - f(x)}{h}$

$$= \frac{2(x+h) + 5 - 2x - 5}{h} = \frac{2h}{h} = 2$$

   b. When $x = 1$:

$h = 0.5 \Rightarrow m_{sec} = 2$

$h = 0.1 \Rightarrow m_{sec} = 2$

$h = 0.01 \Rightarrow m_{sec} = 2$

as $h \to 0$, $m_{sec} \to 2$

   c. Using the point $(1, f(1)) = (1, 7)$ and slope,

$m = 2$, we get the secant line:

$y - 7 = 2(x - 1)$

$y - 7 = 2x - 2$

$y = 2x + 5$

   d. Graphing:

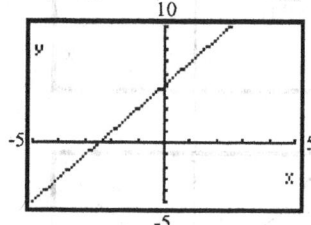

The graph and the secant line coincide.

89. $f(x) = x^2 + 2x$

   a. $m_{sec} = \dfrac{f(x+h) - f(x)}{h}$

$$= \frac{(x+h)^2 + 2(x+h) - (x^2 + 2x)}{h}$$

$$= \frac{x^2 + 2xh + h^2 + 2x + 2h - x^2 - 2x}{h}$$

$$= \frac{2xh + h^2 + 2h}{h}$$

$$= 2x + h + 2$$

   b. When $x = 1$,

$h = 0.5 \Rightarrow m_{sec} = 2 \cdot 1 + 0.5 + 2 = 4.5$

$h = 0.1 \Rightarrow m_{sec} = 2 \cdot 1 + 0.1 + 2 = 4.1$

$h = 0.01 \Rightarrow m_{sec} = 2 \cdot 1 + 0.01 + 2 = 4.01$

as $h \to 0$, $m_{sec} \to 2 \cdot 1 + 0 + 2 = 4$

   c. Using point $(1, f(1)) = (1, 3)$ and

slope $= 4.01$, we get the secant line:

$y - 3 = 4.01(x - 1)$

$y - 3 = 4.01x - 4.01$

$y = 4.01x - 1.01$

   d. Graphing:

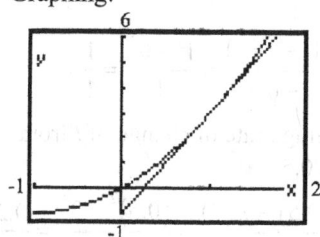

91. $f(x) = 2x^2 - 3x + 1$

   a. $m_{sec} = \dfrac{f(x+h) - f(x)}{h}$

$$= \frac{2(x+h)^2 - 3(x+h) + 1 - (2x^2 - 3x + 1)}{h}$$

$$= \frac{2(x^2 + 2xh + h^2) - 3x - 3h + 1 - 2x^2 + 3x - 1}{h}$$

$$= \frac{2x^2 + 4xh + 2h^2 - 3x - 3h + 1 - 2x^2 + 3x - 1}{h}$$

$$= \frac{4xh + 2h^2 - 3h}{h}$$

$$= 4x + 2h - 3$$

   b. When $x = 1$,

$h = 0.5 \Rightarrow m_{sec} = 4 \cdot 1 + 2(0.5) - 3 = 2$

$h = 0.1 \Rightarrow m_{sec} = 4 \cdot 1 + 2(0.1) - 3 = 1.2$

$h = 0.01 \Rightarrow m_{sec} = 4 \cdot 1 + 2(0.01) - 3 = 1.02$

as $h \to 0$, $m_{sec} \to 4 \cdot 1 + 2(0) - 3 = 1$

   c. Using point $(1, f(1)) = (1, 0)$ and

slope $= 1.02$, we get the secant line:

$y - 0 = 1.02(x - 1)$

$y = 1.02x - 1.02$

   d. Graphing:

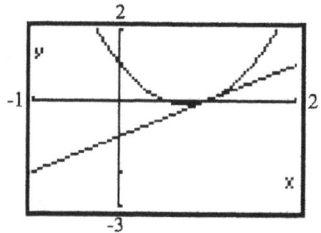

93. $f(x) = \dfrac{1}{x}$

a. $m_{sec} = \dfrac{f(x+h) - f(x)}{h}$

$= \dfrac{\left(\dfrac{1}{x+h} - \dfrac{1}{x}\right)}{h} = \dfrac{\left(\dfrac{x-(x+h)}{(x+h)x}\right)}{h}$

$= \left(\dfrac{x-x-h}{(x+h)x}\right)\left(\dfrac{1}{h}\right) = \left(\dfrac{-h}{(x+h)x}\right)\left(\dfrac{1}{h}\right)$

$= -\dfrac{1}{(x+h)x}$

b. When $x = 1$,

$h = 0.5 \Rightarrow m_{sec} = -\dfrac{1}{(1+0.5)(1)}$

$= -\dfrac{1}{1.5} = -\dfrac{2}{3} \approx -0.667$

$h = 0.1 \Rightarrow m_{sec} = -\dfrac{1}{(1+0.1)(1)}$

$= -\dfrac{1}{1.1} = -\dfrac{10}{11} \approx -0.909$

$h = 0.01 \Rightarrow m_{sec} = -\dfrac{1}{(1+0.01)(1)}$

$= -\dfrac{1}{1.01} = -\dfrac{100}{101} \approx -0.990$

as $h \to 0$, $m_{sec} \to -\dfrac{1}{(1+0)(1)} = -\dfrac{1}{1} = -1$

c. Using point $(1, f(1)) = (1,1)$ and

slope $= -\dfrac{100}{101}$, we get the secant line:

$y - 1 = -\dfrac{100}{101}(x-1)$

$y - 1 = -\dfrac{100}{101}x + \dfrac{100}{101}$

$y = -\dfrac{100}{101}x + \dfrac{201}{101}$

d. Graphing:

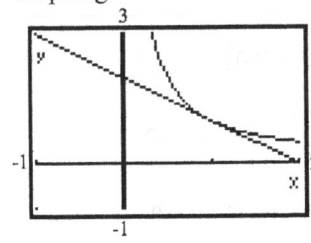

95. Answers will vary. One possibility follows:

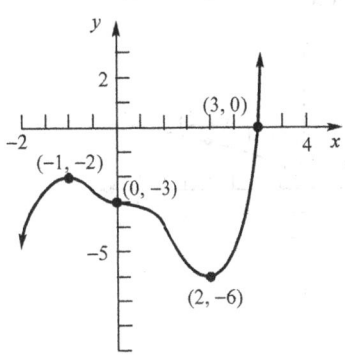

97. A function that is increasing on an interval can have at most one x-intercept on the interval. The graph of f could not "turn" and cross it again or it would start to decrease.

99. To be an even function we need $f(-x) = f(x)$ and to be an odd function we need $f(-x) = -f(x)$. In order for a function be both even and odd, we would need $f(x) = -f(x)$. This is only possible if $f(x) = 0$.

101. Not necessarily. It just means $f(5) > f(2)$. The function could have both increasing and decreasing intervals.

103. a. $0.00000701 = 7.01 \times 10^{-6}$

b. $2,305,000,000 = 2.305 \times 10^{9}$

105. $14 < 5 - 3x \le 29$

$9 < -3x \le 24$

$-3 > x \ge -8$

$-8 \le x < -3$

The solution is: $[-8, -3)$.

145

Section 3.4

1. $y = \sqrt{x}$

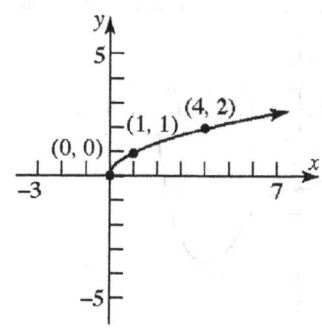

3. $y = x^3 - 8$

y-intercept:

Let $x = 0$, then $y = (0)^3 - 8 = -8$.

x-intercept:

Let $y = 0$, then $0 = x^3 - 8$

$$x^3 = 8$$
$$x = 2$$

The intercepts are $(0, -8)$ and $(2, 0)$.

5. piecewise-defined

7. False; the cube root function is odd and increasing on the interval $(-\infty, \infty)$.

9. b

11. C

13. E

15. B

17. F

19. $f(x) = x$

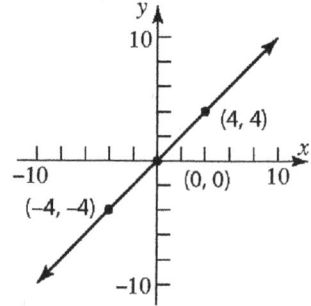

21. $f(x) = x^3$

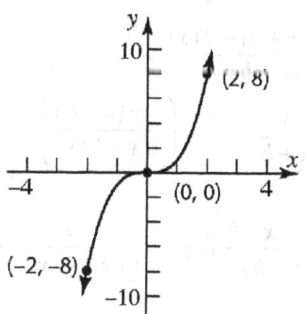

23. $f(x) = \dfrac{1}{x}$

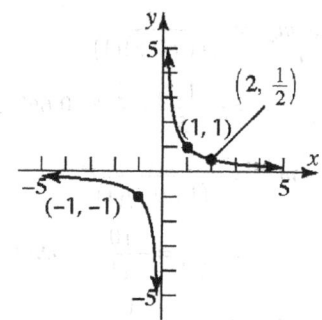

25. $f(x) = \sqrt[3]{x}$

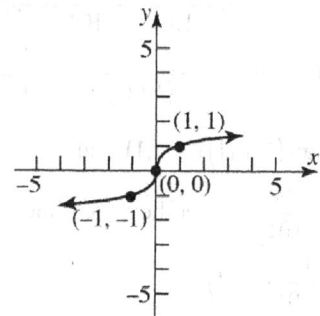

27. a. $f(-2) = (-2)^2 = 4$

b. $f(0) = 2$

c. $f(2) = 2(2) + 1 = 5$

29. a. $f(0) = 2(0) - 4 = -4$

b. $f(1) = 2(1) - 4 = -2$

c. $f(2) = 2(2) - 4 = 0$

d. $f(3) = (3)^3 - 2 = 25$

146

31. $f(x) = \begin{cases} 2x & \text{if } x \neq 0 \\ 1 & \text{if } x = 0 \end{cases}$

   a. Domain: $\{x \mid x \text{ is any real number}\}$

   b. x-intercept: none
      y-intercept:
      $f(0) = 1$

      The only intercept is $(0,1)$.

   c. Graph:

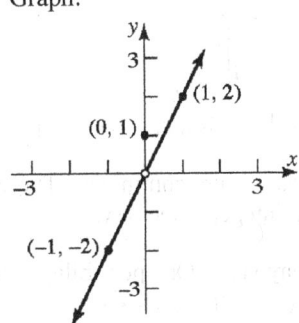

   d. Range: $\{y \mid y \neq 0\}$; $(-\infty, 0) \cup (0, \infty)$

   e. The graph is not continuous. There is a jump at $x = 0$.

33. $f(x) = \begin{cases} -2x+3 & \text{if } x < 1 \\ 3x-2 & \text{if } x \geq 1 \end{cases}$

   a. Domain: $\{x \mid x \text{ is any real number}\}$

   b. x-intercept: none
      y-intercept: $f(0) = -2(0) + 3 = 3$

      The only intercept is $(0,3)$.

   c. Graph:

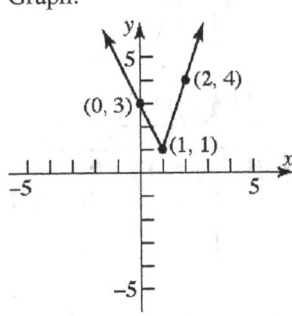

   d. Range: $\{y \mid y \geq 1\}$; $[1, \infty)$

   e. The graph is continuous. There are no holes or gaps.

35. $f(x) = \begin{cases} x+3 & \text{if } -2 \leq x < 1 \\ 5 & \text{if } x = 1 \\ -x+2 & \text{if } x > 1 \end{cases}$

   a. Domain: $\{x \mid x \geq -2\}$; $[-2, \infty)$

   b. $\quad x+3 = 0 \qquad\qquad -x+2 = 0$
      $\quad x = -3 \qquad\qquad\quad -x = -2$
      (not in domain) $\qquad\quad x = 2$
      x-intercept: 2

      y-intercept: $f(0) = 0 + 3 = 3$

      The intercepts are $(2,0)$ and $(0,3)$.

   c. Graph:

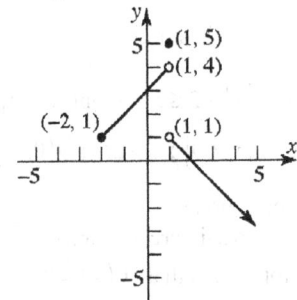

   d. Range: $\{y \mid y < 4, \ y = 5\}$; $(-\infty, 4) \cup \{5\}$

   e. The graph is not continuous. There is a jump at $x = 1$.

37. $f(x) = \begin{cases} 1+x & \text{if } x < 0 \\ x^2 & \text{if } x \geq 0 \end{cases}$

   a. Domain: $\{x \mid x \text{ is any real number}\}$

   b. $\quad 1+x = 0 \qquad\quad x^2 = 0$
      $\qquad x = -1 \qquad\qquad x = 0$
      x-intercepts: $-1, 0$

      y-intercept: $f(0) = 0^2 = 0$

      The intercepts are $(-1,0)$ and $(0,0)$.

147

c. Graph:

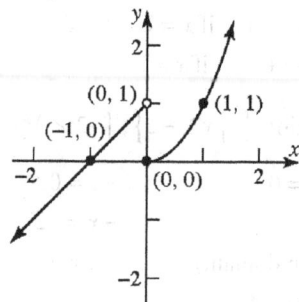

d. Range: $\{y \mid y \text{ is any real number}\}$

e. The graph is not continuous. There is a jump at $x = 0$.

39. $f(x) = \begin{cases} |x| & \text{if } -2 \le x < 0 \\ x^3 & \text{if } x > 0 \end{cases}$

a. Domain: $\{x \mid -2 \le x < 0 \text{ and } x > 0\}$ or $\{x \mid x \ge -2, x \ne 0\}$; $[-2,0) \cup (0,\infty)$.

b. x-intercept: none
There are no x-intercepts since there are no values for x such that $f(x) = 0$.

y-intercept:
There is no y-intercept since $x = 0$ is not in the domain.

c. Graph:

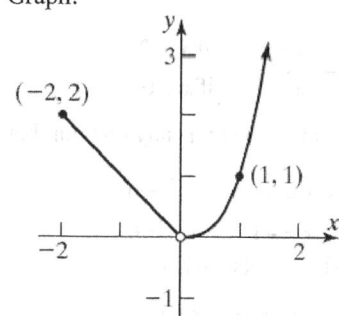

d. Range: $\{y \mid y > 0\}$; $(0, \infty)$

e. The graph is not continuous. There is a hole at $x = 0$.

41. $f(x) = 2\,\text{int}(x)$

a. Domain: $\{x \mid x \text{ is any real number}\}$

b. x-intercepts:
All values for x such that $0 \le x < 1$.

y-intercept: $f(0) = 2\,\text{int}(0) = 0$

The intercepts are all ordered pairs $(x, 0)$ when $0 \le x < 1$.

c. Graph:

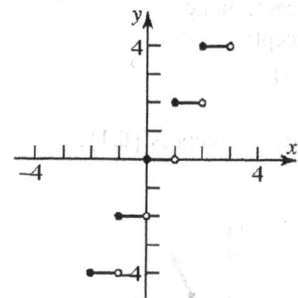

d. Range: $\{y \mid y \text{ is an even integer}\}$

e. The graph is not continuous. There is a jump at each integer value of x.

43. Answers may vary. One possibility follows:
$$f(x) = \begin{cases} -x & \text{if } -1 \le x \le 0 \\ \dfrac{1}{2}x & \text{if } 0 < x \le 2 \end{cases}$$

45. Answers may vary. One possibility follows:
$$f(x) = \begin{cases} -x & \text{if } x \le 0 \\ -x+2 & \text{if } 0 < x \le 2 \end{cases}$$

47. a. $f(1.2) = \text{int}(2(1.2)) = \text{int}(2.4) = 2$

b. $f(1.6) = \text{int}(2(1.6)) = \text{int}(3.2) = 3$

c. $f(-1.8) = \text{int}(2(-1.8)) = \text{int}(-3.6) = -4$

49. a.

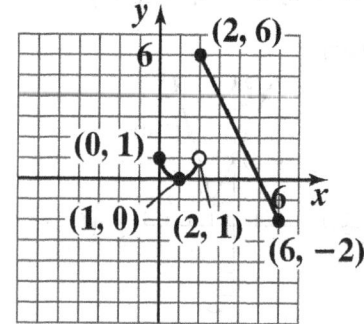

b. The domain is $[0, 6]$.

c. Absolute max: $f(2) = 6$
Absolute min: $f(6) = -2$

51. $C = \begin{cases} 34.99 & \text{if } 0 < x \le 3 \\ 15x - 10.01 & \text{if } x > 3 \end{cases}$

a. $C(2) = \$34.99$

b. $C(5) = 15(5) - 10.01 = \$64.99$

c. $C(13) = 15(13) - 10.01 = \$184.99$

53. a. Charge for 20 therms:
$C = 19.50 + 0.91686(20) + 0.348(20)$
$\quad = \$44.80$

b. Charge for 150 therms:
$C = 19.50 + 0.91686(30) + 0.348(30)$
$\quad\quad + 0.5922(120)$
$\quad = \$128.51$

c. For $0 \le x \le 30$:
$C = 19.50 + 0.91686x + 0.348x$
$\quad = 1.26486x + 19.50$

For $x > 30$:
$C = 19.50 + 0.91686(30) + 0.5922(x - 30)$
$\quad\quad + 0.348(30)$
$\quad = 19.50 + 27.5058 + 0.5922x - 17.766$
$\quad\quad + 10.44$
$\quad = 0.5922x + 39.6798$

The monthly charge function:
$C = \begin{cases} 1.26486x + 19.50 & \text{for } 0 \le x \le 30 \\ 0.5922x + 39.6798 & \text{for } x > 30 \end{cases}$

d. Graph:

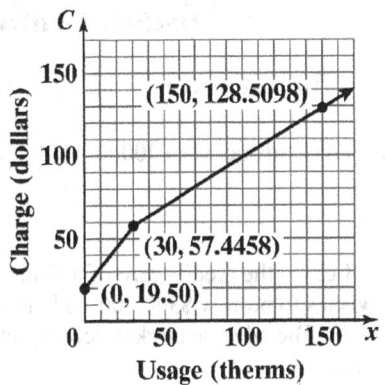

55. For schedule X:

$f(x) = \begin{cases} 0.10x & \text{if } 0 < x \le 9225 \\ 922.50 + 0.15(x - 9225) & \text{if } 9225 < x \le 37,450 \\ 5156.25 + 0.25(x - 37,450) & \text{if } 37,450 < x \le 90,750 \\ 18,481.25 + 0.28(x - 90,750) & \text{if } 90,750 < x \le 189,300 \\ 46,075.25 + 0.33(x - 189,300) & \text{if } 189,300 < x \le 411,500 \\ 119,401.25 + 0.35(x - 411,500) & \text{if } 411,500 < x \le 413,200 \\ 119,996.25 + 0.396(x - 413,200) & \text{if } x > 413,200 \end{cases}$

57. a. Let x represent the number of miles and C be the cost of transportation.

$C(x) = \begin{cases} 0.50x & \text{if } 0 \le x \le 100 \\ 0.50(100) + 0.40(x - 100) & \text{if } 100 < x \le 400 \\ 0.50(100) + 0.40(300) + 0.25(x - 400) & \text{if } 400 < x \le 800 \\ 0.50(100) + 0.40(300) + 0.25(400) + 0(x - 800) & \text{if } 800 < x \le 960 \end{cases}$

$C(x) = \begin{cases} 0.50x & \text{if } 0 \le x \le 100 \\ 10 + 0.40x & \text{if } 100 < x \le 400 \\ 70 + 0.25x & \text{if } 400 < x \le 800 \\ 270 & \text{if } 800 < x \le 960 \end{cases}$

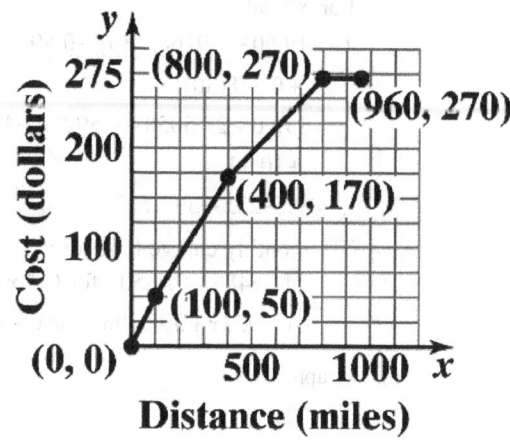

b. For hauls between 100 and 400 miles the cost is: $C(x) = 10 + 0.40x$.

c. For hauls between 400 and 800 miles the cost is: $C(x) = 70 + 0.25x$.

59. a. Let s = the credit score of an individual who wishes to borrow \$300,000 with an 80% LTV ratio. The adverse market delivery charge is given by

$$C(s) = \begin{cases} 9000 & \text{if } s \le 659 \\ 7500 & \text{if } 660 \le s \le 679 \\ 5250 & \text{if } 680 \le s \le 699 \\ 3000 & \text{if } 700 \le s \le 719 \\ 1500 & \text{if } 720 \le s \le 739 \\ 750 & \text{if } s \ge 740 \end{cases}$$

b. 725 is between 720 and 739 so the charge would be \$1500.

c. 670 is between 660 and 679 so the charge would be \$7500.

61. a. $W = 10°C$

b. $W = 33 - \dfrac{(10.45 + 10\sqrt{5} - 5)(33 - 10)}{22.04} \approx 4°C$

c. $W = 33 - \dfrac{(10.45 + 10\sqrt{15} - 15)(33 - 10)}{22.04} \approx -3°C$

d. $W = 33 - 1.5958(33 - 10) = -4°C$

e. When $0 \le v < 1.79$, the wind speed is so small that there is no effect on the temperature.

f. When the wind speed exceeds 20, the wind chill depends only on the air temperature.

63. Let x = the number of ounces and $C(x)$ = the postage due.

For $0 < x \le 1$: $C(x) = \$0.98$

For $1 < x \le 2$: $C(x) = 0.98 + 0.22 = \$1.20$

For $2 < x \le 3$: $C(x) = 0.98 + 2(0.22) = \$1.42$

For $3 < x \le 4$: $C(x) = 0.98 + 3(0.22) = \$1.64$

$\vdots$

For $12 < x \le 13$: $C(x) - 0.98 + 12(0.22) = \$3.62$

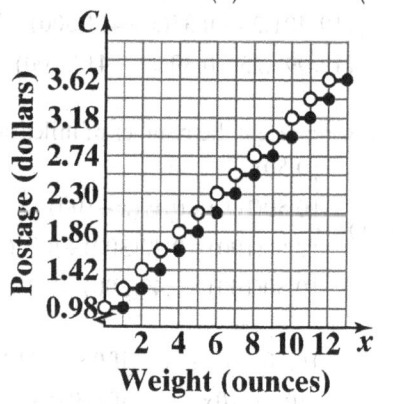

150

65. Each graph is that of $y = x^2$, but shifted horizontally.

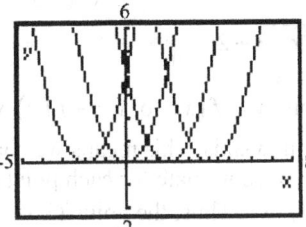

If $y = (x-k)^2$, $k > 0$, the shift is to the right k units; if $y = (x+k)^2$, $k > 0$, the shift is to the left k units. The graph of $y = (x+4)^2$ is the same as the graph of $y = x^2$, but shifted to the left 4 units. The graph of $y = (x-5)^2$ is the graph of $y = x^2$, but shifted to the right 5 units.

67. The graph of $y = -x^2$ is the reflection of the graph of $y = x^2$ about the x-axis.

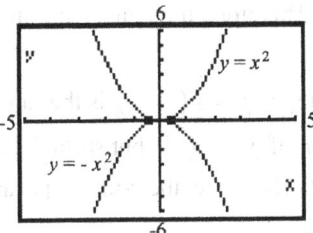

The graph of $y = -|x|$ is the reflection of the graph of $y = |x|$ about the x-axis.

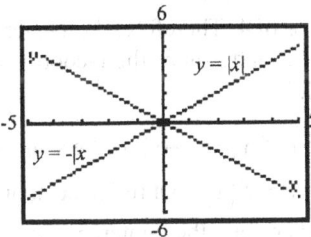

Multiplying a function by -1 causes the graph to be a reflection about the x-axis of the original function's graph.

69. The graph of $y = (x-1)^3 + 2$ is a shifting of the graph of $y = x^3$ one unit to the right and two units up. Yes, the result could be predicted.

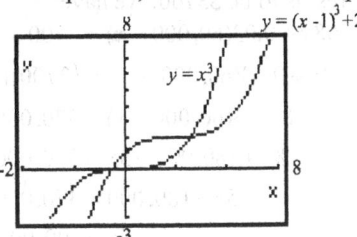

71. The graphs of $y = x^n$, n a positive odd integer, all have the same general shape. All go through the points $(-1, -1)$, $(0, 0)$, and $(1, 1)$. As n increases, the graph of the function increases at a greater rate for $|x| > 1$ and is flatter around 0 for $|x| < 1$.

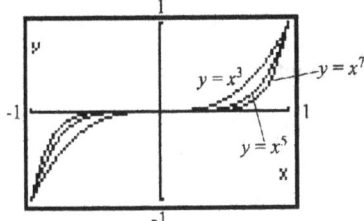

73. For $0 < x < 1$, the graph of $y = x^r$, r rational and $r > 0$, flattens down toward the x-axis as r gets bigger. For $x > 1$, the graph of $y = x^r$ increases at a greater rate as r gets bigger.

75.
$$x^2 + y^2 = 6y + 16$$
$$x^2 + y^2 - 6y = 16$$
$$x^2 + (y^2 - 6y + 9) = 16 + 9$$
$$x^2 + (y-3)^2 = 5^2$$

Center (h,k): (0, 3); Radius = 5

77. Let x represent the amount of money invested in a mutual fund. Then $60,000 - x$ represents the amount of money invested in CD's. Since the total interest is to be $3700, we have:

$$0.08x + 0.03(60,000 - x) = 3700$$
$$(100)(0.08x + 0.03(60,000 - x)) = (3700)(100)$$
$$8x + 3(60,000 - x) = 370,000$$
$$8x + 180,000 - 3x = 370,000$$
$$5x + 180,000 = 370,000$$
$$5x = 190,000$$
$$x = 38,000$$

$38,000 should be invested in a mutual fund at 8% and $22,000 should be invested in CD's at 3%.

Section 3.5

1. horizontal; right

3. False

5. d

7. B

9. H

11. I

13. L

15. F

17. G

19. $y = (x-4)^3$

21. $y = x^3 + 4$

23. $y = (-x)^3 = -x^3$

25. $y = 4x^3$

27. (1) $y = \sqrt{x} + 2$

   (2) $y = -(\sqrt{x} + 2)$

   (3) $y = -(\sqrt{-x} + 2) = -\sqrt{-x} - 2$

29. (1) $y = -\sqrt{x}$

   (2) $y = -\sqrt{x} + 2$

   (3) $y = -\sqrt{x+3} + 2$

31. (c); To go from $y = f(x)$ to $y = -f(x)$ we reflect about the x-axis. This means we change the sign of the y-coordinate for each point on the graph of $y = f(x)$. Thus, the point $(3, 6)$ would become $(3, -6)$.

33. (c); To go from $y = f(x)$ to $y = 2f(x)$, we stretch vertically by a factor of 2. Multiply the y-coordinate of each point on the graph of $y = f(x)$ by 2. Thus, the point $(1, 3)$ would become $(1, 6)$.

35. a. The graph of $y = f(x+2)$ is the same as the graph of $y = f(x)$, but shifted 2 units to the left. Therefore, the x-intercepts are -7 and 1.

   b. The graph of $y = f(x-2)$ is the same as the graph of $y = f(x)$, but shifted 2 units to the right. Therefore, the x-intercepts are -3 and 5.

   c. The graph of $y = 4f(x)$ is the same as the graph of $y = f(x)$, but stretched vertically by a factor of 4. Therefore, the x-intercepts are still -5 and 3 since the y-coordinate of each is 0.

   d. The graph of $y = f(-x)$ is the same as the graph of $y = f(x)$, but reflected about the y-axis. Therefore, the x-intercepts are 5 and -3.

37. a. The graph of $y = f(x+2)$ is the same as the graph of $y = f(x)$, but shifted 2 units to the left. Therefore, the graph of $f(x+2)$ is increasing on the interval $[-3, 3]$.

   b. The graph of $y = f(x-5)$ is the same as the graph of $y = f(x)$, but shifted 5 units to the right. Therefore, the graph of $f(x-5)$ is increasing on the interval $[4,10]$.

c. The graph of $y=-f(x)$ is the same as the graph of $y=f(x)$, but reflected about the x-axis. Therefore, we can say that the graph of $y=-f(x)$ must be *decreasing* on the interval $[-1,5]$.

d. The graph of $y=f(-x)$ is the same as the graph of $y=f(x)$, but reflected about the y-axis. Therefore, we can say that the graph of $y=f(-x)$ must be *decreasing* on the interval $[-5,1]$.

39. $f(x)=x^2-1$

Using the graph of $y=x^2$, vertically shift downward 1 unit.

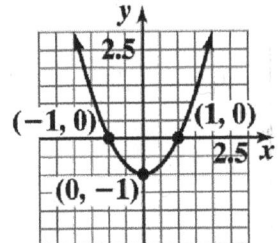

The domain is $(-\infty,\infty)$ and the range is $[-1,\infty)$.

41. $g(x)=x^3+1$

Using the graph of $y=x^3$, vertically shift upward 1 unit.

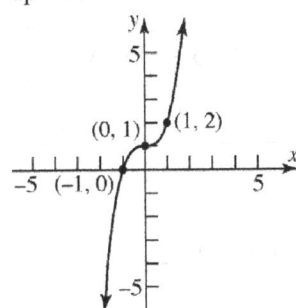

The domain is $(-\infty,\infty)$ and the range is $(-\infty,\infty)$.

43. $h(x)=\sqrt{x+2}$

Using the graph of $y=\sqrt{x}$, horizontally shift to the left 2 units.

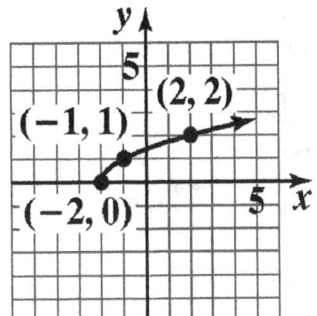

The domain is $[-2,\infty)$ and the range is $[0,\infty)$.

45. $f(x)=(x-1)^3+2$

Using the graph of $y=x^3$, horizontally shift to the right 1 unit $\left[y=(x-1)^3\right]$, then vertically shift up 2 units $\left[y=(x-1)^3+2\right]$.

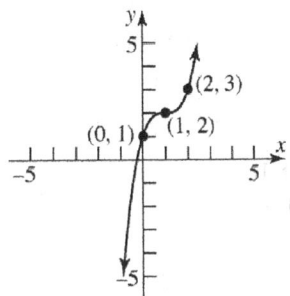

The domain is $(-\infty,\infty)$ and the range is $(-\infty,\infty)$.

47. $g(x)=4\sqrt{x}$

Using the graph of $y=\sqrt{x}$, vertically stretch by a factor of 4.

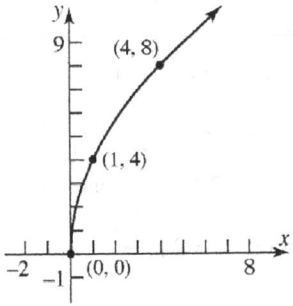

The domain is $[0,\infty)$ and the range is $[0,\infty)$.

49. $f(x) = -\sqrt[3]{x}$

Using the graph of $y = \sqrt[3]{x}$, reflect the graph about the x axis.

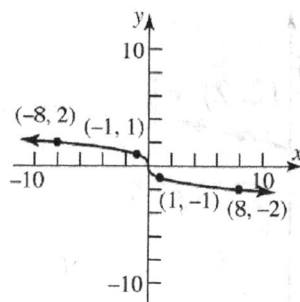

The domain is $(-\infty, \infty)$ and the range is $(-\infty, \infty)$.

51. $f(x) = 2(x+1)^2 - 3$

Using the graph of $y = x^2$, horizontally shift to the left 1 unit $\left[y = (x+1)^2 \right]$, vertically stretch by a factor of 2 $\left[y = 2(x+1)^2 \right]$, and then vertically shift downward 3 units $\left[y = 2(x+1)^2 - 3 \right]$.

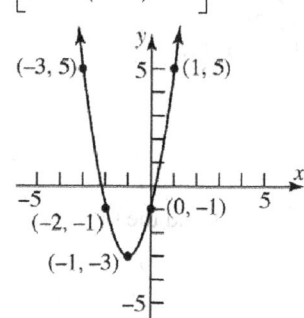

The domain is $(-\infty, \infty)$ and the range is $[-3, \infty)$.

53. $g(x) = 2\sqrt{x-2} + 1$

Using the graph of $y = \sqrt{x}$, horizontally shift to the right 2 units $\left[y = \sqrt{x-2} \right]$, vertically stretch by a factor of 2 $\left[y = 2\sqrt{x-2} \right]$, and vertically shift upward 1 unit $\left[y = 2\sqrt{x-2} + 1 \right]$.

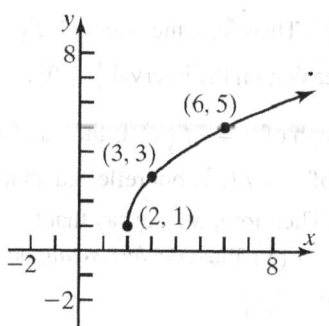

The domain is $[2, \infty)$ and the range is $[1, \infty)$.

55. $h(x) = \sqrt{-x} - 2$

Using the graph of $y = \sqrt{x}$, reflect the graph about the y-axis $\left[y = \sqrt{-x} \right]$ and vertically shift downward 2 units $\left[y = \sqrt{-x} - 2 \right]$.

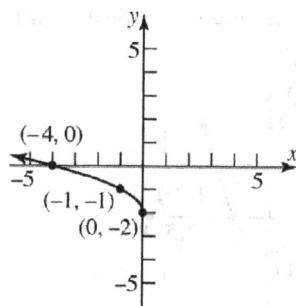

The domain is $(-\infty, 0]$ and the range is $[-2, \infty)$.

57. $f(x) = -(x+1)^3 - 1$

Using the graph of $y = x^3$, horizontally shift to the left 1 unit $\left[y = (x+1)^3 \right]$, reflect the graph about the x-axis $\left[y = -(x+1)^3 \right]$, and vertically shift downward 1 unit $\left[y = -(x+1)^3 - 1 \right]$.

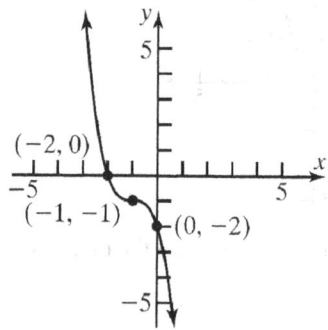

The domain is $(-\infty, \infty)$ and the range is $(-\infty, \infty)$.

59. $g(x) = 2|1-x| = 2|-(-1+x)| = 2|x-1|$

Using the graph of $y = |x|$, horizontally shift to the right 1 unit $\left[y = |x-1|\right]$, and vertically stretch by a factor or 2 $\left[y = 2|x-1|\right]$.

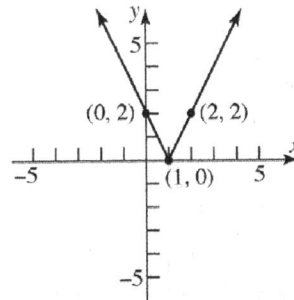

The domain is $(-\infty, \infty)$ and the range is $[0, \infty)$.

61. $h(x) = \dfrac{1}{2x}$

Using the graph of $y = \dfrac{1}{x}$, vertically compress by a factor of $\dfrac{1}{2}$.

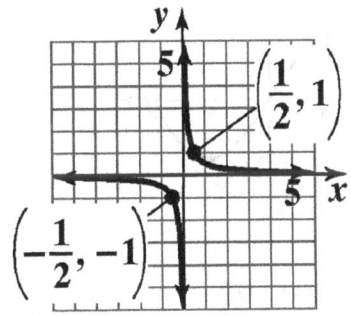

The domain is $(-\infty, 0) \cup (0, \infty)$ and the range is $(-\infty, 0) \cup (0, \infty)$.

63. a. $F(x) = f(x) + 3$

Shift up 3 units.

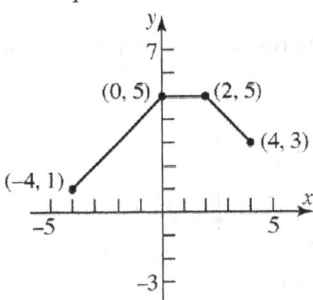

b. $G(x) = f(x+2)$

Shift left 2 units.

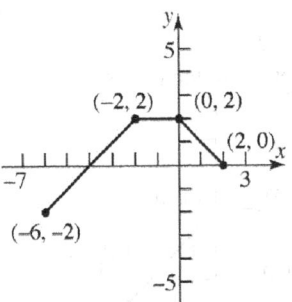

c. $P(x) = -f(x)$

Reflect about the x-axis.

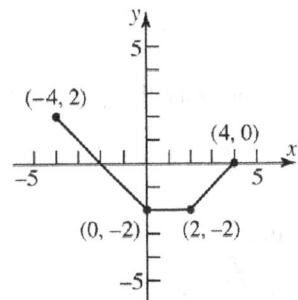

d. $H(x) = f(x+1) - 2$

Shift left 1 unit and shift down 2 units.

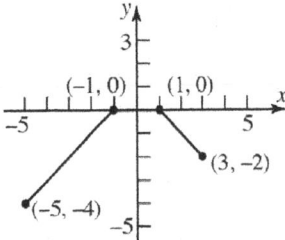

155

e. $Q(x) = \frac{1}{2} f(x)$

Compress vertically by a factor of $\frac{1}{2}$.

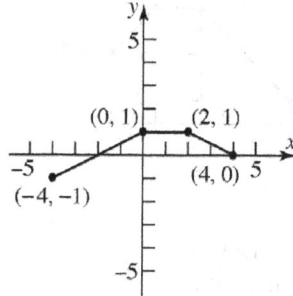

f. $g(x) = f(-x)$

Reflect about the y-axis.

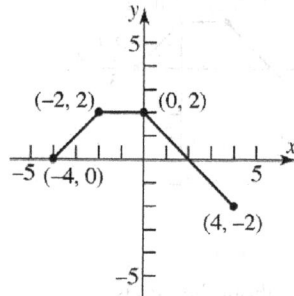

g. $h(x) = f(2x)$

Compress horizontally by a factor of $\frac{1}{2}$.

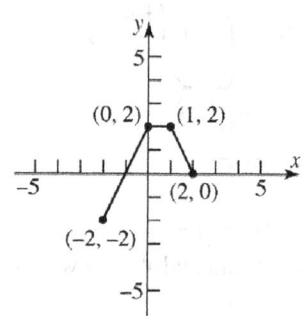

65. a. $F(x) = f(x) + 3$

Shift up 3 units.

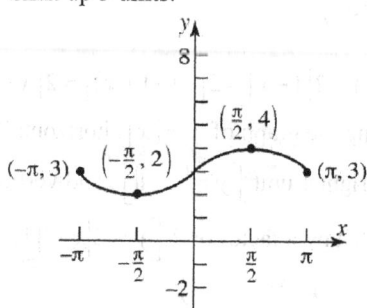

b. $G(x) = f(x+2)$

Shift left 2 units.

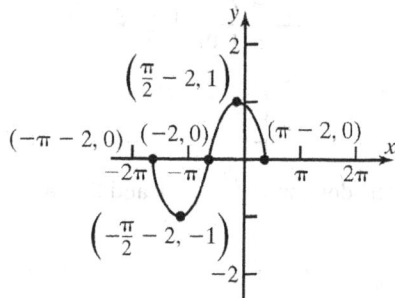

c. $P(x) = -f(x)$

Reflect about the x-axis.

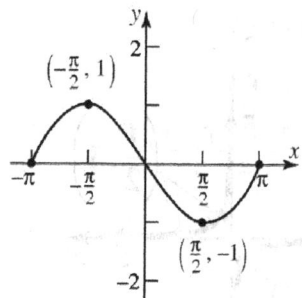

d. $H(x) = f(x+1) - 2$

Shift left 1 unit and shift down 2 units.

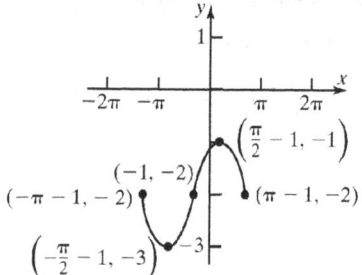

156

e. $Q(x) = \frac{1}{2} f(x)$

Compress vertically by a factor of $\frac{1}{2}$.

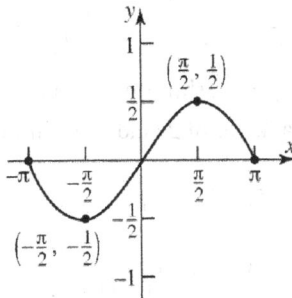

f. $g(x) = f(-x)$

Reflect about the y-axis.

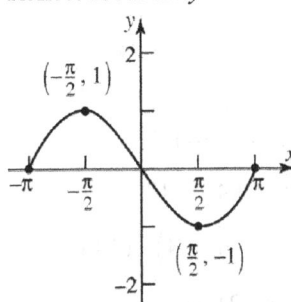

g. $h(x) = f(2x)$

Compress horizontally by a factor of $\frac{1}{2}$.

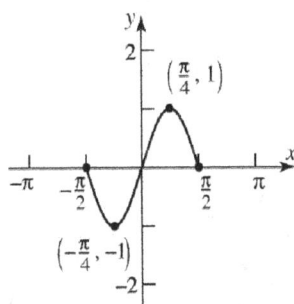

67. a. $f(x) = x^3 - 9x$, $-4 \le x \le 4$

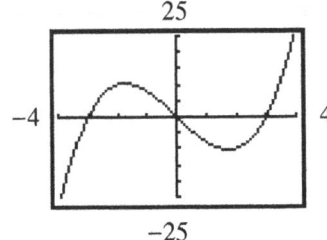

b. $0 = x^3 - 9x$

$0 = x(x^2 - 9)$

$0 = x(x-3)(x+3)$

$x = 0$, $x = 3$, $x = -3$

The x-intercepts are -3, 0, and 3.

c. The local minimum can be found by using the Minimum feature from the CALC menu on a TI-84 Plus graphing calculator.

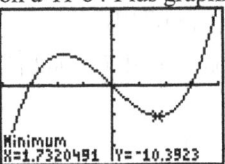

The local minimum is approximately -10.39 when $x \approx 1.73$.

The local maximum can be found by using the Maximum feature from the CALC menu on a TI-84 Plus graphing calculator.

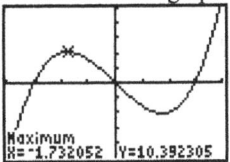

The local maximum is approximately 10.39 when $x \approx -1.73$.

d. From the graph above, we see that f is initially increasing, decreasing between the two extrema, and increasing again at the end. Thus, f is increasing on the interval $[-4, -1.73]$ and on the interval $[1.73, 4]$. It is decreasing on the interval $[-1.73, 1.73]$.

e. $y = f(x+2)$ involves a shift to the left 2 units so we subtract 2 from each x-value. Therefore, we have the following:
x-intercepts: -5, -2, and 1
local minimum: -10.39 when $x \approx -0.27$
local maximum: 10.39 when $x \approx -3.73$
increasing on $[-6, -3.73]$ and $[-0.27, 2]$;
decreasing on $[-3.73, -0.27]$.

f. $y = 2f(x)$ involves a vertical stretch by a factor of 2 so we multiply each y-value by 2. Therefore, we have the following:
x-intercepts: -3, 0, and 3
local minimum: -20.78 when $x \approx 1.73$

local maximum: 20.78 when $x \approx -1.73$
increasing on $[-4, -1.73]$ and $[1.73, 4]$;

decreasing on $[-1.73, 1.73]$.

g. $y = f(-x)$ involves a reflection about the
y-axis. Therefore, we have the following:
x-intercepts: -3, 0, and 3
local minimum: -10.39 when $x \approx -1.73$
local maximum: 10.39 when $x \approx 1.73$
increasing on $[-1.73, 1.73]$;

decreasing on $[-4, -1.73]$ and $[1.73, 4]$.

69. $f(x) = x^2 + 2x$

$f(x) = (x^2 + 2x + 1) - 1$

$f(x) = (x+1)^2 - 1$

Using $f(x) = x^2$, shift left 1 unit and shift down
1 unit.

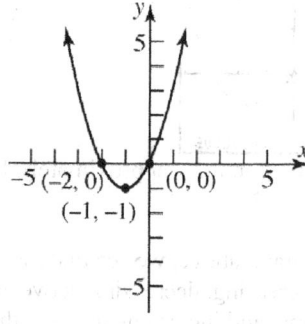

71. $f(x) = x^2 - 8x + 1$

$f(x) = (x^2 - 8x + 16) + 1 - 16$

$f(x) = (x-4)^2 - 15$

Using $f(x) = x^2$, shift right 4 units and shift
down 15 units.

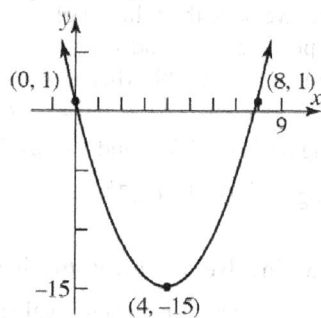

73. $f(x) = 2x^2 - 12x + 19$

$= 2(x^2 - 6x) + 19$

$= 2(x^2 - 6x + 9) + 19 - 18$

$= 2(x-3)^2 + 1$

Using $f(x) = x^2$, shift right 3 units, vertically
stretch by a factor of 2, and then shift up 1 unit.

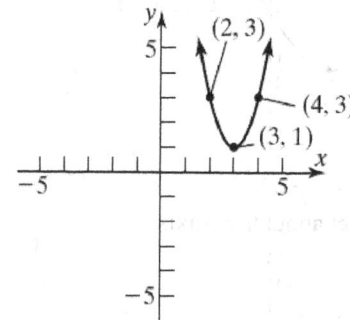

75. $f(x) = -3x^2 - 12x - 17$

$= -3(x^2 + 4x) - 17$

$= -3(x^2 + 4x + 4) - 17 + 12$

$= -3(x+2)^2 - 5$

Using $f(x) = x^2$, shift left 2 units, stretch
vertically by a factor of 3, reflect about the x-
axis, and shift down 5 units.

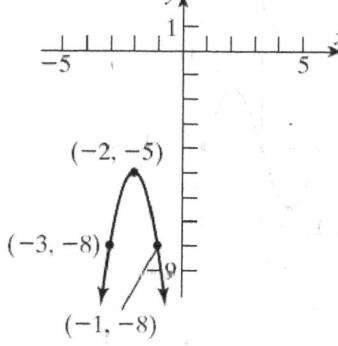

77. a. $y = |f(x)|$

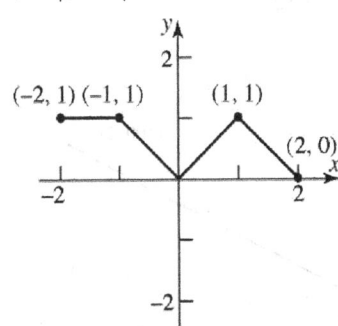

b. $y = f(|x|)$

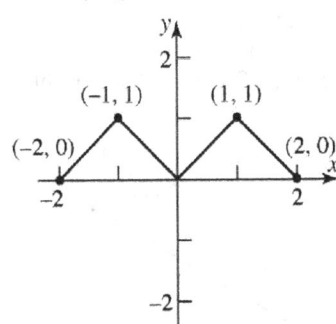

79. a. The graph of $y = f(x+3) - 5$ is the graph of $y = f(x)$ but shifted left 3 units and down 5 units. Thus, the point $(1,3)$ becomes the point $(-2,-2)$.

b. The graph of $y = -2f(x-2) + 1$ is the graph of $y = f(x)$ but shifted right 2 units, stretched vertically by a factor of 2, reflected about the x-axis, and shifted up 1 unit. Thus, the point $(1,3)$ becomes the point $(3,-5)$.

c. The graph of $y = f(2x+3)$ is the graph of $y = f(x)$ but shifted left 3 units and horizontally compressed by a factor of 2. Thus, the point $(1,3)$ becomes the point $(-1,3)$.

81. a. $f(x) = \text{int}(-x)$

Reflect the graph of $y = \text{int}(x)$ about the y-axis.

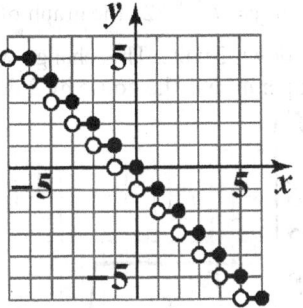

b. $g(x) = -\text{int}(x)$

Reflect the graph of $y = \text{int}(x)$ about the x-axis.

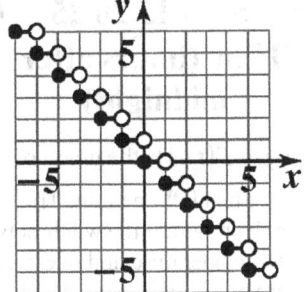

83. a. $f(x) = |x-3| - 3$

Using the graph of $y = |x|$, horizontally shift to the right 3 units $\left[y = |x-3| \right]$ and vertically shift downward 3 units $\left[y = |x-3| - 3 \right]$.

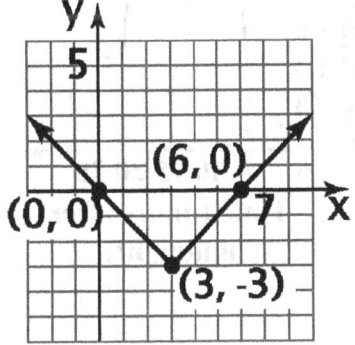

b. $A = \dfrac{1}{2}bh$

$= \dfrac{1}{2}(6)(3) = 9$

The area is 9 square units.

159

85. a. From the graph, the thermostat is set at $72°F$ during the daytime hours. The thermostat appears to be set at $65°F$ overnight.

b. To graph $y = T(t) - 2$, the graph of $T(t)$ is shifted down 2 units. This change will lower the temperature in the house by 2 degrees.

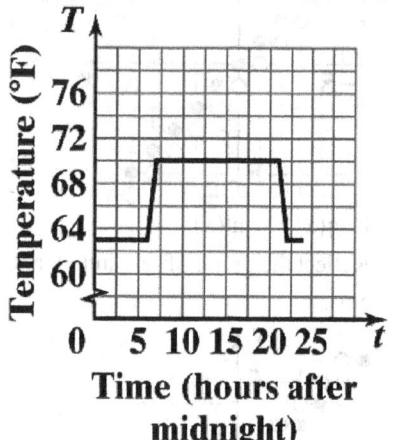

c. To graph $y = T(t+1)$, the graph of $T(t)$ should be shifted left one unit. This change will cause the program to switch between the daytime temperature and overnight temperature one hour sooner. The home will begin warming up at 5am instead of 6am and will begin cooling down at 8pm instead of 9pm.

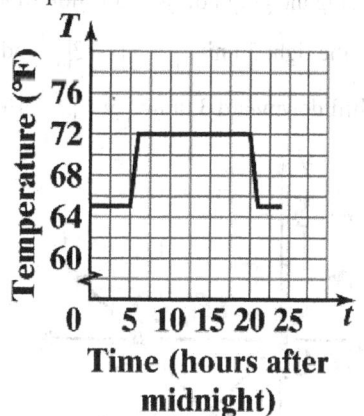

87. $F = \dfrac{9}{5}C + 32$

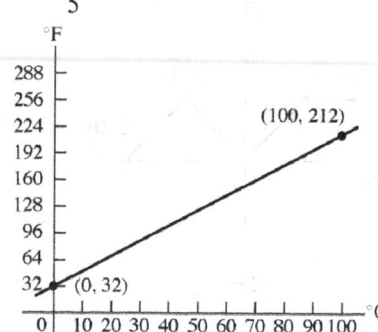

$$F = \frac{9}{5}(K - 273) + 32$$

Shift the graph 273 units to the right.

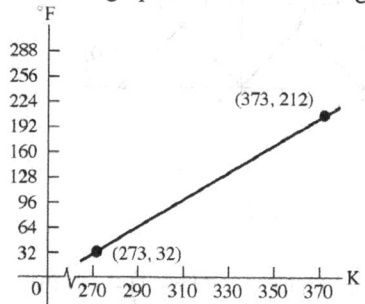

89. $y = (x - c)^2$

If $c = 0$, $y = x^2$.

If $c = 3$, $y = (x-3)^2$; shift right 3 units.

If $c = -2$, $y = (x+2)^2$; shift left 2 units.

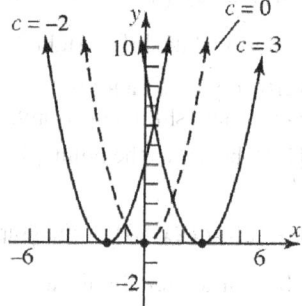

91. The graph of $y = 4f(x)$ is a vertical stretch of the graph of f by a factor of 4, while the graph of $y = f(4x)$ is a horizontal compression of the graph of f by a factor of $\frac{1}{4}$.

93. The graph of $y = \sqrt{-x}$ is the graph of $y = \sqrt{x}$ but reflected about the y-axis. Therefore, our region is simply rotated about the y-axis and does not change shape. Instead of the region being bounded on the right by $x = 4$, it is bounded on the left by $x = -4$. Thus, the area of the second region would also be $\dfrac{16}{3}$ square units.

95. The domain of $g(x) = \sqrt{x}$ is $[0, \infty)$. The graph of $g(x - k)$ is the graph of g shifted k units to the right, so the domaine of g is $[k, \infty)$.

97.
$$\frac{\left(x^{-2}y^3\right)^4}{\left(x^2 y^{-5}\right)^{-2}} = \frac{\left(x^{-8}y^{12}\right)}{\left(x^{-4}y^{10}\right)}$$
$$= \frac{\left(x^4 y^{12}\right)}{\left(x^8 y^{10}\right)} = \frac{y^2}{x^4}$$

99. $y^2 = x + 4$

x-intercepts: y-intercepts:

$(0)^2 = x + 4$ $y^2 = 0 + 4$

$0 = x + 4$ $y^2 = 4$

$x = -4$ $y = \pm 2$

The intercepts are $(-4, 0)$, $(0, -2)$ and $(0, 2)$.

Test x-axis symmetry: Let $y = -y$

$(-y)^2 = x + 4$
$y^2 = x + 4$ same

Test y-axis symmetry: Let $x = -x$

$y^2 = -x + 4$ different

Test origin symmetry: Let $x = -x$ and $y = -y$.

$(-y)^2 = -x + 4$
$y^2 = -x + 4$ different

Therefore, the graph will have x-axis symmetry.

Section 3.6

1. a. The distance d from P to the origin is $d = \sqrt{x^2 + y^2}$. Since P is a point on the graph of $y = x^2 - 8$, we have:
$$d(x) = \sqrt{x^2 + (x^2 - 8)^2} = \sqrt{x^4 - 15x^2 + 64}$$

b. $d(0) = \sqrt{0^4 - 15(0)^2 + 64} = \sqrt{64} = 8$

c. $d(1) = \sqrt{(1)^4 - 15(1)^2 + 64}$
$= \sqrt{1 - 15 + 64} = \sqrt{50} = 5\sqrt{2} \approx 7.07$

d.

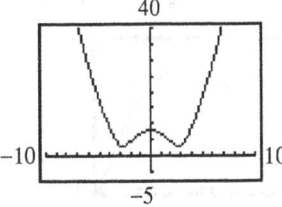

e. d is smallest when $x \approx -2.74$ or when $x \approx 2.74$.

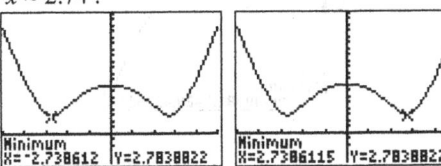

3. a. The distance d from P to the point $(1, 0)$ is $d = \sqrt{(x - 1)^2 + y^2}$. Since P is a point on the graph of $y = \sqrt{x}$, we have:
$$d(x) = \sqrt{(x - 1)^2 + \left(\sqrt{x}\right)^2} = \sqrt{x^2 - x + 1}$$
where $x \geq 0$.

b.

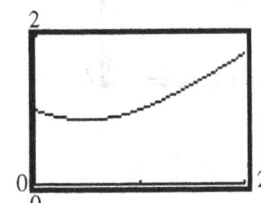

c. d is smallest when $x = \frac{1}{2}$.

161

5. By definition, a triangle has area

$A = \dfrac{1}{2}bh$, b = base, h = height. From the figure, we know that $b = x$ and $h = y$. Expressing the area of the triangle as a function of x, we have:

$A(x) = \dfrac{1}{2}xy = \dfrac{1}{2}x\left(x^3\right) = \dfrac{1}{2}x^4$.

7. a. $A(x) = xy = x\left(16 - x^2\right)$

b. Domain: $\{x \mid 0 < x < 4\}$

c. The area is largest when $x \approx 2.31$.

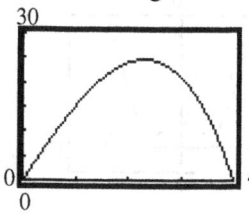

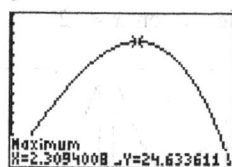

9. a. In Quadrant I, $x^2 + y^2 = 4 \rightarrow y = \sqrt{4 - x^2}$

$A(x) = (2x)(2y) = 4x\sqrt{4 - x^2}$

b. $p(x) = 2(2x) + 2(2y) = 4x + 4\sqrt{4 - x^2}$

c. Graphing the area equation:

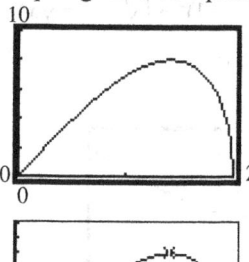

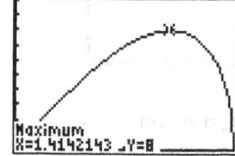

The area is largest when $x \approx 1.41$.

d. Graphing the perimeter equation:

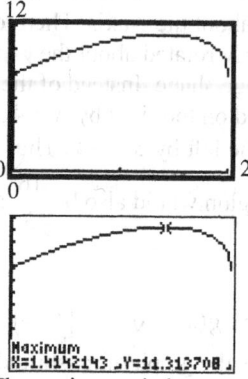

The perimeter is largest when $x \approx 1.41$.

11. a. C = circumference, A = total area,
r = radius, x = side of square

$C = 2\pi r = 10 - 4x \;\Rightarrow\; r = \dfrac{5 - 2x}{\pi}$

Total Area = area$_{\text{square}}$ + area$_{\text{circle}}$ = $x^2 + \pi r^2$

$A(x) = x^2 + \pi\left(\dfrac{5 - 2x}{\pi}\right)^2 = x^2 + \dfrac{25 - 20x + 4x^2}{\pi}$

b. Since the lengths must be positive, we have:
$10 - 4x > 0$ and $x > 0$
$-4x > -10$ and $x > 0$
$x < 2.5$ and $x > 0$
Domain: $\{x \mid 0 < x < 2.5\}$

c. The total area is smallest when $x \approx 1.40$ meters.

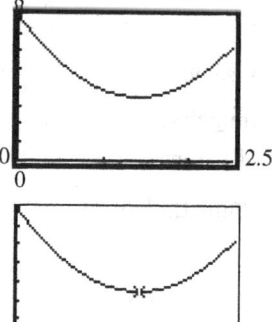

13. a. Since the wire of length x is bent into a circle, the circumference is x. Therefore, $C(x) = x$.

b. Since $C = x = 2\pi r$, $r = \dfrac{x}{2\pi}$.

$$A(x) = \pi r^2 = \pi \left(\frac{x}{2\pi}\right)^2 = \frac{x^2}{4\pi}.$$

15. a. A = area, r = radius; diameter = $2r$

$$A(r) = (2r)(r) = 2r^2$$

b. p = perimeter

$$p(r) = 2(2r) + 2r = 6r$$

17. Area of the equilateral triangle

$$A = \frac{1}{2} x \cdot \frac{\sqrt{3}}{2} x = \frac{\sqrt{3}}{4} x^2$$

From problem 16, we have $r^2 = \dfrac{x^2}{3}$.

Area inside the circle, but outside the triangle:

$$A(x) = \pi r^2 - \frac{\sqrt{3}}{4} x^2$$

$$= \pi \frac{x^2}{3} - \frac{\sqrt{3}}{4} x^2 = \left(\frac{\pi}{3} - \frac{\sqrt{3}}{4}\right) x^2$$

19. a. $d^2 = d_1^2 + d_2^2$

$$d^2 = (2 - 30t)^2 + (3 - 40t)^2$$

$$d(t) = \sqrt{(2 - 30t)^2 + (3 - 40t)^2}$$

$$= \sqrt{4 - 120t + 900t^2 + 9 - 240t + 1600t^2}$$

$$= \sqrt{2500t^2 - 360t + 13}$$

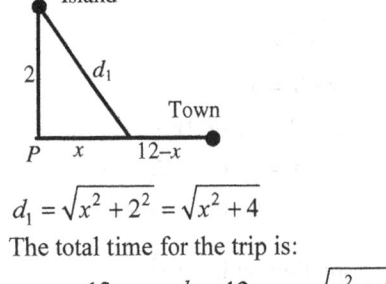

$d_2 = 3 - 40t$

$d_1 = 2 - 30t$

d

b. The distance is smallest at $t \approx 0.07$ hours.

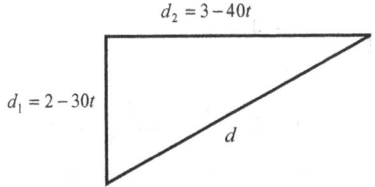
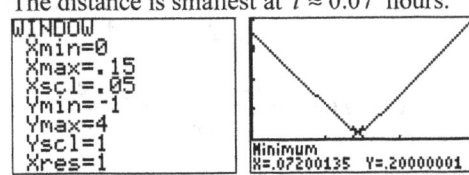

21. r = radius of cylinder, h = height of cylinder, V = volume of cylinder

By similar triangles: $\dfrac{H}{R} = \dfrac{H - h}{r}$

$$Hr = R(H - h)$$

$$Hr = RH - Rh$$

$$Rh = RH - Hr$$

$$h = \frac{RH - Hr}{R} = \frac{H(R - r)}{R}$$

$$V = \pi r^2 h = \pi r^2 \left(\frac{H(R - r)}{R}\right) = \frac{\pi H(R - r) r^2}{R}$$

23. a. The time on the boat is given by $\dfrac{d_1}{3}$. The time on land is given by $\dfrac{12 - x}{5}$.

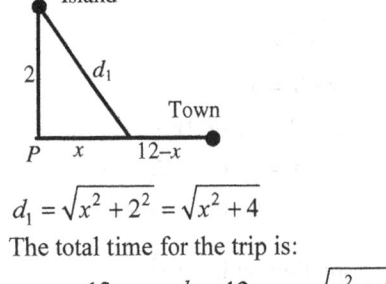

Island

2 d_1

Town

P x $12 - x$

$$d_1 = \sqrt{x^2 + 2^2} = \sqrt{x^2 + 4}$$

The total time for the trip is:

$$T(x) = \frac{12 - x}{5} + \frac{d_1}{3} = \frac{12 - x}{5} + \frac{\sqrt{x^2 + 4}}{3}$$

b. Domain: $\{x \mid 0 \le x \le 12\}$

c. $T(4) = \dfrac{12 - 4}{5} + \dfrac{\sqrt{4^2 + 4}}{3}$

$$= \frac{8}{5} + \frac{\sqrt{20}}{3} \approx 3.09 \text{ hours}$$

d. $T(8) = \dfrac{12 - 8}{5} + \dfrac{\sqrt{8^2 + 4}}{3}$

$$= \frac{4}{5} + \frac{\sqrt{68}}{3} \approx 3.55 \text{ hours}$$

25. a. length = $24 - 2x$; width = $24 - 2x$; height = x

$$V(x) = x(24 - 2x)(24 - 2x) = x(24 - 2x)^2$$

b. $V(3) = 3(24 - 2(3))^2 = 3(18)^2$

$$= 3(324) = 972 \text{ in}^3.$$

c. $V(10) = 10(24 - 2(10))^2 = 10(4)^2$
$= 10(16) = 160 \text{ in}^3.$

d. $y_1 = x(24 - 2x)^2$

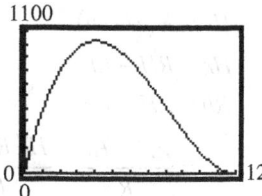

Use MAXIMUM.

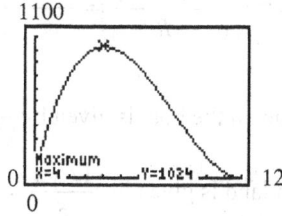

The volume is largest when $x = 4$ inches.

27. $|2x - 3| - 5 = -2$
$|2x - 3| = 3$
$2x - 3 = -3 \text{ or } 2x - 3 = 3$
$2x = 0 \quad \text{or} \quad 2x = 6$
$x = 0 \quad \text{or} \quad x = 3$

The solution set is $\{0, 3\}$.

29. $m = \dfrac{y_2 - y_1}{x_2 - x_1} = \dfrac{6 - (-2)}{1 - 3} = \dfrac{8}{-2} = -4$

Chapter 3 Review Exercises

1. This relation represents a function.
Domain = $\{-1, 2, 4\}$; Range = $\{0, 3\}$.

2. This relation does not represent a function, since 4 is paired with two different values.

3. $f(x) = \dfrac{3x}{x^2 - 1}$

a. $f(2) = \dfrac{3(2)}{(2)^2 - 1} = \dfrac{6}{4-1} = \dfrac{6}{3} = 2$

b. $f(-2) = \dfrac{3(-2)}{(-2)^2 - 1} = \dfrac{-6}{4-1} = \dfrac{-6}{3} = -2$

c. $f(-x) = \dfrac{3(-x)}{(-x)^2 - 1} = \dfrac{-3x}{x^2 - 1}$

d. $-f(x) = -\left(\dfrac{3x}{x^2 - 1}\right) = \dfrac{-3x}{x^2 - 1}$

e. $f(x - 2) = \dfrac{3(x - 2)}{(x - 2)^2 - 1}$
$= \dfrac{3x - 6}{x^2 - 4x + 4 - 1} = \dfrac{3(x - 2)}{x^2 - 4x + 3}$

f. $f(2x) = \dfrac{3(2x)}{(2x)^2 - 1} = \dfrac{6x}{4x^2 - 1}$

4. $f(x) = \sqrt{x^2 - 4}$

a. $f(2) = \sqrt{2^2 - 4} = \sqrt{4 - 4} = \sqrt{0} = 0$

b. $f(-2) = \sqrt{(-2)^2 - 4} = \sqrt{4 - 4} = \sqrt{0} = 0$

c. $f(-x) = \sqrt{(-x)^2 - 4} = \sqrt{x^2 - 4}$

d. $-f(x) = -\sqrt{x^2 - 4}$

e. $f(x - 2) = \sqrt{(x - 2)^2 - 4}$
$= \sqrt{x^2 - 4x + 4 - 4}$
$= \sqrt{x^2 - 4x}$

f. $f(2x) = \sqrt{(2x)^2 - 4} = \sqrt{4x^2 - 4}$
$= \sqrt{4(x^2 - 1)} = 2\sqrt{x^2 - 1}$

5. $f(x) = \dfrac{x^2 - 4}{x^2}$

a. $f(2) = \dfrac{2^2 - 4}{2^2} = \dfrac{4 - 4}{4} = \dfrac{0}{4} = 0$

b. $f(-2) = \dfrac{(-2)^2 - 4}{(-2)^2} = \dfrac{4 - 4}{4} = \dfrac{0}{4} = 0$

c. $f(-x) = \dfrac{(-x)^2 - 4}{(-x)^2} = \dfrac{x^2 - 4}{x^2}$

d. $-f(x) = -\left(\dfrac{x^2 - 4}{x^2}\right) = \dfrac{4 - x^2}{x^2} = -\dfrac{x^2 - 4}{x^2}$

e. $f(x-2) = \dfrac{(x-2)^2 - 4}{(x-2)^2} = \dfrac{x^2 - 4x + 4 - 4}{(x-2)^2}$

$= \dfrac{x^2 - 4x}{(x-2)^2} = \dfrac{x(x-4)}{(x-2)^2}$

f. $f(2x) = \dfrac{(2x)^2 - 4}{(2x)^2} = \dfrac{4x^2 - 4}{4x^2}$

$= \dfrac{4(x^2 - 1)}{4x^2} = \dfrac{x^2 - 1}{x^2}$

6. $f(x) = \dfrac{x}{x^2 - 9}$

The denominator cannot be zero:

$x^2 - 9 \neq 0$

$(x+3)(x-3) \neq 0$

$x \neq -3 \text{ or } 3$

Domain: $\{x \mid x \neq -3, \; x \neq 3\}$

7. $f(x) = \sqrt{2-x}$

The radicand must be non-negative:

$2 - x \geq 0$

$x \leq 2$

Domain: $\{x \mid x \leq 2\}$ or $(-\infty, 2]$

8. $g(x) = \dfrac{|x|}{x}$

The denominator cannot be zero:

$x \neq 0$

Domain: $\{x \mid x \neq 0\}$

9. $f(x) = \dfrac{x}{x^2 + 2x - 3}$

The denominator cannot be zero:

$x^2 + 2x - 3 \neq 0$

$(x+3)(x-1) \neq 0$

$x \neq -3 \text{ or } 1$

Domain: $\{x \mid x \neq -3, \; x \neq 1\}$

10. $f(x) = \dfrac{\sqrt{x+1}}{x^2 - 4}$

The denominator cannot be zero:

$x^2 - 4 \neq 0$

$(x+2)(x-2) \neq 0$

$x \neq -2 \text{ or } 2$

Also, the radicand must be non-negative:

$x + 1 \geq 0$

$x \geq -1$

Domain: $[-1, 2) \cup (2, \infty)$

11. $f(x) = \dfrac{x}{\sqrt{x+8}}$

The radicand must be non-negative and not zero:

$x + 8 > 0$

$x > -8$

Domain: $\{x \mid x > -8\}$

12. $f(x) = 2 - x \quad g(x) = 3x + 1$

$(f+g)(x) = f(x) + g(x)$

$= 2 - x + 3x + 1 = 2x + 3$

Domain: $\{x \mid x \text{ is any real number}\}$

$(f-g)(x) = f(x) - g(x)$

$= 2 - x - (3x + 1)$

$= 2 - x - 3x - 1$

$= -4x + 1$

Domain: $\{x \mid x \text{ is any real number}\}$

$(f \cdot g)(x) = f(x) \cdot g(x)$

$= (2 - x)(3x + 1)$

$= 6x + 2 - 3x^2 - x$

$= -3x^2 + 5x + 2$

Domain: $\{x \mid x \text{ is any real number}\}$

$\left(\dfrac{f}{g}\right)(x) = \dfrac{f(x)}{g(x)} = \dfrac{2-x}{3x+1}$

$3x + 1 \neq 0$

$3x \neq -1 \Rightarrow x \neq -\dfrac{1}{3}$

Domain: $\left\{x \mid x \neq -\dfrac{1}{3}\right\}$

13. $f(x) = 3x^2 + x + 1 \qquad g(x) = 3x$

$(f+g)(x) = f(x) + g(x)$

$= 3x^2 + x + 1 + 3x$

$= 3x^2 + 4x + 1$

Domain: $\{x \mid x \text{ is any real number}\}$

$(f-g)(x)=f(x)-g(x)$

$\qquad = 3x^2 + x + 1 - 3x$

$\qquad = 3x^2 - 2x + 1$

Domain: $\{x \mid x \text{ is any real number}\}$

$(f \cdot g)(x)=f(x) \cdot g(x)$

$\qquad = \left(3x^2 + x + 1\right)(3x)$

$\qquad = 9x^3 + 3x^2 + 3x$

Domain: $\{x \mid x \text{ is any real number}\}$

$\left(\dfrac{f}{g}\right)(x)=\dfrac{f(x)}{g(x)}=\dfrac{3x^2 + x + 1}{3x}$

$3x \neq 0 \Rightarrow x \neq 0$

Domain: $\{x \mid x \neq 0\}$

14. $f(x)=\dfrac{x+1}{x-1} \qquad g(x)=\dfrac{1}{x}$

$(f+g)(x)=f(x)+g(x)$

$\qquad = \dfrac{x+1}{x-1}+\dfrac{1}{x}=\dfrac{x(x+1)+1(x-1)}{x(x-1)}$

$\qquad = \dfrac{x^2 + x + x - 1}{x(x-1)}=\dfrac{x^2 + 2x - 1}{x(x-1)}$

Domain: $\{x \mid x \neq 0, x \neq 1\}$

$(f-g)(x)=f(x)-g(x)$

$\qquad = \dfrac{x+1}{x-1}-\dfrac{1}{\cdot x}=\dfrac{x(x+1)-1(x-1)}{x(x-1)}$

$\qquad = \dfrac{x^2 + x - x + 1}{x(x-1)}=\dfrac{x^2 + 1}{x(x-1)}$

Domain: $\{x \mid x \neq 0, x \neq 1\}$

$(f \cdot g)(x)=f(x) \cdot g(x)=\left(\dfrac{x+1}{x-1}\right)\left(\dfrac{1}{x}\right)=\dfrac{x+1}{x(x-1)}$

Domain: $\{x \mid x \neq 0, x \neq 1\}$

$\left(\dfrac{f}{g}\right)(x)=\dfrac{f(x)}{g(x)}=\dfrac{\dfrac{x+1}{x-1}}{\dfrac{1}{x}}=\left(\dfrac{x+1}{x-1}\right)\left(\dfrac{x}{1}\right)=\dfrac{x(x+1)}{x-1}$

Domain: $\{x \mid x \neq 0, x \neq 1\}$

15. $f(x)=-2x^2 + x + 1$

$\dfrac{f(x+h)-f(x)}{h}$

$=\dfrac{-2(x+h)^2 + (x+h)+1-\left(-2x^2 + x + 1\right)}{h}$

$=\dfrac{-2\left(x^2 + 2xh + h^2\right)+x+h+1+2x^2 - x - 1}{h}$

$=\dfrac{-2x^2 - 4xh - 2h^2 + x + h + 1 + 2x^2 - x - 1}{h}$

$=\dfrac{-4xh - 2h^2 + h}{h}=\dfrac{h(-4x - 2h + 1)}{h}$

$=-4x - 2h + 1$

16. **a.** Domain: $\{x \mid -4 \leq x \leq 3\}$; $[-4, 3]$

Range: $\{y \mid -3 \leq y \leq 3\}$; $[-3, 3]$

b. Intercept: $(0,0)$

c. $f(-2)=-1$

d. $f(x)=-3$ when $x = -4$

e. $f(x)>0$ when $0 < x \leq 3$

$\{x \mid 0 < x \leq 3\}$

f. To graph $y = f(x-3)$, shift the graph of f horizontally 3 units to the right.

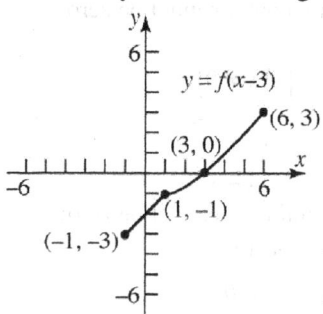

g. To graph $y = f\left(\dfrac{1}{2}x\right)$, stretch the graph of

f horizontally by a factor of 2.

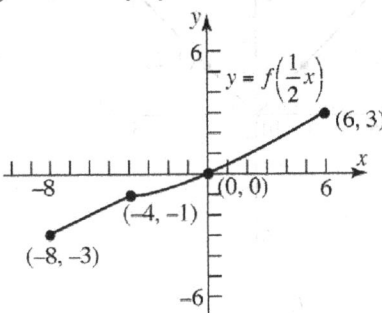

h. To graph $y = -f(x)$, reflect the graph of f

vertically about the y-axis.

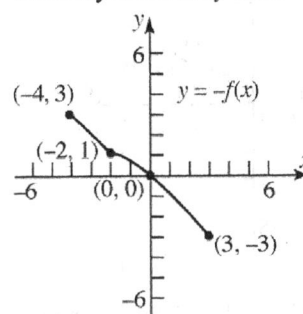

17. a. Domain: $(-\infty, 4]$

Range: $(-\infty, 3]$

b. Increasing: $(-\infty, -2)$ and $(2, 4)$;

Decreasing: $(-2, 2)$

c. Local minimum is -1 at $x = 2$;

Local maximum is 1 at $x = -2$

d. No absolute minimum;

Absolute maximum is 3 at $x = 4$

e. The graph has no symmetry.

f. The function is neither.

g. x-intercepts: $-3, 0, 3$;

y-intercept: 0

18. $f(x) = x^3 - 4x$

$f(-x) = (-x)^3 - 4(-x) = -x^3 + 4x$

$\qquad = -\left(x^3 - 4x\right) = -f(x)$

f is odd.

19. $g(x) = \dfrac{4 + x^2}{1 + x^4}$

$g(-x) = \dfrac{4 + (-x)^2}{1 + (-x)^4} = \dfrac{4 + x^2}{1 + x^4} = g(x)$

g is even.

20. $G(x) = 1 - x + x^3$

$G(-x) = 1 - (-x) + (-x)^3$

$\qquad = 1 + x - x^3 \neq -G(x)$ or $G(x)$

G is neither even nor odd.

21. $f(x) = \dfrac{x}{1 + x^2}$

$f(-x) = \dfrac{-x}{1 + (-x)^2} = \dfrac{-x}{1 + x^2} = -f(x)$

f is odd.

22. $f(x) = 2x^3 - 5x + 1$ on the interval $(-3, 3)$

Use MAXIMUM and MINIMUM on the graph

of $y_1 = 2x^3 - 5x + 1$.

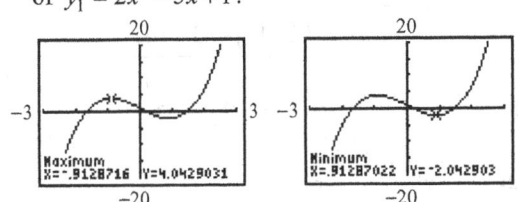

local maximum: 4.04 when $x \approx -0.91$

local minimum: -2.04 when $x = 0.91$

f is increasing on: $[-3, -0.91]$ and $[0.91, 3]$;

f is decreasing on: $[-0.91, 0.91]$.

23. $f(x) = 2x^4 - 5x^3 + 2x + 1$ on the interval $(-2, 3)$

Use MAXIMUM and MINIMUM on the graph

of $y_1 = 2x^4 - 5x^3 + 2x + 1$.

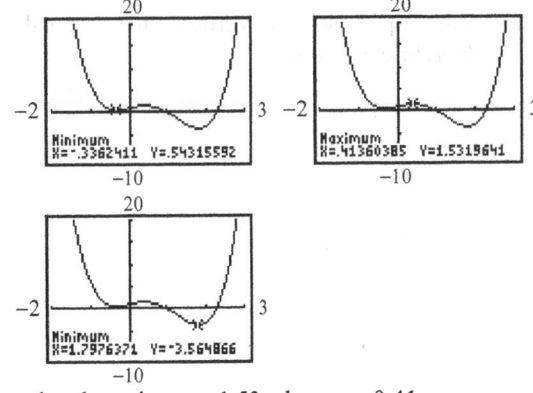

local maximum: 1.53 when $x = 0.41$

167

local minima: 0.54 when $x = -0.34$, -3.56 when $x = 1.80$

f is increasing on: $[-0.34, 0.41]$ and $[1.80, 3]$;

f is decreasing on: $[-2, -0.34]$ and $[0.41, 1.80]$.

24. $f(x) = 8x^2 - x$

a. $\dfrac{f(2) - f(1)}{2 - 1} = \dfrac{8(2)^2 - 2 - [8(1)^2 - 1]}{1}$

$= 32 - 2 - (7) = 23$

b. $\dfrac{f(1) - f(0)}{1 - 0} = \dfrac{8(1)^2 - 1 - [8(0)^2 - 0]}{1}$

$= 8 - 1 - (0) = 7$

c. $\dfrac{f(4) - f(2)}{4 - 2} = \dfrac{8(4)^2 - 4 - [8(2)^2 - 2]}{2}$

$= \dfrac{128 - 4 - (30)}{2} = \dfrac{94}{2} = 47$

25. $f(x) = 2 - 5x$

$\dfrac{f(3) - f(2)}{3 - 2} = \dfrac{[2 - 5(3)] - [2 - 5(2)]}{3 - 2}$

$= \dfrac{(2 - 15) - (2 - 10)}{1}$

$= -13 - (-8) = -5$

26. $f(x) = 3x - 4x^2$

$\dfrac{f(3) - f(2)}{3 - 2} = \dfrac{[3(3) - 4(3)^2] - [3(2) - 4(2)^2]}{3 - 2}$

$= \dfrac{(9 - 36) - (6 - 16)}{1}$

$= -27 + 10 = -17$

27. The graph does not pass the Vertical Line Test and is therefore not a function.

28. The graph passes the Vertical Line Test and is therefore a function.

29. $f(x) = |x|$

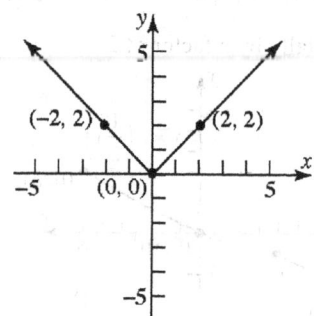

30. $f(x) = \sqrt{x}$

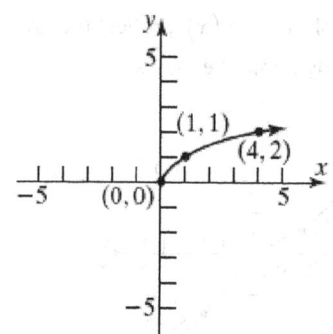

31. $F(x) = |x| - 4$. Using the graph of $y = |x|$, vertically shift the graph downward 4 units.

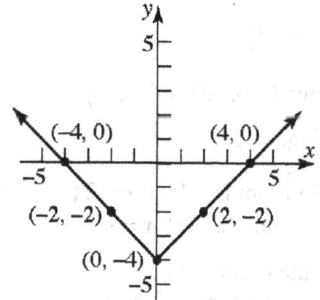

Intercepts: $(-4, 0)$, $(4, 0)$, $(0, -4)$

Domain: $\{x \mid x \text{ is any real number}\}$

Range: $\{y \mid y \geq -4\}$ or $[-4, \infty)$

32. $g(x) = -2|x|$. Reflect the graph of $y = |x|$ about the x-axis and vertically stretch the graph by a factor of 2.

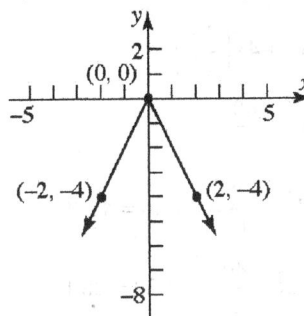

Intercepts: $(0, 0)$

Domain: $\{x \mid x \text{ is any real number}\}$

Range: $\{y \mid y \le 0\}$ or $(-\infty, 0]$

33. $h(x) = \sqrt{x-1}$. Using the graph of $y = \sqrt{x}$, horizontally shift the graph to the right 1 unit.

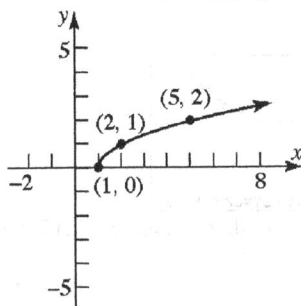

Intercept: $(1, 0)$

Domain: $\{x \mid x \ge 1\}$ or $[1, \infty)$

Range: $\{y \mid y \ge 0\}$ or $[0, \infty)$

34. $f(x) = \sqrt{1-x} = \sqrt{-(x-1)}$. Reflect the graph of $y = \sqrt{x}$ about the y-axis and horizontally shift the graph to the right 1 unit.

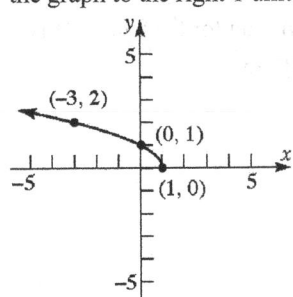

Intercepts: $(1, 0), (0, 1)$

Domain: $\{x \mid x \le 1\}$ or $(-\infty, 1]$

Range: $\{y \mid y \ge 0\}$ or $[0, \infty)$

35. $h(x) = (x-1)^2 + 2$. Using the graph of $y = x^2$, horizontally shift the graph to the right 1 unit and vertically shift the graph up 2 units.

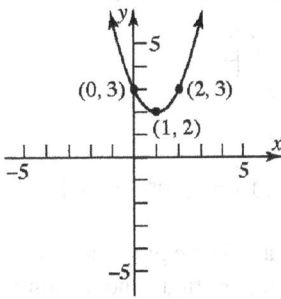

Intercepts: $(0, 3)$

Domain: $\{x \mid x \text{ is any real number}\}$

Range: $\{y \mid y \ge 2\}$ or $[2, \infty)$

36. $g(x) = -2(x+2)^3 - 8$

Using the graph of $y = x^3$, horizontally shift the graph to the left 2 units, vertically stretch the graph by a factor of 2, reflect about the x-axis, and vertically shift the graph down 8 units.

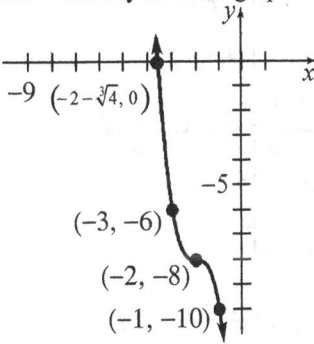

Intercepts: $(0, -24), \left(-2 - \sqrt[3]{4}, 0\right) \approx (-3.6, 0)$

Domain: $\{x \mid x \text{ is any real number}\}$

Range: $\{y \mid y \text{ is any real number}\}$

37. $f(x) = \begin{cases} 3x & \text{if } -2 < x \le 1 \\ x+1 & \text{if } x > 1 \end{cases}$

a. Domain: $\{x \mid x > -2\}$ or $(-2, \infty)$

b. Intercept: $(0, 0)$

c. Graph:

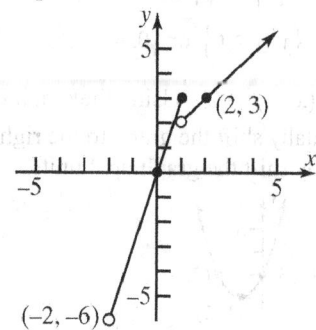

d. Range: $\{y \mid y > -6\}$ or $(-6, \infty)$

e. There is a jump in the graph at $x = 1$. Therefore, the function is not continuous.

38. $f(x) = \begin{cases} x & \text{if } -4 \le x < 0 \\ 1 & \text{if } x = 0 \\ 3x & \text{if } x > 0 \end{cases}$

a. Domain: $\{x \mid x \ge -4\}$ or $[-4, \infty)$

b. Intercept: $(0, 1)$

c. Graph:

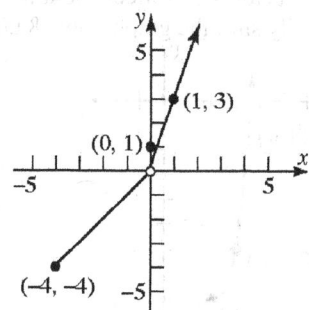

d. Range: $\{y \mid y \ge -4, y \ne 0\}$

e. There is a jump at $x = 0$. Therefore, the function is not continuous.

39. $f(x) = \dfrac{Ax + 5}{6x - 2}$ and $f(1) = 4$

$\dfrac{A(1) + 5}{6(1) - 2} = 4$

$\dfrac{A + 5}{4} = 4$

$A + 5 = 16$

$A = 11$

40. a. $x^2 h = 10 \implies h = \dfrac{10}{x^2}$

$A(x) = 2x^2 + 4xh$

$\quad\quad = 2x^2 + 4x\left(\dfrac{10}{x^2}\right)$

$\quad\quad = 2x^2 + \dfrac{40}{x}$

b. $A(1) = 2 \cdot 1^2 + \dfrac{40}{1} = 2 + 40 = 42 \text{ ft}^2$

c. $A(2) = 2 \cdot 2^2 + \dfrac{40}{2} = 8 + 20 = 28 \text{ ft}^2$

d. Graphing:

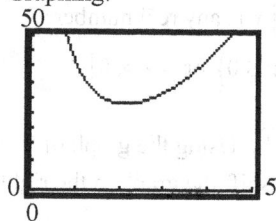

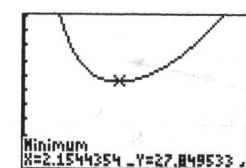

The area is smallest when $x \approx 2.15$ feet.

41. a. Consider the following diagram:

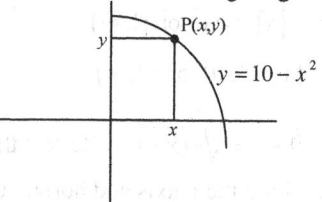

The area of the rectangle is $A = xy$. Thus, the area function for the rectangle is:

$A(x) = x(10 - x^2)$

b. The maximum value occurs at the vertex:

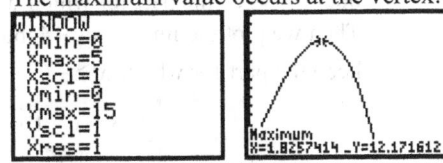

The maximum area is roughly:
$$A(1.83) = -(1.83)^3 + 10(1.83)$$
$$\approx 12.17 \text{ square units}$$

Chapter 3 Test

1. a. $\{(2,5),(4,6),(6,7),(8,8)\}$

This relation is a function because there are no ordered pairs that have the same first element and different second elements.

Domain: $\{2,4,6,8\}$

Range: $\{5,6,7,8\}$

b. $\{(1,3),(4,-2),(-3,5),(1,7)\}$

This relation is not a function because there are two ordered pairs that have the same first element but different second elements.

c. This relation is not a function because the graph fails the vertical line test.

d. This relation is a function because it passes the vertical line test.

Domain: $\{x \mid x \text{ is any real number}\}$

Range: $\{y \mid y \geq 2\}$ or $[2, \infty)$

2. $f(x) = \sqrt{4-5x}$

The function tells us to take the square root of $4-5x$. Only nonnegative numbers have real square roots so we need $4-5x \geq 0$.

$$4-5x \geq 0$$
$$4-5x-4 \geq 0-4$$
$$-5x \geq -4$$
$$\frac{-5x}{-5} \leq \frac{-4}{-5}$$
$$x \leq \frac{4}{5}$$

Domain: $\left\{x \mid x \leq \frac{4}{5}\right\}$ or $\left(-\infty, \frac{4}{5}\right]$

$$f(-1) = \sqrt{4-5(-1)} = \sqrt{4+5} = \sqrt{9} = 3$$

3. $g(x) = \dfrac{x+2}{|x+2|}$

The function tells us to divide $x+2$ by $|x+2|$. Division by 0 is undefined, so the denominator can never equal 0. This means that $x \neq -2$.

Domain: $\{x \mid x \neq -2\}$

$$g(-1) = \frac{(-1)+2}{|(-1)+2|} = \frac{1}{|1|} = 1$$

4. $h(x) = \dfrac{x-4}{x^2+5x-36}$

The function tells us to divide $x-4$ by $x^2+5x-36$. Since division by 0 is not defined, we need to exclude any values which make the denominator 0.

$$x^2+5x-36 = 0$$
$$(x+9)(x-4) = 0$$
$$x = -9 \text{ or } x = 4$$

Domain: $\{x \mid x \neq -9, x \neq 4\}$

(note: there is a common factor of $x-4$ but we must determine the domain prior to simplifying)

$$h(-1) = \frac{(-1)-4}{(-1)^2+5(-1)-36} = \frac{-5}{-40} = \frac{1}{8}$$

5. a. To find the domain, note that all the points on the graph will have an x-coordinate between -5 and 5, inclusive. To find the range, note that all the points on the graph will have a y-coordinate between -3 and 3, inclusive.

Domain: $\{x \mid -5 \leq x \leq 5\}$ or $[-5, 5]$

Range: $\{y \mid -3 \leq y \leq 3\}$ or $[-3, 3]$

b. The intercepts are $(0,2)$, $(-2,0)$, and $(2,0)$.

x-intercepts: $-2, 2$

y-intercept: 2

c. $f(1)$ is the value of the function when $x=1$. According to the graph, $f(1) = 3$.

d. Since $(-5,-3)$ and $(3,-3)$ are the only points on the graph for which

$y = f(x) = -3$, we have $f(x) = -3$ when $x = -5$ and $x = 3$.

e. To solve $f(x) < 0$, we want to find x values such that the graph is below the x-axis. The graph is below the x-axis for values in the domain that are less than -2 and greater than 2. Therefore, the solution set is $\{x \mid -5 \le x < -2 \text{ or } 2 < x \le 5\}$. In interval notation we would write the solution set as $[-5, -2) \cup (2, 5]$.

6. $f(x) = -x^4 + 2x^3 + 4x^2 - 2$

We set Xmin = -5 and Xmax = 5. The standard Ymin and Ymax will not be good enough to see the whole picture so some adjustment must be made.

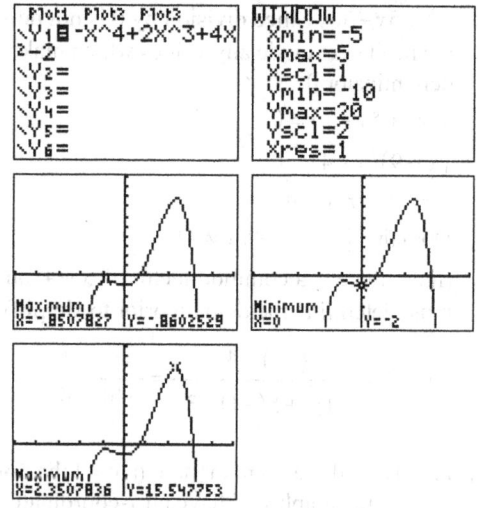

We see that the graph has a local maximum of -0.86 (rounded to two places) when $x = -0.85$ and another local maximum of 15.55 when $x = 2.35$. There is a local minimum of -2 when $x = 0$. Thus, we have

Local maxima: $f(-0.85) \approx -0.86$
$$f(2.35) \approx 15.55$$

Local minima: $f(0) = -2$

The function is increasing on the intervals $[-5, -0.85]$ and $[0, 2.35]$ and decreasing on the intervals $[-0.85, 0]$ and $[2.35, 5]$.

7. a. $f(x) = \begin{cases} 2x + 1 & x < -1 \\ x - 4 & x \ge -1 \end{cases}$

To graph the function, we graph each "piece." First we graph the line $y = 2x + 1$

but only keep the part for which $x < -1$. Then we plot the line $y = x - 4$ but only keep the part for which $x \ge -1$.

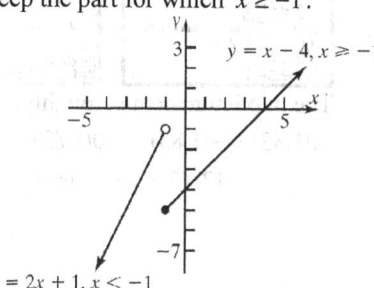

b. To find the intercepts, notice that the only piece that hits either axis is $y = x - 4$.

$y = x - 4$	$y = x - 4$
$y = 0 - 4$	$0 = x - 4$
$y = -4$	$4 = x$

The intercepts are $(0, -4)$ and $(4, 0)$.

c. To find $g(-5)$ we first note that $x = -5$ so we must use the first "piece" because $-5 < -1$.
$$g(-5) = 2(-5) + 1 = -10 + 1 = -9$$

d. To find $g(2)$ we first note that $x = 2$ so we must use the second "piece" because $2 \ge -1$.
$$g(2) = 2 - 4 = -2$$

8. The average rate of change from 3 to 4 is given by
$$\frac{\Delta y}{\Delta x} = \frac{f(4) - f(3)}{4 - 3}$$
$$= \frac{\left(3(4)^2 - 2(4) + 4\right) - \left(3(3)^2 - 2(3) + 4\right)}{4 - 3}$$
$$= \frac{44 - 25}{4 - 3} = \frac{19}{1} = 19$$

9. a. $(f - g)(x) = (2x^2 + 1) - (3x - 2)$
$$= 2x^2 + 1 - 3x + 2 = 2x^2 - 3x + 3$$

b. $(f \cdot g)(x) = (2x^2 + 1)(3x - 2)$
$$= 6x^3 - 4x^2 + 3x - 2$$

c. $f(x+h)-f(x)$

$=\left(2(x+h)^2+1\right)-\left(2x^2+1\right)$

$=\left(2\left(x^2+2xh+h^2\right)+1\right)-\left(2x^2+1\right)$

$=2x^2+4xh+2h^2+1-2x^2-1$

$=4xh+2h^2$

10. a. The basic function is $y=x^3$ so we start with the graph of this function.

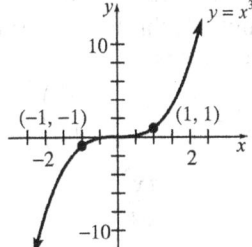

Next we shift this graph 1 unit to the left to obtain the graph of $y=(x+1)^3$.

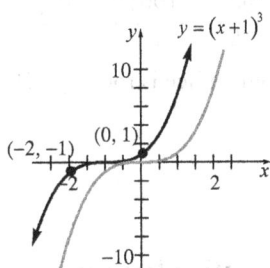

Next we reflect this graph about the x-axis to obtain the graph of $y=-(x+1)^3$.

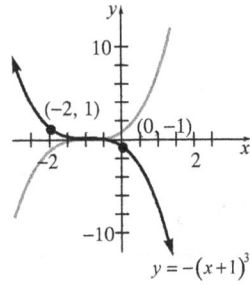

Next we stretch this graph vertically by a factor of 2 to obtain the graph of

$y=-2(x+1)^3$.

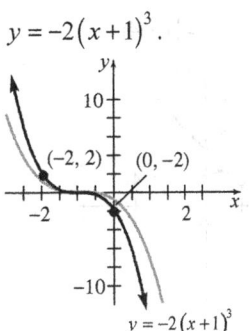

The last step is to shift this graph up 3 units to obtain the graph of $y=-2(x+1)^3+3$.

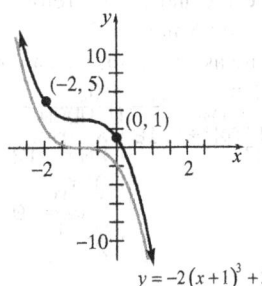

b. The basic function is $y=|x|$ so we start with the graph of this function.

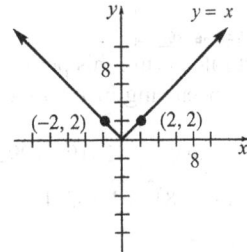

Next we shift this graph 4 units to the left to obtain the graph of $y=|x+4|$.

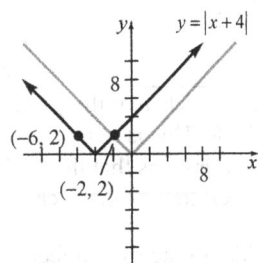

Next we shift this graph up 2 units to obtain

173

the graph of $y = |x+4| + 2$.

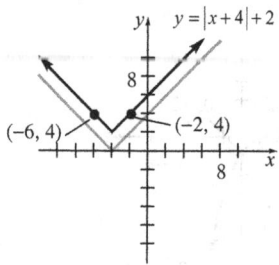

11. **a.** $r(x) = -0.115x^2 + 1.183x + 5.623$

For the years 1992 to 2004, we have values of x between 0 and 12. Therefore, we can let Xmin = 0 and Xmax = 12. Since r is the interest rate as a percent, we can try letting Ymin = 0 and Ymax = 10.

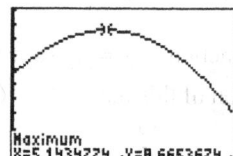

The highest rate during this period appears to be 8.67%, occurring in 1997 ($x \approx 5$).

b. For 2010, we have $x = 2010 - 1992 = 18$.

$r(18) = -0.115(18)^2 + 1.183(18) + 5.623$

$\qquad = -10.343$

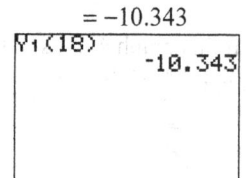

The model predicts that the interest rate will be -10.343%. This is not a reasonable value since it implies that the bank would be paying interest to the borrower.

12. **a.** Let x = width of the rink in feet. Then the length of the rectangular portion is given by $2x - 20$. The radius of the semicircular portions is half the width, or $r = \dfrac{x}{2}$.

To find the volume, we first find the area of the surface and multiply by the thickness of the ice. The two semicircles can be

combined to form a complete circle, so the area is given by

$A = l \cdot w + \pi r^2$

$\qquad = (2x - 20)(x) + \pi \left(\dfrac{x}{2}\right)^2$

$\qquad = 2x^2 - 20x + \dfrac{\pi x^2}{4}$

We have expressed our measures in feet so we need to convert the thickness to feet as well.

$0.75 \text{ in} \cdot \dfrac{1 \text{ ft}}{12 \text{ in}} = \dfrac{0.75}{12} \text{ ft} = \dfrac{1}{16} \text{ ft}$

Now we multiply this by the area to obtain the volume. That is,

$V(x) = \dfrac{1}{16}\left(2x^2 - 20x + \dfrac{\pi x^2}{4}\right)$

$V(x) = \dfrac{x^2}{8} - \dfrac{5x}{4} + \dfrac{\pi x^2}{64}$

b. If the rink is 90 feet wide, then we have $x = 90$.

$V(90) = \dfrac{90^2}{8} - \dfrac{5(90)}{4} + \dfrac{\pi(90)^2}{64} \approx 1297.61$

The volume of ice is roughly 1297.61 ft^3.

Chapter 3 Cumulative Review

1. $\qquad 3x - 8 = 10$

$\qquad 3x - 8 + 8 = 10 + 8$

$\qquad\qquad 3x = 18$

$\qquad\qquad \dfrac{3x}{3} = \dfrac{18}{3}$

$\qquad\qquad x = 6$

The solution set is $\{6\}$.

2. $3x^2 - x = 0$

$x(3x-1) = 0$

$x = 0$ or $3x - 1 = 0$

$\qquad\qquad 3x = 1$

$\qquad\qquad x = \dfrac{1}{3}$

The solution set is $\left\{0, \dfrac{1}{3}\right\}$.

3. $x^2 - 8x - 9 = 0$

$(x-9)(x+1) = 0$

$x - 9 = 0$ or $x + 1 = 0$

$x = 9$ $\qquad\quad x = -1$

The solution set is $\{-1, 9\}$.

4. $6x^2 - 5x + 1 = 0$

$(3x-1)(2x-1) = 0$

$3x - 1 = 0$ or $2x - 1 = 0$

$3x = 1$ $\qquad\quad 2x = 1$

$x = \dfrac{1}{3}$ $\qquad x = \dfrac{1}{2}$

The solution set is $\left\{\dfrac{1}{3}, \dfrac{1}{2}\right\}$.

5. $|2x+3| = 4$

$2x + 3 = -4$ or $2x + 3 = 4$

$2x = -7$ $\qquad\quad 2x = 1$

$x = -\dfrac{7}{2}$ $\qquad x = \dfrac{1}{2}$

The solution set is $\left\{-\dfrac{7}{2}, \dfrac{1}{2}\right\}$.

6. $\sqrt{2x+3} = 2$

$\left(\sqrt{2x+3}\right)^2 = 2^2$

$2x + 3 = 4$

$2x = 1$

$x = \dfrac{1}{2}$

Check:

$\sqrt{2\left(\dfrac{1}{2}\right)+3} \overset{?}{=} 2$

$\sqrt{1+3} \overset{?}{=} 2$

$\sqrt{4} \overset{?}{=} 2$

$2 = 2$ T

The solution set is $\left\{\dfrac{1}{2}\right\}$.

7. $2 - 3x > 6$

$-3x > 4$

$x < -\dfrac{4}{3}$

Solution set: $\left\{x \mid x < -\dfrac{4}{3}\right\}$

Interval notation: $\left(-\infty, -\dfrac{4}{3}\right)$

8. $|2x - 5| < 3$

$-3 < 2x - 5 < 3$

$2 < 2x < 8$

$1 < x < 4$

Solution set: $\{x \mid 1 < x < 4\}$

Interval notation: $(1, 4)$

9. $|4x + 1| \geq 7$

$4x + 1 \leq -7$ or $4x + 1 \geq 7$

$4x \leq -8$ $\qquad\quad 4x \geq 6$

$x \leq -2$ $\qquad\qquad x \geq \dfrac{3}{2}$

Solution set: $\left\{x \mid x \leq -2 \text{ or } x \geq \dfrac{3}{2}\right\}$

Interval notation: $(-\infty, -2] \cup \left[\dfrac{3}{2}, \infty\right)$

10. a. $d = \sqrt{(x_2 - x_1)^2 + (y_2 - y_1)^2}$

$= \sqrt{(3-(-2))^2 + (-5-(-3))^2}$

$= \sqrt{(3+2)^2 + (-5+3)^2}$

$= \sqrt{5^2 + (-2)^2} = \sqrt{25+4}$

$= \sqrt{29}$

b. $M = \left(\dfrac{x_1 + x_2}{2}, \dfrac{y_1 + y_2}{2} \right)$

$= \left(\dfrac{-2+3}{2}, \dfrac{-3+(-5)}{2} \right)$

$= \left(\dfrac{1}{2}, -4 \right)$

c. $m = \dfrac{y_2 - y_1}{x_2 - x_1} = \dfrac{-5-(-3)}{3-(-2)} = \dfrac{-2}{5} = -\dfrac{2}{5}$

11. $3x - 2y = 12$

x-intercept:

$3x - 2(0) = 12$

$3x = 12$

$x = 4$

The point $(4,0)$ is on the graph.

y-intercept:

$3(0) - 2y = 12$

$-2y = 12$

$y = -6$

The point $(0,-6)$ is on the graph.

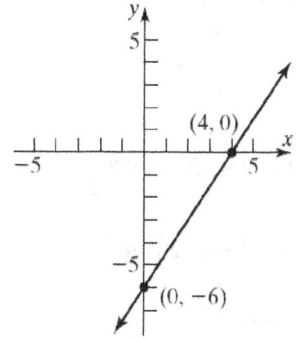

12. $x = y^2$

y	$x = y^2$	(x,y)
-2	$x = (-2)^2 = 4$	$(4,-2)$
-1	$x = (-1)^2 = 1$	$(1,-1)$
0	$x = 0^2 = 0$	$(0,0)$
1	$x = 1^2 = 1$	$(1,1)$
2	$x = 2^2 = 4$	$(4,2)$

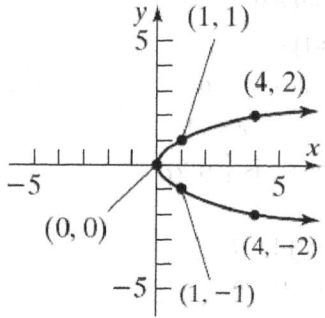

13. $x^2 + (y-3)^2 = 16$

This is the equation of a circle with radius $r = \sqrt{16} = 4$ and center at $(0,3)$. Starting at the center we can obtain some points on the graph by moving 4 units up, down, left, and right. The corresponding points are $(0,7)$, $(0,-1)$, $(-4,3)$, and $(4,3)$, respectively.

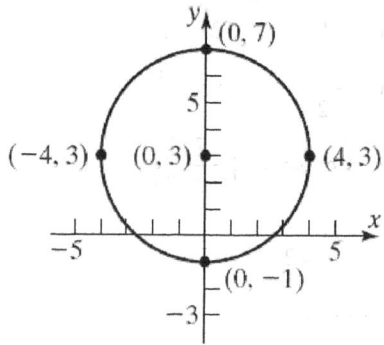

14. $y = \sqrt{x}$

x	$y = \sqrt{x}$	(x, y)
0	$y = \sqrt{0} = 0$	$(0,0)$
1	$y = \sqrt{1} = 1$	$(1,1)$
4	$y = \sqrt{4} = 2$	$(4,2)$

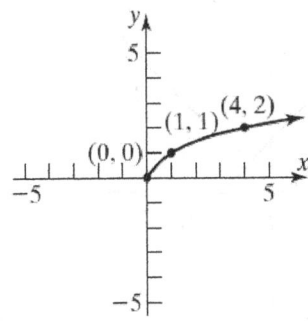

15. $3x^2 - 4y = 12$

x-intercepts:
$3x^2 - 4(0) = 12$
$3x^2 = 12$
$x^2 = 4$
$x = \pm 2$

y-intercept:
$3(0)^2 - 4y = 12$
$-4y = 12$
$y = -3$

The intercepts are $(-2,0)$, $(2,0)$, and $(0,-3)$.

Check x-axis symmetry:
$3x^2 - 4(-y) = 12$
$3x^2 + 4y = 12$ different

Check y-axis symmetry:
$3(-x)^2 - 4y = 12$
$3x^2 - 4y = 12$ same

Check origin symmetry:
$3(-x)^2 - 4(-y) = 12$
$3x^2 + 4y = 12$ different

The graph of the equation has y-axis symmetry.

16. First we find the slope:
$$m = \frac{8-4}{6-(-2)} = \frac{4}{8} = \frac{1}{2}$$

Next we use the slope and the given point $(6,8)$ in the point-slope form of the equation of a line:
$$y - y_1 = m(x - x_1)$$
$$y - 8 = \frac{1}{2}(x - 6)$$
$$y - 8 = \frac{1}{2}x - 3$$
$$y = \frac{1}{2}x + 5$$

17. $f(x) = (x+2)^2 - 3$

Starting with the graph of $y = x^2$, shift the graph 2 units to the left $\left[y = (x+2)^2 \right]$ and down 3 units $\left[y = (x+2)^2 - 3 \right]$.

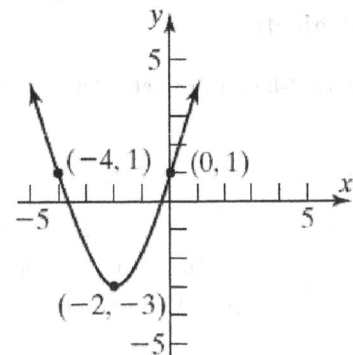

18. $f(x) = \dfrac{1}{x}$

x	$y = \dfrac{1}{x}$	(x, y)
-1	$y = \dfrac{1}{-1} = -1$	$(-1, -1)$
1	$y = \dfrac{1}{1} = 1$	$(1, 1)$
2	$y = \dfrac{1}{2}$	$\left(2, \dfrac{1}{2}\right)$

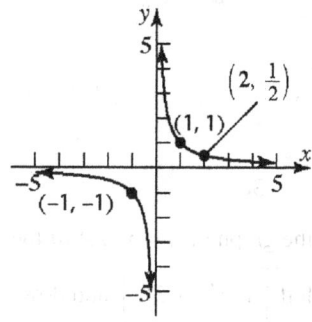

19. $f(x) = \begin{cases} 2 - x & \text{if } x \le 2 \\ |x| & \text{if } x > 2 \end{cases}$

Graph the line $y = 2 - x$ for $x \le 2$. Two points on the graph are $(0, 2)$ and $(2, 0)$.

Graph the line $y = x$ for $x > 2$. There is a hole in the graph at $x = 2$.

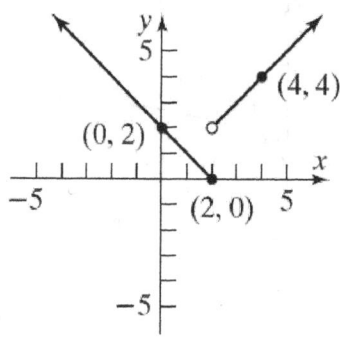

Chapter 3 Projects

Project I – Internet-based Project – Answers will vary

Project II

1. Silver: $C(x) = 20 + 0.16(x - 200) = 0.16x - 12$

$$C(x) = \begin{cases} 20 & 0 \le x \le 200 \\ 0.16x - 12 & x > 200 \end{cases}$$

Gold: $C(x) = 50 + 0.08(x - 1000) = 0.08x - 30$

$$C(x) = \begin{cases} 50.00 & 0 \le x \le 1000 \\ 0.08x - 30 & x > 1000 \end{cases}$$

Platinum: $C(x) = 100 + 0.04(x - 3000)$

$$= 0.04x - 20$$

$$C(x) = \begin{cases} 100.00 & 0 \le x \le 3000 \\ 0.04x - 20 & x > 3000 \end{cases}$$

2.

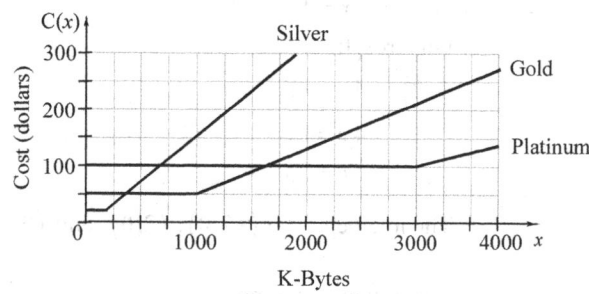

3. Let y = #K-bytes of service over the plan minimum.

Silver: $20 + 0.16y \le 50$

$$0.16y \le 30$$

$$y \le 187.5$$

Silver is the best up to $187.5 + 200 = 387.5$ K-bytes of service.

Gold: $50 + 0.08y \le 100$

$$0.08y \le 50$$

$$y \le 625$$

Gold is the best from 387.5 K-bytes to $625 + 1000 = 1625$ K-bytes of service.

Platinum: Platinum will be the best if more than 1625 K-bytes is needed.

4. Answers will vary.

Project III

1.

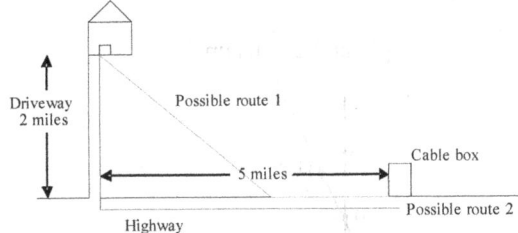

2.

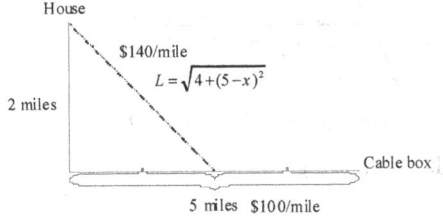

$$C(x) = 100x + 140L$$

$$C(x) = 100x + 140\sqrt{4 + (5-x)^2}$$

3.

x	$C(x)$
0	$100(0) + 140\sqrt{4+25} \approx \$753.92$
1	$100(1) + 140\sqrt{4+16} \approx \$726.10$
2	$100(2) + 140\sqrt{4+9} \approx \$704.78$
3	$100(3) + 140\sqrt{4+4} \approx \$695.98$
4	$100(4) + 140\sqrt{4+1} \approx \$713.05$
5	$100(5) + 140\sqrt{4+0} = \$780.00$

The choice where the cable goes 3 miles down the road then cutting up to the house seems to yield the lowest cost.

4. Since all of the costs are less than $800, there would be a profit made with any of the plans.

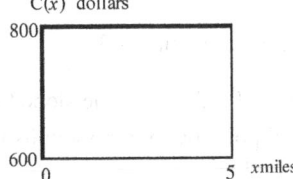

Using the MINIMUM function on a graphing calculator, the minimum occurs at $x \approx 2.96$.

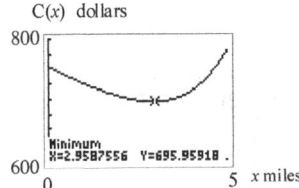

The minimum cost occurs when the cable runs for 2.96 mile along the road.

6. $C(4.5) = 100(4.5) + 140\sqrt{4 + (5 - 4.5)^2}$

$\approx \$738.62$

The cost for the Steven's cable would be $738.62.

7. $5000(738.62) = \$3,693,100$ State legislated
$5000(695.96) = \$3,479,800$ cheapest cost
It will cost the company $213,300 more.

Project IV

1. $A = \pi r^2$

2. $r = 2.2t$

3. $r = 2.2(2) = 4.4$ ft

$r = 2.2(2.5) = 5.5$ ft

4. $A = \pi(4.4)^2 = 60.82$ ft^2

$A = \pi(5.5)^2 = 95.03$ ft^2

5. $A = \pi(2.2t)^2 = 4.84\pi t^2$

6. $A = 4.84\pi(2)^2 = 60.82$ ft^2

$A = 4.84\pi(2.5)^2 = 95.03$ ft^2

7. $\dfrac{A(2.5) - A(2)}{2.5 - 2} = \dfrac{95.03 - 60.82}{0.5} = 68.42$ ft/hr

8. $\dfrac{A(3.5) - A(3)}{3.5 - 3} = \dfrac{186.27 - 136.85}{0.5} = 98.84$ ft/hr

9. The average rate of change is increasing.

10. 150 yds = 450 ft
$r = 2.2t$

$t = \dfrac{450}{2.2} = 204.5$ hours

11. 6 miles = 31680 ft
Therefore, we need a radius of 15,840 ft.

$t = \dfrac{15,840}{2.2} = 7200$ hours

Chapter 4
Linear and Quadratic Functions

Section 4.1

1. From the equation $y = 2x - 3$, we see that the y-intercept is -3. Thus, the point $(0, -3)$ is on the graph. We can obtain a second point by choosing a value for x and finding the corresponding value for y. Let $x = 1$, then $y = 2(1) - 3 = -1$. Thus, the point $(1, -1)$ is also on the graph. Plotting the two points and connecting with a line yields the graph below.

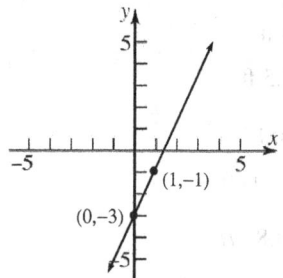

3. $f(2) = 3(2)^2 - 2 = 10$

   $f(4) = 3(4)^2 - 2 = 46$

   $\dfrac{\Delta y}{\Delta x} = \dfrac{f(4) - f(2)}{4 - 2} = \dfrac{46 - 10}{4 - 2} = \dfrac{36}{2} = 18$

5. $f(-2) = (-2)^2 - 4 = 4 - 4 = 0$

7. slope; y-intercept

9. True

11. a

13. $f(x) = 2x + 3$

   a. Slope = 2; y-intercept = 3

   b. Plot the point (0, 3). Use the slope to find an additional point by moving 1 unit to the right and 2 units up.

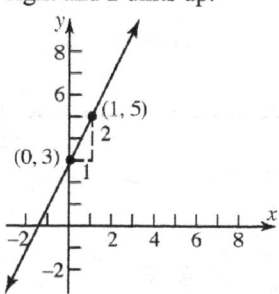

   c. average rate of change = 2

   d. increasing

15. $h(x) = -3x + 4$

   a. Slope = -3; y-intercept = 4

   b. Plot the point (0, 4). Use the slope to find an additional point by moving 1 unit to the right and 3 units down.

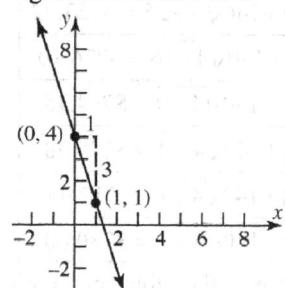

   c. average rate of change = -3

   d. decreasing

17. $f(x) = \dfrac{1}{4}x - 3$

   a. Slope = $\dfrac{1}{4}$; y-intercept = -3

   b. Plot the point $(0, -3)$. Use the slope to find an additional point by moving 4 units to the right and 1 unit up.

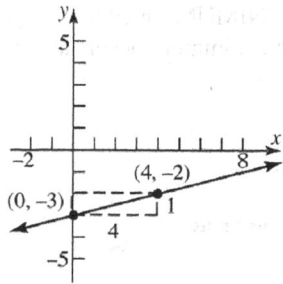

c. average rate of change $= \dfrac{1}{4}$

d. increasing

19. $F(x) = 4$

   a. Slope $= 0$; y-intercept $= 4$

   b. Plot the point $(0, 4)$ and draw a horizontal line through it.

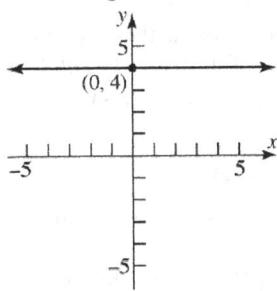

   c. average rate of change $= 0$

   d. constant

21.

x	y	Avg. rate of change $= \dfrac{\Delta y}{\Delta x}$
-2	4	
-1	1	$\dfrac{1-4}{-1-(-2)} = \dfrac{-3}{1} = -3$
0	-2	$\dfrac{-2-1}{0-(-1)} = \dfrac{-3}{1} = -3$
1	-5	$\dfrac{-5-(-2)}{1-0} = \dfrac{-3}{1} = -3$
2	-8	$\dfrac{-8-(-5)}{2-1} = \dfrac{-3}{1} = -3$

Since the average rate of change is constant at -3, this is a linear function with slope $= -3$. The y-intercept is $(0, -2)$, so the equation of the line is $y = -3x - 2$.

23.

x	y	Avg. rate of change $= \dfrac{\Delta y}{\Delta x}$
-2	-8	
-1	-3	$\dfrac{-3-(-8)}{-1-(-2)} = \dfrac{5}{1} = 5$
0	0	$\dfrac{0-(-3)}{0-(-1)} = \dfrac{3}{1} = 3$
1	1	
2	0	

Since the average rate of change is not constant, this is not a linear function.

25.

x	y	Avg. rate of change $= \dfrac{\Delta y}{\Delta x}$
-2	-26	
-1	-4	$\dfrac{-4-(-26)}{-1-(-2)} = \dfrac{22}{1} = 22$
0	2	$\dfrac{2-(-4)}{0-(-1)} = \dfrac{6}{1} = 6$
1	-2	
2	-10	

Since the average rate of change is not constant, this is not a linear function.

27.

x	y	Avg. rate of change $= \dfrac{\Delta y}{\Delta x}$
-2	8	
-1	8	$\dfrac{8-8}{-1-(-2)} = \dfrac{0}{1} = 0$
0	8	$\dfrac{8-8}{0-(-1)} = \dfrac{0}{1} = 0$
1	8	$\dfrac{8-8}{1-0} = \dfrac{0}{1} = 0$
2	8	$\dfrac{8-8}{2-1} = \dfrac{0}{1} = 0$

Since the average rate of change is constant at 0, this is a linear function with slope $= 0$. The y-intercept is $(0, 8)$, so the equation of the line is $y = 0x + 8$ or $y = 8$.

29. $f(x) = 4x - 1$; $g(x) = -2x + 5$

   a. $f(x) = 0$

$$4x - 1 = 0$$
$$x = \frac{1}{4}$$

   b. $f(x) > 0$

$$4x - 1 > 0$$
$$x > \frac{1}{4}$$

The solution set is $\left\{ x \mid x > \dfrac{1}{4} \right\}$ or $\left(\dfrac{1}{4}, \infty \right)$.

c. $f(x) = g(x)$

$4x - 1 = -2x + 5$

$6x = 6$

$x = 1$

d. $f(x) \le g(x)$

$4x - 1 \le -2x + 5$

$6x \le 6$

$x \le 1$

The solution set is $\{x \mid x \le 1\}$ or $(-\infty, 1]$.

e.

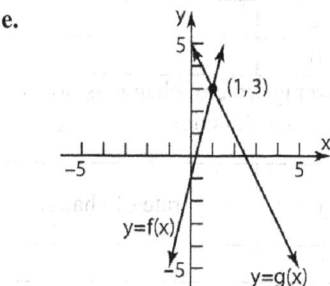

31. a. The point (40, 50) is on the graph of $y = f(x)$, so the solution to $f(x) = 50$ is $x = 40$.

b. The point (88, 80) is on the graph of $y = f(x)$, so the solution to $f(x) = 80$ is $x = 88$.

c. The point (−40, 0) is on the graph of $y = f(x)$, so the solution to $f(x) = 0$ is $x = -40$.

d. The y-coordinates of the graph of $y = f(x)$ are above 50 when the x-coordinates are larger than 40. Thus, the solution to $f(x) > 50$ is $\{x \mid x > 40\}$ or $(40, \infty)$.

e. The y-coordinates of the graph of $y = f(x)$ are below 80 when the x-coordinates are smaller than 88. Thus, the solution to $f(x) \le 80$ is $\{x \mid x \le 88\}$ or $(-\infty, 88]$.

f. The y-coordinates of the graph of $y = f(x)$ are between 0 and 80 when the x-coordinates are between −40 and 88. Thus, the solution to $0 < f(x) < 80$ is $\{x \mid -40 < x < 88\}$ or $(-40, 88)$.

33. a. $f(x) = g(x)$ when their graphs intersect. Thus, $x = -4$.

b. $f(x) \le g(x)$ when the graph of f is above the graph of g. Thus, the solution is $\{x \mid x < -4\}$ or $(-\infty, -4)$.

35. a. $f(x) = g(x)$ when their graphs intersect. Thus, $x = -6$.

b. $g(x) \le f(x) < h(x)$ when the graph of f is above or intersects the graph of g and below the graph of h. Thus, the solution is $\{x \mid -6 \le x < 5\}$ or $[-6, 5)$.

37. $C(x) = 0.35x + 45$

a. $C(40) = 0.35(40) + 45 = \$59$.

b. Solve $C(x) = 0.35x + 45 = 108$

$0.35x + 45 = 108$

$0.35x = 63$

$x = \dfrac{63}{0.35} = 180$ miles

c. Solve $C(x) = 0.35x + 45 \le 150$

$0.35x + 45 \le 150$

$0.35x \le 105$

$x \le \dfrac{105}{0.35} = 300$ miles

d. The number of mile driven cannot be negative, so the implied domain for C is $\{x \mid x \ge 0\}$ or $[0, \infty)$.

e. The cost of renting the car for a day increases $0.35 for each mile driven, or there is a charge of $0.35 per mile to rent the car in addition to a fixed charge of $45.

f. It costs $45 to rent the car if 0 miles are driven, or there is a fixed charge of $45 to rent the car in addition to a charge that depends on mileage.

39. $S(p) = -600 + 50p; \quad D(p) = 1200 - 25p$

a. Solve $S(p) = D(p)$.

$-600 + 50p = 1200 - 25p$

$75p = 1800$

$p = \dfrac{1800}{75} = 24$

$S(16) = -600 + 50(24) = 600$

Thus, the equilibrium price is $24, and the equilibrium quantity is 600 T-shirts.

182

b. Solve $D(p) > S(p)$.

$$1200 - 25p > -600 + 50p$$
$$1800 > 75p$$
$$\frac{1800}{75} > p$$
$$24 > p$$

The demand will exceed supply when the price is less than $16 (but still greater than $0). That is, $0 < p < \$24$.

c. The price will eventually be increased.

41. a. We are told that the tax function T is for adjusted gross incomes x between $9,225 and $37,450 inclusive. Thus, the domain is $\{x \mid 9,225 \le x \le 37,450\}$ or $[9225, 37450]$.

b. $T(20,000) = 0.15(20,000 - 9225) + 922.5$
$$= 2538.75$$
If a single filer's adjusted gross income is $20,000, then his or her tax bill will be $2538.75.

c. The independent variable is adjusted gross income, x. The dependent variable is the tax bill, T.

d. Evaluate T at $x = 9225,\ 20000,$ and 37450.
$$T(9225) = 0.15(9225 - 9225) + 922.5$$
$$= 922.5$$
$$T(20,000) = 0.15(20,000 - 9225) + 922.5$$
$$= 2538.75$$
$$T(37,450) = 0.15(37,450 - 9225) + 922.5$$
$$= 5156.25$$
Thus, the points $(9225, 922.5)$,
$(20000, 2538.75)$, and $(37450, 5456.25)$
are on the graph.

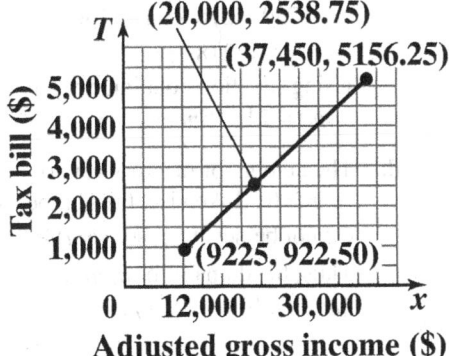

e. We must solve $T(x) = 3663.75$.
$$0.15(x - 9225) + 922.5 = 3663.75$$
$$0.15x - 1383.75 + 922.5 = 3663.75$$
$$0.15x - 461.25 = 3663.75$$
$$0.15x = 4125$$
$$x = 27,500$$
A single filer with an adjusted gross income of $27,500 will have a tax bill of $3663.75.

f. For each additional dollar of taxable income between $9225 and $37,450, the tax bill of a single person in 2015 increased by $0.15.

43. $R(x) = 8x;\ C(x) = 4.5x + 17,500$

a. Solve $R(x) = C(x)$.
$$8x = 4.5x + 17,500$$
$$3.5x = 17,500$$
$$x = 5000$$
The break-even point occurs when the company sells 5000 units.

b. Solve $R(x) > C(x)$
$$8x > 4.5x + 17,500$$
$$3.5x > 17,500$$
$$x > 5000$$
The company makes a profit if it sells more than 5000 units.

45. a. Consider the data points (x, y), where $x =$ the age in years of the computer and $y =$ the value in dollars of the computer. So we have the points $(0, 3000)$ and $(3, 0)$. The slope formula yields:
$$m = \frac{\Delta y}{\Delta x} = \frac{0 - 3000}{3 - 0} = \frac{-3000}{3} = -1000$$
The y-intercept is $(0, 3000)$, so $b = 3000$. Therefore, the linear function is
$V(x) = mx + b = -1000x + 3000$.

b. The age of the computer cannot be negative, and the book value of the computer will be $0 after 3 years. Thus, the implied domain for V is $\{x \mid 0 \le x \le 3\}$ or $[0, 3]$.

c. The graph of $V(x) = -1000x + 3000$

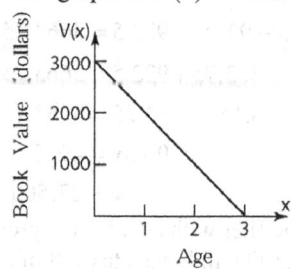

d. $V(2) = -1000(2) + 3000 = 1000$

The computer's book value after 2 years will be $1000.

e. Solve $V(x) = 2000$

$-1000x + 3000 = 2000$

$-1000x = -1000$

$x = 1$

The computer will have a book value of $2000 after 1 year.

47. a. Let x = the number of bicycles manufactured. We can use the cost function $C(x) = mx + b$, with $m = 90$ and $b = 1800$. Therefore $C(x) = 90x + 1800$

b. The graph of $C(x) = 90x + 1800$

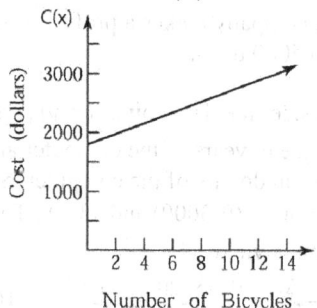

c. The cost of manufacturing 14 bicycles is given by $C(14) = 90(14) + 1800 = \$3060$.

d. Solve $C(x) = 90x + 1800 = 3780$

$90x + 1800 = 3780$

$90x = 1980$

$x = 22$

So 22 bicycles could be manufactured for $3780.

49. a. Let x = number of miles driven, and let C = cost in dollars. Total cost = (cost per mile)(number of miles) + fixed cost
$C(x) = 0.89x + 39.95$

b. $C(110) = (0.89)(110) + 39.95 = \$137.85$

$C(230) = (0.89)(230) + 39.95 = \$244.65$

51. a.

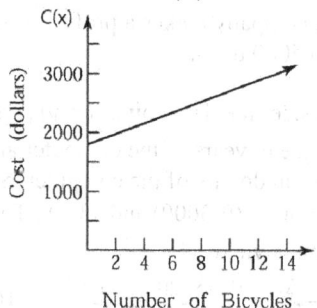

b.

m	n	Avg. rate of change $= \dfrac{\Delta n}{\Delta m}$
8	1750	
16	3500	$\dfrac{3500 - 1750}{16 - 8} = \dfrac{1750}{8} = 218.75$
32	7000	$\dfrac{7000 - 3500}{32 - 16} = \dfrac{3500}{16} = 218.75$
64	14000	$\dfrac{14000 - 7000}{64 - 32} = \dfrac{7000}{32} = 218.75$

Since each input (memory) corresponds to a single output (number of songs), we know that the number of songs is a function of memory. Also, because the average rate of change is constant at 218.75 per gigabyte of memory, the function is linear.

c. From part (b), we know $m = 218.75$. Using $(m_1, n_1) = (8, 1750)$, we get the equation:

$n - n_1 = m(m - m_1)$

$n - 1750 = 218.75(m - 8)$

$n - 1750 = 218.75m - 1750$

$n = 218.75m$

Using function notation, we have $n(m) = 218.75m$.

d. The amount of memory cannot be negative, so $n \geq 0$. Likewise, the number of songs cannot be negative, so, $n(m) \geq 0$. Thus, the domain for $n(m)$ is $\{m \mid m \geq 0\}$ or $[0, \infty)$.

e.

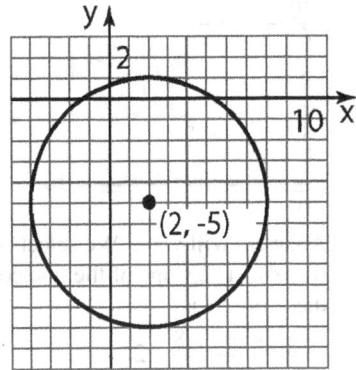

Memory (gigabytes)

f. If memory increases by 1GB, then the number of songs increases by 218.75.

53. The graph shown has a positive slope and a positive y-intercept. Therefore, the function from (d) and (e) might have the graph shown.

55. A linear function $f(x) = mx + b$ will be odd provided $f(-x) = -f(x)$.

That is, provided $m(-x) + b = -(mx + b)$.

$$-mx + b = -mx - b$$
$$b = -b$$
$$2b = 0$$
$$b = 0$$

So a linear function $f(x) = mx + b$ will be odd provided $b = 0$.

A linear function $f(x) = mx + b$ will be even provided $f(-x) = f(x)$.

That is, provided $m(-x) + b = mx + b$.

$$-mx + b = mx + b$$
$$-mxb = mx$$
$$0 = 2mx$$
$$m = 0$$

So, yes, a linear function $f(x) = mx + b$ cab be even provided $m = 0$.

57.
$$x^2 - 4x + y^2 + 10y - 7 = 0$$
$$(x^2 - 4x + 4) + (y^2 + 10y + 25) = 7 + 4 + 25$$
$$(x-2)^2 + (y+5)^2 = 6^2$$

Center: (2, -5); Radius = 6

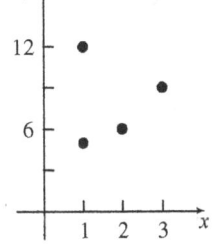

59.
$$\frac{f(3) - f(1)}{3 - 1}$$
$$= \frac{12 - (-2)}{2}$$
$$= \frac{14}{2}$$
$$= 7$$

Section 4.2

1.

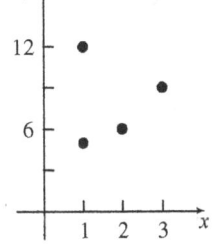

No, the relation is not a function because an input, 1, corresponds to two different outputs, 5 and 12.

3. scatter diagram

5. Linear relation, $m > 0$

7. Linear relation, $m < 0$

9. Nonlinear relation

11. a.

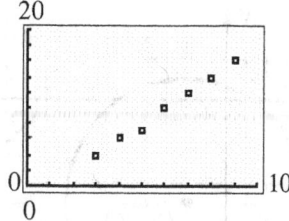

b. Answers will vary. We select (4, 6) and (8, 14). The slope of the line containing these points is:

$$m = \frac{14-6}{8-4} = \frac{8}{4} = 2$$

The equation of the line is:

$$y - y_1 = m(x - x_1)$$
$$y - 6 = 2(x - 4)$$
$$y - 6 = 2x - 8$$
$$y = 2x - 2$$

c.

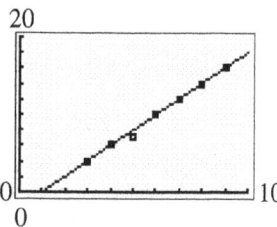

d. Using the LINear REGression program, the line of best fit is:

$$y = 2.0357x - 2.3571$$

e.

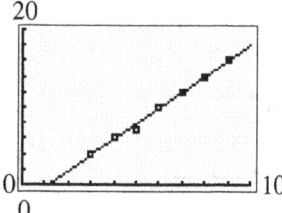

13. a.

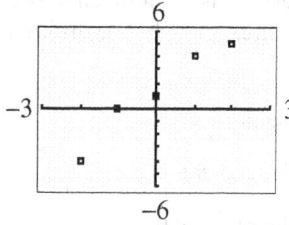

b. Answers will vary. We select (−2,−4) and (2, 5). The slope of the line containing these points is: $m = \dfrac{5-(-4)}{2-(-2)} = \dfrac{9}{4}$.

The equation of the line is:

$$y - y_1 = m(x - x_1)$$
$$y - (-4) = \frac{9}{4}(x - (-2))$$
$$y + 4 = \frac{9}{4}x + \frac{9}{2}$$
$$y = \frac{9}{4}x + \frac{1}{2}$$

c.

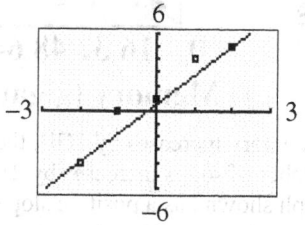

e. Using the LINear REGression program, the line of best fit is:

$$y = 2.2x + 1.2$$

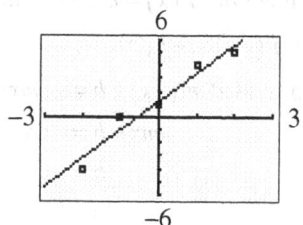

15. a.

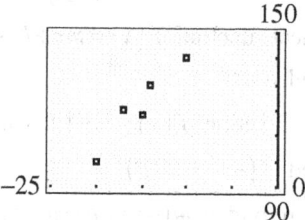

b. Answers will vary. We select (−20,100) and (−10,140). The slope of the line containing these points is:

$$m = \frac{140-100}{-10-(-20)} = \frac{40}{10} = 4$$

The equation of the line is:

$$y - y_1 = m(x - x_1)$$
$$y - 100 = 4(x - (-20))$$
$$y - 100 = 4x + 80$$
$$y = 4x + 180$$

c.

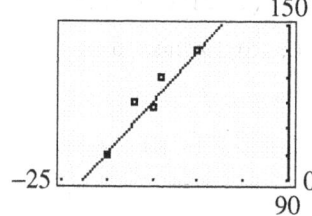

d. Using the LINear REGression program, the line of best fit is:
$$y = 3.8613x + 180.2920$$

e.

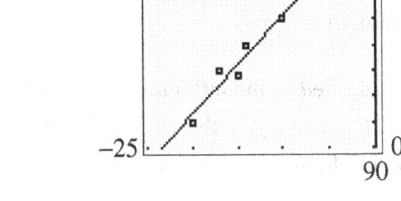

17. a.

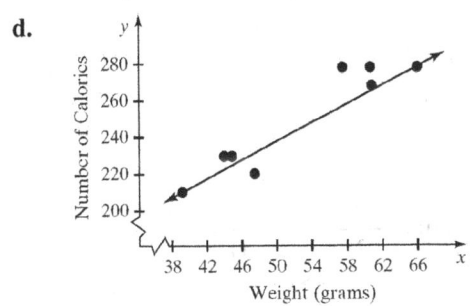

b. Linear.

c. Answers will vary. We will use the points $(39.52, 210)$ and $(66.45, 280)$.

$$m = \frac{280 - 210}{66.45 - 39.52} = \frac{70}{26.93} \approx 2.5993316$$

$$y - 210 = 2.5993316(x - 39.52)$$

$$y - 210 = 2.5993316x - 102.7255848$$

$$y = 2.599x + 107.274$$

d.

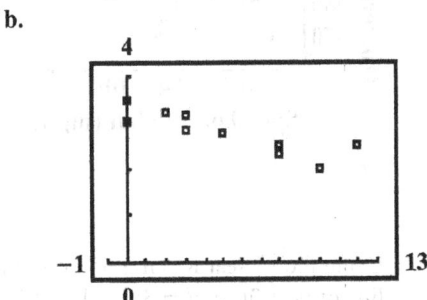

e. $x = 62.3$: $y = 2.599(62.3) + 107.274 \approx 269$

We predict that a candy bar weighing 62.3 grams will contain 269 calories.

f. If the weight of a candy bar is increased by one gram, then the number of calories will increase by 2.599.

19. a. The independent variable is the number of hours spent playing video games and cumulative grade-point average is the dependent variable because we are using number of hours playing video games to predict (or explain) cumulative grade-point average.

b.

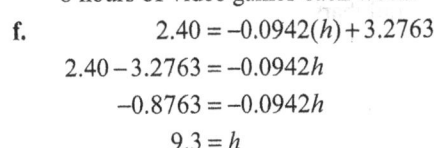

c. Using the LINear REGression program, the line of best fit is: $G(h) = -0.0942h + 3.2763$

d. If the number of hours playing video games in a week increases by 1 hour, the cumulative grade-point average decreases 0.09, on average.

e. $G(8) = -0.0942(8) + 3.2763 = 2.52$

We predict a grade-point average of approximately 2.52 for a student who plays 8 hours of video games each week.

f.
$$2.40 = -0.0942(h) + 3.2763$$
$$2.40 - 3.2763 = -0.0942h$$
$$-0.8763 = -0.0942h$$
$$9.3 = h$$

A student who has a grade-point average of 2.40 will have played approximately 9.3 hours of video games.

21. a. This relation does not represent a function since the values of the input variable s are repeated.

b.

c. Using the LINear REGression program, the line of best fit is: $d = 3.3641s + 51.8233$

d. For each 1-mph increase in the speed off bat, the homerun distance increases by 3.3641 feet, on average.

e. $d(s) = 3.3641s + 51.8233$

f. Since the speed off bat must be greater than 0 the domain is $\{s \mid s > 0\}$.

g. $d(103) = 3.3641(103) + 51.8233 \approx 398$ ft

A hurricane with a wind speed of 85 knots would have a pressure of approximately 967 millibars.

23.

Age of Mother

The data do not follow a linear pattern so it would not make sense to find the line of best fit.

25. A correlation coefficient of 0 implies that the data do not have a linear relationship.

27. $G(0) = -0.0942(0) + 3.2763 = 3.2763$. The approximate grade-point average of a student who plays 0 hours of video games per week would be 3.28.

29. The domain would be all real numbers except those that make the denominator zero.

$x^2 - 25 = 0$

$x^2 = 25 \rightarrow x = \pm 5$

So the domain is: $\{x \mid x \neq 5, -5\}$

31. Since y is shifted to the left 3 units we would use $y = (x+3)^2$. Since y is also shifted down 4 units, we would use $y = (x+3)^2 - 4$.

Section 4.3

1. $y = x^2 - 9$

To find the y-intercept, let $x = 0$:

$y = 0^2 - 9 = -9$.

To find the x-intercept(s), let $y = 0$:

$x^2 - 9 = 0$

$x^2 = 9$

$x = \pm\sqrt{9} = \pm 3$

The intercepts are $(0, -9)$, $(-3, 0)$, and $(3, 0)$.

3. $\left(\frac{1}{2} \cdot (-5)\right)^2 = \frac{25}{4}$

5. parabola

7. $-\dfrac{b}{2a}$

9. True; $-\dfrac{b}{2a} = -\dfrac{4}{2(-1)} = 2$

11. a

13. C

15. F

17. G

19. H

21. $f(x) = \dfrac{1}{4}x^2$

Using the graph of $y = x^2$, compress vertically

by a factor of $\frac{1}{4}$.

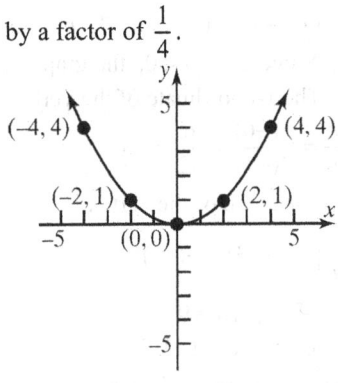

23. $f(x) = (x+2)^2 - 2$

Using the graph of $y = x^2$, shift left 2 units, then shift down 2 units.

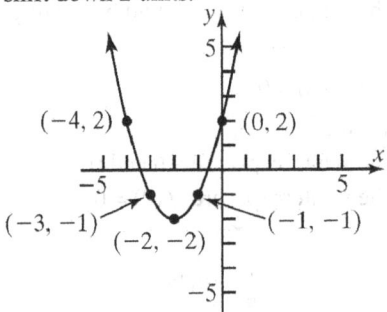

25. $f(x) = x^2 + 4x + 2$

$= (x^2 + 4x + 4) + 2 - 4$

$= (x+2)^2 - 2$

Using the graph of $y = x^2$, shift left 2 units, then shift down 2 units.

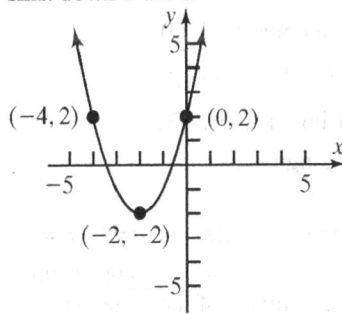

27. $f(x) = 2x^2 - 4x + 1$

$= 2(x^2 - 2x) + 1$

$= 2(x^2 - 2x + 1) + 1 - 2$

$= 2(x-1)^2 - 1$

Using the graph of $y = x^2$, shift right 1 unit, stretch vertically by a factor of 2, then shift

down 1 unit.

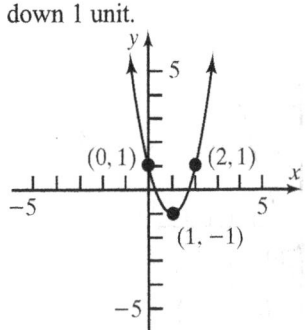

29. $f(x) = -x^2 - 2x$

$= -(x^2 + 2x)$

$= -(x^2 + 2x + 1) + 1$

$= -(x+1)^2 + 1$

Using the graph of $y = x^2$, shift left 1 unit, reflect across the x-axis, then shift up 1 unit.

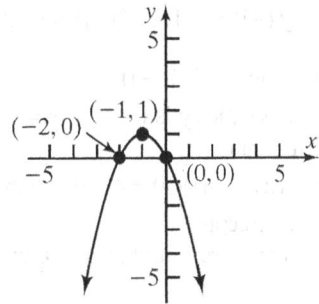

31. $f(x) = \frac{1}{2}x^2 + x - 1$

$= \frac{1}{2}(x^2 + 2x) - 1$

$= \frac{1}{2}(x^2 + 2x + 1) - 1 - \frac{1}{2}$

$= \frac{1}{2}(x+1)^2 - \frac{3}{2}$

Using the graph of $y = x^2$, shift left 1 unit, compress vertically by a factor of $\frac{1}{2}$, then shift down $\frac{3}{2}$ units.

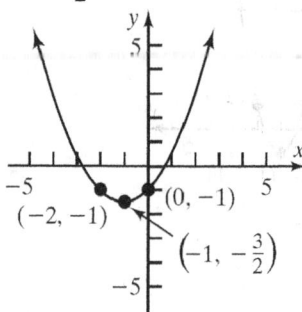

33. a. For $f(x) = x^2 + 2x$, $a = 1$, $b = 2$, $c = 0$.
Since $a = 1 > 0$, the graph opens up.
The x-coordinate of the vertex is
$$x = \frac{-b}{2a} = \frac{-(2)}{2(1)} = \frac{-2}{2} = -1.$$
The y-coordinate of the vertex is
$$f\left(\frac{-b}{2a}\right) = f(-1) = (-1)^2 + 2(-1) = 1 - 2 = -1.$$

Thus, the vertex is $(-1, -1)$.
The axis of symmetry is the line $x = -1$.
The discriminant is
$b^2 - 4ac = (2)^2 - 4(1)(0) = 4 > 0$, so the graph
has two x-intercepts.
The x-intercepts are found by solving:
$$x^2 + 2x = 0$$
$$x(x+2) = 0$$
$$x = 0 \text{ or } x = -2$$
The x-intercepts are -2 and 0.
The y-intercept is $f(0) = 0$.

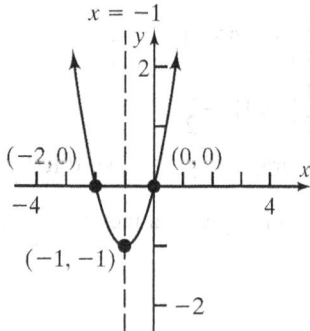

b. The domain is $(-\infty, \infty)$.
The range is $[-1, \infty)$.

c. Decreasing on $(-\infty, -1)$.
Increasing on $(-1, \infty)$.

35. a. For $f(x) = -x^2 - 6x$, $a = -1$, $b = -6$,
$c = 0$. Since $a = -1 < 0$, the graph opens
down. The x-coordinate of the vertex is
$$x = \frac{-b}{2a} = \frac{-(-6)}{2(-1)} = \frac{6}{-2} = -3.$$
The y-coordinate of the vertex is
$$f\left(\frac{-b}{2a}\right) = f(-3) = -(-3)^2 - 6(-3)$$
$$= -9 + 18 = 9.$$
Thus, the vertex is $(-3, 9)$.
The axis of symmetry is the line $x = -3$.
The discriminant is:
$$b^2 - 4ac = (-6)^2 - 4(-1)(0) = 36 > 0,$$
so the graph has two x-intercepts.
The x-intercepts are found by solving:
$$-x^2 - 6x = 0$$
$$-x(x+6) = 0$$
$$x = 0 \text{ or } x = -6.$$
The x-intercepts are -6 and 0.
The y-intercepts are $f(0) = 0$.

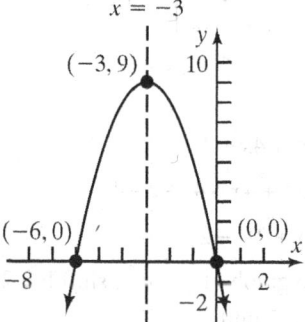

b. The domain is $(-\infty, \infty)$.
The range is $(-\infty, 9]$.

c. Increasing on $(-\infty, -3)$.
Decreasing on $(-3, \infty)$.

37. a. For $f(x) = x^2 + 2x - 8$, $a = 1$, $b = 2$, $c = -8$.
Since $a = 1 > 0$, the graph opens up.
The x-coordinate of the vertex is
$$x = \frac{-b}{2a} = \frac{-2}{2(1)} = \frac{-2}{2} = -1.$$
The y-coordinate of the vertex is
$$f\left(\frac{-b}{2a}\right) = f(-1) = (-1)^2 + 2(-1) - 8$$
$$= 1 - 2 - 8 = -9.$$
Thus, the vertex is $(-1, -9)$.
The axis of symmetry is the line $x = -1$.

The discriminant is:

$b^2 - 4ac = 2^2 - 4(1)(-8) = 4 + 32 = 36 > 0$,

so the graph has two x-intercepts.

The x-intercepts are found by solving:

$x^2 + 2x - 8 = 0$

$(x+4)(x-2) = 0$

$x = -4$ or $x = 2$.

The x-intercepts are -4 and 2.

The y-intercept is $f(0) = -8$.

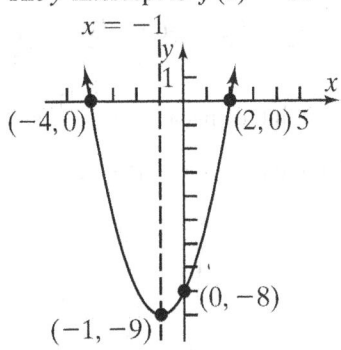

b. The domain is $(-\infty, \infty)$.

The range is $[-9, \infty)$.

c. Decreasing on $(-\infty, -1)$.

Increasing on $(-1, \infty)$.

39. a. For $f(x) = x^2 + 2x + 1$, $a = 1$, $b = 2$, $c = 1$.

Since $a = 1 > 0$, the graph opens up.

The x-coordinate of the vertex is

$x = \dfrac{-b}{2a} = \dfrac{-2}{2(1)} = \dfrac{-2}{2} = -1$.

The y-coordinate of the vertex is

$f\left(\dfrac{-b}{2a}\right) = f(-1)$

$= (-1)^2 + 2(-1) + 1 = 1 - 2 + 1 = 0$.

Thus, the vertex is $(-1, 0)$.

The axis of symmetry is the line $x = -1$.

The discriminant is:

$b^2 - 4ac = 2^2 - 4(1)(1) = 4 - 4 = 0$,

so the graph has one x-intercept.

The x-intercept is found by solving:

$x^2 + 2x + 1 = 0$

$(x+1)^2 = 0$

$x = -1$.

The x-intercept is -1.

The y-intercept is $f(0) = 1$.

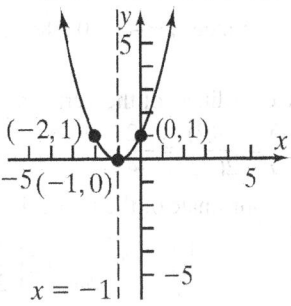

b. The domain is $(-\infty, \infty)$.

The range is $[0, \infty)$.

c. Decreasing on $(-\infty, -1)$.

Increasing on $(-1, \infty)$.

41. a. For $f(x) = 2x^2 - x + 2$, $a = 2$, $b = -1$, $c = 2$.

Since $a = 2 > 0$, the graph opens up.

The x-coordinate of the vertex is

$x = \dfrac{-b}{2a} = \dfrac{-(-1)}{2(2)} = \dfrac{1}{4}$.

The y-coordinate of the vertex is

$f\left(\dfrac{-b}{2a}\right) = f\left(\dfrac{1}{4}\right) = 2\left(\dfrac{1}{4}\right)^2 - \dfrac{1}{4} + 2$

$= \dfrac{1}{8} - \dfrac{1}{4} + 2 = \dfrac{15}{8}$.

Thus, the vertex is $\left(\dfrac{1}{4}, \dfrac{15}{8}\right)$.

The axis of symmetry is the line $x = \dfrac{1}{4}$.

The discriminant is:

$b^2 - 4ac = (-1)^2 - 4(2)(2) = 1 - 16 = -15$,

so the graph has no x-intercepts.

The y-intercept is $f(0) = 2$.

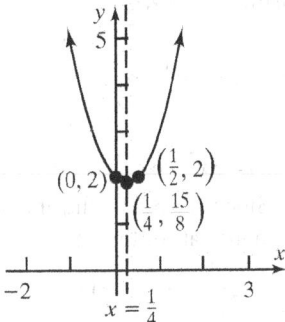

b. The domain is $(-\infty, \infty)$. The range is $\left[\dfrac{15}{8}, \infty\right)$.

c. Decreasing on $\left(-\infty, \dfrac{1}{4}\right)$. Increasing on $\left(\dfrac{1}{4}, \infty\right)$.

43. a. For $f(x) = -2x^2 + 2x - 3$, $a = -2$, $b = 2$, $c = -3$. Since $a = -2 < 0$, the graph opens down.

The x-coordinate of the vertex is

$$x = \frac{-b}{2a} = \frac{-(2)}{2(-2)} = \frac{-2}{-4} = \frac{1}{2}.$$

The y-coordinate of the vertex is

$$f\left(\frac{-b}{2a}\right) = f\left(\frac{1}{2}\right) = -2\left(\frac{1}{2}\right)^2 + 2\left(\frac{1}{2}\right) - 3$$

$$= -\frac{1}{2} + 1 - 3 = -\frac{5}{2}.$$

Thus, the vertex is $\left(\frac{1}{2}, -\frac{5}{2}\right)$.

The axis of symmetry is the line $x = \frac{1}{2}$.

The discriminant is:

$$b^2 - 4ac = 2^2 - 4(-2)(-3) = 4 - 24 = -20,$$

so the graph has no x-intercepts.
The y-intercept is $f(0) = -3$.

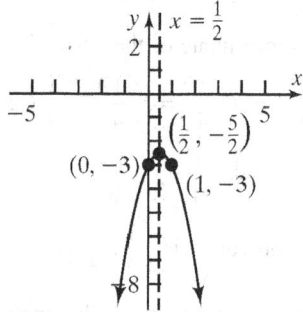

b. The domain is $(-\infty, \infty)$.

The range is $\left(-\infty, -\frac{5}{2}\right]$.

c. Increasing on $\left(-\infty, \frac{1}{2}\right)$.

Decreasing on $\left(\frac{1}{2}, \infty\right)$.

45. a. For $f(x) = 3x^2 + 6x + 2$, $a = 3$, $b = 6$, $c = 2$. Since $a = 3 > 0$, the graph opens up.
The x-coordinate of the vertex is

$$x = \frac{-b}{2a} = \frac{-6}{2(3)} = \frac{-6}{6} = -1.$$

The y-coordinate of the vertex is

$$f\left(\frac{-b}{2a}\right) = f(-1) = 3(-1)^2 + 6(-1) + 2$$

$$= 3 - 6 + 2 = -1.$$

Thus, the vertex is $(-1, -1)$.
The axis of symmetry is the line $x = -1$.
The discriminant is:

$$b^2 - 4ac = 6^2 - 4(3)(2) = 36 - 24 = 12,$$

so the graph has two x-intercepts.
The x-intercepts are found by solving:

$$3x^2 + 6x + 2 = 0$$

$$x = \frac{-b \pm \sqrt{b^2 - 4ac}}{2a}$$

$$= \frac{-6 \pm \sqrt{12}}{6} = \frac{-6 \pm 2\sqrt{3}}{6} = \frac{-3 \pm \sqrt{3}}{3}$$

The x-intercepts are $-1 - \frac{\sqrt{3}}{3}$ and $-1 + \frac{\sqrt{3}}{3}$.

The y-intercept is $f(0) = 2$.

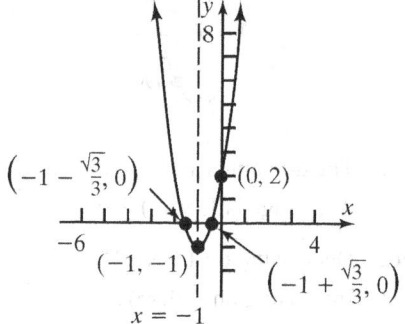

b. The domain is $(-\infty, \infty)$.

The range is $[-1, \infty)$.

c. Decreasing on $(-\infty, -1)$.

Increasing on $(-1, \infty)$.

47. a. For $f(x) = -4x^2 - 6x + 2$, $a = -4$, $b = -6$, $c = 2$. Since $a = -4 < 0$, the graph opens down.
The x-coordinate of the vertex is

$$x = \frac{-b}{2a} = \frac{-(-6)}{2(-4)} = \frac{6}{-8} = -\frac{3}{4}.$$

The y-coordinate of the vertex is

$$f\left(\frac{-b}{2a}\right) = f\left(-\frac{3}{4}\right) = -4\left(-\frac{3}{4}\right)^2 - 6\left(-\frac{3}{4}\right) + 2$$

$$= -\frac{9}{4} + \frac{9}{2} + 2 = \frac{17}{4}.$$

Thus, the vertex is $\left(-\frac{3}{4}, \frac{17}{4}\right)$.

The axis of symmetry is the line $x = -\frac{3}{4}$.

192

The discriminant is:

$b^2 - 4ac = (-6)^2 - 4(-4)(2) = 36 + 32 = 68$,

so the graph has two x-intercepts.

The x-intercepts are found by solving:

$-4x^2 - 6x + 2 = 0$

$x = \dfrac{-b \pm \sqrt{b^2 - 4ac}}{2a} = \dfrac{-(-6) \pm \sqrt{68}}{2(-4)}$

$= \dfrac{6 \pm \sqrt{68}}{-8} = \dfrac{6 \pm 2\sqrt{17}}{-8} = \dfrac{3 \pm \sqrt{17}}{-4}$

The x-intercepts are $\dfrac{-3 + \sqrt{17}}{4}$ and $\dfrac{-3 - \sqrt{17}}{4}$.

The y-intercept is $f(0) = 2$.

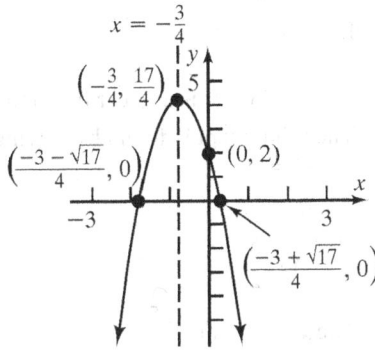

b. The domain is $(-\infty, \infty)$.

The range is $\left(-\infty, \dfrac{17}{4}\right]$.

c. Decreasing on $\left(-\dfrac{3}{4}, \infty\right)$.

Increasing on $\left(-\infty, -\dfrac{3}{4}\right)$.

49. Consider the form $y = a(x-h)^2 + k$. From the graph we know that the vertex is $(-1,-2)$ so we have $h = -1$ and $k = -2$. The graph also passes through the point $(x,y) = (0,-1)$. Substituting these values for x, y, h, and k, we can solve for a:

$-1 = a(0 - (-1))^2 + (-2)$

$-1 = a(1)^2 - 2$

$-1 = a - 2$

$1 = a$

The quadratic function is

$f(x) = (x+1)^2 - 2 = x^2 + 2x - 1$.

51. Consider the form $y = a(x-h)^2 + k$. From the graph we know that the vertex is $(-3,5)$ so we have $h = -3$ and $k = 5$. The graph also passes through the point $(x,y) = (0,-4)$. Substituting these values for x, y, h, and k, we can solve for a:

$-4 = a(0 - (-3))^2 + 5$

$-4 = a(3)^2 + 5$

$-4 = 9a + 5$

$-9 = 9a$

$-1 = a$

The quadratic function is

$f(x) = -(x+3)^2 + 5 = -x^2 - 6x - 4$.

53. Consider the form $y = ax^2 + bx + c$. Substituting the three points from the graph into the general form we have the following three equations.

$5 = a(-1)^2 + b(-1) + c \Rightarrow 5 = a - b + c$

and

$5 = a(3)^2 + b(3) + c \Rightarrow 5 = 9a + 3b + c$

and

$-1 = a(0)^2 + b(0) + c \Rightarrow -1 = c$

Since $-1 = c$, we have the following equations:

$5 = a - b - 1, \quad 5 = 9a + 3b - 1, \quad -1 = c$

Solving the first two simultaneously we have

$\left.\begin{array}{l} 5 = a - b - 1 \\ 5 = 9a + 3b - 1 \end{array}\right\}$

$\left.\begin{array}{l} 6 = a - b \\ 6 = 9a + 3b \end{array}\right\} \rightarrow a = 2, b = -4$

The quadratic function is $f(x) = 2x^2 - 4x - 1$.

55. For $f(x) = 2x^2 + 12x$, $a = 2$, $b = 12$, $c = 0$. Since $a = 2 > 0$, the graph opens up, so the vertex is a minimum point. The minimum occurs at $x = \dfrac{-b}{2a} = \dfrac{-12}{2(2)} = \dfrac{-12}{4} = -3$.

The minimum value is

$f(-3) = 2(-3)^2 + 12(-3) = 18 - 36 = -18$.

57. For $f(x) = 2x^2 + 12x - 3$, $a = 2$, $b = 12$, $c = -3$.
Since $a = 2 > 0$, the graph opens up, so the vertex is a minimum point. The minimum occurs at
$$x = \frac{-b}{2a} = \frac{-12}{2(2)} = \frac{-12}{4} = -3.$$ The minimum value is
$$f(-3) = 2(-3)^2 + 12(-3) - 3 = 18 - 36 - 3 = -21.$$

59. For $f(x) = -x^2 + 10x - 4$, $a = -1$, $b = 10$, $c = -4$.
Since $a = -1 < 0$, the graph opens down, so the vertex is a maximum point. The maximum occurs
at $x = \frac{-b}{2a} = \frac{-10}{2(-1)} = \frac{-10}{-2} = 5$. The maximum value is
$$f(5) = -(5)^2 + 10(5) - 4 = -25 + 50 - 4 = 21.$$

61. For $f(x) = -3x^2 + 12x + 1$, $a = -3$, $b = 12$, $c = 1$.
Since $a = -3 < 0$, the graph opens down, so the vertex is a maximum point. The maximum occurs
at $x = \frac{-b}{2a} = \frac{-12}{2(-3)} = \frac{-12}{-6} = 2$. The maximum value
is $f(2) = -3(2)^2 + 12(2) + 1 = -12 + 24 + 1 = 13$.

63. a. For $f(x) = x^2 - 2x - 15$, $a = 1$, $b = -2$,
$c = -15$. Since $a = 1 > 0$, the graph opens up.
The x-coordinate of the vertex is
$$x = \frac{-b}{2a} = \frac{-(-2)}{2(1)} = \frac{2}{2} = 1.$$
The y-coordinate of the vertex is
$$f\left(\frac{-b}{2a}\right) = f(1) = (1)^2 - 2(1) - 15$$
$$= 1 - 2 - 15 = -16.$$
Thus, the vertex is $(1, -16)$.
The discriminant is:
$$b^2 - 4ac = (-2)^2 - 4(1)(-15) = 4 + 60 = 64 > 0,$$
so the graph has two x-intercepts.
The x-intercepts are found by solving:
$$x^2 - 2x - 15 = 0$$
$$(x + 3)(x - 5) = 0$$
$$x = -3 \text{ or } x = 5$$
The x-intercepts are -3 and 5.
The y-intercept is $f(0) = -15$.

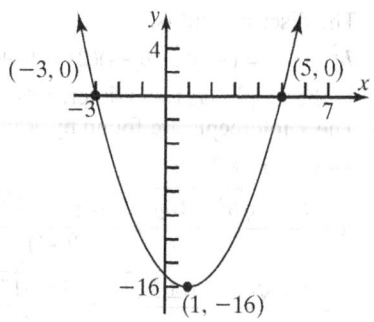

b. The domain is $(-\infty, \infty)$.
The range is $[-16, \infty)$.

c. Decreasing on $(-\infty, 1)$.
Increasing on $(1, \infty)$.

65. a. $F(x) = 2x - 5$ is a linear function.
The x-intercept is found by solving:
$$2x - 5 = 0$$
$$2x = 5$$
$$x = \frac{5}{2}$$
The x-intercept is $\frac{5}{2}$.
The y-intercept is $F(0) = -5$.

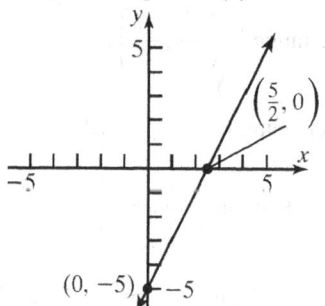

b. The domain is $(-\infty, \infty)$.
The range is $(-\infty, \infty)$.

c. Increasing on $(-\infty, \infty)$.

67. a. $g(x) = -2(x-3)^2 + 2$

Using the graph of $y = x^2$, shift right 3 units, reflect about the x-axis, stretch vertically by a factor of 2, then shift up 2 units.

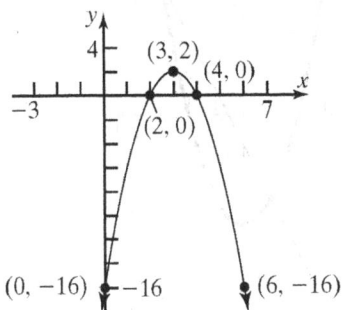

b. The domain is $(-\infty, \infty)$.
The range is $(-\infty, 2]$.

c. Increasing on $(-\infty, 3)$.
Decreasing on $(3, \infty)$.

69. a. For $f(x) = 2x^2 + x + 1$, $a = 2$, $b = 1$, $c = 1$.
Since $a = 2 > 0$, the graph opens up.
The x-coordinate of the vertex is
$$x = \frac{-b}{2a} = \frac{-1}{2(2)} = \frac{-1}{4} = -\frac{1}{4}.$$
The y-coordinate of the vertex is
$$f\left(\frac{-b}{2a}\right) = f\left(-\frac{1}{4}\right) = 2\left(-\frac{1}{4}\right)^2 + \left(-\frac{1}{4}\right) + 1$$
$$= \frac{1}{8} - \frac{1}{4} + 1 = \frac{7}{8}.$$
Thus, the vertex is $\left(-\frac{1}{4}, \frac{7}{8}\right)$.

The discriminant is:
$b^2 - 4ac = 1^2 - 4(2)(1) = 1 - 8 = -7$,
so the graph has no x-intercepts.
The y-intercept is $f(0) = 1$.

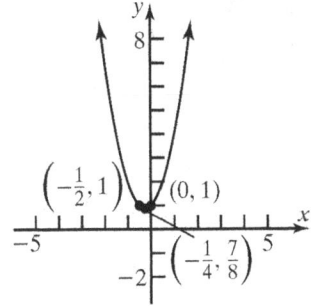

b. The domain is $(-\infty, \infty)$.
The range is $\left[\frac{7}{8}, \infty\right)$.

c. Decreasing on $\left(-\infty, -\frac{1}{4}\right)$.
Increasing on $\left(-\frac{1}{4}, \infty\right)$.

71. a. $h(x) = -\frac{2}{5}x + 4$ is a linear function.
The x-intercept is found by solving:
$$-\frac{2}{5}x + 4 = 0$$
$$-\frac{2}{5}x = -4$$
$$x = -4\left(-\frac{5}{2}\right) = 10$$
The x-intercept is 10.
The y-intercept is $h(0) = 4$.

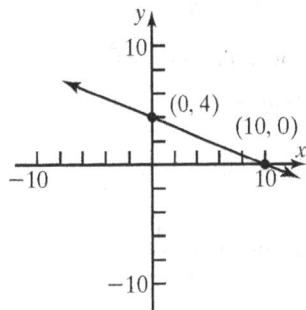

b. The domain is $(-\infty, \infty)$.
The range is $(-\infty, \infty)$.

c. Decreasing on $(-\infty, \infty)$.

73. a. For $H(x) = -4x^2 - 4x - 1$, $a = -4$, $b = -4$, $c = -1$. Since $a = -4 < 0$, the graph opens down. The x-coordinate of the vertex is
$$x = \frac{-b}{2a} = \frac{-(-4)}{2(-4)} = \frac{4}{-8} = -\frac{1}{2}.$$
The y-coordinate of the vertex is
$$H\left(\frac{-b}{2a}\right) = H\left(-\frac{1}{2}\right) = -4\left(-\frac{1}{2}\right)^2 - 4\left(-\frac{1}{2}\right) - 1$$
$$= -1 + 2 - 1 = 0$$
Thus, the vertex is $\left(-\frac{1}{2}, 0\right)$.

The discriminant is:
$b^2 - 4ac = (-4)^2 - 4(-4)(-1) = 16 - 16 = 0$,

195

so the graph has one x-intercept.
The x-intercept is found by solving:

$$-4x^2 - 4x - 1 = 0$$
$$4x^2 + 4x + 1 = 0$$
$$(2x+1)^2 = 0$$
$$2x + 1 = 0$$
$$x = -\frac{1}{2}$$

The x-intercept is $-\frac{1}{2}$.

The y-intercept is $H(0) = -1$.

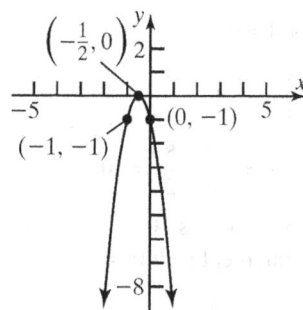

b. The domain is $(-\infty, \infty)$.
The range is $(-\infty, 0]$.

c. Increasing on $\left(-\infty, -\frac{1}{2}\right)$.

Decreasing on $\left(-\frac{1}{2}, \infty\right)$.

75. Use the form $f(x) = a(x-h)^2 + k$.
The vertex is $(0,2)$, so $h = 0$ and $k = 2$.
$f(x) = a(x-0)^2 + 2 = ax^2 + 2$.
Since the graph passes through $(1, 8)$, $f(1) = 8$.

$$f(x) = ax^2 + 2$$
$$8 = a(1)^2 + 2$$
$$8 = a + 2$$
$$6 = a$$
$$f(x) = 6x^2 + 2.$$

$a = 6, b = 0, c = 2$

77. a and d.

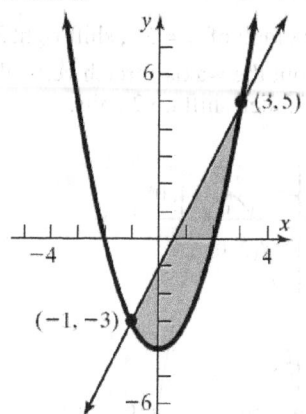

b. $f(x) = g(x)$
$$2x - 1 = x^2 - 4$$
$$0 = x^2 - 2x - 3$$
$$0 = (x+1)(x-3)$$
$$x + 1 = 0 \quad \text{or} \quad x - 3 = 0$$
$$x = -1 \qquad x = 3$$

The solution set is $\{-1, 3\}$.

c. $f(-1) = 2(-1) - 1 = -2 - 1 = -3$
$g(-1) = (-1)^2 - 4 = 1 - 4 = -3$
$f(3) = 2(3) - 1 = 6 - 1 = 5$
$g(3) = (3)^2 - 4 = 9 - 4 = 5$

Thus, the graphs of f and g intersect at the points $(-1, -3)$ and $(3, 5)$.

79. a and d.

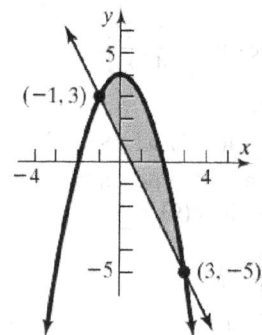

b. $f(x) = g(x)$

$-x^2 + 4 = -2x + 1$

$0 = x^2 - 2x - 3$

$0 = (x+1)(x-3)$

$x + 1 = 0$ or $x - 3 = 0$

$x = -1$ $\qquad x = 3$

The solution set is $\{-1, 3\}$.

c. $f(1) = -(-1)^2 + 4 = -1 + 4 = 3$

$g(1) = -2(-1) + 1 = 2 + 1 = 3$

$f(3) = -(3)^2 + 4 = -9 + 4 = -5$

$g(3) = -2(3) + 1 = -6 + 1 = -5$

Thus, the graphs of f and g intersect at the points $(-1, 3)$ and $(3, -5)$.

81. a and d.

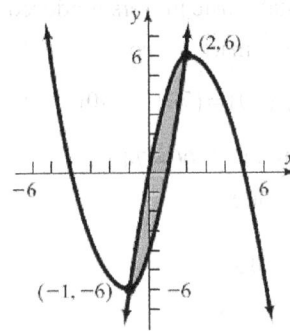

b. $f(x) = g(x)$

$-x^2 + 5x = x^2 + 3x - 4$

$0 = 2x^2 - 2x - 4$

$0 = x^2 - x - 2$

$0 = (x+1)(x-2)$

$x + 1 = 0$ or $x - 2 = 0$

$x = -1$ $\qquad x = 2$

The solution set is $\{-1, 2\}$.

c. $f(-1) = -(-1)^2 + 5(-1) = -1 - 5 = -6$

$g(-1) = (-1)^2 + 3(-1) - 4 = 1 - 3 - 4 = -6$

$f(2) = -(2)^2 + 5(2) = -4 + 10 = 6$

$g(2) = 2^2 + 3(2) - 4 = 4 + 6 - 4 = 6$

Thus, the graphs of f and g intersect at the points $(-1, -6)$ and $(2, 6)$.

83. a. For $a = 1$:

$f(x) = a(x - r_1)(x - r_2)$

$= 1(x - (-3))(x - 1)$

$= (x + 3)(x - 1) = x^2 + 2x - 3$

For $a = 2$:

$f(x) = 2(x - (-3))(x - 1)$

$= 2(x + 3)(x - 1)$

$= 2(x^2 + 2x - 3) = 2x^2 + 4x - 6$

For $a = -2$:

$f(x) = -2(x - (-3))(x - 1)$

$= -2(x + 3)(x - 1)$

$= -2(x^2 + 2x - 3) = -2x^2 - 4x + 6$

For $a = 5$:

$f(x) = 5(x - (-3))(x - 1)$

$= 5(x + 3)(x - 1)$

$= 5(x^2 + 2x - 3) = 5x^2 + 10x - 15$

b. The x-intercepts are not affected by the value of a. The y-intercept is multiplied by the value of a.

c. The axis of symmetry is unaffected by the value of a. For this problem, the axis of symmetry is $x = -1$ for all values of a.

d. The x-coordinate of the vertex is not affected by the value of a. The y-coordinate of the vertex is multiplied by the value of a.

e. The x-coordinate of the vertex is the mean of the x-intercepts.

85. a. $x = -\dfrac{b}{2a} = -\dfrac{4}{2(1)} = -2$

$y = f(-2) = (-2)^2 + 4(-2) - 21 = -25$

The vertex is $(-2, -25)$.

b. $f(x) = 0$

$x^2 + 4x - 21 = 0$

$(x + 7)(x - 3) = 0$

$x + 7 = 0$ or $x - 3 = 0$

$x = -7$ $\qquad x = 3$

The x-intercepts of f are $(-7, 0)$ and $(3, 0)$.

197

c.
$$f(x) = -21$$
$$x^2 + 4x - 21 = -21$$
$$x^2 + 4x - 0$$
$$x(x+4) = 0$$
$$x = 0 \quad \text{or} \quad x + 4 = 0$$
$$x = -4$$

The solutions $f(x) = -21$ are -4 and 0.

Thus, the points $(-4, -21)$ and $(0, -21)$ are on the graph of f.

d.

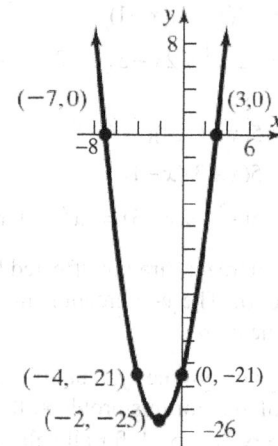

87. Let (x, y) represent a point on the line $y = x$. Then the distance from (x, y) to the point $(3, 1)$ is
$d = \sqrt{(x-3)^2 + (y-1)^2}$. Since $y = x$, we can replace the y variable with x so that we have the distance expressed as a function of x:

$$d(x) = \sqrt{(x-3)^2 + (x-1)^2}$$
$$= \sqrt{x^2 - 6x + 9 + x^2 - 2x + 1}$$
$$= \sqrt{2x^2 - 8x + 10}$$

Squaring both sides of this function, we obtain $[d(x)]^2 = 2x^2 - 8x + 10$.

Now, the expression on the right is quadratic. Since $a = 2 > 0$, it has a minimum. Finding the x-coordinate of the minimum point of $[d(x)]^2$ will also give us the x-coordinate of the minimum of

$d(x)$: $x = \dfrac{-b}{2a} = \dfrac{-(-8)}{2(2)} = \dfrac{8}{4} = 2$. So, 2 is the x-coordinate of the point on the line $y = x$ that is closest to the point $(3, 1)$. Since $y = x$, the y-coordinate is also 2. Thus, the point is $(2, 2)$ is the point on the line $y = x$ that is closest to $(3, 1)$.

89. $R(p) = -4p^2 + 4000p$, $a = -4, b = 4000, c = 0$.

Since $a = -4 < 0$ the graph is a parabola that opens down, so the vertex is a maximum point. The maximum occurs at $p = \dfrac{-b}{2a} = \dfrac{-4000}{2(-4)} = 500$.

Thus, the unit price should be \$500 for maximum revenue. The maximum revenue is
$$R(500) = -4(500)^2 + 4000(500)$$
$$= -1000000 + 2000000$$
$$= \$1,000,000$$

91. a. $C(x) = x^2 - 140x + 7400$,

$a = 1, b = -140, c = 7400$. Since $a = 1 > 0$, the graph opens up, so the vertex is a minimum point. The minimum marginal cost occurs at $x = \dfrac{-b}{2a} = \dfrac{-(-140)}{2(1)} = \dfrac{140}{2} = 70$,

70,000 digital music players produced.

b. The minimum marginal cost is
$$f\left(\frac{-b}{2a}\right) = f(70) = (70)^2 - 140(70) + 7400$$
$$= 4900 - 9800 + 7400$$
$$= \$2500$$

93. a. $R(x) = 75x - 0.2x^2$

$a = -0.2, b = 75, c = 0$

The maximum revenue occurs when
$$x = \frac{-b}{2a} = \frac{-75}{2(-0.2)} = \frac{-75}{-0.4} = 187.5$$

The maximum revenue occurs when $x = 187$ or $x = 188$ watches.
The maximum revenue is:
$$R(187) = 75(187) - 0.2(187)^2 = \$7031.20$$
$$R(188) = 75(188) - 0.2(188)^2 = \$7031.20$$

b. $P(x) = R(x) - C(x)$
$$= 75x - 0.2x^2 - (32x + 1750)$$
$$= -0.2x^2 + 43x - 1750$$

c. $P(x) = -0.2x^2 + 43x - 1750$

$a = -0.2, b = 43, c = -1750$

$$x = \frac{-b}{2a} = \frac{-43}{2(-0.2)} = \frac{-43}{-0.4} = 107.5$$

The maximum profit occurs when $x = 107$ or $x = 108$ watches.

The maximum profit is:

$P(107) = -0.2(107)^2 + 43(107) - 1750$

$\qquad = \$561.20$

$P(108) = -0.2(108)^2 + 43(108) - 1750$

$\qquad = \$561.20$

d. Answers will vary.

95. a. $d(v) = 1.1v + 0.06v^2$

$d(45) = 1.1(45) + 0.06(45)^2$

$\qquad = 49.5 + 121.5 = 171$ ft.

b. $200 = 1.1v + 0.06v^2$

$0 = -200 + 1.1v + 0.06v^2$

$x = \dfrac{-(1.1) \pm \sqrt{(1.1)^2 - 4(0.06)(-200)}}{2(0.06)}$

$= \dfrac{-1.1 \pm \sqrt{49.21}}{0.12}$

$\approx \dfrac{-1.1 \pm 7.015}{0.12}$

$v \approx 49$ or $v \approx -68$

Disregard the negative value since we are talking about speed. So the maximum speed you can be traveling would be approximately 49 mph.

c. The 1.1v term might represent the reaction time.

97. If x is even, then ax^2 and bx are even. When two even numbers are added to an odd number the result is odd. Thus, $f(x)$ is odd. If x is odd, then ax^2 and bx are odd. The sum of three odd numbers is an odd number. Thus, $f(x)$ is odd.

99. $y = x^2 + 2x - 3$; $y = x^2 + 2x + 1$; $y = x^2 + 2x$

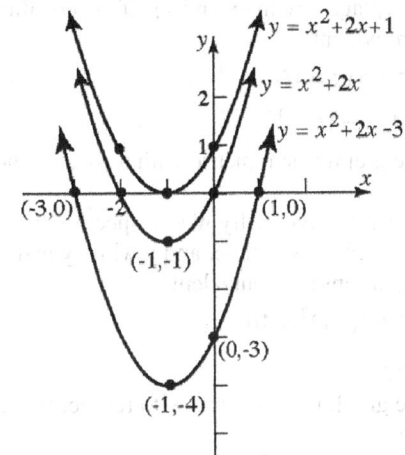

Each member of this family will be a parabola with the following characteristics:

(i) opens upwards since a > 0;

(ii) vertex occurs at $x = -\dfrac{b}{2a} = -\dfrac{2}{2(1)} = -1$;

(iii) There is at least one x-intercept since $b^2 - 4ac \geq 0$.

101. The graph of the quadratic function $f(x) = ax^2 + bx + c$ will not have any x-intercepts whenever $b^2 - 4ac < 0$.

103. No. We know that the graph of a quadratic function $f(x) = ax^2 + bx + c$ is a parabola with vertex $\left(-\frac{b}{2a}, f\left(-\frac{b}{2a}\right)\right)$. If $a > 0$, then the vertex is a minimum point, so the range is $\left[f\left(-\frac{b}{2a}\right), \infty\right)$. If $a < 0$, then the vertex is a maximum point, so the range is $\left(-\infty, f\left(-\frac{b}{2a}\right)\right]$. Therefore, it is impossible for the range to be $(-\infty, \infty)$.

105. $x^2 + 4y^2 = 16$

To check for symmetry with respect to the x-axis, replace y with $-y$ and see if the equations are equivalent.

$x^2 + 4(-y)^2 = 16$

$x^2 + 4y^2 = 16$

So the graph is symmetric with respect to the x-axis.

To check for symmetry with respect to the y-axis, replace x with –x and see if the equations are equivalent.

$(-x)^2 + 4y^2 = 16$

$x^2 + 4y^2 = 16$

So the graph is symmetric with respect to the y-axis.

To check for symmetry with respect to the origin, replace x with –x and y with –y and see if the equations are equivalent.

$(-x)^2 + 4(-y)^2 = 16$

$x^2 + 4y^2 = 16$

So the graph is symmetric with respect to the origin.

107. $x^2 + y^2 - 10x + 4y + 20 = 0$

$$x^2 - 10x + y^2 + 4y = -20$$

$$(x^2 - 10x + 25) + (y^2 + 4y + 4) = -20 + 25 + 4$$

$$(x - 5)^2 + (y + 2)^2 = 3^2$$

Center: (5, -2); Radius = 3

Section 4.4

1. $R = 3x$

3. a. $R(x) = x\left(-\dfrac{1}{6}x + 100\right) = -\dfrac{1}{6}x^2 + 100x$

b. The quantity sold price cannot be negative, so $x \geq 0$. Similarly, the price should be positive, so $p > 0$.

$$-\dfrac{1}{6}x + 100 > 0$$

$$-\dfrac{1}{6}x > -100$$

$$x < 600$$

Thus, the implied domain for R is $\{x \mid 0 \leq x < 600\}$ or $[0, 600)$.

c. $R(200) = -\dfrac{1}{6}(200)^2 + 100(200)$

$$= \dfrac{-20000}{3} + 20000$$

$$= \dfrac{40000}{3} \approx \$13,333.33$$

d. $x = \dfrac{-b}{2a} = \dfrac{-100}{2\left(-\frac{1}{6}\right)} = \dfrac{-100}{\left(-\frac{1}{3}\right)} = \dfrac{300}{1} = 300$

The maximum revenue is

$$R(300) = -\dfrac{1}{6}(300)^2 + 100(300)$$

$$= -15000 + 30000$$

$$= \$15,000$$

e. $p = -\dfrac{1}{6}(300) + 100 = -50 + 100 = \$50$

5. a. If $x = -5p + 100$, then $p = \dfrac{100 - x}{5}$.

$$R(x) = x\left(\dfrac{100 - x}{5}\right) = -\dfrac{1}{5}x^2 + 20x$$

b. $R(15) = -\dfrac{1}{5}(15)^2 + 20(15)$

$$= -45 + 300 = \$255$$

c. $x = \dfrac{-b}{2a} = \dfrac{-20}{2\left(-\frac{1}{5}\right)} = \dfrac{-20}{\left(-\frac{2}{5}\right)} = \dfrac{100}{2} = 50$

The maximum revenue is

$$R(50) = -\dfrac{1}{5}(50)^2 + 20(50)$$

$$= -500 + 1000 = \$500$$

d. $p = \dfrac{100 - 50}{5} = \dfrac{50}{5} = \$10$

e. Graph $R = -\dfrac{1}{5}x^2 + 20x$ and $R = 480$. Find where the graphs intersect by solving $480 = -\dfrac{1}{5}x^2 + 20x$.

Intersection X=40 Y=480   Intersection X=60 Y=480

$$\dfrac{1}{5}x^2 - 20x + 480 = 0$$

$$x^2 - 100x + 2400 = 0$$

$$(x - 40)(x - 60) = 0$$

$$x = 40, x = 60$$

Solve for price.

$x = -5p + 100$

$40 = -5p + 100 \Rightarrow p = \$12$

$60 = -5p + 100 \Rightarrow p = \$8$

The company should charge between $8 and $12.

7. a. Let w = width and l = length of the rectangular area.
Solving $P = 2w + 2l = 400$ for l:

$l = \dfrac{400 - 2w}{2} = 200 - w$.

Then $A(w) = (200 - w)w = 200w - w^2$

$\qquad = -w^2 + 200w$

b. $w = \dfrac{-b}{2a} = \dfrac{-200}{2(-1)} = \dfrac{-200}{-2} = 100$ yards

c. $A(100) = -100^2 + 200(100)$

$\qquad = -10000 + 20000$

$\qquad = 10,000 \text{ yd}^2$

9. Let x = width and y = length of the rectangle.
Solving $P = 2x + y = 4000$ for y:

$y = 4000 - 2x$.

Then $A(x) = (4000 - 2x)x$

$\qquad = 4000x - 2x^2$

$\qquad = -2x^2 + 4000x$

$x = \dfrac{-b}{2a} = \dfrac{-4000}{2(-2)} = \dfrac{-4000}{-4} = 1000 \text{ meters}$

maximizes area.

$A(1000) = -2(1000)^2 + 4000(1000)$.

$\qquad = -2000000 + 4000000$

$\qquad = 2,000,000$

The largest area that can be enclosed is 2,000,000 square meters.

11. $h(x) = \dfrac{-32x^2}{(50)^2} + x + 200 = -\dfrac{8}{625}x^2 + x + 200$

a. $a = -\dfrac{8}{625}, b = 1, c = 200$.

The maximum height occurs when

$x = \dfrac{-b}{2a} = \dfrac{-1}{2(-8/625)} = \dfrac{625}{16} \approx 39$ feet from

base of the cliff.

b. The maximum height is

$h\left(\dfrac{625}{16}\right) = \dfrac{-8}{625}\left(\dfrac{625}{16}\right)^2 + \dfrac{625}{16} + 200$

$\qquad = \dfrac{7025}{32} \approx 219.5$ feet.

c. Solving when $h(x) = 0$:

$-\dfrac{8}{625}x^2 + x + 200 = 0$

$x = \dfrac{-1 \pm \sqrt{1^2 - 4(-8/625)(200)}}{2(-8/625)}$

$x \approx \dfrac{-1 \pm \sqrt{11.24}}{-0.0256}$

$x \approx -91.90$ or $x \approx 170$

Since the distance cannot be negative, the projectile strikes the water approximately 170 feet from the base of the cliff.

d.

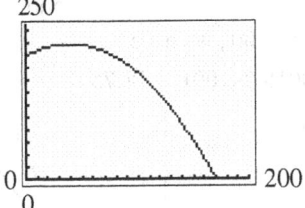

e. Using the MAXIMUM function

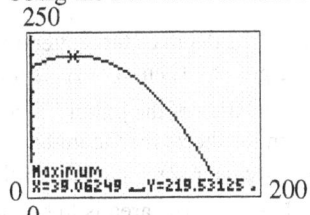

Using the ZERO function

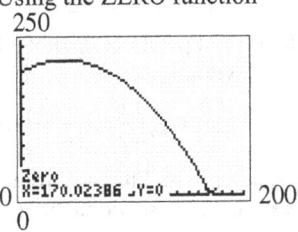

f. $-\dfrac{8}{625}x^2 + x + 200 = 100$

$-\dfrac{8}{625}x^2 + x + 100 = 0$

$x = \dfrac{\sqrt{1^2 - 4(-8/625)(100)}}{2(-8/625)} = \dfrac{-1 \pm \sqrt{6.12}}{-0.0256}$

$x \approx -57.57 \ \text{ or } \ x \approx 135.70$

Since the distance cannot be negative, the projectile is 100 feet above the water when it is approximately 135.7 feet from the base of the cliff.

13. Locate the origin at the point where the cable touches the road. Then the equation of the parabola is of the form: $y = ax^2$, where $a > 0$. Since the point (200, 75) is on the parabola, we can find the constant a:

Since $75 = a(200)^2$, then $a = \dfrac{75}{200^2} = 0.001875$.

When $x = 100$, we have:

$y = 0.001875(100)^2 = 18.75 \ \text{meters}$.

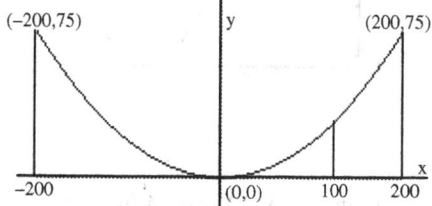

15. a. Let x = the depth of the gutter and y the width of the gutter. Then $A = xy$ is the cross-sectional area of the gutter. Since the aluminum sheets for the gutter are 12 inches wide, we have $2x + y = 12$. Solving for y:

$y = 12 - 2x$. The area is to be maximized, so:

$A = xy = x(12 - 2x) = -2x^2 + 12x$. This equation is a parabola opening down; thus, it has a maximum

when $x = \dfrac{-b}{2a} = \dfrac{-12}{2(-2)} = \dfrac{-12}{-4} = 3$.

Thus, a depth of 3 inches produces a maximum cross-sectional area.

b. Graph $A = -2x^2 + 12x$ and $A = 16$. Find where the graphs intersect by solving

$16 = -2x^2 + 12x$.

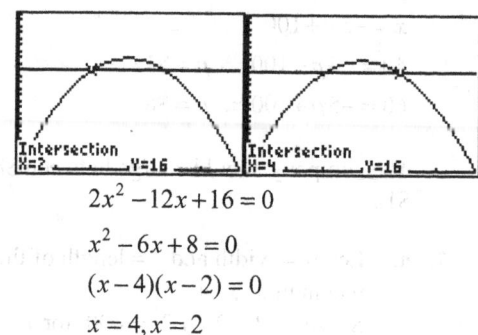

$2x^2 - 12x + 16 = 0$

$x^2 - 6x + 8 = 0$

$(x - 4)(x - 2) = 0$

$x = 4, x = 2$

The graph of $A = -2x^2 + 12x$ is above the graph of $A = 16$ where the depth is between 2 and 4 inches.

17. Let x = the width of the rectangle or the diameter of the semicircle and let y = the length of the rectangle. The perimeter of each semicircle is $\dfrac{\pi x}{2}$.

The perimeter of the track is given

by: $\dfrac{\pi x}{2} + \dfrac{\pi x}{2} + y + y = 1500$.

Solving for x:

$\pi x + 2y = 1500$

$\pi x = 1500 - 2y$

$x = \dfrac{1500 - 2y}{\pi}$

The area of the rectangle is:

$A = xy = \left(\dfrac{1500 - 2y}{\pi}\right) y = \dfrac{-2}{\pi}y^2 + \dfrac{1500}{\pi}y$.

This equation is a parabola opening down; thus, it has a maximum when

$y = \dfrac{-b}{2a} = \dfrac{\dfrac{-1500}{\pi}}{2\left(\dfrac{-2}{\pi}\right)} = \dfrac{-1500}{-4} = 375$.

Thus, $x = \dfrac{1500 - 2(375)}{\pi} = \dfrac{750}{\pi} \approx 238.73$

The dimensions for the rectangle with maximum area are $\dfrac{750}{\pi} \approx 238.73$ meters by 375 meters.

19. We are given: $V(x) = kx(a - x) = -kx^2 + akx$.

The reaction rate is a maximum when:

$x = \dfrac{-b}{2a} = \dfrac{-ak}{2(-k)} = \dfrac{ak}{2k} = \dfrac{a}{2}$.

21. $f(x) = -5x^2 + 8, \ h = 1$

Area $= \dfrac{h}{3}\left(2ah^2 + 6c\right) = \dfrac{1}{3}\left(2(-5)(1)^2 + 6(8)\right)$

$= \dfrac{1}{3}(-10 + 48) = \dfrac{38}{3}$ sq. units

23. $f(x) = x^2 + 3x + 5, \ h = 4$

Area $= \dfrac{h}{3}\left(2ah^2 + 6c\right) = \dfrac{4}{3}\left(2(1)(4)^2 + 6(5)\right)$

$= \dfrac{4}{3}(32 + 30) = \dfrac{248}{3}$ sq. units

25. a.

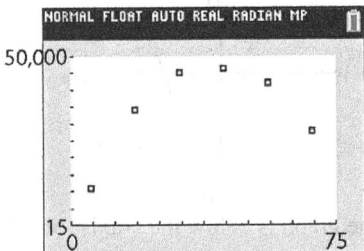

From the graph, the data appear to follow a quadratic relation with $a < 0$.

b. Using the QUADratic REGression program

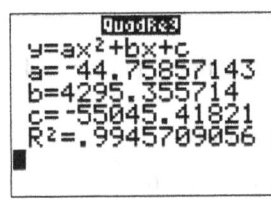

$I(x) = -44.759x^2 + 4295.356x - 55,045.418$

c. $x = \dfrac{-b}{2a} = \dfrac{-4295.356}{2(-44.759)} \approx 48.0$

An individual will earn the most income at about 48.0 years of age.

d. The maximum income will be: $I(48.0) =$

$-44.759(48.0)^2 + 4295.356(48.0) - 55,045.418$

$\approx \$48,007$

e.

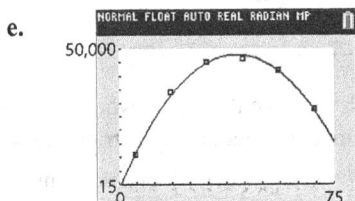

27. a.

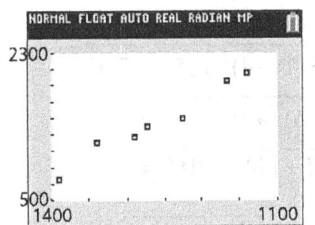

From the graph, the data appear to be linearly related with $m > 0$.

b. Using the LINear REGression program

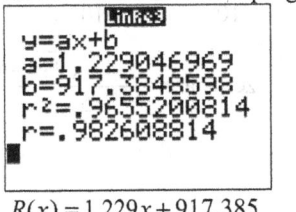

$R(x) = 1.229x + 917.385$

c. $R(875) = 1.229(875) + 917.385 \approx 1993$

The rent for an 875 square-foot apartment in San Diego will be about \$1993 per month.

29. a.

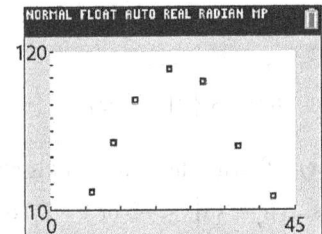

From the graph, the data appear to follow a quadratic relation with $a < 0$.

b. Using the QUADratic REGression program

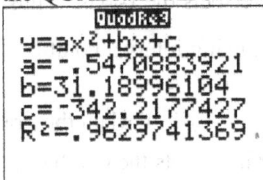

$B(a) = -0.547a^2 + 31.190a - 342.218$

c. $B(35) = -0.547(35)^2 + 31.190(35) - 342.218$

≈ 79.357

The birthrate of 35-year-old women is about 79.4 per 1000.

31. Answers will vary. One possibility follows: If the price is \$140, no one will buy the calculators, thus making the revenue \$0.

33. $d = \sqrt{(x_2 - x_1)^2 + (y_2 - y_1)^2}$

$= \sqrt{((-1) - 4)^2 + (5 - (-7))^2}$

$= \sqrt{(-5)^2 + (12)^2}$

$= \sqrt{25 + 144} = \sqrt{169} = 13$

35. $x = \dfrac{-(8) \pm \sqrt{8^2 - 4(5)(-3)}}{2(5)} = \dfrac{-8 \pm \sqrt{64 + 60}}{10}$

$= \dfrac{-8 \pm \sqrt{124}}{10} = \dfrac{-8 \pm 2\sqrt{31}}{10} = \dfrac{-4 \pm \sqrt{31}}{5}$

So the zeros are: $\dfrac{-4 + \sqrt{31}}{5}, \dfrac{-4 - \sqrt{31}}{5}$

Section 4.5

1. $-3x - 2 < 7$

$-3x < 9$

$x > -3$

The solution set is $\{x \mid x > -3\}$ or $(-3, \infty)$.

3. a. $f(x) > 0$ when the graph of f is above the x-axis. Thus, $\{x \mid x < -2 \text{ or } x > 2\}$ or, using interval notation, $(-\infty, -2) \cup (2, \infty)$.

b. $f(x) \le 0$ when the graph of f is below or intersects the x-axis. Thus, $\{x \mid -2 \le x \le 2\}$ or, using interval notation, $[-2, 2]$.

5. a. $g(x) \ge f(x)$ when the graph of g is above or intersects the graph of f. Thus $\{x \mid -2 \le x \le 1\}$ or, using interval notation, $[-2, 1]$.

b. $f(x) > g(x)$ when the graph of f is above the graph of g. Thus, $\{x \mid x < -2 \text{ or } x > 1\}$ or,

7. $x^2 - 3x - 10 < 0$

We graph the function $f(x) = x^2 - 3x - 10$. The intercepts are

y-intercept: $f(0) = -10$

x-intercepts: $x^2 - 3x - 10 = 0$

$(x - 5)(x + 2) = 0$

$x = 5, x = -2$

The vertex is at $x = \dfrac{-b}{2a} = \dfrac{-(-3)}{2(1)} = \dfrac{3}{2}$. Since

$f\left(\dfrac{3}{2}\right) = -\dfrac{49}{4}$, the vertex is $\left(\dfrac{3}{2}, -\dfrac{49}{4}\right)$.

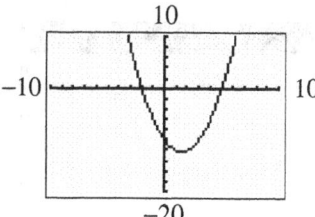

The graph is below the x-axis for $-2 < x < 5$. Since the inequality is strict, the solution set is $\{x \mid -2 < x < 5\}$ or, using interval notation, $(-2, 5)$.

9. $x^2 - 4x > 0$

We graph the function $f(x) = x^2 - 4x$. The intercepts are

y-intercept: $f(0) = 0$

x-intercepts: $x^2 - 4x = 0$

$x(x - 4) = 0$

$x = 0, x = 4$

The vertex is at $x = \dfrac{-b}{2a} = \dfrac{-(-4)}{2(1)} = \dfrac{4}{2} = 2$. Since

$f(2) = -4$, the vertex is $(2, -4)$.

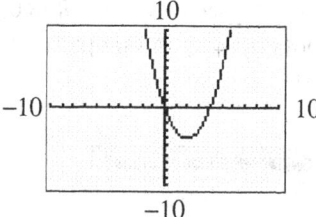

The graph is above the x-axis when $x < 0$ or $x > 4$. Since the inequality is strict, the solution set is $\{x \mid x < 0 \text{ or } x > 4\}$ or, using interval notation, $(-\infty, 0) \cup (4, \infty)$.

11. $x^2 - 9 < 0$

We graph the function $f(x) = x^2 - 9$. The intercepts are

y-intercept: $f(0) = -9$

x-intercepts: $\quad x^2 - 9 = 0$

$$(x+3)(x-3) = 0$$
$$x = -3, x = 3$$

The vertex is at $x = \dfrac{-b}{2a} = \dfrac{-(0)}{2(1)} = 0$. Since $f(0) = -9$, the vertex is $(0, -9)$.

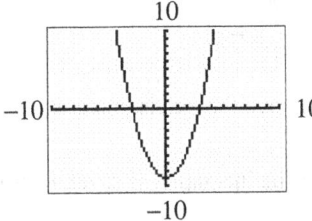

The graph is below the x-axis when $-3 < x < 3$. Since the inequality is strict, the solution set is $\{x \mid -3 < x < 3\}$ or, using interval notation, $(-3, 3)$.

13. $\quad x^2 + x > 12$

$x^2 + x - 12 > 0$

We graph the function $f(x) = x^2 + x - 12$.

y-intercept: $f(0) = -12$

x-intercepts: $\quad x^2 + x - 12 = 0$

$$(x+4)(x-3) = 0$$
$$x = -4, x = 3$$

The vertex is at $x = \dfrac{-b}{2a} = \dfrac{-(1)}{2(1)} = -\dfrac{1}{2}$. Since $f\left(-\dfrac{1}{2}\right) = -\dfrac{49}{4}$, the vertex is $\left(-\dfrac{1}{2}, -\dfrac{49}{4}\right)$.

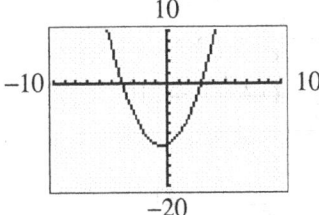

The graph is above the x-axis when $x < -4$ or $x > 3$. Since the inequality is strict, the solution set is $\{x \mid x < -4 \text{ or } x > 3\}$ or, using interval notation, $(-\infty, -4) \cup (3, \infty)$.

15. $\qquad 2x^2 < 5x + 3$

$2x^2 - 5x - 3 < 0$

We graph the function $f(x) = 2x^2 - 5x - 3$. The intercepts are

y-intercept: $f(0) = -3$

x-intercepts: $\quad 2x^2 - 5x - 3 = 0$

$$(2x+1)(x-3) = 0$$
$$x = -\dfrac{1}{2}, x = 3$$

The vertex is at $x = \dfrac{-b}{2a} = \dfrac{-(-5)}{2(2)} = \dfrac{5}{4}$. Since $f\left(\dfrac{5}{4}\right) = -\dfrac{49}{8}$, the vertex is $\left(\dfrac{5}{4}, -\dfrac{49}{8}\right)$.

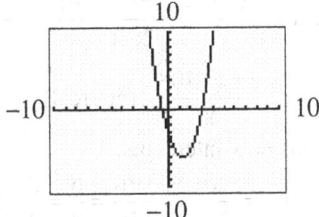

The graph is below the x-axis when $-\dfrac{1}{2} < x < 3$. Since the inequality is strict, the solution set is $\left\{x \mid -\dfrac{1}{2} < x < 3\right\}$ or, using interval notation, $\left(-\dfrac{1}{2}, 3\right)$.

17. $x^2 - x + 1 \le 0$

We graph the function $f(x) = x^2 - x + 1$. The intercepts are

y-intercept: $f(0) = 1$

x-intercepts: $x = \dfrac{-(-1) \pm \sqrt{(-1)^2 - 4(1)(1)}}{2(1)}$

$$= \dfrac{1 \pm \sqrt{-3}}{2} \text{ (not real)}$$

Therefore, f has no x-intercepts.

The vertex is at $x = \dfrac{-b}{2a} = \dfrac{-(-1)}{2(1)} = \dfrac{1}{2}$. Since $f\left(\dfrac{1}{2}\right) = \dfrac{3}{4}$, the vertex is $\left(\dfrac{1}{2}, \dfrac{3}{4}\right)$.

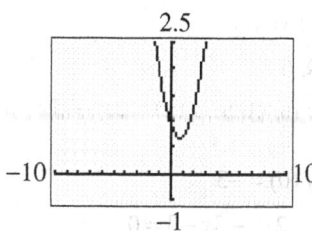

The graph is never below the x-axis. Thus, there is no real solution.

19. $4x^2 + 9 < 6x$

$4x^2 - 6x + 9 < 0$

We graph the function $f(x) = 4x^2 - 6x + 9$.

y-intercept: $f(0) = 9$

x-intercepts: $x = \dfrac{-(-6) \pm \sqrt{(-6)^2 - 4(4)(9)}}{2(4)}$

$= \dfrac{6 \pm \sqrt{-108}}{8}$ (not real)

Therefore, f has no x-intercepts.

The vertex is at $x = \dfrac{-b}{2a} = \dfrac{-(-6)}{2(4)} = \dfrac{6}{8} = \dfrac{3}{4}$. Since

$f\left(\dfrac{3}{4}\right) = \dfrac{27}{4}$, the vertex is $\left(\dfrac{3}{4}, \dfrac{27}{4}\right)$.

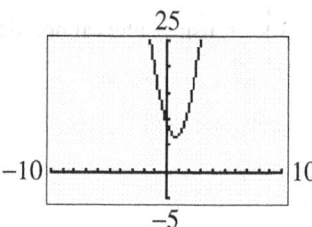

The graph is never below the x-axis. Thus, there is no real solution.

21. $6(x^2 - 1) > 5x$

$6x^2 - 6 > 5x$

$6x^2 - 5x - 6 > 0$

We graph the function $f(x) = 6x^2 - 5x - 6$.

y-intercept: $f(0) = -6$

x-intercepts: $6x^2 - 5x - 6 = 0$

$(3x + 2)(2x - 3) = 0$

$x = -\dfrac{2}{3}, \ x = \dfrac{3}{2}$

The vertex is at $x = \dfrac{-b}{2a} = \dfrac{-(-5)}{2(6)} = \dfrac{5}{12}$. Since

$f\left(\dfrac{5}{12}\right) = -\dfrac{169}{24}$, the vertex is $\left(\dfrac{5}{12}, -\dfrac{169}{24}\right)$.

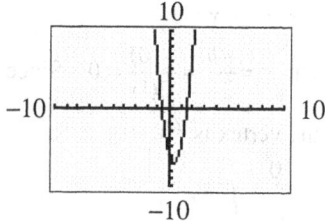

The graph is above the x-axis when $x < -\dfrac{2}{3}$ or

$x > \dfrac{3}{2}$. Since the inequality is strict, solution set

is $\left\{ x \ \middle| \ x < -\dfrac{2}{3} \ \text{or} \ x > \dfrac{3}{2} \right\}$ or, using interval

notation, $\left(-\infty, -\dfrac{2}{3}\right) \cup \left(\dfrac{3}{2}, \infty\right)$.

23. The domain of the expression $f(x) = \sqrt{x^2 - 16}$ includes all values for which $x^2 - 16 \geq 0$.

We graph the function $p(x) = x^2 - 16$. The intercepts of p are

y-intercept: $p(0) = -6$

x-intercepts: $x^2 - 16 = 0$

$(x + 4)(x - 4) = 0$

$x = -4, \ x = 4$

The vertex of p is at $x = \dfrac{-b}{2a} = \dfrac{-(0)}{2(1)} = 0$. Since

$p(0) = -16$, the vertex is $(0, -16)$.

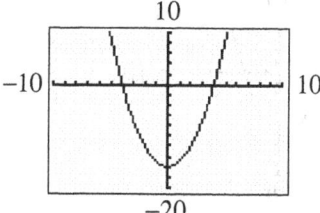

The graph of p is above the x-axis when $x < -4$ or $x > 4$. Since the inequality is not strict, the solution set of $x^2 - 16 \geq 0$ is $\{x \mid x \leq -4 \text{ or } x \geq 4\}$.

Thus, the domain of f is also $\{x \mid x \leq -4 \text{ or } x \geq 4\}$ or, using interval notation, $(-\infty, -4] \cup [4, \infty)$.

25. $f(x) = x^2 - 1; \quad g(x) = 3x + 3$

a. $\qquad f(x) = 0$

$x^2 - 1 = 0$

$(x-1)(x+1) = 0$

$x = 1; \ x = -1$

Solution set: $\{-1, 1\}$.

b. $g(x) = 0$

$3x + 3 = 0$

$3x = -3$

$x = -1$

Solution set: $\{-1\}$.

c. $\qquad f(x) = g(x)$

$x^2 - 1 = 3x + 3$

$x^2 - 3x - 4 = 0$

$(x-4)(x+1) = 0$

$x = 4; \ x = -1$

Solution set: $\{-1, 4\}$.

d. $f(x) > 0$

We graph the function $f(x) = x^2 - 1$.

y-intercept: $f(0) = -1$

x-intercepts: $\qquad x^2 - 1 = 0$

$(x+1)(x-1) = 0$

$x = -1, \ x = 1$

The vertex is at $x = \dfrac{-b}{2a} = \dfrac{-(0)}{2(1)} = 0$. Since

$f(0) = -1$, the vertex is $(0, -1)$.

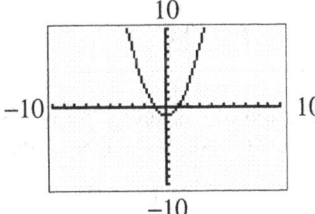

The graph is above the x-axis when $x < -1$ or $x > 1$. Since the inequality is strict, the solution set is $\{x \mid x < -1 \text{ or } x > 1\}$ or, using interval notation, $(-\infty, -1) \cup (1, \infty)$.

e. $g(x) \le 0$

$3x + 3 \le 0$

$3x \le -3$

$x \le -1$

The solution set is $\{x \mid x \le -1\}$ or, using interval notation, $(-\infty, -1]$.

f. $\qquad f(x) > g(x)$

$x^2 - 1 > 3x + 3$

$x^2 - 3x - 4 > 0$

We graph the function $p(x) = x^2 - 3x - 4$.

The intercepts of p are

y-intercept: $p(0) = -4$

x-intercepts: $\quad x^2 - 3x - 4 = 0$

$(x-4)(x+1) = 0$

$x = 4, \ x = -1$

The vertex is at $x = \dfrac{-b}{2a} = \dfrac{-(-3)}{2(1)} = \dfrac{3}{2}$. Since

$p\left(\dfrac{3}{2}\right) = -\dfrac{25}{4}$, the vertex is $\left(\dfrac{3}{2}, -\dfrac{25}{4}\right)$.

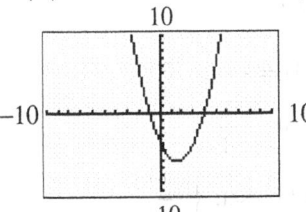

The graph of p is above the x-axis when $x < -1$ or $x > 4$. Since the inequality is strict, the solution set is $\{x \mid x < -1 \text{ or } x > 4\}$ or, using interval notation, $(-\infty, -1) \cup (4, \infty)$.

g. $f(x) \ge 1$

$x^2 - 1 \ge 1$

$x^2 - 2 \ge 0$

We graph the function $p(x) = x^2 - 2$. The intercepts of p are

y-intercept: $p(0) = -2$

x-intercepts: $x^2 - 2 = 0$

$x^2 = 2$

$x = \pm\sqrt{2}$

The vertex is at $x = \dfrac{-b}{2a} = \dfrac{-(0)}{2(1)} = 0$. Since

$p(0) = -2$, the vertex is $(0, -2)$.

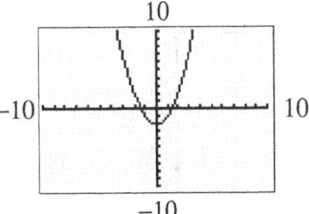

The graph of p is above the x-axis when $x < -\sqrt{2}$ or $x > \sqrt{2}$. Since the inequality is

not strict, the solution set is
$\left\{ x \mid x \le -\sqrt{2} \text{ or } x \ge \sqrt{2} \right\}$ or, using interval
notation, $\left(-\infty, -\sqrt{2} \right] \cup \left[\sqrt{2}, \infty \right)$.

27. $f(x) = -x^2 + 1;$ $g(x) = 4x + 1$

a. $f(x) = 0$
$-x^2 + 1 = 0$
$1 - x^2 = 0$
$(1 - x)(1 + x) = 0$
$x = 1;\ x = -1$
Solution set: $\{-1, 1\}$.

b. $g(x) = 0$
$4x + 1 = 0$
$4x = -1$
$x = -\dfrac{1}{4}$
Solution set: $\left\{ -\dfrac{1}{4} \right\}$.

c. $f(x) = g(x)$
$-x^2 + 1 = 4x + 1$
$0 = x^2 + 4x$
$0 = x(x + 4)$
$x = 0;\ x - 4$
Solution set: $\{-4, 0\}$.

d. $f(x) > 0$
We graph the function $f(x) = -x^2 + 1$.
y-intercept: $f(0) = 1$
x-intercepts: $-x^2 + 1 = 0$
$x^2 - 1 = 0$
$(x + 1)(x - 1) = 0$
$x = -1;\ x = 1$
The vertex is at $x = \dfrac{-b}{2a} = \dfrac{-(0)}{2(-1)} = 0$. Since
$f(0) = 1$, the vertex is $(0, 1)$.

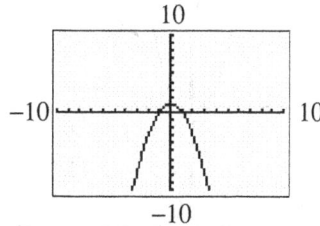

The graph is above the x-axis when
$-1 < x < 1$. Since the inequality is strict, the

solution set is $\left\{ x \mid -1 < x < 1 \right\}$ or, using
interval notation, $(-1, 1)$.

e. $g(x) \le 0$
$4x + 1 \le 0$
$4x \le -1$
$x \le -\dfrac{1}{4}$
The solution set is $\left\{ x \mid x \le -\dfrac{1}{4} \right\}$ or, using
interval notation, $\left(-\infty, -\dfrac{1}{4} \right]$.

f. $f(x) > g(x)$
$-x^2 + 1 > 4x + 1$
$-x^2 - 4x > 0$
We graph the function $p(x) = -x^2 - 4x$.
The intercepts of p are
y-intercept: $p(0) = 0$
x-intercepts: $-x^2 - 4x = 0$
$-x(x + 4) = 0$
$x = 0;\ x = -4$
The vertex is at $x = \dfrac{-b}{2a} = \dfrac{-(-4)}{2(-1)} = \dfrac{4}{-2} = -2$.
Since $p(-2) = 4$, the vertex is $(-2, 4)$.

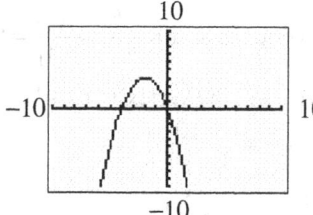

The graph of p is above the x-axis when
$-4 < x < 0$. Since the inequality is strict,
the solution set is $\left\{ x \mid -4 < x < 0 \right\}$ or, using
interval notation, $(-4, 0)$.

g. $f(x) \ge 1$
$-x^2 + 1 \ge 1$
$-x^2 \ge 0$
We graph the function $p(x) = -x^2$. The
vertex is at $x = \dfrac{-b}{2a} = \dfrac{-(0)}{2(-1)} = 0$. Since
$p(0) = 0$, the vertex is $(0, 0)$. Since
$a = -1 < 0$, the parabola opens downward.

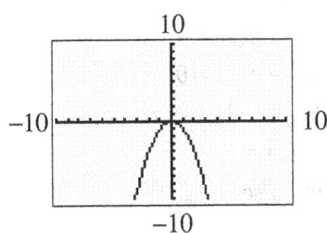

The graph of p is never above the x-axis, but it does touch the x-axis at $x = 0$. Since the inequality is not strict, the solution set is $\{0\}$.

29. $f(x) = x^2 - 4; \quad g(x) = -x^2 + 4$

   a.
   $$f(x) = 0$$
   $$x^2 - 4 = 0$$
   $$(x-2)(x+2) = 0$$
   $$x = 2; x = -2$$
   Solution set: $\{-2, 2\}$.

   b.
   $$g(x) = 0$$
   $$-x^2 + 4 = 0$$
   $$x^2 - 4 = 0$$
   $$(x+2)(x-2) = 0$$
   $$x = -2; x = 2$$
   Solution set: $\{-2, 2\}$.

   c.
   $$f(x) = g(x)$$
   $$x^2 - 4 = -x^2 + 4$$
   $$2x^2 - 8 = 0$$
   $$2(x-2)(x+2) = 0$$
   $$x = 2; x = -2$$
   Solution set: $\{-2, 2\}$.

   d. $f(x) > 0$
   $$x^2 - 4 > 0$$
   We graph the function $f(x) = x^2 - 4$.
   y-intercept: $f(0) = -4$
   x-intercepts: $\quad x^2 - 4 = 0$
   $$(x+2)(x-2) = 0$$
   $$x = -2; x = 2$$
   The vertex is at $x = \dfrac{-b}{2a} = \dfrac{-(0)}{2(-1)} = 0$. Since
   $f(0) = -4$, the vertex is $(0, -4)$.

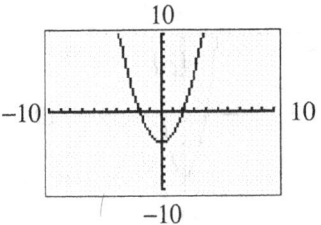

The graph is above the x-axis when $x < -2$ or $x > 2$. Since the inequality is strict, the solution set is $\{ x \mid x < -2 \text{ or } x > 2 \}$ or, using interval notation, $(-\infty, -2) \cup (2, \infty)$.

   e.
   $$g(x) \le 0$$
   $$-x^2 + 4 \le 0$$
   We graph the function $g(x) = -x^2 + 4$.
   y-intercept: $g(0) = 4$
   x-intercepts: $\quad -x^2 + 4 = 0$
   $$x^2 - 4 = 0$$
   $$(x+2)(x-2) = 0$$
   $$x = -2; x = 2$$
   The vertex is at $x = \dfrac{-b}{2a} = \dfrac{-(0)}{2(-1)} = 0$. Since
   $g(0) = 4$, the vertex is $(0, 4)$.

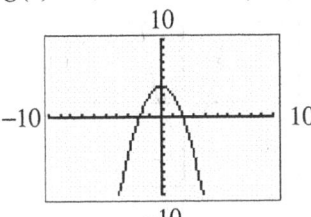

The graph is below the x-axis when $x < -2$ or $x > 2$. Since the inequality is not strict, the solution set is $\{ x \mid x \le -2 \text{ or } x \ge 2 \}$ or, using interval notation, $(-\infty, -2] \cup [2, \infty)$.

   f.
   $$f(x) > g(x)$$
   $$x^2 - 4 > -x^2 + 4$$
   $$2x^2 - 8 > 0$$
   We graph the function $p(x) = 2x^2 - 8$.
   y-intercept: $p(0) = -8$
   x-intercepts: $\quad 2x^2 - 8 = 0$
   $$2(x+2)(x-2) = 0$$
   $$x = -2; x = 2$$
   The vertex is at $x = \dfrac{-b}{2a} = \dfrac{-(0)}{2(2)} = 0$. Since
   $p(0) = -8$, the vertex is $(0, -8)$.

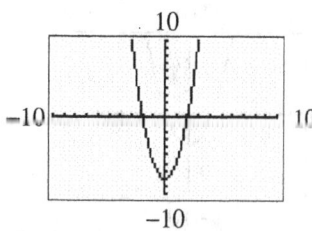

The graph is above the x-axis when $x < -2$ or $x > 2$. Since the inequality is strict, the solution set is $\{x \mid x < -2 \text{ or } x > 2\}$ or, using interval notation, $(-\infty, -2) \cup (2, \infty)$.

g. $f(x) \geq 1$

$x^2 - 4 \geq 1$

$x^2 - 5 \geq 0$

We graph the function $p(x) = x^2 - 5$.

y-intercept: $p(0) = -5$

x-intercepts: $x^2 - 5 = 0$

$x^2 = 5$

$x = \pm\sqrt{5}$

The vertex is at $x = \dfrac{-b}{2a} = \dfrac{-(0)}{2(1)} = 0$. Since $p(0) = -5$, the vertex is $(0, -5)$.

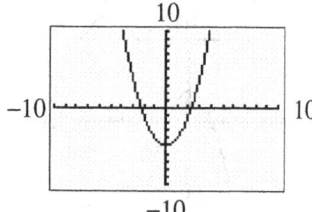

The graph of p is above the x-axis when $x < -\sqrt{5}$ or $x > \sqrt{5}$. Since the inequality is not strict, the solution set is $\{x \mid x \leq -\sqrt{5} \text{ or } x \geq \sqrt{5}\}$ or, using interval notation, $\left(-\infty, -\sqrt{5}\right] \cup \left[\sqrt{5}, \infty\right)$.

31. $f(x) = x^2 - x - 2; \quad g(x) = x^2 + x - 2$

a. $f(x) = 0$

$x^2 - x - 2 = 0$

$(x - 2)(x + 1) = 0$

$x = 2, x = -1$

Solution set: $\{-1, 2\}$.

b. $g(x) = 0$

$x^2 + x - 2 = 0$

$(x + 2)(x - 1) = 0$

$x = -2; x = 1$

Solution set: $\{-2, 1\}$.

c. $f(x) = g(x)$

$x^2 - x - 2 = x^2 + x - 2$

$-2x = 0$

$x = 0$

Solution set: $\{0\}$.

d. $f(x) > 0$

$x^2 - x - 2 > 0$

We graph the function $f(x) = x^2 - x - 2$.

y-intercept: $f(0) = -2$

x-intercepts: $x^2 - x - 2 = 0$

$(x - 2)(x + 1) = 0$

$x = 2; x = -1$

The vertex is at $x = \dfrac{-b}{2a} = \dfrac{-(-1)}{2(1)} = \dfrac{1}{2}$. Since $f\left(\dfrac{1}{2}\right) = -\dfrac{9}{4}$, the vertex is $\left(\dfrac{1}{2}, -\dfrac{9}{4}\right)$.

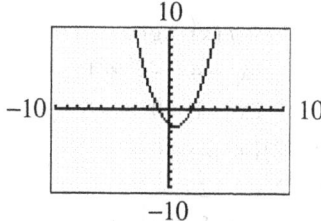

The graph is above the x-axis when $x < -1$ or $x > 2$. Since the inequality is strict, the solution set is $\{x \mid x < -1 \text{ or } x > 2\}$ or, using interval notation, $(-\infty, -1) \cup (2, \infty)$.

e. $g(x) \leq 0$

$x^2 + x - 2 \leq 0$

We graph the function $g(x) = x^2 + x - 2$.

y-intercept: $g(0) = -2$

x-intercepts: $x^2 + x - 2 = 0$

$(x + 2)(x - 1) = 0$

$x = -2; x = 1$

The vertex is at $x = \dfrac{-b}{2a} = \dfrac{-(1)}{2(1)} = -\dfrac{1}{2}$. Since $f\left(-\dfrac{1}{2}\right) = -\dfrac{7}{4}$, the vertex is $\left(-\dfrac{1}{2}, -\dfrac{7}{4}\right)$.

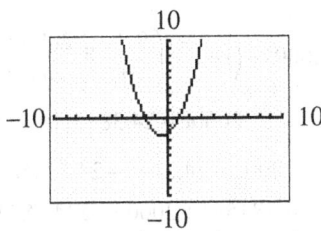

The graph is below the x-axis when $-2 < x < 1$. Since the inequality is not strict, the solution set is $\{x \mid -2 \le x \le 1\}$ or, using interval notation, $[-2, 1]$.

f.
$$f(x) > g(x)$$
$$x^2 - x - 2 > x^2 + x - 2$$
$$-2x > 0$$
$$x < 0$$

The solution set is $\{x \mid x < 0\}$ or, using interval notation, $(-\infty, 0)$.

g.
$$f(x) \ge 1$$
$$x^2 - x - 2 \ge 1$$
$$x^2 - x - 3 \ge 0$$

We graph the function $p(x) = x^2 - x - 3$.

y-intercept: $p(0) = -3$

x-intercepts: $x^2 - x - 3 = 0$

$$x = \frac{-(-1) \pm \sqrt{(-1)^2 - 4(1)(-3)}}{2(1)}$$
$$= \frac{1 \pm \sqrt{1 + 12}}{2} = \frac{1 \pm \sqrt{13}}{2}$$
$$x \approx -1.30 \text{ or } x \approx 2.30$$

The vertex is at $x = \dfrac{-b}{2a} = \dfrac{-(-1)}{2(1)} = \dfrac{1}{2}$. Since $p\left(\dfrac{1}{2}\right) = -\dfrac{13}{4}$, the vertex is $\left(\dfrac{1}{2}, -\dfrac{13}{4}\right)$.

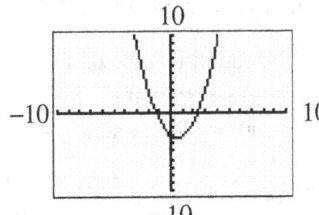

The graph of p is above the x-axis when $x < \dfrac{1 - \sqrt{13}}{2}$ or $x > \dfrac{1 + \sqrt{13}}{2}$. Since the inequality is not strict, the solution set is

$\left\{x \;\middle|\; x \le \dfrac{1 - \sqrt{13}}{2} \text{ or } x \ge \dfrac{1 + \sqrt{13}}{2}\right\}$ or, using

interval notation,

$$\left(-\infty, \frac{1 - \sqrt{13}}{2}\right] \cup \left[\frac{1 + \sqrt{13}}{2}, \infty\right).$$

33. a. The ball strikes the ground when $s(t) = 80t - 16t^2 = 0$.

$$80t - 16t^2 = 0$$
$$16t(5 - t) = 0$$
$$t = 0, t = 5$$

The ball strikes the ground after 5 seconds.

b. Find the values of t for which
$$80t - 16t^2 > 96$$
$$-16t^2 + 80t - 96 > 0$$

We graph the function $f(t) = -16t^2 + 80t - 96$. The intercepts are

y-intercept: $f(0) = -96$

t-intercepts: $-16t^2 + 80t - 96 = 0$
$$-16(t^2 - 5t + 6) = 0$$
$$16(t - 2)(t - 3) = 0$$
$$t = 2, t = 3$$

The vertex is at $t = \dfrac{-b}{2a} = \dfrac{-(80)}{2(-16)} = 2.5$.

Since $f(2.5) = 4$, the vertex is $(2.5, 4)$.

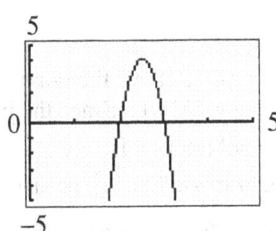

The graph of f is above the t-axis when $2 < t < 3$. Since the inequality is strict, the solution set is $\{t \mid 2 < t < 3\}$ or, using interval notation, $(2, 3)$. The ball is more than 96 feet above the ground for times between 2 and 3 seconds.

35. a.
$$R(p) = -4p^2 + 4000p = 0$$
$$-4p(p - 1000) = 0$$
$$p = 0, p = 1000$$

Thus, the revenue equals zero when the price is $0 or $1000.

b. Find the values of p for which
$$-4p^2 + 4000p > 800,000$$
$$-4p^2 + 4000p - 800,000 > 0$$

We graph $f(p) = -4p^2 + 4000p - 800,000$.
The intercepts are
y-intercept: $f(0) = -800,000$
p-intercepts:
$$-4p^2 + 4000p - 800000 = 0$$
$$p^2 - 1000p + 200000 = 0$$

$$p = \frac{-(-1000) \pm \sqrt{(-1000)^2 - 4(1)(200000)}}{2(1)}$$

$$= \frac{1000 \pm \sqrt{200000}}{2}$$

$$= \frac{1000 \pm 200\sqrt{5}}{2}$$

$$= 500 \pm 100\sqrt{5}$$

$$p \approx 276.39; \quad p \approx 723.61.$$

The vertex is at $p = \dfrac{-b}{2a} = \dfrac{-(4000)}{2(-4)} = 500$.

Since $f(500) = 200,000$, the vertex is
$(500, 200000)$.

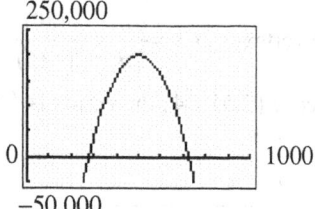

The graph of f is above the p-axis when
$276.39 < p < 723.61$. Since the inequality is
strict, the solution set is
$\{p \mid 276.39 < p < 723.61\}$ or, using interval

notation, $(276.39, 723.61)$. The revenue is
more than \$800,000 for prices between
\$276.39 and \$723.61.

37. $y = cx - (1+c^2)\left(\dfrac{g}{2}\right)\left(\dfrac{x}{v}\right)^2$

a. Since the round must clear a hill 200 meters
high, this mean $y > 200$.

Now $x = 2000$, $v = 897$, and $g = 9.81$.

$$c(2000) - (1+c^2)\left(\frac{9.81}{2}\right)\left(\frac{2000}{897}\right)^2 > 200$$

$$2000c - 24.3845(1+c^2) > 200$$

$$2000c - 24.3845 - 24.3845c^2 > 200$$

$$-24.3845c^2 + 2000c - 224.3845 > 0$$

We graph
$f(c) = -24.3845c^2 + 2000c - 224.3845$.
The intercepts are
y-intercept: $f(0) = -224.3845$
c-intercepts:
$$-24.3845c^2 + 2000c - 224.3845 = 0$$

$$c = \frac{-2000 \pm \sqrt{(2000)^2 - 4(-24.3845)(-224.3845)}}{2(-24.3845)}$$

$$= \frac{-2000 \pm \sqrt{3,978,113.985}}{-48.769}$$

$$c \approx 0.112 \quad \text{or} \quad c \approx 81.907$$

The vertex is at
$$c = \frac{-b}{2a} = \frac{-(2000)}{2(-24.3845)} = 41.010. \text{ Since}$$

$f(41.010) \approx 40,785.273$, the vertex is
$(41.010, 40785.273)$.

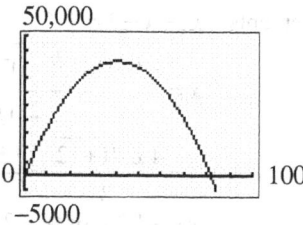

The graph of f is above the c-axis when
$0.112 < c < 81.907$. Since the inequality is
strict, the solution set is
$\{c \mid 0.112 < c < 81.907\}$ or, using interval

notation, $(0.112, 81.907)$.

b. Since the round is to be on the ground $y = 0$.
Note, 75 km $= 75,000$ m. So,
$x = 75,000$, $v = 897$, and $g = 9.81$.

$$c(75,000) - (1+c^2)\left(\frac{9.81}{2}\right)\left(\frac{75,000}{897}\right)^2 = 0$$

$$75,000c - 34,290.724(1+c^2) = 0$$

$$75,000c - 34,290.724 - 34,290.724c^2 = 0$$

$$-34,290.724c^2 + 75,000c - 34,290.724 = 0$$

We graph
$f(c) = -34,290.724c^2 + 75,000c - 34,290.724$.

The intercepts are

y-intercept: $f(0) = -34,290.724$

c-intercepts:

$-34,290.724c^2 + 75,000c - 34,290.724 = 0$

$c = \dfrac{-(75,000) \pm \sqrt{(75,000)^2 - 4(-34,290.724)(-34,290.724)}}{2(-34,290.724)}$

$= \dfrac{-75,000 \pm \sqrt{921,584,990.2}}{-68,581.448}$

$c \approx 0.651$ or $c \approx 1.536$

It is possible to hit the target 75 kilometers away so long as $c \approx 0.651$ or $c \approx 1.536$.

39. $(x-4)^2 \le 0$

We graph the function $f(x) = (x-4)^2$.

y-intercept: $f(0) = 16$

x-intercepts: $(x-4)^2 = 0$

$\qquad\qquad x - 4 = 0$

$\qquad\qquad\quad x = 4$

The vertex is the vertex is $(4, 0)$.

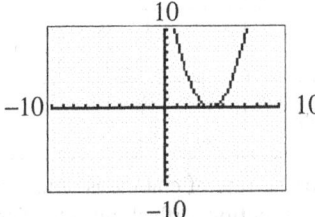

The graph is never below the x-axis. Since the inequality is not strict, the only solution comes from the x-intercept. Therefore, the given inequality has exactly one real solution, namely $x = 4$.

41. Solving $x^2 + x + 1 > 0$

We graph the function $f(x) = x^2 + x + 1$.

y-intercept: $f(0) = 1$

x-intercepts: $b^2 - 4ac = 1^2 - 4(1)(1) = -3$, so f has no x-intercepts.

The vertex is at $x = \dfrac{-b}{2a} = \dfrac{-(1)}{2(1)} = -\dfrac{1}{2}$. Since

$f\left(-\dfrac{1}{2}\right) = \dfrac{3}{4}$, the vertex is $\left(-\dfrac{1}{2}, \dfrac{3}{4}\right)$.

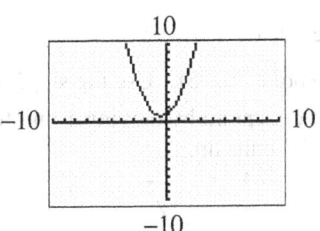

The graph is always above the x-axis. Thus, the solution is the set of all real numbers or, using interval notation, $(-\infty, \infty)$.

43. The x-intercepts are included when the original inequality is not strict (when it contains an equal sign with the inequality).

45. a. $0 = \dfrac{2}{3}x - 6$

$6 = \dfrac{2}{3}x$

$x = 9$

$y = \dfrac{2}{3}(0) - 6$

$\quad = -6$

The intercepts are: $(9,0), (0,-6)$

b.

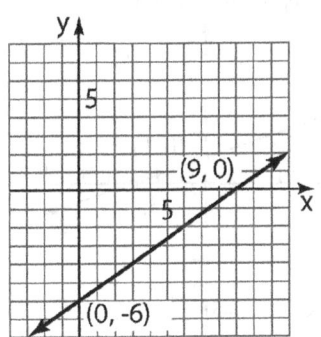

47. $(6-5i)(4-i) = 24 - 6i - 20i + 5i^2$

$= 24 - 26i + 5(-1)$

$= 19 - 26i$

Chapter 4 Review Exercises

1. $f(x) = 2x - 5$

a. Slope $= 2$; y-intercept $= -5$

213

b. average rate of change = 2

c. Plot the point $(0, -5)$. Use the slope to find an additional point by moving 1 unit to the right and 2 units up.

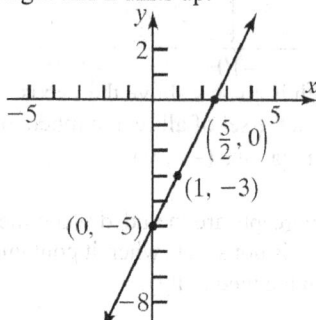

d. increasing

2. $F(x) = -\dfrac{1}{3}x + 1$

a. Slope = $-\dfrac{1}{3}$; y-intercept = 1

b. average rate of change = $-\dfrac{1}{3}$

c. Plot the point $(0, 1)$. Use the slope to find an additional point by moving 3 units to the right and 1 unit down.

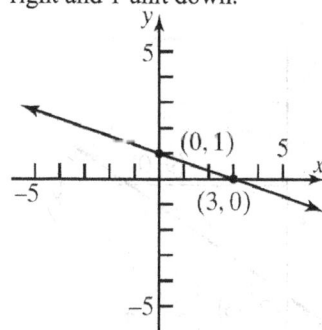

d. decreasing

3. $G(x) = 4$

a. Slope = 0; y-intercept = 4

b. average rate of change = 0

c. Plot the point $(0, 4)$ and draw a horizontal line through it.

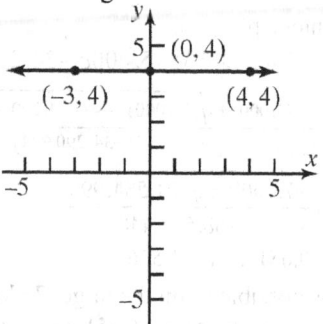

d. constant

4.

x	$y = f(x)$	Avg. rate of change $= \dfrac{\Delta y}{\Delta x}$
-1	-2	
0	3	$\dfrac{3-(-2)}{0-(-1)} = \dfrac{5}{1} = 5$
1	8	$\dfrac{8-3}{1-0} = \dfrac{5}{1} = 5$
2	13	$\dfrac{13-8}{2-1} = \dfrac{5}{1} = 5$
3	18	$\dfrac{18-13}{3-2} = \dfrac{5}{1} = 5$

Since the average rate of change is constant at 5, this is a linear function with slope = 5. The y-intercept is $(0, 3)$, so the equation of the line is $y = 5x + 3$.

5.

x	$y = f(x)$	Avg. rate of change $= \dfrac{\Delta y}{\Delta x}$
-1	-3	
0	4	$\dfrac{4-(-3)}{0-(-1)} = \dfrac{7}{1} = 7$
1	7	$\dfrac{7-4}{1-0} = \dfrac{3}{1} = 3$
2	6	
3	1	

This is not a linear function, since the average rate of change is not constant.

6. $f(x) = (x-2)^2 + 2$

Using the graph of $y = x^2$, shift right 2 units, then shift up 2 units.

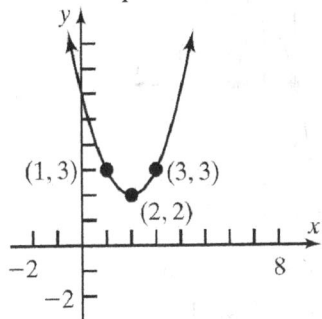

7. $f(x) = -(x-4)^2$

Using the graph of $y = x^2$, shift the graph 4 units right, then reflect about the x-axis.

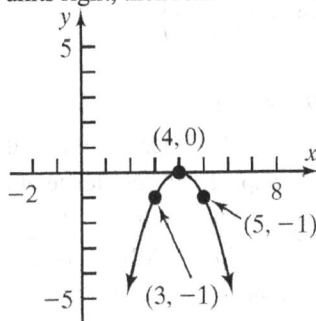

8. $f(x) = 2(x+1)^2 + 4$

Using the graph of $y = x^2$, stretch vertically by a factor of 2, then shift 1 unit left, then shift 4 units up.

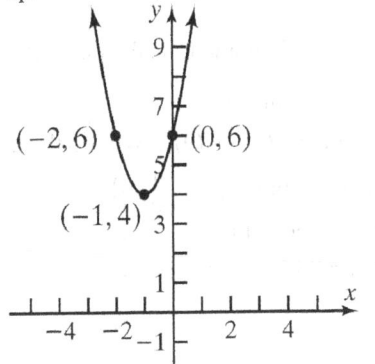

9. a. $f(x) = x^2 - 4x + 6$

$a = 1, b = -4, c = 6$. Since $a = 1 > 0$, the graph opens up. The x-coordinate of the vertex is $x = -\dfrac{b}{2a} = -\dfrac{-4}{2(1)} = \dfrac{4}{2} = 2$.

The y-coordinate of the vertex is

$f\left(-\dfrac{b}{2a}\right) = f(2) = (2)^2 - 4(2) + 6 = 2$.

Thus, the vertex is (2, 2).
The axis of symmetry is the line $x = 2$.
The discriminant is:
$b^2 - 4ac = (-4)^2 - 4(1)(6) = -8 < 0$, so the graph has no x-intercepts.
The y-intercept is $f(0) = 6$.

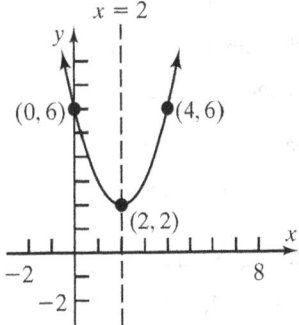

b. The domain is $(-\infty, \infty)$.
The range is $[2, \infty)$.

c. Decreasing on $(-\infty, 2)$.
Increasing on $(2, \infty)$.

10. a. $f(x) = -\dfrac{1}{2}x^2 + 2$

$a = -\dfrac{1}{2}, b = 0, c = 2$. Since $a = -\dfrac{1}{2} < 0$, the graph opens down. The x-coordinate of the vertex is $x = -\dfrac{b}{2a} = -\dfrac{0}{2\left(-\frac{1}{2}\right)} = -\dfrac{0}{-1} = 0$.

The y-coordinate of the vertex is

$f\left(-\dfrac{b}{2a}\right) = f(0) = -\dfrac{1}{2}(0)^2 + 2 = 2$.

The axis of symmetry is the line $x = 0$.
he discriminant is:
$b^2 - 4ac = (0)^2 - 4\left(-\frac{1}{2}\right)(2) = 4 > 0$, so the graph has two x-intercepts.
The x-intercepts are found by solving:

$-\dfrac{1}{2}x^2 + 2 = 0$

$x^2 - 4 = 0$

$x^2 = 4$

$x = -2$ or $x = 2$

The x-intercepts are –2 and 2.
The y-intercept is $f(0) = 2$.

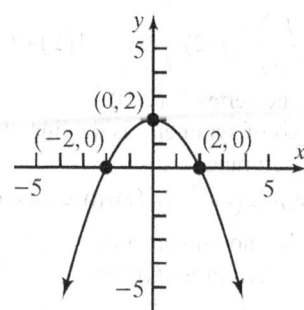

b. The domain is $(-\infty, \infty)$.
The range is $(-\infty, 2]$.

c. Increasing on $(-\infty, 0)$
Decreasing on $(0, \infty)$.

11. a. $f(x) = -4x^2 + 4x$

$a = -4, b = 4, c = 0$. Since $a = -4 < 0$, the graph opens down. The x-coordinate of the vertex is $x = -\dfrac{b}{2a} = -\dfrac{4}{2(-4)} = -\dfrac{4}{-8} = \dfrac{1}{2}$.

The y-coordinate of the vertex is

$f\left(-\dfrac{b}{2a}\right) = f\left(\dfrac{1}{2}\right) = -4\left(\dfrac{1}{2}\right)^2 + 4\left(\dfrac{1}{2}\right)$

$= -1 + 2 = 1$

Thus, the vertex is $\left(\dfrac{1}{2}, 1\right)$.

The axis of symmetry is the line $x = \dfrac{1}{2}$.

The discriminant is:
$b^2 - 4ac = 4^2 - 4(-4)(0) = 16 > 0$, so the graph has two x-intercepts.
The x-intercepts are found by solving:
$-4x^2 + 4x = 0$
$-4x(x-1) = 0 \Rightarrow x = 0$ or $x = 1$
The x-intercepts are 0 and 1.
The y-intercept is $f(0) = -4(0)^2 + 4(0) = 0$.

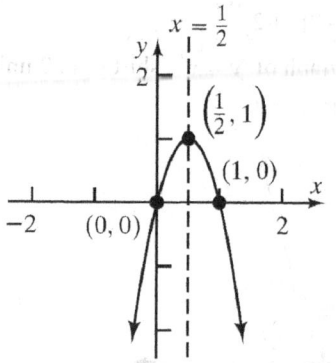

b. The domain is $(-\infty, \infty)$.
The range is $(-\infty, 1]$.

c. Increasing on $\left(-\infty, \dfrac{1}{2}\right)$

Decreasing on $\left(\dfrac{1}{2}, \infty\right)$.

12. a. $f(x) = 9x^2 + 6x + 1$

$a = 9, b = 6, c = 1$. Since $a = 9 > 0$, the graph opens up. The x-coordinate of the vertex is $x = -\dfrac{b}{2a} = -\dfrac{6}{2(9)} = -\dfrac{6}{18} = -\dfrac{1}{3}$.

The y-coordinate of the vertex is

$f\left(-\dfrac{b}{2a}\right) = f\left(-\dfrac{1}{3}\right) = 9\left(-\dfrac{1}{3}\right)^2 + 6\left(-\dfrac{1}{3}\right) + 1$

$= 1 - 2 + 1 = 0$

Thus, the vertex is $\left(-\dfrac{1}{3}, 0\right)$.

The axis of symmetry is the line $x = -\dfrac{1}{3}$.

The discriminant is:
$b^2 - 4ac = (6)^2 - 4(9)(1) = 0$, so the graph has one x-intercept, $\left(-\frac{1}{3}, 0\right)$.
The y-intercept is
$f(0) = 9(0)^2 + 6(0) + 1 = 1$.

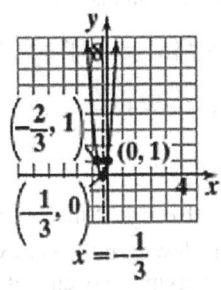

b. The domain is $(-\infty, \infty)$.

The range is $[2, \infty)$.

c. Decreasing on $\left(-\infty, \dfrac{1}{3}\right)$.

Increasing on $\left(\dfrac{1}{3}, \infty\right)$.

13. a. $f(x) = -x^2 + x + \dfrac{1}{2}$

$a = -1, b = 1, c = \dfrac{1}{2}$. Since $a = -1 < 0$, the graph opens down. The x-coordinate of the vertex is $x = -\dfrac{b}{2a} = -\dfrac{1}{2(-1)} = -\dfrac{1}{-2} = \dfrac{1}{2}$.

The y-coordinate of the vertex is

$$f\left(-\dfrac{b}{2a}\right) = f\left(\dfrac{1}{2}\right) = -\left(\dfrac{1}{2}\right)^2 + \left(\dfrac{1}{2}\right) + \dfrac{1}{2}$$

$$= -\dfrac{1}{4} + 1 = \dfrac{3}{4}$$

Thus, the vertex is $\left(\dfrac{1}{2}, \dfrac{3}{4}\right)$. The axis of symmetry is the line $x = \dfrac{1}{2}$. The discriminant is: $b^2 - 4ac = 1^2 - 4(-1)\left(\dfrac{1}{2}\right) = 3 > 0$, so the graph has two x-intercepts. The x-intercepts are found by solving: $-x^2 + x + \dfrac{1}{2} = 0$.

$$x = \dfrac{-b \pm \sqrt{b^2 - 4ac}}{2a} = \dfrac{-1 \pm \sqrt{3}}{2(-1)} = \dfrac{-1 \pm \sqrt{3}}{-2} = \dfrac{1 \pm \sqrt{3}}{2}$$

The x-intercepts are $\dfrac{1 - \sqrt{3}}{2} \approx -0.4$ and $\dfrac{1 + \sqrt{3}}{2} \approx 1.4$.

The y-intercept is $f(0) = -(0)^2 + (0) + \dfrac{1}{2} = \dfrac{1}{2}$.

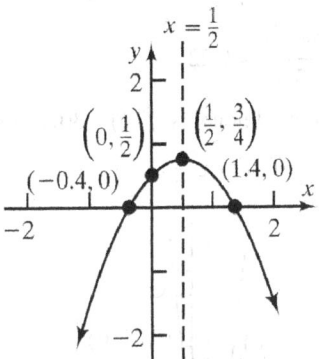

b. The domain is $(-\infty, \infty)$.

The range is $\left(-\infty, \dfrac{3}{4}\right]$.

c. Increasing on $\left(-\infty, \dfrac{1}{2}\right)$.

Decreasing on $\left(\dfrac{1}{2}, \infty\right)$.

14. a. $f(x) = 3x^2 + 4x - 1$

$a = 3, b = 4, c = -1$. Since $a = 3 > 0$, the graph opens up. The x-coordinate of the vertex is $x = -\dfrac{b}{2a} = -\dfrac{4}{2(3)} = -\dfrac{4}{6} = -\dfrac{2}{3}$.

The y-coordinate of the vertex is

$$f\left(-\dfrac{b}{2a}\right) = f\left(-\dfrac{2}{3}\right) = 3\left(-\dfrac{2}{3}\right)^2 + 4\left(-\dfrac{2}{3}\right) - 1$$

$$= \dfrac{4}{3} - \dfrac{8}{3} - 1 = -\dfrac{7}{3}$$

Thus, the vertex is $\left(-\dfrac{2}{3}, -\dfrac{7}{3}\right)$.

The axis of symmetry is the line $x = -\dfrac{2}{3}$.

The discriminant is:
$b^2 - 4ac = (4)^2 - 4(3)(-1) = 28 > 0$, so the graph has two x-intercepts.
The x-intercepts are found by solving:
$3x^2 + 4x - 1 = 0$.

$$x = \dfrac{-b \pm \sqrt{b^2 - 4ac}}{2a} = \dfrac{-4 \pm \sqrt{28}}{2(3)}$$

$$= \dfrac{-4 \pm 2\sqrt{7}}{6} = \dfrac{-2 \pm \sqrt{7}}{3}$$

217

The x-intercepts are $\dfrac{-2-\sqrt{7}}{3} \approx -1.55$ and

$\dfrac{-2+\sqrt{7}}{3} \approx 0.22$.

The y-intercept is $f(0) = 3(0)^2 + 4(0) - 1 = -1$.

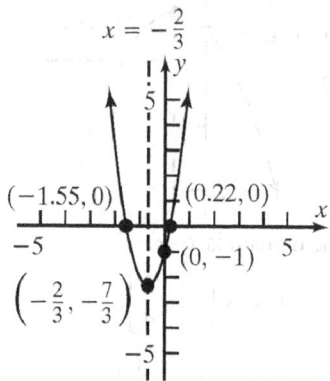

b.   The domain is $(-\infty, \infty)$.

The range is $\left[-\dfrac{7}{3}, \infty \right)$.

c.   Decreasing on $\left(-\infty, -\dfrac{2}{3} \right)$

Increasing on $\left(-\dfrac{2}{3}, \infty \right)$.

15. $f(x) = 3x^2 - 6x + 4$

$a = 3, b = -6, c = 4$. Since $a = 3 > 0$, the graph opens up, so the vertex is a minimum point. The minimum occurs at

$x = -\dfrac{b}{2a} = -\dfrac{-6}{2(3)} = \dfrac{6}{6} = 1$.

The minimum value is

$f\left(-\dfrac{b}{2a} \right) = f(1) = 3(1)^2 - 6(1) + 4$

$= 3 - 6 + 4 = 1$

16. $f(x) = -x^2 + 8x - 4$

$a = -1, b = 8, c = -4$. Since $a = -1 < 0$, the graph opens down, so the vertex is a maximum point. The maximum occurs at

$x = -\dfrac{b}{2a} = -\dfrac{8}{2(-1)} = -\dfrac{8}{-2} = 4$.

The maximum value is

$f\left(-\dfrac{b}{2a} \right) = f(4) = -(4)^2 + 8(4) - 4$

$= -16 + 32 - 4 = 12$

17. $f(x) = -2x^2 + 4$

$a = -2, b = 0, c = 4$. Since $a = -2 < 0$, the graph opens down, so the vertex is a maximum point. The maximum occurs at

$x = -\dfrac{b}{2a} = -\dfrac{0}{2(-2)} = 0$.

The maximum value is

$f\left(-\dfrac{b}{2a} \right) = f(0) = -2(0)^2 + 4 = 4$.

18. $x^2 + 6x - 16 < 0$

We graph the function $f(x) = x^2 + 6x - 16$. The intercepts are

y-intercept: $f(0) = -16$

x-intercepts: $x^2 + 6x - 16 = 0$

$(x+8)(x-2) = 0$

$x = -8, x = 2$

The vertex is at $x = \dfrac{-b}{2a} = \dfrac{-(6)}{2(1)} = -3$. Since $f(-3) = -25$, the vertex is $(-3, -25)$.

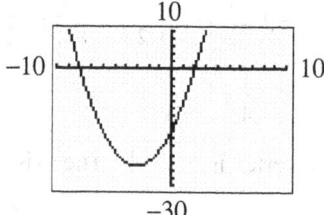

The graph is below the x-axis when $-8 < x < 2$. Since the inequality is strict, the solution set is $\{x \mid -8 < x < 2\}$ or, using interval notation, $(-8, 2)$.

19. $3x^2 \geq 14x + 5$

$3x^2 - 14x - 5 \geq 0$

We graph the function $f(x) = 3x^2 - 14x - 5$. The intercepts are

y-intercept: $f(0) = -5$

x-intercepts: $3x^2 - 14x - 5 = 0$

$(3x+1)(x-5) = 0$

$$x = -\frac{1}{3}, x = 5$$

The vertex is at $x = \dfrac{-b}{2a} = \dfrac{-(-14)}{2(3)} = \dfrac{14}{6} = \dfrac{7}{3}$.

Since $f\left(\dfrac{7}{3}\right) = -\dfrac{64}{3}$, the vertex is $\left(\dfrac{7}{3}, -\dfrac{64}{3}\right)$.

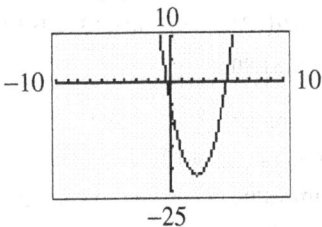

The graph is above the x-axis when $x < -\dfrac{1}{3}$ or $x > 5$. Since the inequality is not strict, the solution set is $\left\{x \mid x \le -\dfrac{1}{3} \text{ or } x \ge 5\right\}$ or, using interval notation, $\left(-\infty, -\dfrac{1}{3}\right] \cup [5, \infty)$.

20. Use the form $f(x) = a(x-h)^2 + k$.

The vertex is $(-1, 2)$, so $h = -1$ and $k = 2$.

$f(x) = a(x+1)^2 + 2$.

Since the graph passes through $(1, 6)$, $f(1) = 6$.

$6 = a(1+1)^2 + 2$

$6 = a(2)^2 + 2$

$6 = 4a + 2$

$4 = 4a$

$1 = a$

$f(x) = 1(x+1)^2 + 2 = (x^2 + 2x + 1) + 2$

$\qquad = x^2 + 2x + 3$

21. Consider the form $y = ax^2 + bx + c$. Substituting the three points from the graph into the general form we have the following three equations.

$2 = a(1)^2 + b(1) + c \Rightarrow 2 = a + b + c$

and

$2 = a(3)^2 + b(3) + c \Rightarrow 2 = 9a + 3b + c$

and

$5 = a(0)^2 + b(0) + c \Rightarrow 5 = c$

Since $5 = c$, we have the following equations:

$2 = a + b + 5, \quad 2 = 9a + 3b + 5, \quad 5 = c$

Solving the first two simultaneously we have

$\left.\begin{aligned} 2 &= a + b + 5 \\ 2 &= 9a + 3b + 5 \end{aligned}\right)$

$\left.\begin{aligned} -3 &= a + b \\ -3 &= 9a + 3b \end{aligned}\right) \rightarrow a = 1, b = -4$

The quadratic function is $f(x) = x^2 - 4x + 5$.

22. a. $S(x) = 0.01x + 25,000$

b. $S(1,000,000) = 0.01(1,000,000) + 25,000$

$\qquad\qquad\qquad = 10,000 + 25,000 = 35,000$

In 2005, Bill's salary was $35,000.

c. $0.01x + 25,000 = 100,000$

$\qquad 0.01x = 75,000$

$\qquad x = 7,500,000$

Bill's sales would have to be $7,500,000 in order to earn $100,000.

d. $0.01x + 25,000 > 150,000$

$\qquad 0.01x > 125,000$

$\qquad x > 12,500,000$

Bill's sales would have to be more than $12,500,000 in order for his salary to exceed $150,000.

23. a. The revenue will equal the quantity x sold times the price p. That is, $R = xp$. Thus,

$$R(x) = x\left(-\frac{1}{10}x + 150\right) = -\frac{1}{10}x^2 + 150x$$

b. $R(100) = -\dfrac{1}{10}(100)^2 + 150(100) = 14,000$

The revenue is $14,000 if 100 units are sold.

c. $a = -\dfrac{1}{10}, b = 150, c = 0$. Since $a = -\dfrac{1}{10} < 0$, the graph opens down, so the vertex is a maximum point. The maximum occurs at $x = \dfrac{-b}{2a} = \dfrac{-(150)}{2(-1/10)} = \dfrac{-150}{-1/5} = 750$. Thus, the quantity that maximizes revenue is 750 units. The maximum revenue is

$R(750) = -\dfrac{1}{10}(750)^2 + 150(750)$

$\qquad\quad = -56,250 + 112,500$

$\qquad\quad = \$56,250$

d. From part (c), we know revenue is maximizes when $x = 750$ units are sold.

The price that should be charged for this is
$$p = -\frac{1}{10}(750) + 150 = \$75.$$

24. Consider the diagram

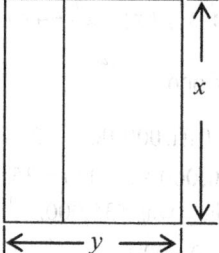

Total amount of fence = $3x + 2y = 10,000$

$$y = \frac{10,000 - 3x}{2} = 5000 - \frac{3}{2}x$$

Total area enclosed = $(x)(y) = (x)\left(5000 - \frac{3}{2}x\right)$

$A(x) = 5000x - \frac{3}{2}x^2 = -\frac{3}{2}x^2 + 5000x$ is a

quadratic function with $a = -\frac{3}{2} < 0$.

So the vertex corresponds to the maximum value for this function. The vertex occurs when

$$x = -\frac{b}{2a} = -\frac{5000}{2(-3/2)} = \frac{5000}{3}.$$

The maximum area is:

$$A\left(\frac{5000}{3}\right) = -\frac{3}{2}\left(\frac{5000}{3}\right)^2 + 5000\left(\frac{5000}{3}\right)$$

$$= -\frac{3}{2}\left(\frac{25,000,000}{9}\right) + \frac{25,000,000}{3}$$

$$= -\frac{12,500,000}{3} + \frac{25,000,000}{3}$$

$$= \frac{12,500,000}{3}$$

$$\approx 4,166,666.67 \text{ square meters}$$

25. $C(x) = 4.9x^2 - 617.4x + 19,600$;

$a = 4.9$, $b = -617.4$, $c = 19,600$. Since
$a = 4.9 > 0$, the graph opens up, so the vertex is a minimum point.

a. The minimum marginal cost occurs at
$$x = -\frac{b}{2a} = -\frac{-617.40}{2(4.9)} = \frac{617.40}{9.8} = 63.$$
Thus, 63 golf clubs should be manufactured in order to minimize the marginal cost.

b. The minimum marginal cost is
$$C(63) = 4.9(63)^2 - (617.40)(63) + 19600$$
$$= \$151.90$$

26. The area function is:
$$A(x) = x(10 - x) = -x^2 + 10x$$
The maximum value occurs at the vertex:
$$x = -\frac{b}{2a} = -\frac{10}{2(-1)} = -\frac{10}{-2} = 5$$
The maximum area is:
$$A(5) = -(5)^2 + 10(5)$$
$$= -25 + 50 = 25 \text{ square units}$$

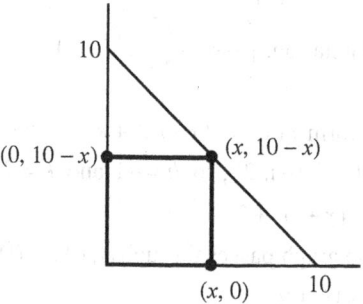

27. Locate the origin at the point directly under the highest point of the arch. Then the equation is in the form: $y = -ax^2 + k$, where $a > 0$. Since the maximum height is 10 feet, when $x = 0$, $y = k = 10$. Since the point $(10, 0)$ is on the parabola, we can find the constant:
$$0 = -a(10)^2 + 10$$
$$a = \frac{10}{10^2} = \frac{1}{10} = 0.10$$
The equation of the parabola is:
$$y = -\frac{1}{10}x^2 + 10$$
At $x = 8$:
$$y = -\frac{1}{10}(8)^2 + 10 = -6.4 + 10 = 3.6 \text{ feet}$$

28. a.

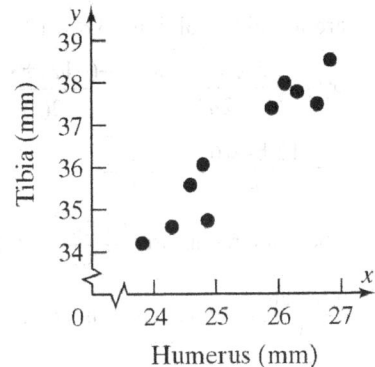

Humerus (mm)

b. Yes, the two variables appear to have a linear relationship.

c. Using the LINear REGression program, the line of best fit is: $y = 1.3902x + 1.1140$

```
LinReg
y=ax+b
a=1.390171918
b=1.113952697
r²=.9050023758
r=.9513161282
```

d. $y = 1.39017(26.5) + 1.11395 \approx 37.95$ mm

29. a.

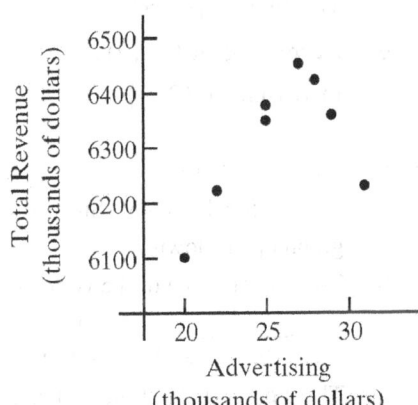

Advertising
(thousands of dollars)

The data appear to be quadratic with $a < 0$.

b. The maximum revenue occurs at

$$A = \frac{-b}{2a} = \frac{-(411.88)}{2(-7.76)}$$

$$= \frac{-411.88}{-15.52} \approx \$26.5 \text{ thousand}$$

c. The maximum revenue is

$$R\left(\frac{-b}{2a}\right) = R(26.53866)$$

$$= -7.76(26.5)^2 + (411.88)(26.5) + 942.72$$

$$\approx \$6408 \text{ thousand}$$

d. Using the QUADratic REGression program, the quadratic function of best fit is:

$$y = -7.76x^2 + 411.88x + 942.72 .$$

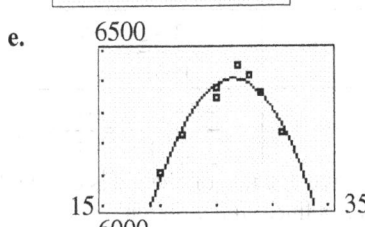

```
QuadReg
y=ax²+bx+c
a=-7.759570754
b=411.8750353
c=942.721091
```

e.

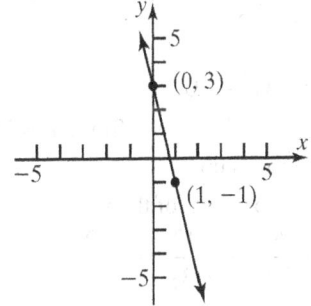

Chapter 4 Test

1. $f(x) = -4x + 3$

a. Slope $= -4$; y-intercept $= 3$.

b. The slope is negative, so the graph is decreasing.

c. Plot the point $(0, 3)$. Use the slope to find an additional point by moving 1 unit to the right and 4 units down.

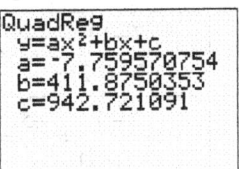

221

2.

x	y	Avg. rate of change $= \dfrac{\Delta y}{\Delta x}$
-2	12	
-1	7	$\dfrac{7-12}{-1-(-2)}=\dfrac{-5}{1}=-5$
0	2	$\dfrac{2-7}{0-(-1)}=\dfrac{-5}{1}=-5$
1	-3	$\dfrac{-3-2}{1-0}=\dfrac{-5}{1}=-5$
2	-8	$\dfrac{-8-(-3)}{2-1}=\dfrac{-5}{1}=-5$

Since the average rate of change is constant at -5, this is a linear function with slope $=-5$. The y-intercept is $(0,2)$, so the equation of the line is $y=-5x+2$.

3. $f(x)=(x-3)^2-2$

Using the graph of $y=x^2$, shift right 3 units, then shift down 2 units.

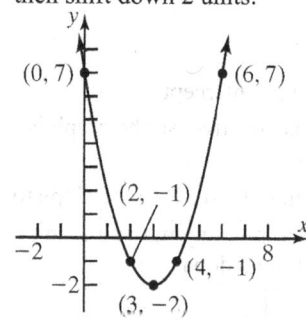

4. a. $f(x)=3x^2-12x+4$

$a=3, b=-12, c=4$. Since $a=3>0$, the graph opens up.

b. The x-coordinate of the vertex is
$$x=-\frac{b}{2a}=-\frac{-12}{2(3)}=-\frac{-12}{6}=2.$$
The y-coordinate of the vertex is
$$f\left(-\frac{b}{2a}\right)=f(2)=3(2)^2-12(2)+4$$
$$=12-24+4=-8$$
Thus, the vertex is $(2,-8)$.

c. The axis of symmetry is the line $x=2$.

d. The discriminant is:
$b^2-4ac=(-12)^2-4(3)(4)=96>0$, so the graph has two x-intercepts. The x-intercepts

are found by solving: $3x^2-12x+4=0$.
$$x=\frac{-b\pm\sqrt{b^2-4ac}}{2a}=\frac{-(-12)\pm\sqrt{96}}{2(3)}$$
$$=\frac{12\pm4\sqrt{6}}{6}=\frac{6\pm2\sqrt{6}}{3}$$
The x-intercepts are $\dfrac{6-2\sqrt{6}}{3}\approx0.37$ and $\dfrac{6\pm2\sqrt{6}}{3}\approx3.63$. The y-intercept is $f(0)=3(0)^2-12(0)+4=4$.

e.

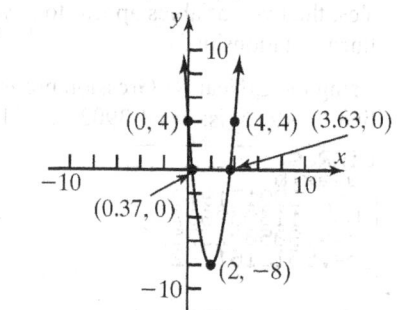

f. The domain is $(-\infty,\infty)$.
The range is $[-8,\infty)$.

g. Decreasing on $(-\infty,2)$.
Increasing on $(2,\infty)$.

5. a. $g(x)=-2x^2+4x-5$
$a=-2, b=4, c=-5$. Since $a=-2<0$, the graph opens down.

b. The x-coordinate of the vertex is
$$x=-\frac{b}{2a}=-\frac{4}{2(-2)}=-\frac{4}{-4}=1.$$
The y-coordinate of the vertex is
$$g\left(-\frac{b}{2a}\right)=g(1)=-2(1)^2+4(1)-5$$
$$=-2+4-5=-3$$
Thus, the vertex is $(1,-3)$.

c. The axis of symmetry is the line $x=1$.

d. The discriminant is:
$b^2-4ac=(4)^2-4(-2)(-5)=-24<0$, so the graph has no x-intercepts. The y-intercept is $g(0)=-2(0)^2+4(0)-5=-5$.

e.

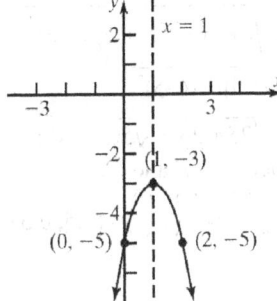

f. The domain is $(-\infty, \infty)$.

The range is $(-\infty, -3]$.

g. Increasing on $(-\infty, 1)$.

Decreasing on $(1, \infty)$.

6. Consider the form $y = a(x-h)^2 + k$. From the graph we know that the vertex is $(1, -32)$ so we have $h = 1$ and $k = -32$. The graph also passes through the point $(x, y) = (0, -30)$. Substituting these values for x, y, h, and k, we can solve for a:

$$-30 = a(0-1)^2 + (-32)$$
$$-30 = a(-1)^2 - 32$$
$$-30 = a - 32$$
$$2 = a$$

The quadratic function is

$$f(x) = 2(x-1)^2 - 32 = 2x^2 - 4x - 30.$$

7. $f(x) = -2x^2 + 12x + 3$

$a = -2$, $b = 12$, $c = 3$. Since $a = -2 < 0$, the graph opens down, so the vertex is a maximum point. The maximum occurs at

$$x = -\frac{b}{2a} = -\frac{12}{2(-2)} = -\frac{12}{-4} = 3.$$

The maximum value is

$$f(3) = -2(3)^2 + 12(3) + 3 = -18 + 36 + 3 = 21.$$

8. $x^2 - 10x + 24 \geq 0$

We graph the function $f(x) = x^2 - 10x + 24$.

The intercepts are

y-intercept: $f(0) = 24$

x-intercepts: $x^2 - 10x + 24 = 0$

$$(x-4)(x-6) = 0$$
$$x = 4, \; x = 6$$

The vertex is at $x = \dfrac{-b}{2a} = \dfrac{-(-10)}{2(1)} = \dfrac{10}{2} = 5$.

Since $f(5) = -1$, the vertex is $(5, -1)$.

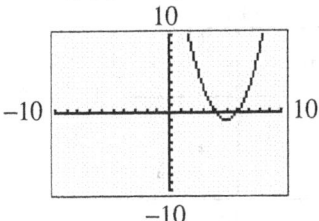

The graph is above the x-axis when $x < 4$ or $x > 6$. Since the inequality is not strict, the solution set is $\{x \mid x \leq 4 \text{ or } x \geq 6\}$ or, using interval notation, $(-\infty, 4] \cup [6, \infty)$.

9. **a.** $C(m) = 0.15m + 129.50$

   b. $C(860) = 0.15(860) + 129.50$
   $$= 129 + 129.50 = 258.50$$
   If 860 miles are driven, the rental cost is $258.50.

   c. $C(m) = 213.80$
   $$0.15m + 129.50 = 213.80$$
   $$0.15m = 84.30$$
   $$m = 562$$
   The rental cost is $213.80 if 562 miles were driven.

10. **a.** $R(x) = x\left(-\dfrac{1}{10}x + 1000\right) = -\dfrac{1}{10}x^2 + 1000x$

    b. $R(400) = -\dfrac{1}{10}(400)^2 + 1000(400)$
    $$= -16,000 + 400,000$$
    $$= \$384,000$$

    c. $x = \dfrac{-b}{2a} = \dfrac{-1000}{2\left(-\frac{1}{10}\right)} = \dfrac{-1000}{\left(-\frac{1}{5}\right)} = 5000$
    The maximum revenue is
    $$R(5000) = -\frac{1}{10}(5000)^2 + 1000(5000)$$
    $$= -250,000 + 5,000,000$$
    $$= \$2,500,000$$
    Thus, 5000 units maximizes revenue at $2,500,000.

    d. $p = -\dfrac{1}{10}(5000) + 1000$
    $$= -500 + 1000$$
    $$= \$500$$

11. a. Set A:

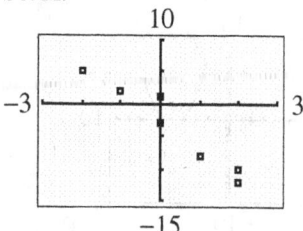

The data appear to be linear with a negative slope.

Set B:

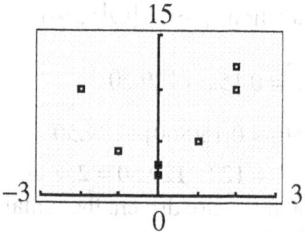

The data appear to be quadratic and opens up.

b. Using the LINear REGression program, the linear function of best fit is:

$y = -4.234x - 2.362$.

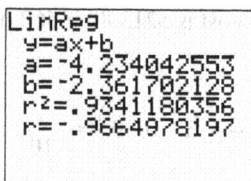

```
LinReg
y=ax+b
a=-4.234042553
b=-2.361702128
r²=.9341180356
r=-.9664978197
```

c. Using the QUADratic REGression program, the quadratic function of best fit is:

$y = 1.993x^2 + 0.289x + 2.503$.

```
QuadReg
y=ax²+bx+c
a=1.992842536
b=.2893660532
c=2.503067485
```

Chapter 4 Cumulative Review

1. $P = (-1,3); Q = (4,-2)$

Distance between P and Q:

$$d(P,Q) = \sqrt{(4-(-1))^2 + (-2-3)^2}$$
$$= \sqrt{(5)^2 + (-5)^2}$$
$$= \sqrt{25+25} = \sqrt{50} = 5\sqrt{2}$$

Midpoint between P and Q:

$$\left(\frac{-1+4}{2}, \frac{3-2}{2}\right) = \left(\frac{3}{2}, \frac{1}{2}\right) = (1.5, 0.5)$$

2. $y = x^3 - 3x + 1$

a. $(-2,-1): \quad -1 = (-2)^3 - 3(-2) + 1$
$$-1 = -8 + 6 + 1$$
$$-1 = -1$$

Yes, $(-2,-1)$ is on the graph.

b. $(2,3): \quad 3 = (2)^3 - 3(2) + 1$
$$3 = 8 - 6 + 1$$
$$3 = 3$$

Yes, $(2,3)$ is on the graph.

c. $(3,1): \quad 1 = (3)^3 - 3(3) + 1$
$$1 = -27 - 9 + 1$$
$$1 \neq -35$$

No, $(3,1)$ is not on the graph.

3. $5x + 3 \geq 0$
$$5x \geq -3$$
$$x \geq -\frac{3}{5}$$

The solution set is $\left\{ x \mid x \geq -\frac{3}{5} \right\}$ or $\left[-\frac{3}{5}, \infty \right)$.

4. $(-1,4)$ and $(2,-2)$ are points on the line.

$$\text{Slope} = \frac{-2-4}{2-(-1)} = \frac{-6}{3} = -2$$

$$y - y_1 = m(x - x_1)$$
$$y - 4 = -2(x - (-1))$$
$$y - 4 = -2(x + 1)$$
$$y - 4 = -2x - 2$$
$$y = -2x + 2$$

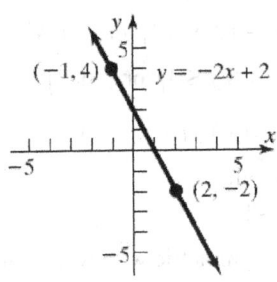

5. Perpendicular to $y = 2x + 1$;
Containing $(3,5)$

Slope of perpendicular $= -\dfrac{1}{2}$

$$y - y_1 = m(x - x_1)$$

$$y - 5 = -\frac{1}{2}(x - 3)$$

$$y - 5 = -\frac{1}{2}x + \frac{3}{2}$$

$$y = -\frac{1}{2}x + \frac{13}{2}$$

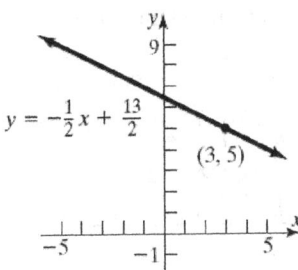

6. $x^2 + y^2 - 4x + 8y - 5 = 0$

$$x^2 - 4x + y^2 + 8y = 5$$

$$(x^2 - 4x + 4) + (y^2 + 8y + 16) = 5 + 4 + 16$$

$$(x - 2)^2 + (y + 4)^2 = 25$$

$$(x - 2)^2 + (y + 4)^2 = 5^2$$

Center: $(2,-4)$ Radius $= 5$

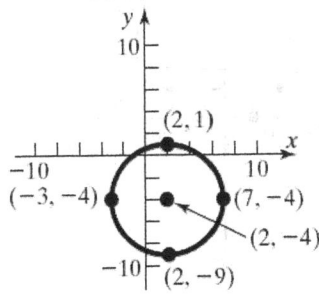

7. Yes, this is a function since each x-value is paired with exactly one y-value.

8. $f(x) = x^2 - 4x + 1$

   a. $f(2) = 2^2 - 4(2) + 1 = 4 - 8 + 1 = -3$

   b. $f(x) + f(2) = x^2 - 4x + 1 + (-3)$

$$= x^2 - 4x - 2$$

   c. $f(-x) = (-x)^2 - 4(-x) + 1 = x^2 + 4x + 1$

   d. $-f(x) = -\left(x^2 - 4x + 1\right) = -x^2 + 4x - 1$

   e. $f(x + 2) = (x + 2)^2 - 4(x + 2) + 1$

$$= x^2 + 4x + 4 - 4x - 8 + 1$$

$$= x^2 - 3$$

   f. $\dfrac{f(x + h) - f(x)}{h}$

$$= \frac{(x + h)^2 - 4(x + h) + 1 - \left(x^2 - 4x + 1\right)}{h}$$

$$= \frac{x^2 + 2xh + h^2 - 4x - 4h + 1 - x^2 + 4x - 1}{h}$$

$$= \frac{2xh + h^2 - 4h}{h}$$

$$= \frac{h(2x + h - 4)}{h} = 2x + h - 4$$

9. $h(z) = \dfrac{3z - 1}{6z - 7}$

The denominator cannot be zero:

$$6z - 7 \neq 0$$

$$6z \neq 7$$

$$z \neq \frac{7}{6}$$

Domain: $\left\{ z \mid z \neq \dfrac{7}{6} \right\}$

10. Yes, the graph represents a function since it passes the Vertical Line Test.

11. $f(x) = \dfrac{x}{x + 4}$

   a. $f(1) = \dfrac{1}{1 + 4} = \dfrac{1}{5} \neq \dfrac{1}{4}$, so $\left(1, \dfrac{1}{4}\right)$ is not on the graph of f.

   b. $f(-2) = \dfrac{-2}{-2 + 4} = \dfrac{-2}{2} = -1$, so $(-2, -1)$ is a point on the graph of f.

c. Solve for x:

$$2 = \frac{x}{x+4}$$

$$2x+8 = x$$

$$x = -8$$

So, $(-8, 2)$ is a point on the graph of f.

12. $f(x) = \dfrac{x^2}{2x+1}$

$$f(-x) = \frac{(-x)^2}{2(-x)+1} = \frac{x^2}{-2x+1} \neq f(x) \text{ or } -f(x)$$

Therefore, f is neither even nor odd.

13. $f(x) = x^3 - 5x + 1$ on the interval $(-4, 4)$

Use MAXIMUM and MINIMUM on the graph of $y_1 = x^3 - 5x + 4$.

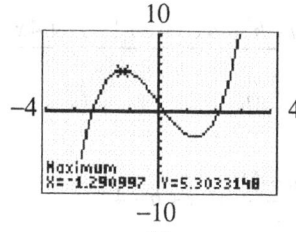

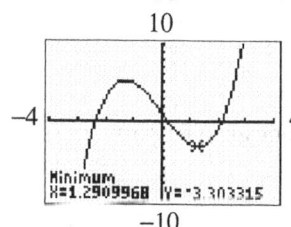

Local maximum is 5.30 and occurs at $x \approx -1.29$;

Local minimum is -3.30 and occurs at $x \approx 1.29$;

f is increasing on $(-4, -1.29)$ or $(1.29, 4)$;

f is decreasing on $(-1.29, 1.29)$.

14. $f(x) = 3x+5;\quad g(x) = 2x+1$

a. $f(x) = g(x)$

$$3x+5 = 2x+1$$

$$3x+5 = 2x+1$$

$$x = -4$$

b. $f(x) > g(x)$

$$3x+5 > 2x+1$$

$$3x+5 > 2x+1$$

$$x > -4$$

The solution set is $\{x \mid x > -4\}$ or $(-4, \infty)$.

15. a. Domain: $\{x \mid -4 \le x \le 4\}$ or $[-4, 4]$

Range: $\{y \mid -1 \le y \le 3\}$ or $[-1, 3]$

b. Intercepts: $(-1, 0)$, $(0, -1)$, $(1, 0)$

x-intercepts: $-1, 1$

y-intercept: -1

c. The graph is symmetric with respect to the y-axis.

d. When $x = 2$, the function takes on a value of 1. Therefore, $f(2) = 1$.

e. The function takes on the value 3 at $x = -4$ and $x = 4$.

f. $f(x) < 0$ means that the graph lies below the x-axis. This happens for x values between -1 and 1. Thus, the solution set is $\{x \mid -1 < x < 1\}$ or $(-1, 1)$.

g. The graph of $y = f(x) + 2$ is the graph of $y = f(x)$ but shifted up 2 units.

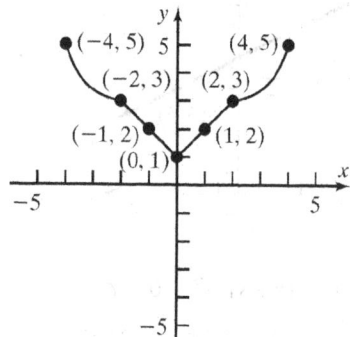

h. The graph of $y = f(-x)$ is the graph of $y = f(x)$ but reflected about the y-axis.

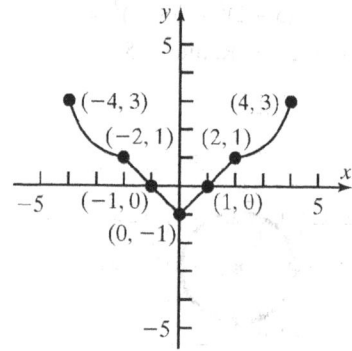

i. The graph of $y = 2f(x)$ is the graph of $y = f(x)$ but stretched vertically by a factor of 2. That is, the coordinate of each point is multiplied by 2.

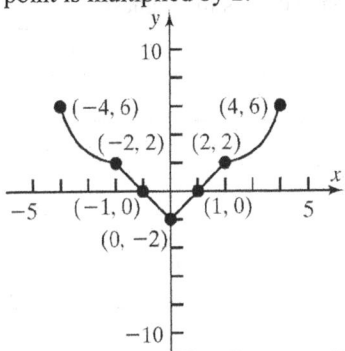

j. Since the graph is symmetric about the y-axis, the function is even.

k. The function is increasing on the open interval $(0, 4)$.

Chapter 4 Projects

Project I – Internet-based Project

Answers will vary.

Project II

a.

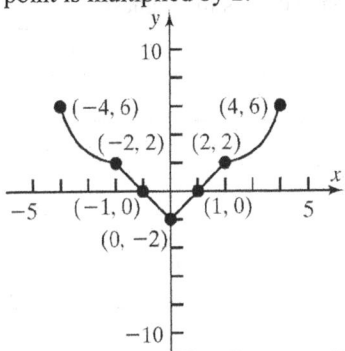

b. The data would be best fit by a quadratic function.

```
QuadReg
y=ax²+bx+c
a=.0851846811
b=-14.46460932
c=1069.518992
```

$$y = 0.085x^2 - 14.46x + 1069.52$$

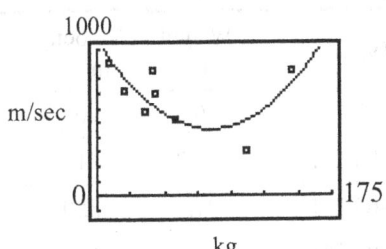

These results seem reasonable since the function fits the data well.

c. $s_0 = 0$m

Type	Weight kg	Velocity m/sec	Equation in the form: $s(t) = -4.9t^2 + \dfrac{\sqrt{2}}{2}v_0 t + s_0$
MG 17	10.2	905	$s(t) = -4.9t^2 + 639.93t$ Best. (It goes the highest)
MG 131	19.7	710	$s(t) = -4.9t^2 + 502.05t$
MG 151	41.5	850	$s(t) = -4.9t^2 + 601.04t$
MG 151/20	42.3	695	$s(t) = -4.9t^2 + 491.44t$
MG/FF	35.7	575	$s(t) = -4.9t^2 + 406.59t$
MK 103	145	860	$s(t) = -4.9t^2 + 608.11t$
MK 108	58	520	$s(t) = -4.9t^2 + 367.70t$
WGr 21	111	315	$s(t) = -4.9t^2 + 222.74t$

$s_0 = 200$m

Type	Weight kg	Velocity m/sec	Equation in the form: $s(t) = -4.9t^2 + \dfrac{\sqrt{2}}{2}v_0 t + s_0$
MG 17	10.2	905	$s(t) = -4.9t^2 + 639.93t + 200$ Best. (It goes the highest)
MG 131	19.7	710	$s(t) = -4.9t^2 + 502.05t + 200$
MG 151	41.5	850	$s(t) = -4.9t^2 + 601.04t + 200$
MG 151/20	42.3	695	$s(t) = -4.9t^2 + 491.44t + 200$
MG/FF	35.7	575	$s(t) = -4.9t^2 + 406.59t + 200$
MK 103	145	860	$s(t) = -4.9t^2 + 608.11t + 200$
MK 108	58	520	$s(t) = -4.9t^2 + 367.70t + 200$
WGr 21	111	315	$s(t) = 4.9t^2 + 222.74t + 200$

$s_0 = 30$m

Type	Weight kg	Velocity m/sec	Equation in the form: $s(t) = -4.9t^2 + \dfrac{\sqrt{2}}{2}v_0 t + s_0$
MG 17	10.2	905	$s(t) = -4.9t^2 + 639.93t + 30$ Best. (It goes the highest)
MG 131	19.7	710	$s(t) = -4.9t^2 + 502.05t + 30$
MG 151	41.5	850	$s(t) = -4.9t^2 + 601.04t + 30$
MG 151/20	42.3	695	$s(t) = -4.9t^2 + 491.44t + 30$
MG/FF	35.7	575	$s(t) = -4.9t^2 + 406.59t + 30$
MK 103	145	860	$s(t) = -4.9t^2 + 608.11t + 30$
MK 108	58	520	$s(t) = -4.9t^2 + 367.70t + 30$
WGr 21	111	315	$s(t) = -4.9t^2 + 222.74t + 30$

Notice that the gun is what makes the difference, not how high it is mounted necessarily. The only way to change the true maximum height that the projectile can go is to change the angle at which it fires.

Project III

a.

x	1	2	3	4	5
$y = -2x + 5$	3	1	-1	-3	-5

b.
$$\frac{\Delta y}{\Delta x} = \frac{y_2 - y_1}{x_2 - x_1} = \frac{1-3}{1} = -2$$

$$\frac{\Delta y}{\Delta x} = \frac{y_2 - y_1}{x_2 - x_1} = \frac{-1-1}{1} = -2$$

$$\frac{\Delta y}{\Delta x} = \frac{y_2 - y_1}{x_2 - x_1} = \frac{-3-(-1)}{1} = -2$$

$$\frac{\Delta y}{\Delta x} = \frac{y_2 - y_1}{x_2 - x_1} = \frac{-5-(-3)}{1} = -2$$

All of the values of $\frac{\Delta y}{\Delta x}$ are the same.

c.

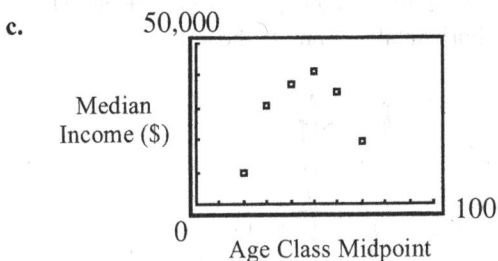

d.
$$\frac{\Delta I}{\Delta x} = \frac{30633 - 9548}{10} = 2108.50$$

$$\frac{\Delta I}{\Delta x} = \frac{37088 - 30633}{10} = 645.50$$

$$\frac{\Delta I}{\Delta x} = \frac{41072 - 37088}{10} = 398.40$$

$$\frac{\Delta I}{\Delta x} = \frac{34414 - 41072}{10} = -665.80$$

$$\frac{\Delta I}{\Delta x} = \frac{19167 - 34414}{10} = -1524.70$$

These $\frac{\Delta I}{\Delta x}$ values are not all equal. The data are not linearly related.

e.

x	-2	-1	0	1	2	3	4
y	23	9	3	5	15	33	59
$\frac{\Delta y}{\Delta x}$		-14	-6	2	10	18	26

As x increases, $\frac{\Delta y}{\Delta x}$ increases. This makes sense because the parabola is increasing (going up) steeply as x increases.

f.

x	-2	-1	0	1	2	3	4
y	23	9	3	5	15	33	59
$\frac{\Delta^2 y}{\Delta x^2}$			8	8	8	8	8

The second differences are all the same.

g. The paragraph should mention at least two observations:

1. The first differences for a linear function are all the same.

2. The second differences for a quadratic function are the same.

Project IV

a. – i. Answers will vary, depending on where the CBL is located above the bouncing ball.

j. The ratio of the heights between bounces will be the same.

Chapter 5
Polynomial and Rational Functions

Section 5.1

1. $(-2,0)$, $(2,0)$, and $(0,9)$

   x-intercepts: let $y = 0$ and solve for x
   $$9x^2 + 4(0) = 36$$
   $$9x^2 = 36$$
   $$x^2 = 4$$
   $$x = \pm 2$$

   y-intercepts: let $x = 0$ and solve for y
   $$9(0)^2 + 4y = 36$$
   $$4y = 36$$
   $$y = 9$$

3. down; 4

5. True: $f(x) = 0$ indicates that $y = 0$ which indicates that the point is an x-intercept.

7. smooth; continuous

9. $(-1,1)$, $(0,0)$, and $(1,1)$

11. turning points

13. ∞; $-\infty$

15. b

17. $f(x) = 4x + x^3$ is a polynomial function of degree 3.

19. $g(x) = \dfrac{1-x^2}{2} = \dfrac{1}{2} - \dfrac{1}{2}x^2$ is a polynomial function of degree 2.

21. $f(x) = 1 - \dfrac{1}{x} = 1 - x^{-1}$ is not a polynomial function because it contains a negative exponent.

23. $g(x) = x^{3/2} - x^2 + 2$ is not a polynomial function because it contains a fractional exponent.

25. $F(x) = 5x^4 - \pi x^3 + \dfrac{1}{2}$ is a polynomial function of degree 4.

27. $G(x) = 2(x-1)^2(x^2+1) = 2(x^2 - 2x + 1)(x^2 + 1)$
    $$= 2(x^4 + x^2 - 2x^3 - 2x + x^2 + 1)$$
    $$= 2(x^4 - 2x^3 + 2x^2 - 2x + 1)$$
    $$= 2x^4 - 4x^3 + 4x^2 - 4x + 2$$
    is a polynomial function of degree 4.

29. $f(x) = (x+1)^4$

    Using the graph of $y = x^4$, shift the graph horizontally, 1 unit to the left.

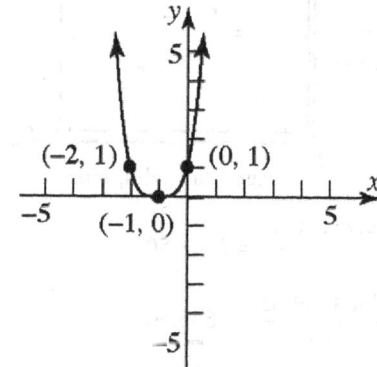

31. $f(x) = x^5 - 3$

    Using the graph of $y = x^5$, shift the graph vertically, 3 units down.

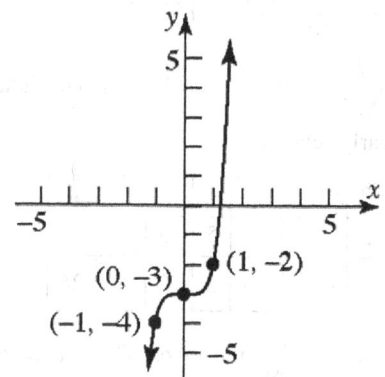

230

33. $f(x) = \frac{1}{2}x^4$

Using the graph of $y = x^4$, compress the graph vertically by a factor of $\frac{1}{2}$.

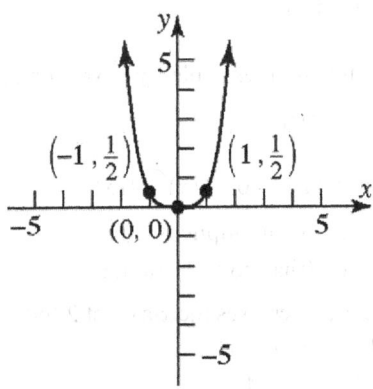

35. $f(x) = -x^5$

Using the graph of $y = x^5$, reflect the graph about the x-axis.

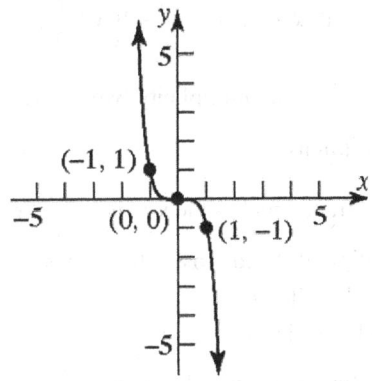

37. $f(x) = (x-1)^5 + 2$

Using the graph of $y = x^5$, shift the graph horizontally, 1 unit to the right, and shift vertically 2 units up.

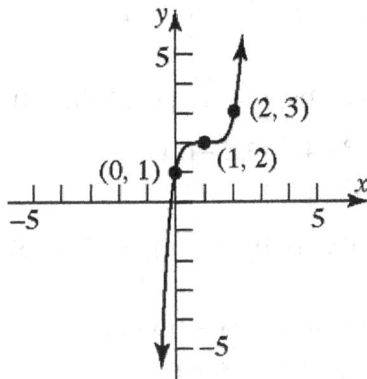

39. $f(x) = 2(x+1)^4 + 1$

Using the graph of $y = x^4$, shift the graph horizontally, 1 unit to the left, stretch vertically by a factor of 2, and shift vertically 1 unit up.

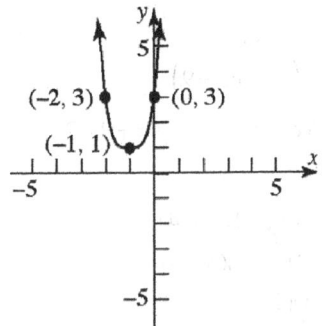

41. $f(x) = 4 - (x-2)^5 = -(x-2)^5 + 4$

Using the graph of $y = x^5$, shift the graph horizontally, 2 units to the right, reflect about the x-axis, and shift vertically 4 units up.

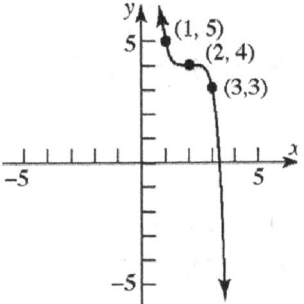

43. $f(x) = a\big(x - (-1)\big)(x-1)(x-3)$

For $a = 1$:

$$f(x) = (x+1)(x-1)(x-3) = \left(x^2 - 1\right)(x-3)$$
$$= x^3 - 3x^2 - x + 3$$

45. $f(x) = a(x-(-3))(x-0)(x-4)$

For $a = 1$:

$f(x) = (x+3)(x)(x-4) = (x^2 + 3x)(x-4)$

$\quad = x^3 - 4x^2 + 3x^2 - 12x$

$\quad = x^3 - x^2 - 12x$

47. $f(x) = a(x-(-4))(x-(-1))(x-2)(x-3)$

For $a = 1$:

$f(x) = (x+4)(x+1)(x-2)(x-3)$

$= (x^2 + 5x + 4)(x^2 - 5x + 6)$

$= x^4 - 5x^3 + 6x^2 + 5x^3 - 25x^2 + 30x + 4x^2 - 20x + 24$

$= x^4 - 15x^2 + 10x + 24$

49. $f(x) = a(x-(-1))(x-3)^2$

For $a = 1$:

$f(x) = (x+1)(x-3)^2$

$\quad = (x+1)(x^2 - 6x + 9)$

$\quad = x^3 - 6x^2 + 9x + x^2 - 6x + 9$

$\quad = x^3 - 5x^2 + 3x + 9$

51. $f(x) = a(x+3)(x-1)(x-4)$

$180 = a(6+3)(6-1)(6-4)$

$180 = a(9)(5)(2)$

$180 = 90a$

$\quad a = 2$

$f(x) = 2(x+3)(x-1)(x-4)$

$\quad = 2x^3 - 4x^2 - 22x + 24$

53. $f(x) = ax(x+1)(x-2)(x-4)$

$63 = a\left(\frac{1}{2}\right)\left(\frac{1}{2}+1\right)\left(\frac{1}{2}-2\right)\left(\frac{1}{2}-4\right)$

$63 = \frac{63}{16}a$

$\quad a = 16$

$f(x) = 16x(x+1)(x-2)(x-4)$

$\quad = 16x^4 - 80x^3 + 32x^2 + 128x$

55. $f(x) = a(x+1)^2(x-1)^2$

$45 = a(-2+1)^2(-2-1)^2$

$45 = 9a$

$\quad a = 5$

$f(x) = 5(x+1)^2(x-1)^2$

$\quad = 5x^4 - 10x^2 + 5$

57. a. The real zeros of $f(x) = 3(x-7)(x+3)^2$ are: 7, with multiplicity one; and −3, with multiplicity two.

b. The graph crosses the x-axis at 7 (odd multiplicity) and touches it at −3 (even multiplicity).

c. $n-1 = 3-1 = 2$

d. The function resembles $y = 3x^3$ for large values of $|x|$.

59. a. The real zeros of $f(x) = 4(x^2 + 1)(x-2)^3$ is: 2, with multiplicity three.

$x^2 + 1 = 0$ has no real solution.

b. The graph crosses the x-axis at 2 (odd multiplicity).

c. $n-1 = 5-1 = 4$

d. The function resembles $y = 4x^5$ for large values of $|x|$.

61. a. The real zero of $f(x) = -2\left(x+\frac{1}{2}\right)^2(x+4)^3$

are: $-\frac{1}{2}$, with multiplicity two; -4 with multiplicity 3.

b. The graph touches the x-axis at $-\frac{1}{2}$ (even multiplicity) and crosses the x-axis at -4 (odd multiplicity).

c. $n-1 = 5-1 = 4$

d. The function resembles $y = -2x^5$ for large values of $|x|$.

63. a. The real zeros of $f(x) = (x-5)^3(x+4)^2$ are: 5, with multiplicity three; and −4, with multiplicity two.

b. The graph crosses the x-axis at 5 (odd multiplicity) and touches it at −4 (even multiplicity).

c. $n-1 = 5-1 = 4$

d. The function resembles $y = x^5$ for large values of $|x|$.

65. a. $f(x) = 3(x^2 + 8)(x^2 + 9)^2$ has no real zeros.

$x^2 + 8 = 0$ and $x^2 + 9 = 0$ have no real solutions.

b. The graph neither touches nor crosses the x-axis.

c. $n - 1 = 6 - 1 = 5$

d. The function resembles $y = 3x^6$ for large values of $|x|$.

67. a. The real zeros of $f(x) = -2x^2(x^2 - 2)$ are: $-\sqrt{2}$ and $\sqrt{2}$ with multiplicity one; and 0, with multiplicity two.

b. The graph touches the x-axis at 0 (even multiplicity) and crosses the x-axis at $-\sqrt{2}$ and $\sqrt{2}$ (odd multiplicities).

c. $n - 1 = 4 - 1 = 3$

d. The function resembles $y = -2x^4$ for large values of $|x|$.

69. The graph could be the graph of a polynomial function.; zeros: $-1, 1, 2$; min degree = 3

71. The graph cannot be the graph of a polynomial.; not continuous at $x = -1$

73. The graph crosses the x-axis at $x = 0$, $x = 1$, and $x = 2$. Thus, each of these zeros has an odd multiplicity. Using one for each of these multiplicities, a possible function is $f(x) = ax(x-1)(x-2)$. Since the y-intercept is 0, we know $f(0) = 0$. Thus, a can be any positive constant. Using $a = 1$, the function is $f(x) = x(x-1)(x-2)$.

75. The graph crosses the x-axis at $x = -1$ and $x = 2$ and touches it at $x = 1$. Thus, -1 and 2 each have odd multiplicities while 1 has an even multiplicity. Using one for each odd multiplicity and two for the even multiplicity, a possible function is $f(x) = ax(x-1)^2(x-2)$. Since the y-intercept is 1, we know $f(0) = 1$. Thus,

$a(0+1)(0-1)(0-2) = 1$

$a(1)(-1)(-2) = 1$

$2a = 1$

$a = \dfrac{1}{2}$

The function is $f(x) = -\dfrac{1}{2}(x+1)(x-1)^2(x-2)$.

77. The graph crosses the x-axis at $x = -4$ and $x = 3$ and touches it at $x = -1$. Thus, -4 and 3 each have odd multiplicities while -1 has an even multiplicity. Using one for each odd multiplicity and two for the even multiplicity, a possible function is $f(x) = a(x+4)(x+1)^2(x-3)$. We know $f(1) = -8$. Thus,

$a(1+4)(1+1)^2(1-3) = -8$

$a(5)(2)^2(-2) = -8$

$-40a = -8$

$a = 0.2$

The function is $f(x) = 0.2(x+4)(x+1)^2(x-3)$.

79. The graph crosses the x-axis at $x = 0$ and touches it at $x = -3$ and $x = 3$. Thus, -3 and 3 each have even multiplicities while 0 has an odd multiplicity. Using one for each odd multiplicity and two for the even multiplicity, a possible function is $f(x) = ax(x+3)^2(x-3)^2$. We know $f(2) = -50$. Thus,

$a(2)(2+3)^2(2-3)^2 = -50$

$a(2)(5)^2(-1)^2 = -50$

$50a = -50$

$a = -1$

The function is $f(x) = -x(x+3)^2(x-3)^2$.

81. $f(x) = x^2(x-3)$

Step 1: Degree is 3. The function resembles $y = x^3$ for large values of $|x|$.

Step 2: y-intercept: $f(0) = 0^2(0-3) = 0$
x-intercepts: solve $f(x) = 0$

$0 = x^2(x-3)$

$x = 0, \ x = 3$

Step 3: Real zeros: 0 with multiplicity two, 3 with multiplicity one. The graph touches the x-axis at $x = 0$ and crosses the x-axis at $x = 3$.

Step 4: Graphing utility:

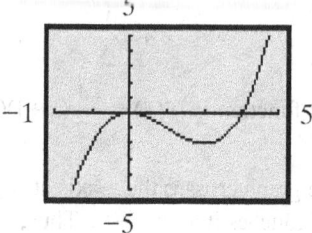

Step 5: 2 turning points; local maximum: $(0, 0)$; local minimum: $(2, -4)$

Step 6: Graphing by hand:

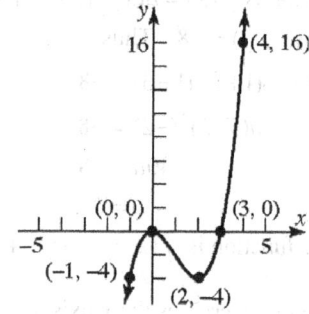

Step 7: Domain: $(-\infty, \infty)$; Range: $(-\infty, \infty)$

Step 8: Increasing on $(-\infty, 0]$ and $[2, \infty)$; decreasing on $[0, 2]$

83. $f(x) = (x + 4)^2(1 - x)$

Step 1: Degree is 3. The function resembles $y = -x^3$ for large values of $|x|$.

Step 2: y-intercept: $f(0) = (0 + 4)^2(1 - 0) = 16$

x-intercepts: solve $f(x) = 0$

$0 = (x + 4)^2(1 - x)$

$x = -4, 1$

Step 3: Real zeros: -4 with multiplicity two, 1 with multiplicity one. The graph touches the x-axis at $x = -4$ and crosses the x-axis at $x = 1$

Step 4: Graphing utility:

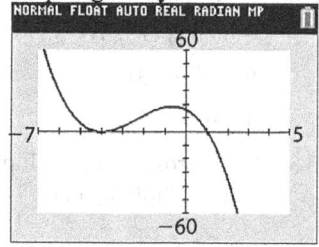

Step 5: 2 turning points; local maximum: $(-0.67, 18.52)$

local minimum: $(-4, 0)$

Step 6: Graphing by hand;

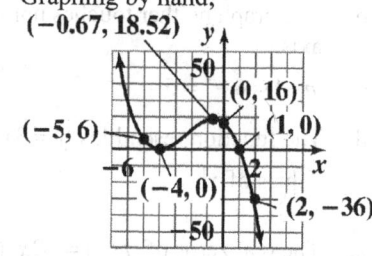

Step 7: Domain: $(-\infty, \infty)$; Range: $(-\infty, \infty)$

Step 8: Increasing on $[-4, -0.67]$; decreasing on $(-\infty, -4]$ and $[-0.67, \infty)$

85. $f(x) = -2(x + 2)(x - 2)^3$

Step 1: Degree is 4. The function resembles $y = -2x^4$ for large values of $|x|$.

Step 2: y-intercept: $f(0) = -2(0 + 2)(0 - 2)^3 = 32$

x-intercepts: solve $f(x) = 0$

$0 = -2(x + 2)(x - 2)^3$

$x = -2, 2$

Step 3: Real zeros: -2 with multiplicity one, 2 with multiplicity three. The graph crosses the x-axis at $x = -2$ and $x = 2$.

Step 4: Graphing utility;

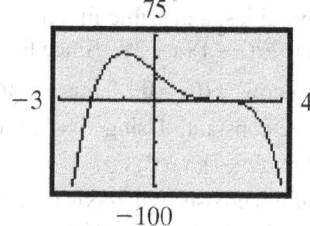

Step 5: 1 turning point; local maximum: $(-1, 54)$

Step 6: Graphing by hand;

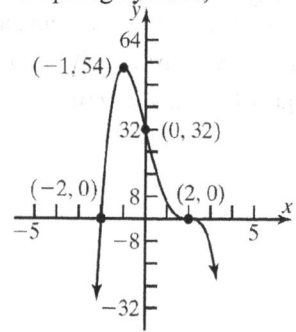

Step 7: Domain: $(-\infty,\infty)$; Range: $(-\infty,54]$

Step 8: Increasing on $(-\infty,-1]$; decreasing on $[-1,\infty)$

87. $f(x) = (x+1)(x-2)(x+4)$

Step 1: Degree is 3. The function resembles $y = x^3$ for large values of $|x|$.

Step 2: y-intercept: $f(0) = (0+1)(0-2)(0+4)$
$= -8$
x-intercepts: solve $f(x) = 0$
$0 = (x+1)(x-2)(x+4)$
$x = -1, 2, -4$

Step 3: Real zeros: -4 with multiplicity one, -1 with multiplicity one, 2 with multiplicity one. The graph crosses the x-axis at $x = -4, -1, 2$.

Step 4: Graphing utility:

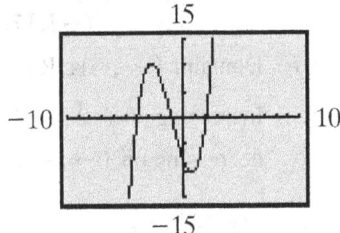

Step 5: 2 turning points;
local maximum: $(-2.73, 10.39)$;
local minimum: $(0.73, -10.39)$

Step 6: Graphing by hand;

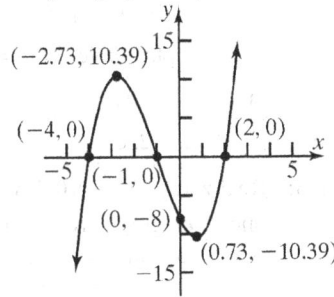

Step 7: Domain: $(-\infty,\infty)$; Range: $(-\infty,\infty)$

Step 8: Increasing on $(-\infty, -2.73]$ and $[0.73,\infty)$; decreasing on $[-2.73, 0.73]$

89. $f(x) = x^2(x-2)(x+2)$

Step 1: Degree is 4. The function resembles $y = x^4$ for large values of $|x|$.

Step 2: y-intercept: $f(x) = 0^2(0-2)(0+2) = 0$
x-intercepts: solve $f(x) = 0$
$0 = x^2(x-2)(x+2)$
$x = 0, 2, -2$

Step 3: Real zeros: -2 with multiplicity one, 0 with multiplicity two, 2 with multiplicity one. The graph crosses the x-axis at $x = -2$ and $x = -2$, and touches it at $x = 0$.

Step 4: Graphing utility:

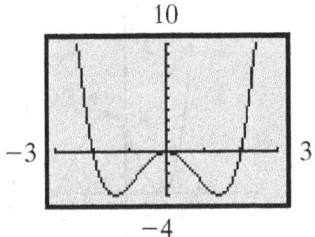

Step 5: 3 turning points;
local maximum: $(0, 0)$;
local minima: $(-1.41, -4), (1.41, -4)$

Step 6: graphing by hand

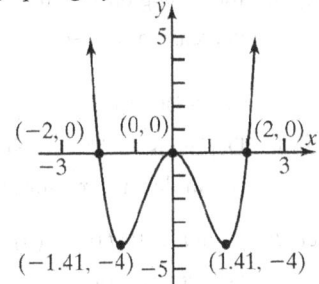

Step 7: Domain: $(-\infty,\infty)$; Range: $[-4,\infty)$

Step 8: Increasing on $[-1.41, 0]$ and $[1.41,\infty)$; decreasing on $(-\infty, -1.41]$ and $[0, 1.41]$.

91. $f(x) = (x+1)^2(x-2)^2$

Step 1: Degree is 4. The graph of the function resembles $y = x^4$ for large values of $|x|$.

Step 2: y-intercept: $f(0) = (0+1)^2(0-2)^2 = 4$
x-intercepts: solve $f(x) = 0$
$(x+1)^2(x-2)^2 = 0$
$x = -1$ or $x = 2$

Step 3: Real zeros: -1 with multiplicity two, 2 with multiplicity two. The graph touches the x-axis at $x = -1$ and $x = 2$.

Step 4: Graphing utility:

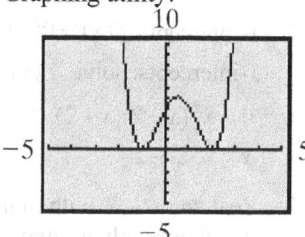

Step 5: 3 turning points; local maximum: $(0.5, 5.06)$; local minima: $(-1, 0)$, $(2, 0)$

Step 6: Graphing by hand:

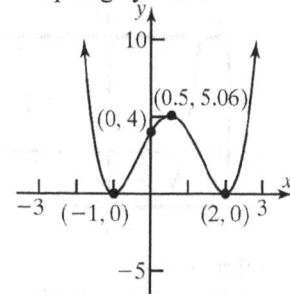

Step 7: Domain: $(-\infty, \infty)$; Range: $[0, \infty)$

Step 8: Increasing on $[-1, 0.5]$ and $[2, \infty)$; decreasing on $(-\infty, -1]$ and $[0.5, 2]$.

93. $f(x) = x^2(x+3)(x+1)$

Step 1: Degree is 4. The graph of the function resembles $y = x^4$ for large values of $|x|$.

Step 2: y-intercept: $f(0) = (0)^2(0+3)(0+1) = 0$
x-intercept: solve $f(x) = 0$
$$x^2(x+3)(x+1) = 0$$
$x = 0$ or $x = -3$ or $x = -1$

Step 3: Real zeros: 0 with multiplicity two, -3 with multiplicity one, -1 with multiplicity one. The graph touches the x-axis at $x = 0$, and crosses it at $x = -3$ and $x = -1$.

Step 4: Graphing utility:

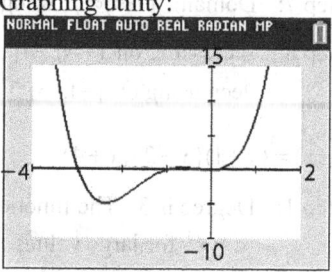

Step 5: 3 turning points; local maximum: $(0, 0)$; local minima: $(-0.69, -0.54)$, $(2.19, -12.39)$

Step 6: Graphing by hand;

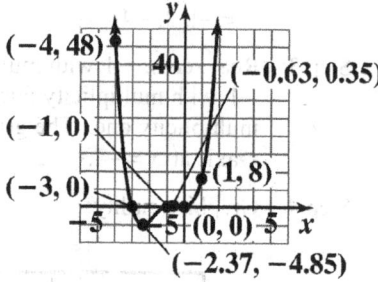

Step 7: Domain: $(-\infty, \infty)$; Range: $[-4.85, \infty)$

Step 8: Increasing on $(-2.37, -0.63)$ and $[0, \infty)$; decreasing on $(-\infty, -2.37]$ and $[-0.63, \infty)$

95. $f(x) = 5x(x^2 - 4)(x+3)$

Step 1: Degree is 4. The function resembles $y = x^4$ for large values of $|x|$.

Step 2: y-intercept:
$$f(x) = 5(0)(0^2 - 4)(0 + 3) = 0$$
x-intercepts: solve $f(x) = 0$
$$0 = 5x(x^2 - 4)(x+3)$$
$$x = 0, 2, -2, -3$$

Step 3: Real zeros: $-3, -2, 0, 2$ with multiplicity one. The graph crosses the x-axis at $x = -3$ and $x = -2$, $x = 0$ and $x = 2$.

Step 4:   Graphing utility:

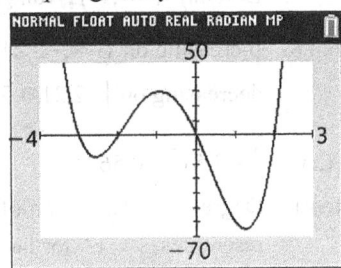

Step 5:   3 turning points;
local maximum: $[-0.93, 30.18]$
local minima:
$[-2.57, -14.39]$ and $[1.25, -64.75]$

Step 6:   graphing by hand

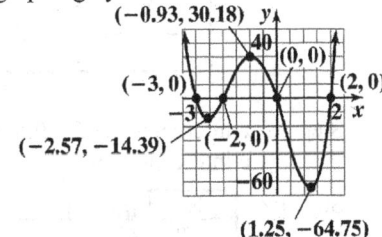

Step 7:   Domain: $(-\infty, \infty)$; Range: $[-64.75, \infty)$

Step 8:   Increasing on
$[-2.57, -0.93]$ and $[1.25, \infty)$;
decreasing on
$(-\infty, -2.57]$ and $[-0.93, 1.25]$.

97. $f(x) = x^2(x-2)(x^2+3)$

Step 1:   Degree is 5. The graph of the function
resembles $y = x^5$ for large values of $|x|$.

Step 2: y-intercept: $f(0) = 0^2(0-2)(0^2+3) = 0$
x-intercept: solve $f(x) = 0$
$x^2(x-2)(x^2+3) = 0$
$x = 0$ or $x = 2$
Note: $x^2 + 3 = 0$ has no real solution.

Step 3:   Real zeros: 0 with multiplicity two,
2 with multiplicity one. The graph
touches the x-axis at $x = 0$ and crosses
it $x = 2$.

Step 4:   Graphing utility:

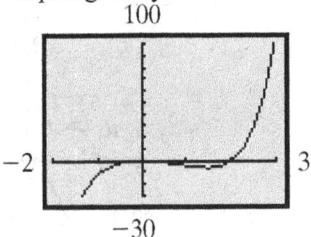

Step 5:   2 turning points;
local maximum: $(0, 0)$;
local minimum: $(1.48, -5.91)$

Step 6:   Graphing by hand:

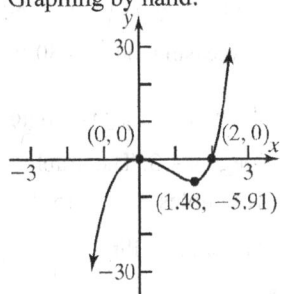

Step 7:   Domain: $(-\infty, \infty)$; Range: $(-\infty, \infty)$

Step 8:   Increasing on $(-\infty, 0]$ and $[1.48, \infty)$;
decreasing on $[0, 1.48)$

99. $f(x) = x^3 + 0.2x^2 - 1.5876x - 0.31752$

Step 1:   Degree = 3; The graph of the function
resembles $y = x^3$ for large values of $|x|$.

Step 2:   Graphing utility

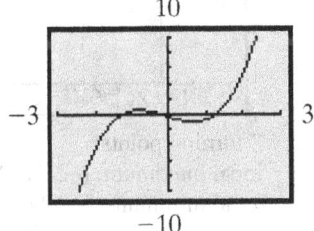

Step 3: x-intercepts: $-1.26, -0.20, 1.26$;
y-intercept: -0.31752

Step 4:

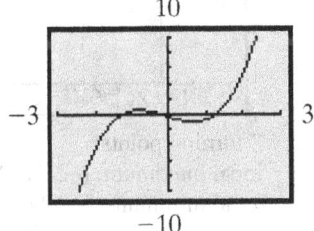

Step 5:   2 turning points;
local maximum: $(-0.80, 0.57)$;
local minimum: $(0.66, -0.99)$

Step 6: Graphing by hand

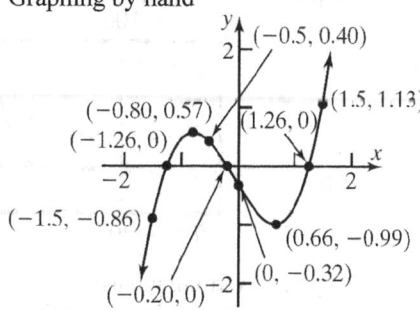

Step 7: Domain: $(-\infty, \infty)$; Range: $(-\infty, \infty)$

Step 8: Increasing on $(-\infty, -0.80]$ and $[0.66, \infty)$;
decreasing on $[-0.80, 0.66]$

101. $f(x) = x^3 + 2.56x^2 - 3.31x + 0.89$

Step 1: Degree = 3; The graph of the function resembles $y = x^3$ for large values of $|x|$.

Step 2: Graphing utility

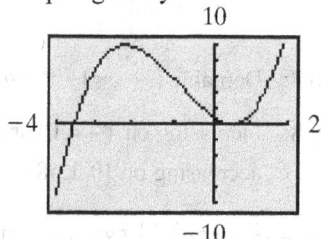

Step 3: x-intercepts: $-3.56, 0.50$;
y-intercept: 0.89

Step 4:

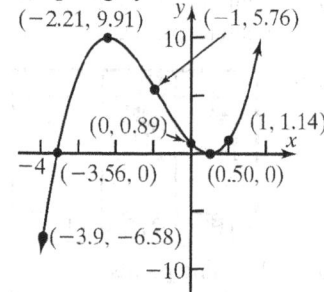

Step 5: 2 turning points;
local maximum: $(-2.21, 9.91)$;
local minimum: $(0.50, 0)$

Step 6: Graphing by hand

Step 7: Domain: $(-\infty, \infty)$; Range: $(-\infty, \infty)$

Step 8: Increasing on $(-\infty, -2.21]$ and $[0.50, \infty)$;
decreasing on $[2.21, 0.50]$.

103. $f(x) = x^4 - 2.5x^2 + 0.5625$

Step 1: Degree = 4; The graph of the function resembles $y = x^4$ for large values of $|x|$.

Step 2: Graphing utility

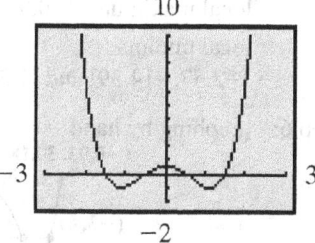

Step 3: x-intercepts: $-1.5, -0.5, 0.5, 1.5$;
y-intercept: 0.5625

Step 4:

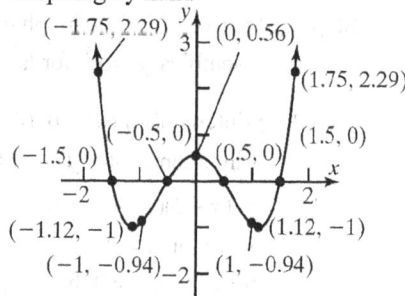

Step 5: 3 turning points:
local maximum: $(0, 0.5625)$;
local minima: $(-1.12, -1)$, $(1.12, -1)$

Step 6: Graphing by hand

Step 7: Domain: $(-\infty, \infty)$; Range: $[-1, \infty)$

Step 8: Increasing on $[-1.12, 0]$ and $[1.12, \infty)$;
decreasing on $(-\infty, -1.12]$ and $[0, 1.12]$

238

105. $f(x) = 2x^4 - \pi x^3 + \sqrt{5}x - 4$

Step 1: Degree = 4; The graph of the function resembles $y = 2x^4$ for large values of $|x|$.

Step 2: Graphing utility:

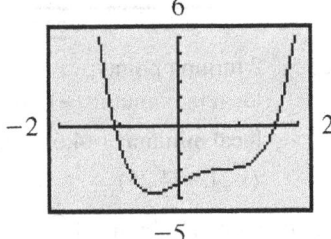

Step 3: x-intercepts: $-1.07, 1.62$; y-intercept: -4

Step 4:

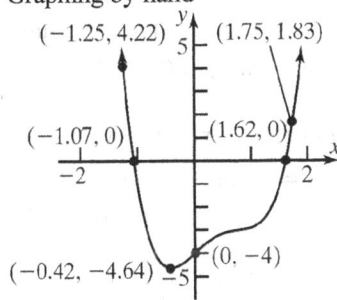

Step 5: 1 turning point; local minimum: $(-0.42, -4.64)$

Step 6: Graphing by hand

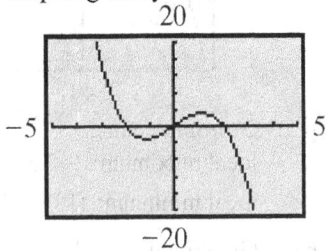

Step 7: Domain: $(-\infty, \infty)$; Range: $[-4.64, \infty)$

Step 8: Increasing on $[-0.42, \infty)$; decreasing on $(-\infty, -0.42]$

107. $f(x) = 4x - x^3 = -x(x^2 - 4) = -x(x+2)(x-2)$

Step 1: Degree is 3. The function resembles $y = -x^3$ for large values of $|x|$.

Step 2: y-intercept: $f(0) = 4(0) - 0^3 = 0$
x-intercepts: Solve $f(x) = 0$
$0 = -x(x+2)(x-2)$
$x = 0, -2, 2$

Step 3: Real zeros: 0 with multiplicity one, -2 with multiplicity one, 2 with multiplicity one. The graph crosses the x-axis at $x = 0$, $x = -2$, and $x = 2$.

Step 4: Graphing utility:

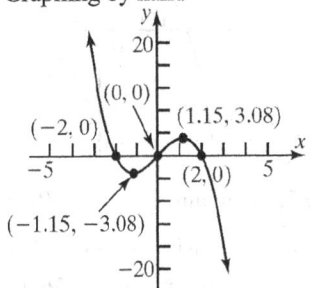

Step 5: 2 turning points; local maximum: $(1.15, 3.08)$; local minimum: $(-1.15, -3.08)$

Step 6: Graphing by hand

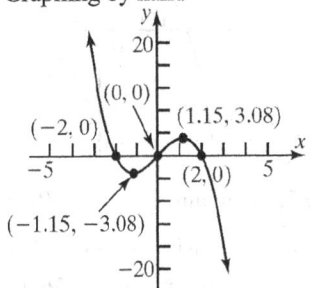

Step 7: Domain: $(-\infty, \infty)$; Range: $(-\infty, \infty)$

Step 8: Increasing on $[-1.15, 1.15]$; decreasing on $(-\infty, -1.15]$ and $[1.15, \infty)$.

109. $f(x) = x^3 + x^2 - 12x$
$= x(x^2 + x - 12)$
$= x(x+4)(x-3)$

Step 1: Degree is 3. The function resembles $y = -x^3$ for large values of $|x|$.

Step 2: y-intercept: $f(0) = 0^3 + 0^2 - 12(0) = 0$
x-intercepts: Solve $f(x) = 0$
$0 = x(x+4)(x-3)$
$x = 0, -4, 3$

Step 3: Real zeros: 0 with multiplicity one, -4 with multiplicity one, 3 with multiplicity one. The graph crosses the x-axis at $x = 0$, $x = -4$, and $x = 3$.

Step 4: Graphing utility:

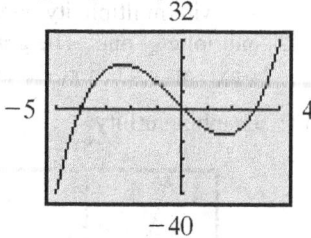

Step 5: 2 turning points;
local maximum: $(-2.36, 20.75)$;
local minimum: $(1.69, -12.60)$

Step 6: Graphing by hand

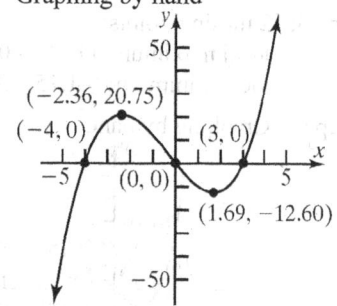

Step 7: Domain: $(-\infty, \infty)$; Range: $(-\infty, \infty)$

Step 8: Increasing on $[-\infty, -2.36]$ and $[1.69, \infty)$;
decreasing on $[-2.36, 1.69]$.

111. $f(x) = 2x^4 + 12x^3 - 8x^2 - 48x$
$= 2x(x^3 + 6x^2 - 4x - 24)$
$= 2x\left[x^2(x+6) - 4(x+6)\right]$
$= 2x(x+6)(x^2 - 4)$
$= 2x(x+6)(x-2)(x+2)$

Step 1: Degree is 3. The function resembles
$y = -x^3$ for large values of $|x|$.

Step 2: y-intercept:
$f(0) = 2(0)^4 + 12(0)^3 - 8(0)^2 - 48(0) = 0$
x-intercepts: Solve $f(x) = 0$
$0 = 2x(x+6)(x-2)(x+2)$
$x = 0, -6, 2, -2$

Step 3: Real zeros: 0 with multiplicity one,
-6 with multiplicity one, 2 with
multiplicity one, -2 with multiplicity
one. The graph crosses the x-axis at
$x = 0$, $x = -6$, $x = 2$, and $x = -2$.

Step 4: Graphing utility:

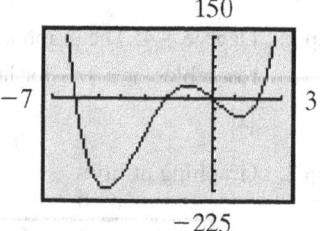

Step 5: 2 turning points;
local maximum: $(-1.06, 30.12)$;
local minima: $(-4.65, -221.25)$,
$(1.21, -44.25)$

Step 6: Graphing by hand

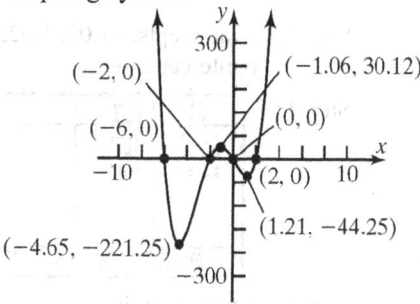

Step 7: Domain: $(-\infty, \infty)$; Range: $[-221.25, \infty)$

Step 8: Increasing on $[-4.65, -1.06]$ and
$(-1.21, \infty)$; decreasing on $(-\infty, -4.65)$
and $(-1.06, 1.21)$.

113. $f(x) = -x^5 - x^4 + x^3 + x^2$
$= -x^2(x^3 + x^2 - x - 1)$
$= -x^2\left[x^2(x+1) - 1(x+1)\right]$
$= -x^2(x+1)(x^2 - 1)$
$= -x^2(x+1)(x+1)(x-1)$
$= -x^2(x+1)^2(x-1)$

Step 1: Degree is 5. The graph of the function
resembles $y = -x^5$ for large values of $|x|$.

Step 2: y-intercept:
$f(0) = -(0)^5 - (0)^4 + (0)^3 + (0)^2 = 0$
x-intercept: Solve $f(x) = 0$
$-x^2(x+1)^2(x-1) = 0$
$x = 0, -1, 1$

Step 3: Real zeros: 0 with multiplicity two,
-1 with multiplicity two, 1 with
multiplicity one. The graph touches the

x-axis at $x=0$ and $x=-1$, and crosses it at $x=1$.

Step 4: Graphing utility:

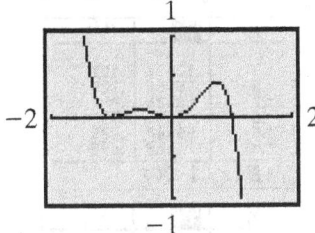

Step 5: 4 turning points;
local maxima: $(-0.54, 0.10), (0.74, 0.43)$;
local minima: $(-1, 0)$, $(0, 0)$

Step 6: Graphing by hand:

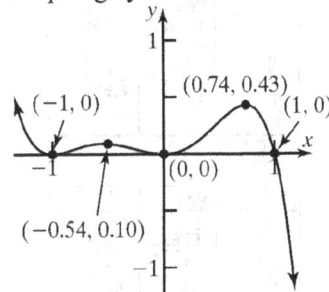

Step 7: Domain: $(-\infty, \infty)$; Range: $(-\infty, \infty)$

Step 8: Increasing on $[-1, -0.54]$ and $[0, 0.74]$;
decreasing on $(-\infty, -1]$, $[-0.54, 0]$, and $[0.74, \infty)$

115. $f(x) = a(x+3)(x-1)(x-4)$
$36 = a(0+3)(0-1)(0-4)$
$36 = 12a$
$a = 3$
$f(x) = 3(x+3)(x-1)(x-4)$

117. $f(x) = a(x+5)^2(x-2)(x-4)$
$128 = a(3+5)^2(3-2)(3-4)$
$128 = -64a$
$a = -2$
$f(x) = -2(x+5)^2(x-2)(x-4)$

119. a. $0 = (x+3)^2(x-2) \Rightarrow x=-3, x=2$

b. The graph is shifted to the left 3 units so the x-intercepts would be
$x = -3 - 3 = -6$
and $x = 2 - 3 = -1$

121. a. Graphing, we see that the graph may be a cubic relation.

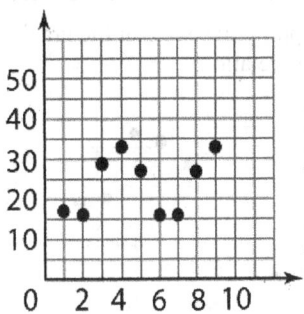

b. The cubic function of best fit is $H(x) = 0.3948x^3 - 5.9563x^2 + 26.1965x - 7.4127$

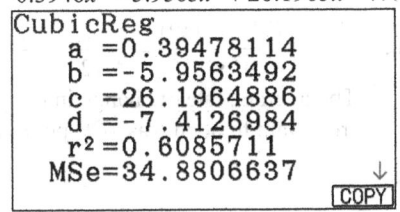

c. For the decade 1961-1970, we have $x = 5$.
$H(5) = 0.3948(5)^3 - 5.9563(5)^2 + 26.1695(5) - 7.4127$
≈ 24
The model predicts that about 24 major hurricanes struck the Atlantic Basin between 1961 and 1970.

d.

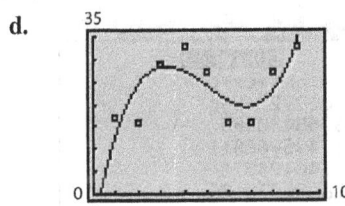

e. For the decade 2011 to 2020 we have $x = 10$.
$H(10) = 0.3948(10)^3 - 5.9563(10)^2 + 26.1695(10) - 7.4127$
≈ 54
The model predicts that approximately 54 major hurricanes will strike the Atlantic Basin between 2011 and 2020. The prediction does not seem to be reasonable. It appears to be too high. The end behavior of the model indicates that as time goes on, the number of major hurricanes will continue to increase each decade without limit. This is unrealistic. End behavior

should not be used to make predictions too far outside the data used to create the model.

123. a. Graphing, we see that the graph may be a cubic relation.

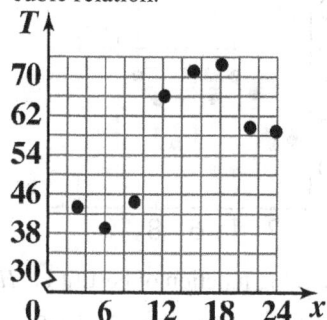

b. $\dfrac{\Delta T}{\Delta x} = \dfrac{T(12) - T(9)}{12 - 9} = \dfrac{62.1 - 44.1}{12 - 9} = \dfrac{18}{3} = 6$

The average rate of change in temperature from 9am to noon was $6°F$ per hour.

c.

$\dfrac{\Delta T}{\Delta x} = \dfrac{T(18) - T(15)}{18 - 15} = \dfrac{71.6 - 71.1}{18 - 15} = \dfrac{0.5}{3} = 0.17$

The average rate of change in temperature from 3pm to 6pm was $0.17°F$ per hour.

d. The cubic function of best fit is $T(x) =$

$-0.01992x^3 + 0.6745x^2 - 4.4360x + 48.4643$

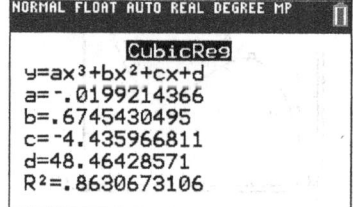

At 5pm we have $x = 17$.

$T(17) = -0.01992(17)^3 + 0.6745(17)^2$

$\qquad\qquad - 4.4360(17) + 48.4643$

$\qquad = 70.1$

The predicted temperature at 5pm is $\approx 70.1°F$.

e.

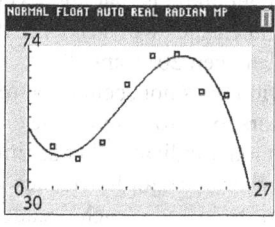

f. The y-intercept $(x = 0)$ is approximately $48.5°F$. The model predicts that the midnight temperature was $48.5°F$.

125. a.

X	Y1	Y2
-1	.5	0
-.9	.52632	.181
-.8	.55556	.328
-.7	.58824	.447
-.6	.625	.544
-.5	.66667	.625
-.4	.71429	.696

Y₁☐1/(1-X)

X	Y1	Y2
-.3	.76923	.763
-.2	.83333	.832
-.1	.90909	.909
0	1	1
.1	1.1111	1.111
.2	1.25	1.248
.3	1.4286	1.417

Y₁☐1/(1-X)

X	Y1	Y2
.4	1.6667	1.624
.5	2	1.875
.6	2.5	2.176
.7	3.3333	2.533
.8	5	2.952
.9	10	3.439
1	ERR:	4

Y₁☐1/(1-X)

b.

X	Y1	Y2
-1	.5	1
-.9	.52632	.8371
-.8	.55556	.7376
-.7	.58824	.6871
-.6	.625	.6736
-.5	.66667	.6875
-.4	.71429	.7216

Y₁☐1/(1-X)

X	Y1	Y2
-.3	.76923	.7711
-.2	.83333	.8336
-.1	.90909	.9091
0	1	1
.1	1.1111	1.1111
.2	1.25	1.2496
.3	1.4286	1.4251

Y₁☐1/(1-X)

X	Y1	Y2
.4	1.6667	1.6496
.5	2	1.9375
.6	2.5	2.3056
.7	3.3333	2.7731
.8	5	3.3616
.9	10	4.0951
1	ERR:	5

Y₁☐1/(1-X)

c.

X	Y1	Y2
-1	.5	0
-.9	.52632	.24661
-.8	.55556	.40992
-.7	.58824	.51903
-.6	.625	.59584
-.5	.66667	.65625
-.4	.71429	.71136

Y₁☐1/(1-X)

X	Y1	Y2
-.3	.76923	.76867
-.2	.83333	.83328
-.1	.90909	.90909
0	1	1
.1	1.1111	1.1111
.2	1.25	1.2499
.3	1.4286	1.4275

Y₁☐1/(1-X)

d. The values of the polynomial function get closer to the values of f. The approximations near 0 are better than those near -1 or 1.

127. Answers will vary.

129. Answers will vary, $f(x) = (x+2)(x-1)^2$ and $g(x) = (x+2)^3(x-1)^2$ are two such polynomials.

131. $f(x) = x^3 + bx^2 + cx + d$

a. True since every polynomial function has exactly one y-intercept, in this case $(0,d)$.

b. True, a third degree polynomial will have at most 3 x-intercepts since the equation $x^3 + bx^2 + cx + d = 0$ will have at most 3 real solutions.

c. True, a third degree polynomial will have at least one x-intercept since the equation $x^3 + bx^2 + cx + d = 0$ will have at least one real solution.

d. True, since f has degree 3 and the leading coefficient 1.

e. False, since

$$f(-x) = (-x)^3 + b(-x)^2 + c(-x) + d$$
$$= -x^3 + bx^2 - cx + d$$
$$\neq -f(x). \text{ (unless } b = d = 0)$$

f. True only if $d = 0$, otherwise the statement is false.

133. Answers will vary. One possibility:

$$f(x) = -5(x-1)^3(x-2)\left(x - \frac{1}{2}\right)\left(x + \frac{3}{5}\right)$$

135. We need to put the equation in standard form.

$$5x - 2y = 6$$
$$-2y = -5x + 6$$
$$y = \frac{5}{2}x - 3$$

Since we are looking for a perpendicular line, the new slope must be $m = -\frac{2}{5}$.

$$y - y_1 = -\frac{2}{5}(x - x_1)$$
$$y + 3 = -\frac{2}{5}(x - 2)$$
$$y + 3 = -\frac{2}{5}x + \frac{4}{5}$$
$$y = -\frac{2}{5}x - \frac{11}{5}$$

137. $x = \dfrac{-b \pm \sqrt{b^2 - 4ac}}{2a}$

$$= \frac{-(8) \pm \sqrt{(8)^2 - 4(4)(-3)}}{2(4)}$$

$$= \frac{-8 \pm \sqrt{64 + 48}}{8} = \frac{-8 \pm \sqrt{112}}{8}$$

$$= \frac{-8 \pm 4\sqrt{7}}{8} = \frac{-2 \pm \sqrt{7}}{2}$$

So the zeros are: $\dfrac{-2 - \sqrt{7}}{2}, \dfrac{-2 + \sqrt{7}}{2}$

243

Section 5.2

1. $f(-1) = 2(-1)^2 - (-1) = 2 + 1 = 3$

3. Using synthetic division:

$$3\overline{)3 \ \ -5 \ \ \ 0 \ \ \ \ 7 \ \ \ -4}$$
$$\ \ \ \ \ \ \ \ 9 \ \ 12 \ \ 36 \ \ 129$$
$$\overline{\ 3 \ \ \ 4 \ \ 12 \ \ 43 \ \ 125}$$

Quotient: $3x^3 + 4x^2 + 12x + 43$
Remainder: 125

5. a

7. b

9. 0.

11. $f(x) = 4x^3 - 3x^2 - 8x + 4; \ \ c = 2$

$f(2) = 4(2)^3 - 3(2)^2 - 8(2) + 4$
$\ \ \ \ \ \ = 32 - 12 - 16 + 4 = 8 \ne 0$

Thus, 2 is not a zero of f and $x - 2$ is not a factor of f.

13. $f(x) = 3x^4 - 6x^3 - 5x + 10; \ \ c = 2$

$f(2) = 3(2)^4 - 6(2)^3 - 5(2) + 10$
$\ \ \ \ \ \ = 48 - 48 - 10 + 10 = 0$

Thus, 2 is a zero of f and $x - 2$ is a factor of f.

15. $f(x) = 3x^6 + 82x^3 + 27; \ \ c = -3$

$f(-3) = 3(-3)^6 + 82(-3)^3 + 27$
$\ \ \ \ \ \ \ = 2187 - 2214 + 27 = 0$

Thus, -3 is a zero of f and $x + 3$ is a factor of f.

17. $f(x) = 4x^6 - 64x^4 + x^2 - 15; \ \ c = -4$

$f(-4) = 4(-4)^6 - 64(-4)^4 + (-4)^2 - 15$
$\ \ \ \ \ \ \ = 16,384 - 16,384 + 16 - 15 = 1 \ne 0$

Thus, -4 is not a zero of f and $x + 4$ is not a factor of f.

19. $f(x) = 2x^4 - x^3 + 2x - 1; \ \ c = \dfrac{1}{2}$

$f\left(\dfrac{1}{2}\right) = 2\left(\dfrac{1}{2}\right)^4 - \left(\dfrac{1}{2}\right)^3 + 2\left(\dfrac{1}{2}\right) - 1$

$\ \ \ \ \ \ \ = \dfrac{1}{8} - \dfrac{1}{8} + 1 - 1 = 0$

Thus, $\dfrac{1}{2}$ is a zero of f and $x - \dfrac{1}{2}$ is a factor of f.

21. $f(x) = -4x^7 + x^3 - x^2 + 2$

Examining $f(x) = -4x^7 + x^3 - x^2 + 2$, there are three variations in sign; thus, there are three positive real zeros or there is one positive real zero.
Examining

$f(-x) = -4(-x)^7 + (-x)^3 - (-x)^2 + 2$,
$\ \ \ \ \ \ \ = 4x^7 - x^3 - x^2 + 2$

there are two variations in sign; thus, there are two negative real zeros or no negative real zeros.

23. $f(x) = 2x^6 - 3x^2 - x + 1$

Examining $f(x) = 2x^6 - 3x^2 - x + 1$, there are two variations in sign; thus, there are two positive real zeros or no positive real zeros.
Examining

$f(-x) = 2(-x)^6 - 3(-x)^2 - (-x) + 1$,
$\ \ \ \ \ \ \ = 2x^6 - 3x^2 + x + 1$

there are two variations in sign; thus, there are two negative real zeros or no negative real zeros.

25. $f(x) = 3x^3 - 2x^2 + x + 2$

Examining $f(x) = 3x^3 - 2x^2 + x + 2$, there are two variations in sign; thus, there are two positive real zeros or no positive real zeros.
Examining

$f(-x) = 3(-x)^3 - 2(-x)^2 + (-x) + 2$,
$\ \ \ \ \ \ \ = -3x^3 - 2x^2 - x + 2$

there is one variation in sign; thus, there is one negative real zero.

27. $f(x) = -x^4 + x^2 - 1$

Examining $f(x) = -x^4 + x^2 - 1$, there are two variations in sign; thus, there are two positive real zeros or no positive real zeros.

Examining $f(-x) = -(-x)^4 + (-x)^2 - 1$

$= -x^4 + x^2 - 1$, there are two variations in sign; thus, there are two negative real zeros or no negative real zeros.

29. $f(x) = x^5 + x^4 + x^2 + x + 1$

Examining $f(x) = x^5 + x^4 + x^2 + x + 1$, there are no variations in sign; thus, there are no positive real zeros.

Examining

$f(-x) = (-x)^5 + (-x)^4 + (-x)^2 + (-x) + 1$,

$= -x^5 + x^4 + x^2 - x + 1$

there are three variations in sign; thus, there are three negative real zeros or there is one negative real zero.

31. $f(x) = x^6 - 1$

Examining $f(x) = x^6 - 1$, there is one variation in sign; thus, there is one positive real zero.

Examining $f(-x) = (-x)^6 - 1 = x^6 - 1$, there is one variation in sign; thus, there is one negative real zero.

33. $f(x) = 3x^4 - 3x^3 + x^2 - x + 1$

The maximum number of zeros is the degree of the polynomial, which is 4.

p must be a factor of 1: $p = \pm 1$

q must be a factor of 3: $q = \pm 1, \pm 3$

The possible rational zeros are: $\dfrac{p}{q} = \pm 1, \pm \dfrac{1}{3}$

35. $f(x) = x^5 - 6x^2 + 9x - 3$

The maximum number of zeros is the degree of the polynomial, which is 5.

p must be a factor of –3: $p = \pm 1, \pm 3$

q must be a factor of 1: $q = \pm 1$

The possible rational zeros are: $\dfrac{p}{q} = \pm 1, \pm 3$

37. $f(x) = -4x^3 - x^2 + x + 2$

The maximum number of zeros is the degree of the polynomial, which is 3.

p must be a factor of 2: $p = \pm 1, \pm 2$

q must be a factor of –4: $q = \pm 1, \pm 2, \pm 4$

The possible rational zeros are:

$\dfrac{p}{q} = \pm 1, \pm 2, \pm \dfrac{1}{2}, \pm \dfrac{1}{4}$

39. $f(x) = 6x^4 - x^2 + 9$

The maximum number of zeros is the degree of the polynomial, which is 4.

p must be a factor of 9: $p = \pm 1, \pm 3, \pm 9$

q must be a factor of 6: $q = \pm 1, \pm 2, \pm 3, \pm 6$

The possible rational zeros are:

$\dfrac{p}{q} = \pm 1, \pm \dfrac{1}{2}, \pm \dfrac{1}{3}, \pm \dfrac{1}{6}, \pm 3, \pm \dfrac{3}{2}, \pm 9, \pm \dfrac{9}{2}$

41. $f(x) = 2x^5 - x^3 + 2x^2 + 12$

The maximum number of zeros is the degree of the polynomial, which is 5.

p must be a factor of 12:

$p = \pm 1, \pm 2, \pm 3, \pm 4, \pm 6, \pm 12$

q must be a factor of 2: $q = \pm 1, \pm 2$

The possible rational zeros are:

$\dfrac{p}{q} = \pm 1, \pm 2, \pm 4, \pm \dfrac{1}{2}, \pm 3, \pm \dfrac{3}{2}, \pm 6, \pm 12$

43. $f(x) = 6x^4 + 2x^3 - x^2 + 20$

The maximum number of zeros is the degree of the polynomial, which is 4.

p must be a factor of 20:

$p = \pm 1, \pm 2, \pm 4, \pm 5, \pm 10, \pm 20$

q must be a factor of 6: $q = \pm 1, \pm 2, \pm 3, \pm 6$

The possible rational zeros are:

$\dfrac{p}{q} = \pm 1, \pm 2, \pm \dfrac{1}{2}, \pm \dfrac{1}{3}, \pm \dfrac{2}{3}, \pm \dfrac{1}{6}, \pm 4, \pm \dfrac{4}{3}, \pm 5, \pm \dfrac{5}{2},$

$\pm \dfrac{5}{3}, \pm \dfrac{5}{6}, \pm 10, \pm \dfrac{10}{3}, \pm 20, \pm \dfrac{20}{3}$

45. $f(x) = 2x^3 + x^2 - 1$ p must be a factor of 1: $p = \pm 1$

q must be a factor of 2: $q = \pm 1, \pm 2$

$\dfrac{p}{q} = \pm 1, \pm \dfrac{1}{2}$

r	coeff of q(x)			remainder
1	2	3	3	2
-1	2	-1	1	-2

For r = 1, the last row of synthetic division contains only numbers that are positive or 0, so we know there are no zeros greater than 1. For r = -1, the last row of synthetic division results in alternating positive (or 0) and negative (or 0) values, so we know that there are no zeros less than -1. The upper bound is 1 and the lower bound is -1.

We graph using the bounds and ZOOM-FIT.

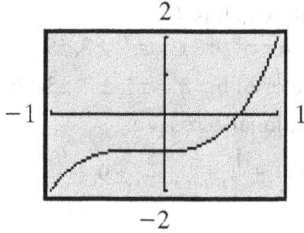

47. $f(x) = x^3 - 5x^2 - 11x + 11$ p must be a factor of 11: $p = \pm 1, \pm 11$

q must be a factor of 1: $q = \pm 1$

$\dfrac{p}{q} = \pm 1, \pm 11$

r	coeff of q(x)			remainder
1	1	-4	-15	-4
11	1	-3	-17	-23
...				
-1	1	-6	5	6
-11	1	-16	165	-1804

For r = 11, the last row of synthetic division contains only numbers that are positive or 0, so we know there are no zeros greater than 11. For r = -11, the last row of synthetic division results in alternating positive (or 0) and negative (or 0) values, so we know that there are no zeros less than -11. The upper bound is 11 and the lower bound is -11.

We graph using the bounds and ZOOM-FIT, and adjust the viewing window to improve the graph.

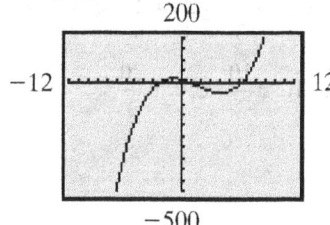

49. $f(x) = x^4 + 3x^3 - 5x^2 + 9$ p must be a factor of 9: $p = \pm 1, \pm 3, \pm 9$

q must be a factor of 1: $q = \pm 1$

$\dfrac{p}{q} = \pm 1, \pm 3, \pm 9$

r		coeff of q(x)			remainder
1	1	4	−1	−1	8
3	1	6	13	39	126
−1	1	2	−7	7	2
−3	1	0	−5	15	−36
−9	1	−6	49	−441	3978

For r = 3, the last row of synthetic division contains only numbers that are positive or 0, so we know there are no zeros greater than 3. For r = -9, the last row of synthetic division results in alternating positive (or 0) and negative (or 0) values, so we know that there are no zeros less than -9. The upper bound is 3 and the lower bound is -9.

We graph using the bounds and ZOOM-FIT, and adjust the viewing window to improve the graph.

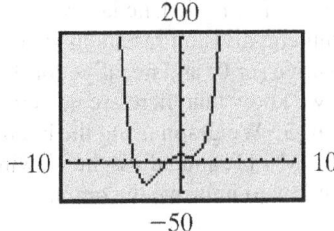

51. $f(x) = x^3 + 2x^2 - 5x - 6$

Step 1: $f(x)$ has at most 3 real zeros.

Step 2: Possible rational zeros:
$p = \pm 1, \pm 2, \pm 3, \pm 6; \quad q = \pm 1;$

$\dfrac{p}{q} = \pm 1, \pm 2, \pm 3, \pm 6$

Step 3: Using the Bounds on Zeros Theorem:

r		coeff of q(x)		remainder
2	1	4	3	0
−6	1	−4	19	−120

Referring to the Theorem, try the possible rational (integer) zeros from r = 1 and r = -1 until the desired result is obtained. For r = 2, the last row of synthetic division contains only numbers that are positive or 0, so we know there are no zeros greater than 2. For r = -6, the last row of synthetic division results in alternating positive (or 0) and negative (or 0) values, so we know that there are no zeros less than -6. We graph using the bounds and ZOOM-FIT, and adjust the viewing window to improve the graph.

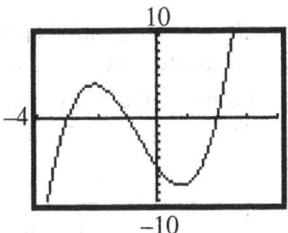

Step 4: From the synthetic division one of the x-intercepts is 2.

Since the remainder is 0, $x - 2$ is a factor.

Thus, $f(x) = (x - 2)\left(x^2 + 4x + 3\right)$

$= (x - 2)(x + 3)(x + 1)$

The zeros are −3, −1, and 2.

53. $f(x) = 2x^3 - 13x^2 + 24x - 9$

Step 1: $f(x)$ has at most 3 real zeros.

Step 2: Possible rational zeros:
$p = \pm 1, \pm 3, \pm 9; \quad q = \pm 1, \pm 2;$

$\dfrac{p}{q} = \pm 1, \pm 3, \pm 9, \pm \dfrac{1}{2}, \pm \dfrac{3}{2}, \pm \dfrac{9}{2}$

247

Step 3: Using the Bounds on Zeros Theorem:

r	coeff of q(x)			remainder
9	2	5	69	612
−1	2	−15	39	−48

Referring to the Theorem, try the possible rational (integer) zeros from r = 1 and r = -1 until the desired result is obtained. For r = 9, the last row of synthetic division contains only numbers that are positive or 0, so we know there are no zeros greater than 9. For r = -1, the last row of synthetic division results in alternating positive (or 0) and negative (or 0) values, so we know that there are no zeros less than -1. We graph using the bounds and ZOOM-FIT, and adjust the viewing window to improve the graph.

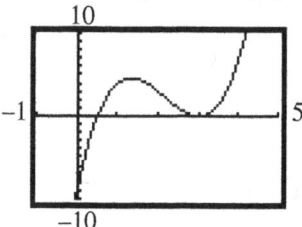

Step 4: From the graph it appears that there are x-intercepts at 0.5 and 3 .

Using synthetic division:

$$3{\overline{\smash{\big)}\,2\quad -13\quad 24\quad -9}}$$
$$\underline{\quad 6\quad -21\quad 9}$$
$$2\quad -7\quad 3\quad 0$$

Since the remainder is 0, $x - 3$ is a factor. Thus,

$$f(x) = (x-3)\left(2x^2 - 7x + 3\right)$$
$$= (x-3)(2x-1)(x-3)$$
$$= (2x-1)(x-3)^2$$

The zeros are $\frac{1}{2}$ and 3 (multiplicity 2).

55. $f(x) = 3x^3 + 4x^2 + 4x + 1$

Step 1: $f(x)$ has at most 3 real zeros..

Step 2: Possible rational zeros:

$$p = \pm 1; \quad q = \pm 1, \pm 3; \quad \frac{p}{q} = \pm 1, \pm \frac{1}{3}$$

Step 3: Using the Bounds on Zeros Theorem:

r	coeff of q(x)			remainder
1	3	7	11	12
−3	3	−5	19	−56

Referring to the Theorem, try the possible rational (integer) zeros from r = 1 and r = -1 until the desired result is obtained. For r = 1, the last row of synthetic division contains only numbers that are positive or 0, so we know there are no zeros greater than 1. For r = -3, the last row of synthetic division results in alternating positive (or 0) and negative (or 0) values, so we know there are no zeros less than -3. We graph using the bounds and ZOOM-FIT, and adjust the viewing window to improve the graph.

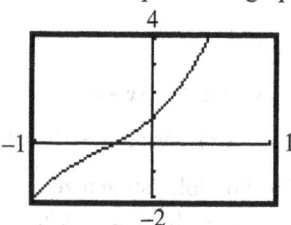

Step 4: From the graph it appears that there is an x-intercepts at $-\frac{1}{3}$.

Using synthetic division:

$$-\frac{1}{3}{\overline{\smash{\big)}\,3\quad 4\quad 4\quad 1}}$$
$$\underline{\quad -1\quad -1\quad -1}$$
$$3\quad 3\quad 3\quad 0$$

The remainder is 0, so $x - \left(-\frac{1}{3}\right) = x + \frac{1}{3}$

Thus, $f(x) = \left(x + \frac{1}{3}\right)(3x^2 + 3x + 3)$.

$$= 3\left(x + \frac{1}{3}\right)(x^2 + x + 1)$$
$$= (3x+1)(x^2 + x + 1)$$

Note that $x^2 + x + 1 = 0$ has no real solution. The only real zero is $-\frac{1}{3}$.

57. $f(x) = x^3 - 10x^2 + 28x - 16$

Step 1: $f(x)$ has at most 3 real zeros.

Step 2: Possible rational zeros:
$p = \pm 1, \pm 2, \pm 4, \pm 8, \pm 16;$

$q = \pm 1;$

$\dfrac{p}{q} = \pm 1, \pm 2, \pm 4, \pm 8, \pm 16$

Step 3: Using the Bounds on Zeros Theorem:

r	coeff of q(x)			remainder
16	1	6	124	1968
−1	1	−11	39	−55

Referring to the Theorem, try the possible rational (integer) zeros from r = 1 and r = -1 until the desired result is obtained. For r = 16, the last row of synthetic division contains only numbers that are positive or 0, so we know there are no zeros greater than 16. For r = -1, the last row of synthetic division results in alternating positive (or 0) and negative (or 0) values, so we know that there are no zeros less than -1. We graph using the bounds and ZOOM-FIT, and adjust the viewing window to improve the graph.

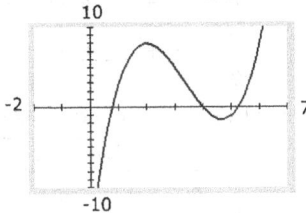

Step 4: From the graph it appears that there is an x-intercept at 4.
Using synthetic division:

```
4)1  −10   28   −16
        4  −24    16
   ─────────────────
   1   −6    4     0
```

Since the remainder is 0, $x - 4$ is a factor. Thus,

$f(x) = (x-4)(x^2 - 6x + 4)$.

Using the quadratic formula to find the solutions of the depressed equation $x^2 - 6x + 4 = 0$:

$x = \dfrac{-(-6) \pm \sqrt{(-6)^2 - 4(1)(4)}}{2(1)} = \dfrac{6 \pm \sqrt{20}}{2}$

$= \dfrac{6 \pm 2\sqrt{5}}{2} = 3 \pm \sqrt{5}$

Thus, $f(x) =$

$(x-4)\left(x + (3 - \sqrt{5})\right)\left(x + (3 + \sqrt{5})\right)$.

The zeros are 4, $3 - \sqrt{5}$, and $3 + \sqrt{5}$.

59. $f(x) = x^4 + x^3 - 3x^2 - x + 2$

Step 1: $f(x)$ has at most 4 real zeros.

Step 2: Possible rational zeros:

$p = \pm 1, \pm 2; \quad q = \pm 1; \quad \dfrac{p}{q} = \pm 1, \pm 2$

Step 3: Using the Bounds on Zeros Theorem:

r	coeff of q(x)				remainder
2	1	3	3	5	12
−3	1	−2	3	−10	32

Referring to the Theorem, try the possible rational (integer) zeros from r = 1 and r = -1 until the desired result is obtained. For r = 2, the last row of synthetic division contains only numbers that are positive or 0, so we know there are no zeros greater than 2. For r = -3, the last row of synthetic division results in alternating positive (or 0) and negative (or 0) values, so we know that there are no zeros less than -3. We graph using the bounds and ZOOM-FIT, and adjust the viewing window to improve the graph.

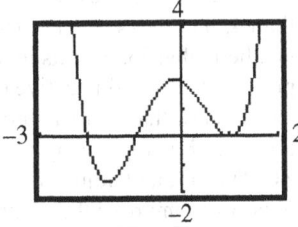

Step 4: From the graph it appears that there are x-intercepts at −2, −1, and 1.
Using synthetic division:

```
−2)1   1   −3   −1    2
       −2    2    2   −2
   ────────────────────
   1  −1   −1    1    0
```

Since the remainder is 0, $x + 2$ is a factor. Using synthetic division again:

249

Copyright © 2017 Pearson Education, Inc.

$$-1)\overline{\begin{array}{rrr} 1 & -1 & -1 & 1 \\ & -1 & 2 & -1 \end{array}}$$
$$\overline{\begin{array}{rrrr} 1 & -2 & 1 & 0 \end{array}}$$

Since the remainder is 0, $x+1$ is also a factor. Thus,

$$f(x) = (x+2)(x+1)(x^2 - 2x+1)$$
$$= (x+2)(x+1)(x-1)^2$$

The zeros are $-2, -1$, and 1 (multiplicity 2).

61. $f(x) = 21x^4 + 22x^3 - 99x^2 - 72x + 28$

Step 1: $f(x)$ has at most 4 real zeros.

Step 2: Possible rational zeros:
$p = \pm 1, \pm 2, \pm 4, \pm 7, \pm 14, \pm 28;$
$q = \pm 1, \pm 3, \pm 7, \pm 21;$

$\dfrac{p}{q} = \pm 1, \pm 2, \pm 4, \pm 7, \pm 14, \pm 28,$

$\pm \frac{1}{3}, \pm \frac{1}{7}, \pm \frac{1}{21}, \pm \frac{2}{3}, \pm \frac{2}{7}, \pm \frac{2}{21},$

$\pm \frac{4}{3}, \pm \frac{4}{7}, \pm \frac{4}{21}, \pm \frac{7}{3}, \pm \frac{14}{3}, \pm \frac{28}{3}$

Step 3: Using the Bounds on Zeros Theorem:

r		coeff of q(x)			remainder
4	21	106	325	1228	4940
-4	21	-62	149	-668	2700

Referring to the Theorem, try the possible rational (integer) zeros from r = 1 and r = -1 until the desired result is obtained. For r = 4, the last row of synthetic division contains only numbers that are positive or 0, so we know there are no zeros greater than 4. For r = -4, the last row of synthetic division results in alternating positive (or 0) and negative (or 0) values, so we know that there are no zeros less than -4. We graph using the bounds and ZOOM-FIT, and adjust the viewing window to improve the graph.

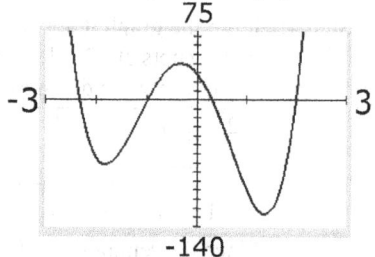

Step 4: From the graph it appears that there are x-intercepts at $-\dfrac{7}{3}, \dfrac{2}{7}, -1,$ and 2.

Using synthetic division:

$$2)\overline{\begin{array}{rrrrr} 21 & 22 & -99 & -72 & 28 \\ & 42 & 128 & 28 & -28 \end{array}}$$
$$\overline{\begin{array}{rrrrr} 21 & 64 & 29 & -14 & 0 \end{array}}$$

Since the remainder is 0, $x-2$ is a factor. Using synthetic division again:

$$-1)\overline{\begin{array}{rrrr} 21 & 64 & 29 & -14 \\ & -21 & -43 & -14 \end{array}}$$
$$\overline{\begin{array}{rrrr} 21 & 43 & -14 & 0 \end{array}}$$

Since the remainder is 0, $x+1$ is also a factor. Thus,

$$f(x) = (x-2)(x+1)(21x^2 + 43x - 14)$$
$$= (x-2)(x+1)(3x+7)(7x-2)$$

The zeros are $-\dfrac{7}{3}, \dfrac{2}{7}, -1,$ and 2.

63. $f(x) = x^3 - 8x^2 + 17x - 6$

Step 1: $f(x)$ has at most 3 real zeros.

Step 2: Possible rational zeros:
$p = \pm 1, \pm 2, \pm 3, \pm 6;$ $q = \pm 1;$

$\dfrac{p}{q} = \pm 1, \pm 2, \pm 3, \pm 6$

Step 3: Using the Bounds on Zeros Theorem:

r		coeff of q(x)		remainder
8	1	0	17	130
-1	1	-9	26	-32

Referring to the Theorem, try the possible rational (integer) zeros from r = 1 and r = -1 until the desired result is obtained. For r = 8, the last row of synthetic division contains only numbers that are positive or 0, so we know there are no zeros greater than 8. For r = -1, the last row of synthetic division results in alternating positive (or 0) and negative (or 0) values, so we know that there are no zeros less than -1. We graph using the bounds and ZOOM-FIT, and adjust the viewing window to improve the graph.

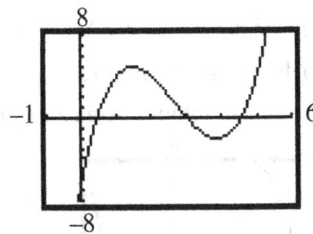

Step 4: From the graph it appears that there are x-intercepts at 0.5, 3, and 4.5.
Using synthetic division:

$$3\overline{)\begin{array}{cccc} 1 & -8 & 17 & -6 \\ & 3 & -15 & 6 \end{array}}$$
$$\;\; 1 \;\; -5 \;\;\;\; 2 \;\;\;\;\; 0$$

Since the remainder is 0, $x - 3$ is a factor. Thus,

$$f(x) = (x-3)\left(x^2 - 5x + 2\right).$$

Using the quadratic formula to find the solutions of the depressed equation $x^2 - 5x + 2 = 0$:

$$x = \frac{-(-5) \pm \sqrt{(-5)^2 - 4(1)(2)}}{2(1)} = \frac{5 \pm \sqrt{17}}{2}$$

Thus,

$$f(x) = (x-3)\left(x - \left(\frac{5+\sqrt{17}}{2}\right)\right)\left(x - \left(\frac{5-\sqrt{17}}{2}\right)\right).$$

The zeros are 3, $\dfrac{5+\sqrt{17}}{2}$, and $\dfrac{5-\sqrt{17}}{2}$.

65. $f(x) = 4x^4 + 7x^2 - 2$

Step 1: $f(x)$ has at most 4 real zeros.

Step 2: Possible rational zeros:
$p = \pm 1, \pm 2; \quad q = \pm 1, \pm 2, \pm 4;$
$$\frac{p}{q} = \pm 1, \pm 2, \pm\frac{1}{2}, \pm\frac{1}{4}$$

Step 3: Using the Bounds on Zeros Theorem:

r		coeff of q(x)			remainder
1	4	4	11	11	9
−1	4	−4	11	−11	9

Referring to the Theorem, try the possible rational (integer) zeros from r = 1 and r = -1 until the desired result is obtained. For r = 1, the last row of synthetic division contains only numbers that are positive or 0, so we know there are no zeros greater than 1. For r = -1, the last row of synthetic division results in alternating

positive (or 0) and negative (or 0) values, so we know that there are no zeros less than -1. We graph using the bounds and ZOOM-FIT, and adjust the viewing window to improve the graph.

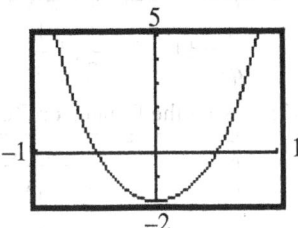

Step 4: From the graph it appears that there are x-intercepts at $-\dfrac{1}{2}$ and $\dfrac{1}{2}$.
Using synthetic division:

$$-\frac{1}{2}\overline{)\begin{array}{ccccc} 4 & 0 & 7 & 0 & -2 \\ & -2 & 1 & -4 & 2 \end{array}}$$
$$\phantom{-\frac{1}{2})}\;\; 4 \;\; -2 \;\;\; 8 \;\; -4 \;\;\;\; 0$$

Since the remainder is 0, $x + \dfrac{1}{2}$ is a factor. Using synthetic division again:

$$\frac{1}{2}\overline{)\begin{array}{cccc} 4 & -2 & 8 & -4 \\ & 2 & 0 & 4 \end{array}}$$
$$\phantom{\frac{1}{2})}\;\; 4 \;\;\;\; 0 \;\;\; 8 \;\;\;\; 0$$

Since the remainder is 0, $x - \dfrac{1}{2}$ is also a factor. Thus,

$$f(x) = \left(x + \frac{1}{2}\right)\left(x - \frac{1}{2}\right)\left(4x^2 + 8\right)$$
$$= 4\left(x + \frac{1}{2}\right)\left(x - \frac{1}{2}\right)\left(x^2 + 2\right)$$
$$= (2x+1)(2x-1)\left(x^2 + 2\right)$$

The depressed equation has no real zeros. The real zeros are $-\dfrac{1}{2}$ and $\dfrac{1}{2}$.

67. $f(x) = 4x^5 - 8x^4 - x + 2$

    Step 1: $f(x)$ has at most 5 real zeros.

    Step 2: Possible rational zeros:
    $p = \pm 1, \pm 2; \quad q = \pm 1, \pm 2, \pm 4;$

    $\dfrac{p}{q} = \pm 1, \pm 2, \pm \dfrac{1}{2}, \pm \dfrac{1}{4}$

    Step 3: Using the Bounds on Zeros Theorem:

r		coeff of q(x)				remainder
3	4	4	12	36	107	323
−1	4	−12	12	−12	11	−9

    Referring to the Theorem, try the possible rational (integer) zeros from r = 1 and r = -1 until the desired result is obtained. For r = 3, the last row of synthetic division contains only numbers that are positive or 0, so we know there are no zeros greater than 3. For r = -1, the last row of synthetic division results in alternating positive (or 0) and negative (or 0) values, so we know that there are no zeros less than -1. We graph using the bounds and ZOOM-FIT, and adjust the viewing window to improve the graph.

    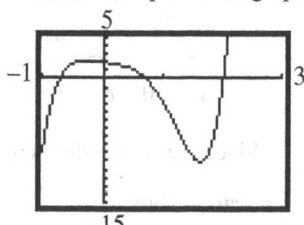

    Step 4: From the graph it appears that there are x-intercepts at −0.7, 0.7 and 2. Using synthetic division:

    ```
    2)4  −8   0   0  −1   2
         8   0   0   0  −2
      ─────────────────────
      4   0   0   0  −1   0
    ```

    Since the remainder is 0, $x - 2$ is a factor. Thus,

    $f(x) = (x-2)(4x^4 - 1)$

    $\qquad = (x-2)(2x^2 - 1)(2x^2 + 1)$

    $\qquad = (x-2)\left(\sqrt{2}x - 1\right)\left(\sqrt{2}x + 1\right)\left(2x^2 + 1\right)$

    Note that the depressed equation $2x^2 + 1 = 0$ has no real solution. The real zeros are $-\dfrac{\sqrt{2}}{2}, \dfrac{\sqrt{2}}{2}$, and 2.

69. $f(x) = x^3 + 3.2x^2 - 16.83x - 5.31$

    $f(x)$ has at most 3 real zeros.

    Solving by graphing (using ZERO):

    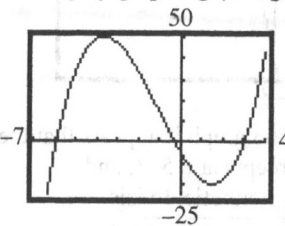

    The zeros are approximately −5.9, −0.3, and 3.

71. $f(x) = x^4 - 1.4x^3 - 33.71x^2 + 23.94x + 292.41$

    $f(x)$ has at most 4 real zeros.

    Solving by graphing (using ZERO):

    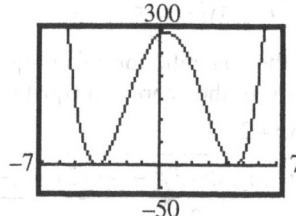

    The zeros are approximately −3.8 and 4.5. These zeros are each of multiplicity 2.

73. $f(x) = x^3 + 19.5x^2 - 1021x + 1000.5$

    $f(x)$ has at most 3 real zeros.

    Solving by graphing (using ZERO):

    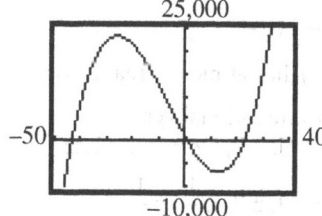

    The zeros are approximately −43.5, 1, and 23.

75. $x^4 - x^3 + 2x^2 - 4x - 8 = 0$

    The solutions of the equation are the zeros of $f(x) = x^4 - x^3 + 2x^2 - 4x - 8$.

    Step 1: $f(x)$ has at most 4 real zeros.

    Step 2: Possible rational zeros:
    $p = \pm 1, \pm 2, \pm 4, \pm 8; \quad q = \pm 1;$

    $\dfrac{p}{q} = \pm 1, \pm 2, \pm 4, \pm 8$

Step 3: Using the Bounds on Zeros Theorem:

r		coeff of q(x)		remainder
2	1	2	4	4	0
−1	1	−2	4	−8	0

Referring to the Theorem, try the possible rational (integer) zeros from r = 1 and r = -1 until the desired result is obtained. For r = 2, the last row of synthetic division contains only numbers that are positive or 0, so we know there are no zeros greater than 2. For r = -1, the last row of synthetic division results in alternating positive (or 0) and negative (or 0) values, so we know that there are no zeros less than -1. We graph using the bounds and ZOOM-FIT, and adjust the viewing window to improve the graph.

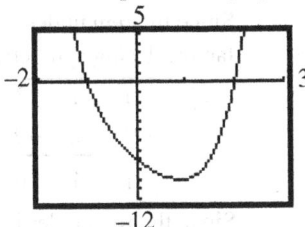

Step 4: From the graph it appears that there are x-intercepts at -1 and 2.
Using synthetic division:

$$-1 \overline{)\,1 \quad -1 \quad 2 \quad -4 \quad -8}$$
$$\underline{\quad\quad -1 \quad 2 \quad -4 \quad 8}$$
$$\quad\quad 1 \quad -2 \quad 4 \quad -8 \quad 0$$

Since the remainder is 0, $x+1$ is a factor. Using synthetic division again:

$$2 \overline{)\,1 \quad -2 \quad 4 \quad -8}$$
$$\underline{\quad\quad 2 \quad 0 \quad 8}$$
$$\quad\quad 1 \quad 0 \quad 4 \quad 0$$

Since the remainder is 0, $x-2$ is also a factor. Thus,
$$f(x) = (x+1)(x-2)(x^2+4).$$
The real solutions of the equation are -1 and 2. (Note that $x^2 + 4 = 0$ has no real solutions.)

77. $3x^3 + 4x^2 - 7x + 2 = 0$
The solutions of the equation are the zeros of
$f(x) = 3x^3 + 4x^2 - 7x + 2$.

Step 1: $f(x)$ has at most 3 real zeros.

Step 2: Possible rational zeros:
$p = \pm 1, \pm 2; \quad q = \pm 1, \pm 3;$
$$\frac{p}{q} = \pm 1, \pm 2, \pm \frac{1}{3}, \pm \frac{2}{3}$$

Step 3: Using the Bounds on Zeros Theorem:

r		coeff of q(x)		remainder
1	3	7	0	2
−3	3	−5	8	−22

Referring to the Theorem, try the possible rational (integer) zeros from r = 1 and r = -1 until the desired result is obtained. For r = 1, the last row of synthetic division contains only numbers that are positive or 0, so we know there are no zeros greater than 1. For r = -3, the last row of synthetic division results in alternating positive (or 0) and negative (or 0) values, so we know that there are no zeros less than -3. We graph using the bounds and ZOOM-FIT, and adjust the viewing window to improve the graph.

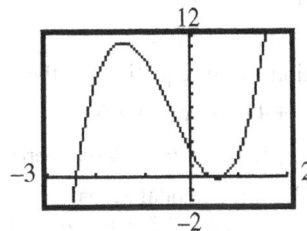

Step 4: From the graph it appears that there are x-intercepts at $\frac{1}{3}, \frac{2}{3}$, and -2.4.
Using synthetic division:

$$\frac{2}{3} \overline{)\,3 \quad 4 \quad -7 \quad 2}$$
$$\underline{\quad\quad 2 \quad 4 \quad -2}$$
$$\quad\quad 3 \quad 6 \quad -3 \quad 0$$

Since the remainder is 0, $x - \frac{2}{3}$ is a factor. The other factor is the quotient:
$3x^2 + 6x - 3$.

$$f(x) = \left(x - \frac{2}{3}\right)\left(3x^2 + 6x - 3\right)$$

$$= 3\left(x - \frac{2}{3}\right)\left(x^2 + 2x - 1\right)$$

Using the quadratic formula to solve
$x^2 + 2x - 1 = 0$:

$$x = \frac{-2 \pm \sqrt{4 - 4(1)(-1)}}{2(1)} = \frac{-2 \pm \sqrt{8}}{2}$$

$$= \frac{-2 \pm 2\sqrt{2}}{2} = -1 \pm \sqrt{2}$$

The real solutions of the equation are
$\frac{2}{3}, -1 + \sqrt{2},$ and $-1 - \sqrt{2}$.

79. $3x^3 - x^2 - 15x + 5 = 0$
Solving by factoring:

$$x^2(3x - 1) - 5(3x - 1) = 0$$

$$(3x - 1)(x^2 - 5) = 0$$

$$(3x - 1)(x - \sqrt{5})(x + \sqrt{5}) = 0$$

The solutions of the equation are $\frac{1}{3}, \sqrt{5},$ and
$-\sqrt{5}$.

81. $x^4 + 4x^3 + 2x^2 - x + 6 = 0$
The solutions of the equation are the zeros of
$f(x) = x^4 + 4x^3 + 2x^2 - x + 6$.

Step 1: $f(x)$ has at most 4 real zeros.

Step 2: Possible rational zeros:
$p = \pm 1, \pm 2, \pm 3, \pm 6; \quad q = \pm 1;$

$$\frac{p}{q} = \pm 1, \pm 2, \pm 3, \pm 6$$

Step 3: Using the Bounds on Zeros Theorem:

r		coeff of q(x)			remainder
1	1	5	7	6	12
-6	1	-2	14	-85	516

Referring to the Theorem, try the
possible rational (integer) zeros from r
= 1 and r = -1 until the desired result is
obtained. For r = 2, the last row of
synthetic division contains only
numbers that are positive or 0, so we
know there are no zeros greater than 2.
For r = -6, the last row of synthetic
division results in alternating positive
(or 0) and negative (or 0) values, so we

know that there are no zeros less than
-6. We graph using the bounds and
ZOOM-FIT, and adjust the viewing
window to improve the graph.

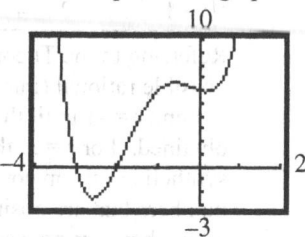

Step 4: From the graph it appears that there are
x-intercepts at -3 and -2.
Using synthetic division:

$$-3\overline{)\begin{array}{ccccc} 1 & 4 & 2 & -1 & 6 \\ & -3 & -3 & 3 & -6 \\ \hline 1 & 1 & -1 & 2 & 0 \end{array}}$$

Since the remainder is 0, $x + 3$ is a
factor. Using synthetic division again:

$$-2\overline{)\begin{array}{cccc} 1 & 1 & -1 & 2 \\ & -2 & 2 & -2 \\ \hline 1 & -1 & 1 & 0 \end{array}}$$

Since the remainder is 0, $x + 2$ is also a
factor. Thus,

$$f(x) = x^4 + 4x^3 + 2x^2 - x + 6$$

$$= (x + 2)(x + 3)(x^2 - x + 1)$$

The real solutions of the equation are -3
and -2. (Note that $x^2 - x + 1 = 0$ has no
real solutions.)

83. $x^3 - \frac{2}{3}x^2 + \frac{8}{3}x + 1 = 0$

The solutions of the equation are the zeros of

$$f(x) = x^3 - \frac{2}{3}x^2 + \frac{8}{3}x + 1.$$

Step 1: $f(x)$ has at most 3 real zeros.

Step 2: Use the equivalent equation
$3x^3 - 2x^2 + 8x + 3 = 0$ to find the possible
rational zeros:
$p = \pm 1, \pm 3; \quad q = \pm 1, \pm 3;$

$$\frac{p}{q} = \pm 1, \pm 3, \pm \frac{1}{3}$$

Step 3: Using the Bounds on Zeros Theorem:

r		coeff of q(x)		remainder
1	1	$\frac{1}{1}$	3	4
-1	1	$-\frac{5}{3}$	$\frac{13}{3}$	$-\frac{10}{3}$

Referring to the Theorem, try the possible rational (integer) zeros from $r = 1$ and $r = -1$ until the desired result is obtained. For $r = 1$, the last row of synthetic division contains only numbers that are positive or 0, so we know there are no zeros greater than 1. For $r = -1$, the last row of synthetic division results in alternating positive (or 0) and negative (or 0) values, so we know that there are no zeros less than -1. We graph using the bounds and ZOOM-FIT, and adjust the viewing window to improve the graph.

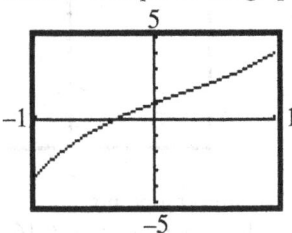

Step 4: From the graph it appears that there is an x-intercept at $-\frac{1}{3}$.

Using synthetic division:

$$-\frac{1}{3}\overline{)\begin{array}{cccc} 1 & -\frac{2}{3} & \frac{8}{3} & 1 \\ & -\frac{1}{3} & \frac{1}{3} & -1 \\ \hline 1 & -1 & 3 & 0 \end{array}}$$

Since the remainder is 0, $x + \frac{1}{3}$ is a factor. Thus,

$$f(x) = x^3 - \frac{2}{3}x^2 + \frac{8}{3}x + 1$$

$$= \left(x + \frac{1}{3}\right)(x^2 - x + 3)$$

The only real solution to the equation is $-\frac{1}{3}$. (Note that $x^2 - x + 3 = 0$ has no

85. Using the TABLE feature to show that there is a zero in the interval:

$$f(x) = 8x^4 - 2x^2 + 5x - 1; \quad [0,1]$$

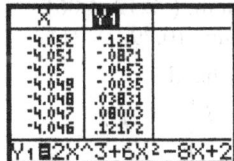

$$f(0) = -1 < 0 \text{ and } f(1) = 10 > 0$$

Since one is positive and one is negative, there is a zero in the interval. Using the TABLE feature to approximate the zero to two decimal places:

The zero is approximately 0.22.

87. Using the TABLE feature to show that there is a zero in the interval:

$$f(x) = 2x^3 + 6x^2 - 8x + 2; \quad [-5, -4]$$

$$f(-5) = -58 < 0 \text{ and } f(-4) = 2 > 0$$

Since one is positive and one is negative, there is a zero in the interval. Using the TABLE feature to approximate the zero to two decimal places:

The zero is approximately −4.05.

89. Using the TABLE feature to show that there is a zero in the interval:

$$f(x) = x^5 - x^4 + 7x^3 - 7x^2 - 18x + 18; \quad [1.4, 1.5]$$

$$f(1.4) = -0.1754 < 0 \text{ and } f(1.5) = 1.4063 > 0$$

Since one is positive and one is negative, there is

a zero in the interval. Using the TABLE feature to approximate the zero to two decimal places:

X	Y1
1.411	-.041
1.412	-.0283
1.413	-.0156
1.414	-.0028
1.415	.01016
1.416	.02315
1.417	.03621

Y1☰X^5-X^4+7X^3...

The zero is approximately 1.41.

91. $f(x) = x^3 + 2x^2 - 5x - 6$

Step 1: Degree = 3; The graph of the function resembles $y = x^3$ for large values of $|x|$.

Step 2: y-intercept:
$$f(0) = (0)^3 + 2(0)^2 - 5(0) - 6 = -6$$
x-intercepts: Solve $f(x) = 0$.
From Problem 39, we found
$x = -3, -1,$ and 2

Step 3: Real zeros: -3 with multiplicity one, -1 with multiplicity one, 2 with multiplicity one. The graph crosses the x-axis at $x = -3, -1, 2$.

Step 4: Graphing utility:

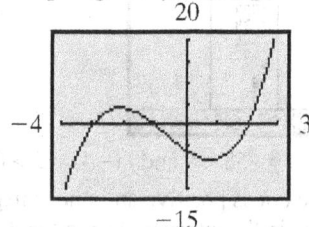

Step 5: 2 turning points;
local maximum: $(-2.12, 4.06)$;
local minimum: $(0.79, -8.21)$

Step 6: Graphing by hand

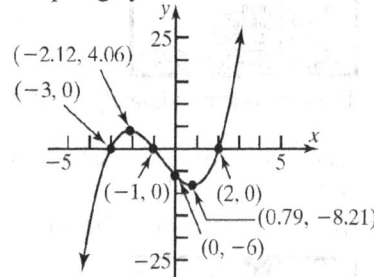

Step 7: Domain: $(-\infty, \infty)$; Range: $(-\infty, \infty)$

Step 8: Increasing on $(-\infty, -2.12]$ and $[0.79, \infty)$;
decreasing on $[-2.12, 0.79]$

93. $f(x) = x^4 + x^3 - 3x^2 - x + 2$

Step 1: Degree = 4; The graph of the function resembles $y = x^4$ for large values of $|x|$.

Step 2: y-intercept:
$$f(0) = (0)^4 + (0)^3 - 3(0)^2 - 0 + 2 = 2$$
x-intercepts: Solve $f(x) = 0$.
From Problem 47, we found
$x = -2, -1,$ and 1

Step 3: Real zeros: -2 with multiplicity one, -1 with multiplicity one, 1 with multiplicity two. The graph crosses the x-axis at $x = -2$ and $x = -1$, and touches it at $x = 1$.

Step 4: Graphing utility:

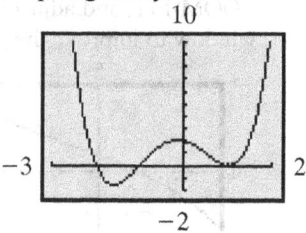

Step 5: 2 turning points;
local maximum: $(-0.16, 2.08)$;
local minima: $(-1.59, -1.62)$ and $(1, 0)$

Step 6: Graphing by hand

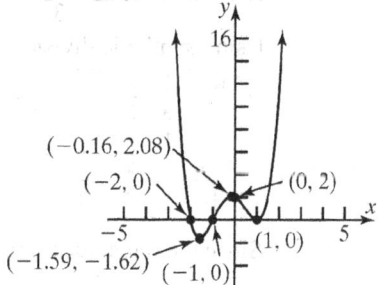

Step 7: Domain: $(-\infty, \infty)$; Range: $[-1.62, \infty)$

Step 8: Increasing on $[-1.59, -0.16]$ and $[1, \infty)$;
decreasing on $(-\infty, -1.59]$ and $[-0.16, 1]$

95. $f(x) = 4x^5 - 8x^4 - x + 2$

Step 1: Degree = 5; The graph of the function resembles $y = 4x^5$ for large values of $|x|$.

Step 2: y-intercept:
$$f(0) = 4(0)^5 - 8(0)^4 - 0 + 2 = 2$$
x-intercepts: Solve $f(x) = 0$.
From Problem 55, we found
$$x = -\frac{\sqrt{2}}{2}, \frac{\sqrt{2}}{2}, \text{ and } 2.$$

Step 3: Real zeros: $-\frac{\sqrt{2}}{2}$ with multiplicity one, $\frac{\sqrt{2}}{2}$ with multiplicity one, 2 with multiplicity one. The graph crosses the x-axis at $x = -\frac{\sqrt{2}}{2}$, $x = \frac{\sqrt{2}}{2}$, and $x = 2$.

Step 4: Graphing utility:

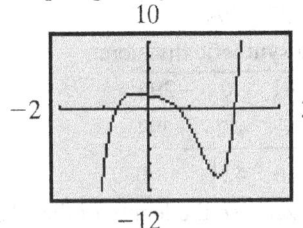

Step 5: 2 turning points;
local maximum: $(-0.30, 2.23)$;
local minimum: $(1.61, -10.09)$

Step 6: Graphing by hand

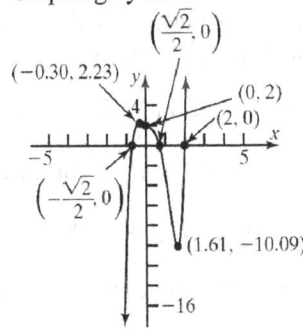

Step 7: Domain: $(-\infty, \infty)$; Range: $(-\infty, \infty)$

Step 8: Increasing on $(-\infty, -0.30]$ and $(1.61, \infty)$; decreasing on $[-0.30, 1.61]$

97. $f(x) = 6x^4 - 37x^3 + 58x^2 + 3x - 18$

Step 1: Degree = 4; The graph of the function resembles $y = 6x^4$ for large values of $|x|$.

Step 2: y-intercept:
$$f(0) = 6(0)^4 - 37(0)^3 + 58(0)^2 + 3(0) - 18$$
$$= -18$$
x-intercepts: Solve $f(x) = 0$.
The possible rational zeroes are:
$$\pm 1, \pm 2, \pm 3, \pm 6, \pm 9, \pm 18,$$
$$\pm \frac{1}{2}, \pm \frac{1}{3}, \pm \frac{3}{2}, \pm \frac{9}{2}, \pm \frac{2}{3}$$
We find the factoring to be:
$$f(x) = (x-3)^2(2x+1)(3x-2) \text{ so the x}$$
intercepts are: $3, -\frac{1}{2}, \frac{2}{3}$

Step 3: Real zeros: $-\frac{1}{2}$ and $\frac{2}{3}$ with multiplicity one, 3 with multiplicity 2. The graph crosses the x-axis at $-\frac{1}{2}$ and $\frac{2}{3}$ and touches at $x = 3$.

Step 4: Graphing utility:

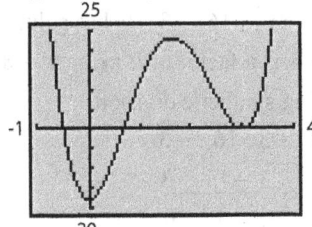

Step 5: 3 turning points;
local minimum: $(-0.03, -18.04), (3, 0)$;
local maximum: $(1.65, 23.12)$

Step 6: Graphing by hand

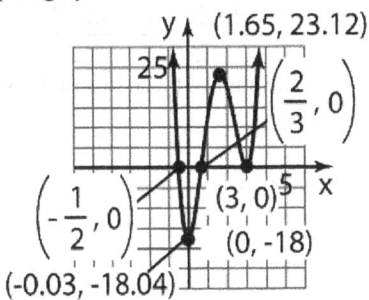

Step 7: Domain: $(-\infty, \infty)$; Range: $(-18.04, \infty)$

Step 8: Increasing on $[-0.03, 1.65]$ and $[3, \infty)$;

decreasing on $(-\infty, -0.03], [1.65, 3)$

99. From the Remainder and Factor Theorems,

$x - 2$ is a factor of f if $f(2) = 0$.

$$(2)^3 - k(2)^2 + k(2) + 2 = 0$$
$$8 - 4k + 2k + 2 = 0$$
$$-2k + 10 = 0$$
$$-2k = -10$$
$$k = 5$$

101. From the Remainder Theorem, we know that the remainder is

$$f(1) = 2(1)^{20} - 8(1)^{10} + 1 - 2 = 2 - 8 + 1 - 2 = -7$$

The remainder is -7.

103. We want to prove that $x - c$ is a factor of

$x^n - c^n$, for any positive integer n. By the Factor Theorem, $x - c$ will be a factor of $f(x)$

provided $f(c) = 0$. Here, $f(x) = x^n - c^n$, so

that $f(c) = c^n - c^n = 0$. Therefore, $x - c$ is a

factor of $x^n - c^n$.

105. $x^3 - 8x^2 + 16x - 3 = 0$ has solution $x = 3$, so

$x - 3$ is a factor of $f(x) = x^3 - 8x^2 + 16x - 3$.

Using synthetic division

```
3)1  -8   16   -3
      3  -15    3
   1  -5    1    0
```

Thus,

$$f(x) = x^3 - 8x^2 + 16x - 3 = (x - 3)(x^2 - 5x + 1).$$

Solving $x^2 - 5x + 1 = 0$

$$x = \frac{5 \pm \sqrt{25 - 4}}{2} = \frac{5 \pm \sqrt{21}}{2}$$

The sum of these two roots is

107. Let x be the length of a side of the original cube. After removing the 1-inch slice, one dimension will be $x - 1$. The volume of the new solid will be: $(x - 1) \cdot x \cdot x$.

Solve the volume equation:

$$(x - 1) \cdot x \cdot x = 294$$
$$x^3 - x^2 = 294$$
$$x^3 - x^2 - 294 = 0$$

The solutions to this equation are the same as the real zeros of $f(x) = x^3 - x^2 - 294$.

By Descartes' Rule of Signs, we know that there is one positive real zero.

$p = \pm 1, \pm 2, \pm 3, \pm 6, \pm 7, \pm 14, \pm 21, \pm 42, \pm 49, \pm 98,$

$\quad \pm 147, \pm 294$

$q = \pm 1$

The possible rational zeros are the same as the values for p.

$\frac{p}{q} = \pm 1, \pm 2, \pm 3, \pm 6, \pm 7, \pm 14, \pm 21, \pm 42, \pm 49, \pm 98,$

$\quad \pm 147, \pm 294$

Using synthetic division:

```
7)1  -1    0   -294
      7   42    294
   1   6   42     0
```

7 is a zero, so the length of the edge of the original cube was 7 inches.

109. $f(x) = x^n + a_{n-1}x^{n-1} + a_{n-2}x^{n-2} + \dots + a_1 x + a_0$;

where $a_{n-1}, a_{n-2}, \dots a_1, a_0$ are integers. If r is a real zero of f, then r is either rational or irrational. We know that the rational roots of f

must be of the form $\frac{p}{q}$ where p is a divisor of

a_0 and q is a divisor of 1. This means that

$q = \pm 1$. So if r is rational, then $r = \frac{p}{q} = \pm p$.

Therefore, r is an integer or r is irrational.

111. $f(x) = 8x^4 - 2x^2 + 5x - 1 \quad 0 \le r \le 1$

We begin with the interval [0,1].

$f(0) = -1; \quad f(1) = 10$

Let m_i = the midpoint of the interval being considered.

So $m_1 = 0.5$

n	m_n	$f(m_n)$	New interval
1	0.5	$f(0.5) = 1.5 > 0$	[0,0.5]
2	0.25	$f(0.25) = 0.15625 > 0$	[0,0.25]
3	0.125	$f(0.125) \approx -0.4043 < 0$	[0.125,0.25]
4	0.1875	$f(0.1875) \approx -0.1229 < 0$	[0.1875,0.25]
5	0.21875	$f(0.21875) \approx 0.0164 > 0$	[0.1875,0.21875]
6	0.203125	$f(0.203125) \approx -0.0533 < 0$	[0.203125,0.21875]
7	0.2109375	$f(0.2109375) \approx -0.0185 < 0$	[0.2109375,0.21875]
8	0.21484375	$f(0.21484375) \approx -0.0011 < 0$	[0.21484375,0.21875]
9	0.216796875	$f(0.216796875) \approx 0.0077 > 0$	[0.21484375,0.216796875]
10	0.2158203125	$f(0.2158203125) \approx 0.0033 > 0$	[0.21484375,0.2158203125]
11	0.2153320313	$f(0.2153320313) \approx 0.0011 > 0$	[0.21484375,0.2153320313]

If rounded to three decimal places, both endpoints of the new interval at Step 11 agree at 0.215. Therefore, $r \approx 0.215$, correct to three decimal places.

---

113. $f(x) = 4x^3 - 5x^2 - 3x + 1$

By the Rational Zero Theorem, the only possible rational zeros are: $\dfrac{p}{q} = \pm 1, \pm \dfrac{1}{2}, \pm \dfrac{1}{4}$.

Since $\dfrac{1}{3}$ is not in the list of possible rational zeros, it is not a zero of f.

115. $f(x) = x^7 + 6x^5 - x^4 + x + 2$

By the Rational Zero Theorem, the only possible rational zeros are: $\dfrac{p}{q} = \pm 1, \pm 2$.

Since $\dfrac{2}{3}$ is not in the list of possible rational zeros, it is not a zero of f.

117. $[3,8)$

119. Increasing: $[-3,2]$ and $[5,\infty)$

Section 5.3

1. $(3-2i) + (-3+5i) = 3 - 3 - 2i + 5i$
$$= 3i$$
$(3-2i)(-3+5i) = -9 + 15i + 6i - 10i^2$
$$= -9 + 21i - 10(-1)$$
$$= 1 + 21i$$

3. one

5. True

7. Since complex zeros appear in conjugate pairs, $4+i$, the conjugate of $4-i$, is the remaining zero of f.

9. Since complex zeros appear in conjugate pairs, $-i$, the conjugate of i, and $1-i$, the conjugate of $1+i$, are the remaining zeros of f.

11. Since complex zeros appear in conjugate pairs, $-i$, the conjugate of i, and $-2i$, the conjugate of $2i$, are the remaining zeros of f.

13. Since complex zeros appear in conjugate pairs, $-i$, the conjugate of i, is the remaining zero.

15. Since complex zeros appear in conjugate pairs, $2-i$, the conjugate of $2+i$, and $-3+i$, the conjugate of $-3-i$, are the remaining zeros.

For 17–22, we will use $a_n = 1$ as the lead coefficient of the polynomial. Also note that

$$(x-(a+bi))(x-(a-bi)) = ((x-a)-bi)((x-a)+bi)$$
$$= (x-a)^2 - (bi)^2$$

17. Since $3+2i$ is a zero, its conjugate $3-2i$ is also a zero of f.

$$f(x) = (x-4)(x-4)(x-(3+2i))(x-(3-2i))$$
$$= (x^2 - 8x + 16)((x-3)-2i)((x-3)+2i)$$
$$= (x^2 - 8x + 16)(x^2 - 6x + 9 - 4i^2)$$
$$= (x^2 - 8x + 16)(x^2 - 6x + 13)$$
$$= x^4 - 6x^3 + 13x^2 - 8x^3 + 48x^2$$
$$\quad - 104x + 16x^2 - 96x + 208$$
$$= x^4 - 14x^3 + 77x^2 - 200x + 208$$

19. Since $-i$ is a zero, its conjugate i is also a zero, and since $1+i$ is a zero, its conjugate $1-i$ is also a zero of f.

$$f(x) = (x-2)(x+i)(x-i)(x-(1+i))(x-(1-i))$$
$$= (x-2)(x^2 - i^2)((x-1)-i)((x-1)+i)$$
$$= (x-2)(x^2+1)(x^2 - 2x + 1 - i^2)$$
$$= (x^3 - 2x^2 + x - 2)(x^2 - 2x + 2)$$
$$= x^5 - 2x^4 + 2x^3 - 2x^4 + 4x^3 - 4x^2$$
$$\quad + x^3 - 2x^2 + 2x - 2x^2 + 4x - 4$$
$$= x^5 - 4x^4 + 7x^3 - 8x^2 + 6x - 4$$

21. Since $-i$ is a zero, its conjugate i is also a zero.
$$f(x) = (x-3)(x-3)(x+i)(x-i)$$
$$= (x^2 - 6x + 9)(x^2 - i^2)$$
$$= (x^2 - 6x + 9)(x^2 + 1)$$
$$= x^4 + x^2 - 6x^3 - 6x + 9x^2 + 9$$
$$= x^4 - 6x^3 + 10x^2 - 6x + 9$$

23. Since $2i$ is a zero, its conjugate $-2i$ is also a zero of f. $x-2i$ and $x+2i$ are factors of f. Thus, $(x-2i)(x+2i) = x^2 + 4$ is a factor of f. Using division to find the other factor:

$$
\begin{array}{r}
x - 4 \\
x^2 + 4 \overline{) x^3 - 4x^2 + 4x - 16} \\
\underline{x^3 \qquad + 4x} \\
-4x^2 \qquad -16 \\
\underline{-4x^2 \qquad -16}
\end{array}
$$

$x-4$ is a factor, so the remaining zero is 4. The zeros of f are $4, 2i, -2i$.

25. Since $-2i$ is a zero, its conjugate $2i$ is also a zero of f. $x-2i$ and $x+2i$ are factors of f.

Thus, $(x-2i)(x+2i) = x^2 + 4$ is a factor of f.

Using division to find the other factor:

$$
\begin{array}{r}
2x^2 + 5x - 3 \\
x^2 + 4 \overline{\smash{\big)}\ 2x^4 + 5x^3 + 5x^2 + 20x - 12} \\
\underline{2x^4 \qquad\quad + 8x^2} \\
5x^3 - 3x^2 + 20x \\
\underline{5x^3 \qquad\quad + 20x} \\
-3x^2 \qquad\quad -12 \\
\underline{-3x^2 \qquad\quad -12}
\end{array}
$$

$2x^2 + 5x - 3 = (2x-1)(x+3)$

The remaining zeros are $\dfrac{1}{2}$ and -3.

The zeros of f are $2i, -2i, -3, \dfrac{1}{2}$.

27. Since $3-2i$ is a zero, its conjugate $3+2i$ is also a zero of h. $x-(3-2i)$ and $x-(3+2i)$ are factors of h.

Thus,

$(x-(3-2i))(x-(3+2i)) = ((x-3)+2i)((x-3)-2i)$
$\qquad\qquad = x^2 - 6x + 9 - 4i^2$
$\qquad\qquad = x^2 - 6x + 13$

is a factor of h.

Using division to find the other factor:

$$
\begin{array}{r}
x^2 - 3x - 10 \\
x^2 - 6x + 13 \overline{\smash{\big)}\ x^4 - 9x^3 + 21x^2 + 21x - 130} \\
\underline{x^4 - 6x^3 + 13x^2} \\
-3x^3 + 8x^2 + 21x \\
\underline{-3x^3 + 18x^2 - 39x} \\
-10x^2 + 60x - 130 \\
\underline{-10x^2 + 60x - 130}
\end{array}
$$

$x^2 - 3x - 10 = (x+2)(x-5)$

The remaining zeros are -2 and 5.

The zeros of h are $3-2i, 3+2i, -2, 5$.

29. Since $-4i$ is a zero, its conjugate $4i$ is also a zero of h. $x-4i$ and $x+4i$ are factors of h.

Thus, $(x-4i)(x+4i) = x^2 + 16$ is a factor of h.

Using division to find the other factor:

$$
\begin{array}{r}
3x^3 + 2x^2 - 33x - 22 \\
x^2 + 16 \overline{\smash{\big)}\ 3x^5 + 2x^4 + 15x^3 + 10x^2 - 528x - 352} \\
\underline{3x^5 \qquad\quad + 48x^3} \\
2x^4 - 33x^3 + 10x^2 \\
\underline{2x^4 \qquad\quad + 32x^2} \\
-33x^3 - 22x^2 - 528x \\
\underline{-33x^3 \qquad\quad - 528x} \\
-22x^2 \qquad\quad - 352 \\
\underline{-22x^2 \qquad\quad - 352}
\end{array}
$$

$3x^3 + 2x^2 - 33x - 22 = x^2(3x+2) - 11(3x+2)$
$\qquad\qquad\qquad\qquad = (3x+2)(x^2 - 11)$

$\qquad\qquad\qquad\qquad = (3x+2)\left(x - \sqrt{11}\right)\left(x + \sqrt{11}\right)$

The remaining zeros are $-\dfrac{2}{3}, \sqrt{11}$, and $-\sqrt{11}$.

The zeros of h are $4i, -4i, -\sqrt{11}, \sqrt{11}, -\dfrac{2}{3}$.

31. $f(x) = x^3 - 1 = (x-1)\left(x^2 + x + 1\right)$ The solutions of $x^2 + x + 1 = 0$ are:

$$x = \frac{-1 \pm \sqrt{1^2 - 4(1)(1)}}{2(1)} = \frac{-1 \pm \sqrt{-3}}{2}$$

$$= -\frac{1}{2} + \frac{\sqrt{3}}{2}i \ \text{ and } \ -\frac{1}{2} - \frac{\sqrt{3}}{2}i$$

The zeros are: $1, \ -\dfrac{1}{2} + \dfrac{\sqrt{3}}{2}i, \ -\dfrac{1}{2} - \dfrac{\sqrt{3}}{2}i$.

$$f(x) = (x-1)\left(x + \frac{1}{2} - \frac{\sqrt{3}}{2}i\right)\left(x + \frac{1}{2} + \frac{\sqrt{3}}{2}i\right)$$

33. $f(x) = x^3 - 8x^2 + 25x - 26$

Step 1: $f(x)$ has 3 complex zeros.

Step 2: By Descartes Rule of Signs, there are three positive real zeros or there is one positive real zero.

$f(-x) = (-x)^3 - 8(-x)^2 + 25(-x) - 26$, thus,

$\quad = -x^3 - 8x^2 - 25x - 26$

there are no negative real zeros.

Step 3: Possible rational zeros:

$p = \pm 1, \pm 2, \pm 13, \pm 26; \quad q = \pm 1;$

$\dfrac{p}{q} = \pm 1, \pm 2, \pm 13, \pm 26$

Step 4: Using synthetic division:

We try $x - 2$:

$$\begin{array}{r|rrrr} 2 & 1 & -8 & 25 & -26 \\ & & 2 & -12 & 26 \\ \hline & 1 & -6 & 13 & 0 \end{array}$$

$x - 2$ is a factor. The other factor is the quotient: $x^2 - 6x + 13$.

The solutions of $x^2 - 6x + 13 = 0$ are:

$x = \dfrac{-(-6) \pm \sqrt{(-6)^2 - 4(1)(13)}}{2(1)}$.

$\quad = \dfrac{6 \pm \sqrt{-16}}{2} = \dfrac{6 \pm 4i}{2} = 3 \pm 2i$

The zeros are 2, $3 - 2i$, $3 + 2i$.

$f(x) = (x - 2)(x - 3 + 2i)(x - 3 - 2i)$

35. $f(x) = x^4 + 5x^2 + 4 = \left(x^2 + 4\right)\left(x^2 + 1\right)$

$\quad = (x + 2i)(x - 2i)(x + i)(x - i)$

The zeros are: $-2i, \ -i, \ i, \ 2i$.

37. $f(x) = x^4 + 2x^3 + 22x^2 + 50x - 75$

Step 1: $f(x)$ has 4 complex zeros.

Step 2: By Descartes Rule of Signs, there is 1 positive real zero.

$f(-x) = (-x)^4 + 2(-x)^3 + 22(-x)^2 + 50(-x) - 75$

$\quad = x^4 - 2x^3 + 22x^2 - 50x - 75$

Thus, there are three negative real zeros or there is one negative real zero.

Step 3: Possible rational zeros:

$p = \pm 1, \pm 3, \pm 5, \pm 15, \pm 25, \pm 75; \quad q = \pm 1;$

$\dfrac{p}{q} = \pm 1, \pm 3, \pm 5, \pm 15, \pm 25, \pm 75$

Step 4: Using synthetic division:

We try $x + 3$:

$$\begin{array}{r|rrrrr} -3 & 1 & 2 & 22 & 50 & -75 \\ & & -3 & 3 & -75 & 75 \\ \hline & 1 & -1 & 25 & -25 & 0 \end{array}$$

$x + 3$ is a factor. The other factor is the quotient: $x^3 - x^2 + 25x - 25$.

$x^3 - x^2 + 25x - 25 = x^2(x - 1) + 25(x - 1)$

$\quad\quad\quad\quad\quad\quad = (x - 1)\left(x^2 + 25\right)$

$\quad\quad\quad\quad\quad\quad = (x - 1)(x + 5i)(x - 5i)$

The zeros are $-3, \ 1, \ -5i, \ 5i$.

$f(x) = (x + 3)(x - 1)(x + 5i)(x - 5i)$

39. $f(x) = 3x^4 - x^3 - 9x^2 + 159x - 52$

Step 1: $f(x)$ has 4 complex zeros.

Step 2: By Descartes Rule of Signs, there are three positive real zeros or there is one positive real zero.

$f(-x) = 3(-x)^4 - (-x)^3 - 9(-x)^2 + 159(-x) - 52$

$\quad = 3x^4 + x^3 - 9x^2 - 159x - 52$

Thus, there is 1 negative real zero.

Step 3: Possible rational zeros:

$p = \pm 1, \pm 2, \pm 4, \pm 13, \pm 26, \pm 52;$

$q = \pm 1, \pm 3;$

$\dfrac{p}{q} = \pm 1, \pm 2, \pm 4, \pm 13, \pm 26, \pm 52,$

$\pm\dfrac{1}{3}, \pm\dfrac{2}{3}, \pm\dfrac{4}{3}, \pm\dfrac{13}{3}, \pm\dfrac{26}{3}, \pm\dfrac{52}{3}$

Step 4: Using synthetic division:

We try $x+4$:

$$
\begin{array}{r|rrrrr}
-4 & 3 & -1 & -9 & 159 & -52 \\
   &   & -12 & 52 & -172 & 52 \\
\hline
   & 3 & -13 & 43 & -13 & 0
\end{array}
$$

$x+4$ is a factor and the quotient is
$3x^3 - 13x^2 + 43x - 13$.

We try $x - \dfrac{1}{3}$ on $3x^3 - 13x^2 + 43x - 13$:

$$
\begin{array}{r|rrrr}
\frac{1}{3} & 3 & -13 & 43 & -13 \\
   &   & 1 & -4 & 13 \\
\hline
   & 3 & -12 & 39 & 0
\end{array}
$$

$x - \dfrac{1}{3}$ is a factor and the quotient is

$3x^2 - 12x + 39$.

$3x^2 - 12x + 39 = 3\left(x^2 - 4x + 13\right)$

The solutions of $x^2 - 4x + 13 = 0$ are:

$x = \dfrac{-(-4) \pm \sqrt{(-4)^2 - 4(1)(13)}}{2(1)}$.

$= \dfrac{4 \pm \sqrt{-36}}{2} = \dfrac{4 \pm 6i}{2} = 2 \pm 3i$

The zeros are $-4,\ \dfrac{1}{3},\ 2 - 3i,\ 2 + 3i$.

$f(x) = 3(x+4)\left(x - \dfrac{1}{3}\right)(x - 2 + 3i)(x - 2 - 3i)$

41. $f(x) = 2x^3 - 14x^2 + bx - 3$

$0 = 2(2)^3 - 14(2)^2 + b(2) - 3$

$0 = 16 - 26 + 2b - 3$

$b = \dfrac{43}{2}$

so

$f(x) = 2x^3 - 14x^2 + \dfrac{43}{2}x - 3$

$g(x) = (3-i)^3 + (3-i)^2 c - 8(3-i) + 30$

$0 = (18 - 26i) + (8 - 6i)c - (24 - 8i) + 30$

$0 = 18 - 26i + 8 - 6ic - 24 + 8i + 30$

$0 = 18 - 24 + 30 + 8c - 26i - 6ic + 8i$

$c = -3$

$g(x) = x^3 - 3x^2 - 8x + 30$

$f(1) = 2 - 14 + \dfrac{43}{2} - 3 = \dfrac{13}{2}$

$g(1) = 1 - 3 - 8 + 30 = 20$

$(f \cdot g)(1) = \dfrac{13}{2}\ 20 = 130$

43. a. $f(x) = x^4 + 1$

$= x^4 + 2x^2 + 1 - 2x^2$

$= \left(x^2 + 1\right)^2 - 2x^2$

$= \left(x^2 + 1 - \sqrt{2}x\right)\left(x^2 + 1 + \sqrt{2}x\right)$

$= \left(x^2 - \sqrt{2}x + 1\right)\left(x^2 + \sqrt{2}x + 1\right)$

b. $x = \dfrac{-\sqrt{2} \pm \sqrt{\left(\sqrt{2}\right)^2 - 4(1)(1)}}{2(1)}$

$= \dfrac{-\sqrt{2} \pm \sqrt{-2}}{2}$

$x = -\dfrac{\sqrt{2}}{2} - \dfrac{\sqrt{-2}}{2},\ -\dfrac{\sqrt{2}}{2} + \dfrac{\sqrt{-2}}{2}$

and

$$x = \frac{\sqrt{2} \pm \sqrt{\left(\sqrt{2}\right)^2 - 4(1)(1)}}{2(1)}$$

$$= \frac{\sqrt{2} \pm \sqrt{-2}}{2}$$

$$x = \frac{\sqrt{2}}{2} - \frac{\sqrt{-2}}{2}, \frac{\sqrt{2}}{2} + \frac{\sqrt{-2}}{2}$$

45. Three zeros are given. If the coefficients are real numbers, then the complex zeros would also have their conjugates as zeros. This would mean that there are 5 zeros, which would require a polynomial of degree 5.

47. One of the remaining zeros must be $4+i$, the conjugate of $4-i$. The third zero is a real number. Thus, the fourth zero must also be a real number in order to have a degree 4 polynomial.

49.

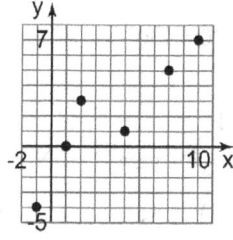

51. $(2x-5)(3x^2 + x - 4) = (2x)3x^2 + (2x)x - (2x)4$
$$- 5(3x^2) - 5(x) - 5(-4)$$
$$= 6x^3 + 2x^2 - 8x - 15x^2 - 5x + 20$$
$$= 6x^3 - 13x^2 - 13x + 20$$

Section 5.4

1. True

3. $y = \dfrac{1}{x}$

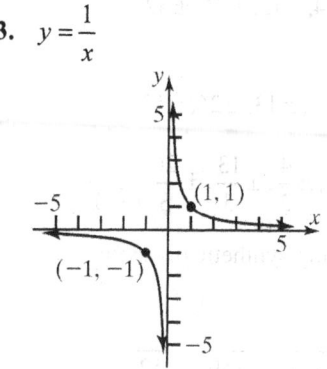

5. False

7. vertical asymptote

9. True

11. $y = 0$

13. d

15. In $R(x) = \dfrac{4x}{x-3}$, the denominator, $q(x) = x-3$, has a zero at 3. Thus, the domain of $R(x)$ is all real numbers except 3. $\{x \mid x \ne 3\}$

17. In $H(x) = \dfrac{-4x^2}{(x-2)(x+4)}$, the denominator, $q(x) = (x-2)(x+4)$, has zeros at 2 and –4. Thus, the domain of $H(x)$ is all real numbers except –4 and 2. $\{x \mid x \ne -4, x \ne 2\}$

19. In $F(x) = \dfrac{3x(x-1)}{2x^2 - 5x - 3}$, the denominator, $q(x) = 2x^2 - 5x - 3 = (2x+1)(x-3)$, has zeros at $-\dfrac{1}{2}$ and 3. Thus, the domain of $F(x)$ is all real numbers except $-\dfrac{1}{2}$ and 3. $\left\{x \mid x \ne -\dfrac{1}{2}, x \ne 3\right\}$

21. In $R(x) = \dfrac{x}{x^3 - 8}$, the denominator, $q(x) = x^3 - 8 = (x-2)(x^2 + 2x + 4)$, has a zero at 2 ($x^2 + 2x + 4$ has no real zeros). Thus, the domain of $R(x)$ is all real numbers except 2. $\{x \mid x \ne 2\}$

23. In $H(x) = \dfrac{3x^2 + x}{x^2 + 4}$, the denominator,

$q(x) = x^2 + 4$, has no real zeros. Thus, the domain of $H(x)$ is all real numbers.

25. In $R(x) = \dfrac{3(x^2 - x - 6)}{4(x^2 - 9)}$, the denominator,

$q(x) = 4(x^2 - 9) = 4(x - 3)(x + 3)$, has zeros at 3 and -3. Thus, the domain of $R(x)$ is all real numbers except -3 and 3. $\{x \mid x \neq -3, x \neq 3\}$

27. a. Domain: $\{x \mid x \neq 2\}$; Range: $\{y \mid y \neq 1\}$

b. Intercept: $(0, 0)$

c. Horizontal Asymptote: $y = 1$

d. Vertical Asymptote: $x = 2$

e. Oblique Asymptote: none

29. a. Domain: $\{x \mid x \neq 0\}$;

Range: all real numbers

b. Intercepts: $(-1, 0)$ and $(1, 0)$

c. Horizontal Asymptote: none

d. Vertical Asymptote: $x = 0$

e. Oblique Asymptote: $y = 2x$

31. a. Domain: $\{x \mid x \neq -2, x \neq 2\}$;

Range: $\{y \mid y \leq 0 \text{ or } y > 1\}$

b. Intercept: $(0, 0)$

c. Horizontal Asymptote: $y = 1$

d. Vertical Asymptotes: $x = -2, x = 2$

e. Oblique Asymptote: none

33. a. $F(x) = 2 + \dfrac{1}{x}$; Using the function, $y = \dfrac{1}{x}$, shift the graph vertically 2 units up.

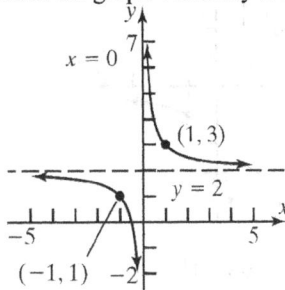

b. Domain: $\{x \mid x \neq 0\}$

Range: $\{y \mid y \neq 2\}$

c. Vertical asymptote: $x = 0$

Horizontal asymptote: $y = 2$

35. a. $R(x) = \dfrac{1}{(x-1)^2}$; Using the function, $y = \dfrac{1}{x^2}$, shift the graph horizontally 1 unit to the right.

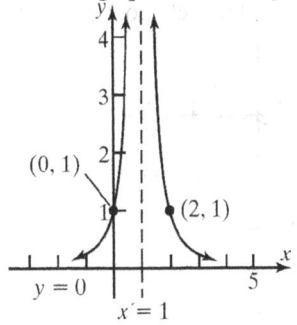

b. Domain: $\{x \mid x \neq 1\}$

Range: $\{y \mid y > 0\}$

c. Vertical asymptote: $x = 1$

Horizontal asymptote: $y = 0$

37. a. $H(x) = \dfrac{-2}{x+1} = -2\left(\dfrac{1}{x+1}\right)$; Using the function

$y = \dfrac{1}{x}$, shift the graph horizontally 1 unit to the left, reflect about the x-axis, and stretch vertically by a factor of 2.

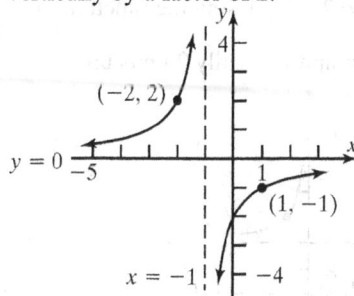

b. Domain: $\{x \mid x \neq -1\}$

Range: $\{y \mid y \neq 0\}$

c. Vertical asymptote: $x = -1$

Horizontal asymptote: $y = 0$

39. a. $R(x) = \dfrac{-1}{x^2 + 4x + 4} = -\dfrac{1}{(x+2)^2}$; Using the

function $y = \dfrac{1}{x^2}$, shift the graph horizontally 2

units to the left, and reflect about the x-axis.

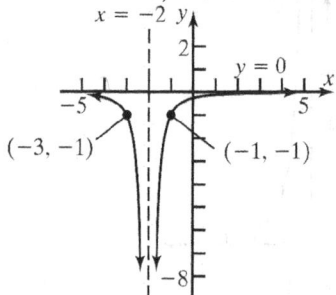

b. Domain: $\{x \mid x \neq -2\}$

Range: $\{y \mid y < 0\}$

c. Vertical asymptote: $x = -2$

Horizontal asymptote: $y = 0$

41. a. $G(x) = 1 + \dfrac{2}{(x-3)^2} = \dfrac{2}{(x-3)^2} + 1$;

$= 2\left[\dfrac{1}{(x-3)^2}\right] + 1$

Using the function $y = \dfrac{1}{x^2}$, shift the graph right 3

units, stretch vertically by a factor of 2, and shift

vertically 1 unit up.

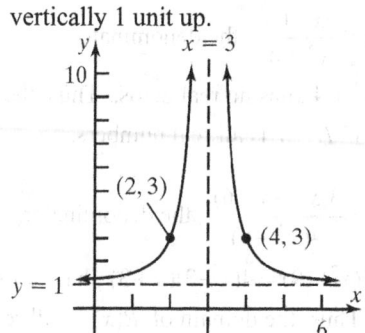

b. Domain: $\{x \mid x \neq 3\}$

Range: $\{y \mid y > 1\}$

c. Vertical asymptote: $x = 3$

Horizontal asymptote: $y = 1$

43. a. $R(x) = \dfrac{x^2 - 4}{x^2} = 1 - \dfrac{4}{x^2} = -4\left(\dfrac{1}{x^2}\right) + 1$; Using

the function $y = \dfrac{1}{x^2}$, reflect about the x-axis,

stretch vertically by a factor of 4, and shift

vertically 1 unit up.

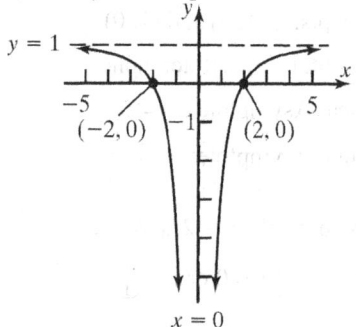

b. Domain: $\{x \mid x \neq 0\}$

Range: $\{y \mid y < 1\}$

c. Vertical asymptote: $x = 0$

Horizontal asymptote: $y = 1$

45. $R(x) = \dfrac{3x}{x+4}$; The degree of the numerator,

$p(x) = 3x$, is $n = 1$. The degree of the

denominator, $q(x) = x + 4$, is $m = 1$. Since

$n = m$, the line $y = \dfrac{3}{1} = 3$ is a horizontal

asymptote. The denominator is zero at $x = -4$,

so $x = -4$ is a vertical asymptote.

47. $H(x) = \dfrac{x^3 - 8}{x^2 - 5x + 6} = \dfrac{(x-2)(x^2 + 2x + 4)}{(x-2)(x-3)}$

$= \dfrac{x^2 + 2x + 4}{x - 3}$, where $x \neq 2, 3$

The degree of the numerator in lowest terms is

$n = 2$. The degree of the denominator in lowest

terms is $m = 1$. Since $n = m + 1$, there is an

oblique asymptote.

Dividing:

$$
\begin{array}{r}
x + 5 \\
x - 3 \overline{\smash{\big)}\, x^2 + 2x + 4} \\
\underline{-(x^2 - 3x)} \\
5x + 4 \\
\underline{-(5x - 15)} \\
19
\end{array}
$$

$H(x) = x + 5 + \dfrac{19}{x - 3}$, $x \neq 2, 3$

Thus, the oblique asymptote is $y = x + 5$.

The denominator in lowest terms is zero at $x = 3$

so $x = 3$ is a vertical asymptote.

49. $T(x) = \dfrac{x^3}{x^4 - 1}$; The degree of the numerator,

$p(x) = x^3$, is $n = 3$. The degree of the

denominator, $q(x) = x^4 - 1$ is $m = 4$. Since

$n < m$, the line $y = 0$ is a horizontal asymptote.

The denominator is zero at $x = -1$ and $x = 1$, so

$x = -1$ and $x = 1$ are vertical asymptotes.

51. $Q(x) = \dfrac{2x^2 - 5x - 12}{3x^2 - 11x - 4} = \dfrac{(2x+3)(x-4)}{(3x+1)(x-4)}$

$= \dfrac{2x + 3}{3x + 1}$, where $x \neq -\dfrac{1}{3}, 4$

The degree of the numerator in lowest terms is

$n = 1$. The degree of the denominator in lowest

terms is $m = 1$. Since $n = m$, the line $y = \dfrac{2}{3}$ is a

horizontal asymptote. The denominator in

lowest terms is zero at $x = -\dfrac{1}{3}$, so $x = -\dfrac{1}{3}$ is a

vertical asymptote.

53. $R(x) = \dfrac{6x^2 + 7x - 5}{3x + 5} = \dfrac{(3x+5)(2x-1)}{3x + 5}$

$= 2x - 1$, where $x \neq -\dfrac{5}{3}$

The degree of the numerator in lowest terms is

$n = 1$. The degree of the denominator in lowest

terms is $m = 0$. Since $n = m + 1$, there is an

oblique asymptote. From the simplification

shown above, the oblique asymptote is

$y = 2x - 1$. The denominator of $R(x)$ in lowest

terms is 1, so there is no vertical asymptote.

55. $G(x) = \dfrac{x^4 - 1}{x^2 - x} = \dfrac{(x^2+1)(x^2-1)}{x(x-1)}$

$= \dfrac{(x^2+1)(x+1)(x-1)}{x(x-1)}$

$= \dfrac{(x^2+1)(x+1)}{x}$, where $x \neq 0, 1$

The degree of the numerator in lowest terms is

$n = 3$. The degree of the denominator in lowest

terms is $m = 1$. Since $n \geq m + 2$, there is no

horizontal asymptote or oblique asymptote. The

denominator in lowest terms is zero at $x = 0$, so

$x = 0$ is a vertical asymptote.

57. $g(h) = \dfrac{3.99 \times 10^{14}}{\left(6.374 \times 10^6 + h\right)^2}$

a. $g(0) = \dfrac{3.99 \times 10^{14}}{\left(6.374 \times 10^6 + 0\right)^2} \approx 9.8208 \text{ m/s}^2$

b. $g(443) = \dfrac{3.99 \times 10^{14}}{\left(6.374 \times 10^6 + 443\right)^2}$

$\approx 9.8195 \text{ m/s}^2$

c. $g(8848) = \dfrac{3.99 \times 10^{14}}{\left(6.374 \times 10^6 + 8848\right)^2}$

$\approx 9.7936 \text{ m/s}^2$

d. $g(h) = \dfrac{3.99 \times 10^{14}}{\left(6.374 \times 10^6 + h\right)^2}$

$\approx \dfrac{3.99 \times 10^{14}}{h^2} \to 0 \text{ as } h \to \infty$

Thus, the h-axis is the horizontal asymptote. This indicates that as the height gets further away from the Earth the acceleration due to gravity approaches zero.

e. $g(h) = \dfrac{3.99 \times 10^{14}}{\left(6.374 \times 10^6 + h\right)^2} = 0$, to solve this

equation would require that $3.99 \times 10^{14} = 0$, which is impossible. Therefore, there is no height above sea level at which $g = 0$. In other words, there is no point in the entire universe that is unaffected by the Earth's gravity!

59. a. $R_{tot} = \dfrac{10 R_2}{10 + R_2}$

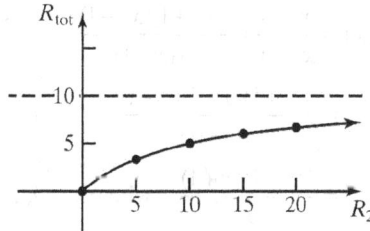

b. Horizontal asymptote: $y = R_{tot} = 10$

As the value of R_2 increases without bound, the total resistance approaches 10 ohms, the resistance of R_1.

c. $R_{tot} = \dfrac{R_1 R_2}{R_1 + R_2}$

$17 = \dfrac{R_1 \cdot 2\sqrt{R_1}}{R_1 + 2\sqrt{R_1}}$

Solving graphically, let $Y_1 = 17$ and

$Y_2 = 2x\sqrt{x}\big/\left(x + 2\sqrt{x}\right)$.

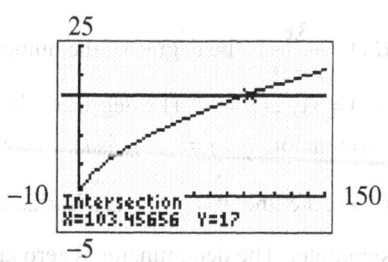

We would need $R_1 \approx 103.5$ ohms.

61. a. Dividing:

$$x - 1 \overline{\smash{\big)}\, 2x + 3} \atop $$

$$\begin{array}{r} 2 \\ x-1 \overline{\smash{)}2x+3} \\ \underline{-(2x-2)} \\ 5 \end{array}$$

$R(x) = 2 + \dfrac{5}{x-1} = 5\,\dfrac{1}{x-1} + 2$

b.

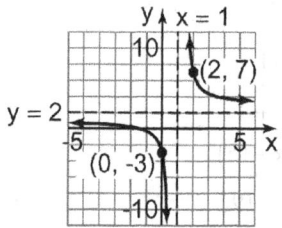

c. HA: $y = 2$. VA: $x = 1$

63. Answers will vary. If $x = 4$ is a vertical asymptote, then $x = 4$ is a zero of the denominator. If $x = 4$ is a zero of a polynomial, then $(x - 4)$ is a factor of the polynomial. Therefore, if $x = 4$ is a vertical asymptote of a rational function, then $(x - 4)$ must be a factor of the denominator.

65. A rational function cannot have both a horizontal and oblique asymptote. To have an oblique asymptote, the degree of the numerator must be exactly one larger than the degree of the denominator. However, if the numerator has a higher degree, there is no horizontal asymptote.

67. The equation of a vertical line through the point $(5, -3)$ is $x = 5$.

69. $2x^3 - xy^2 = 4$

Test x-axis symmetry: Let $y = -y$

$2x^3 - x(-y)^2 = 4$

$2x^3 - xy^2 = 4$ same

Test y-axis symmetry: Let $x = -x$

$2(-x)^3 - (-x)y^2 = 4$

$-2x^3 + xy^2 = 4$ not the same

Test origin symmetry: Let $x = -x$ and $y = -y$

$2(-x)^3 - (-x)(-y)^2 = 4$

$-2x^3 + xy^2 = 4$ not the same

Therefore, the graph will have x-axis symmetry.

Section 5.5

1. a. y-intercept:

$$f(0) = \frac{0^2 - 1}{0^2 - 4} = \frac{-1}{-4} = \frac{1}{4}$$

x-intercepts:
Set the numerator equal to 0 and solve for x.

$x^2 - 1 = 0$

$(x+1)(x-1) = 0$

$x = -1$ or $x = 1$

The intercepts are $\left(0, \frac{1}{4}\right)$, $(-1, 0)$, and $(1, 0)$.

   b. $f(-x) = \frac{(-x)^2 - 1}{(-x)^2 - 4} = \frac{x^2 - 1}{x^2 - 4} = f(x)$.

   Thus, f is an even function.

3. c

5. a. range: $\{x \mid x \neq 2\}$ since 2 makes the denominator 0.

   b. $0 = \frac{x(x-2)^2}{x-2}$

   $0 = x(x-2)^2$

   $x = 0$ or $x = 2$

   Since R(x) in not defined at x=2, the only x-intercept is $x = 0$.

269

In problems 7–44, we will use the terminology: $R(x) = \dfrac{p(x)}{q(x)}$, *where the degree of* $p(x) = n$ *and the degree of*

$q(x) = m$.

7. $R(x) = \dfrac{x+1}{x(x+4)}$ $p(x) = x+1;$ $q(x) = x(x+4) = x^2 + 4x;$ $n = 1;$ $m = 2$

Step 1: Domain: $\{x \mid x \neq -4, x \neq 0\}$

Since 0 is not in the domain, there is no y-intercept.

Step 2 & 3: The function is in lowest terms. The x-intercept is the zero of $p(x)$: $x = -1$

The x-intercept is −1. Near −1, $R(x) \approx -\dfrac{1}{3}(x+1)$. Plot the point $(-1, 0)$ and show a line with negative slope there.

Step 4: $R(x) = \dfrac{x+1}{x(x+4)}$ is in lowest terms.

The vertical asymptotes are the zeros of $q(x)$: $x = -4$ and $x = 0$. Plot these lines using dashes.

Step 5: Since $n < m$, the line $y = 0$ is the horizontal asymptote. Solve $R(x) = 0$ to find intersection points:

$$\frac{x+1}{x(x+4)} = 0$$

$$x + 1 = 0$$

$$x = -1$$

$R(x)$ intersects $y = 0$ at $(-1, 0)$. Plot the point $(-1, 0)$ and the line $y = 0$ using dashes.

Steps 6 & 7: Graphing

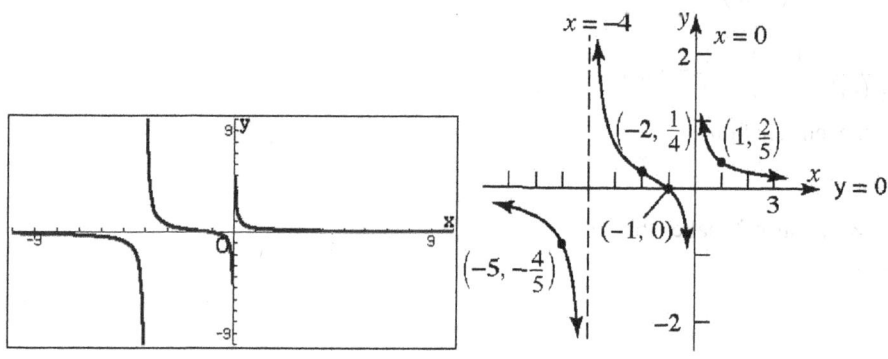

9. $R(x) = \dfrac{3x+3}{2x+4}$ $p(x) = 3x+3; \; q(x) = 2x+4; \; n=1; \; m=1$

Step 1: Domain: $\{x \mid x \neq -2\}$

The y-intercept is $R(0) = \dfrac{3(0)+3}{2(0)+4} = \dfrac{3}{4}$. Plot the point $\left(0, \dfrac{3}{4}\right)$.

Step 2 & 3: $R(x) = \dfrac{3x+3}{2x+4} = \dfrac{3(x+1)}{2(x+2)}$ is in lowest terms. The x-intercept is the zero of $p(x)$, $x=-1$.

Near -1, $R(x) \approx \dfrac{3}{2}(x+1)$. Plot the point $(-1,0)$ and show a line with positive slope there.

Step 4: $R(x) = \dfrac{3x+3}{2x+4} = \dfrac{3(x+1)}{2(x+2)}$ is in lowest terms.

The vertical asymptote is the zero of $q(x)$: $x=-2$. Graph this asymptote using a dashed line.

Step 5: Since $n=m$, the line $y=\dfrac{3}{2}$ is the horizontal asymptote.

Solve to find intersection points:

$$\dfrac{3x+3}{2x+4} = \dfrac{3}{2}$$
$$2(3x+3) = 3(2x+4)$$
$$6x+6 = 6x+4$$
$$0 \neq 2$$

$R(x)$ does not intersect $y=\dfrac{3}{2}$. Plot the line $y=\dfrac{3}{2}$ with dashes.

Steps 6 & 7: Graphing:

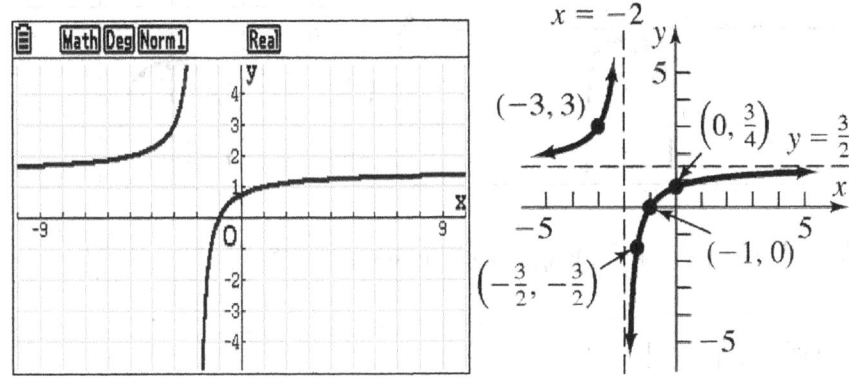

11. $R(x) = \dfrac{3}{x^2 - 4} = \dfrac{3}{(x-2)(x+2)}$ $\quad p(x) = 3;\ q(x) = x^2 - 4;\ n = 0;\ m = 2$

Step 1: Domain: $\{x \mid x \neq -2,\ x \neq 2\}$

The y-intercept is $R(0) = \dfrac{3}{0^2 - 4} = \dfrac{3}{-4} = -\dfrac{3}{4}$. Plot the point $\left(0, -\dfrac{3}{4}\right)$.

Step 2 & 3: R is in lowest terms. The x-intercepts are the zeros of $p(x)$. Since $p(x)$ is a constant, there are no x-intercepts.

Step 4: $R(x) = \dfrac{3}{x^2 - 4}$ is in lowest terms. The vertical asymptotes are the zeros of $q(x)$: $x = -2$ and $x = 2$.

Graph each of these asymptotes using dashed lines.

Step 5: Since $n < m$, the line $y = 0$ is the horizontal asymptote. Solve to find intersection points:

$$\frac{3}{x^2 - 4} = 0$$

$$3 = 0\left(x^2 - 4\right)$$

$$3 \neq 0$$

$R(x)$ does not intersect $y = 0$. Plot the line $y = 0$ with dashes.

Steps 6 & 7: Graphing:

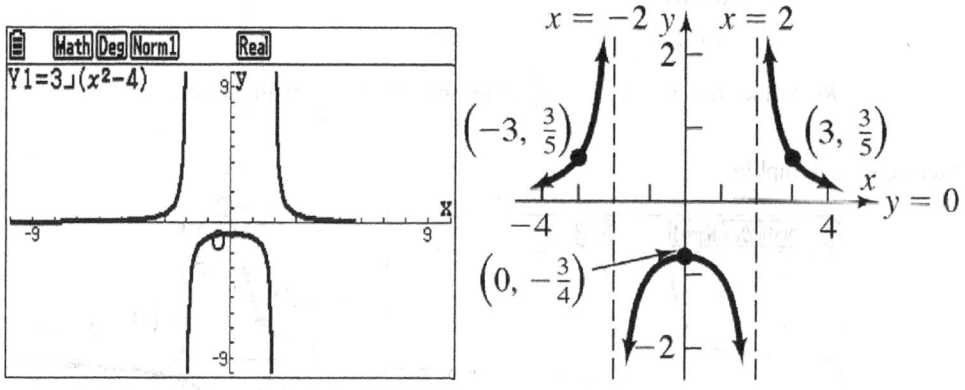

13. $P(x) = \dfrac{x^4 + x^2 + 1}{x^2 - 1}$ $\quad p(x) = x^4 + x^2 + 1;\ q(x) = x^2 - 1;\ n = 4;\ m = 2$

Step 1: Domain: $\{x \mid x \neq -1,\ x \neq 1\}$

The y-intercept is $P(0) = \dfrac{0^4 + 0^2 + 1}{0^2 - 1} = \dfrac{1}{-1} = -1$. Plot the point $(0, -1)$.

Step 2 & 3: $P(x) = \dfrac{x^4 + x^2 + 1}{x^2 - 1}$ is in lowest terms. The x-intercept is the zero of $p(x)$. Since $p(x)$ is never 0, there are no x-intercepts.

Step 4: $P(x) = \dfrac{x^4 + x^2 + 1}{x^2 - 1}$ is in lowest terms. The vertical asymptotes are the zeros of $q(x)$: $x = -1$ and $x = 1$.

Graph each of these asymptotes using dashed lines.

Step 5: Since $n > m + 1$, there is no horizontal or oblique asymptote.

Steps 6 & 7: Graphing:

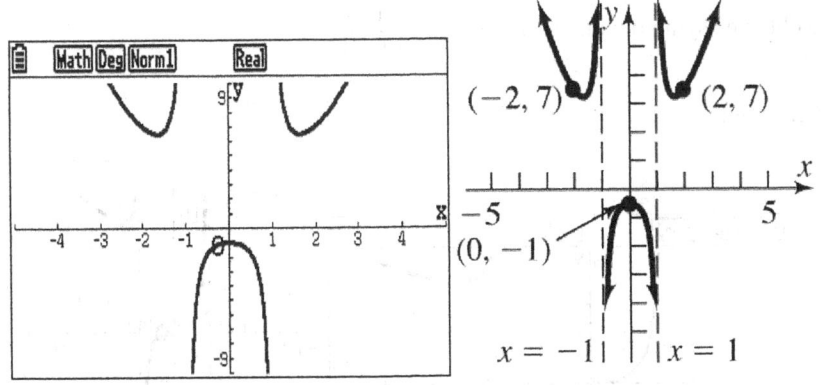

15. $H(x) = \dfrac{x^3 - 1}{x^2 - 9} = \dfrac{(x-1)(x^2 + x + 1)}{(x+3)(x-3)}$ $p(x) = x^3 - 1;\ q(x) = x^2 - 9;\ n = 3;\ m = 2$

Step 1: Domain: $\{x \mid x \neq -3, x \neq 3\}$

The y-intercept is $H(0) = \dfrac{0^3 - 1}{0^2 - 9} = \dfrac{-1}{-9} = \dfrac{1}{9}$. Plot the point $\left(0, \dfrac{1}{9}\right)$.

Step 2 & 3: $H(x)$ is in lowest terms. The x-intercept is the zero of $p(x)$: 1.

Near 1, $H(x) \approx -\dfrac{3}{8}(x-1)$. Plot the point $(1,0)$ and indicate a line with negative slope there.

Step 4: $H(x)$ is in lowest terms. The vertical asymptotes are the zeros of $q(x)$: $x = -3$ and $x = 3$. Graph each of these asymptotes using dashed lines.

Step 5: Since $n = m + 1$, there is an oblique asymptote. Dividing:

$$x^2 - 9 \overline{) x^3 + 0x^2 + 0x - 1} \qquad H(x) = x + \dfrac{9x - 1}{x^2 - 9}$$

$$\underline{x^3 \qquad - 9x}$$
$$9x - 1$$

with quotient x.

The oblique asymptote is $y = x$. Graph this asymptote with a dashed line. Solve to find intersection points:

$$\frac{x^3-1}{x^2-9}=x$$

$$x^3-1=x^3-9x$$

$$-1=-9x$$

$$x=\frac{1}{9}$$

The oblique asymptote intersects $H(x)$ at $\left(\frac{1}{9},\frac{1}{9}\right)$.

Steps 6 & 7: Graphing:

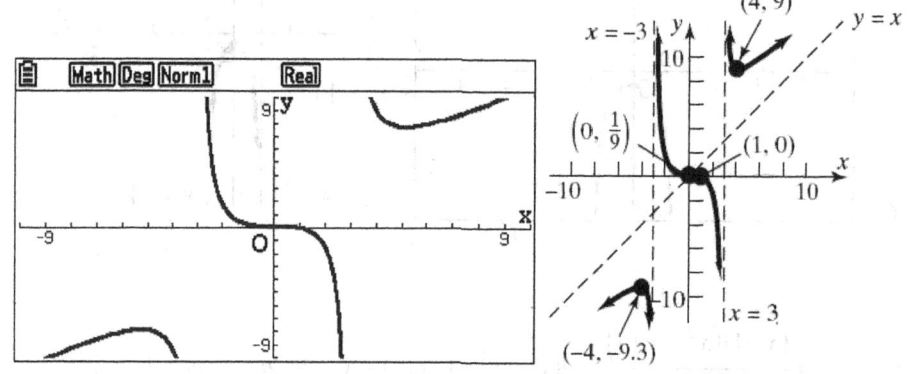

17. $R(x)=\dfrac{x^2}{x^2+x-6}=\dfrac{x^2}{(x+3)(x-2)}$ $p(x)=x^2$; $q(x)=x^2+x-6$; $n=2$; $m=2$

Step 1:   Domain: $\{x\,|\,x\neq-3,\,x\neq2\}$

The y-intercept is $R(0)=\dfrac{0^2}{0^2+0-6}=\dfrac{0}{-6}=0$. Plot the point $(0,0)$.

Step 2 & 3: $R(x)=\dfrac{x^2}{x^2+x-6}$ is in lowest terms. The x-intercept is the zero of $p(x)$: 0

Near 0, $R(x)\approx-\dfrac{1}{6}x^2$. Plot the point $(0,0)$ and indicate a parabola opening down there.

Step 4: $R(x)=\dfrac{x^2}{x^2+x-6}$ is in lowest terms. The vertical asymptotes are the zeros of $q(x)$:

$x=-3$ and $x=2$. Graph each of these asymptotes using dashed lines.

Step 5:   Since $n=m$, the line $y=1$ is the horizontal asymptote. Graph this asymptote with a dashed line.
Solve to find intersection points:

$$\frac{x^2}{x^2+x-6}=1$$
$$x^2 = x^2+x-6$$
$$0 = x-6$$
$$x = 6$$

$R(x)$ intersects $y=1$ at $(6, 1)$.

Steps 6 & 7: Graphing:

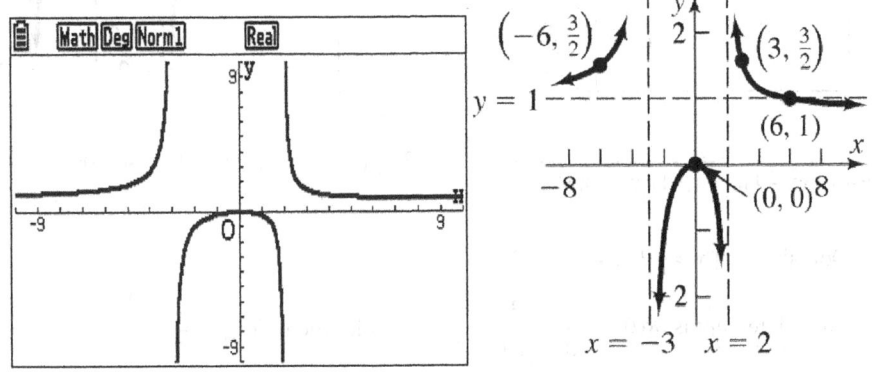

19. $G(x) = \dfrac{x}{x^2-4} = \dfrac{x}{(x+2)(x-2)}$ $\quad p(x)=x;\ q(x)=x^2-4;\ n=1;\ m=2$

Step 1: Domain: $\{x \mid x \neq -2,\ x \neq 2\}$

The y-intercept is $G(0) = \dfrac{0}{0^2-4} = \dfrac{0}{-4} = 0$. Plot the point $(0,0)$.

Step 2 & 3: $G(x) = \dfrac{x}{x^2-4}$ is in lowest terms. The x-intercept is the zero of $p(x)$: 0

Near 0, $G(x) \approx -\dfrac{1}{4}x$. Plot the point $(0,0)$ and indicate a line with negative slope there.

Step 4: $G(x) = \dfrac{x}{x^2-4}$ is in lowest terms. The vertical asymptotes are the zeros of $q(x)$: $x=-2$ and $x=2$.
Graph each of these asymptotes using a dashed line.

Step 5: Since $n < m$, the line $y=0$ is the horizontal asymptote. Graph this asymptote using a dashed line.
Solve to find intersection points:

$$\frac{x}{x^2-4} = 0$$
$$x = 0$$

$G(x)$ intersects $y=0$ at $(0, 0)$.

Steps 6 & 7: Graphing:

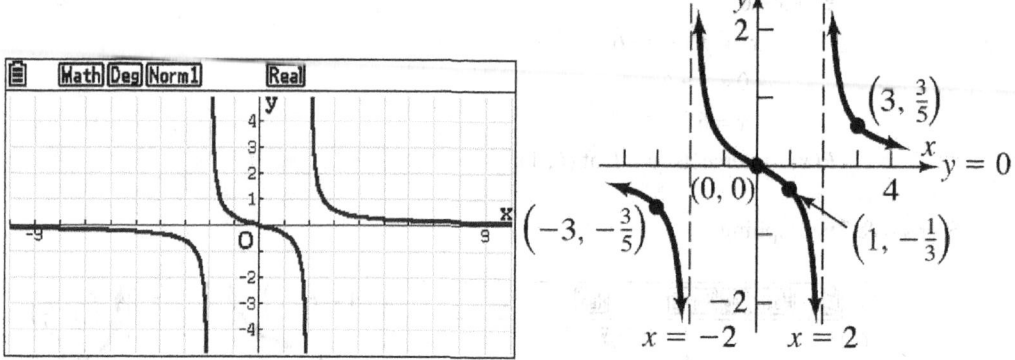

21. $R(x) = \dfrac{3}{(x-1)(x^2-4)} = \dfrac{3}{(x-1)(x+2)(x-2)}$ $p(x) = 3;\ q(x) = (x-1)(x^2-4);\ n=0;\ m=3$

Step 1: Domain: $\{x \mid x \neq -2,\ x \neq 1,\ x \neq 2\}$

The y-intercept is $R(0) = \dfrac{3}{(0-1)(0^2-4)} = \dfrac{3}{4}$. Plot the point $\left(0, \dfrac{3}{4}\right)$.

Step 2 & 3: $R(x) = \dfrac{3}{(x-1)(x^2-4)}$ is in lowest terms. There is no x-intercept.

Step 4: $R(x) = \dfrac{3}{(x-1)(x^2-4)}$ is in lowest terms.

The vertical asymptotes are the zeros of $q(x)$: $x = -2$, $x = 1$, and $x = 2$.

Graph each of these asymptotes using a dashed line.

Step 5: Since $n < m$, the line $y = 0$ is the horizontal asymptote. Graph this asymptote with a dashed line.

Solve to find intersection points:

$$\frac{3}{(x-1)(x^2-4)} = 0$$
$$3 \neq 0$$

$R(x)$ does not intersect $y = 0$.

Steps 6 & 7: Graphing:

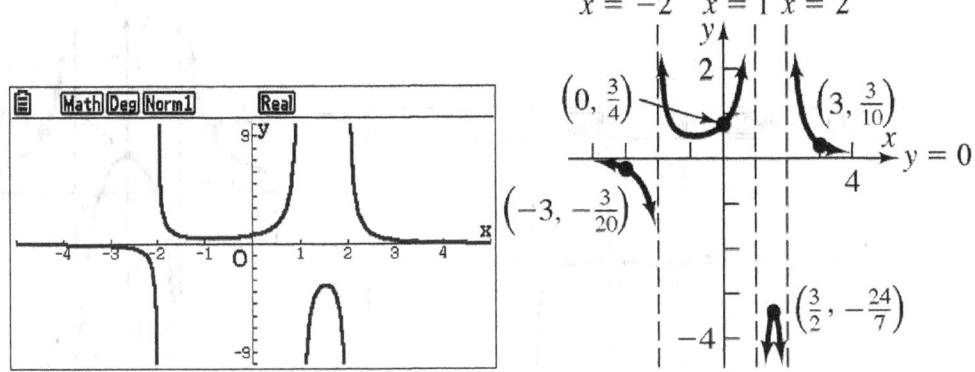

23. $H(x) = \dfrac{x^2-1}{x^4-16} = \dfrac{(x-1)(x+1)}{(x^2+4)(x+2)(x-2)}$ $p(x) = x^2-1;\ q(x) = x^4-16;\ n=2;\ m=4$

Step 1:   Domain: $\{x \mid x \neq -2,\ x \neq 2\}$

The y-intercept is $H(0) = \dfrac{0^2-1}{0^4-16} = \dfrac{-1}{-16} = \dfrac{1}{16}$. Plot the point $\left(0, \dfrac{1}{16}\right)$.

Step 2 & 3: $H(x) = \dfrac{x^2-1}{x^4-16}$ is in lowest terms. The x-intercepts are the zeros of $p(x)$: -1 and 1

Near -1, $H(x) \approx \dfrac{2}{15}(x+1)$; Near 1, $H(x) \approx -\dfrac{2}{15}(x-1)$

Plot $(-1,0)$ and indicate a line with positive slope there.

Plot $(1,0)$ and indicate a line with negative slope there.

Step 4: $H(x) = \dfrac{x^2-1}{x^4-16}$ is in lowest terms. The vertical asymptotes are the zeros of $q(x)$: $x=-2$ and $x=2$

Graph each of these asymptotes using a dashed line.

Step 5:   Since $n < m$, the line $y=0$ is the horizontal asymptote. Graph this asymptote using a dashed line.

Solve to find intersection points:

$$\dfrac{x^2-1}{x^4-16} = 0$$
$$x^2-1 = 0$$
$$x = \pm 1$$

$H(x)$ intersects $y=0$ at $(-1, 0)$ and $(1, 0)$.

277

Steps 6 & 7: Graphing:

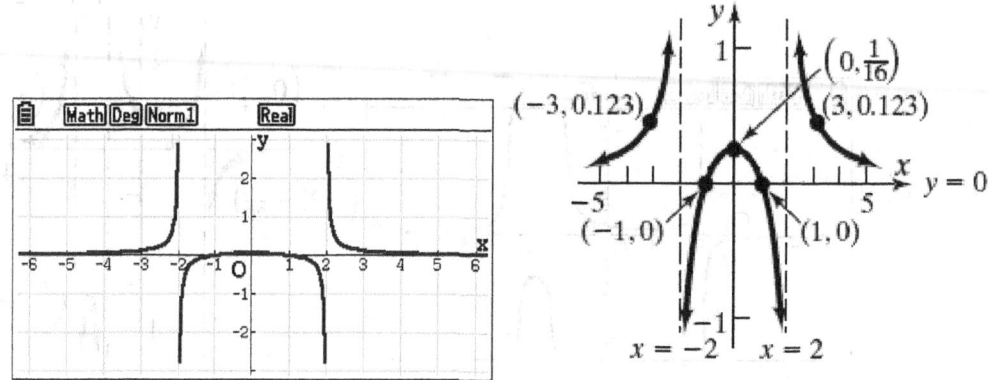

25. $F(x) = \dfrac{x^2 - 3x - 4}{x+2} = \dfrac{(x+1)(x-4)}{x+2}$ $p(x) = x^2 - 3x - 4;\ q(x) = x + 2;\ n = 2;\ m = 1$

Step 1:   Domain: $\{x \mid x \neq -2\}$

The y-intercept is $F(0) = \dfrac{0^2 - 3(0) - 4}{0 + 2} = \dfrac{-4}{2} = -2$. Plot the point $(0, -2)$.

Step 2 & 3: $F(x) = \dfrac{x^2 - 3x - 4}{x+2}$ is in lowest terms. The x-intercepts are the zeros of $p(x)$: -1 and 4.

Near -1, $F(x) \approx -5(x+1)$;   Near 4, $F(x) \approx \dfrac{5}{6}(x-4)$.

Plot $(-1, 0)$ and indicate a line with negative slope there.

Plot $(4, 0)$ and indicate a line with positive slope there.

Step 4: $F(x) = \dfrac{x^2 - 3x - 4}{x+2}$ is in lowest terms. The vertical asymptote is the zero of $q(x)$: $x = -2$

Graph this asymptote using a dashed line.

Step 5:   Since $n = m + 1$, there is an oblique asymptote. Dividing:

$$\begin{array}{r} x - 5 \\ x+2\overline{)x^2 - 3x - 4} \\ \underline{x^2 + 2x} \\ -5x - 4 \\ \underline{-5x - 10} \\ 6 \end{array} \qquad\qquad F(x) = x - 5 + \dfrac{6}{x+2}$$

The oblique asymptote is $y = x - 5$. Graph this asymptote using a dashed line. Solve to find intersection points:

$$\frac{x^2 - 3x - 4}{x + 2} = x - 5$$

$$x^2 - 3x - 4 = x^2 - 3x - 10$$

$$-4 \neq -10$$

The oblique asymptote does not intersect $F(x)$.

Steps 6 & 7: Graphing:

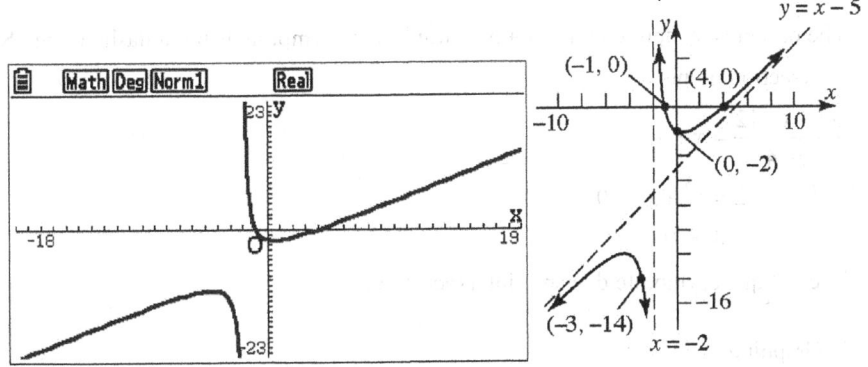

27. $R(x) = \dfrac{x^2 + x - 12}{x - 4} = \dfrac{(x+4)(x-3)}{x-4}$ $p(x) = x^2 + x - 12;$ $q(x) = x - 4;$ $n = 2;$ $m = 1$

Step 1:   Domain: $\{x \mid x \neq 4\}$

The y-intercept is $R(0) = \dfrac{0^2 + 0 - 12}{0 - 4} = \dfrac{-12}{-4} = 3$. Plot the point $(0, 3)$.

Step 2 & 3: $R(x) = \dfrac{x^2 + x - 12}{x - 4}$ is in lowest terms. The x-intercepts are the zeros of $p(x)$: -4 and 3.

Near -4, $R(x) \approx \dfrac{7}{8}(x + 4)$; Near 3, $R(x) \approx -7(x - 3)$.

Plot $(-4, 0)$ and indicate a line with positive slope there.

Plot $(3, 0)$ and indicate a line with negative slope there.

Step 4: $R(x) = \dfrac{x^2 + x - 12}{x - 4}$ is in lowest terms. The vertical asymptote is the zero of $q(x)$: $x = 4$

Graph this asymptote using a dashed line.

Step 5: Since $n = m + 1$, there is an oblique asymptote. Dividing:

$$x - 4 \overline{\smash{\big)}\, x^2 + x - 12} \quad \begin{array}{c} x + 5 \end{array}$$
$$\underline{x^2 - 4x}$$
$$5x - 12$$
$$\underline{5x - 20}$$
$$8$$

$$R(x) = x + 5 + \frac{8}{x - 4}$$

The oblique asymptote is $y = x + 5$. Graph this asymptote using a dashed line. Solve to find intersection points:

$$\frac{x^2 + x - 12}{x - 4} = x + 5$$
$$x^2 + x - 12 = x^2 + x - 20$$
$$-12 \neq -20$$

The oblique asymptote does not intersect $R(x)$.

Steps 6 & 7: Graphing:

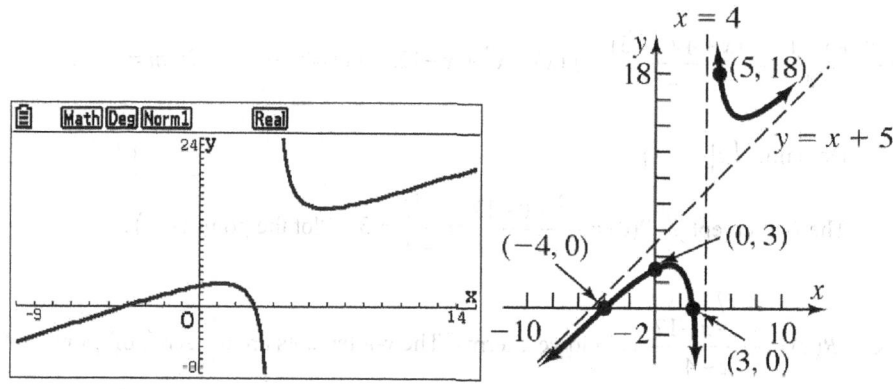

29. $F(x) = \dfrac{x^2 + x - 12}{x + 2} = \dfrac{(x+4)(x-3)}{x+2}$ $\quad p(x) = x^2 + x - 12; \ q(x) = x + 2; \ n = 2; \ m = 1$

Step 1: Domain: $\{x \mid x \neq -2\}$

The y-intercept is $F(0) = \dfrac{0^2 + 0 - 12}{0 + 2} = \dfrac{-12}{2} = -6$. Plot the point $(0, -6)$.

Step 2 & 3: $F(x) = \dfrac{x^2 + x - 12}{x + 2}$ is in lowest terms. The x-intercepts are the zeros of $p(x)$: -4 and 3.

Near -4, $F(x) \approx \dfrac{7}{2}(x + 4)$; Near 3, $F(x) \approx \dfrac{7}{5}(x - 3)$.

Plot $(-4, 0)$ and indicate a line with positive slope there.

Plot $(3, 0)$ and indicate a line with positive slope there.

Step 4: $F(x) = \dfrac{x^2 + x - 12}{x+2}$ is in lowest terms. The vertical asymptote is the zero of $q(x)$: $x = -2$

Graph this asymptote using a dashed line.

Step 5: Since $n = m+1$, there is an oblique asymptote. Dividing:

$$x+2 \overline{\smash{)}\begin{array}{r} x-1 \\ x^2 + x - 12 \\ \end{array}}$$
$$\underline{x^2 + 2x}$$
$$-x - 12$$
$$\underline{-x-2}$$
$$-10$$

$$F(x) = x - 1 + \dfrac{-10}{x+2}$$

The oblique asymptote is $y = x - 1$. Graph this asymptote using a dashed line. Solve to find intersection points:

$$\dfrac{x^2 + x - 12}{x+2} = x - 1$$
$$x^2 + x - 12 = x^2 + x - 2$$
$$-12 \neq -2$$

The oblique asymptote does not intersect $F(x)$.

Steps 6 & 7: Graphing:

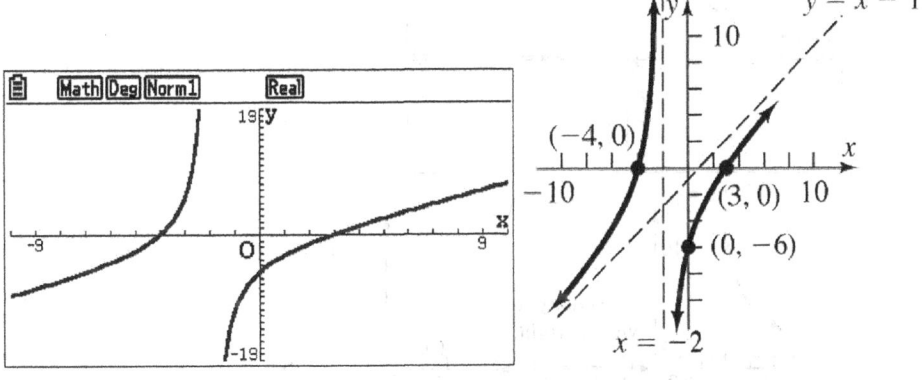

31. $R(x) = \dfrac{x(x-1)^2}{(x+3)^3}$ $p(x) = x(x-1)^2$; $q(x) = (x+3)^3$; $n = 3$; $m = 3$

Step 1: Domain: $\{x \mid x \neq -3\}$

The y-intercept is $R(0) = \dfrac{0(0-1)^2}{(0+3)^3} = \dfrac{0}{27} = 0$. Plot the point $(0,0)$.

Step 2 & 3: $R(x) = \dfrac{x(x-1)^2}{(x+3)^3}$ is in lowest terms. The x-intercepts are the zeros of $p(x)$: 0 and 1

Near 0, $R(x) \approx \dfrac{1}{27}x$; Near 1, $R(x) \approx \dfrac{1}{64}(x-1)^2$.

Plot $(0,0)$ and indicate a line with positive slope there.

Plot $(1,0)$ and indicate a parabola that opens up there.

Step 4: $R(x) = \dfrac{x(x-1)^2}{(x+3)^3}$ is in lowest terms. The vertical asymptote is the zero of $q(x)$: $x = -3$

Graph this asymptote with a dashed line.

Step 5: Since $n = m$, the line $y = 1$ is the horizontal asymptote. Graph this asymptote with a dashed line.

Solve to find intersection points:

$$\frac{x(x-1)^2}{(x+3)^3} = 1$$

$$x^3 - 2x^2 + x = x^3 + 9x^2 + 27x + 27$$

$$0 = 11x^2 + 26x + 27$$

$$b^2 - 4ac = 26^2 - 4(11)(27) = -512$$

no real solution

$R(x)$ does not intersect $y = 1$.

Steps 6 & 7: Graphing:

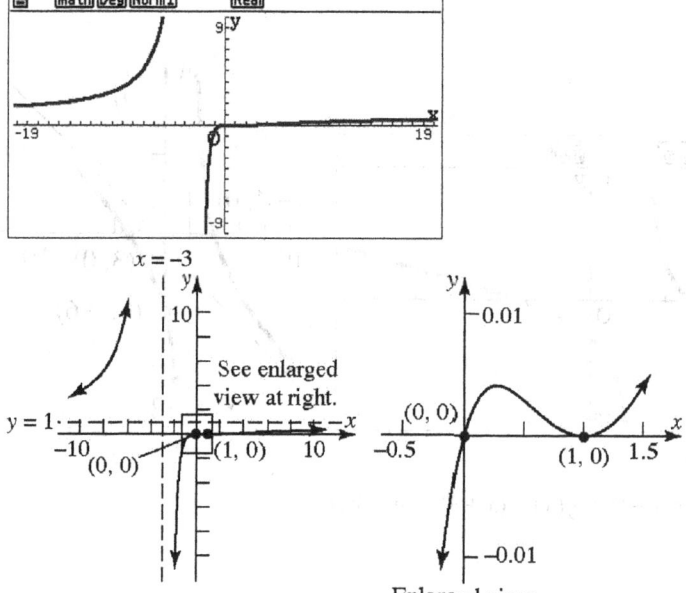

33. $R(x) = \dfrac{x^2 + x - 12}{x^2 - x - 6} = \dfrac{(x+4)(x-3)}{(x-3)(x+2)} = \dfrac{x+4}{x+2}$ $p(x) = x^2 + x - 12;\ \ q(x) = x^2 - x - 6;\ \ n = 2;\ m = 2$

Step 1:   Domain: $\{x \mid x \neq -2,\ x \neq 3\}$

The y-intercept is $R(0) = \dfrac{0^2 + 0 - 12}{0^2 - 0 - 6} = \dfrac{-12}{-6} = 2$. Plot the point $(0, 2)$.

Step 2 & 3: In lowest terms, $R(x) = \dfrac{x+4}{x+2},\ x \neq 3$. Note: $R(x)$ is still undefined at both 3 and -2.

The x-intercept is the zero of $y = x + 4$: -4

Near -4, $R(x) \approx -\dfrac{1}{2}(x+4)$. Plot $(-4, 0)$ and indicate a line with negative slope there.

Step 4:   In lowest terms, $R(x) = \dfrac{x+4}{x+2},\ x \neq 3$. The vertical asymptote is the zero of $f(x) = x + 2$: $x = -2$;

Graph this asymptote using a dashed line. Note: $x = 3$ is not a vertical asymptote because the reduced

form must be used to find the asymptotes. The graph has a hole at $\left(3, \dfrac{7}{5}\right)$.

Step 5:   Since $n = m$, the line $y = 1$ is the horizontal asymptote. Graph this asymptote using a dashed line.

Solve to find intersection points:

$$\frac{x^2 + x - 12}{x^2 - x - 6} = 1$$

$$x^2 + x - 12 = x^2 - x - 6$$

$$2x = 6$$

$$x = 3$$

$R(x)$ does not intersect $y = 1$ because $R(x)$ is not defined at $x = 3$.

Steps 6 & 7: Graphing:

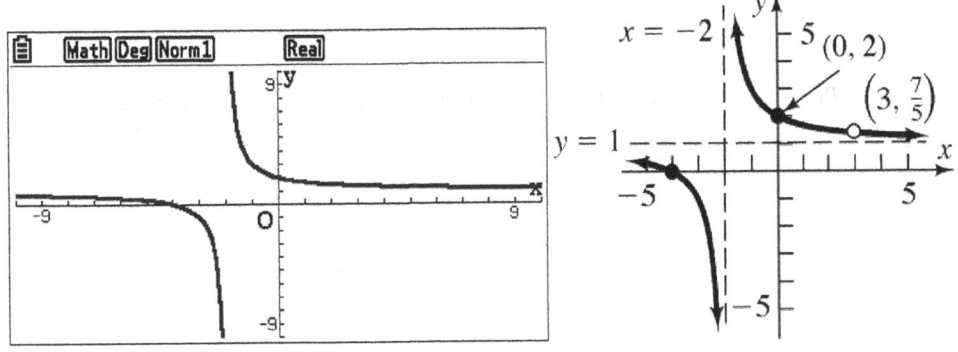

35. $R(x) = \dfrac{6x^2 - 7x - 3}{2x^2 - 7x + 6} = \dfrac{(3x+1)(2x-3)}{(2x-3)(x-2)} = \dfrac{3x+1}{x-2}$ $p(x) = 6x^2 - 7x - 3;$ $q(x) = 2x^2 - 7x + 6;$ $n = 2;$ $m = 2$

Step 1: Domain: $\left\{ x \mid x \neq \dfrac{3}{2}, x \neq 2 \right\}$

The y-intercept is $R(0) = \dfrac{6(0)^2 - 7(0) - 3}{2(0)^2 - 7(0) + 6} = \dfrac{-3}{6} = -\dfrac{1}{2}$. Plot the point $\left(0, -\dfrac{1}{2} \right)$.

Step 2 & 3: In lowest terms, $R(x) = \dfrac{3x+1}{x-2}$, $x \neq \dfrac{3}{2}$. The x-intercept is the zero of $y = 3x + 1$: $-\dfrac{1}{3}$;

Note: $x = \dfrac{3}{2}$ is not a zero because reduced form must be used to find the zeros.

Near $-\dfrac{1}{3}$, $R(x) \approx -\dfrac{3}{7}(3x+1)$. Plot the point $\left(-\dfrac{1}{3}, 0 \right)$ and indicate a line with negative slope there.

Step 4: In lowest terms, $R(x) = \dfrac{3x+1}{x-2}$, $x \neq \dfrac{3}{2}$. The vertical asymptote is the zero of $f(x) = x - 2$: $x = 2$; Graph this asymptote using a dashed line.

Note: $x = \dfrac{3}{2}$ is not a vertical asymptote because reduced form must be used to find the asymptotes.

The graph has a hole at $\left(\dfrac{3}{2}, -11 \right)$.

Step 5: Since $n = m$, the line $y = 3$ is the horizontal asymptote. Graph this asymptote using a dashed line. Solve to find intersection points:

$$\dfrac{6x^2 - 7x - 3}{2x^2 - 7x + 6} = 3$$
$$6x^2 - 7x - 3 = 6x^2 - 21x + 18$$
$$14x = 21$$
$$x = \dfrac{3}{2}$$

$R(x)$ does not intersect $y = 3$ because $R(x)$ is not defined at $x = \dfrac{3}{2}$.

Steps 6 & 7: Graphing:

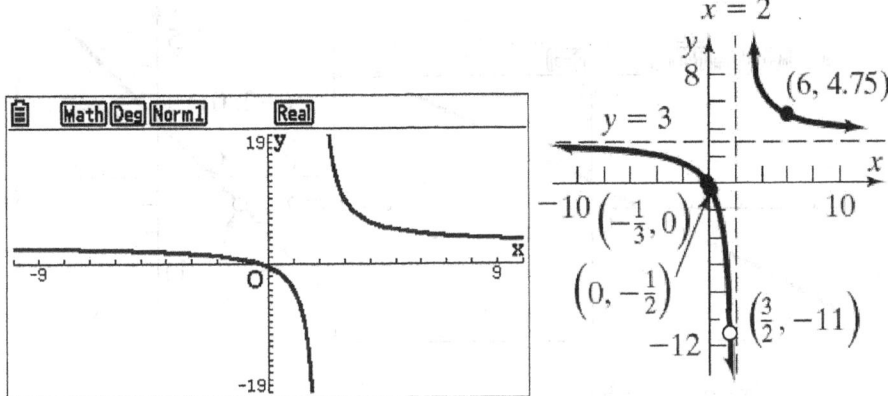

37. $R(x) = \dfrac{x^2+5x+6}{x+3} = \dfrac{(x+2)(x+3)}{x+3} = x+2 \qquad p(x) = x^2+5x+6; \quad q(x) = x+3; \quad n=2; \ m=1$

Step 1: Domain: $\{x \mid x \neq -3\}$

The y-intercept is $R(0) = \dfrac{0^2+5(0)+6}{0+3} = \dfrac{6}{3} = 2$. Plot the point $(0,2)$.

Step 2 & 3: In lowest terms, $R(x) = x+2$, $x \neq -3$. The x-intercept is the zero of $y = x+2$: -2;

Note: -3 is not a zero because reduced form must be used to find the zeros.

Near -2, $R(x) = x+2$. Plot the point $(0,-2)$ and indicate the line $y = x+2$ there.

Step 4: In lowest terms, $R(x) = x+2$, $x \neq -3$. There are no vertical asymptotes. Note: $x = -3$ is not a vertical asymptote because reduced form must be used to find the asymptotes. The graph has a hole at $(-3,-1)$.

Step 5: Since $n = m+1$ there is an oblique asymptote. The line $y = x+2$ is the oblique asymptote. Solve to find intersection points:

$$\dfrac{x^2+5x+6}{x+3} = x+2$$
$$x^2+5x+6 = (x+2)(x+3)$$
$$x^2+5x+6 = x^2+5x+6$$
$$0 = 0$$

The oblique asymptote intersects $R(x)$ at every point of the form $(x, x+2)$ except $(-3,-1)$.

Steps 6 & 7: Graphing:

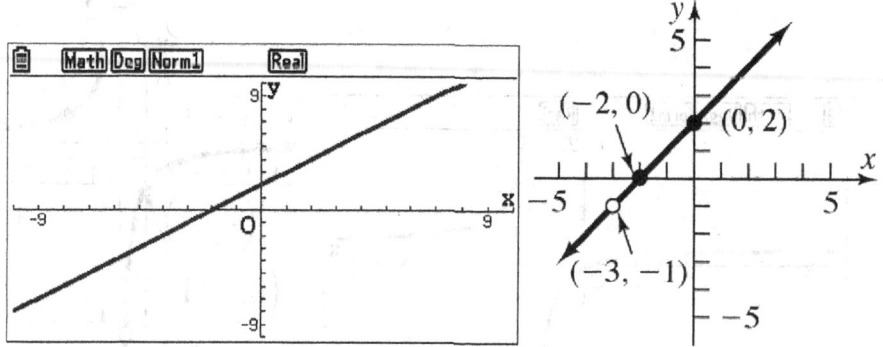

39. $H(x) = \dfrac{3x-6}{4-x^2} = \dfrac{3x-6}{-(x^2-4)} = \dfrac{-3(x-2)}{(x+2)(x-2)}$ $p(x) = 3x-6;\ q(x) = 4-x^2;\ n = 1;\ m = 2$

Step 1:   Domain: $\{x \mid x \neq -2,\ x \neq 2\}$

The y-intercept is $H(0) = \dfrac{3(0)-6}{4-0^2} = \dfrac{-6}{4} = -\dfrac{3}{2}$. Plot the point $\left(0, -\dfrac{3}{2}\right)$.

Step 2 & 3: $H(x) = \dfrac{3x-6}{4-x^2} = \dfrac{-3}{x+2}$ is in lowest terms. The x-intercept is the zero of $p(x)$: none

Step 4: $H(x) = \dfrac{3x-6}{4-x^2} = \dfrac{-3}{x+2}$ is in lowest terms so $R(x) = x+2$. The vertical asymptotes are the zeros of

$R(x):\ x = -2$.

Graph the asymptote using a dashed line. The multiplicity of -2 is odd so the graph will approach plus or minus infinity on either side of the asymptote.

Step 5:   Since $n < m$, the line $y = 0$ is the horizontal asymptote. Graph this asymptote using a dashed line.
Solve to find intersection points:

$\dfrac{3x-6}{4-x^2} = 0$

$3x-6 = 0$

$x = 2$

The function is not defined at $x = 2$ so there is no interection.

Steps 6 & 7: Graphing:

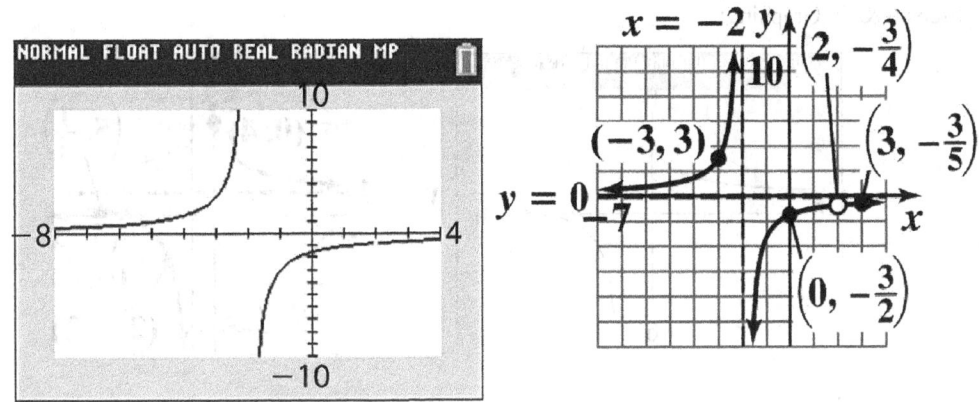

41. $F(x) = \dfrac{x^2 - 5x + 4}{x^2 - 2x + 1} = \dfrac{(x-1)(x-4)}{(x-1)^2} = \dfrac{x-4}{x-1}$ $p(x) = x^2 - 5x + 4;$ $q(x) = x^2 - 2x + 1;$ $n = 2;$ $m = 2$

Step 1:   Domain: $\{x \mid x \neq 1\}$

The y-intercept is $R(0) = \dfrac{0^2 - 5(0) + 4}{0^2 - 2(0) + 1} = \dfrac{4}{1} = 4$. Plot the point $(0, 4)$.

Step 2 & 3: In lowest terms, $F(x) = \dfrac{x-4}{x-1}$, $x \neq 1$. The x-intercept is the zero of $y = x - 4$: 4 with odd multiplicity.

Note: -5 is not a zero because reduced form must be used to find the zeros.

Plot the point $(4, 0)$. The graph crosses the x-axis at this point.

Step 4:   In lowest terms, $F(x) = \dfrac{x-4}{x-1}$, $x \neq 1$. The vertical asymptote is the zero of $f(x) = x - 1$: $x = 1$; Graph this asymptote using a dashed line. The multiplicity of 1 is odd so the graph will approach plus or minus infinity on either side of the asymptote.

Step 5:   Since $n = m$, the line $y = 1$ is the horizontal asymptote. Graph this asymptote using a dashed line. Solve to find intersection points:

$$\dfrac{x^2 - 5x + 4}{x^2 - 2x + 1} = 1$$
$$x^2 - 5x + 4 = x^2 - 2x + 1$$
$$-3x = -3$$
$$x = 1$$

$F(x)$ does not intersect $y = 1$ because $R(x)$ is not defined at $x = 1$.

Steps 6 & 7: Graphing:

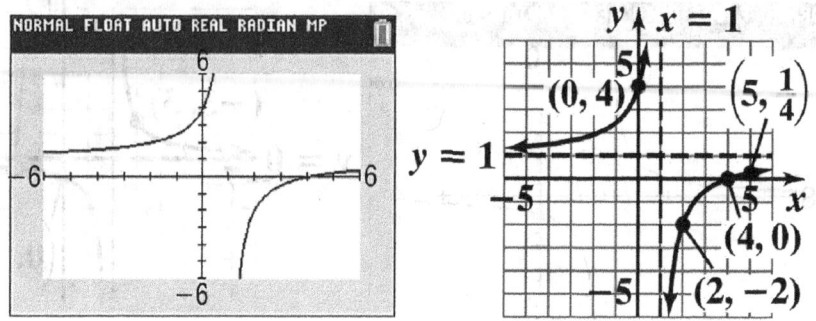

43. $G(x) = \dfrac{x}{(x+2)^2}$ $p(x) = x$; $q(x) = (x+2)^2$; $n = 1$; $m = 2$

Step 1:   Domain: $\{x \mid x \neq -2\}$.

Step 2: $G(x) = \dfrac{x}{(x+2)^2}$ is in lowest terms.

Step 3:   The y-intercept is $G(0) = \dfrac{(0)}{(0+2)^2} = \dfrac{0}{4} = 0$. Plot the point $(0,0)$.

The x-intercept is the zero of $p(x)$: 0 with odd multiplicity.
Plot the point $(0,0)$. The graph crosses the x-axis at this point.

Step 4: $G(x) = \dfrac{x}{(x+2)^2}$ is in lowest terms. The vertical asymptote is the zero of $q(x)$: $x = -2$. Graph this
asymptote. The multiplicity of -2 is even so the graph will approach the same infinity on both sides of
the asymptote.

Step 5:   Since $n < m$, the line $y = 0$ is the horizontal asymptote. Graph this asymptote using a dashed line.
Solve to find intersection points:
$$\dfrac{x}{(x+2)^2} = 0$$
$$x = 0$$
$G(x)$ intersects $y = 0$ at $(0,0)$.

Steps 6 & 7: Graphing:

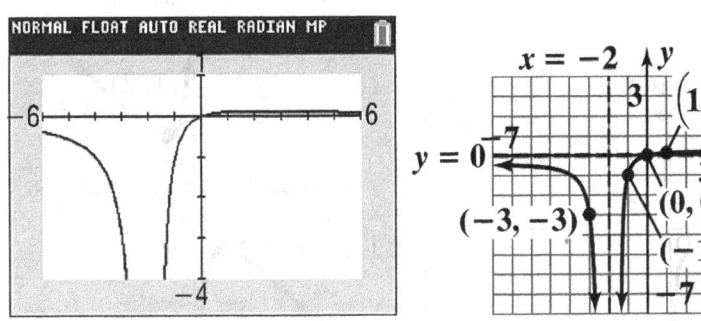

45. $f(x) = x + \dfrac{1}{x} = \dfrac{x^2 + 1}{x}$ $p(x) = x^2 + 1;\ q(x) = x;\ n = 2;\ m = 1$

Step 1:   Domain: $\{x \mid x \neq 0\}$

There is no y-intercept because 0 is not in the domain.

Step 2 & 3: $f(x) = \dfrac{x^2 + 1}{x}$ is in lowest terms. There are no x-intercepts since $x^2 + 1 = 0$ has no real solutions.

Step 4: $f(x) = \dfrac{x^2 + 1}{x}$ is in lowest terms. The vertical asymptote is the zero of $q(x)$: $x = 0$ Graph this asymptote using a dashed line.

Step 5:   Since $n = m + 1$, there is an oblique asymptote.

Dividing:
$$\begin{array}{r} x \\ x{\overline{\smash{\big)}\,x^2 + 1}} \\ \underline{x^2} \\ 1 \end{array}$$

$f(x) = x + \dfrac{1}{x}$

The oblique asymptote is $y = x$.

Graph this asymptote using a dashed line. Solve to find intersection points:

$$\dfrac{x^2 + 1}{x} = x$$
$$x^2 + 1 = x^2$$
$$1 \neq 0$$

The oblique asymptote does not intersect $f(x)$.

Steps 6 & 7: Graphing :

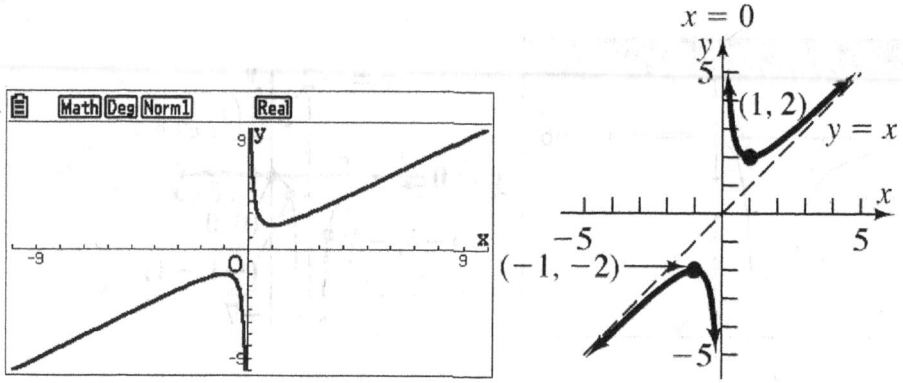

47. $f(x) = x^2 + \dfrac{1}{x} = \dfrac{x^3 + 1}{x} = \dfrac{(x+1)(x^2 - x + 1)}{x}$ $p(x) = x^3 + 1;$ $q(x) = x;$ $n = 3;$ $m = 1.$

Step 1:   Domain: $\{x \mid x \neq 0\}$

There is no y-intercept because 0 is not in the domain.

Step 2 & 3: $f(x) = \dfrac{x^3 + 1}{x}$ is in lowest terms. The x-intercept is the zero of $p(x)$: -1

Near -1, $f(x) \approx -3(x+1)$. Plot the point $(-1,0)$ and indicate a line with negative slope there.

Step 4: $f(x) = \dfrac{x^3 + 1}{x}$ is in lowest terms. The vertical asymptote is the zero of $q(x)$: $x = 0$

Graph this asymptote using a dashed line.

Step 5:   Since $n > m + 1$, there is no horizontal or oblique asymptote.

Steps 6 & 7: Graphing:

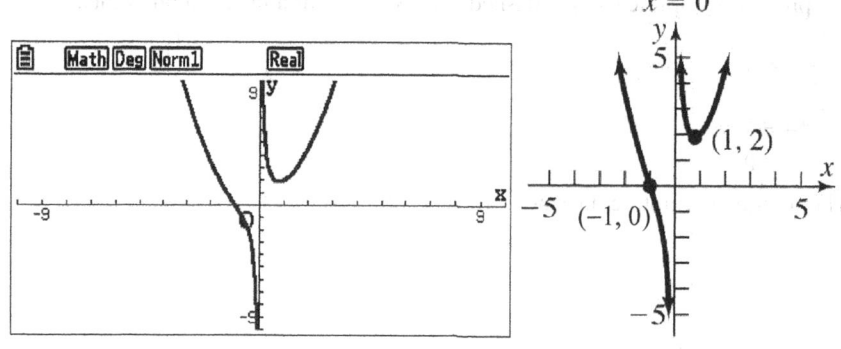

49. $f(x) = x + \dfrac{1}{x^3} = \dfrac{x^4+1}{x^3}$ $\qquad$ $p(x) = x^4+1;\ q(x) = x^3;\ n = 4;\ m = 3$

Step 1: $\quad$ Domain: $\{x \mid x \neq 0\}$

There is no y-intercept because 0 is not in the domain.

Step 2 & 3: $f(x) = \dfrac{x^4+1}{x^3}$ is in lowest terms. There are no x-intercepts since $x^4+1 = 0$ has no real solutions.

Step 4: $\quad$ $f(x) = \dfrac{x^4+1}{x^3}$ is in lowest terms. The vertical asymptote is the zero of $q(x)$: $x = 0$

Graph this asymptote using a dashed line.

Step 5: $\quad$ Since $n = m+1$, there is an oblique asymptote. Dividing:

$$x^3\overline{\smash{)}\,x^4+1} \quad \begin{array}{c} x \\ \underline{x^4} \\ 1 \end{array}$$

$\qquad\qquad\qquad f(x) = x + \dfrac{1}{x^3}$

The oblique asymptote is $y = x$. Graph this asymptote using a dashed line. Solve to find intersection points:

$$\frac{x^4+1}{x^3} = x$$

$$x^4+1 = x^4$$

$$1 \neq 0$$

The oblique asymptote does not intersect $f(x)$.

Steps 6 & 7: Graphing:

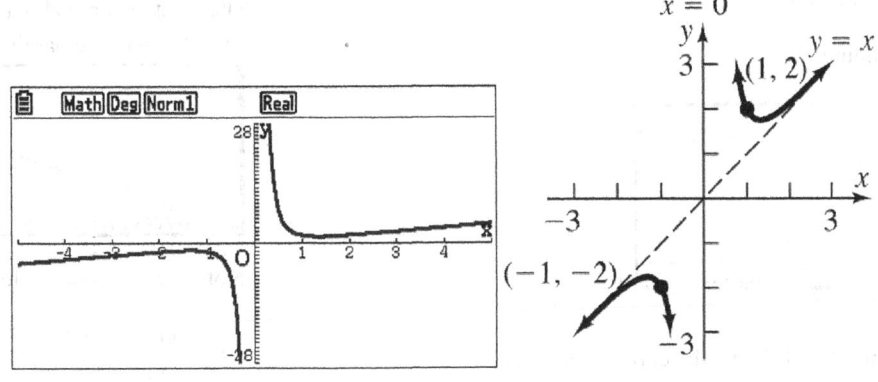

51. One possibility: $R(x) = \dfrac{x^2}{x^2 - 4}$

53. One possibility: $R(x) = \dfrac{(x-1)(x-3)(x^2 + a)}{(x+1)^2 (x-2)^2}$

(Using the point $(0,1)$ leads to $a = 4/3$.) Thus,

$$R(x) = \dfrac{(x-1)(x-3)\left(x^2 + \frac{4}{3}\right)}{(x+1)^2 (x-2)^2}.$$

55.

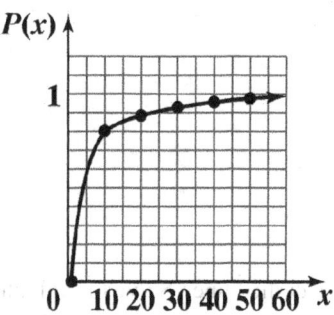

The likelihood your ball is chosen decreases very quickly and approaches 0 as the number of attendees, x, increases.

57. a. The degree of the numerator is 1 and the degree of the denominator is 2. Thus, the horizontal asymptote is $y = 0$. The concentration of the drug decreases to 0 as time increases.

b. Graphing:

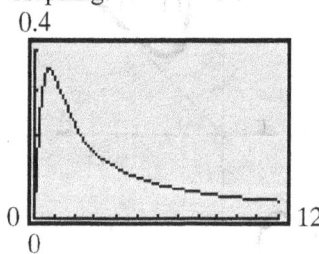

c. Using MAXIMUM, the concentration is highest after $t \approx 0.71$ hours.

59. a. The cost of the project is the sum of the cost for the parallel side, the two other sides, and the posts.

$A = xy$

$1000 = xy$

$y = \dfrac{1000}{x}$

If the length of a perpendicular side is x feet, the length of the parallel side is

$y = \dfrac{1000}{x}$ feet. Thus,

$$C(x) = 2 \cdot 8 \cdot x + 5 \cdot \dfrac{1000}{x} + 4(25)$$

$$= 16x + \dfrac{5000}{x} + 100$$

b. The domain is $x > 0$. Note that x is a length so it cannot be negative. In addition, if $x = 0$, there is no rectangle (that is, the area is 0 square feet).

c. $C(x) = 16x + \dfrac{5000}{x} + 100$

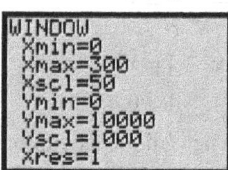 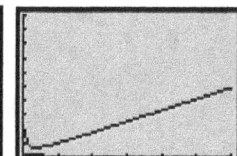

d. Using MINIMUM, the dimensions of cheapest cost are about 17.7 feet by 56.6 feet (longer side parallel to river).

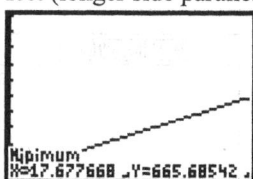

Note: $x = 17\dfrac{2}{3} = \dfrac{53}{3}$ feet and

$y = \dfrac{1000}{53/3} = \dfrac{3000}{53}$ feet .

61. a. The surface area is the sum of the areas of the six sides.

$$S = xy + xy + xy + xy + x^2 + x^2 = 4xy + 2x^2$$

The volume is $x \cdot x \cdot y = x^2 y = 10,000$.

Thus, $y = \dfrac{10,000}{x^2}$, so

$$S(x) = 4x\left(\dfrac{10,000}{x^2}\right) + 2x^2$$

$$= 2x^2 + \dfrac{40,000}{x}$$

$$= \dfrac{2x^3 + 40,000}{x}$$

b. Graphing:
10,000

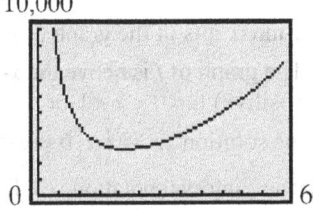

0   60
0

c. Using MINIMUM, the minimum surface area (amount of cardboard) is about 2784.95 square inches.

d. The surface area is a minimum when $x \approx 21.54$ inches.

$$y = \dfrac{10,000}{(21.544)^2} \approx 21.54 \text{ inches}$$

The dimensions of the box are: 21.54 in. by 21.54 in. by 21.54 in.

e. Answers will vary. One possibility is to save costs or reduce weight by minimizing the material needed to construct the box.

63. a. $500 = \pi r^2 h$

$$h = \dfrac{500}{\pi r^2}$$

$$C(r) = 6(2\pi r^2) + 4(2\pi r h)$$

$$= 12\pi r^2 + 8\pi r\left(\dfrac{500}{\pi r^2}\right)$$

$$= 12\pi r^2 + \dfrac{4000}{r}$$

b. Graphing:
6000

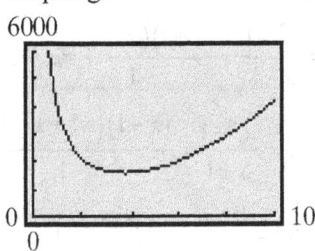

0   10
0

Using MINIMUM, the cost is least for $r \approx 3.76$ cm.

65. $y = \dfrac{x^2 - 1}{x - 1}$

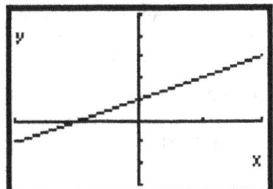

$y = \dfrac{x^3 - 1}{x - 1}$

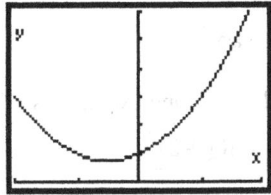

$y = \dfrac{x^4 - 1}{x - 1}$

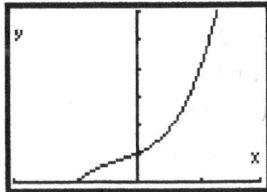

$y = \dfrac{x^5 - 1}{x - 1}$

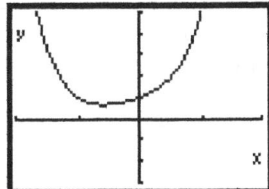

$x = 1$ is not a vertical asymptote because of the following behavior:

When $x \neq 1$:

$$y = \frac{x^2-1}{x-1} = \frac{(x+1)(x-1)}{x-1} = x+1$$

$$y = \frac{x^3-1}{x-1} = \frac{(x-1)(x^2+x+1)}{x-1} = x^2+x+1$$

$$y = \frac{x^4-1}{x-1} = \frac{(x^2+1)(x^2-1)}{x-1}$$
$$= \frac{(x^2+1)(x-1)(x+1)}{x-1}$$
$$= x^3+x^2+x+1$$

$$y = \frac{x^5-1}{x-1} = \frac{(x^4+x^3+x^2+x+1)(x-1)}{x-1}$$
$$= x^4+x^3+x^2+x+1$$

In general, the graph of

$$y = \frac{x^n-1}{x-1}, \ n \geq 1, \text{ an integer,}$$

will have a "hole" with coordinates $(1, n)$.

67. Answers will vary.

69. Answers will vary. One example is

$$R(x) = \frac{2(x-3)(x+2)^2}{(x-1)^3}.$$

71. Answers will vary.

73. $\dfrac{3x}{3x+1} = \dfrac{x-2}{x+5}$

$3x(x+5) = (3x+1)(x-2)$

$3x^2+15x = 3x^2-5x-2$

$20x = -2$

$x = -\dfrac{1}{10}$

75. $\dfrac{\sqrt{5}-3}{\sqrt{7}+2} \approx -0.164$

Section 5.6

1. $3-4x > 5$

$-4x > 2$

$x < -\dfrac{1}{2}$

The solution set is $\left\{ x \middle| x < -\dfrac{1}{2} \right\}$ or, using interval

notation, $\left(-\infty, -\dfrac{1}{2} \right)$.

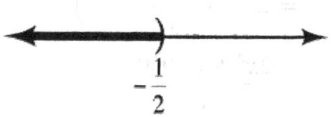

$-\dfrac{1}{2}$

3. c

5. The x-intercepts of the graph of f are 0, 1, and 2.

   a. The graph of f is above the x-axis (so f is positive) for $0 < x < 1$ or $x > 2$. Therefore, the solution set is $\left\{ x \middle| 0 < x < 1 \text{ or } x > 2 \right\}$ or, using interval notation, $(0, 1) \cup (2, \infty)$.

   b. The graph of f is below the x-axis (so f is negative) for $x < 0$ or $1 < x < 2$. Since the inequality is not strict, we include 0, 1, and 2 in the solution set. Therefore, the solution set is $\left\{ x \middle| x \leq 0 \text{ or } 1 \leq x \leq 2 \right\}$ or, using interval notation $(-\infty, 0] \cup [1, 2]$.

7. The x intercept of the graph of f is 0.

   a. The graph of f is below the x-axis (so f is negative) for $-1 < x < 0$ or $x > 1$. Therefore, the solution set is $\left\{ x \middle| -1 < x < 0 \text{ or } x > 1 \right\}$ or, using interval notation, $(-1, 0) \cup (1, \infty)$.

   b. The graph of f is above the x-axis (so f is positive) for $x < -1$ or $0 < x < 1$. Since the inequality is not strict, we include 0 in the solution set. Therefore, the solution set is $\left\{ x \middle| x < -1 \text{ or } 0 \leq x < 1 \right\}$ or, using interval notation $(-\infty, -1) \cup [0, 1)$.

9. We graphed $f(x) = x^2(x-3)$ in Problem 81 of Section 5.1. The graph is reproduced below.

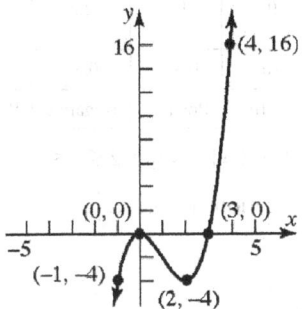

From the graph, we see that f is below the x-axis (so f is negative) for $x < 0$ or $0 < x < 3$. Thus, the solution set is $\{ x \mid x < 0 \text{ or } 0 < x < 3 \}$ or, using interval notation $(-\infty, 0) \cup (0, 3)$.

11. We graphed $f(x) = (x+4)^2(1-x)$ in Problem 83 of Section 5.1. The graph is reproduced below.

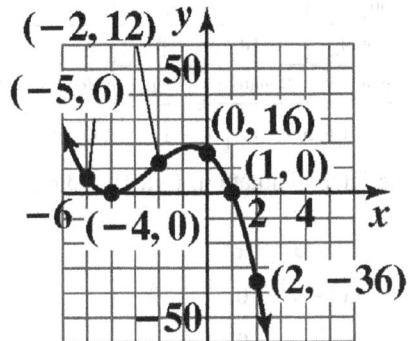

From the graph, we that f is above the x-axis (so f is positive) for $x \le 1$. Since the inequality is not strict, we include 1 in the solution set. Therefore, the solution set is $\{ x \mid x \le 1 \}$ or, using interval notation $(-\infty, 1]$.

13. We graphed $f(x) = -2(x+2)(x-2)^3$ in Problem 85 of Section 5.1. The graph is reproduced below.

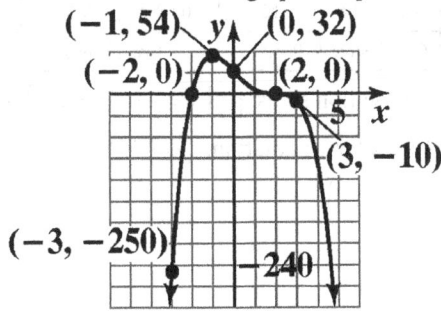

From the graph, we see that f is below the x-axis (so f is negative) for $x < -2$ or $x > 2$. Since the inequality is not strict, we include -2 and 2 in the solution set. Therefore, the solution set is $\{ x \mid x \le -2 \text{ or } x \ge 2 \}$ or, using interval notation $(-\infty, -2] \cup [2, \infty)$.

15. We graphed $R(x) = \dfrac{x+1}{x(x+4)}$ in Problem 7 of Section 5.5. The graph is reproduced below.

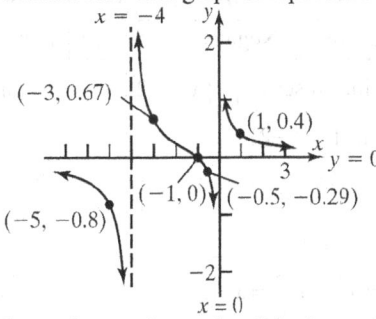

From the graph, we that R is above the x-axis (so R is positive) for $-4 < x < -1$ or $x > 0$. Therefore, the solution set is $\{ x \mid -4 < x < -1 \text{ or } x > 0 \}$ or, using interval notation $(-4, -1) \cup (0, \infty)$.

17. We graphed $R(x) = \dfrac{3x+3}{2x+4}$ in Problem 9 of Section 5.5. The graph is reproduced below.

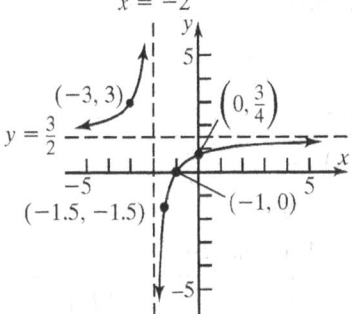

From the graph, we that R is below the x-axis (so R is negative) for $-2 < x < -1$. Since the inequality is not strict, we include -1 in the solution set. Therefore, the solution set is $\{ x \mid -2 < x \le -1 \}$ or, using interval notation $(-2, -1]$.

19. $(x-5)^2(x+2) < 0$

$f(x) = (x-5)^2(x+2)$

$x = 5, x = -2$ are the zeros of f.

Interval	$(-\infty, -2)$	$(-2, 5)$	$(5, \infty)$
Number Chosen	-3	0	6
Value of f	-64	50	8
Conclusion	Negative	Positive	Positive

The solution set is $\{ x \mid x < -2 \}$ or, using interval notation, $(-\infty, -2)$.

21. $x^3 - 4x^2 > 0$

$x^2(x-4) > 0$

$f(x) = x^3 - 4x^2 = x^2(x-4)$

$x = 0, x = 4$ are the zeros of f.

Interval	$(-\infty, 0)$	$(0, 4)$	$(4, \infty)$
Number Chosen	-1	1	5
Value of f	-5	-3	25
Conclusion	Negative	Negative	Positive

The solution set is $\{ x \mid x > 4 \}$ or, using interval notation, $(4, \infty)$.

23. $2x^3 > -8x^2$

$2x^3 + 8x^2 > 0$

$2x^2(x+4) > 0$

$f(x) = 2x^3 + 8x^2$

$x = 0, x = -4$ are the zeros of f.

Interval	$(-\infty, -4)$	$(-4, 0)$	$(0, \infty)$
Number Chosen	-5	-1	1
Value of f	-50	6	10
Conclusion	Negative	Positive	Positive

The solution set is $\{ x \mid -4 < x < 0 \text{ or } x > 0 \}$ or, using interval notation, $(-4, 0) \cup (0, \infty)$.

25. $(x-1)(x-2)(x-3) \le 0$

$f(x) = (x-1)(x-2)(x-3)$

$x = 1, x = 2, x = 3$ are the zeros of f.

Interval	$(-\infty, 1)$	$(1, 2)$	$(2, 3)$	$(3, \infty)$
Number Chosen	0	1.5	2.5	4
Value of f	-6	0.375	-0.375	6
Conclusion	Negative	Positive	Negative	Positive

The solution set is $\{ x \mid x \le 1 \text{ or } 2 \le x \le 3 \}$ or, using interval notation, $(-\infty, 1] \cup [2, 3]$.

27. $x^3 - 2x^2 - 3x > 0$

$x(x^2 - 2x - 3) > 0$

$x(x+1)(x-3) > 0$

$f(x) = x^3 - 2x^2 - 3x$

$x = -1, x = 0, x = 3$ are the zeros of f.

Interval	$(-\infty, -1)$	$(-1, 0)$	$(0, 3)$	$(3, \infty)$
Number Chosen	-2	-0.5	1	4
Value of f	-10	0.875	-4	20
Conclusion	Negative	Positive	Negative	Positive

The solution set is $\{ x \mid -1 < x < 0 \text{ or } x > 3 \}$ or, using interval notation, $(-1, 0) \cup (3, \infty)$.

29. $x^4 > x^2$

$x^4 - x^2 > 0$

$x^2(x^2 - 1) > 0$

$x^2(x-1)(x+1) > 0$

$f(x) = x^2(x-1)(x+1)$

$x = -1, x = 0, x = 1$ are the zeros of f

Interval	$(-\infty, -1)$	$(-1, 0)$	$(0, 1)$	$(1, \infty)$
Number Chosen	-2	-0.5	0.5	2
Value of f	12	-0.1875	-0.1875	12
Conclusion	Positive	Negative	Negative	Positive

The solution set is $\{ x \mid x < -1 \text{ or } x > 1 \}$ or, using interval notation, $(-\infty, -1) \cup (1, \infty)$.

31.
$$x^4 > 1$$
$$x^4 - 1 > 0$$
$$(x^2 - 1)(x^2 + 1) > 0$$
$$(x - 1)(x + 1)(x^2 + 1) > 0$$
$$f(x) = (x - 1)(x + 1)(x^2 + 1)$$

$x = 1$, $x = -1$ are the zeros of f ; $x^2 + 1$ has no real solution

Interval	$(-\infty, -1)$	$(-1, 1)$	$(1, \infty)$
Number Chosen	-2	0	2
Value of f	15	-1	15
Conclusion	Positive	Negative	Positive

The solution set is $\{ x \mid x < -1 \text{ or } x > 1 \}$ or, using interval notation, $(-\infty, -1) \cup (1, \infty)$.

33. $\dfrac{x+1}{x-1} > 0$

$$f(x) = \frac{x+1}{x-1}$$

The zeros and values where f is undefined are $x = -1$ and $x = 1$.

Interval	$(-\infty, -1)$	$(-1, 1)$	$(1, \infty)$
Number Chosen	-2	0	2
Value of f	$\dfrac{1}{3}$	-1	3
Conclusion	Positive	Negative	Positive

The solution set is $\{ x \mid x < -1 \text{ or } x > 1 \}$ or,

35. $\dfrac{(x-1)(x+1)}{x} \le 0$

$$f(x) = \frac{(x-1)(x+1)}{x}$$

The zeros and values where f is undefined are $x = -1$, $x = 0$ and $x = 1$.

Interval	$(-\infty, -1)$	$(-1, 0)$	$(0, 1)$	$(1, \infty)$
Number Chosen	-2	-0.5	0.5	2
Value of f	-1.5	1.5	-1.5	1.5
Conclusion	Negative	Positive	Negative	Positive

The solution set is $\{ x \mid x \le -1 \text{ or } 0 < x \le 1 \}$ or, using interval notation, $(-\infty, -1] \cup (0, 1]$.

37. $\dfrac{(x-2)^2}{x^2 - 1} \ge 0$

$$\frac{(x-2)^2}{(x+1)(x-1)} \ge 0$$

$$f(x) = \frac{(x-2)^2}{x^2 - 1}$$

The zeros and values where f is undefined are $x = -1$, $x = 1$ and $x = 2$.

Interval	$(-\infty, -1)$	$(-1, 1)$	$(1, 2)$	$(2, \infty)$
Number Chosen	-2	0	1.5	3
Value of f	$\dfrac{16}{3}$	-4	0.2	0.125
Conclusion	Positive	Negative	Positive	Positive

The solution set is $\{ x \mid x < -1 \text{ or } x > 1 \}$ or, using interval notation, $(-\infty, -1) \cup (1, \infty)$.

39.
$$\frac{x+4}{x-2} \le 1$$
$$\frac{x+4}{x-2} - 1 \le 0$$
$$\frac{x+4-(x-2)}{x-2} \le 0$$
$$\frac{6}{x-2} \le 0$$
$$f(x) = \frac{6}{x-2}$$

The value where f is undefined is $x = 2$.

Interval	$(-\infty, 2)$	$(2, \infty)$
Number Chosen	0	3
Value of f	-3	6
Conclusion	Negative	Positive

The solution set is $\{ x \mid x < 2 \}$ or, using interval notation, $(-\infty, 2)$.

41.
$$\frac{3x-5}{x+2} \le 2$$

$$\frac{3x-5}{x+2} - 2 \le 0$$

$$\frac{3x-5-2(x+2)}{x+2} \le 0$$

$$\frac{x-9}{x+2} \le 0$$

$$f(x) = \frac{x-9}{x+2}$$

The zeros and values where f is undefined are
$x = -2$ and $x = 9$.

Interval	$(-\infty, -2)$	$(-2, 9)$	$(9, \infty)$
Number Chosen	-3	0	10
Value of f	12	-4.5	$\dfrac{1}{12}$
Conclusion	Positive	Negative	Positive

The solution set is $\left\{ x \mid -2 < x \le 9 \right\}$ or, using
interval notation, $(-2, 9]$.

43.
$$\frac{1}{x-2} < \frac{2}{3x-9}$$

$$\frac{1}{x-2} - \frac{2}{3x-9} < 0$$

$$\frac{3x-9-2(x-2)}{(x-2)(3x-9)} < 0$$

$$\frac{x-5}{(x-2)(3x-9)} < 0$$

$$f(x) = \frac{x-5}{(x-2)(3x-9)}$$

The zeros and values where f is undefined are
$x = 2$, $x = 3$, and $x = 5$.

Interval	$(-\infty, 2)$	$(2, 3)$	$(3, 5)$	$(5, \infty)$
Number Chosen	0	2.5	4	6
Value of f	$-\dfrac{5}{18}$	$\dfrac{10}{3}$	$-\dfrac{1}{6}$	$\dfrac{1}{36}$
Conclusion	Negative	Positive	Negative	Positive

The solution set is $\left\{ x \mid x < 2 \text{ or } 3 < x < 5 \right\}$ or,
using interval notation, $(-\infty, 2) \cup (3, 5)$.

45.
$$\frac{x^2(3+x)(x+4)}{(x+5)(x-1)} \ge 0$$

$$f(x) = \frac{x^2(3+x)(x+4)}{(x+5)(x-1)}$$

The zeros and values where f is undefined are
$x = -5$, $x = -4$, $x = -3$, $x = 0$ and $x = 1$.

Interval	Number Chosen	Value of f	Conclusion
$(-\infty, -5)$	-6	$\dfrac{216}{7}$	Positive
$(-5, -4)$	-4.5	$-\dfrac{243}{44}$	Negative
$(-4, -3)$	-3.5	$\dfrac{49}{108}$	Positive
$(-3, 0)$	-1	-0.75	Negative
$(0, 1)$	0.5	$-\dfrac{63}{44}$	Negative
$(1, \infty)$	2	$\dfrac{120}{7}$	Positive

The solution set is
$\left\{ x \mid x < -5 \text{ or } -4 \le x \le -3 \text{ or } x = 0 \text{ or } x > 1 \right\}$
or, using interval notation,
$(-\infty, -5) \cup [-4, -3] \cup \{0\} \cup (1, \infty)$.

47.
$$\frac{(3-x)^3(2x+1)}{x^3-1} < 0$$

$$\frac{(3-x)^3(2x+1)}{(x-1)(x^2+x+1)} < 0$$

$$f(x) = \frac{(3-x)^3(2x+1)}{(x-1)(x^2+x+1)}$$

The zeros and values where f is undefined are
$x = 3$, $x = -\dfrac{1}{2}$, and $x = 1$.

Interval	Number Chosen	Value of f	Conclusion
$\left(-\infty, -\dfrac{1}{2}\right)$	-1	32	Positive
$\left(-\dfrac{1}{2}, 1\right)$	0	-27	Negative
$(1, 3)$	2	$5/7$	Positive
$(3, \infty)$	4	$-1/7$	Negative

The solution set is $\left\{ x \mid -\frac{1}{2} < x < 1 \text{ or } x > 3 \right\}$ or,

using interval notation, $\left(-\frac{1}{2}, 1\right) \cup (3, \infty)$.

49. $(x+1)(x-3)(x-5) > 0$

$f(x) = (x+1)(x-3)(x-5)$

$x = -1, x = 3, x = 5$ are the zeros of f.

Interval	$(-\infty, -1)$	$(-1, 3)$	$(3, 5)$	$(5, \infty)$
Number Chosen	-2	0	4	6
Value of f	-35	15	-5	21
Conclusion	Negative	Positive	Negative	Positive

We want to know where $f(x) > 0$, so the solution set is $\left\{ x \mid -1 < x < 3 \text{ or } x > 5 \right\}$ or, using interval notation, $(-1, 3) \cup (5, \infty)$.

51. $\qquad 7x - 4 \geq -2x^2$

$2x^2 + 7x - 4 \geq 0$

$(2x-1)(x+4) \geq 0$

$f(x) = 2x^2 + 7x - 4$

$x = \frac{1}{2}, x = -4$ are the zeros of f.

Interval	$(-\infty, -4)$	$\left(-4, \frac{1}{2}\right)$	$\left(\frac{1}{2}, \infty\right)$
Number Chosen	-5	0	1
Value of f	11	-4	5
Conclusion	Positive	Negative	Positive

We want to know where $f(x) \geq 0$, so the

solution set is $\left\{ x \mid x \leq -4 \text{ or } x \geq \frac{1}{2} \right\}$ or, using

interval notation, $\left(-\infty, -4\right] \cup \left[\frac{1}{2}, \infty\right)$.

53. $\qquad \frac{x+1}{x-3} \leq 2$

$\frac{x+1}{x-3} - 2 \leq 0$

$\frac{x+1 - 2(x-3)}{x-3} \leq 0$

$\frac{x+1 - 2x + 6)}{x-3} \leq 0$

$\frac{-x+7}{x-3} \leq 0$

$f(x) = \frac{-x+7}{x-3}$

The zeros and values where f is undefined are $x = 3$ and $x = 7$.

Interval	$(-\infty, 3)$	$(3, 7)$	$(7, \infty)$
Number Chosen	1	5	8
Value of f	-3	1	$-\frac{1}{5}$
Conclusion	Negative	Positive	Negative

We want to know where $f(x) \leq 0$, so the

solution set is $\left\{ x \mid x < 3 \text{ or } x \geq 7 \right\}$ or, using

interval notation, $(-\infty, 3) \cup [7, \infty)$. Note that 3 is not in the solution set because 3 is not in the domain of f.

55. $3(x^2 - 2) < 2(x-1)^2 + x^2$

$3x^2 - 6 < 2(x^2 - 2x + 1) + x^2$

$3x^2 - 6 < 2x^2 - 4x + 2 + x^2$

$3x^2 - 6 < 3x^2 - 4x + 2$

$-6 < -4x + 2$

$4x - 6 < 2$

$4x < 8$

$x < 2$

The solution set is $\left\{ x \mid x < 2 \right\}$ or, using interval notation, $(-\infty, 2)$.

57.
$$6x - 5 < \frac{6}{x}$$

$$6x - 5 - \frac{6}{x} < 0$$

$$\frac{6x^2 - 5x - 6}{x} < 0$$

$$\frac{(2x-3)(3x+2)}{x} < 0$$

$$f(x) = \frac{(2x-3)(3x+2)}{x}$$

The zeros and values where f is undefined are

$$x = -\frac{2}{3}, x = 0 \text{ and } x = \frac{3}{2}.$$

Interval	$\left(-\infty, -\frac{2}{3}\right)$	$\left(-\frac{2}{3}, 0\right)$	$\left(0, \frac{3}{2}\right)$	$\left(\frac{3}{2}, \infty\right)$
Number Chosen	-1	-0.5	1	2
Value of f	-5	4	-5	4
Conclusion	Negative	Positive	Negative	Positive

We want to know where $f(x) < 0$, so the

solution set is $\left\{ x \mid x < -\frac{2}{3} \text{ or } 0 < x < \frac{3}{2} \right\}$ or,

59.
$$x^3 - 9x \le 0$$
$$x(x-3)(x+3) \le 0$$
$$f(x) = x^3 - 9x$$

$x = -3, x = 0, x = 3$ are the zeros of f.

Interval	$(-\infty, -3)$	$(-3, 0)$	$(0, 3)$	$(3, \infty)$
Number Chosen	-4	-1	1	4
Value of f	-28	8	-8	28
Conclusion	Negative	Positive	Negative	Positive

We want to know where $f(x) \le 0$, so the

solution set is $\{x \mid x \le -3 \text{ or } 0 \le x \le 3\}$ or, using

interval notation, $(-\infty, -3] \cup [0, 3]$.

61. $f(x) = 2x^4 + 11x^3 - 11x^2 - 104x - 48$

   a. Step 1: $f(x)$ has at most 4 real zeros.

Step 2: Possible rational zeros:
$$p : \pm 1, \pm 2, \pm 3, \pm 4, \pm 6, \pm 8,$$
$$\pm 12, \pm 24, \pm 48$$
$$q : \pm 1, \pm 2$$

$$\frac{p}{q} = \pm 1, \pm 2, \pm 3, \pm 4, \pm 6, \pm 8, \pm 12,$$
$$\pm 24, \pm \frac{1}{2}, \pm \frac{3}{2}, \pm 48$$

Step 3: From the graph it appears that there are x-intercepts at $-4, -0.5,$ and 3. Using synthetic division:

$$-4 \overline{)2 \quad 11 \quad -11 \quad -104 \quad -48}$$
$$\underline{ \quad -8 \quad -12 \quad 92 \quad 48}$$
$$2 \quad 3 \quad -23 \quad -12 \quad 0$$

Since the remainder is 0, $x + 4$ is a factor. Using synthetic division again:

$$-4 \overline{)2 \quad 3 \quad -23 \quad -12}$$
$$\underline{ \quad -8 \quad 20 \quad 12}$$
$$2 \quad -5 \quad -3 \quad 0$$

Since the remainder is 0, $x + 4$ is also a factor again. Thus,

$$f(x) = (x+4)^2 (2x^2 - 5x - 3)$$
$$= (x-4)^2 (2x+1)(x-3)$$

The zeros are $-4, -\frac{1}{2},$ and 3.

   b. The factoring is:
$$f(x) = (x-4)^2 (2x+1)(x-3)$$

   c.

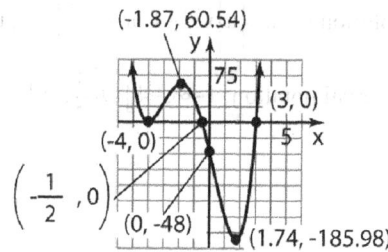

   d. Looking at the graph we have $f(x) < 0$ at
$$\left(-\frac{1}{2}, 3\right)$$

63. a. $R(x) = \dfrac{x^2 + 5x - 6}{x^2 - 4x + 4} = \dfrac{(x+6)(x-1)}{(x-2)(x-2)}$ $p(x) = x^2 + 5x - 6;$ $q(x) = x^2 - 4;$ $n = 2;$ $m = 2$

Step 1: Domain: $\{x \mid x \neq 2\}$

The y-intercept is $R(0) = \dfrac{(0)^2 + 5(0) - 6}{(0)^2 - 4(0) + 4} = \dfrac{-6}{4} = -\dfrac{3}{2}$. Plot the point $\left(0, -\dfrac{3}{2}\right)$.

Step 2 & 3: In lowest terms, $R(x) = \dfrac{(x+6)(x-1)}{(x-2)(x-2)}$, $x \neq 2$. The x-intercepts are the zeros of $y = x + 6$ and $y = x - 1$: $-6, 1$;

Step 4: In lowest terms, $R(x) = \dfrac{(x+6)(x-1)}{(x-2)(x-2)}$, $x \neq 2$. The vertical asymptote is the zero of $f(x) = x - 2$:

$x = 2$;
Graph this asymptote using a dashed line.

Step 5: Since $n = m$, the line $y = 1$ is the horizontal asymptote. Graph this asymptote using a dashed line.
Solve to find intersection points:

$$\frac{x^2 + 5x - 6}{x^2 - 4x + 4} = 1$$

$$x^2 + 5x - 6 = x^2 - 4x + 4$$

$$5x - 6 = -4x + 4$$

$$9x = 10$$

$$x = \frac{10}{9}$$

Steps 6 & 7: Graphing:

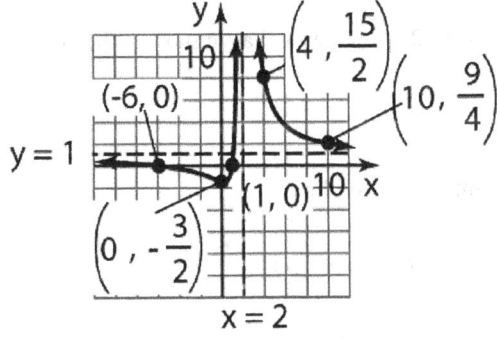

b. $\dfrac{(x+6)(x-1)}{(x-2)(x-2)} \geq 0$

The zeros and values where f is undefined are $x = -6$, $x = 1$, and $x = 2$.

Interval	Number Chosen	Value of f	Conclusion
$(-\infty, -6)$	-6	$\dfrac{216}{7}$	Positive
$(-6, -1)$	-4.5	$-\dfrac{243}{44}$	Negative
$(-1, 2)$	-3.5	$\dfrac{49}{108}$	Positive
$(2, \infty)$	2	$\dfrac{120}{7}$	Positive

The solution set is $\left\{ x \mid x < -6 \text{ or } 1 \le x \le 2 \text{ or } x > 2 \right\}$ or, using interval notation,

$$(-\infty, -6] \cup [1, 2) \cup (2, \infty)$$

65. a. $R(x) = \dfrac{x^3 + 2x^2 - 11x - 12}{x^2 - x - 6} = \dfrac{(x-3)(x+4)(x+1)}{(x-3)(x+2)} = \dfrac{(x+4)(x+1)}{(x+2)}$ $\quad p(x) = x^3 + 2x^2 - 11x - 12;$

$q(x) = x^2 - x - 6; \quad n = 3; \quad m = 2$

Step 1: Domain: $\left\{ x \mid x \ne -2, 3 \right\}$

The y-intercept is $R(0) = \dfrac{(0)^3 + 2(0)^2 - 11(0) - 12}{(0)^2 - (0) - 6} = \dfrac{-12}{-6} = 2$. Plot the point $(0, 2)$.

Step 2 & 3: In lowest terms, $R(x) = \dfrac{(x+4)(x+1)}{(x+2)}$, $x \ne -2$. The x-intercepts are the zeros of $y = x + 4$ and

$y = x + 1$: $-4, -1$;

Note: $x = 3$ is not a zero because reduced form must be used to find the zeros.

Step 4: In lowest terms, $R(x) = \dfrac{(x+4)(x+1)}{(x+2)}$, $x \ne 2$. The vertical asymptote is the zero of $f(x) = x + 2$:

$x = -2$;

Graph this asymptote using a dashed line.

Step 5: Since $n = m + 1$, there is an oblique asymptote. Dividing:

$$x^2 - x - 6 \overline{\smash{\big)}\, x^3 + 2x^2 - 11x - 12} \qquad\qquad G(x) = x + 3 + \dfrac{-2x + 6}{x^2 - x - 6}$$

with quotient $x + 3$ and working:

$$\begin{array}{r} x + 3 \\ x^2 - x - 6 \overline{\smash{\big)}\, x^3 + 2x^2 - 11x - 12} \\ \underline{x^3 - x^2 - 6x} \\ 3x^2 - 5x - 12 \\ \underline{3x^2 - 3x - 18} \\ -2x + 6 \end{array}$$

The oblique asymptote is $y = x + 3$. Graph this asymptote with a dashed line.

Steps 6 & 7: Graphing:

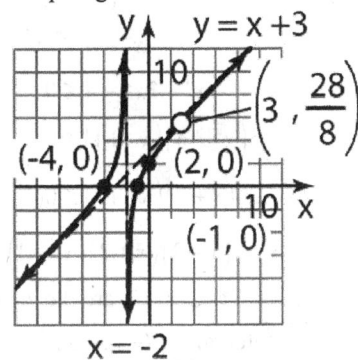

b. $\dfrac{(x-3)(x+4)(x+1)}{(x-3)(x+2)} \geq 0$

The zeros and values where f is undefined are $x=-4$, $x=-2$, and $x=-1$.

Interval	Number Chosen	Value of f	Conclusion
$(-\infty, -4)$	-5	$-\dfrac{4}{3}$	Negative
$(-4, -2)$	-3	2	Positive
$(-2, -1)$	-1.5	$-\dfrac{5}{2}$	Negative
$(-1, \infty)$	0	2	Positive

The solution set is $\left\{ x \mid -4 \leq x < -2 \text{ or } -1 \leq x < 3 \text{ or } x > 3 \right\}$ or, using interval notation,

$[-4, -2) \cup [-1, 3) \cup (3, \infty)$

67. Let x be the positive number. Then

$x^3 > 4x^2$

$x^3 - 4x^2 > 0$

$x^2(x-4) > 0$

$f(x) = x^2(x-4)$

$x = 0$ and $x = 4$ are the zeros of f.

Interval	$(-\infty, 0)$	$(0, 4)$	$(4, \infty)$
Number Chosen	-1	1	5
Value of f	-5	-3	25
Conclusion	Negative	Negative	Positive

Since x must be positive, all real numbers greater than 4 satisfy the condition. The solution set is $\left\{ x \mid x > 4 \right\}$ or, using interval notation, $(4, \infty)$.

69. The domain of $f(x) = \sqrt{x^4 - 16}$ consists of all real numbers x for which

$x^4 - 16 \geq 0$

$(x^2 + 4)(x^2 - 4) \geq 0$

$\left(x^2 + 4\right)(x-2)(x+2) \geq 0$

$p(x) = \left(x^2 + 4\right)(x-2)(x+2)$

$x = -2$ and $x = 2$ are the zeros of p.

Interval	$(-\infty, -2)$	$(-2, 2)$	$(2, \infty)$
Number Chosen	-3	0	3
Value of p	65	-16	65
Conclusion	Positive	Negative	Positive

The domain of f will be where $p(x) \geq 0$. Thus, the domain of f is $\left\{ x \mid x \leq -2 \text{ or } x \geq 2 \right\}$ or, using interval notation, $(-\infty, -2] \cup [2, \infty)$.

71. The domain of $f(x) = \sqrt{\dfrac{x-2}{x+4}}$ includes all

values for which $\dfrac{x-2}{x+4} \geq 0$.

$R(x) = \dfrac{x-2}{x+4}$

The zeros and values where R is undefined are
$x = -4$ and $x = 2$.

Interval	$(-\infty, -4)$	$(-4, 2)$	$(2, \infty)$
Number Chosen	-5	0	3
Value of R	7	$-\frac{1}{2}$	$\frac{1}{7}$
Conclusion	Positive	Negative	Positive

The domain of f will be where $R(x) \geq 0$. Thus,
the domain of f is $\{x \mid x < -4 \text{ or } x \geq 2\}$ or,
using interval notation, $(-\infty, -4) \cup [2, \infty)$.

73. $$f(x) \leq g(x)$$
$$x^4 - 1 \leq -2x^2 + 2$$
$$x^4 + 2x^2 - 3 \leq 0$$
$$(x^2 + 3)(x^2 - 1) \leq 0$$
$$(x^2 + 3)(x - 1)(x + 1) \leq 0$$
$$h(x) = (x^2 + 3)(x - 1)(x + 1)$$

$x = -1$ and $x = 1$ are the zeros of h.

Interval	$(-\infty, -1)$	$(-1, 1)$	$(1, \infty)$
Number Chosen	-2	0	2
Value of h	21	-3	21
Conclusion	Positive	Negative	Positive

$f(x) \leq g(x)$ if $-1 \leq x \leq 1$. That is, on the
interval $[-1, 1]$.

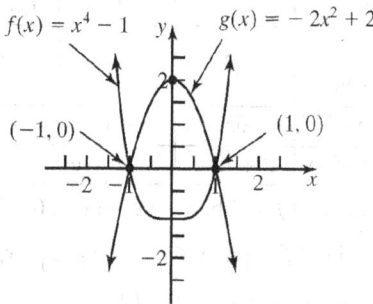

75. $$f(x) \leq g(x)$$
$$x^4 - 4 \leq 3x^2$$
$$x^4 - 3x^2 - 4 \leq 0$$
$$(x^2 - 4)(x^2 + 1) \leq 0$$
$$(x - 2)(x + 2)(x^2 + 1) \leq 0$$
$$h(x) = (x - 2)(x + 2)(x^2 + 1)$$

$x = -2$ and $x = 2$ are the zeros of h.

Interval	$(-\infty, -2)$	$(-2, 2)$	$(2, \infty)$
Number Chosen	-3	0	3
Value of h	50	-4	50
Conclusion	Positive	Negative	Positive

$f(x) \leq g(x)$ if $-2 \leq x \leq 2$. That is, on the
interval $[-2, 2]$.

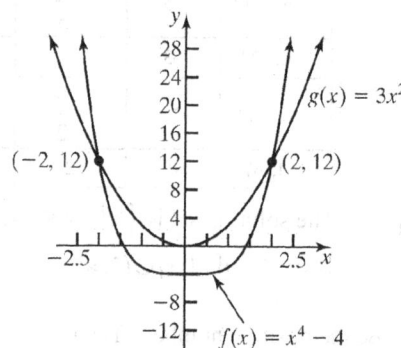

77. We need to solve $\overline{C}(x) \leq 100$.
$$\frac{80x + 5000}{x} \leq 100$$
$$\frac{80x + 5000}{x} - \frac{100x}{x} \leq 0$$
$$\frac{5000 - 20x}{x} \leq 0$$
$$\frac{20(250 - x)}{x} \leq 0$$
$$f(x) = \frac{20(250 - x)}{x}$$

The zeros and values where the expression is
undefined are $x = 0$ and $x = 250$.

Interval	$(-\infty, 0)$	$(0, 250)$	$(250, \infty)$
Number Chosen	-1	1	260
Value of f	-5020	4980	$-10/13$
Conclusion	Negative	Positive	Negative

The number of bicycles produced cannot be negative, so the solution is $\{x \mid x \geq 250\}$ or, using interval notation, $[250, \infty)$. The company must produce at least 250 bicycles each day to keep average costs to no more than $100.

79. a.
$$K \geq 16$$
$$\frac{2(150)(S+42)}{S^2} \geq 16$$
$$\frac{300S+12,600}{S^2} \geq 16$$
$$\frac{300S+12,600}{S^2} - 16 \geq 0$$
$$\frac{300S+12,600-16S^2}{S^2} \geq 0$$

Solve $-16S^2 + 300S + 12,600 = 0$ and $S^2 = 0$. The zeros and values where the left-hand side is undefined are $S = 0$, $S \approx 39$, $S \approx -20$. Since the stretch cannot be negative, we only consider cases where $S \geq 0$.

Interval	$(0, 39)$	$(39, \infty)$
Number Chosen	1	40
Value of left side	12884	-0.625
Conclusion	Positive	Negative

The cord will stretch less than 39 feet.

b. The safe height is determined by the minimum clearance (3 feet), the free length of the cord (42 feet), and the stretch in the cord (39 feet). Therefore, the platform must be at least $3 + 42 + 39 = 84$ feet above the ground for a 150-pound jumper.

81. Let x represent the number of student that attend the play. Then the discounted price per ticket is $40 - 0.20x$. Each student's share of the bus cost is $\frac{500}{x}$. Thus, each student's total cost will be
$$C(x) = 40 - 0.20x + \frac{500}{x}.$$
We need to solve $C(x) \leq 40$.
$$40 - 0.20x + \frac{500}{x} \leq 40$$
$$-0.20x + \frac{500}{x} \leq 0$$
$$0.20x - \frac{500}{x} \geq 0$$
$$\frac{0.20x^2 - 500}{x} \geq 0$$
$$\frac{0.20(x^2 - 2500)}{x} \geq 0$$
$$\frac{0.20(x+50)(x-50)}{x} \geq 0$$
$$f(x) = \frac{0.20(x+50)(x-50)}{x}$$

The zeros and values where f is undefined are $x = -50$, $x = 0$ and $x = 50$. Since the number of students cannot be negative, we only consider cases where $x \geq 0$.

Interval	$(0, 50)$	$(50, \infty)$
Number Chosen	25	100
Value of f	-15	15
Conclusion	Negative	Positive

We are looking for where $f(x) \geq 0$. Thus, the solution is $\{x \mid x \geq 50\}$ or, using interval notation, $[50, \infty)$. If at least 50 students attend the play, the price per student will be at or below $40.

83. $x^4 + 1 < -5$ has no solution because the quantity $x^4 + 1$ is never negative. ($x^4 + 1 \geq 1$)

85. Answers will vary. One example:
$$\frac{x-5}{x+3} \leq 0$$

87.
$$f(x) = x^2 + 3x - 2$$
$$f(x-2) = (x-2)^2 + 3(x-2) - 2$$
$$= x^2 - 4x + 4 + 3x - 6 - 2$$
$$= x^2 - x - 4$$

89. Since we have y varying directly with $\sqrt{x}$ then
$y = k\sqrt{x}$. When x = 9, then y = 2 so we have

$$2 = k\sqrt{9}$$

$$2 = 3k \quad \text{so} \quad k = \frac{2}{3}$$

So $y = \frac{2}{3}\sqrt{x}$

6. $f(x) = -(x-1)^4$

Using the graph of $y = x^4$, shift right 1 unit,
then reflect about the x-axis.

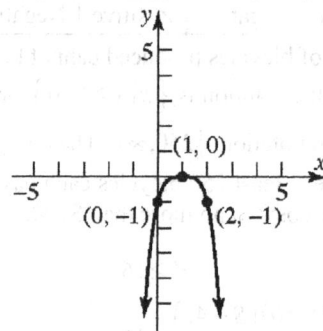

Chapter 5 Review Exercises

1. $f(x) = 4x^5 - 3x^2 + 5x - 2$ is a polynomial of
degree 5.

2. $f(x) = \frac{3x^5}{2x+1}$ is a rational function. It is not a
polynomial because there are variables in the
denominator.

3. $f(x) = 3x^2 + 5x^{1/2} - 1$ is not a polynomial

because the variable x is raised to the $\frac{1}{2}$ power,

which is not a nonnegative integer.

4. $f(x) = 3$ is a polynomial of degree 0.

5. $f(x) = (x+2)^3$

Using the graph of $y = x^3$, shift left 2 units.

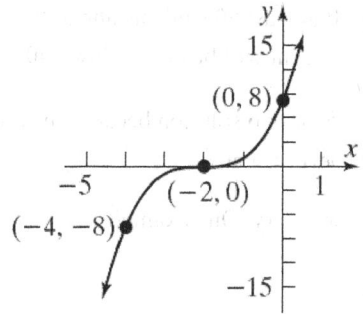

7. $f(x) = (x-1)^4 + 2$

Using the graph of $y = x^4$, shift right 1 unit,
then shift up 2 units.

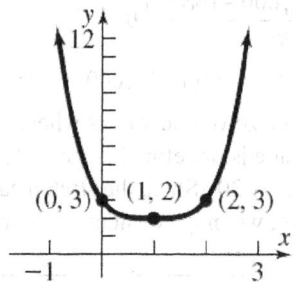

8. $f(x) = x(x+2)(x+4)$

Step 1: Degree is 3. The function resembles $y = x^3$ for large values of $|x|$.

Step 2: y-intercept: $f(0) = 0(0+2)(0+4) = 0$

x-intercepts: solve $f(x) = 0$

$x(x+2)(x+4) = 0$

$x = 0$ or $x = -2$ or $x = -4$

Step 3: Real zeros: -4 with multiplicity one, -2 with multiplicity one, 0 with multiplicity one. The graph crosses the x-axis at $x = -4$, $x = -2$, and $x = 0$.

Step 4: Graphing utility:

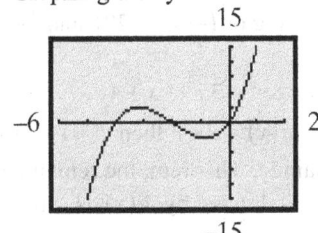

Step 5: 2 turning points; local maximum: $(-3.15, 3.08)$; local minimum: $(-0.85, -3.08)$

Step 6: Graphing by hand:

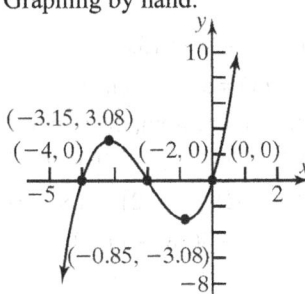

Step 7: Domain: $(-\infty, \infty)$; Range: $(-\infty, \infty)$

Step 8: Increasing on $(-\infty, -3.15]$ and $[-0.85, \infty)$; decreasing on $[-3.15, -0.85]$

9. $f(x) = (x-2)^2(x+4)$

Step 1: Degree is 3. The function resembles $y = x^3$ for large values of $|x|$.

Step 2: y-intercept: $f(0) = (0-2)^2(0+4) = 16$

x-intercepts: solve $f(x) = 0$

$(x-2)^2(x+4) = 0$

$x = 2$ or $x = -4$

Step 3: Real zeros: -4 with multiplicity one, 2 with multiplicity two. The graph crosses the x-axis at $x = -4$ and touches it at $x = 2$.

Step 4: Graphing utility:

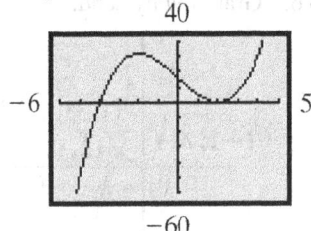

Step 5: 2 turning points; local maximum: $(-2, 32)$; local minimum: $(2, 0)$

Step 6: Graphing by hand:

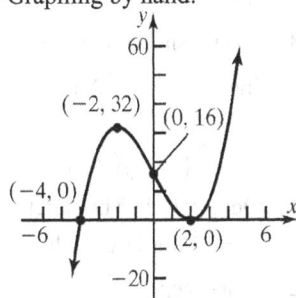

Step 7: Domain: $(-\infty, \infty)$; Range: $(-\infty, \infty)$

Step 8: Increasing on $(-\infty, -2]$ and $[2, \infty)$; decreasing on $[-2, 2]$

10. $f(x) = -2x^3 + 4x^2$

Step 1: Degree is 3. The function resembles $y = -2x^3$ for large values of $|x|$.

Step 2: y-intercept: $f(0) = -2(0)^3 + 4(0)^2 = 0$

x-intercepts: solve $f(x) = 0$

$-2x^3 + 4x^2 = 0$

$-2x^2(x-2) = 0$

$x = 0$ or $x = 2$

Step 3: Real zeros: 0 with multiplicity two, 2 with multiplicity one. The graph touches the x-axis at $x = 0$ and crosses it at $x = 2$.

Step 4: Graphing utility:

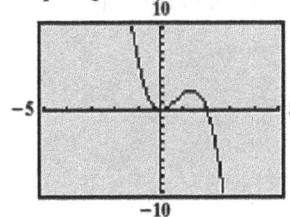

Step 5: 2 turning points; local minimum: $(0,0)$; local maximum: $(1.33, 2.37)$

Step 6: Graphing by hand:

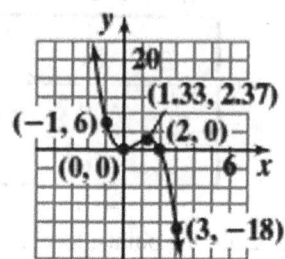

Step 7: Domain: $(-\infty, \infty)$; Range: $(-\infty, \infty)$

Step 8: Decreasing on $(-\infty, 0]$ and $[1.33, \infty)$; increasing on $[0, 1.33]$

11. $f(x) = (x-1)^2 (x+3)(x+1)$

Step 1: Degree is 4. The function resembles $y = x^4$ for large values of $|x|$.

Step 2: y-intercept:
$$f(0) = (0-1)^2 (0+3)(0+1) = 3$$
x-intercepts: solve $f(x) = 0$
$$(x-1)^2 (x+3)(x+1) = 0$$
$$x = 1 \text{ or } x = -3 \text{ or } x = -1$$

Step 3: Real zeros: -3 with multiplicity one, -1 with multiplicity one, 1 with multiplicity two. The graph crosses the x-axis at $x = -3$ and $x = -1$, and touches it at $x = 1$.

Step 4: Graphing utility:

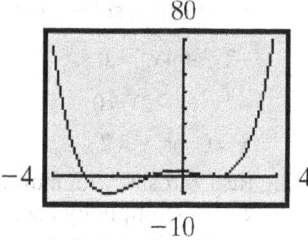

Step 5: 3 turning points;
local maximum: $(-0.22, 3.23)$;
local minima: $(-2.28, -9.91)$ and $(1, 0)$

Step 6: Graphing by hand:

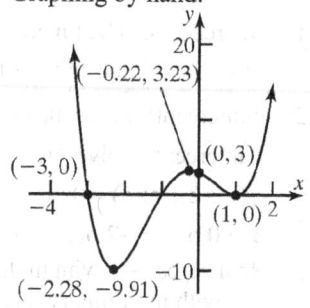

Step 7: Domain: $(-\infty, \infty)$; Range: $[-9.91, \infty)$

Step 8: Increasing on $[-2.28, -0.22]$ and $[1, \infty)$; decreasing on $(-\infty, -2.28]$ and $[-0.22, 1]$

12. $f(x) = 8x^3 - 3x^2 + x + 4$

Since $g(x) = x - 1$ then $c = 1$. From the Remainder Theorem, the remainder R when $f(x)$ is divided by $g(x)$ is $f(c)$:
$$f(1) = 8(1)^3 - 3(1)^2 + 1 + 4$$
$$= 8 - 3 + 1 + 4$$
$$= 10$$
So $R = 10$ and g is not a factor of f.

13. $f(x) = x^4 - 2x^3 + 15x - 2$

Since $g(x) = x + 2$ then $c = -2$. From the Remainder Theorem, the remainder R when $f(x)$ is divided by $g(x)$ is $f(c)$:
$$f(-2) = (-2)^4 - 2(-2)^3 + 15(-2) - 2$$
$$= 16 - 2(-8) - 30 - 2$$
$$= 0$$
So $R = 0$ and g is a factor of f.

14.

	12	0	-8	0	0	0	1
4)		48	192	736	2944	11,776	47,104
	12	48	184	736	2944	11,776	47,105

$f(4) = 47,105$

15. $f(x) = 12x^8 - x^7 + 8x^4 - 2x^3 + x + 3$

The maximum number of zeros is the degree of the polynomial, which is 8.

Examining $f(x) = 12x^8 - x^7 + 8x^4 - 2x^3 + x + 3$,

there are four variations in signs; thus, there are four, two or no positive real zeros.

Examining

$f(-x) = 12(-x)^8 - (-x)^7 + 8(-x)^4 - 2(-x)^3 + (-x) + 3$

$= 12x^8 + x^7 + 8x^4 + 2x^3 - x + 3$

,

there are two variations in sign; thus, there are two negative real zeros or no negative real zeros.

16. $f(x) = -6x^5 + x^4 + 5x^3 + x + 1$

The maximum number of zeros is the degree of the polynomial, which is 5.

Examining $f(x) = -6x^5 + x^4 + 5x^3 + x + 1$, there is one variation in sign; thus, there is one positive real zero.

Examining

$f(-x) = -6(-x)^5 + (-x)^4 + 5(-x)^3 + (-x) + 1$,

$= 6x^5 + x^4 - 5x^3 - x + 1$

there are two variations in sign; thus, there are two negative real zeros or no negative real zeros.

17. $a_0 = -3$, $a_8 = 12$

$p = \pm 1, \pm 3$

$q = \pm 1, \pm 2, \pm 3, \pm 4, \pm 6, \pm 12$

$\dfrac{p}{q} = \pm 1, \pm 3, \pm \dfrac{1}{2}, \pm \dfrac{3}{2}, \pm \dfrac{1}{3}, \pm \dfrac{1}{4}, \pm \dfrac{3}{4}, \pm \dfrac{1}{6}, \pm \dfrac{1}{12}$

18. $f(x) = x^3 - 3x^2 - 6x + 8$

Possible rational zeros:

$p = \pm 1, \pm 2, \pm 4, \pm 8; \quad q = \pm 1;$

$\dfrac{p}{q} = \pm 1, \pm 2, \pm 4, \pm 8$

Using synthetic division:

We try $x + 2$:

$$
\begin{array}{r|rrrr}
-2 & 1 & -3 & -6 & 8 \\
   &   & -2 & 10 & -8 \\
\hline
   & 1 & -5 & 4 & 0 \\
\end{array}
$$

$x + 2$ is a factor. The other factor is the quotient: $x^2 - 5x + 4$.

Thus, $f(x) = (x+2)(x^2 - 5x + 4)$.

$= (x+2)(x-1)(x-4)$

The zeros are –2, 1, and 4, each of multiplicity 1.

19. $f(x) = 4x^3 + 4x^2 - 7x + 2$

Possible rational zeros:

$p = \pm 1, \pm 2; \quad q = \pm 1, \pm 2, \pm 4;$

$\dfrac{p}{q} = \pm 1, \pm 2, \pm \dfrac{1}{2}, \pm \dfrac{1}{4}$

Using synthetic division:

We try $x + 2$:

$$
\begin{array}{r|rrrr}
-2 & 4 & 4 & -7 & 2 \\
   &   & -8 & 8 & -2 \\
\hline
   & 4 & -4 & 1 & 0 \\
\end{array}
$$

$x + 2$ is a factor. The other factor is the quotient: $4x^2 - 4x + 1$.

Thus, $f(x) = (x+2)(4x^2 - 4x + 1)$.

$= (x+2)(2x-1)(2x-1)$

The zeros are –2, of multiplicity 1 and $\dfrac{1}{2}$, of multiplicity 2.

20. $f(x) = x^4 - 4x^3 + 9x^2 - 20x + 20$

Possible rational zeros:

$p = \pm 1, \pm 2, \pm 4, \pm 5, \pm 10, \pm 20; \quad q = \pm 1;$

$\dfrac{p}{q} = \pm 1, \pm 2, \pm 4, \pm 5, \pm 10, \pm 20$

Using synthetic division:

We try $x - 2$:

$$
\begin{array}{r|rrrrr}
2 & 1 & -4 & 9 & -20 & 20 \\
  &   & 2 & -4 & 10 & -20 \\
\hline
  & 1 & -2 & 5 & -10 & 0 \\
\end{array}
$$

$x - 2$ is a factor and the quotient is

$x^3 - 2x^2 + 5x - 10 = x^2(x-2) + 5(x-2)$

$= (x-2)(x^2 + 5)$

Thus, $f(x) = (x-2)(x-2)(x^2 + 5)$

$= (x-2)^2(x^2 + 5)$

Since $x^2 + 5 = 0$ has no real solutions, the only zero is 2, of multiplicity 2.

21. $2x^4 + 2x^3 - 11x^2 + x - 6 = 0$

The solutions of the equation are the zeros of
$f(x) = 2x^4 + 2x^3 - 11x^2 + x - 6$.

Possible rational zeros:
$p = \pm 1, \pm 2, \pm 3, \pm 6; \quad q = \pm 1, \pm 2;$

$\dfrac{p}{q} = \pm 1, \pm 2, \pm 3, \pm 6, \pm \dfrac{1}{2}, \pm \dfrac{3}{2}$

Using synthetic division:
We try $x + 3$:

$$
\begin{array}{r|rrrrr}
-3 & 2 & 2 & -11 & 1 & -6 \\
   &   & -6 & 12 & -3 & 6 \\
\hline
   & 2 & -4 & 1 & -2 & 0
\end{array}
$$

$x + 3$ is a factor and the quotient is
$2x^3 - 4x^2 + x - 2 = 2x^2(x - 2) + 1(x - 2)$
$\qquad\qquad = (x - 2)(2x^2 + 1)$

Thus, $f(x) = (x + 3)(x - 2)(2x^2 + 1)$.

Since $2x^2 + 1 = 0$ has no real solutions, the
solution set is $\{-3, 2\}$.

22. $2x^4 + 7x^3 + x^2 - 7x - 3 = 0$

The solutions of the equation are the zeros of
$f(x) = 2x^4 + 7x^3 + x^2 - 7x - 3$.

Possible rational zeros:
$p = \pm 1, \pm 3; \quad q = \pm 1, \pm 2;$

$\dfrac{p}{q} = \pm 1, \pm 3, \pm \dfrac{1}{2}, \pm \dfrac{3}{2}$

Using synthetic division:
We try $x + 3$:

$$
\begin{array}{r|rrrrr}
-3 & 2 & 7 & 1 & -7 & -3 \\
   &   & -6 & -3 & 6 & 3 \\
\hline
   & 2 & 1 & -2 & -1 & 0
\end{array}
$$

$x + 3$ is a factor and the quotient is
$2x^3 + x^2 - 2x - 1 = x^2(2x + 1) - 1(2x + 1)$
$\qquad\qquad = (2x + 1)(x^2 - 1)$

Thus, $\begin{aligned} f(x) &= (x + 3)(2x + 1)(x^2 - 1) \\ &= (x + 3)(2x + 1)(x - 1)(x + 1) \end{aligned}$

The solution set is $\left\{-3, -1, -\dfrac{1}{2}, 1\right\}$.

23. $f(x) = x^3 - x^2 - 4x + 2$

r	coeff of q(x)			remainder
1	1	0	-4	-2
2	1	1	-2	-2
3	1	2	2	8
-1	1	-2	-2	4
-2	1	-3	2	-2

For r = 3, the last row of synthetic division
contains only numbers that are positive or 0, so
we know there are no zeros greater than 3. For r
= -2, the last row of synthetic division results in
alternating positive (or 0) and negative (or 0)
values, so we know that there are no zeros less
than -2. The upper bound is 3 and the lower
bound is -2.

We graph using the bounds and ZOOM-FIT, and
adjust the viewing window to improve the graph.

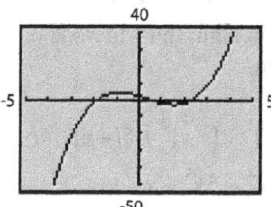

24. $f(x) = 2x^3 - 7x^2 - 10x + 35$

$p = \pm 1, \pm 5, \pm 7, \pm 35; \quad q = \pm 1, \pm 2;$

$\dfrac{p}{q} = \pm 1, \pm 5, \pm 7, \pm 35, \pm \dfrac{1}{2}, \pm \dfrac{5}{2}, \pm \dfrac{7}{2}, \pm \dfrac{35}{2}$

r	coeff of q(x)			remainder
1	2	-5	-15	20
5	2	3	5	60
-1	2	-9	-1	36
-5	2	-17	75	-340

For r = 5, the last row of synthetic division
contains only numbers that are positive or 0, so
we know there are no zeros greater than 5. For r
= -5, the last row of synthetic division results in
alternating positive (or 0) and negative (or 0)
values, so we know that there are no zeros less
than -5. The upper bound is 5 and the lower
bound is -5.

We graph using the bounds and ZOOM-FIT, and
adjust the viewing window to improve the graph.

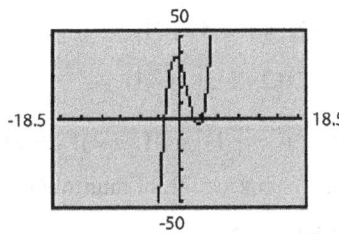

25. $f(x) = 3x^3 - x - 1; \quad [0,1]$

$f(0) = -1 < 0$ and $f(1) = 1 > 0$

The value of the function is positive at one endpoint and negative at the other. Since the function is continuous, the Intermediate Value Theorem guarantees at least one zero in the given interval.

26. $f(x) = 8x^4 - 4x^3 - 2x - 1; \quad [0,1]$

$f(0) = -1 < 0$ and $f(1) = 1 > 0$

The value of the function is positive at one endpoint and negative at the other. Since the function is continuous, the Intermediate Value Theorem guarantees at least one zero in the given interval.

27. $f(x) = x^3 - x - 2$

$f(1) = -2; \quad f(2) = 4$

So by the Intermediate Value Theorem, f has a zero on the interval $[1,2]$.
Subdivide the interval $[1,2]$ into 10 equal subintervals:
$[1,1.1]$; $[1.1,1.2]$; $[1.2,1.3]$; $[1.3,1.4]$; $[1.4,1.5]$; $[1.5,1.6]$; $[1.6,1.7]$; $[1.7,1.8]$; $[1.8,1.9]$; $[1.9,2]$

$f(1) = -2; f(1.1) = -1.769$

$f(1.1) = -1.769; f(1.2) = -1.472$

$f(1.2) = -1.472; f(1.3) = -1.103$

$f(1.3) = -1.103; f(1.4) = -0.656$

$f(1.4) = -0.656; f(1.5) = -0.125$

$f(1.5) = -0.125; f(1.6) = 0.496$

So f has a real zero on the interval $[1.5,1.6]$.

Subdivide the interval $[1.5,1.6]$ into 10 equal subintervals:
$[1.5,1.51]$; $[1.51,1.52]$; $[1.52,1.53]$; $[1.53,1.54]$; $[1.54,1.55]$; $[1.55,1.56]$; $[1.56,1.57]$; $[1.57,1.58]$; $[1.58,1.59]$; $[1.59,1.6]$

$f(1.5) = -0.125; f(1.51) \approx -0.0670$

$f(1.51) \approx -0.0670; f(1.52) \approx -0.0082$

$f(1.52) \approx -0.0082; f(1.53) \approx 0.0516$

So f has a real zero on the interval $[1.52,1.53]$, therefore the zero is 1.52, correct to two decimal places.

28. $f(x) = 8x^4 - 4x^3 - 2x - 1$

$f(0) = -1; \quad f(1) = 1$,

So by the Intermediate Value Theorem, f has a zero on the interval $[0,1]$.
Subdivide the interval $[0,1]$ into 10 equal subintervals:
$[0,0.1]$; $[0.1,0.2]$; $[0.2,0.3]$; $[0.3,0.4]$; $[0.4,0.5]$; $[0.5,0.6]$; $[0.6,0.7]$; $[0.7,0.8]$; $[0.8,0.9]$; $[0.9,1]$

$f(0) = -1; f(0.1) = -1.2032$

$f(0.1) = -1.2032; f(0.2) = -1.4192$

$f(0.2) = -1.4192; f(0.3) = -1.6432$

$f(0.3) = -1.6432; f(0.4) = -1.8512$

$f(0.4) = -1.8512; f(0.5) = -2$

$f(0.5) = -2; f(0.6) = -2.0272$

$f(0.6) = -2.0272; f(0.7) = -1.8512$

$f(0.7) = -1.8512; f(0.8) = -1.3712$

$f(0.8) = -1.3712; f(0.9) = -0.4672$

$f(0.9) = -0.4672; f(1) = 1$

So f has a real zero on the interval $[0.9,1]$.

Subdivide the interval $[0.9,1]$ into 10 equal subintervals:
$[0.9,0.91]$; $[0.91,0.92]$; $[0.92,0.93]$; $[0.93,0.94]$; $[0.94,0.95]$; $[0.95,0.96]$; $[0.96,0.97]$; $[0.97,0.98]$; $[0.98,0.99]$; $[0.99,1]$

$f(0.9) = -0.4672; f(0.91) \approx -0.3483$

$f(0.91) \approx -0.3483; f(0.92) \approx -0.2236$

$f(0.92) \approx -0.2236; f(0.93) \approx -0.0930$

$f(0.93) \approx -0.0930; f(0.94) \approx 0.0437$

So f has a real zero on the interval $[0.93,0.94]$, therefore the zero is 0.93, correct to two decimal places.

29. Since complex zeros appear in conjugate pairs, $4-i$, the conjugate of $4+i$, is the remaining zero of f.

$$f(x) = (x-6)(x-4-i)(x-4+i)$$
$$= x^3 - 14x^2 + 65x - 102$$

30. Since complex zeros appear in conjugate pairs, $-i$, the conjugate of i, and $1-i$, the conjugate of $1+i$, are the remaining zeros of f.

$$f(x) = (x-i)(x+i)(x-1-i)(x-1+i)$$
$$= x^4 - 2x^3 + 3x^2 - 2x + 2$$

31. $f(x) = x^3 - 3x^2 - 6x + 8$.

Possible rational zeros:
$$p = \pm1, \pm2, \pm4, \pm8; \quad q = \pm1;$$
$$\frac{p}{q} = \pm1, \pm2, \pm4, \pm8$$

Using synthetic division:
We try $x-1$:

```
1)1  -3  -6   8
      1  -2  -8
   1  -2  -8   0
```

$x-1$ is a factor and the quotient is $x^2 - 2x - 8$
Thus,
$$f(x) = (x-1)(x^2 - 2x - 8) = (x-1)(x-4)(x+2).$$

The complex zeros are 1, 4, and –2, each of multiplicity 1.

32. $f(x) = 4x^3 + 4x^2 - 7x + 2$.

Possible rational zeros:
$$p = \pm1, \pm2; \quad q = \pm1, \pm2, \pm4;$$
$$\frac{p}{q} = \pm1, \pm\frac{1}{2}, \pm\frac{1}{4}, \pm2$$

Using synthetic division:
We try $x+2$:

```
-2)4   4  -7   2
      -8   8  -2
   4  -4   1   0
```

$x+2$ is a factor and the quotient is $4x^2 - 4x + 1$.
Thus,

$$f(x) = (x+2)(4x^2 - 4x + 1)$$
$$= (x+2)(2x-1)(2x-1)$$
$$= (x+2)(2x-1)^2 = 4(x+2)\left(x+\frac{1}{2}\right)^2$$

The complex zeros are –2, of multiplicity 1, and $\frac{1}{2}$, of multiplicity 2.

33. $f(x) = x^4 - 4x^3 + 9x^2 - 20x + 20$.

Possible rational zeros:
$$p = \pm1, \pm2, \pm4, \pm5, \pm10, \pm20; \quad q = \pm1;$$
$$\frac{p}{q} = \pm1, \pm2, \pm4, \pm5, \pm10, \pm20$$

Using synthetic division:
We try $x-2$:

```
2)1  -4   9  -20   20
      2  -4   10  -20
   1  -2   5  -10    0
```

$x-2$ is a factor and the quotient is $x^3 - 2x^2 + 5x - 10$.

Thus, $f(x) = (x-2)(x^3 - 2x^2 + 5x - 10)$.

We can factor $x^3 - 2x^2 + 5x - 10$ by grouping.
$$x^3 - 2x^2 + 5x - 10 = x^2(x-2) + 5(x-2)$$
$$= (x-2)(x^2 + 5)$$
$$= (x-2)(x+\sqrt{5}i)(x-\sqrt{5}i)$$
$$f(x) = (x-2)^2(x+\sqrt{5}i)(x-\sqrt{5}i)$$

The complex zeros are 2, of multiplicity 2, and $\sqrt{5}i$ and $-\sqrt{5}i$, each of multiplicity 1.

34. $f(x) = 2x^4 + 2x^3 - 11x^2 + x - 6$.

Possible rational zeros:
$$p = \pm1, \pm2, \pm3, \pm6; \quad q = \pm1, \pm2;$$
$$\frac{p}{q} = \pm1, \pm\frac{1}{2}, \pm2, \pm3, \pm\frac{3}{2}, \pm6$$

Using synthetic division:
We try $x-2$:

```
2)2   2  -11   1  -6
      4   12   2   6
   2   6    1   3   0
```

$x-2$ is a factor and the quotient is $2x^3 + 6x^2 + x + 3$.

Thus, $f(x) = (x-2)(2x^3 + 6x^2 + x + 3)$.

We can factor $2x^3 + 6x^2 + x + 3$ by grouping.

$2x^3 + 6x^2 + x + 3 = 2x^2(x+3) + (x+3)$

$= (x+3)(2x^2 + 1)$

$= (x+3)(\sqrt{2}x + i)(\sqrt{2}x - i)$

$f(x) = (x-2)(x+3)(\sqrt{2}x + i)(\sqrt{2}x - i)$

$= 2(x-2)(x+3)\left(x + \frac{\sqrt{2}}{2}i\right)\left(x - \frac{\sqrt{2}}{2}i\right)$

The complex zeros are $2, -3, -\frac{\sqrt{2}}{2}i$, and $\frac{\sqrt{2}}{2}i$,
each of multiplicity 1.

35. $R(x) = \frac{x+2}{x^2 - 9} = \frac{x+2}{(x+3)(x-3)}$ is in lowest terms.

The denominator has zeros at -3 and 3. Thus,
the domain is $\{x \mid x \neq -3, x \neq 3\}$. The degree of
the numerator, $p(x) = x + 2$, is $n = 1$. The

degree of the denominator
$q(x) = x^2 - 9$, is $m = 2$. Since $n < m$, the line
$y = 0$ is a horizontal asymptote. Since the
denominator is zero at -3 and 3, $x = -3$ and $x = 3$
are vertical asymptotes.

36. $R(x) = \frac{x^2 + 3x + 2}{(x+2)^2} = \frac{(x+2)(x+1)}{(x+2)^2} = \frac{x+1}{x+2}$ is in

lowest terms. The denominator has a zero at -2.
Thus, the domain is $\{x \mid x \neq -2\}$. The degree of

the numerator, $p(x) = x^2 + 3x + 2$, is $n = 2$. The
degree of the denominator,

$q(x) = (x+2)^2 = x^2 + 4x + 4$, is $m = 2$. Since

$n = m$, the line $y = \frac{1}{1} = 1$ is a horizontal

asymptote. Since the denominator of $y = \frac{x+1}{x+2}$

is zero at -2, $x = -2$ is a vertical asymptote.

37. $R(x) = \frac{2x-6}{x}$ $\quad p(x) = 2x - 6;\ q(x) = x;\ n = 1;\ m = 1$

Step 1: Domain: $\{x \mid x \neq 0\}$

There is no y-intercept because 0 is not in the domain.

Step 2: $R(x) = \frac{2x-6}{x} = \frac{2(x-3)}{x}$ is in lowest terms.

Step 3: The x-intercept is the zero of $p(x)$: 3

Near 3: $R(x) \approx \frac{2}{3}(x-3)$. Plot the point $(3,0)$ and show a line with positive slope there.

Step 4: $R(x) = \frac{2x-6}{x} = \frac{2(x-3)}{x}$ is in lowest terms. The vertical asymptote is the zero of $q(x)$: $x = 0$.

Graph this asymptote using a dashed line.

Step 5: Since $n = m$, the line $y = \frac{2}{1} = 2$ is the horizontal asymptote. Solve to find intersection points:

$\frac{2x-6}{x} = 2$

$2x - 6 = 2x$

$-6 \neq 0$

$R(x)$ does not intersect $y = 2$. Plot the line $y = 2$ with dashes.

Steps 6 & 7: Graphing:

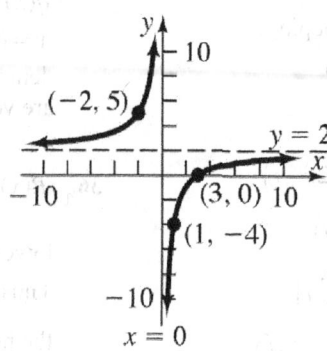

38. $H(x) = \dfrac{x+2}{x(x-2)}$ $p(x) = x+2$; $q(x) = x(x-2) = x^2 - 2x$; $n = 1$; $m = 2$

Step 1:   Domain: $\{x \mid x \neq 0, x \neq 2\}$.

Step 2: $H(x) = \dfrac{x+2}{x(x-2)}$ is in lowest terms.

Step 3:   There is no y-intercept because 0 is not in the domain.

The x-intercept is the zero of $p(x)$: -2

Near -2: $H(x) \approx \dfrac{1}{8}(x+2)$. Plot the point $(-2, 0)$ and show a line with positive slope there.

Step 4: $H(x) = \dfrac{x+2}{x(x-2)}$ is in lowest terms. The vertical asymptotes are the zeros of $q(x)$: $x = 0$ and $x = 2$.

Graph these asymptotes using dashed lines.

Step 5:   Since $n < m$, the line $y = 0$ is the horizontal asymptote. Solve to find intersection points:

$$\frac{x+2}{x(x-2)} = 0$$
$$x+2 = 0$$
$$x = -2$$

$H(x)$ intersects $y = 0$ at $(-2, 0)$. Plot the line $y = 0$ using dashes.

Steps 6 & 7: Graphing:

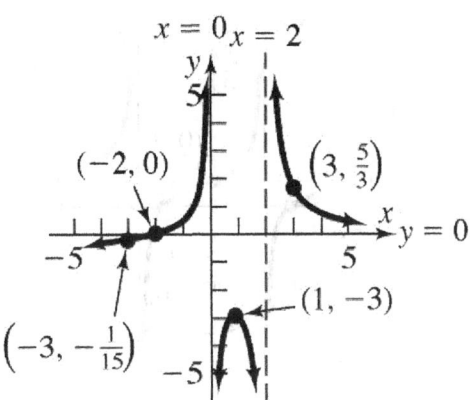

39. $R(x) = \dfrac{x^2 + x - 6}{x^2 - x - 6} = \dfrac{(x+3)(x-2)}{(x-3)(x+2)}$ $p(x) = x^2 + x - 6;\ q(x) = x^2 - x - 6;$

Step 1:   Domain: $\{x \mid x \neq -2,\ x \neq 3\}$.

Step 2: $R(x) = \dfrac{x^2 + x - 6}{x^2 - x - 6}$ is in lowest terms.

Step 3:   The y-intercept is $R(0) = \dfrac{0^2 + 0 - 6}{0^2 - 0 - 6} = \dfrac{-6}{-6} = 1$. Plot the point $(0,1)$.

The x-intercepts are the zeros of $p(x)$: -3 and 2.

Near -3: $R(x) \approx -\dfrac{5}{6}(x+3)$. Plot the point $(-3,0)$ and show a line with negative slope there.

Near 2: $R(x) \approx -\dfrac{5}{4}(x-2)$. Plot the point $(2,0)$ and show a line with negative slope there.

Step 4: $R(x) = \dfrac{x^2 + x - 6}{x^2 - x - 6}$ is in lowest terms. The vertical asymptotes are the zeros of $q(x)$:

$x = -2$ and $x = 3$. Graph these asymptotes with dashed lines.

Step 5:   Since $n = m$, the line $y = \dfrac{1}{1} = 1$ is the horizontal asymptote. Solve to find intersection points:

$$\dfrac{x^2 + x - 6}{x^2 - x - 6} = 1$$
$$x^2 + x - 6 = x^2 - x - 6$$
$$2x = 0$$
$$x = 0$$

$R(x)$ intersects $y = 1$ at $(0, 1)$. Plot the line $y = 1$ using dashes.

Steps 6 & 7: Graphing:

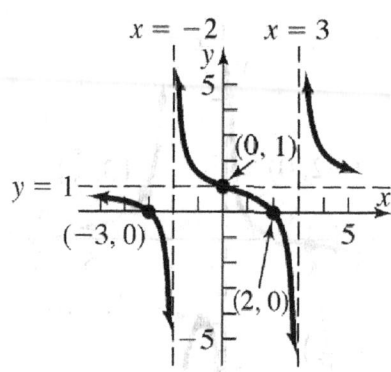

40. $F(x) = \dfrac{x^3}{x^2-4} = \dfrac{x^3}{(x+2)(x-2)}$ $p(x)=x^3$; $q(x)=x^2-4$; $n=3$; $m=2$

Step 1:   Domain: $\{x \mid x \neq -2, x \neq 2\}$.

Step 2: $F(x) = \dfrac{x^3}{x^2-4}$ is in lowest terms.

Step 3:   The y-intercept is $F(0) = \dfrac{0^3}{0^2-4} = \dfrac{0}{-4} = 0$. Plot the point $(0,0)$.

The x-intercept is the zero of $p(x)$: 0.

Near 0: $F(x) \approx -\dfrac{1}{4}x^3$. Plot the point $(0,0)$ and indicate a cubic function there (left tail up and right tail down).

Step 4: $F(x) = \dfrac{x^3}{x^2-4}$ is in lowest terms. The vertical asymptotes are the zeros of $q(x)$: $x=-2$ and $x=2$.
Graph these asymptotes using dashed lines.

Step 5:   Since $n=m+1$, there is an oblique asymptote. Dividing:

$$x^2-4 \overline{\smash{\big)}\,x^3 } \qquad \dfrac{x^3}{x^2-4} = x + \dfrac{4x}{x^2-4}$$

$$\begin{array}{r} x \\ x^2-4 \overline{\smash{\big)}\,x^3 } \\ \underline{x^3 -4x} \\ 4x \end{array}$$

The oblique asymptote is $y=x$. Solve to find intersection points:

$$\dfrac{x^3}{x^2-4} = x$$
$$x^3 = x^3 - 4x$$
$$4x = 0$$
$$x = 0$$

$F(x)$ intersects $y=x$ at $(0,0)$. Plot the line $y=x$ using dashed lines.

Steps 6 & 7: Graphing:

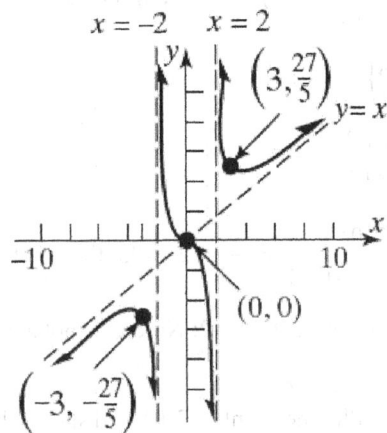

41. $R(x) = \dfrac{2x^4}{(x-1)^2}$ $p(x) = 2x^4$; $q(x) = (x-1)^2$; $n = 4$; $m = 2$

Step 1:   Domain: $\{x \mid x \neq 1\}$.

Step 2: $R(x) = \dfrac{2x^4}{(x-1)^2}$ is in lowest terms.

Step 3:   The y-intercept is $R(0) = \dfrac{2(0)^4}{(0-1)^2} = \dfrac{0}{1} = 0$. Plot the point $(0,0)$.

The x-intercept is the zero of $p(x)$: 0.

Near 0: $R(x) \approx 2x^4$. Plot the point $(0,0)$ and show the graph of a quartic opening up there.

Step 4: $R(x) = \dfrac{2x^4}{(x-1)^2}$ is in lowest terms. The vertical asymptote is the zero of $q(x)$: $x = 1$.

Graph this asymptote using a dashed line.

Step 5:   Since $n > m+1$, there is no horizontal asymptote and no oblique asymptote.

Steps 6 & 7: Graphing:

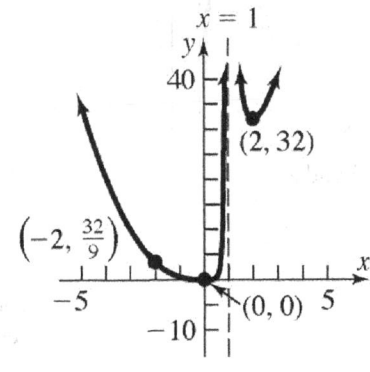

317

42. $G(x) = \dfrac{x^2-4}{x^2-x-2} = \dfrac{(x+2)(x-2)}{(x-2)(x+1)} = \dfrac{x+2}{x+1}$ $p(x) = x^2-4;\ q(x) = x^2-x-2;$

Step 1:   Domain: $\{x \mid x \neq -1, x \neq 2\}$.

Step 2:   In lowest terms, $G(x) = \dfrac{x+2}{x+1},\ x \neq 2$.

Step 3:   The y-intercept is $G(0) = \dfrac{0^2-4}{0^2-0-2} = \dfrac{-4}{-2} = 2$. Plot the point $(0,2)$.

The x-intercept is the zero of $y = x+2$: -2; Note: 2 is not a zero because reduced form must be used to find the zeros.

Near -2: $G(x) \approx -x-2$. Plot the point $(-2,0)$ and show a line with negative slope there.

Step 4:   In lowest terms, $G(x) = \dfrac{x+2}{x+1},\ x \neq 2$. The vertical asymptote is the zero of $f(x) = x+1$: $x = -1$;
Graph this asymptote using a dashed line. Note: $x = 2$ is not a vertical asymptote because reduced form must be used to find the asymptotes. The graph has a hole at $\left(2, \dfrac{4}{3}\right)$.

Step 5:   Since $n = m$, the line $y = \dfrac{1}{1} = 1$ is the horizontal asymptote. Solve to find intersection points:

$$\frac{x^2-4}{x^2-x-2} = 1$$
$$x^2-4 = x^2-x-2$$
$$x = 2$$

$G(x)$ does not intersect $y = 1$ because $G(x)$ is not defined at $x = 2$. Plot the line $y = 1$ using dashes.

Steps 6 & 7:   Graphing:

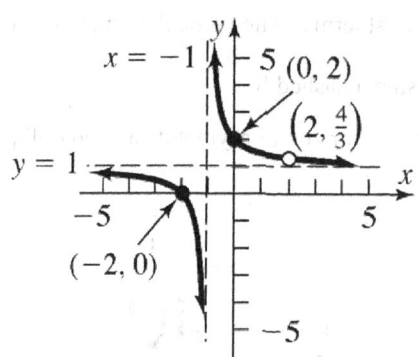

43. The x-intercepts of the graph of f are -3 and 2.

a. $f(x) = 0$ for $x = -3$ and 2.

b.   The graph of f is above the x-axis (so f is positive) for $-3 < x < 2$ or $x > 2$.

Therefore, the solution set is
$\{x \mid -3 < x < 2 \text{ or } x > 2\}$ or, using interval notation, $(-3, 2) \cup (2, \infty)$.

c.   The graph of f is below the x-axis (so f is negative) for $x < -3$. Since the inequality is not strict, we include -3 and 2 in the solution set. Therefore, the solution set is $\{x \mid x \le -3 \text{ or } x = 2\}$ or, using interval notation $(-\infty, -3] \cup \{2\}$.

d. The graph crosses the x-axis at $x = -3$ and touches at $x = 2$. Thus, -3 has odd multiplicity while 2 has an even multiplicity. Using one for the odd multiplicity and two for the even multiplicity, a possible function is $f(x) = a(x-2)^2(x+3)$. Since the y-intercept is 12, we know $f(0) = 12$. Thus, $a=1$. Using $a=1$, the function is $f(x) = (x-2)^2(x+3)$.

44. The x-intercepts of the graph of f are -3 and -1.

a. The horizontal asymptote is $y = 0.25$.

b. The vertical asymptotes are $x = -2$ and $x = 2$

c. The graph of f is below the x-axis (so f is negative) for $-3 < x < -2$ or $-1 < x < 2$. Therefore, the solution set is $\{x \mid -3 < x < -2 \text{ or } -1 < x < 2\}$ or, using interval notation, $(-3, -2) \cup (-1, 2)$.

d. The graph of f is above the x-axis (so f is positive) for $-\infty < x < -3$ or $-2 < x < -1$ or $x > 2$. Since the inequality is not strict, we include -3 and -1 in the solution set. Therefore, the solution set is $\{x \mid -\infty < x < -3 \text{ or } -2 < x < -1 \text{ or } x > 2\}$ or, using interval notation $(-\infty, -3] \cup (-2, -1] \cup (2, \infty)$.

e. One possibility: $R(x) = a\dfrac{(x+3)(x+1)}{(x+2)(x-2)}$

(Using the point $\left(0, -\dfrac{3}{16}\right)$ leads to $a = 1/4$.) Thus,

$R(x) = \dfrac{(x+3)(x+1)}{4(x+2)(x-2)} = \dfrac{x^2+4x+3}{4x^2-16}$.

45.
$$x^3 + x^2 < 4x + 4$$
$$x^3 + x^2 - 4x - 4 < 0$$
$$x^2(x+1) - 4(x+1) < 0$$
$$(x^2-4)(x+1) < 0$$
$$(x-2)(x+2)(x+1) < 0$$
$$f(x) = (x-2)(x+2)(x+1)$$
$x = -2$, $x = -1$, and $x = 2$ are the zeros of f.

Interval	$(-\infty, -2)$	$(-2, -1)$	$(-1, 2)$	$(2, \infty)$
Number Chosen	-3	$-3/2$	0	3
Value of f	-10	0.875	-4	20
Conclusion	Negative	Positive	Negative	Positive

The solution set is $\{x \mid x < -2 \text{ or } -1 < x < 2\}$, or, using interval notation, $(-\infty, -2) \cup (-1, 2)$.

46.
$$x^3 + 4x^2 \ge x + 4$$
$$x^3 + 4x^2 - x - 4 \ge 0$$
$$x^2(x+4) - 1(x+4) \ge 0$$
$$(x^2-1)(x+4) \ge 0$$
$$(x-1)(x+1)(x+4) \ge 0$$
$$f(x) = (x-1)(x+1)(x+4)$$
$x = -4$, $x = -1$, and $x = 1$ are the zeros of f.

Interval	$(-\infty, -4)$	$(-4, -1)$	$(-1, 1)$	$(1, \infty)$
Number Chosen	-5	-2	0	2
Value of f	-24	6	-4	18
Conclusion	Negative	Positive	Negative	Positive

The solution set is $\{x \mid -4 \le x \le -1 \text{ or } x \ge 1\}$, or, using interval notation, $[-4, -1] \cup [1, \infty)$.

47. $\dfrac{2x-6}{1-x} < 2$

$$\dfrac{2x-6}{1-x} - 2 < 0$$

$$\dfrac{2x-6-2(1-x)}{1-x} < 0$$

$$\dfrac{4x-8}{1-x} < 0$$

$$f(x) = \dfrac{4(x-2)}{1-x}$$

The zeros and values where the expression is undefined are $x = 1$, and $x = 2$.

Interval	$(-\infty, 1)$	$(1, 2)$	$(2, \infty)$
Number Chosen	0	1.5	3
Value of f	-8	4	-2
Conclusion	Negative	Positive	Negative

The solution set is $\{x \mid x < 1 \text{ or } x > 2\}$, or, using interval notation, $(-\infty, 1) \cup (2, \infty)$.

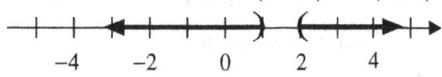

48. $\dfrac{(x-2)(x-1)}{x-3} \geq 0$

$$f(x) = \dfrac{(x-2)(x-1)}{x-3}$$

The zeros and values where the expression is undefined are $x = 1$, $x = 2$, and $x = 3$.

Interval	$(-\infty, 1)$	$(1, 2)$	$(2, 3)$	$(3, \infty)$
Number Chosen	0	1.5	2.5	4
Value of f	$-\dfrac{2}{3}$	$\dfrac{1}{6}$	$-\dfrac{3}{2}$	6
Conclusion	Negative	Positive	Negative	Positive

The solution set is $\{x \mid 1 \leq x \leq 2 \text{ or } x > 3\}$, or, using interval notation, $[1, 2] \cup (3, \infty)$.

49. $\dfrac{x^2 - 8x + 12}{x^2 - 16} > 0$

$$f(x) = \dfrac{x^2 - 8x + 12}{x^2 - 16}$$

$$\dfrac{(x-2)(x-6)}{(x+4)(x-4)} > 0$$

The zeros and values where the expression is undefined are $x = -4$, $x = 2$, $x = 4$, and $x = 6$.

Interval	Number Chosen	Value of f	Conclusion
$(-\infty, -4)$	-5	$\dfrac{77}{9}$	Positive
$(-4, 2)$	0	$-\dfrac{3}{4}$	Negative
$(2, 4)$	3	$\dfrac{3}{7}$	Positive
$(4, 6)$	5	$-\dfrac{1}{3}$	Negative
$(6, \infty)$	7	$\dfrac{5}{33}$	Positive

The solution set is
$\{x \mid x < -4 \text{ or } 2 < x < 4 \text{ or } x > 6\}$, or, using interval notation, $(-\infty, -4) \cup (2, 4) \cup (6, \infty)$.

50. a. $250 = \pi r^2 h \;\Rightarrow\; h = \dfrac{250}{\pi r^2}$;

$$A(r) = 2\pi r^2 + 2\pi r h = 2\pi r^2 + 2\pi r \left(\dfrac{250}{\pi r^2}\right)$$

$$= 2\pi r^2 + \dfrac{500}{r}$$

b. $A(3) = 2\pi \cdot 3^2 + \dfrac{500}{3}$

$$= 18\pi + \dfrac{500}{3} \approx 223.22 \text{ square cm}$$

c. $A(5) = 2\pi \cdot 5^2 + \dfrac{500}{5}$

$$= 50\pi + 100 \approx 257.08 \text{ square cm}$$

320

d. Use MINIMUM on the graph of

$$y_1 = 2\pi x^2 + \frac{500}{x}$$

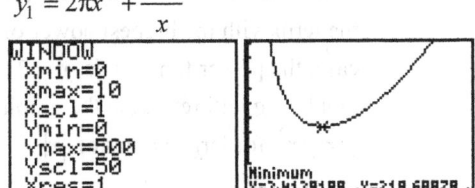

The area is smallest when the radius is approximately 3.41 cm.

51. a.   **c.**

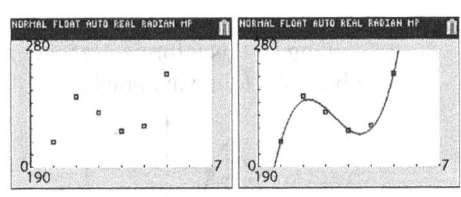

b. $P(t) = 4.4926t^3 - 45.5294t^2 + 136.1209t$
$+ 115.4667$

$P(9) = 4.4926(9)^3 - 45.5294(9)^2$
$+ 136.1209(9) + 115.4667 \approx 928$

The predicted new home price for January 2020 is approximately \$928,000.

52. a. Since the graph points upward for large values of $|x|$, the degree is even.

b. Likewise, the leading coefficient is positive.

c. The graph is symmetric about the y-axis, so the function is even.

d. The graph touches the x-axis at $x = 0$, which means 0 is a zero of the function with even multiplicity. Thus, x^2 must be a factor.

e. The graph has 7 turning points, so the degree of the polynomial must be at least 8.

f. Answers will vary.

Chapter 5 Test

1. $f(x) = (x-3)^4 - 2$

Using the graph of $y = x^4$, shift right 3 units, then shift down 2 units.

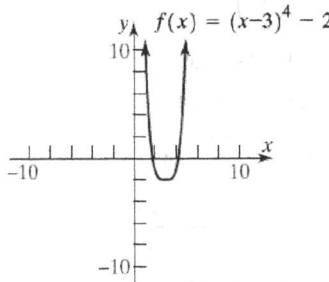

2. a. The maximum number of real zeros is the degree, $n = 3$.

b. $g(x) = 2x^3 + 5x^2 - 28x - 15$

We list all integers p that are factors of $a_0 = -15$ and all the integers q that are factors of $a_3 = 2$.

$p : \pm 1, \pm 3, \pm 5, \pm 15$

$q : \pm 1, \pm 2$

Now we form all possible ratios $\frac{p}{q}$:

$\frac{p}{q} : \pm \frac{1}{2}, \pm 1, \pm \frac{3}{2}, \pm \frac{5}{2}, \pm 3, \pm 5, \pm \frac{15}{2}, \pm 15$

If g has a rational zero, it must be one of the 16 possibilities listed.

c. We can find the rational zeros by using the fact that if r is a zero of g, then $g(r) = 0$.

That is, we evaluate the function for different values from our list of rational zeros. If we get $g(r) = 0$, we have a zero.

Then we use long division to reduce the polynomial and start again on the reduced polynomial.

We will start with the positive integers:

$g(1) = 2(1)^3 + 5(1)^2 - 28(1) - 15$,
$= 2 + 5 - 28 - 15$
$= -36$

$g(3) = 2(3)^3 + 5(3)^2 - 28(3) - 15$
$= 54 + 45 - 84 - 15$
$= 0$

So, we know that 3 is a zero. This means

that $(x-3)$ must be a factor of g. Using long division we get

$$x-3 \overline{)2x^3+5x^2-28x-15}$$
$$\underline{-(2x^3-6x^2)}$$
$$11x^2-28x$$
$$\underline{-(11x^2-33x)}$$
$$5x-15$$
$$\underline{-(5x-15)}$$
$$0$$

with quotient $2x^2+11x+5$.

Thus, we can now write

$$g(x)=(x-3)(2x^2+11x+5)$$

The quadratic factor can be factored so we get:

$$g(x)=(x-3)(2x+1)(x+5)$$

To find the remaining zeros of g, we set the last two factors equal to 0 and solve.

$$2x+1=0 \qquad x+5=0$$
$$2x=-1 \qquad x=-5$$
$$x=-\frac{1}{2}$$

Therefore, the zeros are -5, $-\frac{1}{2}$, and 3.

Notice how these rational zeros were all in the list of potential rational zeros.

d. The x-intercepts of a graph are the same as the zeros of the function. In the previous part, we found the zeros to be -5, $-\frac{1}{2}$, and 3. Therefore, the x-intercepts are -5, $-\frac{1}{2}$, and 3.

To find the y-intercept, we simply find $g(0)$.

$$g(0)=2(0)^3+5(0)^2-28(0)-15=-15$$

So, the y-intercept is -15.

e. Whether the graph crosses or touches at an x-intercept is determined by the multiplicity. Each factor of the polynomial occurs once, so the multiplicity of each zero is 1. For odd multiplicity, the graph will cross the x-axis at the zero. Thus, the graph crosses the x-axis at each of the three x-intercepts.

f. The power function that the graph of g resembles for large values of $|x|$ is given by the term with the highest power of x. In this case, the power function is $y=2x^3$. So, the graph of g will resemble the graph of $y=2x^3$ for large values of $|x|$.

g. 2 turning points;
local maximum: $(-3.15, 60.30)$;
local minima: $(1.48, -39.00)$

h. We could first evaluate the function at several values for x to help determine the scale.
Putting all this information together, we obtain the following graph:

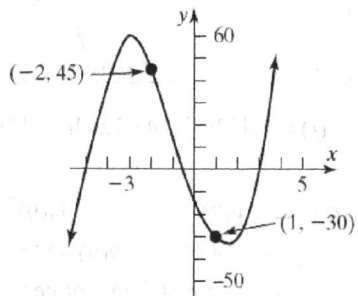

3. $x^3-4x^2+25x-100=0$
$$x^2(x-4)+25(x-4)=0$$
$$(x-4)(x^2+25)=0$$
$$x-4=0 \quad \text{or} \quad x^2+25=0$$
$$x=4 \qquad x^2=-25$$
$$x=\pm\sqrt{-25}$$
$$x=\pm 5i$$
The solution set is $\{4, -5i, 5i\}$.

4. $\qquad 3x^3+2x-1=8x^2-4$
$$3x^3-8x^2+2x+3=0$$
If we let the left side of the equation be $f(x)$, then we are simply finding the zeros of f.
We list all integers p that are factors of $a_0=3$ and all the integers q that are factors of $a_3=3$.
$$p:\pm 1, \pm 3; \quad q:\pm 1, \pm 3$$

Now we form all possible ratios $\frac{p}{q}$:

$\dfrac{p}{q}: \pm\dfrac{1}{3}, \pm 1, \pm 3$

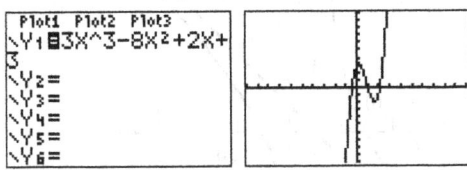

It appears that there is a zero near $x = 1$.

$f(1) = 3(1)^3 - 8(1)^2 + 2(1) + 3 = 0$

Therefore, x=1 is a zero and $(x-1)$ is a factor of $f(x)$. We can reduce the polynomial expression by using synthetic division.

$$\begin{array}{r|rrrr} 1 & 3 & -8 & 2 & 3 \\ & & 3 & -5 & -3 \\ \hline & 3 & -5 & -3 & 0 \end{array}$$

Thus, $f(x) = (x-1)\left(3x^2 - 5x - 3\right)$. We can find the remaining zeros by using the quadratic formula.

$3x^2 - 5x - 3 = 0$

$a = 3, b = -5, c = -3$

$x = \dfrac{-(-5) \pm \sqrt{(-5)^2 - 4(3)(-3)}}{2(3)}$

$= \dfrac{5 \pm \sqrt{25 + 36}}{6} = \dfrac{5 \pm \sqrt{61}}{6}$

Thus, the solution set is $\left\{1, \dfrac{5 - \sqrt{61}}{6}, \dfrac{5 + \sqrt{61}}{6}\right\}$.

5. We start by factoring the numerator and denominator.

$g(x) = \dfrac{2x^2 - 14x + 24}{x^2 + 6x - 40} = \dfrac{2(x-3)(x-4)}{(x+10)(x-4)}$

The domain of f is $\{x \mid x \neq -10, x \neq 4\}$.

In lowest terms, $g(x) = \dfrac{2(x-3)}{x+10}$ with $x \neq 4$.

The graph has one vertical asymptote, $x = -10$, since $x + 10$ is the only factor of the denominator of g in lowest terms. The graph is still undefined at $x = 4$, but there is a hole in the graph there instead of an asymptote.

Since the degree of the numerator is the same as the degree of the denominator, the graph has a horizontal asymptote equal to the quotient of the leading coefficients. The leading coefficient in the numerator is 2 and the leading coefficient in the denominator is 1. Therefore, the graph has the horizontal asymptote $y = \dfrac{2}{1} = 2$.

6. $r(x) = \dfrac{x^2 + 2x - 3}{x + 1}$

Start by factoring the numerator.

$r(x) = \dfrac{(x+3)(x-1)}{x+1}$

The domain of the function is $\{x \mid x \neq -1\}$.

Asymptotes:
Since the function is in lowest terms, the graph has one vertical asymptote, $x = -1$.
The degree of the numerator is one more than the degree of the denominator so the graph will have an oblique asymptote. To find it, we need to use long division (note: we could also use synthetic division in this case because the dividend is linear).

$$\begin{array}{r} x+1 \\ x+1\overline{\smash{)}\,x^2 + 2x - 3} \\ \underline{-\left(x^2 + x\right)} \\ x - 3 \\ \underline{-(x+1)} \\ -4 \end{array}$$

The oblique asymptote is $y = x + 1$.

7. From problem 6 we know that the domain is $\{x \mid x \neq -1\}$ and that the graph has one vertical asymptote, $x = -1$, and one oblique asymptote, $y = x + 1$.

x-intercepts:
To find the x-intercepts, we need to set the numerator equal to 0 and solve the resulting equation.

$(x+3)(x-1) = 0$

$x + 3 = 0 \quad$ or $\quad x - 1 = 0$

$x = -3 \qquad\qquad x = 1$

The x-intercepts are -3 and 1.

The points $(-3, 0)$ and $(1, 0)$ are on the graph.

y-intercept:

$r(0) = \dfrac{0^2 + 2(0) - 3}{0 + 1} = -3$

The y-intercept is -3. The point $(0,-3)$ is on the graph.

Test for symmetry:

$$r(-x) = \frac{(-x)^2 + 2(-x) - 3}{(-x) + 1} = \frac{x^2 - 2x - 3}{-x + 1}$$

Since $r(-x) \neq r(x)$, the graph is not symmetric with respect to the y-axis.
Since $r(-x) \neq -r(x)$, the graph is not symmetric with respect to the origin.

Behavior near the asymptotes:
To determine if the graph crosses the oblique asymptote, we solve the equation

$$r(x) = x + 1$$

$$\frac{x^2 + 2x - 3}{x + 1} = x + 1, \quad x \neq -1$$

$$x^2 + 2x - 3 = x^2 + 2x + 1$$

$$-3 = 1 \quad \text{false}$$

The result is a contradiction so the graph does not cross the oblique asymptote.

The zeros of the numerator and denominator, -3, -1, and 1, divide the x-axis into four subintervals.

$$(-\infty, -3), (-3, -1), (-1, 1), (1, \infty)$$

We can check a point in each subinterval to determine if the graph is above or below the x-axis.

Interval	$(-\infty,-3)$	$(-3,-1)$	$(-1,1)$	$(1,\infty)$
Number	-5	-2	0	3
Value of r	-3	3	-3	3
Location	below	above	below	above
Point	$(-5,-3)$	$(-2,3)$	$(0,-3)$	$(3,3)$

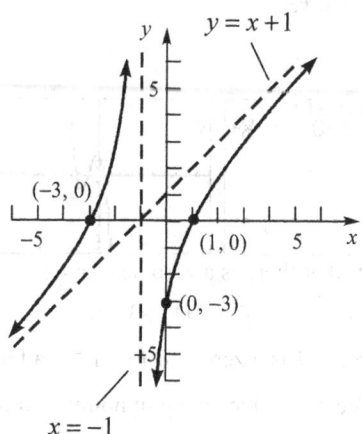

8. Since the polynomial has real coefficients, we can apply the Conjugate Pairs Theorem to find the remaining zero. If $3+i$ is a zero, then its conjugate, $3-i$, must also be a zero. Thus, the four zeros are $-2, 0, 3-i$, and $3+i$. The Factor Theorem says that if $f(c) = 0$, then $(x-c)$ is a factor of the polynomial. This allows us to write the following function:

$$f(x) = a(x-(-2))(x-0)(x-(3-i))(x-(3+i))$$

where a is any real number. If we let $a = 1$, we get

$$f(x) = (x+2)(x)(x-3+i)(x-3-i)$$

$$= (x^2 + 2x)(x-3+i)(x-3-i)$$

$$= (x^2 + 2x)(x^2 - 6x + 10)$$

$$= x^4 - 6x^3 + 10x^2 + 2x^3 - 12x^2 + 20x$$

$$= x^4 - 4x^3 - 2x^2 + 20x$$

9. Since the domain excludes 4 and 9, the denominator must contain the factors $(x-4)$ and $(x-9)$. However, because there is only one vertical asymptote, $x = 4$, the numerator must also contain the factor $(x-9)$.

The horizontal asymptote, $y = 2$, indicates that the degree of the numerator must be the same as the degree of the denominator and that the ratio of the leading coefficients needs to be 2. We can accomplish this by including another factor in the numerator, $(x-a)$, where $a \neq 4$, along with a factor of 2.

Therefore, we have $r(x) = \dfrac{2(x-9)(x-a)}{(x-4)(x-9)}$.

If we let $a = 1$, we get

$$r(x) = \frac{2(x-9)(x-1)}{(x-4)(x-9)} = \frac{2x^2 - 20x + 18}{x^2 - 13x + 36}.$$

10. Since we have a polynomial function and polynomials are continuous, we simply need to show that $f(a)$ and $f(b)$ have opposite signs (where a and b are the endpoints of the interval).

$$f(0) = -2(0)^2 - 3(0) + 8 = 8$$

$$f(4) = -2(4)^2 - 3(4) + 8 = -36$$

Since $f(0) = 8 > 0$ and $f(4) = -36 < 0$, the Intermediate Value Theorem guarantees that there is at least one real zero between 0 and 4.

11. $\dfrac{x+2}{x-3} < 2$

We note that the domain of the variable consists of all real numbers except 3.

Rearrange the terms so that the right side is 0.

$$\frac{x+2}{x-3} - 2 < 0$$

For $f(x) = \dfrac{x+2}{x-3} - 2$, we find the zeros of f and the values of x at which f is undefined. To do this, we need to write f as a single rational expression.

$$f(x) = \frac{x+2}{x-3} - 2$$

$$= \frac{x+2}{x-3} - 2 \cdot \frac{x-3}{x-3}$$

$$= \frac{x+2-2x+6}{x-3}$$

$$= \frac{-x+8}{x-3}$$

The zero of f is $x = 8$ and f is undefined at $x = 3$. We use these two values to divide the real number line into three subintervals.

Interval	$(-\infty, 3)$	$(3, 8)$	$(8, \infty)$
Num. chosen	0	4	9
Value of f	$-\dfrac{8}{3}$	4	$-\dfrac{1}{6}$
Conclusion	negative	positive	negative

Since we want to know where $f(x)$ is negative, we conclude that values of x for which $x < 3$ or $x > 8$ are solutions. The inequality is strict so

the solution set is $\{x \mid x < 3 \text{ or } x > 8\}$. In interval notation we write $(-\infty, 3)$ or $(8, \infty)$.

Chapter 5 Cumulative Review

1. $P = (1, 3)$, $Q = (-4, 2)$

$$d_{P,Q} = \sqrt{(-4-1)^2 + (2-3)^2}$$

$$= \sqrt{(-5)^2 + (-1)^2} = \sqrt{25+1}$$

$$= \sqrt{26}$$

2. $x^2 \geq x$

$$x^2 - x \geq 0$$

$$x(x-1) \geq 0$$

$$f(x) = x^2 - x$$

$x = 0$, $x = 1$ are the zeros of f.

Interval	$(-\infty, 0)$	$(0, 1)$	$(1, \infty)$
Number Chosen	-1	0.5	2
Value of f	2	-0.25	2
Conclusion	Positive	Negative	Positive

The solution set is $\{x \mid x \leq 0 \text{ or } x \geq 1\}$ or $(-\infty, 0]$ or $[1, \infty)$ in interval notation.

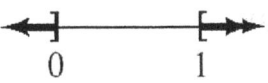

3. $x^2 - 3x < 4$

$$x^2 - 3x - 4 < 0$$

$$(x-4)(x+1) < 0$$

$$f(x) = x^2 - 3x - 4$$

$x = -1$, $x = 4$ are the zeros of f.

Interval	$(-\infty, -1)$	$(-1, 4)$	$(4, \infty)$
Number Chosen	-2	0	5
Value of f	6	-4	6
Conclusion	Positive	Negative	Positive

325

The solution set is $\{x|-1 < x < 4\}$ or $(-1,4)$ in interval notation.

4. Slope -3, Containing the point $(-1, 4)$

   Using the point-slope formula yields:
   $$y - y_1 = m(x - x_1)$$
   $$y - 4 = -3(x - (-1))$$
   $$y - 4 = -3x - 3$$
   $$y = -3x + 1$$
   Thus, $f(x) = -3x + 1$.

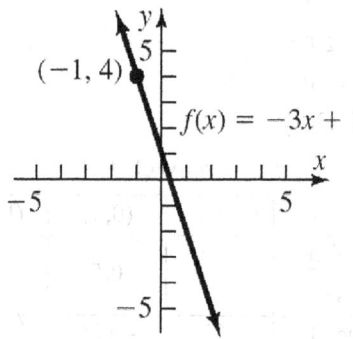

5. Parallel to $y = 2x + 1$; Slope 2, Containing the point $(3, 5)$

   Using the point-slope formula yields:
   $$y - y_1 = m(x - x_1)$$
   $$y - 5 = 2(x - 3)$$
   $$y - 5 = 2x - 6$$
   $$y = 2x - 1$$

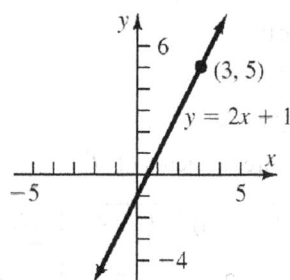

6. $y = x^3$

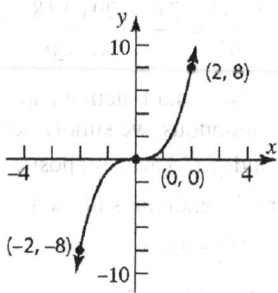

7. This relation is not a function because the ordered pairs (3, 6) and (3, 8) have the same first element, but different second elements.

8. $$x^3 - 6x^2 + 8x = 0$$
   $$x(x^2 - 6x + 8) = 0$$
   $$x(x - 4)(x - 2) = 0$$
   $$x = 0 \text{ or } x = 4 \text{ or } x = 2$$

   The solution set is $\{0, 2, 4\}$.

9. $$3x + 2 \le 5x - 1$$
   $$3 \le 2x$$
   $$\frac{3}{2} \le x$$
   $$x \ge \frac{3}{2}$$

   The solution set is $\left\{x \middle| x \ge \frac{3}{2}\right\}$ or $\left[\frac{3}{2}, \infty\right)$ in interval notation.

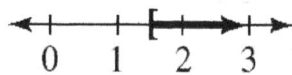

10. $$x^2 + 4x + y^2 - 2y - 4 = 0$$
    $$(x^2 + 4x + 4) + (y^2 - 2y + 1) = 4 + 4 + 1$$
    $$(x + 2)^2 + (y - 1)^2 = 9$$
    $$(x + 2)^2 + (y - 1)^2 = 3^2$$
    Center: $(-2, 1)$
    Radius 3

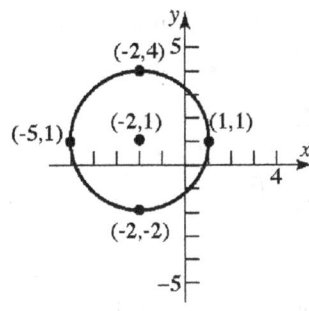

11. $y = x^3 - 9x$

x-intercepts: $0 = x^3 - 9x$

$$0 = x(x^2 - 9)$$
$$0 = x(x+3)(x-3)$$
$$x = 0, -3, \text{ and } 3$$
$$(0,0), \ (-3,0), \ (3,0)$$

y-intercepts: $y = 0^3 - 9(0) = 0 \Rightarrow (0,0)$

Test for symmetry:

x-axis: Replace y by $-y$: $-y = x^3 - 9x$, which is not equivalent to $y = x^3 - 9x$.

y-axis: Replace x by $-x$: $y = (-x)^3 - 9(-x)$
$$= -x^3 + 9x$$

which is not equivalent to $y = x^3 - 9x$.

Origin: Replace x by $-x$ and y by $-y$:
$$-y = (-x)^3 - 9(-x)$$
$$y = -x^3 + 9x$$

which is equivalent to $y = x^3 - 9x$. Therefore, the graph is symmetric with respect to origin.

12. $3x - 2y = 7$
$$-2y = -3x + 7$$
$$y = \frac{3}{2}x - \frac{7}{2}$$

The given line has slope $\frac{3}{2}$. Every line that is perpendicular to the given line will have slope $-\frac{2}{3}$. Using the point $(1,5)$ and the point-slope

formula yields:
$$y - y_1 = m(x - x_1)$$
$$y - 5 = -\frac{2}{3}(x-1)$$
$$y - 5 = -\frac{2}{3}x + \frac{2}{3}$$
$$y = -\frac{2}{3}x + \frac{17}{3}$$

13. Not a function, since the graph fails the Vertical Line Test, for example, when $x = 0$.

14. $f(x) = x^2 + 5x - 2$

a. $f(3) = 3^2 + 5(3) - 2 = 9 + 15 - 2 = 22$

b. $f(-x) = (-x)^2 + 5(-x) - 2 = x^2 - 5x - 2$

c. $-f(x) = -(x^2 + 5x - 2) = -x^2 - 5x + 2$

d. $f(3x) = (3x)^2 + 5(3x) - 2 = 9x^2 + 15x - 2$

e. $\dfrac{f(x+h) - f(x)}{h}$

$$= \frac{(x+h)^2 + 5(x+h) - 2 - (x^2 + 5x - 2)}{h}$$
$$= \frac{x^2 + 2xh + h^2 + 5x + 5h - 2 - x^2 - 5x + 2}{h}$$
$$= \frac{2xh + h^2 + 5h}{h}$$
$$= 2x + h + 5$$

15. $f(x) = \dfrac{x+5}{x-1}$

a. Domain $\{x \mid x \neq 1\}$.

b. $f(2) = \dfrac{2+5}{2-1} = \dfrac{7}{1} = 7 \neq 6$;

$(2,6)$ is not on the graph of f.

The point $(2,7)$ is on the graph.

c. $f(3) = \dfrac{3+5}{3-1} = \dfrac{8}{2} = 4$;

$(3,4)$ is on the graph of f.

d. Solve for x

$$\frac{x+5}{x-1}=9$$
$$x+5=9(x-1)$$
$$x+5=9x-9$$
$$14=8x$$
$$x=\frac{14}{8}=\frac{7}{4}$$

Therefore, $\left(\frac{7}{4},9\right)$ is on the graph of f.

e. $f(x)$ is a rational function since it is in the form $\frac{p(x)}{q(x)}$.

16. $f(x)=-3x+7$

The graph is a line with slope -3 and y-intercept $(0, 7)$.

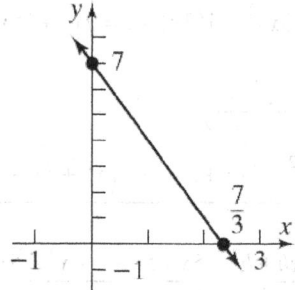

17. $f(x)=2x^2-4x+1$

$a=2$, $b=-4$, $c=1$. Since $a=2>0$, the graph opens up.

The x-coordinate of the vertex is

$$x=-\frac{b}{2a}=-\frac{-4}{2(2)}=1.$$

The y-coordinate of the vertex is

$$f\left(-\frac{b}{2a}\right)=f(1)=2(1)^2-4(1)+1=-1.$$

Thus, the vertex is $(1,-1)$.
The axis of symmetry is the line $x=1$.

The discriminant is:
$b^2-4ac=(-4)^2-4(2)(1)=8>0$, so the graph has two x-intercepts.

The x-intercepts are found by solving:
$$2x^2-4x+1=0$$
$$x=\frac{-(-4)\pm\sqrt{8}}{2(2)}$$
$$=\frac{4\pm2\sqrt{2}}{4}=\frac{2\pm\sqrt{2}}{2}$$

The x-intercepts are $\frac{2-\sqrt{2}}{2}$ and $\frac{2+\sqrt{2}}{2}$.

The y-intercept is $f(0)=1$.

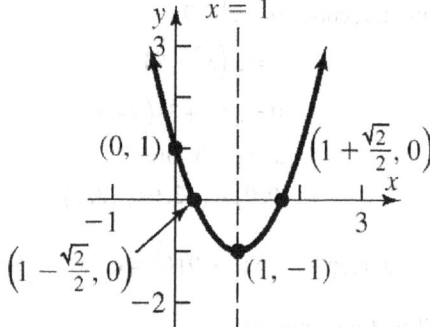

18. $f(x)=x^2+3x+1$

average rate of change of f from 1 to 2:
$$\frac{f(2)-f(1)}{2-1}=\frac{11-5}{1}=6=m_{sec}$$

$f(2)=11$ so the point $(2,11)$ is on the graph. Using this point and the slope $m=6$, we can obtain the equation of the secant line:
$$y-y_1=m(x-x_1)$$
$$y-11=6(x-2)$$
$$y-11=6x-12$$
$$y=6x-1$$

19. a. x-intercepts: $(-5,0);(-1,0);(5,0)$; y-intercept: $(0,-3)$

b. The graph is not symmetric with respect to the origin, x-axis or y-axis.

c. The function is neither even nor odd.

d. f is increasing on $(-\infty,-3]$ and $[2,\infty)$; f is decreasing on $[-3,2]$;

e. f has a local maximum at $x = -3$, and the local maximum is $f(-3) = 5$.

f. f has a local minimum at $x = 2$, and the local minimum is $f(2) = -6$.

20. $f(x) = \dfrac{5x}{x^2 - 9}$

$f(-x) = \dfrac{5(-x)}{(-x)^2 - 9} = \dfrac{-5x}{x^2 - 9} = -f(x)$, therefore

f is an odd function.

21. $f(x) = \begin{cases} 2x+1 & \text{if } -3 < x < 2 \\ -3x+4 & \text{if } x \geq 2 \end{cases}$

a. Domain: $\{x \mid x > -3\}$ or $(-3, \infty)$

b. x-intercept: $\left(-\dfrac{1}{2}, 0\right)$

y-intercept: $(0,1)$

c.

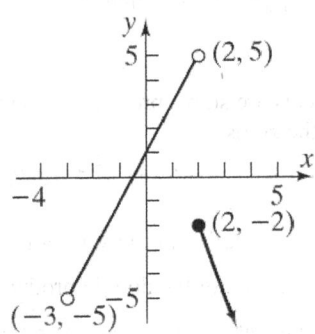

d. Range: $\{y \mid y < 5\}$ or $(-\infty, 5)$

22. $f(x) = -3(x+1)^2 + 5$

Using the graph of $y = x^2$, shift left 1 unit, vertically stretch by a factor of 3, reflect about the x-axis, then shift up 5 units.

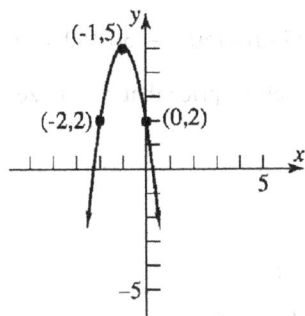

23. $f(x) = x^2 - 5x + 1 \qquad g(x) = -4x - 7$

a. $(f+g)(x) = x^2 - 5x + 1 + (-4x - 7)$

$= x^2 - 9x - 6$

The domain is: $\{x \mid x \text{ is a real number}\}$.

b. $\left(\dfrac{f}{g}\right)(x) = \dfrac{f(x)}{g(x)} = \dfrac{x^2 - 5x + 1}{-4x - 7}$

The domain is: $\left\{x \mid x \neq -\dfrac{7}{4}\right\}$.

24. a. $R(x) = x \cdot p$

$= x\left(-\dfrac{1}{10}x + 150\right)$

$= -\dfrac{1}{10}x^2 + 150x$

b. $R(100) = -\dfrac{1}{10}(100)^2 + 150(100)$

$= -1000 + 15,000 = \$14,000$

c. Since $R(x) = -\dfrac{1}{10}x^2 + 150x$ is a quadratic

function with $a = -\dfrac{1}{10} < 0$, the vertex will

be a maximum point. The vertex occurs

when $x = -\dfrac{b}{2a} = -\dfrac{150}{2(-1/10)} = 750$.

Thus, the revenue is maximized when $x = 750$ units sold.

The maximum revenue is given by

$R(750) = -\dfrac{1}{10}(750)^2 + 150(750)$

$= -56,250 + 112,500 = \$56,250$

d. $p = -\dfrac{1}{10}(750) + 150 = -75 + 150 = \$75$ is

the selling price that maximizes the revenue.

Chapter 5 Projects

Project I – Internet-based Project

Answers will vary

Project II

a. $x^2 + 8x - 9 = 0$

$(x+9)(x-1) = 0$

sum $= -9 + 1 = -8$, product $= (-9)(1) = -9$

$x = -9$ or $x = 1$

b. $x^2 + bx + c = 0$

$(x - r_1)(x - r_2) = 0$

$x^2 - r_1 x - r_2 x + r_1 r_2 = 0$

$x^2 - (r_1 + r_2)x + r_1 r_2 = 0$

$b = -(r_1 + r_2)$

$c = r_1 r_2$

c. $f(x) = x^3 - x^2 - 10x - 8$

$f(x) = (x+2)(x^2 - 3x - 4)$

$f(x) = (x+2)(x-4)(x+1)$

zeros: -2, 4, -1

sum $= -2 + 4 - 1 = 1$, product $= (-2)(4)(-1) = 8$

sum of double products

$= -2(4) + (-2)(-1) + 4(-1) = -8 + 2 - 4 = -10$

The coefficient of x^2 is the negative sum. The coefficient of x is the sum of the double products. The constant term is the negative product.

d. $f(x) = x^3 + bx^2 + cx + d$

$f(x) = (x - r_1)(x - r_2)(x - r_3)$

$f(x) = (x^2 - (r_1 + r_2)x + r_1 r_2)(x - r_3)$

$f(x) = x^3 - (r_1 + r_2 + r_3)x^2$

$\qquad + (r_1 r_2 + r_1 r_3 + r_2 r_3)x - r_1 r_2 r_3$

$b = -(r_1 + r_2 + r_3)$

$c = r_1 r_2 + r_1 r_3 + r_2 r_3$

$d = -r_1 r_2 r_3$

e. $f(x) = x^4 + bx^3 + cx^2 + dx + e$

$f(x) = (x - r_1)(x - r_2)(x - r_3)(x - r_4)$

$f(x) = (x^3 - (r_1 + r_2 + r_3)x^2$

$\qquad + (r_1 r_2 + r_1 r_3 + r_2 r_3)x - r_1 r_2 r_3)(x - r_4)$

$f(x) = x^4 - (r_1 + r_2 + r_3 + r_4)x^3$

$\qquad + (r_1 r_2 + r_1 r_3 + r_2 r_3 + r_1 r_4 + r_2 r_4 + r_3 r_4)x^2$

$\qquad - (r_1 r_2 r_4 + r_1 r_3 r_4 + r_2 r_3 r_4 + r_1 r_2 r_3)x + r_1 r_2 r_3 r_4$

$b = -(r_1 + r_2 + r_3 + r_4)$

$c = r_1 r_2 + r_1 r_3 + r_2 r_3 + r_1 r_4 + r_2 r_4 + r_3 r_4$

$d = -(r_1 r_2 r_4 + r_1 r_3 r_4 + r_2 r_3 r_4 + r_1 r_2 r_3)$

$e = r_1 r_2 r_3 r_4$

f. The coefficients are sums, products, or sums of products of the zeros.

If $f(x) = x^n + a_{n-1}x^{n-1} + a_{n-2}x^{n-2} + ... + a_1 x + a_0$, then:

a_{n-1} will be the negative of the sum of the zeros.

a_{n-2} will be the sum of the double products.

a_1 will be the negative (if n is even) or positive (if n is odd) of the sum of (n-1) products.

a_0 will be the negative (if n is odd) or positive (if n is even) product of the zeros.

These will always hold. These would be useful if you needed to multiply a number of binomials in $x - c$ form together and you did not want to have to do the multiplication out. These formulas would help same time.

Chapter 6
Exponential and Logarithmic Functions

Section 6.1

1. $f(3) = -4(3)^2 + 5(3)$
$= -4(9) + 15$
$= -36 + 15$
$= -21$

3. $f(x) = \dfrac{x^2 - 1}{x^2 - 25}$
$x^2 - 25 \neq 0$
$(x + 5)(x - 5) \neq 0$
$x \neq -5, \quad x \neq 5$
Domain: $\{x \mid x \neq -5, \ x \neq 5\}$

5. False: $(f \circ g)(x) = f(g(4))$
$= f(\sqrt{4 + 9}) = f(\sqrt{13})$
$= (\sqrt{13})^2 = 13$

7. a

9. a. $(f \circ g)(1) = f(g(1)) = f(0) = -1$

b. $(f \circ g)(-1) = f(g(-1)) = f(0) = -1$

c. $(g \circ f)(-1) = g(f(-1)) = g(-3) = 8$

d. $(g \circ f)(0) = g(f(0)) = g(-1) = 0$

e. $(g \circ g)(-2) = g(g(-2)) = g(3) = 8$

f. $(f \circ f)(-1) = f(f(-1)) = f(-3) = -7$

11. a. $(g \circ f)(-1) = g(f(-1)) = g(1) = 4$

b. $(g \circ f)(0) = g(f(0)) = g(0) = 5$

c. $(f \circ g)(-1) = f(g(-1)) = f(3) = -1$

d. $(f \circ g)(4) = f(g(4)) = f(2) = -2$

13. $f(x) = 2x \qquad g(x) = 3x^2 + 1$

a. $(f \circ g)(4) = f(g(4))$
$= f\left(3(4)^2 + 1\right)$
$= f(49)$
$= 2(49)$
$= 98$

b. $(g \circ f)(2) = g(f(2))$
$= g(2 \cdot 2)$
$= g(4)$
$= 3(4)^2 + 1$
$= 48 + 1$
$= 49$

c. $(f \circ f)(1) = f(f(1))$
$= f(2(1))$
$= f(2)$
$= 2(2)$
$= 4$

d. $(g \circ g)(0) = g(g(0))$
$= g\left(3(0)^2 + 1\right)$
$= g(1)$
$= 3(1)^2 + 1$
$= 4$

15. $f(x) = 4x^2 - 3 \qquad g(x) = 3 - \dfrac{1}{2}x^2$

a. $(f \circ g)(4) = f(g(4))$
$= f\left(3 - \dfrac{1}{2}(4)^2\right)$
$= f(-5)$
$= 4(-5)^2 - 3$
$= 97$

b. $(g \circ f)(2) = g(f(2))$
$$= g(4(2)^2 - 3)$$
$$= g(13)$$
$$= 3 - \frac{1}{2}(13)^2$$
$$= 3 - \frac{169}{2}$$
$$= -\frac{163}{2}$$

c. $(f \circ f)(1) = f(f(1))$
$$= f(4(1)^2 - 3)$$
$$= f(1)$$
$$= 4(1)^2 - 3$$
$$= 1$$

d. $(g \circ g)(0) = g(g(0))$
$$= g\left(3 - \frac{1}{2}(0)^2\right)$$
$$= g(3)$$
$$= 3 - \frac{1}{2}(3)^2$$
$$= 3 - \frac{9}{2}$$
$$= -\frac{3}{2}$$

17. $f(x) = \sqrt{x} \qquad g(x) = 2x$

a. $(f \circ g)(4) = f(g(4))$
$$= f(2(4))$$
$$= f(8)$$
$$= \sqrt{8}$$
$$= 2\sqrt{2}$$

b. $(g \circ f)(2) = g(f(2))$
$$= g\left(\sqrt{2}\right)$$
$$= 2\sqrt{2}$$

c. $(f \circ f)(1) = f(f(1))$
$$= f\left(\sqrt{1}\right)$$
$$= f(1)$$
$$= \sqrt{1}$$
$$= 1$$

d. $(g \circ g)(0) = g(g(0))$
$$= g(2(0))$$
$$= g(0)$$
$$= 2(0)$$
$$= 0$$

19. $f(x) = |x| \qquad g(x) = \frac{1}{x^2 + 1}$

a. $(f \circ g)(4) = f(g(4))$
$$= f\left(\frac{1}{4^2 + 1}\right)$$
$$= f\left(\frac{1}{17}\right)$$
$$= \left|\frac{1}{17}\right|$$
$$= \frac{1}{17}$$

b. $(g \circ f)(2) = g(f(2))$
$$= g(|2|)$$
$$= g(2)$$
$$= \frac{1}{2^2 + 1}$$
$$= \frac{1}{5}$$

c. $(f \circ f)(1) = f(f(1))$
$$= f(|1|)$$
$$= f(1)$$
$$= |1|$$
$$= 1$$

d. $(g \circ g)(0) = g(g(0))$
$$= g\left(\frac{1}{0^2 + 1}\right)$$
$$= g(1)$$
$$= \frac{1}{1^2 + 1}$$
$$= \frac{1}{2}$$

21. $f(x) = \dfrac{3}{x+1}$ $\qquad g(x) = \sqrt[3]{x}$

   a. $(f \circ g)(4) = f(g(4))$

$$= f\left(\sqrt[3]{4}\right)$$

$$= \dfrac{3}{\sqrt[3]{4}+1}$$

   b. $(g \circ f)(2) = g(f(2))$

$$= g\left(\dfrac{3}{2+1}\right)$$

$$= g\left(\dfrac{3}{3}\right)$$

$$= g(1)$$

$$= \sqrt[3]{1}$$

$$= 1$$

   c. $(f \circ f)(1) = f(f(1))$

$$= f\left(\dfrac{3}{1+1}\right)$$

$$= f\left(\dfrac{3}{2}\right)$$

$$= \dfrac{3}{\dfrac{3}{2}+1}$$

$$= \dfrac{3}{\dfrac{5}{2}}$$

$$= \dfrac{6}{5}$$

   d. $(g \circ g)(0) = g(g(0))$

$$= g\left(\sqrt[3]{0}\right)$$

$$= g(0)$$

$$= \sqrt[3]{0}$$

$$= 0$$

23. $f(x) = 2x+3 \qquad g(x) = 3x$

The domain of f is $\{x \mid x$ is any real number$\}$. The domain of g is $\{x \mid x$ is any real number$\}$.

   a. $(f \circ g)(x) = f(g(x))$

$$= f(3x)$$

$$= 2(3x)+3$$

$$= 6x+3$$

Domain: $\{x \mid x$ is any real number$\}$.

   b. $(g \circ f)(x) = g(f(x))$

$$= g(2x+3)$$

$$= 3(2x+3)$$

$$= 6x+9$$

Domain: $\{x \mid x$ is any real number$\}$.

   c. $(f \circ f)(x) = f(f(x))$

$$= f(2x+3)$$

$$= 2(2x+3)+3$$

$$= 4x+6+3$$

$$= 4x+9$$

Domain: $\{x \mid x$ is any real number$\}$.

   d. $(g \circ g)(x) = g(g(x))$

$$= g(3x)$$

$$= 3(3x)$$

$$= 9x$$

Domain: $\{x \mid x$ is any real number$\}$.

25. $f(x) = 3x+1 \qquad g(x) = x^2$

The domain of f is $\{x \mid x$ is any real number$\}$.

The domain of g is $\{x \mid x$ is any real number$\}$.

   a. $(f \circ g)(x) = f(g(x))$

$$= f\left(x^2\right)$$

$$= 3x^2+1$$

Domain: $\{x \mid x$ is any real number$\}$.

   b. $(g \circ f)(x) = g(f(x))$

$$= g(3x+1)$$

$$= (3x+1)^2$$

$$= 9x^2+6x+1$$

Domain: $\{x \mid x$ is any real number$\}$.

333

c. $(f \circ f)(x) = f(f(x))$
$= f(3x+1)$
$= 3(3x+1)+1$
$= 9x+3+1$
$= 9x+4$
Domain: $\{x \mid x \text{ is any real number}\}$.

d. $(g \circ g)(x) = g(g(x))$
$= g\left(x^2\right)$
$= \left(x^2\right)^2$
$= x^4$
Domain: $\{x \mid x \text{ is any real number}\}$.

27. $f(x) = x^2 \qquad g(x) = x^2 + 4$
The domain of f is $\{x \mid x \text{ is any real number}\}$.
The domain of g is $\{x \mid x \text{ is any real number}\}$.

a. $(f \circ g)(x) = f(g(x))$
$= f\left(x^2+4\right)$
$= \left(x^2+4\right)^2$
$= x^4 + 8x^2 + 16$
Domain: $\{x \mid x \text{ is any real number}\}$.

b. $(g \circ f)(x) = g(f(x))$
$= g\left(x^2\right)$
$= \left(x^2\right)^2 + 4$
$= x^4 + 4$
Domain: $\{x \mid x \text{ is any real number}\}$.

c. $(f \circ f)(x) = f(f(x))$
$= f\left(x^2\right)$
$= \left(x^2\right)^2$
$= x^4$
Domain: $\{x \mid x \text{ is any real number}\}$.

d. $(g \circ g)(x) = g(g(x))$
$= g\left(x^2+4\right)$
$= \left(x^2+4\right)^2 + 4$
$= x^4 + 8x^2 + 16 + 4$
$= x^4 + 8x^2 + 20$
Domain: $\{x \mid x \text{ is any real number}\}$.

29. $f(x) = \dfrac{3}{x-1} \qquad g(x) = \dfrac{2}{x}$
The domain of f is $\{x \mid x \neq 1\}$. The domain of g is $\{x \mid x \neq 0\}$.

a. $(f \circ g)(x) = f(g(x))$
$= f\left(\dfrac{2}{x}\right)$
$= \dfrac{3}{\dfrac{2}{x}-1}$
$= \dfrac{3}{\dfrac{2-x}{x}}$
$= \dfrac{3x}{2-x}$
Domain $\{x \mid x \neq 0,\ x \neq 2\}$.

b. $(g \circ f)(x) = g(f(x))$
$= g\left(\dfrac{3}{x-1}\right)$
$= \dfrac{2}{\dfrac{3}{x-1}}$
$= \dfrac{2(x-1)}{3}$
Domain $\{x \mid x \neq 1\}$

c. $(f \circ f)(x) = f(f(x))$
$= f\left(\dfrac{3}{x-1}\right)$
$= \dfrac{3}{\dfrac{3}{x-1}-1} = \dfrac{3}{\dfrac{3-(x-1)}{x-1}}$
$= \dfrac{3(x-1)}{4-x}$
Domain $\{x \mid x \neq 1,\ x \neq 4\}$.

d. $(g \circ g)(x) = g(g(x)) = g\left(\dfrac{2}{x}\right) = \dfrac{2}{\dfrac{2}{x}} = \dfrac{2x}{2} = x$
Domain $\{x \mid x \neq 0\}$.

31. $f(x) = \dfrac{x}{x-1}$ $\qquad g(x) = -\dfrac{4}{x}$

The domain of f is $\{x \mid x \neq 1\}$. The domain of g is $\{x \mid x \neq 0\}$.

a. $(f \circ g)(x) = f(g(x))$

$= f\left(-\dfrac{4}{x}\right)$

$= \dfrac{-\dfrac{4}{x}}{-\dfrac{4}{x}-1} = \dfrac{-\dfrac{4}{x}}{\dfrac{-4-x}{x}} = \dfrac{-4}{-4-x}$

$= \dfrac{4}{4+x}$

Domain $\{x \mid x \neq -4, x \neq 0\}$.

b. $(g \circ f)(x) = g(f(x))$

$= g\left(\dfrac{x}{x-1}\right)$

$= -\dfrac{4}{\dfrac{x}{x-1}}$

$= \dfrac{-4(x-1)}{x}$

Domain $\{x \mid x \neq 0, x \neq 1\}$.

c. $(f \circ f)(x) = f(f(x))$

$= f\left(\dfrac{x}{x-1}\right)$

$= \dfrac{\dfrac{x}{x-1}}{\dfrac{x}{x-1}-1} = \dfrac{\dfrac{x}{x-1}}{\dfrac{x-(x-1)}{x-1}} = \dfrac{\dfrac{x}{x-1}}{\dfrac{1}{x-1}}$

$= x$

Domain $\{x \mid x \neq 1\}$.

d. $(g \circ g)(x) = g(g(x))$

$= g\left(\dfrac{-4}{x}\right)$

$= -\dfrac{4}{\dfrac{-4}{x}} = \dfrac{-4x}{-4}$

$= x$

Domain $\{x \mid x \neq 0\}$.

33. $f(x) = \sqrt{x}$ $\qquad g(x) = 2x+3$

The domain of f is $\{x \mid x \geq 0\}$. The domain of g is $\{x \mid x \text{ is any real number}\}$.

a. $(f \circ g)(x) = f(g(x)) = f(2x+3) = \sqrt{2x+3}$

Domain $\left\{x \mid x \geq -\dfrac{3}{2}\right\}$.

b. $(g \circ f)(x) = g(f(x)) = g\left(\sqrt{x}\right) = 2\sqrt{x}+3$

Domain $\{x \mid x \geq 0\}$.

c. $(f \circ f)(x) = f(f(x))$

$= f\left(\sqrt{x}\right)$

$= \sqrt{\sqrt{x}}$

$= \left(x^{1/2}\right)^{1/2}$

$= x^{1/4}$

$= \sqrt[4]{x}$

Domain $\{x \mid x \geq 0\}$.

d. $(g \circ g)(x) = g(g(x))$

$= g(2x+3)$

$= 2(2x+3)+3$

$= 4x+6+3$

$= 4x+9$

Domain $\{x \mid x \text{ is any real number}\}$.

35. $f(x) = x^2+1$ $\qquad g(x) = \sqrt{x-1}$

The domain of f is $\{x \mid x \text{ is any real number}\}$.

The domain of g is $\{x \mid x \geq 1\}$.

a. $(f \circ g)(x) = f(g(x))$

$= f\left(\sqrt{x-1}\right)$

$= \left(\sqrt{x-1}\right)^2 + 1$

$= x-1+1$

$= x$

Domain $\{x \mid x \geq 1\}$.

b. $(g \circ f)(x) = g(f(x))$

$= g\left(x^2 + 1\right)$

$= \sqrt{x^2 + 1 - 1}$

$= \sqrt{x^2}$

$= |x|$

Domain $\{x \mid x \text{ is any real number}\}$.

c. $(f \circ f)(x) = f(f(x))$

$= f\left(x^2 + 1\right)$

$= \left(x^2 + 1\right)^2 + 1$

$= x^4 + 2x^2 + 1 + 1$

$= x^4 + 2x^2 + 2$

Domain $\{x \mid x \text{ is any real number}\}$.

d. $(g \circ g)(x) = g(g(x))$

$= g\left(\sqrt{x-1}\right) = \sqrt{\sqrt{x-1} - 1}$

Now, $\sqrt{x-1} - 1 \geq 0$

$\sqrt{x-1} \geq 1$

$x - 1 \geq 1$

$x \geq 2$

Domain $\{x \mid x \geq 2\}$.

37. $f(x) = \dfrac{x-5}{x+1} \qquad g(x) = \dfrac{x+2}{x-3}$

The domain of f is $\{x \mid x \neq -1\}$. The domain of g is $\{x \mid x \neq 3\}$.

a. $(f \circ g)(x) = f(g(x)) = f\left(\dfrac{x+2}{x-3}\right)$

$= \dfrac{\dfrac{x+2}{x-3} - 5}{\dfrac{x+2}{x-3} + 1} = \dfrac{\left(\dfrac{x+2}{x-3} - 5\right)(x-3)}{\left(\dfrac{x+2}{x-3} + 1\right)(x-3)}$

$= \dfrac{x+2-5(x-3)}{x+2+1(x-3)} = \dfrac{x+2-5x+15}{x+2+x-3}$

$= \dfrac{-4x+17}{2x-1}$ or $-\dfrac{4x-17}{2x-1}$

Now, $2x - 1 \neq 0$, so $x \neq \dfrac{1}{2}$. Also, from the

domain of g, we know $x \neq 3$.

Domain of $f \circ g : \left\{x \mid x \neq \dfrac{1}{2}, x \neq 3\right\}$.

b. $(g \circ f)(x) = g(f(x)) = g\left(\dfrac{x-5}{x+1}\right)$

$= \dfrac{\dfrac{x-5}{x+1} + 2}{\dfrac{x-5}{x+1} - 3} = \dfrac{\left(\dfrac{x-5}{x+1} + 2\right)(x+1)}{\left(\dfrac{x-5}{x+1} - 3\right)(x+1)}$

$= \dfrac{x-5+2(x+1)}{x-5-3(x+1)} = \dfrac{x-5+2x+2}{x-5-3x-3}$

$= \dfrac{3x-3}{-2x-8}$ or $-\dfrac{3x-3}{2x+8}$

Now, $-2x - 8 \neq 0$, so $x \neq -4$. Also, from

the domain of f, we know $x \neq -1$.

Domain of $g \circ f : \{x \mid x \neq -4, x \neq -1\}$.

c. $(f \circ f)(x) = f(f(x)) = f\left(\dfrac{x-5}{x+1}\right)$

$= \dfrac{\dfrac{x-5}{x+1} - 5}{\dfrac{x-5}{x+1} + 1} = \dfrac{\left(\dfrac{x-5}{x+1} - 5\right)(x+1)}{\left(\dfrac{x-5}{x+1} + 1\right)(x+1)}$

$= \dfrac{x-5-5(x+1)}{x-5+1(x+1)} = \dfrac{x-5-5x-5}{x-5+x+1}$

$= \dfrac{-4x-10}{2x-4} = \dfrac{-2(2x+5)}{2(x-2)} = -\dfrac{2x+5}{x-2}$

Now, $x - 2 \neq 0$, so $x \neq 2$. Also, from the

domain of f, we know $x \neq -1$.

Domain of $f \circ f : \{x \mid x \neq -1, x \neq 2\}$.

d. $(g \circ g)(x) = g(g(x)) = g\left(\dfrac{x+2}{x-3}\right)$

$= \dfrac{\dfrac{x+2}{x-3} + 2}{\dfrac{x+2}{x-3} - 3} = \dfrac{\left(\dfrac{x+2}{x-3} + 2\right)(x-3)}{\left(\dfrac{x+2}{x-3} - 3\right)(x-3)}$

$= \dfrac{x+2+2(x-3)}{x+2-3(x-3)} = \dfrac{x+2+2x-6}{x+2-3x+9}$

$= \dfrac{3x-4}{-2x+11}$ or $-\dfrac{3x-4}{2x-11}$

Now, $-2x + 11 \neq 0$, so $x \neq \dfrac{11}{2}$. Also, from the

domain of g, we know $x \neq 3$.

Domain of $g \circ g : \left\{x \mid x \neq \dfrac{11}{2}, x \neq 3\right\}$.

39. $(f \circ g)(x) = f(g(x)) = f\left(\dfrac{1}{2}x\right) = 2\left(\dfrac{1}{2}x\right) = x$

$(g \circ f)(x) = g(f(x)) = g(2x) = \dfrac{1}{2}(2x) = x$

41. $(f \circ g)(x) = f(g(x)) = f\left(\sqrt[3]{x}\right) = \left(\sqrt[3]{x}\right)^3 = x$

$(g \circ f)(x) = g(f(x)) = g\left(x^3\right) = \sqrt[3]{x^3} = x$

43. $(f \circ g)(x) = f(g(x))$

$\qquad = f\left(\dfrac{1}{2}(x+6)\right)$

$\qquad = 2\left(\dfrac{1}{2}(x+6)\right) - 6$

$\qquad = x + 6 - 6$

$\qquad = x$

$(g \circ f)(x) = g(f(x))$

$\qquad = g(2x-6)$

$\qquad = \dfrac{1}{2}\left((2x-6)+6\right)$

$\qquad = \dfrac{1}{2}(2x)$

$\qquad = x$

45. $(f \circ g)(x) = f(g(x))$

$\qquad = f\left(\dfrac{1}{a}(x-b)\right)$

$\qquad = a\left(\dfrac{1}{a}(x-b)\right) + b$

$\qquad = x - b + b$

$\qquad = x$

$(g \circ f)(x) = g(f(x))$

$\qquad = g(ax+b)$

$\qquad = \dfrac{1}{a}\left((ax+b)-b\right)$

$\qquad = \dfrac{1}{a}(ax)$

$\qquad = x$

47. $H(x) = (2x+3)^4$
Answers may vary. One possibility is
$f(x) = x^4, \quad g(x) = 2x+3$

49. $H(x) = \sqrt{x^2+1}$
Answers may vary. One possibility is
$f(x) = \sqrt{x}, \quad g(x) = x^2+1$

51. $H(x) = \left|2x+1\right|$
Answers may vary. One possibility is
$f(x) = \left|x\right|, \quad g(x) = 2x+1$

53. $f(x) = 2x^3 - 3x^2 + 4x - 1 \qquad g(x) = 2$

$(f \circ g)(x) = f(g(x))$

$\qquad = f(2)$

$\qquad = 2(2)^3 - 3(2)^2 + 4(2) - 1$

$\qquad = 16 - 12 + 8 - 1$

$\qquad = 11$

$(g \circ f)(x) = g(f(x)) = g\left(2x^3 - 3x^2 + 4x - 1\right) = 2$

55. $f(x) = 2x^2 + 5 \qquad g(x) = 3x + a$

$(f \circ g)(x) = f(g(x)) = f(3x+a) = 2(3x+a)^2 + 5$

When $x = 0$, $(f \circ g)(0) = 23$.

Solving: $2(3 \cdot 0 + a)^2 + 5 = 23$

$\qquad\qquad 2a^2 + 5 = 23$

$\qquad\qquad 2a^2 - 18 = 0$

$\qquad\qquad 2(a+3)(a-3) = 0$

$\qquad\qquad a = -3 \ \text{ or } \ a = 3$

57. a. $(f \circ g)(x) = f(g(x))$

$\qquad\qquad = f(cx+d)$

$\qquad\qquad = a(cx+d) + b$

$\qquad\qquad = acx + ad + b$

b. $(g \circ f)(x) = g(f(x))$

$\qquad\qquad = g(ax+b)$

$\qquad\qquad = c(ax+b) + d$

$\qquad\qquad = acx + bc + d$

c. Since the domain of f is the set of all real numbers and the domain of g is the set of all real numbers, the domains of both $f \circ g$ and $g \circ f$ are all real numbers.

d. $(f \circ g)(x) = (g \circ f)(x)$

$acx + ad + b = acx + bc + d$

$ad + b = bc + d$

Thus, $f \circ g = g \circ f$ when $ad + b = bc + d$.

59. $S(r) = 4\pi r^2 \qquad r(t) = \frac{2}{3}t^3,\ t \ge 0$

$$S(r(t)) = S\left(\frac{2}{3}t^3\right)$$

$$= 4\pi\left(\frac{2}{3}t^3\right)^2$$

$$= 4\pi\left(\frac{4}{9}t^6\right)$$

$$= \frac{16}{9}\pi t^6$$

Thus, $S(t) = \frac{16}{9}\pi t^6$.

61. $N(t) = 100t - 5t^2,\ 0 \le t \le 10$

$C(N) = 15,000 + 8000N$

$$C(N(t)) = C\left(100t - 5t^2\right)$$

$$= 15,000 + 8000\left(100t - 5t^2\right)$$

$$= 15,000 + 800,000t - 40,000t^2$$

Thus, $C(t) = 15,000 + 800,000t - 40,000t^2$.

63. $p = -\frac{1}{4}x + 100,\quad 0 \le x \le 400$

$$\frac{1}{4}x = 100 - p$$

$$x = 4(100 - p)$$

$$C = \frac{\sqrt{x}}{25} + 600$$

$$= \frac{\sqrt{4(100 - p)}}{25} + 600$$

$$= \frac{2\sqrt{100 - p}}{25} + 600,\quad 0 \le p \le 100$$

65. $V = \pi r^2 h \qquad h = 2r$

$$V(r) = \pi r^2(2r) = 2\pi r^3$$

67. $f(x)$ = the number of Euros bought for x dollars;

$g(x)$ = the number of yen bought for x Euros

a. $f(x) = 0.9428x$

b. $g(x) = 126.457x$

c. $(g \circ f)(x) = g(f(x))$

$$= g(0.9428x)$$

$$= 126.457(0.9428x)$$

$$= 119.2236596x$$

d. $(g \circ f)(1000) = 119.2236596(1000)$

$$= 119,223.6596 \text{ yen}$$

69. a. $f(p) = p - 200$

b. $g(p) = 0.80p$

c. $(f \circ g)(p) = f(g(p))$

$$= (0.80p) - 200$$

$$= 0.80p - 200$$

This represents the final price when the rebate is issued on the sale price.

$$(g \circ f)(p) = g(f(p))$$

$$= 0.80(p - 200)$$

$$= 0.80p - 160$$

This represents the final price when the sale price is calculated after the rebate is given.

Appling the 20% first is a better deal since a larger portion will be removed up front.

71. $f(x) = ax + b;\ g(x) = bx + a$

$f(1) = 8;\ f(g(20)) - g(f(20)) = -14$

We will solve as a system of equations. The first equation is $f(1) = a(1) + b = a + b = 8$. The second is:

$$\left[f(20b + a)\right] - \left[g(20a + b)\right] = -14$$

$$a(20b + a) + b - b(20a + b) - a = -14$$

$$20ab + a^2 + b - 20ab - b^2 - a = -14$$

$$a^2 + b - b^2 - a = -14$$

Now we substitute from the first equation, $a = 8 - b$.

$$(8 - b)^2 + b - b^2 - (8 - b) = -14$$

$$64 - 16b + b^2 + b - b^2 - 8 + b = -14$$

$$-14b = -70$$

$$b = 5$$

Substituting back into the first equation to solve for a gives $a = 3$. So the product is:

$$ab = (3)(5) = 15$$

338

73. Given that f is odd and g is even, we know that
$f(-x) = -f(x)$ and $g(-x) = g(x)$ for all x in
the domain of f and g, respectively. The composite
function $(f \circ g)(x) = f(g(x))$ has the following
property:
$$(f \circ g)(-x) = f(g(-x))$$
$$= f(g(x)) \quad \text{since } g \text{ is even}$$
$$= (f \circ g)(x)$$
Thus, $f \circ g$ is an even function.

The composite function $(g \circ f)(x) = g(f(x))$ has
the following property:
$$(g \circ f)(-x) = g(f(-x))$$
$$= g(-f(x)) \quad \text{since } f \text{ is odd}$$
$$= g(f(x)) \quad \text{since } g \text{ is even}$$
$$= (g \circ f)(x)$$
Thus, $g \circ f$ is an even function.

75. $2x - 5\sqrt{x} + 2 = 0$
$(2\sqrt{x} - 1)(\sqrt{x} - 2) = 0$
$2\sqrt{x} - 1 = 0$ or $\sqrt{x} - 2 = 0$
$\sqrt{x} = \dfrac{1}{2}$ $\qquad \sqrt{x} = 2$
$x = \dfrac{1}{4}$ $\qquad x = 4$

The solution set is $\dfrac{1}{4}, 4$

77. $R(x) = \dfrac{x^2 + 6x + 5}{x - 3} = \dfrac{(x + 5)(x + 1)}{x - 3}$
where the domain is $\{x \mid x \neq 3\}$
The degree of the numerator in lowest terms is
$n = 2$. The degree of the denominator in lowest
terms is $m = 1$. Since $n > m$, There is no
horizontal asymptote. The denominator in lowest
terms is zero at $x = 3$, so $x = 3$ is a vertical
asymptote.

$$\begin{array}{r} x + 9 \\ x - 3 \overline{\smash{)} x^2 + 6x + 5} \\ \underline{-(x^2 - 3x)} \\ 9x + 5 \\ \underline{-(9x - 27)} \\ 32 \end{array}$$

$R(x) = x + 9 + \dfrac{32}{x - 3}, \ x \neq 3$
Thus, the oblique asymptote is $y = x + 9$.

Section 6.2

1. The set of ordered pairs is a function because there
are no ordered pairs with the same first element
and different second elements.

3. The function is not defined when $x^2 + 3x - 18 = 0$.
Solve: $x^2 + 3x - 18 = 0$
$(x + 6)(x - 3) = 0$
$x = -6$ or $x = 3$
The domain is $\{x \mid x \neq -6, \ x \neq 3\}$.

5. $f(x_1) \neq f(x_2)$

7. 3

9. $[4, \infty)$

11. a

13. The function is one-to-one because there are no
two distinct inputs that correspond to the same
output.

15. The function is not one-to-one because there are
two different inputs, 20 Hours and 50 Hours, that
correspond to the same output, $200.

17. The function is not one-to-one because there are
two distinct inputs, 2 and -3, that correspond to
the same output.

19. The function is one-to-one because there are no
two distinct inputs that correspond to the same
output.

21. The function f is one-to-one because every horizontal line intersects the graph at exactly one point.

23. The function f is not one-to-one because there are horizontal lines (for example, $y=1$) that intersect the graph at more than one point.

25. The function f is one-to-one because every horizontal line intersects the graph at exactly one point.

27. To find the inverse, interchange the elements in the domain with the elements in the range:

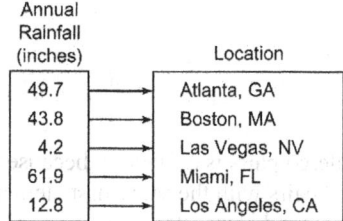

Domain: {49.7, 43.8, 4.2, 61.9, 12.8}

Range: {Atlanta, Boston, Las Vegas, Miami, Los Angeles}

29. To find the inverse, interchange the elements in the domain with the elements in the range:

Monthly Cost of Life Insurance → Age

$7.09 → 30
$8.40 → 40
$11.29 → 45

Domain: {$10.59, $12.52, $15.94}

Range: {30, 40, 45}

31. Interchange the entries in each ordered pair:
{(5,−3), (9,−2), (2,−1), (11,0), (−5,1)}
Domain: {5, 9, 2, 11,−5}
Range: {−3,−2,−1, 0, 1}

33. Interchange the entries in each ordered pair:
{(1,−2), (2,−3), (0,−10), (9,1), (4,2)}
Domain: {1, 2, 0, 9, 4}
Range: {−2,−3,−10, 1, 2}

35. $f(x)=3x+4;\qquad g(x)=\dfrac{1}{3}(x-4)$

$$f(g(x))=f\left(\frac{1}{3}(x-4)\right)$$
$$=3\left(\frac{1}{3}(x-4)\right)+4$$
$$=(x-4)+4$$
$$=x$$

$$g(f(x))=g(3x+4)$$
$$=\frac{1}{3}((3x+4)-4)$$
$$=\frac{1}{3}(3x)=x$$

37. $f(x)=4x-8;\qquad g(x)=\dfrac{x}{4}+2$

$$f(g(x))=f\left(\frac{x}{4}+2\right)$$
$$=4\left(\frac{x}{4}+2\right)-8$$
$$=x+8-8$$
$$=x$$

$$g(f(x))=g(4x-8)$$
$$=\frac{4x-8}{4}+2$$
$$=x-2+2$$
$$=x$$

Thus, f and g are inverses of each other.

39. $f(x)=x^3-8;\qquad g(x)=\sqrt[3]{x+8}$

$$f(g(x))=f\left(\sqrt[3]{x+8}\right)$$
$$=\left(\sqrt[3]{x+8}\right)^3-8$$
$$=x+8-8$$
$$=x$$

$$g(f(x))=g(x^3-8)$$
$$=\sqrt[3]{(x^3-8)+8}$$
$$=\sqrt[3]{x^3}$$
$$=x$$

Thus, f and g are inverses of each other.

41. $f(x) = \dfrac{1}{x}; \qquad g(x) = \dfrac{1}{x}$

$f(g(x)) = f\left(\dfrac{1}{x}\right) = \dfrac{1}{\frac{1}{x}} = 1 \cdot \dfrac{x}{1} = x$

$g(f(x)) = g\left(\dfrac{1}{x}\right) = \dfrac{1}{\frac{1}{x}} = 1 \cdot \dfrac{x}{1} = x$

Thus, f and g are inverses of each other.

43. $f(x) = \dfrac{2x+3}{x+4}; \quad g(x) = \dfrac{4x-3}{2-x}$

$f(g(x)) = f\left(\dfrac{4x-3}{2-x}\right), \ x \neq 2$

$= \dfrac{2\left(\dfrac{4x-3}{2-x}\right)+3}{\dfrac{4x-3}{2-x}+4}$

$= \dfrac{\left(2\left(\dfrac{4x-3}{2-x}\right)+3\right)(2-x)}{\left(\dfrac{4x-3}{2-x}+4\right)(2-x)}$

$= \dfrac{2(4x-3)+3(2-x)}{4x-3+4(2-x)}$

$= \dfrac{8x-6+6-3x}{4x-3+8-4x}$

$= \dfrac{5x}{5}$

$= x$

$g(f(x)) = g\left(\dfrac{2x+3}{x+4}\right), \ x \neq -4$

$= \dfrac{4\left(\dfrac{2x+3}{x+4}\right)-3}{2-\dfrac{2x+3}{x+4}}$

$= \dfrac{\left(4\left(\dfrac{2x+3}{x+4}\right)-3\right)(x+4)}{\left(2-\dfrac{2x+3}{x+4}\right)(x+4)}$

$= \dfrac{4(2x+3)-3(x+4)}{2(x+4)-(2x+3)}$

$= \dfrac{8x+12-3x-12}{2x+8-2x-3}$

$= \dfrac{5x}{5}$

$= x$

Thus, f and g are inverses of each other.

45. Graphing the inverse:

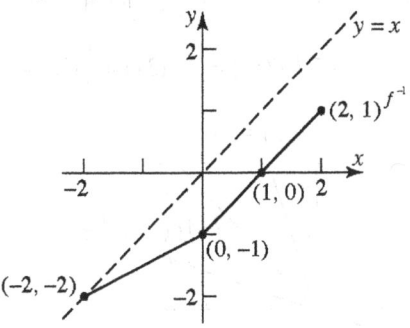

47. Graphing the inverse:

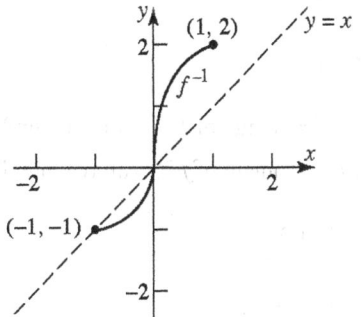

49. Graphing the inverse:

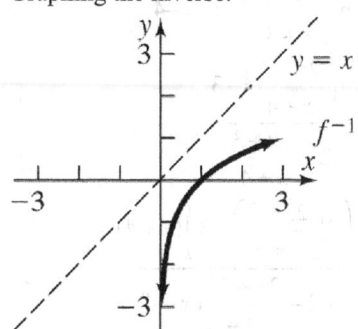

51. a. $f(x) = 3x$

$y = 3x$

$x = 3y$ Inverse

$y = \dfrac{x}{3}$

$f^{-1}(x) = \dfrac{1}{3}x$

Verifying: $f\left(f^{-1}(x)\right) = f\left(\dfrac{1}{3}x\right) = 3\left(\dfrac{1}{3}x\right) = x$

$f^{-1}(f(x)) = f^{-1}(3x) = \dfrac{1}{3}(3x) = x$

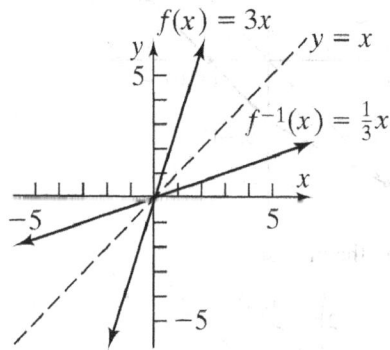

b. domain of f = range of f^{-1} = all real numbers

range of f = domain of f^{-1} = all real numbers

53. a. $f(x) = 4x + 2$

$y = 4x + 2$

$x = 4y + 2$ Inverse

$4y = x - 2$

$y = \dfrac{x-2}{4}$

$y = \dfrac{x}{4} - \dfrac{1}{2}$

$f^{-1}(x) = \dfrac{x}{4} - \dfrac{1}{2}$

Verifying:

$f\left(f^{-1}(x)\right) = f\left(\dfrac{x}{4} - \dfrac{1}{2}\right) = 4\left(\dfrac{x}{4} - \dfrac{1}{2}\right) + 2$

$= x - 2 + 2 = x$

$f^{-1}(f(x)) = f^{-1}(4x + 2) = \dfrac{4x+2}{4} - \dfrac{1}{2}$

$= x + \dfrac{1}{2} - \dfrac{1}{2} = x$

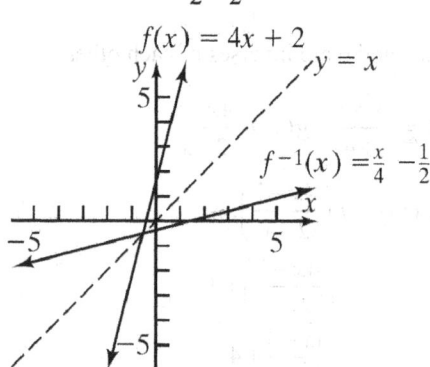

b. domain of f = range of f^{-1} = all real numbers

range of f = domain of f^{-1} = all real numbers

55. a. $f(x) = x^3 - 1$

$y = x^3 - 1$

$x = y^3 - 1$ Inverse

$y^3 = x + 1$

$y = \sqrt[3]{x+1}$

$f^{-1}(x) = \sqrt[3]{x+1}$

Verifying: $f\left(f^{-1}(x)\right) = f\left(\sqrt[3]{x+1}\right) = \left(\sqrt[3]{x+1}\right)^3 - 1$

$= x + 1 - 1 = x$

$f^{-1}(f(x)) = f^{-1}\left(x^3 - 1\right) = \sqrt[3]{\left(x^3 - 1\right) + 1}$

$= \sqrt[3]{x^3} = x$

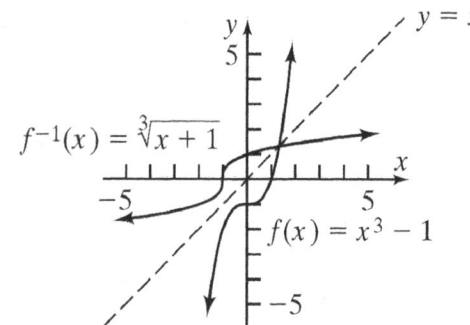

342

b. domain of f = range of f^{-1} = all real numbers

range of f = domain of f^{-1} = all real numbers

57. a. $f(x) = x^2 + 4, \ x \geq 0$

$y = x^2 + 4, \ x \geq 0$

$x = y^2 + 4, \ y \geq 0$ Inverse

$y^2 = x - 4, \ x \geq 4$

$y = \sqrt{x-4}, \ x \geq 4$

$f^{-1}(x) = \sqrt{x-4}, \ x \geq 4$

Verifying: $f\left(f^{-1}(x)\right) = f\left(\sqrt{x-4}\right)$

$= \left(\sqrt{x-4}\right)^2 + 4$

$= x - 4 + 4 = x$

$f^{-1}\left(f(x)\right) = f^{-1}\left(x^2 + 4\right)$

$= \sqrt{\left(x^2 + 4\right) - 4}$

$= \sqrt{x^2} = |x|$

$= x, \ x \geq 0$

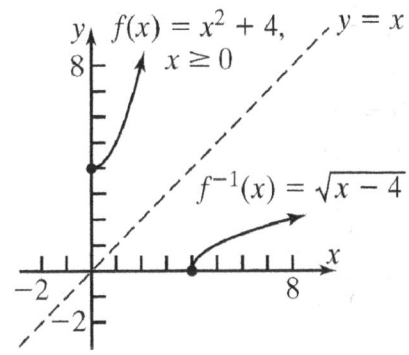

b. domain of f = range of $f^{-1} = \{x \mid x \geq 0\}$

range of f = domain of $f^{-1} = \{x \mid x \geq 4\}$

59. a. $f(x) = \dfrac{4}{x}$

$y = \dfrac{4}{x}$

$x = \dfrac{4}{y}$ Inverse

$xy = 4$

$y = \dfrac{4}{x}$

$f^{-1}(x) = \dfrac{4}{x}$

Verifying:

$f\left(f^{-1}(x)\right) = f\left(\dfrac{4}{x}\right) = \dfrac{4}{\frac{4}{x}} = 4 \cdot \left(\dfrac{x}{4}\right) = x$

$f^{-1}\left(f(x)\right) = f^{-1}\left(\dfrac{4}{x}\right) = \dfrac{4}{\frac{4}{x}} = 4 \cdot \left(\dfrac{x}{4}\right) = x$

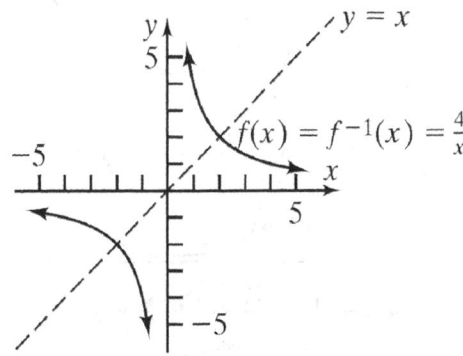

b. domain of f = range of $f^{-1} = \{x \mid x \neq 0\}$

range of f = domain of $f^{-1} = \{x \mid x \neq 0\}$

61. a. $f(x) = \dfrac{1}{x-2}$

$y = \dfrac{1}{x-2}$

$x = \dfrac{1}{y-2}$ Inverse

$xy - 2x = 1$

$xy = 2x + 1$

$y = \dfrac{2x+1}{x}$

$f^{-1}(x) = \dfrac{2x+1}{x}$

Verifying:

$f\left(f^{-1}(x)\right) = f\left(\dfrac{2x+1}{x}\right) = \dfrac{1}{\dfrac{2x+1}{x} - 2}$

$= \dfrac{1 \cdot x}{\left(\dfrac{2x+1}{x} - 2\right)x} = \dfrac{x}{2x+1-2x}$

$= \dfrac{x}{1} = x$

$$f^{-1}\left(f(x)\right) = f^{-1}\left(\dfrac{1}{x-2}\right) = \dfrac{2\left(\dfrac{1}{x-2}\right)+1}{\dfrac{1}{x-2}}$$

$$= \dfrac{\left(2\left(\dfrac{1}{x-2}\right)+1\right)(x-2)}{\left(\dfrac{1}{x-2}\right)(x-2)}$$

$$= \dfrac{2+(x-2)}{1} = \dfrac{x}{1} = x$$

$$f^{-1}(x) = \dfrac{2x+1}{x}$$
$$f(x) = \dfrac{1}{x-2}$$

b. domain of f = range of f^{-1} = $\{x \mid x \neq 2\}$
   range of f = domain of f^{-1} = $\{x \mid x \neq 0\}$

63. a. $f(x) = \dfrac{2}{3+x}$

$$y = \dfrac{2}{3+x}$$

$$x = \dfrac{2}{3+y} \quad \text{Inverse}$$

$$x(3+y) = 2$$
$$3x + xy = 2$$
$$xy = 2 - 3x$$
$$y = \dfrac{2-3x}{x}$$
$$f^{-1}(x) = \dfrac{2-3x}{x}$$

Verifying:

$$f\left(f^{-1}(x)\right) = f\left(\dfrac{2-3x}{x}\right) = \dfrac{2}{3+\dfrac{2-3x}{x}}$$

$$= \dfrac{2 \cdot x}{\left(3+\dfrac{2-3x}{x}\right)x} = \dfrac{2x}{3x+2-3x}$$

$$= \dfrac{2x}{2} = x$$

$$f^{-1}\left(f(x)\right) = f^{-1}\left(\dfrac{2}{3+x}\right) = \dfrac{2-3\left(\dfrac{2}{3+x}\right)}{\dfrac{2}{3+x}}$$

$$= \dfrac{\left(2-3\left(\dfrac{2}{3+x}\right)\right)(3+x)}{\left(\dfrac{2}{3+x}\right)(3+x)}$$

$$= \dfrac{2(3+x)-3(2)}{2} = \dfrac{6+2x-6}{2}$$

$$= \dfrac{2x}{2} = x$$

b. domain of f = range of f^{-1} = $\{x \mid x \neq -3\}$
   range of f = domain of f^{-1} = $\{x \mid x \neq 0\}$

65. a. $f(x) = \dfrac{3x}{x+2}$

$$y = \dfrac{3x}{x+2}$$

$$x = \dfrac{3y}{y+2} \quad \text{Inverse}$$

$$x(y+2) = 3y$$
$$xy + 2x = 3y$$
$$xy - 3y = -2x$$
$$y(x-3) = -2x$$
$$y = \dfrac{-2x}{x-3}$$
$$f^{-1}(x) = \dfrac{-2x}{x-3}$$

Verifying:

$$f\left(f^{-1}(x)\right) = f\left(\dfrac{-2x}{x-3}\right)$$

$$= \dfrac{3\left(\dfrac{-2x}{x-3}\right)}{\dfrac{-2x}{x-3}+2} = \dfrac{\left(3\left(\dfrac{-2x}{x-3}\right)\right)(x-3)}{\left(\dfrac{-2x}{x-3}+2\right)(x-3)}$$

$$= \dfrac{-6x}{-2x+2x-6} = \dfrac{-6x}{-6} = x$$

$$f^{-1}\left(f(x)\right) = f^{-1}\left(\frac{3x}{x+2}\right)$$

$$= \frac{-2\left(\frac{3x}{x+2}\right)}{\frac{3x}{x+2} - 3} = \frac{\left(-2\left(\frac{3x}{x+2}\right)\right)(x+2)}{\left(\frac{3x}{x+2} - 3\right)(x+2)}$$

$$= \frac{-6x}{3x - 3x - 6} = \frac{-6x}{-6} = x$$

b. domain of f = range of f^{-1} = $\{x \mid x \neq -2\}$

range of f = domain of f^{-1} = $\{x \mid x \neq 3\}$

67. a. $f(x) = \dfrac{2x}{3x-1}$

$$y = \frac{2x}{3x-1}$$

$$x = \frac{2y}{3y-1} \quad \text{Inverse}$$

$$3xy - x = 2y$$

$$3xy - 2y = x$$

$$y(3x-2) = x$$

$$y = \frac{x}{3x-2}$$

$$f^{-1}(x) = \frac{x}{3x-2}$$

Verifying:

$$f\left(f^{-1}(x)\right) = f\left(\frac{x}{3x-2}\right) = \frac{2\left(\frac{x}{3x-2}\right)}{3\left(\frac{x}{3x-2}\right) - 1}$$

$$= \frac{\left(2\left(\frac{x}{3x-2}\right)\right)(3x-2)}{\left(3\left(\frac{x}{3x-2}\right) - 1\right)(3x-2)}$$

$$= \frac{2x}{3x - (3x-2)} = \frac{2x}{2} = x$$

$$f^{-1}\left(f(x)\right) = f\left(\frac{2x}{3x-1}\right) = \frac{\frac{2x}{3x-1}}{3\left(\frac{2x}{3x-1}\right) - 2}$$

$$= \frac{\left(\frac{2x}{3x-1}\right)(3x-1)}{\left(3\left(\frac{2x}{3x-1}\right) - 2\right)(3x-1)}$$

$$= \frac{2x}{3(2x) - 2(3x-1)}$$

$$= \frac{2x}{6x - 6x + 2} = \frac{2x}{2} = x$$

b. domain of f = range of f^{-1} = $x \mid x \neq \dfrac{1}{3}$

range of f = domain of f^{-1} = $x \mid x \neq \dfrac{2}{3}$

69. a. $f(x) = \dfrac{3x+4}{2x-3}$

$$y = \frac{3x+4}{2x-3}$$

$$x = \frac{3y+4}{2y-3} \quad \text{Inverse}$$

$$x(2y-3) = 3y+4$$

$$2xy - 3x = 3y + 4$$

$$2xy - 3y = 3x + 4$$

$$y(2x-3) = 3x + 4$$

$$y = \frac{3x+4}{2x-3}$$

$$f^{-1}(x) = \frac{3x+4}{2x-3}$$

Verifying:

$$f\left(f^{-1}(x)\right) = f\left(\frac{3x+4}{2x-3}\right) = \frac{3\left(\frac{3x+4}{2x-3}\right) + 4}{2\left(\frac{3x+4}{2x-3}\right) - 3}$$

$$= \frac{\left(3\left(\frac{3x+4}{2x-3}\right) + 4\right)(2x-3)}{\left(2\left(\frac{3x+4}{2x-3}\right) - 3\right)(2x-3)}$$

$$= \frac{3(3x+4) + 4(2x-3)}{2(3x+4) - 3(2x-3)}$$

$$= \frac{9x + 12 + 8x - 12}{6x + 8 - 6x + 9} = \frac{17x}{17} = x$$

$$f^{-1}\left(f(x)\right) = f^{-1}\left(\frac{3x+4}{2x-3}\right) = \frac{3\left(\dfrac{3x+4}{2x-3}\right)+4}{2\left(\dfrac{3x+4}{2x-3}\right)-3}$$

$$= \frac{\left(3\left(\dfrac{3x+4}{2x-3}\right)+4\right)(2x-3)}{\left(2\left(\dfrac{3x+4}{2x-3}\right)-3\right)(2x-3)}$$

$$= \frac{3(3x+4)+4(2x-3)}{2(3x+4)-3(2x-3)}$$

$$= \frac{9x+12+8x-12}{6x+8-6x+9} = \frac{17x}{17} = x$$

b. domain of f = range of f^{-1} = $\left\{x \mid x \neq \dfrac{3}{2}\right\}$

range of f = domain of f^{-1} = $\left\{x \mid x \neq \dfrac{3}{2}\right\}$

71. a. $f(x) = \dfrac{2x+3}{x+2}$

$y = \dfrac{2x+3}{x+2}$

$x = \dfrac{2y+3}{y+2}$ Inverse

$xy + 2x = 2y + 3$

$xy - 2y = -2x + 3$

$y(x-2) = -2x + 3$

$y = \dfrac{-2x+3}{x-2}$

$f^{-1}(x) = \dfrac{-2x+3}{x-2}$

Verifying:

$$f\left(f^{-1}(x)\right) = f\left(\frac{-2x+3}{x-2}\right) = \frac{2\left(\dfrac{-2x+3}{x-2}\right)+3}{\dfrac{-2x+3}{x-2}+2}$$

$$= \frac{\left(2\left(\dfrac{-2x+3}{x-2}\right)+3\right)(x-2)}{\left(\dfrac{-2x+3}{x-2}+2\right)(x-2)}$$

$$= \frac{2(-2x+3)+3(x-2)}{-2x+3+2(x-2)}$$

$$= \frac{-4x+6+3x-6}{-2x+3+2x-4} = \frac{-x}{-1} = x$$

$$f^{-1}\left(f(x)\right) = f^{-1}\left(\frac{2x+3}{x+2}\right) = \frac{-2\left(\dfrac{2x+3}{x+2}\right)+3}{\dfrac{2x+3}{x+2}-2}$$

$$= \frac{\left(-2\left(\dfrac{2x+3}{x+2}\right)+3\right)(x+2)}{\left(\dfrac{2x+3}{x+2}-2\right)(x+2)}$$

$$= \frac{-2(2x+3)+3(x+2)}{2x+3-2(x+2)}$$

$$= \frac{-4x-6+3x+6}{2x+3-2x-4} = \frac{-x}{-1} = x$$

b. domain of f = range of f^{-1} = $\left\{x \mid x \neq -2\right\}$

range of f = domain of f^{-1} = $\left\{x \mid x \neq 2\right\}$

73. a. $f(x) = \dfrac{x^2-4}{2x^2},\ x > 0$

$y = \dfrac{x^2-4}{2x^2},\quad x > 0$

$x = \dfrac{y^2-4}{2y^2},\quad y > 0$ Inverse

$2xy^2 = y^2 - 4,\quad x < \dfrac{1}{2}$

$2xy^2 - y^2 = -4,\quad x < \dfrac{1}{2}$

$y^2(2x-1) = -4,\quad x < \dfrac{1}{2}$

$y^2(1-2x) = 4,\quad x < \dfrac{1}{2}$

$y^2 = \dfrac{4}{1-2x},\quad x < \dfrac{1}{2}$

$y = \sqrt{\dfrac{4}{1-2x}},\quad x < \dfrac{1}{2}$

$y = \dfrac{2}{\sqrt{1-2x}},\quad x < \dfrac{1}{2}$

$f^{-1}(x) = \dfrac{2}{\sqrt{1-2x}},\quad x < \dfrac{1}{2}$

Verifying:

$$f\left(f^{-1}(x)\right) = f\left(\frac{2}{\sqrt{1-2x}}\right) = \frac{\left(\frac{2}{\sqrt{1-2x}}\right)^2 - 4}{2\left(\frac{2}{\sqrt{1-2x}}\right)^2}$$

$$= \frac{\frac{4}{1-2x} - 4}{2\left(\frac{4}{1-2x}\right)} = \frac{\left(\frac{4}{1-2x} - 4\right)(1-2x)}{\left(2\left(\frac{4}{1-2x}\right)\right)(1-2x)}$$

$$= \frac{4 - 4(1-2x)}{2(4)} = \frac{4 - 4 + 8x}{8} = \frac{8x}{8} = x$$

$$f^{-1}\left(f(x)\right) = f^{-1}\left(\frac{x^2-4}{2x^2}\right) = \frac{2}{\sqrt{1-2\left(\frac{x^2-4}{2x^2}\right)}}$$

$$= \frac{2}{\sqrt{1-\frac{x^2-4}{x^2}}} = \frac{2}{\sqrt{1-1+\frac{4}{x^2}}}$$

$$= \frac{2}{\sqrt{\frac{4}{x^2}}} = \frac{2}{\frac{2}{|x|}} = 2 \cdot \frac{|x|}{2}$$

$$= |x| = x, \quad x > 0$$

b. domain of f = range of $f^{-1} = \{x \mid x > 0\}$

range of f = domain of $f^{-1} = \left\{x \mid x < \frac{1}{2}\right\}$

75. a. Because the ordered pair $(-1,0)$ is on the graph, $f(-1) = 0$.

b. Because the ordered pair $(1,2)$ is on the graph, $f(1) = 2$.

c. Because the ordered pair $(0,1)$ is on the graph, $f^{-1}(1) = 0$.

d. Because the ordered pair $(1,2)$ is on the graph, $f^{-1}(2) = 1$.

77. Since $f(7) = 13$, we have $f^{-1}(13) = 7$; the input of the function is the output of the inverse when the output of the function is the input of the inverse.

79. Since the domain of a function is the range of the inverse, and the range of the function is the domain of the inverse, we get the following for f^{-1}:
Domain: $[-2, \infty)$ Range: $[5, \infty)$

81. Since the domain of a function is the range of the inverse, and the range of the function is the domain of the inverse, we get the following for g^{-1}:
Domain: $[0, \infty)$ Range: $(-\infty, 0]$

83. Since $f(x)$ is increasing on the interval $(0,5)$, it is one-to-one on the interval and has an inverse, $f^{-1}(x)$. In addition, we can say that $f^{-1}(x)$ is increasing on the interval $(f(0), f(5))$.

85.
$$f(x) = mx + b, \quad m \neq 0$$
$$y = mx + b$$
$$x = my + b \quad \text{Inverse}$$
$$x - b = my$$
$$y = \frac{1}{m}(x-b)$$
$$f^{-1}(x) = \frac{1}{m}(x-b), \quad m \neq 0$$

87. If (a,b) is on the graph of f, then (b,a) is on the graph of f^{-1}. Since the graph of f^{-1} lies in Quadrant I, both coordinates of (a,b) are positive, which means that both coordinates of (b,a) are positive. Thus, the graph of f^{-1} must lie in Quadrant I.

89. Answers may vary. One possibility follows:
$f(x) = |x|, x \geq 0$ is one-to-one.
Thus, $f(x) = x, x \geq 0$
$$y = x, x \geq 0$$
$$f^{-1}(x) = x, x \geq 0$$

91. a.
$$d = 6.97r - 90.39$$
$$d + 90.39 = 6.97r$$
$$\frac{d + 90.39}{6.97} = r$$
Therefore, we would write
$$r(d) = \frac{d + 90.39}{6.97}$$

b. $r(d(r)) = \dfrac{(6.97r - 90.39) + 90.39}{6.97}$
$$= \frac{6.97r + 90.39 - 90.39}{6.97} = \frac{6.97r}{6.97}$$
$$= r$$

$$d(r(d)) = 6.97\left(\frac{d+90.39}{6.97}\right) - 90.39$$
$$= d + 90.39 - 90.39$$
$$= d$$

c. $r(300) = \frac{300+90.39}{6.97} \approx 56.01$

If the distance required to stop was 300 feet, the speed of the car was roughly 56 miles per hour.

93. a. 6 feet = 72 inches
$$W(72) = 50 + 2.3(72-60)$$
$$= 50 + 2.3(12) = 50 + 27.6 = 77.6$$

The ideal weight of a 6-foot male is 77.6 kilograms.

b.
$$W = 50 + 2.3(h-60)$$
$$W - 50 = 2.3h - 138$$
$$W + 88 = 2.3h$$
$$\frac{W+88}{2.3} = h$$

Therefore, we would write
$$h(W) = \frac{W+88}{2.3}$$

c. $h(W(h)) = \dfrac{(50+2.3(h-60))+88}{2.3}$
$$= \frac{50+2.3h-138+88}{2.3} = \frac{2.3h}{2.3} = h$$

$$W(h(W)) = 50 + 2.3\left(\frac{W+88}{2.3} - 60\right)$$
$$= 50 + W + 88 - 138 = W$$

d. $h(80) = \dfrac{80+88}{2.3} = \dfrac{168}{2.3} \approx 73.04$

The height of a male who is at his ideal weight of 80 kg is roughly 73 inches.

95. a. From the restriction given in the problem statement, the domain is
$$\{g \mid 37,450 \le g \le 90,750\} \text{ or } [37450, 90750].$$

b. $T(37,450) = 5156.25 + 0.25(37,450 - 37,450)$
$$= 5154.25$$
$$T(90,450) = 5156.25 + 0.25(90,750 - 37,450)$$
$$= 18,481.25$$

Since T is linear and increasing, we have that

the range is $\{T \mid 5156.25 \le T \le 18,481.25\}$ or $[5156.25, 18481.25]$.

c. $T = 5156.25 + 0.25(g - 37,450)$
$$T - 5156.25 = 0.25(g - 37,450)$$
$$\frac{T-5156.25}{0.25} = g - 37,450$$
$$\frac{T-5156.25}{0.25} + 37,450 = g$$

Therefore, we would write
$$g(T) = \frac{T-5156.25}{0.25} + 37,450$$

Domain: $\{T \mid 5156.25 \le T \le 18,481.25\}$

Range: $\{g \mid 37,450 \le g \le 90,750\}$

97. a. The graph of H is symmetric about the y-axis. Since t represents the number of seconds *after* the rock begins to fall, we know that $t \ge 0$. The graph is strictly decreasing over its domain, so it is one-to-one.

b.
$$H = 100 - 4.9t^2$$
$$H + 4.9t^2 = 100$$
$$4.9t^2 = 100 - H$$
$$t^2 = \frac{100-H}{4.9}$$
$$t = \sqrt{\frac{100-H}{4.9}}$$

Therefore, we would write $t(H) = \sqrt{\dfrac{100-H}{4.9}}$.

(Note: we only need the principal square root since we know $t \ge 0$)

$$H(t(H)) = 100 - 4.9\left(\sqrt{\frac{100-H}{4.9}}\right)^2$$
$$= 100 - 4.9\left(\frac{100-H}{4.9}\right)$$
$$= 100 - 100 + H$$
$$= H$$

$$t(H(t)) = \sqrt{\frac{100 - (100 - 4.9t^2)}{4.9}}$$
$$= \sqrt{\frac{4.9t^2}{4.9}} = \sqrt{t^2} = t \quad \text{(since } t \ge 0)$$

c. $t(80) = \sqrt{\dfrac{100-80}{4.9}} \approx 2.02$

It will take the rock about 2.02 seconds to fall 80 meters.

99. $f(x) = \dfrac{ax+b}{cx+d}$

$y = \dfrac{ax+b}{cx+d}$

$x = \dfrac{ay+b}{cy+d}$ Inverse

$x(cy+d) = ay+b$

$cxy+dx = ay+b$

$cxy-ay = b-dx$

$y(cx-a) = b-dx$

$y = \dfrac{b-dx}{cx-a}$

$f^{-1}(x) = \dfrac{-dx+b}{cx-a}$

Now, $f = f^{-1}$ provided that $\dfrac{ax+b}{cx+d} = \dfrac{-dx+b}{cx-a}$.

This is only true if $a = -d$.

101. Answers will vary.

103. No, not every odd function is one-to-one. For example, $f(x) = x^3 - x$ is an odd function, but it is not one-to-one.

105. If a horizontal line passes through two points on a graph of a function, then the y value associated with that horizontal line will be assigned to two different x values which violates the definition of one-to-one.

107. We start with the graph of $y = |x|$. The graph will be shifted horizontally to the left by 2 units. The graph will be reflected on the x-axis. Then the graph will be shifted vertically by 3 units upward.

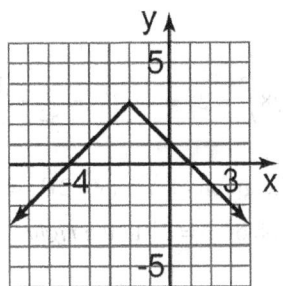

109. $R(x) = \dfrac{6x^2 - 11x - 2}{2x^2 - x - 6} = \dfrac{(x-2)(6x+1)}{(x-2)(2x+3)}$

$= \dfrac{6x+1}{2x+3},$ where $x \neq -\dfrac{3}{2}, 2$

Domain: $x \mid x \neq -\dfrac{3}{2}, 2$

The degree of the numerator in lowest terms is $n = 1$. The degree of the denominator in lowest terms is $m = 1$. Since $n = m$, the line $y = \dfrac{6}{2} = 3$ is a horizontal asymptote. The denominator in lowest terms is zero at $x = -\dfrac{3}{2}$, so $x = -\dfrac{3}{2}$ is a vertical asymptote.

Section 6.3

1. $4^3 = 64$; $8^{2/3} = \left(\sqrt[3]{8}\right)^2 = 2^2 = 4$; $3^{-2} = \dfrac{1}{3^2} = \dfrac{1}{9}$

3. False. To obtain the graph of $y = (x-2)^3$, we would shift the graph of $y = x^3$ to the *right* 2 units.

5. True

7. a

9. True

11. 4

13. b

15. a. $2^{3.14} \approx 8.815$

    b. $2^{3.141} \approx 8.821$

    c. $2^{3.1415} \approx 8.824$

    d. $2^{\pi} \approx 8.825$

17. a. $3.1^{2.7} \approx 21.217$

    b. $3.14^{2.71} \approx 22.217$

    c. $3.141^{2.718} \approx 22.440$

    d. $\pi^e \approx 22.459$

19. $(1+0.04)^6 \approx 1.265$

21. $8.4 \left(\dfrac{1}{3}\right)^{2.9} \approx 0.347$

23. $e^{1.2} \approx 3.320$

25. $125e^{0.025(7)} \approx 149.952$

27.

x	$y = f(x)$	$\dfrac{\Delta y}{\Delta x}$	$\dfrac{f(x+1)}{f(x)}$
-1	3		$\dfrac{6}{3} = 2$
0	6	$\dfrac{6-3}{0-(-1)} = 3$	$\dfrac{12}{6} = 2$
1	12	$\dfrac{12-6}{1-0} = 6$	$\dfrac{18}{12} = \dfrac{3}{2}$
2	18		
3	30		

Not a linear function since the average rate of change is not constant.

Not an exponential function since the ratio of consecutive terms is not constant.

29.

x	$y = H(x)$	$\dfrac{\Delta y}{\Delta x}$	$\dfrac{H(x+1)}{H(x)}$
-1	$\dfrac{1}{4}$		$\dfrac{1}{(1/4)} = 4$
0	1	$\dfrac{1-\frac{1}{4}}{0-(-1)} = \dfrac{3}{4}$	$\dfrac{4}{1} = 4$
1	4	$\dfrac{4-1}{1-0} = 3$	$\dfrac{16}{4} = 4$
2	16		$\dfrac{64}{16} = 4$
3	64		

Not a linear function since the average rate of change is not constant.

The ratio of consecutive outputs is a constant, 4. This is an exponential function with growth factor $a = 4$. The initial value of the exponential function is $C = 1$. Therefore, the exponential function that models the data is $H(x) = Ca^x = 1 \cdot (4)^x = 4^x$.

31.

x	$y = f(x)$	$\dfrac{\Delta y}{\Delta x}$	$\dfrac{f(x+1)}{f(x)}$
-1	$\dfrac{3}{2}$		$\dfrac{3}{(3/2)} = 2$
0	3	$\dfrac{3-\frac{3}{2}}{0-(-1)} = \dfrac{3}{2}$	$\dfrac{6}{3} = 2$
1	6	$\dfrac{6-3}{1-0} = 3$	$\dfrac{12}{6} = 2$
2	12		$\dfrac{24}{12} = 2$
3	24		

Not a linear function since the average rate of change is not constant.

The ratio of consecutive outputs is a constant, 2. This is an exponential function with growth factor $a = 2$. The initial value of the exponential function is $C = 3$. Therefore, the exponential function that models the data is $f(x) = Ca^x = 3 \cdot (2)^x = 3 \cdot 2^x$.

33.

x	$y = H(x)$	$\dfrac{\Delta y}{\Delta x}$	$\dfrac{H(x+1)}{H(x)}$
-1	2		$\dfrac{4}{2} = 2$
0	4	$\dfrac{4-2}{0-(-1)} = 2$	$\dfrac{6}{4} = \dfrac{3}{2}$
1	6	$\dfrac{6-4}{1-0} = 2$	
2	8	$\dfrac{8-6}{2-1} = 2$	
3	10	$\dfrac{10-8}{3-2} = 2$	

Not an exponential function since the ratio of consecutive terms is not constant.

The average rate of change is a constant, 2. Therefore, this is a linear function. In a linear function the average rate of change is the slope m. So, $m = 2$. When $x = 0$ we have $y = 4$ so the y-intercept is $b = 4$. The linear function that models this data is $H(x) = mx + b = 2x + 4$.

35. B

37. D

39. A

41. E

43. $f(x) = 2^x + 1$

Using the graph of $y = 2^x$, shift the graph up 1 unit.
Domain: All real numbers
Range: $\{y \mid y > 1\}$ or $(1, \infty)$
Horizontal Asymptote: $y = 1$

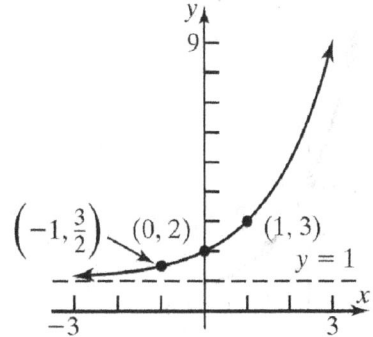

45. $f(x) = 3^{x-1}$

Using the graph of $y = 3^x$, shift the graph right 1 unit.
Domain: All real numbers
Range: $\{y \mid y > 0\}$ or $(0, \infty)$
Horizontal Asymptote: $y = 0$

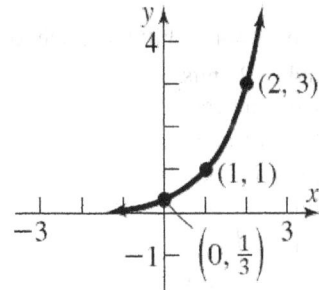

351

47. $f(x) = 3 \cdot \left(\dfrac{1}{2}\right)^x$

Using the graph of $y = \left(\dfrac{1}{2}\right)^x$, vertically stretch the graph by a factor of 3. That is, for each point on the graph, multiply the y-coordinate by 3.
Domain: All real numbers
Range: $\{y \mid y > 0\}$ or $(0, \infty)$
Horizontal Asymptote: $y = 0$

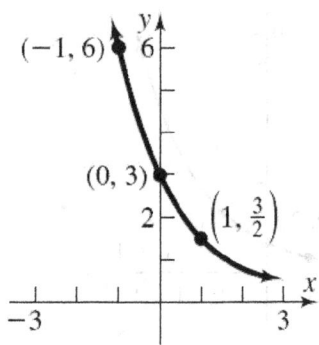

49. $f(x) = 3^{-x} - 2$

Using the graph of $y = 3^x$, reflect the graph about the y-axis, and shift down 2 units.
Domain: All real numbers
Range: $\{y \mid y > -2\}$ or $(-2, \infty)$
Horizontal Asymptote: $y = -2$

51. $f(x) = 2 + 4^{x-1}$

Using the graph of $y = 4^x$, shift the graph to the right one unit and up 2 units.
Domain: All real numbers
Range: $\{y \mid y > 2\}$ or $(2, \infty)$
Horizontal Asymptote: $y = 2$

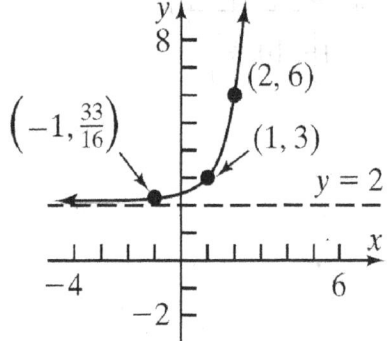

53. $f(x) = 2 + 3^{x/2}$

Using the graph of $y = 3^x$, stretch the graph horizontally by a factor of 2, and shift up 2 units.
Domain: All real numbers
Range: $\{y \mid y > 2\}$ or $(2, \infty)$
Horizontal Asymptote: $y = 2$

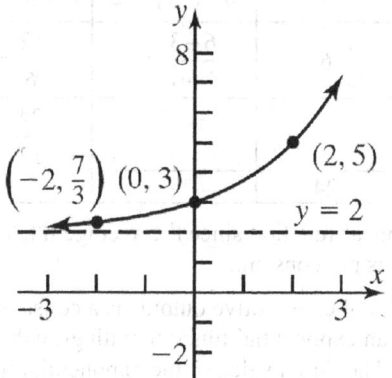

55. $f(x) = e^{-x}$

Using the graph of $y = e^x$, reflect the graph about the y-axis.
Domain: All real numbers
Range: $\{y \mid y > 0\}$ or $(0, \infty)$
Horizontal Asymptote: $y = 0$

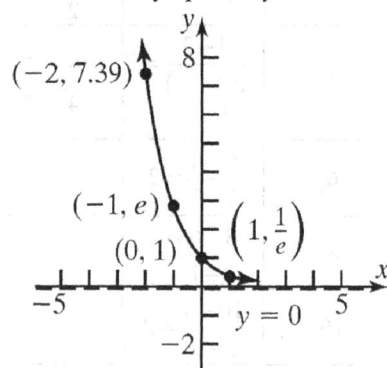

57. $f(x) = e^{x+2}$

Using the graph of $y = e^x$, shift the graph 2 units to the left.
Domain: All real numbers
Range: $\{y \mid y > 0\}$ or $(0, \infty)$
Horizontal Asymptote: $y = 0$

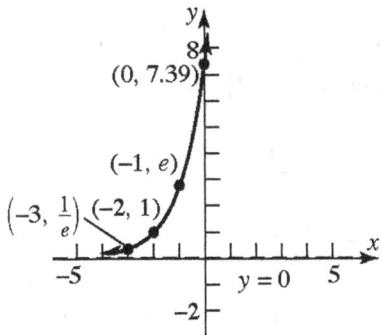

59. $f(x) = 5 - e^{-x}$

Using the graph of $y = e^x$, reflect the graph about the y-axis, reflect about the x-axis, and shift up 5 units.
Domain: All real numbers
Range: $\{y \mid y < 5\}$ or $(-\infty, 5)$
Horizontal Asymptote: $y = 5$

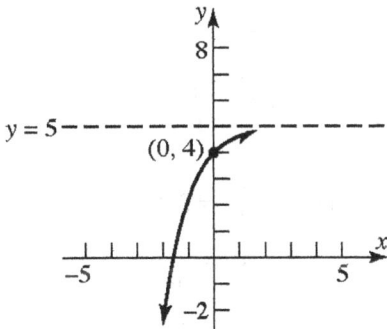

61. $f(x) = 2 - e^{-x/2}$

Using the graph of $y = e^x$, reflect the graph about the y-axis, stretch horizontally by a factor of 2, reflect about the x-axis, and shift up 2 units.
Domain: All real numbers
Range: $\{y \mid y < 2\}$ or $(-\infty, 2)$
Horizontal Asymptote: $y = 2$

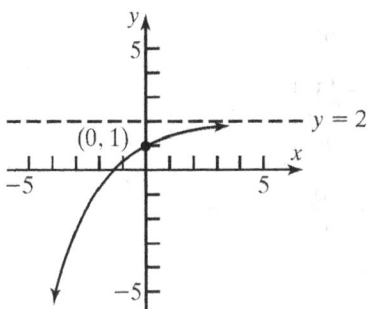

63. $7^x = 7^3$

We have a single term with the same base on both sides of the equation. Therefore, we can set the exponents equal to each other: $x = 3$.
The solution set is $\{3\}$.

65. $2^{-x} = 16$

$2^{-x} = 2^4$
$-x = 4$
$x = -4$
The solution set is $\{-4\}$.

67. $\left(\dfrac{1}{5}\right)^x = \dfrac{1}{25}$

$\left(\dfrac{1}{5}\right)^x = \dfrac{1}{5^2}$

$\left(\dfrac{1}{5}\right)^x = \left(\dfrac{1}{5}\right)^2$

$x = 2$
The solution set is $\{2\}$.

69. $2^{2x-1} = 4$

$2^{2x-1} = 2^2$
$2x - 1 = 2$
$2x = 3$
$x = \dfrac{3}{2}$

The solution set is $\left\{\dfrac{3}{2}\right\}$.

71.
$$3^{x^3} = 9^x$$
$$3^{x^3} = \left(3^2\right)^x$$
$$3^{x^3} = 3^{2x}$$
$$x^3 = 2x$$
$$x^3 - 2x = 0$$
$$x\left(x^2 - 2\right) = 0$$
$$x = 0 \text{ or } x^2 - 2 = 0$$
$$x^2 = 2$$
$$x = \pm\sqrt{2}$$
The solution set is $\left\{-\sqrt{2}, 0, \sqrt{2}\right\}$.

73.
$$8^{-x+14} = 16^x$$
$$\left(2^3\right)^{-x+14} = \left(2^4\right)^x$$
$$2^{-3x+42} = 2^{4x}$$
$$-3x + 42 = 4x$$
$$42 = 7x$$
$$6 = x$$
The solution set is $\{6\}$.

75.
$$3^{x^2-7} = 27^{2x}$$
$$3^{x^2-7} = \left(3^3\right)^{2x}$$
$$3^{x^2-7} = 3^{6x}$$
$$x^2 - 7 = 6x$$
$$x^2 - 6x - 7 = 0$$
$$(x-7)(x+1) = 0$$
$$x - 7 = 0 \text{ or } x + 1 = 0$$
$$x = 7 \qquad x = -1$$
The solution set is $\{-1, 7\}$.

77.
$$4^x \cdot 2^{x^2} = 16^2$$
$$\left(2^2\right)^x \cdot 2^{x^2} = \left(2^4\right)^2$$
$$2^{2x} \cdot 2^{x^2} = 2^8$$
$$2^{2x+x^2} = 2^8$$
$$x^2 + 2x = 8$$
$$x^2 + 2x - 8 = 0$$
$$(x+4)(x-2) = 0$$

$$x + 4 = 0 \text{ or } x - 2 = 0$$
$$x = -4 \qquad x = 2$$
The solution set is $\{-4, 2\}$.

79.
$$e^x = e^{3x+8}$$
$$x = 3x + 8$$
$$-2x = 8$$
$$x = -4$$
The solution set is $\{-4\}$.

81.
$$e^{x^2} = e^{3x} \cdot \frac{1}{e^2}$$
$$e^{x^2} = e^{3x} \cdot e^{-2}$$
$$e^{x^2} = e^{3x-2}$$
$$x^2 = 3x - 2$$
$$x^2 - 3x + 2 = 0$$
$$(x-2)(x-1) = 0$$
$$x - 2 = 0 \text{ or } x - 1 = 0$$
$$x = 2 \qquad x = 1$$
The solution set is $\{1, 2\}$.

83. If $4^x = 7$, then $\left(4^x\right)^{-2} = 7^{-2}$
$$4^{-2x} = \frac{1}{7^2}$$
$$4^{-2x} = \frac{1}{49}$$

85. If $3^{-x} = 2$, then $\left(3^{-x}\right)^{-2} = 2^{-2}$
$$3^{2x} = \frac{1}{2^2}$$
$$3^{2x} = \frac{1}{4}$$

87. If $9^x = 25$, then $\left(3^2\right)^x = 5^2$
$$\left(3^x\right)^2 = 5^2$$
$$3^x = 5$$

89. We need a function of the form $f(x) = k \cdot a^{p \cdot x}$, with $a > 0$, $a \neq 1$. The graph contains the points $\left(-1, \frac{1}{3}\right)$, $(0,1)$, $(1,3)$, and $(2,9)$. In other words,

$f(-1) = \frac{1}{3}$, $f(0) = 1$, $f(1) = 3$, and $f(2) = 9$.

Therefore, $f(0) = k \cdot a^{p \cdot (0)}$

$$1 = k \cdot a^0$$
$$1 = k \cdot 1$$
$$1 = k$$

and $f(1) = a^{p \cdot (1)}$

$$3 = a^p$$

Let's use $a = 3$, $p = 1$. Then $f(x) = 3^x$. Now we need to verify that this function yields the other known points on the graph. $f(-1) = 3^{-1} = \frac{1}{3}$;

$f(2) = 3^2 = 9$

So we have the function $f(x) = 3^x$.

91. We need a function of the form $f(x) = k \cdot a^{p \cdot x}$, with $a > 0$, $a \neq 1$. The graph contains the points $\left(-1, -\frac{1}{6}\right)$, $(0,-1)$, $(1,-6)$, and $(2,-36)$. In other words, $f(-1) = -\frac{1}{6}$, $f(0) = -1$, $f(1) = -6$, and $f(2) = -36$.

Therefore, $f(0) = k \cdot a^{p \cdot (0)}$ and $f(1) = -a^{p \cdot (1)}$.

$$-1 = k \cdot a^0 \qquad\qquad -6 = -a^p$$
$$-1 = k \cdot 1 \qquad\qquad 6 = a^p$$
$$-1 = k$$

Let's use $a = 6$, $p = 1$. Then $f(x) = -6^x$.

Now we need to verify that this function yields the other known points on the graph.

$f(-1) = -6^{-1} = -\frac{1}{6}$; $f(2) = -6^2 = -36$

So we have the function $f(x) = -6^x$.

93. We need a function of the form $f(x) = k \cdot a^{p \cdot x} + b$, with $a > 0$, $a \neq 1$ and b is the vertical shift of 2 units upward. The graph contains the points $(0,3)$, and $(1,5)$. In other words, $f(0) = 1$ and $f(1) = 3$. We can assume the graph has the same shape as the graph of $f(x) = k \cdot a^{p \cdot x}$. The reference (unshifted) graph would contain the points $(0,1)$, and $(1,3)$.

Therefore, $f(0) = k \cdot a^{p \cdot (0)}$ and $f(1) = a^{p \cdot (1)}$

$$1 = k \cdot a^0 \qquad\qquad 3 = a^p$$
$$1 = k \cdot 1$$
$$1 = k$$

Let's use $a = 3$, $p = 1$. Then $f(x) = 3^x$. To shift the graph up by 2 units we would have $f(x) = 3^x + 2$. Now we need to verify that this function yields the other known points on the graph.

$f(0) = 3^0 + 2 = 3$

$f(1) = 3^1 + 2 = 5$

So we have the function $f(x) = 3^x + 2$.

95. a. $f(4) = 2^4 = 16$

The point $(4,16)$ is on the graph of f.

b. $f(x) = \frac{1}{16}$

$$2^x = \frac{1}{16}$$
$$2^x = \frac{1}{2^4}$$
$$2^x = 2^{-4}$$
$$x = -4$$

The point $\left(-4, \frac{1}{16}\right)$ is on the graph of f.

97. a. $g(-1) = 4^{-1} + 2 = \frac{1}{4} + 2 = \frac{9}{4}$

The point $\left(-1, \frac{9}{4}\right)$ is on the graph of g.

b. $g(x) = 66$

$4^x + 2 = 66$

$4^x = 64$

$4^x = 4^3$

$x = 3$

The point $(3, 66)$ is on the graph of g.

99. a. $H(-6) = \left(\frac{1}{2}\right)^{-6} - 4 = (2)^6 - 4 = 60$

The point $(-6, 60)$ is on the graph of H.

b. $H(x) = 12$

$\left(\frac{1}{2}\right)^x - 4 = 12$

$\left(\frac{1}{2}\right)^x = 16$

$(2)^{-x} = 2^4$

$-x = 4$

$x = -4$

The point $(-4, 12)$ is on the graph of H.

c. $\left(\frac{1}{2}\right)^x - 4 = 0$

$\left(\frac{1}{2}\right)^x = 4$

$\left(2^{-1}\right)^x = 2^2$

$2^{-x} = 2^2$

$-x = 2$

$x = -2$

The zero of H is $x = -2$.

101. $f(x) = \begin{cases} e^{-x} & \text{if } x < 0 \\ e^x & \text{if } x \geq 0 \end{cases}$

Domain: $(-\infty, \infty)$

Range: $\{y \mid y \geq 1\}$ or $[1, \infty)$

Intercept: $(0, 1)$

103. $f(x) = \begin{cases} -e^x & \text{if } x < 0 \\ -e^{-x} & \text{if } x \geq 0 \end{cases}$

Domain: $(-\infty, \infty)$

Range: $\{y \mid -1 \leq y < 0\}$ or $[-1, 0)$

Intercept: $(0, -1)$

105. $p(n) = 100(0.97)^n$

a. $p(10) = 100(0.97)^{10} \approx 74\%$ of light

b. $p(25) = 100(0.97)^{25} \approx 47\%$ of light

107. $p(x) = 22,265(0.90)^x$

a. $p(3) = 22,265(0.90)^3 \approx \$16,231$

b. $p(9) = 22,265(0.90)^9 \approx \$8,626$

c. As each year passes, the sedan is worth 90% of its value the previous year.

109. $P(t) = 100(0.3)^t$

  a. $0.3 = 30\%$

  b. $P(2) = 100(0.3)^2 = 9\%$

  c. As each year passes, only 30% of the previous survivors survive again.

111. $D(h) = 5e^{-0.4h}$

$D(1) = 5e^{-0.4(1)} = 5e^{-0.4} \approx 3.35$

After 1 hours, 3.35 milligrams will be present.

$D(6) = 5e^{-0.4(6)} = 5e^{-2.4} \approx 0.45$ milligrams

After 6 hours, 0.45 milligrams will be present.

113. $F(t) = 1 - e^{-0.1t}$

  a. $F(10) = 1 - e^{-0.1(10)} = 1 - e^{-1} \approx 0.632$

The probability that a car will arrive within 10 minutes of 12:00 PM is 0.632.

  b. $F(40) = 1 - e^{-0.1(40)} = 1 - e^{-4} \approx 0.982$

The probability that a car will arrive within 40 minutes of 12:00 PM is 0.982.

  c. As $t \to \infty$, $F(t) = 1 - e^{-0.1t} \to 1 - 0 = 1$

  d. Graphing the function:

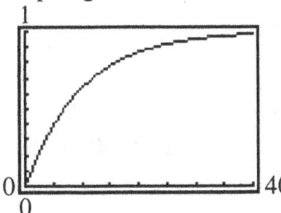

  e. $F(7) \approx 0.50$, so about 7 minutes are needed for the probability to reach 50%.

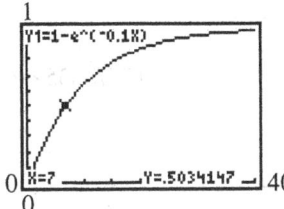

115. $P(x) = \dfrac{20^x e^{-20}}{x!}$

  a. $P(15) = \dfrac{20^{15} e^{-20}}{15!} \approx 0.0516$ or 5.16%

The probability that 15 cars will arrive between 5:00 PM and 6:00 PM is 5.16%.

  b. $P(20) = \dfrac{20^{20} e^{-20}}{20!} \approx 0.0888$ or 8.88%

The probability that 20 cars will arrive between 5:00 PM and 6:00 PM is 8.88%.

117. $R = 10^{\left(\frac{4221}{T+459.4} - \frac{4221}{D+459.4} + 2\right)}$

  a. $R = 10^{\left(\frac{4221}{50+459.4} - \frac{4221}{41+459.4} + 2\right)} \approx 70.95\%$

  b. $R = 10^{\left(\frac{4221}{68+459.4} - \frac{4221}{59+459.4} + 2\right)} \approx 72.62\%$

  c. $R = 10^{\left(\frac{4221}{T+459.4} - \frac{4221}{T+459.4} + 2\right)} = 10^2 = 100\%$

119. $I = \dfrac{E}{R}\left[1 - e^{-\left(\frac{R}{L}\right)t}\right]$

  a. $I_1 = \dfrac{120}{10}\left[1 - e^{-\left(\frac{10}{5}\right)0.3}\right] = 12\left[1 - e^{-0.6}\right] \approx 5.414$

amperes after 0.3 second

$I_1 = \dfrac{120}{10}\left[1 - e^{-\left(\frac{10}{5}\right)0.5}\right] = 12\left[1 - e^{-1}\right] \approx 7.585$

amperes after 0.5 second

$I_1 = \dfrac{120}{10}\left[1 - e^{-\left(\frac{10}{5}\right)1}\right] = 12\left[1 - e^{-2}\right] \approx 10.376$

amperes after 1 second

  b. As $t \to \infty$, $e^{-\left(\frac{10}{5}\right)t} \to 0$. Therefore, as,

$t \to \infty$, $I_1 = \dfrac{120}{10}\left[1 - e^{-\left(\frac{10}{5}\right)t}\right] \to 12[1-0] = 12$,

which means the maximum current is 12 amperes.

  c. See the graph at the end of the solution.

  d. $I_2 = \dfrac{120}{5}\left[1 - e^{-\left(\frac{5}{10}\right)0.3}\right] = 24\left[1 - e^{-0.15}\right]$

≈ 3.343 amperes after 0.3 second

$I_2 = \dfrac{120}{5}\left[1 - e^{-\left(\frac{5}{10}\right)0.5}\right] = 24\left[1 - e^{-0.25}\right]$

≈ 5.309 amperes after 0.5 second

$$I_2 = \frac{120}{5}\left[1 - e^{-\left(\frac{5}{10}\right)1}\right] = 24\left[1 - e^{-0.5}\right]$$

≈ 9.443 amperes after 1 second

which means the maximum current is 24 amperes.

f. See the graph that follows.

e. As $t \to \infty$, $e^{-\left(\frac{5}{10}\right)t} \to 0$. Therefore, as,

$$t \to \infty,\; I_1 = \frac{120}{5}\left[1 - e^{-\left(\frac{10}{5}\right)t}\right] \to 24[1-0] = 24,$$

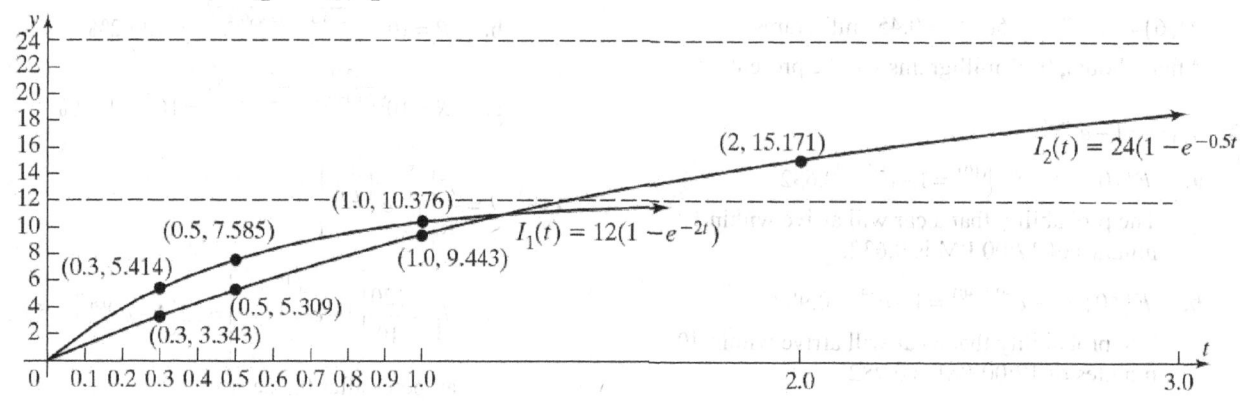

121. Since the growth rate is 3 then $a = 3$. So we have

$$f(x) = C \cdot 3^x$$
$$f(6) = C \cdot 3^6$$
$$12 = C \cdot 3^6$$
$$\frac{12}{3^6} = C$$

So $f(7) = \frac{12}{3^6} \cdot 3^7$
$$= 12 \cdot 3$$
$$= 36$$

So $f(7) = 36$

123. $2 + 1 = 3$

$$2 + \frac{1}{1+1} = 2.5 < e$$

$$2 + \cfrac{1}{1 + \cfrac{1}{2+2}} = 2.8 > e$$

$$2 + \cfrac{1}{1 + \cfrac{1}{2 + \cfrac{2}{3+3}}} = 2.7 < e$$

$$2 + \cfrac{1}{1 + \cfrac{1}{2 + \cfrac{2}{3 + \cfrac{3}{4+4}}}} \approx 2.721649485 > e$$

$$2 + \cfrac{1}{1 + \cfrac{1}{2 + \cfrac{2}{3 + \cfrac{3}{4 + \cfrac{4}{5+5}}}}} \approx 2.717770035 < e$$

$$2+\cfrac{1}{1+\cfrac{1}{2+\cfrac{2}{3+\cfrac{3}{4+\cfrac{4}{5+\cfrac{5}{6+6}}}}}} \approx 2.718348855 > e$$

$$e \approx 2.718281828$$

125. $f(x) = a^x$

$f(A+B) = a^{A+B} = a^A \cdot a^B = f(A) \cdot f(B)$

127. $f(x) = a^x$

$f(\alpha x) = a^{\alpha x} = \left(a^x\right)^{\alpha} = [f(x)]^{\alpha}$

129. $\cosh x = \dfrac{1}{2}\left(e^x + e^{-x}\right)$

a. $f(-x) = \cosh(-x)$

$= \dfrac{1}{2}\left(e^{-x} + e^x\right)$

$= \dfrac{1}{2}\left(e^x + e^{-x}\right)$

$= \cosh x$

$= f(x)$

Thus, $f(x) = \cosh x$ is an even function.

b. Let $Y_1 = \dfrac{1}{2}\left(e^x + e^{-x}\right)$.

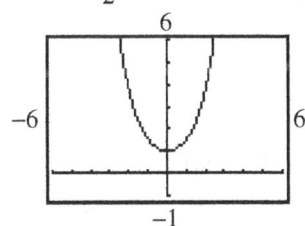

c. $(\cosh x)^2 - (\sinh x)^2$

$= \left(\dfrac{e^x + e^{-x}}{2}\right)^2 - \left(\dfrac{e^x - e^{-x}}{2}\right)^2$

$= \dfrac{e^{2x} + 2 + e^{-2x}}{4} - \dfrac{e^{2x} - 2 + e^{-2x}}{4}$

$= \dfrac{e^{2x} + 2 + e^{-2x} - e^{2x} + 2 - e^{-2x}}{4}$

$= \dfrac{4}{4}$

$= 1$

131. Since the number of bacteria doubles every minute, half of the container is full one minute before it is full. Thus, it takes 59 minutes to fill the container.

133. Answers will vary.

135. Using the laws of exponents, we have:

$a^{-x} = \dfrac{1}{a^x} = \left(\dfrac{1}{a}\right)^x$. So $y = a^{-x}$ and

$y = \left(\dfrac{1}{a}\right)^x$ will have the same graph.

137. $\dfrac{x+1}{x-2} \geq 1$

$\dfrac{x+1}{x-2} - 1 \geq 0$

$\dfrac{x+1-x+2}{x-2} \geq 0$

$\dfrac{3}{x-2} \geq 0$

$f(x) = \dfrac{3}{x-2}$

The value where f is undefined is $x = 2$.

Interval	$(-\infty, 2)$	$(2,\infty)$
Number Chosen	0	5
Value of f	-1.5	1
Conclusion	Negative	Positive

The solution set is $\{x \mid x > 2\}$ or, $(2,\infty)$.

139. a. $f(x) = x^2 + 2x - 3$

$a = 1, b = 2, c = -3$. Since $a = 1 > 0$, the graph opens up. The x-coordinate of the vertex is

$x = -\dfrac{b}{2a} = -\dfrac{2}{2(1)} = -1$.

The y-coordinate of the vertex is

$f\left(-\dfrac{b}{2a}\right) = f(-1) = (-1)^2 + 2(-1) - 3 = -4$.

The discriminant is:

$b^2 - 4ac = (2)^2 - 4(1)(-3) = 16 > 0$, so the graph has two x-intercepts.

The x-intercepts are found by solving:

$x^2 + 2x - 3 = 0$

$(x+3)(x-1) = 0$

$x = -3$ or $x = 1$

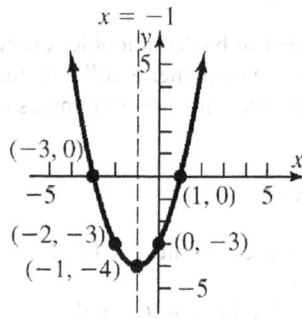

b. Domain: $(-\infty, \infty)$. Range: $[-4, \infty)$.

c. Decreasing on $(-\infty, -1]$; increasing on $[-1, \infty)$.

Section 6.4

1. a. $3x - 7 \le 8 - 2x$

   $5x \le 15$

   $x \le 3$

   The solution set is $\{x \mid x \le 3\}$.

   b. $x^2 - x - 6 > 0$

   We graph the function $f(x) = x^2 - x - 6$.

   The intercepts are

   y-intercept: $f(0) = -6$

   x-intercepts: $x^2 - x - 6 = 0$

   $(x+2)(x-3) = 0$

   $x = -2, x = 3$

   The vertex is at $x = \dfrac{-b}{2a} = \dfrac{-(-1)}{2(1)} = \dfrac{1}{2}$. Since

   $f\left(\dfrac{1}{2}\right) = -\dfrac{25}{4}$, the vertex is $\left(\dfrac{1}{2}, -\dfrac{25}{4}\right)$.

Thus, the vertex is $(-1, -4)$.

The axis of symmetry is the line $x = -1$.

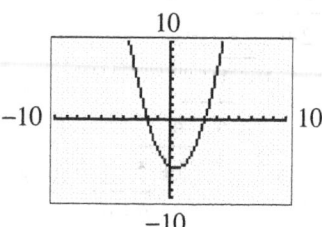

The graph is above the x-axis when $x < -2$ or $x > 3$. Since the inequality is strict, the solution set is $\{x \mid x < -2 \text{ or } x > 3\}$ or, using interval notation, $(-\infty, -2) \cup (3, \infty)$.

2. $\dfrac{x-1}{x+4} > 0$

   $f(x) = \dfrac{x-1}{x+4}$

   f is zero or undefined when $x = 1$ or $x = -4$.

Interval	$(-\infty, -4)$	$(-4, 1)$	$(1, \infty)$
Test Value	-5	0	2
Value of f	6	$-\dfrac{1}{4}$	$\dfrac{1}{6}$
Conclusion	positive	negative	positive

The solution set is $\{x \mid x < -4 \text{ or } x > 1\}$ or, using interval notation, $(-\infty, -4) \cup (1, \infty)$.

3. $2x + 3 = 9$

   $2x = 6$

   $x = 3$

5. $\left(\dfrac{1}{a}, -1\right)$, $(1, 0)$, $(a, 1)$

7. False. If $y = \log_a x$, then $x = a^y$.

9. a

11. $9 = 3^2$ is equivalent to $2 = \log_3 9$.

13. $a^2 = 1.6$ is equivalent to $2 = \log_a 1.6$.

15. $2^x = 7.2$ is equivalent to $x = \log_2 7.2$.

17. $e^x = 8$ is equivalent to $x = \ln 8$.

19. $\log_2 8 = 3$ is equivalent to $2^3 = 8$.

21. $\log_a 3 = 6$ is equivalent to $a^6 = 3$.

23. $\log_3 2 = x$ is equivalent to $3^x = 2$.

25. $\ln 4 = x$ is equivalent to $e^x = 4$.

27. $\log_2 1 = 0$ since $2^0 = 1$.

29. $\log_5 25 = 2$ since $5^2 = 25$.

31. $\log_{1/2} 16 = -4$ since $\left(\dfrac{1}{2}\right)^{-4} = 2^4 = 16$.

33. $\log_{10} \sqrt{10} = \dfrac{1}{2}$ since $10^{1/2} = \sqrt{10}$.

35. $\log_{\sqrt{2}} 4 = 4$ since $\left(\sqrt{2}\right)^4 = 4$.

37. $\ln \sqrt{e} = \dfrac{1}{2}$ since $e^{1/2} = \sqrt{e}$.

39. $f(x) = \ln(x-3)$ requires $x - 3 > 0$.
$$x - 3 > 0$$
$$x > 3$$
The domain of f is $\{x \mid x > 3\}$ or $(3, \infty)$.

41. $F(x) = \log_2 x^2$ requires $x^2 > 0$.
$$x^2 > 0 \text{ for all } x \neq 0.$$
The domain of F is $\{x \mid x \neq 0\}$.

43. $f(x) = 3 - 2\log_4\left[\dfrac{x}{2} - 5\right]$ requires $\dfrac{x}{2} - 5 > 0$.
$$\frac{x}{2} - 5 > 0$$
$$\frac{x}{2} > 5$$
$$x > 10$$
The domain of f is $\{x \mid x > 10\}$ or $(10, \infty)$.

45. $f(x) = \ln\left(\dfrac{1}{x+1}\right)$ requires $\dfrac{1}{x+1} > 0$.

$p(x) = \dfrac{1}{x+1}$ is undefined when $x = -1$.

Interval	$(-\infty, -1)$	$(-1, \infty)$
Test Value	-2	0
Value of p	-1	1
Conclusion	negative	positive

The domain of f is $\{x \mid x > -1\}$ or $(-1, \infty)$.

47. $g(x) = \log_5\left(\dfrac{x+1}{x}\right)$ requires $\dfrac{x+1}{x} > 0$.

$p(x) = \dfrac{x+1}{x}$ is zero or undefined when $x = -1$ or $x = 0$.

Interval	$(-\infty, -1)$	$(-1, 0)$	$(0, \infty)$
Test Value	-2	$-\dfrac{1}{2}$	1
Value of p	$\dfrac{1}{2}$	-1	2
Conclusion	positive	negative	positive

The domain of g is $\{x \mid x < -1 \text{ or } x > 0\}$; $(-\infty, -1) \cup (0, \infty)$.

49. $f(x) = \sqrt{\ln x}$ requires $\ln x \geq 0$ and $x > 0$
$$\ln x \geq 0$$
$$x \geq e^0$$
$$x \geq 1$$
The domain of h is $\{x \mid x \geq 1\}$ or $[1, \infty)$.

51. $\ln\left(\dfrac{5}{3}\right) \approx 0.511$

53. $\dfrac{\ln\dfrac{10}{3}}{0.04} \approx 30.099$

55. $\dfrac{\ln 4 + \ln 2}{\log 4 + \log 2} \approx 2.303$

57. $\dfrac{2\ln 5 + \log 50}{\log 4 - \ln 2} \approx -53.991$

59. If the graph of $f(x) = \log_a x$ contains the point $(2, 2)$, then $f(2) = \log_a 2 = 2$. Thus,

$$\log_a 2 = 2$$
$$a^2 = 2$$
$$a = \pm\sqrt{2}$$

Since the base a must be positive by definition, we have that $a = \sqrt{2}$.

61.

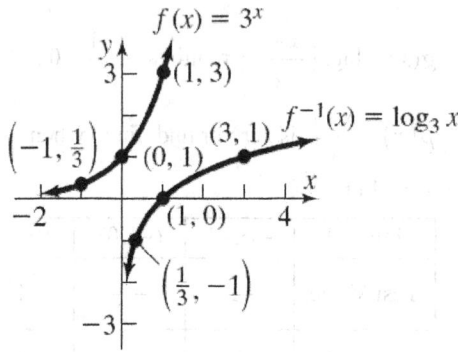

63.

$f(x) = \left(\frac{1}{2}\right)^x$

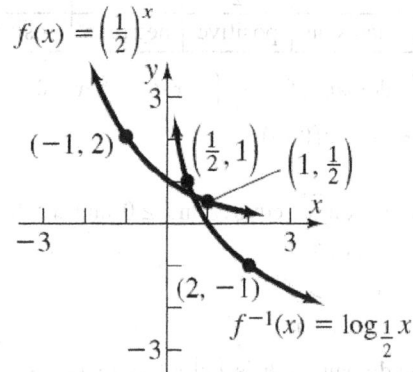

65. B

67. D

69. A

71. E

73. $f(x) = \ln(x+4)$

a. Domain: $(-4, \infty)$

b. Using the graph of $y = \ln x$, shift the graph 4 units to the left.

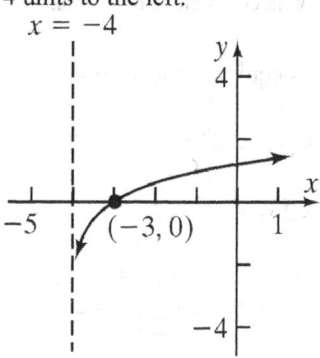

c. Range: $(-\infty, \infty)$
Vertical Asymptote: $x = -4$

d.
$$f(x) = \ln(x+4)$$
$$y = \ln(x+4)$$
$$x = \ln(y+4) \quad \text{Inverse}$$
$$y+4 = e^x$$
$$y = e^x - 4$$
$$f^{-1}(x) = e^x - 4$$

e. The domain of the inverse found in part (d) is all real numbers.

Since the domain of f is the range of f^{-1}, we can use the result from part (a) to say that the range of f^{-1} is $(-4, \infty)$.

f. Shift the graph of $y = e^x$ down 4 units.

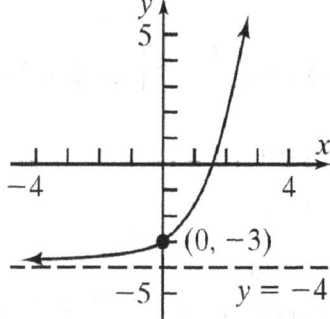

75. $f(x) = 2 + \ln x$

    a. Domain: $(0, \infty)$

    b. Using the graph of $y = \ln x$, shift up 2 units.

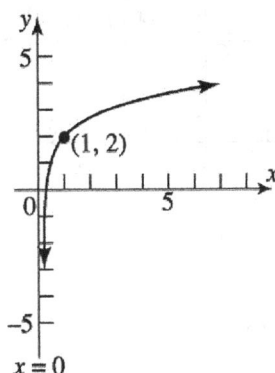

    c. Range: $(-\infty, \infty)$

        Vertical Asymptote: $x = 0$

    d. $\quad f(x) = 2 + \ln x$

$$y = 2 + \ln x$$
$$x = 2 + \ln y \quad \text{Inverse}$$
$$x - 2 = \ln y$$
$$y = e^{x-2}$$
$$f^{-1}(x) = e^{x-2}$$

    e. The domain of the inverse found in part (d) is all real numbers.

        Since the domain of f is the range of f^{-1}, we can use the result from part (a) to say that the range of f^{-1} is $(0, \infty)$.

    f. Using the graph of $y = e^x$, shift the graph 2 units to the right.

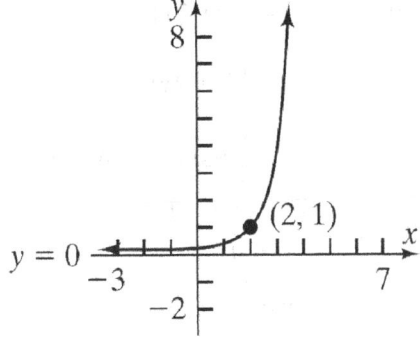

77. $f(x) = \ln(2x) - 3$

    a. Domain: $(0, \infty)$

    b. Using the graph of $y = \ln x$, compress the graph horizontally by a factor of $\frac{1}{2}$, and shift down 3 units.

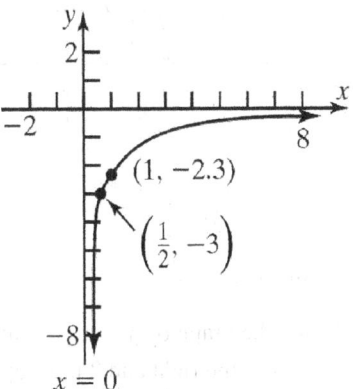

    c. Range: $(-\infty, \infty)$

        Vertical Asymptote: $x = 0$

    d. $\quad f(x) = \ln(2x) - 3$

$$y = \ln(2x) - 3$$
$$x = \ln(2y) - 3 \quad \text{Inverse}$$
$$x + 3 = \ln(2y)$$
$$2y = e^{x+3}$$
$$y = \frac{1}{2} e^{x+3}$$
$$f^{-1}(x) = \frac{1}{2} e^{x+3}$$

    e. The domain of the inverse found in part (d) is all real numbers.

        Since the domain of f is the range of f^{-1}, we can use the result from part (a) to say that the range of f^{-1} is $(0, \infty)$.

f. Using the graph of $y = e^x$, reflect the graph about the y-axis, and reflect about the x-axis.

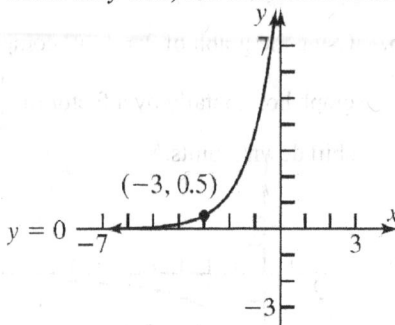

79. $f(x) = \log(x-4) + 2$

    a. Domain: $(4, \infty)$

    b. Using the graph of $y = \log x$, shift the graph 4 units to the right and 2 units up.

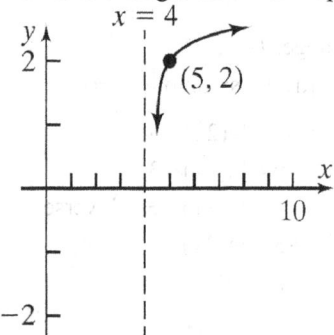

    c. Range: $(-\infty, \infty)$
        Vertical Asymptote: $x = 4$

    d. $\quad f(x) = \log(x-4) + 2$
$$y = \log(x-4) + 2$$
$$x = \log(y-4) + 2 \quad \text{Inverse}$$
$$x - 2 = \log(y-4)$$
$$y - 4 = 10^{x-2}$$
$$y = 10^{x-2} + 4$$
$$f^{-1}(x) = 10^{x-2} + 4$$

    e. The domain of the inverse found in part (d) is all real numbers.

        Since the domain of f is the range of f^{-1}, we can use the result from part (a) to say that the range of f^{-1} is $(4, \infty)$.

f. Using the graph of $y = 10^x$, shift the graph 2 units to the right and 4 units up.

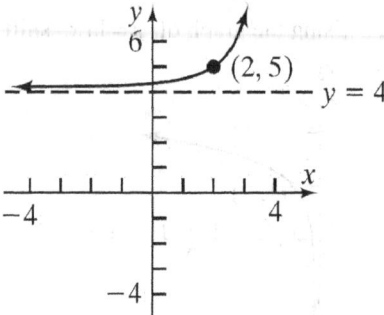

81. $f(x) = \dfrac{1}{2}\log(2x)$

    a. Domain: $(0, \infty)$

    b. Using the graph of $y = \log x$, compress the graph horizontally by a factor of $\dfrac{1}{2}$, and compress vertically by a factor of $\dfrac{1}{2}$.

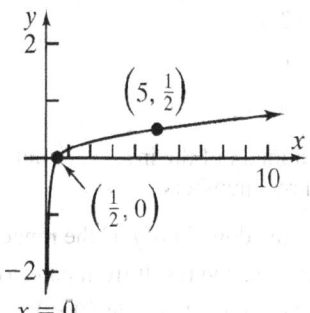

    c. Range: $(-\infty, \infty)$
        Vertical Asymptote: $x = 0$

    d. $\quad f(x) = \dfrac{1}{2}\log(2x)$
$$y = \dfrac{1}{2}\log(2x)$$
$$x = \dfrac{1}{2}\log(2y) \quad \text{Inverse}$$
$$2x = \log(2y)$$
$$2y = 10^{2x}$$
$$y = \dfrac{1}{2} \cdot 10^{2x}$$
$$f^{-1}(x) = \dfrac{1}{2} \cdot 10^{2x}$$

    e. The domain of the inverse found in part (d) is all real numbers.

Since the domain of f is the range of f^{-1}, we can use the result from part (a) to say that the range of f^{-1} is $(0,\infty)$.

f. Using the graph of $y = 10^x$, compress the graph horizontally by a factor of $\frac{1}{2}$, and compress vertically by a factor of $\frac{1}{2}$.

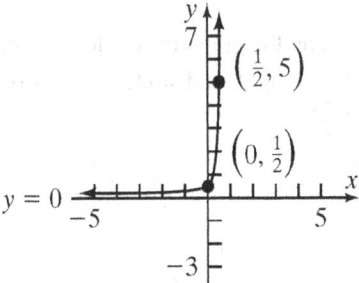

83. $f(x) = 3 + \log_3(x+2)$

a. Domain: $(-2, \infty)$

b. Using the graph of $y = \log_3 x$, shift 2 units to the left, and shift up 3 units.

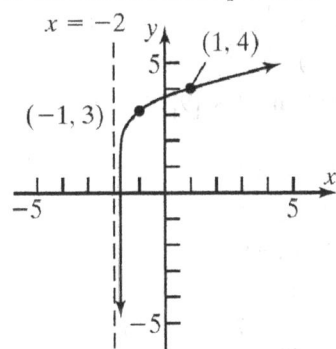

c. Range: $(-\infty, \infty)$
Vertical Asymptote: $x = -2$

d. $f(x) = 3 + \log_3(x+2)$
$y = 3 + \log_3(x+2)$
$x = 3 + \log_3(y+2)$ Inverse
$x - 3 = \log_3(y+2)$
$y + 2 = 3^{x-3}$
$y = 3^{x-3} - 2$
$f^{-1}(x) = 3^{x-3} - 2$

e. The domain of the inverse found in part (d) is all real numbers.

Since the domain of f is the range of f^{-1}, we can use the result from part (a) to say that the range of f^{-1} is $(-2,\infty)$.

f. Using the graph of $y = 3^x$, shift 3 units to the right, and shift down 2 units.

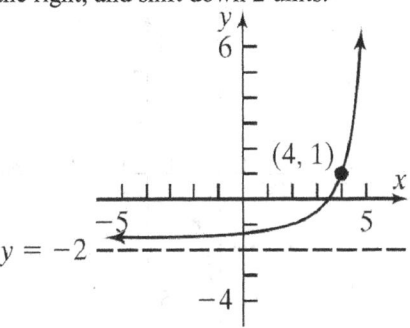

85. $f(x) = e^{x+2} - 3$

a. Domain: $(-\infty, \infty)$

b. Using the graph of $y = e^x$, shift the graph two units to the left, and shift 3 units down.

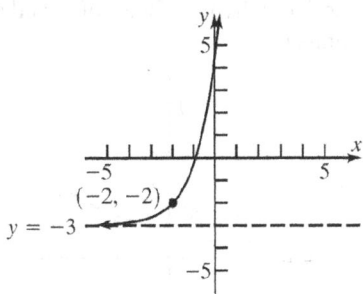

c. Range: $(-3, \infty)$
Horizontal Asymptote: $y = -3$

d. $f(x) = e^{x+2} - 3$
$y = e^{x+2} - 3$
$x = e^{y+2} - 3$ Inverse
$x + 3 = e^{y+2}$
$y + 2 = \ln(x+3)$
$y = \ln(x+3) - 2$
$f^{-1}(x) = \ln(x+3) - 2$

e. For the domain of f^{-1} we need
$x + 3 > 0$
$x > -3$
So the domain of the inverse found in part (d) is $(-3,\infty)$.

Since the domain of f is the range of f^{-1}, we can use the result from part (a) to say that the range of f^{-1} is $(-\infty, \infty)$.

f. Using the graph of $y = \ln x$, shift 3 units to the left, and shift down 2 units.

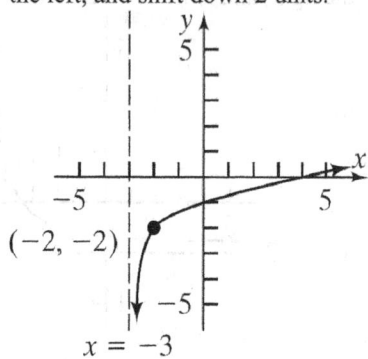

87. $f(x) = 2^{x/3} + 4$

a. Domain: $(-\infty, \infty)$

b. Using the graph of $y = 2^x$, stretch the graph horizontally by a factor of 3, and shift 4 units up.

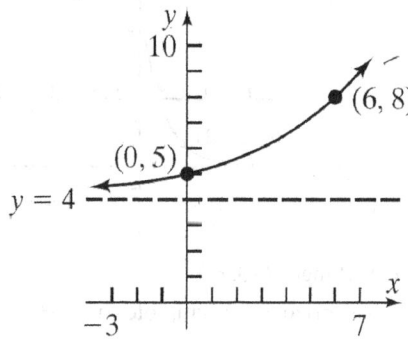

c. Range: $(4, \infty)$

Horizontal Asymptote: $y = 4$

d.
$$f(x) = 2^{x/3} + 4$$
$$y = 2^{x/3} + 4$$
$$x = 2^{y/3} + 4 \quad \text{Inverse}$$
$$x - 4 = 2^{y/3}$$
$$\frac{y}{3} = \log_2 (x - 4)$$
$$y = 3\log_2 (x - 4)$$
$$f^{-1}(x) = 3\log_2 (x - 4)$$

e. For the domain of f^{-1} we need
$$x - 4 > 0$$
$$x > 4$$
The domain of the inverse found in part (d) is $(4, \infty)$.

Since the domain of f is the range of f^{-1}, we can use the result from part (a) to say that the range of f^{-1} is $(-\infty, \infty)$.

f. Using the graph of $y = \log_2 x$, shift 4 units to the right, and stretch vertically by a factor of 3.

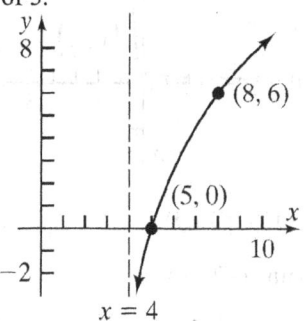

89. $\log_3 x = 2$
$$x = 3^2$$
$$x = 9$$
The solution set is $\{9\}$.

91. $\log_2 (2x + 1) = 3$
$$2x + 1 = 2^3$$
$$2x + 1 = 8$$
$$2x = 7$$
$$x = \frac{7}{2}$$
The solution set is $\left\{\frac{7}{2}\right\}$.

93. $\log_x 4 = 2$
$$x^2 = 4$$
$$x = 2 \quad (x \neq -2, \text{ base is positive})$$
The solution set is $\{2\}$.

95. $\ln e^x = 5$
$$e^x = e^5$$
$$x = 5$$
The solution set is $\{5\}$.

97. $\log_4 64 = x$

$4^x = 64$

$4^x = 4^3$

$x = 3$

The solution set is $\{3\}$.

99. $\log_3 243 = 2x+1$

$3^{2x+1} = 243$

$3^{2x+1} = 3^5$

$2x+1 = 5$

$2x = 4$

$x = 2$

The solution set is $\{2\}$.

101. $e^{3x} = 10$

$3x = \ln 10$

$x = \dfrac{\ln 10}{3}$

The solution set is $\left\{\dfrac{\ln 10}{3}\right\}$.

103. $e^{2x+5} = 8$

$2x+5 = \ln 8$

$2x = -5 + \ln 8$

$x = \dfrac{-5 + \ln 8}{2}$

The solution set is $\left\{\dfrac{-5 + \ln 8}{2}\right\}$.

105. $\log_3\left(x^2 + 1\right) = 2$

$x^2 + 1 = 3^2$

$x^2 + 1 = 9$

$x^2 = 8$

$x = \pm\sqrt{8} = \pm 2\sqrt{2}$

The solution set is $\left\{-2\sqrt{2},\ 2\sqrt{2}\right\}$.

107. $\log_2 8^x = -3$

$8^x = 2^{-3}$

$\left(2^3\right)^x = 2^{-3}$

$2^{3x} = 2^{-3}$

$3x = -3$

$x = -1$

The solution set is $\{-1\}$.

109. $5e^{0.2x} = 7$

$e^{0.2x} = \dfrac{7}{5}$

$0.2x = \ln\dfrac{7}{5}$

$5(0.2x) = 5\left(\ln\dfrac{7}{5}\right)$

$x = 5\ln\dfrac{7}{5}$

The solution set is $\left\{5\ln\dfrac{7}{5}\right\}$.

111. $2 \cdot 10^{2-x} = 5$

$10^{2-x} = \dfrac{5}{2}$

$2 - x = \log\dfrac{5}{2}$

$-x = -2 + \log\dfrac{5}{2}$

$x = 2 - \log\dfrac{5}{2}$

The solution set is $\left\{2 - \log\dfrac{5}{2}\right\}$.

113. a. $G(x) = \log_3\left(2x+1\right) - 2$

We require that $2x+1$ be positive.

$2x+1 > 0$

$2x > -1$

$x > -\dfrac{1}{2}$

Domain: $\left\{x \mid x > -\dfrac{1}{2}\right\}$ or $\left(-\dfrac{1}{2}, \infty\right)$

b. $G(40) = \log_3(2 \cdot 40 + 1) - 2$

$\qquad = \log_3 81 - 2$

$\qquad = 4 - 2$

$\qquad = 2$

The point $(40, 2)$ is on the graph of G.

c. $\qquad G(x) = 3$

$\log_3(2x+1) - 2 = 3$

$\qquad \log_3(2x+1) = 5$

$\qquad 2x + 1 = 3^5$

$\qquad 2x + 1 = 243$

$\qquad 2x = 242$

$\qquad x = 121$

The point $(121, 3)$ is on the graph of G.

d. $\qquad G(x) = 0$

$\log_3(2x+1) - 2 = 0$

$\qquad \log_3(2x+1) = 2$

$\qquad 2x + 1 = 3^2$

$\qquad 2x + 1 = 9$

$\qquad 2x = 8$

$\qquad x = 4$

The zero of G is $x = 4$.

115. $f(x) = \begin{cases} \ln(-x) & \text{if } x < 0 \\ \ln x & \text{if } x > 0 \end{cases}$

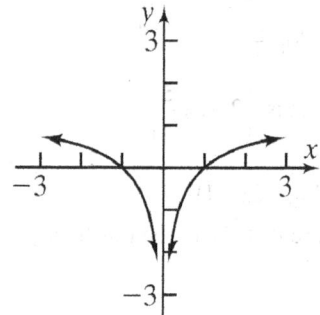

Domain: $\{x \mid x \neq 0\}$

Range: $(-\infty, \infty)$

Intercepts: $(-1, 0), (1, 0)$

117. $f(x) = \begin{cases} -\ln x & \text{if } 0 < x < 1 \\ \ln x & \text{if } x \geq 1 \end{cases}$

Domain: $\{x \mid x > 0\}$; $(0, \infty)$

Range: $\{y \mid y \geq 0\}$; $[0, \infty)$

Intercept: $(1, 0)$

119. $\text{pH} = -\log_{10}\left[H^+\right]$

a. $\text{pH} = -\log_{10}[0.1] = -(-1) = 1$

b. $\text{pH} = -\log_{10}[0.01] = -(-2) = 2$

c. $\text{pH} = -\log_{10}[0.001] = -(-3) = 3$

d. As the H^+ decreases, the pH increases.

e. $\qquad 3.5 = -\log_{10}\left[H^+\right]$

$\qquad -3.5 = \log_{10}\left[H^+\right]$

$\qquad \left[H^+\right] = 10^{-3.5}$

$\qquad \approx 3.16 \times 10^{-4}$

$\qquad = 0.000316$

f. $\qquad 7.4 = -\log_{10}\left[H^+\right]$

$\qquad -7.4 = \log_{10}\left[H^+\right]$

$\qquad \left[H^+\right] = 10^{-7.4}$

$\qquad \approx 3.981 \times 10^{-8}$

$\qquad = 0.00000003981$

121. $p = 760e^{-0.145h}$

a. $320 = 760e^{-0.145h}$

$$\frac{320}{760} = e^{-0.145h}$$

$$\ln\left(\frac{320}{760}\right) = -0.145h$$

$$h = \frac{\ln\left(\dfrac{320}{760}\right)}{-0.145} \approx 5.97$$

Approximately 5.97 kilometers.

b. $667 = 760e^{-0.145h}$

$$\frac{667}{760} = e^{-0.145h}$$

$$\ln\left(\frac{667}{760}\right) = -0.145h$$

$$h = \frac{\ln\left(\dfrac{667}{760}\right)}{-0.145} \approx 0.90$$

Approximately 0.90 kilometers.

123. $F(t) = 1 - e^{-0.1t}$

a. $0.5 = 1 - e^{-0.1t}$

$-0.5 = -e^{-0.1t}$

$0.5 = e^{-0.1t}$

$\ln(0.5) = -0.1t$

$$t = \frac{\ln(0.5)}{-0.1} \approx 6.93$$

Approximately 6.93 minutes.

b. $0.8 = 1 - e^{-0.1t}$

$-0.2 = -e^{-0.1t}$

$0.2 = e^{-0.1t}$

$\ln(0.2) = -0.1t$

$$t = \frac{\ln(0.2)}{-0.1} \approx 16.09$$

Approximately 16.09 minutes.

c. It is impossible for the probability to reach 100% because $e^{-0.1t}$ will never equal zero; thus, $F(t) = 1 - e^{-0.1t}$ will never equal 1.

125. $D = 5e^{-0.4h}$

$2 = 5e^{-0.4h}$

$0.4 = e^{-0.4h}$

$\ln(0.4) = -0.4h$

$$h = \frac{\ln(0.4)}{-0.4} \approx 2.29$$

Approximately 2.29 hours, or 2 hours and 17 minutes.

127. $I = \dfrac{E}{R}\left[1 - e^{-(R/L)t}\right]$

Substituting $E = 12$, $R = 10$, $L = 5$, and $I = 0.5$, we obtain:

$$0.5 = \frac{12}{10}\left[1 - e^{-(10/5)t}\right]$$

$$\frac{5}{12} = 1 - e^{-2t}$$

$$e^{-2t} = \frac{7}{12}$$

$$-2t = \ln(7/12)$$

$$t = \frac{\ln(7/12)}{-2} \approx 0.2695$$

It takes approximately 0.2695 second to obtain a current of 0.5 ampere.

Substituting $E = 12$, $R = 10$, $L = 5$, and $I = 1.0$, we obtain:

$$1.0 = \frac{12}{10}\left[1 - e^{-(10/5)t}\right]$$

$$\frac{10}{12} = 1 - e^{-2t}$$

$$e^{-2t} = \frac{1}{6}$$

$$-2t = \ln(1/6)$$

$$t = \frac{\ln(1/6)}{-2} \approx 0.8959$$

It takes approximately 0.8959 second to obtain a current of 0.5 ampere.

Graphing:

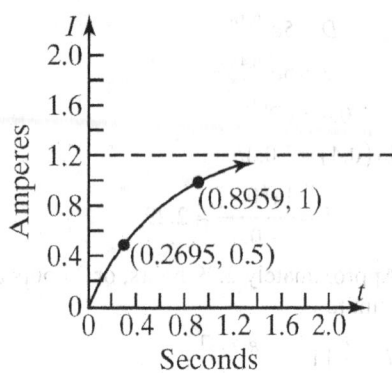

129. $L\left(10^{-7}\right)=10\log\left(\dfrac{10^{-7}}{10^{-12}}\right)$

$=10\log\left(10^{5}\right)$

$=10\cdot5$

$=50$ decibels

131. $L\left(10^{-3}\right)=10\log\left(\dfrac{10^{-3}}{10^{-12}}\right)$

$=10\log\left(10^{9}\right)$

$=10\cdot9$

$=90$ decibels

133. $M(125,892)=\log\left(\dfrac{125,892}{10^{-3}}\right)\approx8.1$

135. $R=e^{kx}$

a. $1.4=e^{k(0.03)}$

$1.4=e^{0.03k}$

$\ln(1.4)=0.03k$

$k=\dfrac{\ln(1.4)}{0.03}\approx11.216$

b. $R=e^{11.216(0.17)}=e^{1.90672}\approx6.73$

c. $100=e^{11.216x}$

$100=e^{11.216x}$

$\ln(100)=11.216x$

$x=\dfrac{\ln(100)}{11.216}\approx0.41$ percent

d. $5=e^{11.216x}$

$\ln5=11.216x$

$x=\dfrac{\ln5}{11.216}\approx0.14$ percent

At a percent concentration of 0.14 or higher, the driver should be charged with a DUI.

e. Answers will vary.

137. If the base of a logarithmic function equals 1, we would have the following:

$f(x)=\log_{1}(x)$

$f^{-1}(x)=1^{x}=1$ for every real number x.

In other words, f^{-1} would be a constant function and, therefore, f^{-1} would not be one-to-one.

139. $g(x)=4x^{4}-37x^{2}+9$

$=(4x^{2}-1)(x^{2}-9)$

$=(2x-1)(2x+1)(x-3)(x+3)$

$0=(2x-1)(2x+1)(x-3)(x+3)$

Setting each factor to 0, we get:

$x=\dfrac{1}{2},x=-\dfrac{1}{2},x=3,x=-3$

So the zeros of the function and the x-intercepts

are: $\dfrac{1}{2},-\dfrac{1}{2},3,-3$

141. $f(1)=4(1)^{3}-2(1)^{2}-7$

$=4-2-7$

$=-5$

$f(2)=4(2)^{3}-2(2)^{2}-7$

$=4(8)-2(4)-7$

$=32-8-7$

$=17$

Since f(b) and f(a) are of opposite signs there is at least one real zero of f between a and b.

Section 6.5

1. 0

3. r

5. $\log_a M \, ; \log_a N$

7. 7

9. False: $\log_2\left(3x^4\right) = \log_2 3 + 4\log_2 x$

11. b

13. $\log_3 3^{71} = 71$

15. $\ln e^{-4} = -4$

17. $2^{\log_2 7} = 7$

19. $\log_8 2 + \log_8 4 = \log_8\left(4\cdot 2\right) = \log_8 8 = 1$

21. $\log_6 18 - \log_6 3 = \log_6 \dfrac{18}{3} = \log_6 6 = 1$

23. $\log_2 6 \cdot \log_6 8 = \log_6 8^{\log_2 6}$
$$= \log_6\left(2^3\right)^{\log_2 6}$$
$$= \log_6 2^{3\log_2 6}$$
$$= \log_6 2^{\log_2 6^3}$$
$$= \log_6 6^3$$
$$= 3$$

25. $3^{\log_3 5 - \log_3 4} = 3^{\log_3 \frac{5}{4}} = \dfrac{5}{4}$

27. $e^{\log_{e^2} 16}$
Let $a = \log_{e^2} 16$, then $\left(e^2\right)^a = 16$.
$$e^{2a} = 16$$
$$e^{2a} = 4^2$$
$$\left(e^{2a}\right)^{1/2} = \left(4^2\right)^{1/2}$$
$$e^a = 4$$
$$a = \ln 4$$
Thus, $e^{\log_{e^2} 16} = e^{\ln 4} = 4$.

29. $\ln 6 = \ln(2\cdot 3) = \ln 2 + \ln 3 = a + b$

31. $\ln 1.5 = \ln\dfrac{3}{2} = \ln 3 - \ln 2 = b - a$

33. $\ln 8 = \ln 2^3 = 3 \cdot \ln 2 = 3a$

35. $\ln \sqrt[5]{6} = \ln 6^{1/5}$
$$= \frac{1}{5}\ln 6$$
$$= \frac{1}{5}\ln\left(2\cdot 3\right)$$
$$= \frac{1}{5}\left(\ln 2 + \ln 3\right)$$
$$= \frac{1}{5}\left(a + b\right)$$

37. $\log_5\left(25x\right) = \log_5 25 + \log_5 x = 2 + \log_5 x$

39. $\log_2 z^3 = 3\log_2 z$

41. $\ln\left(ex\right) = \ln e + \ln x = 1 + \ln x$

43. $\ln\left(\dfrac{x}{e^x}\right) = \ln x - \ln e^x = \ln x - x$

45. $\log_a\left(u^2 v^3\right) = \log_a u^2 + \log_a v^3$
$$= 2\log_a u + 3\log_a v$$

47. $\ln\left(x^2\sqrt{1-x}\right) = \ln x^2 + \ln\sqrt{1-x}$
$$= \ln x^2 + \ln(1-x)^{1/2}$$
$$= 2\ln x + \frac{1}{2}\ln(1-x)$$

49. $\log_2\left(\dfrac{x^3}{x-3}\right) = \log_2 x^3 - \log_2(x-3)$
$$= 3\log_2 x - \log_2(x-3)$$

51. $\log\left[\dfrac{x(x+2)}{(x+3)^2}\right] = \log\left[x(x+2)\right] - \log(x+3)^2$
$$= \log x + \log(x+2) - 2\log(x+3)$$

53. $\ln\left[\dfrac{x^2-x-2}{(x+4)^2}\right]^{1/3}$

$= \dfrac{1}{3}\ln\left[\dfrac{(x-2)(x+1)}{(x+4)^2}\right]$

$= \dfrac{1}{3}\left[\ln(x-2)(x+1)-\ln(x+4)^2\right]$

$= \dfrac{1}{3}\left[\ln(x-2)+\ln(x+1)-2\ln(x+4)\right]$

$= \dfrac{1}{3}\ln(x-2)+\dfrac{1}{3}\ln(x+1)-\dfrac{2}{3}\ln(x+4)$

55. $\ln\dfrac{5x\sqrt{1+3x}}{(x-4)^3}$

$= \ln\left(5x\sqrt{1+3x}\right)-\ln(x-4)^3$

$= \ln 5 + \ln x + \ln\sqrt{1+3x}-3\ln(x-4)$

$= \ln 5 + \ln x + \ln(1+3x)^{1/2}-3\ln(x-4)$

$= \ln 5 + \ln x + \dfrac{1}{2}\ln(1+3x)-3\ln(x-4)$

57. $3\log_5 u + 4\log_5 v = \log_5 u^3 + \log_5 v^4$

$\qquad\qquad\qquad = \log_5\left(u^3 v^4\right)$

59. $\log_3\sqrt{x}-\log_3 x^3 = \log_3\left(\dfrac{\sqrt{x}}{x^3}\right)$

$\qquad\qquad\qquad = \log_3\left(\dfrac{x^{1/2}}{x^3}\right)$

$\qquad\qquad\qquad = \log_3 x^{-5/2}$

$\qquad\qquad\qquad = \log_3\left(\dfrac{1}{x^{5/2}}\right)$

61. $\log_4\left(x^2-1\right)-5\log_4\left(x+1\right)$

$= \log_4\left(x^2-1\right)-\log_4\left(x+1\right)^5$

$= \log_4\left[\dfrac{x^2-1}{(x+1)^5}\right]$

$= \log_4\left[\dfrac{(x+1)(x-1)}{(x+1)^5}\right]$

$= \log_4\left[\dfrac{x-1}{(x+1)^4}\right]$

63. $\ln\left(\dfrac{x}{x-1}\right)+\ln\left(\dfrac{x+1}{x}\right)-\ln\left(x^2-1\right)$

$= \ln\left[\dfrac{x}{x-1}\cdot\dfrac{x+1}{x}\right]-\ln\left(x^2-1\right)$

$= \ln\left[\dfrac{x+1}{x-1}\div\left(x^2-1\right)\right]$

$= \ln\left[\dfrac{x+1}{(x-1)\left(x^2-1\right)}\right]$

$= \ln\left[\dfrac{x+1}{(x-1)(x-1)(x+1)}\right]$

$= \ln\left(\dfrac{1}{(x-1)^2}\right)$

$= \ln(x-1)^{-2}$

$= -2\ln(x-1)$

65. $8\log_2\sqrt{3x-2}-\log_2\left(\dfrac{4}{x}\right)+\log_2 4$

$= \log_2\left(\sqrt{3x-2}\right)^8-\left(\log_2 4-\log_2 x\right)+\log_2 4$

$= \log_2(3x-2)^4-\log_2 4+\log_2 x+\log_2 4$

$= \log_2(3x-2)^4+\log_2 x$

$= \log_2\left[x(3x-2)^4\right]$

67. $2\log_a\left(5x^3\right)-\dfrac{1}{2}\log_a(2x+3)$

$= \log_a\left(5x^3\right)^2-\log_a(2x+3)^{1/2}$

$= \log_a\left(25x^6\right)-\log_a\sqrt{2x+3}$

$= \log_a\left[\dfrac{25x^6}{\sqrt{2x+3}}\right]$

69. $2\log_2(x+1)-\log_2(x+3)-\log_2(x-1)$

$= \log_2(x+1)^2-\log_2(x+3)-\log_2(x-1)$

$= \log_2\dfrac{(x+1)^2}{(x+3)}-\log_2(x-1)$

$= \log_2\left[\dfrac{(x+1)^2}{(x+3)(x-1)}\right]$

71. $\log_3 21 = \dfrac{\log 21}{\log 3}\approx 2.771$

73. $\log_{1/3} 71 = \dfrac{\log 71}{\log(1/3)} = \dfrac{\log 71}{-\log 3} \approx -3.880$

75. $\log_{\sqrt{2}} 7 = \dfrac{\log 7}{\log \sqrt{2}} \approx 5.615$

77. $\log_{\pi} e = \dfrac{\ln e}{\ln \pi} \approx 0.874$

79. $y = \log_4 x = \dfrac{\ln x}{\ln 4}$ or $y = \dfrac{\log x}{\log 4}$

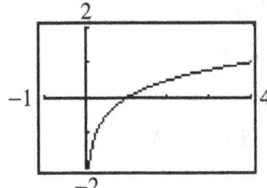

81. $y = \log_2(x+2) = \dfrac{\ln(x+2)}{\ln 2}$ or $y = \dfrac{\log(x+2)}{\log 2}$

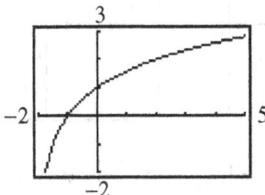

83. $y = \log_{x-1}(x+1) = \dfrac{\ln(x+1)}{\ln(x-1)}$ or $y = \dfrac{\log(x+1)}{\log(x-1)}$

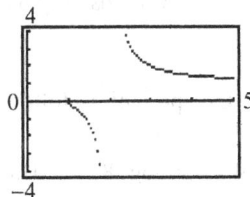

85. $f(x) = \ln x$; $g(x) = e^x$; $h(x) = x^2$

   a. $(f \circ g)(x) = f(g(x)) = \ln(e^x) = x$

     Domain: $\{x \mid x \text{ is any real number}\}$ or

     $(-\infty, \infty)$

   b. $(g \circ f)(x) = g(f(x)) = e^{\ln x} = x$

     Domain: $\{x \mid x > 0\}$ or $(0, \infty)$

     (Note: the restriction on the domain is due to the domain of $\ln x$)

   c. $(f \circ g)(5) = 5$ [from part (a)]

   d. $(f \circ h)(x) = f(h(x)) = \ln(x^2)$

     Domain: $\{x \mid x \neq 0\}$ or $(-\infty, 0) \cup (0, \infty)$

   e. $(f \circ h)(e) = \ln(e^2) = 2\ln e = 2 \cdot 1 = 2$

87. $\ln y = \ln x + \ln C$

   $\ln y = \ln(xC)$

    $y = Cx$

89. $\ln y = \ln x + \ln(x+1) + \ln C$

   $\ln y = \ln(x(x+1)C)$

    $y = Cx(x+1)$

91. $\ln y = 3x + \ln C$

   $\ln y = \ln e^{3x} + \ln C$

   $\ln y = \ln(Ce^{3x})$

    $y = Ce^{3x}$

93. $\ln(y-3) = -4x + \ln C$

   $\ln(y-3) = \ln e^{-4x} + \ln C$

   $\ln(y-3) = \ln(Ce^{-4x})$

    $y - 3 = Ce^{-4x}$

    $y = Ce^{-4x} + 3$

95. $3\ln y = \dfrac{1}{2}\ln(2x+1) - \dfrac{1}{3}\ln(x+4) + \ln C$

   $\ln y^3 = \ln(2x+1)^{1/2} - \ln(x+4)^{1/3} + \ln C$

   $\ln y^3 = \ln\left[\dfrac{C(2x+1)^{1/2}}{(x+4)^{1/3}}\right]$

   $y^3 = \dfrac{C(2x+1)^{1/2}}{(x+4)^{1/3}}$

   $y = \left[\dfrac{C(2x+1)^{1/2}}{(x+4)^{1/3}}\right]^{1/3}$

   $y = \dfrac{\sqrt[3]{C}(2x+1)^{1/6}}{(x+4)^{1/9}}$

373

97. $\log_2 3 \cdot \log_3 4 \cdot \log_4 5 \cdot \log_5 6 \cdot \log_6 7 \cdot \log_7 8$

$= \dfrac{\log 3}{\log 2} \cdot \dfrac{\log 4}{\log 3} \cdot \dfrac{\log 5}{\log 4} \cdot \dfrac{\log 6}{\log 5} \cdot \dfrac{\log 7}{\log 6} \cdot \dfrac{\log 8}{\log 7}$

$= \dfrac{\log 8}{\log 2} = \dfrac{\log 2^3}{\log 2}$

$= \dfrac{3\log 2}{\log 2}$

$= 3$

99. $\log_2 3 \cdot \log_3 4 \cdots \log_n (n+1) \cdot \log_{n+1} 2$

$= \dfrac{\log 3}{\log 2} \cdot \dfrac{\log 4}{\log 3} \cdots \dfrac{\log(n+1)}{\log n} \cdot \dfrac{\log 2}{\log(n+1)}$

$= \dfrac{\log 2}{\log 2}$

$= 1$

101. $\log_a\left(x+\sqrt{x^2-1}\right) + \log_a\left(x-\sqrt{x^2-1}\right):$

$= \log_a\left[\left(x+\sqrt{x^2-1}\right)\left(x-\sqrt{x^2-1}\right)\right]$

$= \log_a\left[x^2-\left(x^2-1\right)\right]$

$= \log_a\left[x^2-x^2+1\right]$

$= \log_a 1$

$= 0$

103. $2x+\ln\left(1+e^{-2x}\right) = \ln e^{2x} + \ln\left(1+e^{-2x}\right)$

$= \ln\left(e^{2x}\left(1+e^{-2x}\right)\right)$

$= \ln\left(e^{2x}+e^0\right)$

$= \ln\left(e^{2x}+1\right)$

105. $f(x)=\log_a x$ means that $x=a^{f(x)}$.
Now, raising both sides to the -1 power, we

obtain $x^{-1} = \left(a^{f(x)}\right)^{-1} = \left(a^{-1}\right)^{f(x)} = \left(\dfrac{1}{a}\right)^{f(x)}$.

$x^{-1} = \left(\dfrac{1}{a}\right)^{f(x)}$ means that $\log_{1/a} x^{-1} = f(x)$.

Thus, $\log_{1/a} x^{-1} = f(x)$

$-\log_{1/a} x = f(x)$

$-f(x) = \log_{1/a} x$

107. $f(x)=\log_a x$

$f\left(\dfrac{1}{x}\right) = \log_a\left(\dfrac{1}{x}\right)$

$= \log_a 1 - \log_a x$

$= -\log_a x$

$= -f(x)$

109. If $A=\log_a M$ and $B=\log_a N$, then $a^A=M$ and $a^B=N$.

$\log_a\left(\dfrac{M}{N}\right) = \log_a\left(\dfrac{a^A}{a^B}\right)$

$= \log_a a^{A-B}$

$= A-B$

$= \log_a M - \log_a N$

111. $Y_1 = \log x^2$ $\qquad Y_2 = 2\log x$

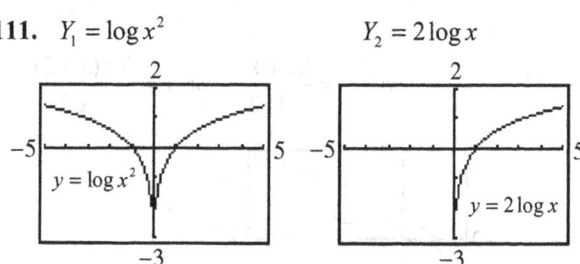

The domain of $Y_1 = \log_a x^2$ is $\{x\,|\,x\neq 0\}$. The domain of $Y_2 = 2\log_a x$ is $\{x\,|\,x>0\}$. These two domains are different because the logarithm property $\log_a x^n = n\cdot\log_a x$ holds only when $\log_a x$ exists.

113. Answers may vary. One possibility follows:
Let $x=4$ and $y=4$. Then
$\log_2(x+y) = \log_2(4+4) = \log_2 8 = 3$. But
$\log_2 x + \log_2 y = \log_2 4 + \log_2 4 = 2+2 = 4$.
Thus, $\log_2(4+4) \neq \log_2 4 + \log_2 4$ and, in general, $\log_2(x+y) \neq \log_2 x + \log_2 y$.

115. Using a graphing utility set
$Y1 = x^3 - 3x^2 - 4x + 8$ and $Y2 = 0$. Graph both and use the graphing utility to find the intersection points. You can also use the SOLVE function on the graphing utility to find the zeros.

The zeros are: $\{-1.78, 1.29, 3.49\}$.

117. $f(x) = 5x^5 + 44x^4 + 116x^3 + 95x^2 - 4x - 4$.

Possible rational zeros:
$p = \pm 1, \pm 2, \pm 4; \quad q = \pm 1, \pm 5;$

$$\frac{p}{q} = \pm 1, \pm 2, \pm 4, \pm \frac{1}{5}, \pm \frac{2}{5}, \pm \frac{4}{5}$$

Using synthetic division:
We try -2:

$$\begin{array}{r|rrrrrr}
-2 & 5 & 44 & 116 & 95 & -4 & -4 \\
& & -10 & -68 & -96 & -2 & 4 \\
\hline
& 5 & 34 & 48 & -1 & -2 & 0
\end{array}$$

-2 is a zero and we try it again:

$$\begin{array}{r|rrrrr}
-2 & 5 & 34 & 48 & -1 & -2 \\
& & -10 & -48 & 0 & 2 \\
\hline
& 5 & 24 & 0 & -1 & 0
\end{array}$$

So -2 is a repeated root. We now try $\frac{1}{5}$

$$\begin{array}{r|rrrr}
\frac{1}{5} & 5 & 24 & 0 & -1 \\
& & 1 & 5 & 1 \\
\hline
& 5 & 25 & 5 & 0
\end{array}$$

We can use the quadratic formula on
$5x^2 + 25x + 5 = 0$

$x^2 + 5x + 1 = 0$

$x = \dfrac{-b \pm \sqrt{b^2 - 4ac}}{2a}$

$= \dfrac{-5 \pm \sqrt{5^2 - 4(1)(1)}}{2(1)}$

$= \dfrac{-5 \pm \sqrt{25 - 4}}{2}$

$= \dfrac{-5 \pm \sqrt{21}}{2}$

The zeros are -2, of multiplicity 2, $\dfrac{1}{5}$,

$\dfrac{-5 \pm \sqrt{21}}{2}$, each of multiplicity 1.

Section 6.6

1. $x^2 - 7x - 30 = 0$

$(x + 3)(x - 10) = 0$

$x + 3 = 0 \quad$ or $\quad x - 10 = 0$

$x = -3 \quad$ or $\qquad x = 10$

The solution set is $\{-3, 10\}$.

3. $x^3 = x^2 - 5$

Using INTERSECT to solve:

$y_1 = x^3; \quad y_2 = x^2 - 5$

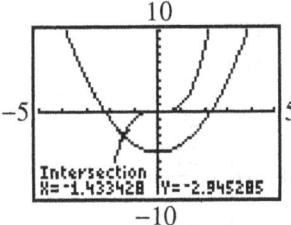

Thus, $x \approx -1.43$, so the solution set is $\{-1.43\}$.

5. $\log_4 x = 2$

$x = 4^2$

$x = 16$

The solution set is $\{16\}$.

7. $\log_2(5x) = 4$

$5x = 2^4$

$5x = 16$

$x = \dfrac{16}{5}$

The solution set is $\left\{\dfrac{16}{5}\right\}$.

9. $\log_4(x + 2) = \log_4 8$

$x + 2 = 8$

$x = 6$

The solution set is $\{6\}$.

11. $\dfrac{1}{2}\log_3 x = 2\log_3 2$

$\log_3 x^{1/2} = \log_3 2^2$

$x^{1/2} = 4$

$x = 16$

The solution set is $\{16\}$.

13. $3\log_2 x = -\log_2 27$

$\log_2 x^3 = \log_2 27^{-1}$

$x^3 = 27^{-1}$

$x^3 = \dfrac{1}{27}$

$x = \dfrac{1}{3}$

The solution set is $\left\{\dfrac{1}{3}\right\}$.

15. $3\log_2(x-1) + \log_2 4 = 5$

$\log_2(x-1)^3 + \log_2 4 = 5$

$\log_2\left(4(x-1)^3\right) = 5$

$4(x-1)^3 = 2^5$

$(x-1)^3 = \dfrac{32}{4}$

$(x-1)^3 = 8$

$x - 1 = 2$

$x = 3$

The solution set is $\{3\}$.

17. $\log x + \log(x+15) = 2$

$\log\left(x(x+15)\right) = 2$

$x(x+15) = 10^2$

$x^2 + 15x - 100 = 0$

$(x+20)(x-5) = 0$

$x = -20$ or $x = 5$

Since $\log(-20)$ is undefined, the solution set is $\{5\}$.

19. $\log(2x+1) = 1 + \log(x-2)$

$\log(2x+1) - \log(x-2) = 1$

$\log\left(\dfrac{2x+1}{x-2}\right) = 1$

$\dfrac{2x+1}{x-2} = 10^1$

$2x+1 = 10(x-2)$

$2x+1 = 10x - 20$

$-8x = -21$

$x = \dfrac{-21}{-8} = \dfrac{21}{8}$

The solution set is $\left\{\dfrac{21}{8}\right\}$.

21. $\log_2(x+7) + \log_2(x+8) = 1$

$\log_2\left[(x+7)(x+8)\right] = 1$

$(x+7)(x+8) = 2^1$

$x^2 + 8x + 7x + 56 = 2$

$x^2 + 15x + 54 = 0$

$(x+9)(x+6) = 0$

$x = -9$ or $x = -6$

Since $\log_2(-9+7) = \log_2(-2)$ is undefined, the solution set is $\{-6\}$.

23. $\log_8(x+6) = 1 - \log_8(x+4)$

$\log_8(x+6) + \log_8(x+4) = 1$

$\log_8\left[(x+6)(x+4)\right] = 1$

$(x+6)(x+4) = 8^1$

$x^2 + 4x + 6x + 24 = 8$

$x^2 + 10x + 16 = 0$

$(x+8)(x+2) = 0$

$x = -8$ or $x = -2$

Since $\log_8(-8+6) = \log_8(-2)$ is undefined, the solution set is $\{-2\}$.

25. $\ln x + \ln(x+2) = 4$

$\ln\left(x(x+2)\right) = 4$

$x(x+2) = e^4$

$x^2 + 2x - e^4 = 0$

$x = \dfrac{-2 + \sqrt{2^2 - 4(1)(-e^4)}}{2(1)}$

$= \dfrac{-2 \pm \sqrt{4 + 4e^4}}{2}$

$= \dfrac{-2 \pm 2\sqrt{1+e^4}}{2}$

$= -1 \pm \sqrt{1+e^4}$

$x = -1 - \sqrt{1+e^4}$ or $x = -1 + \sqrt{1+e^4}$

≈ -8.456 ≈ 6.456

Since $\ln(-8.456)$ is undefined, the solution set is $\left\{-1 + \sqrt{1+e^4}\right\} \approx \{6.456\}$.

27. $\log_3(x+1)+\log_3(x+4)=2$

$$\log_3\left[(x+1)(x+4)\right]=2$$
$$(x+1)(x+4)=3^2$$
$$x^2+4x+x+4=9$$
$$x^2+5x-5=0$$
$$x=\frac{-5\pm\sqrt{5^2-4(1)(-5)}}{2(1)}$$
$$=\frac{-5\pm\sqrt{45}}{2}$$
$$=\frac{-5\pm3\sqrt{5}}{2}$$
$$x=\frac{-5-3\sqrt{5}}{2}\ \text{or}\ x=\frac{-5+3\sqrt{5}}{2}$$
$$\approx-5.854\qquad\qquad\approx0.854$$

Since $\log_3(-8.854+1)=\log_3(-7.854)$ is undefined, the solution set is

$$\left\{\frac{-5+3\sqrt{5}}{2}\right\}\approx\{0.854\}.$$

29. $\log_{1/3}(x^2+x)-\log_{1/3}(x^2-x)=-1$

$$\log_{1/3}\left(\frac{x^2+x}{x^2-x}\right)=-1$$
$$\frac{x^2+x}{x^2-x}=\left(\frac{1}{3}\right)^{-1}$$
$$\frac{x^2+x}{x^2-x}=3$$
$$x^2+x=3\left(x^2-x\right)$$
$$x^2+x=3x^2-3x$$
$$-2x^2+4x=0$$
$$-2x(x-2)=0$$
$$-2x=0\ \text{ or }\ x-2=0$$
$$x=0\ \text{ or }\qquad x=2$$

Since each of the original logarithms are not defined for $x=0$, but are defined for $x=2$, the solution set is $\{2\}$.

31. $\log_a(x-1)-\log_a(x+6)=\log_a(x-2)-\log_a(x+3)$

$$\log_a\left(\frac{x-1}{x+6}\right)=\log_a\left(\frac{x-2}{x+3}\right)$$
$$\frac{x-1}{x+6}=\frac{x-2}{x+3}$$
$$(x-1)(x+3)=(x-2)(x+6)$$
$$x^2+2x-3=x^2+4x-12$$
$$2x-3=4x-12$$
$$9=2x$$
$$x=\frac{9}{2}$$

Since each of the original logarithms is defined for $x=\frac{9}{2}$, the solution set is $\left\{\frac{9}{2}\right\}$.

33. $2\log_5(x-3)-\log_5 8=\log_5 2$

$$\log_5(x-3)^2-\log_5 8=\log_5 2$$
$$\log_5\frac{(x-3)^2}{8}=\log_5 2$$
$$\frac{(x-3)^2}{8}=2$$
$$(x-3)^2=16$$
$$x^2-6x+9=16$$
$$x^2-6x-7=0$$
$$(x-7)(x+1)=0$$
$$x=7\ \text{ or }\ x=-1$$

Since $\log_5(-4)$ is undefined, the solution set is $\{7\}$

35. $2\log_6(x+2)=3\log_6 2+\log_6 4$

$$\log_6(x+2)^2=\log_6 2^3+\log_6 4$$
$$\log_6(x+2)^2=\log_6\left[(8)(4)\right]$$
$$(x+2)^2=32$$
$$x^2+4x+4=32$$
$$x^2+4x-28=0$$

$$x = \frac{-4 \pm \sqrt{4^2 - 4(1)(-28)}}{2(1)}$$

$$= \frac{-4 \pm \sqrt{128}}{2}$$

$$= \frac{-4 \pm 8\sqrt{2}}{2}$$

$$= -2 \pm 4\sqrt{2}$$

$$x = -2 - 4\sqrt{2} \text{ or } x = -2 + 4\sqrt{2}$$

$$\approx \cancel{-8.123} \qquad \approx 0.123$$

37. $2\log_{13}(x+2) = \log_{13}(4x+7)$

$$\log_{13}(x+2)^2 = \log_{13}(4x+7)$$

$$(x+2)^2 = (4x+7)$$

$$x^2 + 4x + 4 = 4x + 7$$

$$x^2 - 3 = 0$$

$$x^2 = 3$$

$$x = \pm\sqrt{3} \approx \pm 1.732$$

The solution set is $\left\{ -\sqrt{3}, \sqrt{3} \right\}$.

39. $\quad (\log_3 x)^2 - 5(\log_3 x) = 6$

$$(\log_3 x)^2 - 5(\log_3 x) - 6 = 0$$

$$(\log_3 x - 6)(\log_3 x + 1) = 0$$

$$\log_3 x = 6 \text{ or } \log_3 x = -1$$

$$x = 3^6 \qquad x = 3^{-1}$$

$$x = 729 \qquad x = \frac{1}{3}$$

The solution set is $\left\{ \frac{1}{3}, 729 \right\}$.

41. $2^{x-5} = 8$

$$2^{x-5} = 2^3$$

$$x - 5 = 3$$

$$x = 8$$

The solution set is $\{8\}$.

43. $2^x = 10$

$$x = \log_2 10 = \frac{\ln 10}{\ln 2} \approx 3.322$$

The solution set is

$$\{\log_2 10\} = \left\{ \frac{\ln 10}{\ln 2} \right\} \approx \{3.322\}.$$

45. $8^{-x} = 1.2$

$$-x = \log_8 1.2$$

$$x = -\log_8 1.2 = -\frac{\log(1.2)}{\log 8} \approx -0.088$$

The solution set is

$$\{-\log_8 1.2\} = \left\{ \frac{\log(1.2)}{-\log 8} \right\} \approx \{-0.088\}.$$

47. $5\left(2^{3x}\right) = 8$

$$2^{3x} = \frac{8}{5}$$

$$3x = \log_2\left(\frac{8}{5}\right)$$

$$x = \frac{1}{3}\log_2\left(\frac{8}{5}\right) = \frac{\ln(8/5)}{3\ln 2} \approx 0.226$$

The solution set is

$$\left\{ \frac{1}{3}\log_2\left(\frac{8}{5}\right) \right\} = \left\{ \frac{\ln(8/5)}{3\ln 2} \right\} \approx \{0.226\}.$$

49. $\qquad 3^{1-2x} = 4^x$

$$\ln\left(3^{1-2x}\right) = \ln\left(4^x\right)$$

$$(1-2x)\ln 3 = x\ln 4$$

$$\ln 3 - 2x\ln 3 = x\ln 4$$

$$\ln 3 = 2x\ln 3 + x\ln 4$$

$$\ln 3 = x(2\ln 3 + \ln 4)$$

$$x = \frac{\ln 3}{2\ln 3 + \ln 4} \approx 0.307$$

The solution set is $\left\{ \frac{\ln 3}{2\ln 3 + \ln 4} \right\} \approx \{0.307\}.$

51.
$$\left(\frac{3}{5}\right)^x = 7^{1-x}$$
$$\ln\left(\frac{3}{5}\right)^x = \ln\left(7^{1-x}\right)$$
$$x\ln(3/5) = (1-x)\ln 7$$
$$x\ln(3/5) = \ln 7 - x\ln 7$$
$$x\ln(3/5) + x\ln 7 = \ln 7$$
$$x\left(\ln(3/5) + \ln 7\right) = \ln 7$$
$$x = \frac{\ln 7}{\ln(3/5) + \ln 7} \approx 1.356$$
The solution set is $\left\{\dfrac{\ln 7}{\ln(3/5) + \ln 7}\right\} \approx \{1.356\}$.

53.
$$1.2^x = (0.5)^{-x}$$
$$\ln 1.2^x = \ln(0.5)^{-x}$$
$$x\ln(1.2) = -x\ln(0.5)$$
$$x\ln(1.2) + x\ln(0.5) = 0$$
$$x\left(\ln(1.2) + \ln(0.5)\right) = 0$$
$$x = 0$$
The solution set is $\{0\}$.

55.
$$\pi^{1-x} = e^x$$
$$\ln \pi^{1-x} = \ln e^x$$
$$(1-x)\ln \pi = x$$
$$\ln \pi - x\ln \pi = x$$
$$\ln \pi = x + x\ln \pi$$
$$\ln \pi = x(1 + \ln \pi)$$
$$x = \frac{\ln \pi}{1 + \ln \pi} \approx 0.534$$
The solution set is $\left\{\dfrac{\ln \pi}{1 + \ln \pi}\right\} \approx \{0.534\}$.

57.
$$2^{2x} + 2^x - 12 = 0$$
$$\left(2^x\right)^2 + 2^x - 12 = 0$$
$$\left(2^x - 3\right)\left(2^x + 4\right) = 0$$
$$2^x - 3 = 0 \quad \text{or} \quad 2^x + 4 = 0$$
$$2^x = 3 \quad \text{or} \quad 2^x = -4$$
$$\ln\left(2^x\right) = \ln 3 \qquad \text{No solution}$$
$$x\ln 2 = \ln 3$$
$$x = \frac{\ln 3}{\ln 2} \approx 1.585$$
The solution set is $\left\{\dfrac{\ln 3}{\ln 2}\right\} \approx \{1.585\}$.

59.
$$3^{2x} + 3^{x+1} - 4 = 0$$
$$\left(3^x\right)^2 + 3 \cdot 3^x - 4 = 0$$
$$\left(3^x - 1\right)\left(3^x + 4\right) = 0$$
$$3^x - 1 = 0 \quad \text{or} \quad 3^x + 4 = 0$$
$$3^x = 1 \quad \text{or} \quad 3^x = -4$$
$$x = 0 \qquad \text{No solution}$$
The solution set is $\{0\}$.

61.
$$16^x + 4^{x+1} - 3 = 0$$
$$\left(4^2\right)^x + 4 \cdot 4^x - 3 = 0$$
$$\left(4^x\right)^2 + 4 \cdot 4^x - 3 = 0$$
Let $u = 4^x$.
$$u^2 + 4u - 3 = 0$$
$$a = 1, b = 4, c = -3$$
$$u = \frac{-4 \pm \sqrt{4^2 - 4(1)(-3)}}{2(1)} = \frac{-4 \pm \sqrt{28}}{2}$$
$$= \frac{-4 \pm 2\sqrt{7}}{2} = -2 \pm \sqrt{7}$$
Therefore, we get
$$\cancel{4^x = -2 - \sqrt{7}} \quad \text{or} \quad 4^x = -2 + \sqrt{7}$$
$$x = \log_4\left(-2 + \sqrt{7}\right)$$
(we ignore the first solution since 4^x is never negative)
The solution set is $\left\{\log_4\left(-2 + \sqrt{7}\right)\right\} \approx \{-0.315\}$.

63. $25^x - 8 \cdot 5^x = -16$

$\left(5^2\right)^x - 8 \cdot 5^x = -16$

$\left(5^x\right)^2 - 8 \cdot 5^x = -16$

Let $u = 5^x$.

$u^2 - 8u = -16$

$u^2 - 8u + 16 = 0$

$(u-4)^2 = 0$

$u = 4$

Therefore, we get

$5^x = 4$

$x = \log_5 4$

The solution set is $\{\log_5 4\} \approx \{0.861\}$.

65. $3 \cdot 4^x + 4 \cdot 2^x + 8 = 0$

$3 \cdot \left(2^2\right)^x + 4 \cdot 2^x + 8 = 0$

$3 \cdot \left(2^x\right)^2 + 4 \cdot 2^x + 8 = 0$

Let $u = 2^x$.

$3u^2 + 4u + 8 = 0$

$a = 3, b = 4, c = 8$

$u = \dfrac{-4 \pm \sqrt{4^2 - 4(3)(8)}}{2(3)}$

$= \dfrac{-4 \pm \sqrt{-80}}{6} = $ not real

The equation has no real solution.

67. $4^x - 10 \cdot 4^{-x} = 3$

Multiply both sides of the equation by 4^x.

$\left(4^x\right)^2 - 10 \cdot 4^{-x} \cdot 4^x = 3 \cdot 4^x$

$\left(4^x\right)^2 - 10 = 3 \cdot 4^x$

$\left(4^x\right)^2 - 3 \cdot 4^x - 10 = 0$

$\left(4^x - 5\right)\left(4^x + 2\right) = 0$

$4^x - 5 = 0$ or $4^x + 2 = 0$

$4^x = 5$ $\cancel{4^x = -2}$

$x = \log_4 5$

The solution set is $\{\log_4 5\} \approx \{1.161\}$.

69. $\log_5(x+1) - \log_4(x-2) = 1$

Using INTERSECT to solve:

$y_1 = \ln(x+1)/\ln(5) - \ln(x-2)/\ln(4)$

$y_2 = 1$

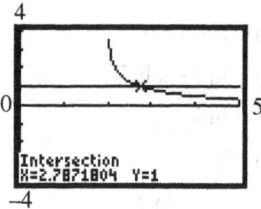

Thus, $x \approx 2.79$, so the solution set is $\{2.79\}$.

71. $e^x = -x$

Using INTERSECT to solve: $y_1 = e^x$; $y_2 = -x$

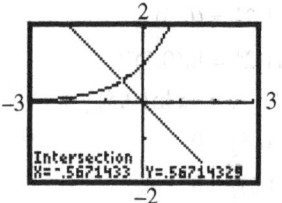

Thus, $x \approx -0.57$, so the solution set is $\{-0.57\}$.

73. $e^x = x^2$

Using INTERSECT to solve:

$y_1 = e^x$; $y_2 = x^2$

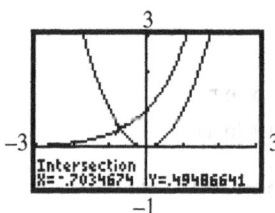

Thus, $x \approx -0.70$, so the solution set is $\{-0.70\}$.

75. $\ln x = -x$

Using INTERSECT to solve:

$y_1 = \ln x$; $y_2 = -x$

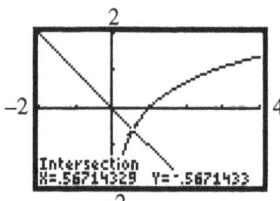

Thus, $x \approx 0.57$, so the solution set is $\{0.57\}$.

77. $\ln x = x^3 - 1$

Using INTERSECT to solve:

$y_1 = \ln x;\ y_2 = x^3 - 1$

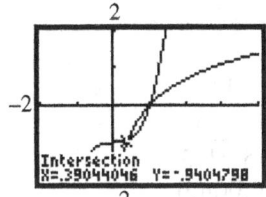 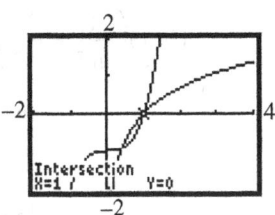

Thus, $x \approx 0.39$ or $x = 1$, so the solution set is $\{0.39,\ 1\}$.

79. $e^x + \ln x = 4$

Using INTERSECT to solve:

$y_1 = e^x + \ln x;\ y_2 = 4$

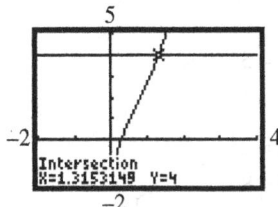

Thus, $x \approx 1.32$, so the solution set is $\{1.32\}$.

81. $e^{-x} = \ln x$

Using INTERSECT to solve:

$y_1 = e^{-x};\ y_2 = \ln x$

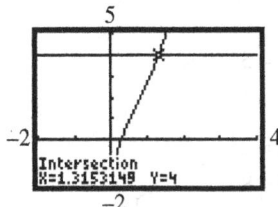

Thus, $x \approx 1.31$, so the solution set is $\{1.31\}$.

83. $\log_2(x+1) - \log_4 x = 1$

$$\log_2(x+1) - \frac{\log_2 x}{\log_2 4} = 1$$

$$\log_2(x+1) - \frac{\log_2 x}{2} = 1$$

$$2\log_2(x+1) - \log_2 x = 2$$

$$\log_2(x+1)^2 - \log_2 x = 2$$

$$\log_2\left(\frac{(x+1)^2}{x}\right) = 2$$

$$\frac{(x+1)^2}{x} = 2^2$$

$$x^2 + 2x + 1 = 4x$$

$$x^2 - 2x + 1 = 0$$

$$(x-1)^2 = 0$$

$$x - 1 = 0$$

$$x = 1$$

Since each of the original logarithms is defined for $x = 1$, the solution set is $\{1\}$.

85. $\log_{16} x + \log_4 x + \log_2 x = 7$

$$\frac{\log_2 x}{\log_2 16} + \frac{\log_2 x}{\log_2 4} + \log_2 x = 7$$

$$\frac{\log_2 x}{4} + \frac{\log_2 x}{2} + \log_2 x = 7$$

$$\log_2 x + 2\log_2 x + 4\log_2 x = 28$$

$$7\log_2 x = 28$$

$$\log_2 x = 4$$

$$x = 2^4 = 16$$

Since each of the original logarithms is defined for $x = 16$, the solution set is $\{16\}$.

87. $\left(\sqrt[3]{2}\right)^{2-x} = 2^{x^2}$

$$\left(2^{1/3}\right)^{2-x} = 2^{x^2}$$

$$2^{\frac{1}{3}(2-x)} = 2^{x^2}$$

$$\frac{1}{3}(2-x) = x^2$$

$$2 - x = 3x^2$$

$$3x^2 + x - 2 = 0$$

$$(3x - 2)(x + 1) = 0$$

$$x = \frac{2}{3} \text{ or } x = -1$$

The solution set is $\left\{ -1, \frac{2}{3} \right\}$.

89.
$$\frac{e^x + e^{-x}}{2} = 1$$
$$e^x + e^{-x} = 2$$
$$e^x (e^x + e^{-x}) = 2e^x$$
$$e^{2x} + 1 = 2e^x$$
$$(e^x)^2 - 2e^x + 1 = 0$$
$$(e^x - 1)^2 = 0$$
$$e^x - 1 = 0$$
$$e^x = 1$$
$$x = 0$$

The solution set is $\{0\}$.

91.
$$\frac{e^x - e^{-x}}{2} = 2$$
$$e^x - e^{-x} = 4$$
$$e^x (e^x - e^{-x}) = 4e^x$$
$$e^{2x} - 1 = 4e^x$$
$$(e^x)^2 - 4e^x - 1 = 0$$

$$e^x = \frac{-(-4) \pm \sqrt{(-4)^2 - 4(1)(-1)}}{2(1)}$$

$$= \frac{4 \pm \sqrt{20}}{2} = \frac{4 \pm 2\sqrt{5}}{2} = 2 \pm \sqrt{5}$$

$$x = \ln(2 - \sqrt{5}) \text{ or } x = \ln(2 + \sqrt{5})$$
$$x \approx \ln(-0.236) \text{ or } x \approx 1.444$$

Since $\ln(-0.236)$ is undefined, the solution set is $\left\{ \ln(2 + \sqrt{5}) \right\} \approx \{1.444\}$.

93.
$$\log_5 x + \log_3 x = 1$$
$$\frac{\ln x}{\ln 5} + \frac{\ln x}{\ln 3} = 1$$
$$(\ln x) \left(\frac{1}{\ln 5} + \frac{1}{\ln 3} \right) = 1$$
$$\ln x = \frac{1}{\dfrac{1}{\ln 5} + \dfrac{1}{\ln 3}}$$
$$\ln x = \frac{(\ln 5)(\ln 3)}{\ln 3 + \ln 5}$$
$$\ln x = \frac{(\ln 5)(\ln 3)}{\ln 15}$$
$$x = e^{\left(\frac{(\ln 5)(\ln 3)}{\ln 15} \right)} \approx 1.921$$

The solution set is $\left\{ e^{\left(\frac{(\ln 5)(\ln 3)}{\ln 15} \right)} \right\} \approx \{1.921\}$.

95. a.
$$f(x) = 3$$
$$\log_2 (x + 3) = 3$$
$$x + 3 = 2^3$$
$$x + 3 = 8$$
$$x = 5$$

The solution set is $\{5\}$. The point $(5, 3)$ is on the graph of f.

b.
$$g(x) = 4$$
$$\log_2 (3x + 1) = 4$$
$$3x + 1 = 2^4$$
$$3x + 1 = 16$$
$$3x = 15$$
$$x = 5$$

The solution set is $\{5\}$. The point $(5, 4)$ is on the graph of g.

c.
$$f(x) = g(x)$$
$$\log_2 (x + 3) = \log_2 (3x + 1)$$
$$x + 3 = 3x + 1$$
$$2 = 2x$$
$$1 = x$$

The solution set is $\{1\}$, so the graphs intersect when $x = 1$. That is, at the point $(1, 2)$.

d.
$$(f+g)(x)=7$$
$$\log_2(x+3)+\log_2(3x+1)=7$$
$$\log_2\left[(x+3)(3x+1)\right]=7$$
$$(x+3)(3x+1)=2^7$$
$$3x^2+10x+3=128$$
$$3x^2+10x-125=0$$
$$(3x+25)(x-5)=0$$
$$3x+25=0 \quad \text{or} \quad x-5=0$$
$$3x=-25 \qquad\qquad x=5$$
$$x=-\frac{25}{3}$$
The solution set is $\{5\}$.

e.
$$(f-g)(x)=2$$
$$\log_2(x+3)-\log_2(3x+1)=2$$
$$\log_2\frac{x+3}{3x+1}=2$$
$$\frac{x+3}{3x+1}=2^2$$
$$x+3=4(3x+1)$$
$$x+3=12x+4$$
$$-1=11x$$
$$-\frac{1}{11}=x$$
The solution set is $\left\{-\frac{1}{11}\right\}$.

97. a.

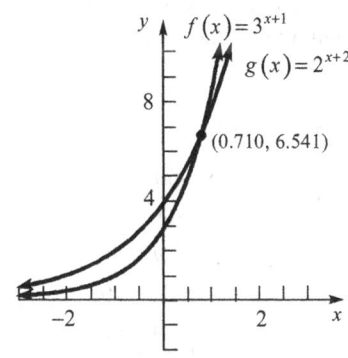

b.
$$f(x)=g(x)$$
$$3^{x+1}=2^{x+2}$$
$$\ln\left(3^{x+1}\right)=\ln\left(2^{x+2}\right)$$
$$(x+1)\ln 3=(x+2)\ln 2$$
$$x\ln 3+\ln 3=x\ln 2+2\ln 2$$
$$x\ln 3-x\ln 2=2\ln 2-\ln 3$$
$$x(\ln 3-\ln 2)=2\ln 2-\ln 3$$
$$x=\frac{2\ln 2-\ln 3}{\ln 3-\ln 2}\approx 0.710$$
$$f\left(\frac{2\ln 2-\ln 3}{\ln 3-\ln 2}\right)\approx 6.541$$
The intersection point is roughly $(0.710, 6.541)$.

c. Based on the graph, $f(x)>g(x)$ for $x>0.710$. The solution set is $\{x\,|\,x>0.710\}$ or $(0.710,\infty)$.

99. a., b.

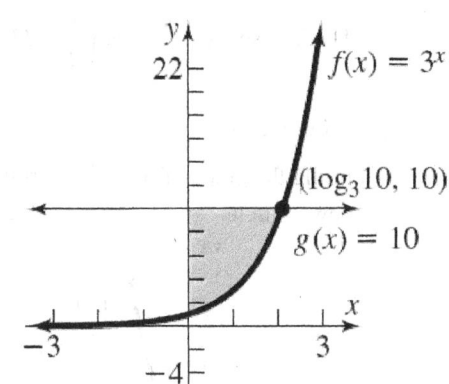

c.
$$f(x)=g(x)$$
$$3^x=10$$
$$x=\log_3 10$$
The intersection point is $\left(\log_3 10, 10\right)$.

383

101. a., b.

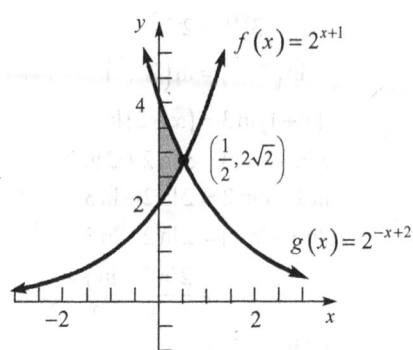

c. $f(x) = g(x)$

$2^{x+1} = 2^{-x+2}$

$x+1 = -x+2$

$2x = 1$

$x = \dfrac{1}{2}$

$f\left(\dfrac{1}{2}\right) = 2^{1/2+1} = 2^{3/2} = 2\sqrt{2}$

The intersection point is $\left(\dfrac{1}{2}, 2\sqrt{2}\right)$.

103. a. $f(x) = 2^x - 4$

Using the graph of $y = 2^x$, shift the graph down 4 units.

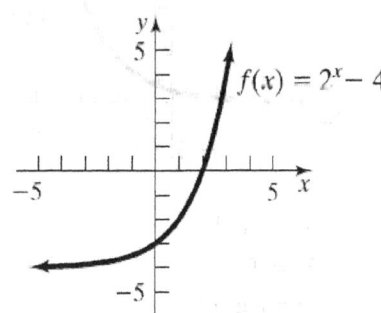

b. $f(x) = 0$

$2^x - 4 = 0$

$2^x = 4$

$2^x = 2^2$

$x = 2$

The zero of f is $x = 2$.

c. Based on the graph, $f(x) < 0$ when $x < 2$.

The solution set is $\{x \mid x < 2\}$ or $(-\infty, 2)$.

105. a. $320(1.007)^{t-2015} = 400$

$(1.007)^{t-2015} = \dfrac{400}{320}$

$\ln(1.007)^{t-2015} = \ln\dfrac{400}{320}$

$(t-2015)\ln(1.007) = \ln\dfrac{400}{320}$

$t - 2015 = \dfrac{\ln(400/320)}{\ln(1.007)}$

$t = \dfrac{\ln(400/320)}{\ln(1.007)} + 2015$

≈ 2047

According to the model, the population of the U.S. will reach 400 million people around the beginning of the year 2047.

b. $320(1.007)^{t-2015} = 435$

$(1.007)^{t-2015} = \dfrac{435}{320}$

$\ln(1.007)^{t-2015} = \ln\dfrac{435}{320}$

$(t-2015)\ln(1.007) = \ln\dfrac{435}{320}$

$t - 2015 = \dfrac{\ln(435/320)}{\ln(1.007)}$

$t = \dfrac{\ln(435/320)}{\ln(1.007)} + 2015$

≈ 2059

According to the model, the population of the U.S. will reach 435 million people in the beginning of the year 2059.

107. a. $18,700(0.84)^t = 9,000$

$(0.84)^t = \dfrac{9,000}{18,700}$

$\log(0.84)^t = \log\dfrac{9,000}{18,700}$

$t\log(0.84) = \log\dfrac{9,000}{18,700}$

$t = \dfrac{\log(9,000/18,700)}{\log(0.84)}$

≈ 4.2

According to the model, the car will be worth $9,000 after about 4.2 years.

b. $18,700(0.84)^t = 6,000$

$$(0.84)^t = \frac{6,000}{18,700}$$

$$\log(0.84)^t = \log\frac{6,000}{18,700}$$

$$t\log(0.84) = \log\frac{6,000}{18,700}$$

$$t = \frac{\log(6,000/18,700)}{\log(0.84)}$$

$$\approx 6.5$$

According to the model, the car will be worth $6,000 after about 6.5 years.

c. $18,700(0.84)^t = 2,000$

$$(0.84)^t = \frac{2,000}{18,700}$$

$$\log(0.84)^t = \log\frac{2,000}{18,700}$$

$$t\log(0.84) = \log\frac{2,000}{18,700}$$

$$t = \frac{\log(2,000/18,700)}{\log(0.84)}$$

$$\approx 12.8$$

According to the model, the car will be worth $2,000 after about 12.8 years.

109. Solution A: change to exponential expression; square root method; meaning of $\pm$; solve.

Solution B: $\log_a M^r = r\log_a M$; divide by 2; change to exponential expression; solve.

The power rule $\log_a M^r = r\log_a M$ only applies when $M > 0$. In this equation, $M = x-1$. Now, $x = -2$ causes $M = -2-1 = -3$. Thus, if we use the power rule, we lose the valid solution $x = -2$.

111. $f(x) = \sqrt{x+3} + \sqrt{x-1}$

The domain of $\sqrt{x+3}$ is $\{x \mid x \geq -3\}$ and the domain of $\sqrt{x-1}$ is $\{x \mid x \geq 1\}$. The domain of the function would be the most restrictive of these so the domain of $f(x)$ is $\{x \mid x \geq 1\}$.

113. Since the x elements are not repeated and the y elements are not repeated the ordered pairs represent a function that is one-to-one.

Section 6.7

1. $P = \$500$, $r = 0.06$, $t = 6$ months $= 0.5$ year

$I = Prt = (500)(0.06)(0.5) = \$15.00$

3. principal

5. 4

7. $P = \$100$, $r = 0.04$, $n = 4$, $t = 2$

$$A = P\left(1+\frac{r}{n}\right)^{nt} = 100\left(1+\frac{0.04}{4}\right)^{(4)(2)} \approx \$108.29$$

9. $P = \$500$, $r = 0.08$, $n = 4$, $t = 2.5$

$$A = P\left(1+\frac{r}{n}\right)^{nt} = 500\left(1+\frac{0.08}{4}\right)^{(4)(2.5)} \approx \$609.50$$

11. $P = \$600$, $r = 0.05$, $n = 365$, $t = 3$

$$A = P\left(1+\frac{r}{n}\right)^{nt} = 600\left(1+\frac{0.05}{365}\right)^{(365)(3)} \approx \$697.09$$

13. $P = \$1000$, $r = 0.11$, $t = 2$

$A = Pe^{rt} = 1000e^{(0.11)(2)} \approx \$1246.08$

15. $A = \$100$, $r = 0.06$, $n = 12$, $t = 2$

$$P = A\left(1+\frac{r}{n}\right)^{-nt} = 100\left(1+\frac{0.06}{12}\right)^{(-12)(2)} \approx \$88.72$$

17. $A = \$1000$, $r = 0.06$, $n = 365$, $t = 2.5$

$$P = A\left(1+\frac{r}{n}\right)^{-nt}$$

$$= 1000\left(1+\frac{0.06}{365}\right)^{(-365)(2.5)} \approx \$860.72$$

19. $A = \$600$, $r = 0.04$, $n = 4$, $t = 2$

$$P = A\left(1+\frac{r}{n}\right)^{-nt} = 600\left(1+\frac{0.04}{4}\right)^{(-4)(2)} \approx \$554.09$$

21. $A = \$80$, $r = 0.09$, $t = 3.25$

$P = Ae^{-rt} = 80e^{(-0.09)(3.25)} \approx \$59.71$

23. Suppose P dollars are invested for 1 year at 5%.

Compounded quarterly yields:

$A = P\left(1+\dfrac{0.05}{4}\right)^{(4)(1)} \approx 1.05095P$.

The interest earned is
$I = 1.05095P - P = 0.05095P$
Thus, $I = Prt$

$0.05095P = P \cdot r \cdot 1$

$0.05095 = r$

The effective interest rate is 5.095%.

25. Suppose P dollars are invested for 1 year at 5%.

Compounded continuously yields:
$A = Pe^{(0.05)(1)} \approx 1.05127P$
The interest earned is
$I = 1.05127P - P = 0.05127P$
Thus, $I = Prt$

$0.05127P = P \cdot r \cdot 1$

$.05127 = r$

The effective interest rate is 5.127%.

27. 6% compounded quarterly:

$A = 10,000\left(1+\dfrac{0.06}{4}\right)^{(4)(1)} = \$10,613.64$

$6\frac{1}{4}\%$ compounded annually:

$A = 10,000(1+0.0625)^1 = \$10,625$

$6\frac{1}{4}\%$ compounded annually is the better deal.

29. 9% compounded monthly:

$A = 10,000\left(1+\dfrac{0.09}{12}\right)^{(12)(1)} = \$10,938.07$

8.8% compounded daily:

$A = 10,000\left(1+\dfrac{0.088}{365}\right)^{365} = \$10,919.77$

31. $2P = P\left(1+\dfrac{r}{1}\right)^{3(1)}$

$2P = P(1+r)^3$

$2 = (1+r)^3$

$\sqrt[3]{2} = 1+r$

$r = \sqrt[3]{2} - 1 \approx 0.25992$

The required rate is 25.992%.

33. $3P = P\left(1+\dfrac{r}{1}\right)^{5(1)}$

$3P = P(1+r)^5$

$3 = (1+r)^5$

$\sqrt[5]{3} = 1+r$

$r = \sqrt[5]{3} - 1 \approx 0.24573$

The required rate is 24.573%.

35. a. $2P = P\left(1+\dfrac{0.08}{12}\right)^{12t}$

$2 = \left(1+\dfrac{0.08}{12}\right)^{12t}$

$\ln 2 = \ln\left(1+\dfrac{0.08}{12}\right)^{12t}$

$\ln 2 = 12t \ln\left(1+\dfrac{0.08}{12}\right)$

$t = \dfrac{\ln 2}{12 \ln\left(1+\dfrac{0.08}{12}\right)} \approx 8.69$

It will take about 8.69 years to double.

b. $2P = Pe^{0.08t}$

$2 = e^{0.08t}$

$\ln 2 = 0.08t$

$t = \dfrac{\ln 2}{0.08} \approx 8.66$

It will take about 8.66 years to double.

37. Since the effective interest rate is 7%, we have:
$I = Prt$
$I = P \cdot 0.07 \cdot 1$
$I = 0.07P$
Thus, the amount in the account is
$A = P + 0.07P = 1.07P$

Let x be the required interest rate. Then,

$1.07P = P\left(1+\dfrac{r}{4}\right)^{(4)(1)}$

$1.07 = \left(1+\dfrac{r}{4}\right)^4$

$\sqrt[4]{1.07} = 1+\dfrac{r}{4}$

$\sqrt[4]{1.07} - 1 = \dfrac{r}{4}$

$r = 4\left(\sqrt[4]{1.07} - 1\right) \approx 0.06823$

Thus, an interest rate of 6.823% compounded quarterly has an effective interest rate of 7%.

39. $150 = 100\left(1+\dfrac{0.04}{12}\right)^{12t}$

$1.5 \approx (1.003333)^{12t}$

$\ln 1.5 \approx 12t \ln(1.003333)$

$t \approx \dfrac{\ln 1.5}{12 \ln(1.003333)} \approx 10.15$

Compounded monthly, it will take about 10.15 years.

$150 = 100e^{0.04t}$

$1.5 = e^{0.04t}$

$\ln 1.5 = 0.04t$

$t = \dfrac{\ln 1.5}{0.04} \approx 10.14$

Compounded continuously, it will take about 10.14 years.

41. $25,000 = 10,000e^{0.06t}$

$2.5 = e^{0.06t}$

$\ln 2.5 = 0.06t$

$t = \dfrac{\ln 2.5}{0.06} \approx 15.27$

It will take about 15.27 years (or 15 years, 3 months).

43. $A = 90,000(1+0.03)^5 = \$104,335$

The house will cost \$104,335 in five years.

45. $P = 15,000e^{(-0.05)(3)} \approx \$12,910.62$

Jerome should ask for \$12,910.62.

47. $A = 15(1+0.15)^5 = 15(1.15)^5 \approx \$30.17$ per share for a total of about \$3017.

49. 5.6% compounded continuously:

$A = 1000e^{(0.056)(1)} = \$1057.60$

Jim will not have enough money to buy the computer.

5.9% compounded monthly:

$A = 1000\left(1+\dfrac{0.059}{12}\right)^{12} = \$1060.62$

The second bank offers the better deal.

51. Will: 9% compounded semiannually:

$A = 2000\left(1+\dfrac{0.09}{2}\right)^{(2)(20)} = \$11,632.73$

Henry: 8.5% compounded continuously:

$A = 2000e^{(0.085)(20)} = \$10,947.89$

Will has more money after 20 years.

53. a. Let x = the year, then the average annual cost C of a 4-year private college is by the function $C(x) = 31,231(1.037)^{x-2014}$.

$C(2034) = 31,231(1.037)^{2034-2014}$

$\qquad\quad = 31,231(1.037)^{20}$

$\qquad\quad \approx 64,589$

In 2034, the average annual cost at a 4-year private college will be about \$64,589.

b. $A = Pe^{rt}$

$64,589 = Pe^{0.02(18)}$

$P = \dfrac{64,589}{e^{0.02(18)}} \approx \$45,062$

An investment of \$45,062 in 2016 would pay for the cost of college at a 4-year private college in 2034.

55. $A = P\left(1+\dfrac{r}{n}\right)^{nt}$

$A = 787\left(1+\dfrac{0.013}{2}\right)^{2(20)} = 787(1.0065)^{40} \approx 1019$

The government would have to pay back approximately \$1019 billion in 2029. The amount of interest would be $1019 - 787 = \$232$ billion.

57. $P = 1000, r = 0.03, n = 2$

$A = 1000(1-0.03)^2 = \$940.90$

59. $P = 1000, A = 950, n = 2$

$950 = 1000(1-r)^2$

$0.95 = (1-r)^2$

$\pm\sqrt{0.95} = 1-r$

$r = 1 \pm \sqrt{0.95}$

$r \approx 0.0253$ or $r \approx 1.9747$

Disregard $r \approx 1.9747$. The inflation rate was 2.53%.

61. $r = 0.02$

$$\frac{1}{2}P = P(1 - 0.02)^t$$

$$0.5P = P(0.98)^t$$

$$0.5 = (0.98)^t$$

$$t = \log_{0.98}(0.5)$$

$$= \frac{\ln 0.5}{\ln 0.98} \approx 34.31$$

The purchasing power will be half in 34.31 years.

63. a. $A = \$10,000,\ r = 0.05,\ n = 12,\ t = 20$

$$P = 10,000\left(1 + \frac{0.05}{12}\right)^{(-12)(20)} \approx \$3686.45$$

b. $A = \$10,000,\ r = 0.05,\ t = 20$

$$P = 10,000e^{(-0.05)(20)} \approx \$3678.79$$

65. $A = \$10,000,\ r = 0.045,\ n = 1,\ t = 10$

$$P = 10,000\left(1 + \frac{0.045}{1}\right)^{(-1)(10)} \approx \$6439.28$$

67. a.
$$t = \frac{\ln 2}{1 \cdot \ln\left(1 + \frac{0.06}{1}\right)}$$

$$= \frac{\ln 2}{\ln(1.06)} \approx 11.90 \text{ years}$$

b.
$$t = \frac{\ln 3}{4 \cdot \ln\left(1 + \frac{0.05}{4}\right)}$$

$$= \frac{\ln 3}{4\ln(1.0125)} \approx 22.11 \text{ years}$$

c.
$$mP = P\left(1 + \frac{r}{n}\right)^{nt}$$

$$m = \left(1 + \frac{r}{n}\right)^{nt}$$

$$\ln m = nt \cdot \ln\left(1 + \frac{r}{n}\right)$$

$$t = \frac{\ln m}{n \cdot \ln\left(1 + \frac{r}{n}\right)}$$

69. a. $CPI_0 = 214.5,\ CPI = 236.7,$
$n = 2014 - 2009 = 5$

$$236.7 = 214.5\left(1 + \frac{r}{100}\right)^5$$

$$\frac{236.7}{214.5} = \left(1 + \frac{r}{100}\right)^5$$

$$1 + \frac{r}{100} = \sqrt[5]{\frac{236.7}{214.5}}$$

$$\frac{r}{100} = \sqrt[5]{\frac{236.7}{214.5}} - 1$$

$$r = 100\left(\sqrt[5]{\frac{236.7}{214.5}} - 1\right) \approx 1.99\%$$

b. $CPI_0 = 236.7,\ CPI = 300,\ r = 1.99$

$$300 = 236.7\left(1 + \frac{1.99}{100}\right)^n$$

$$\frac{300}{236.7} = \left(1 + \frac{1.99}{100}\right)^n$$

$$\ln\left(\frac{300}{236.7}\right) = \ln\left(1 + \frac{1.99}{100}\right)^n$$

$$\ln\left(\frac{300}{236.7}\right) = n\ln\left(1 + \frac{1.99}{100}\right)$$

$$n = \frac{\ln\left(\frac{300}{236.7}\right)}{\ln\left(1 + \frac{1.99}{100}\right)} \approx 12.0 \text{ years}$$

The CPI will reach 300 about 12 years after 2014, or in the year 2026.

71. $r = 3.1\%$

$$2 \cdot CPI_0 = CPI_0\left(1 + \frac{3.1}{100}\right)^n$$

$$2 = (1.031)^n$$

$$n = \log_{1.031} 2 = \frac{\ln 2}{\ln 1.031} \approx 22.7$$

It will take about 22.7 years for the CPI index to double.

73. Answers will vary.

75. Answers will vary.

77. $f(x) = \dfrac{x}{x-2}$

$x = \dfrac{y}{y-2}$

$x(y-2) = y$

$xy - 2x = y$

$xy - y = 2x$

$y(x-1) = 2x$

$y = \dfrac{2x}{x-1}$

$f^{-1}(x) = \dfrac{2x}{x-1}$

79. $\log_2(x+3) = 2\log_2(x-3)$

$\log_2(x+3) = \log_2(x-3)^2$

$(x+3) = (x-3)^2$

$x+3 = x^2 - 6x + 9$

$x^2 - 7x + 6 = 0$

$(x-6)(x-1) = 0$

$x = 6$ or $x = 1$

But x = 1 will not work since we cannot take the log of a negative number so the solution set is
$\{6\}$

Section 6.8

1. $P(t) = 500e^{0.02t}$

a. $P(0) = 500e^{(0.02)\cdot(0)} = 500$ insects

b. growth rate $= 2\%$

c.

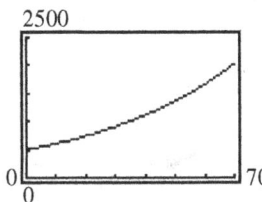

d. $P(10) = 500e^{(0.02)\cdot(10)} \approx 611$ insects

e. Find t when $P = 800$:

$800 = 500e^{0.02t}$

$1.6 = e^{0.02t}$

$\ln 1.6 = 0.02t$

$t = \dfrac{\ln 1.6}{0.02} \approx 23.5$ days

f. Find t when $P = 1000$:

$1000 = 500e^{0.02t}$

$2 = e^{0.02t}$

$\ln 2 = 0.02t$

$t = \dfrac{\ln 2}{0.02} \approx 34.7$ days

3. $A(t) = A_0 e^{-0.0244t} = 500e^{-0.0244t}$

a. decay rate $= -2.44\%$

b.

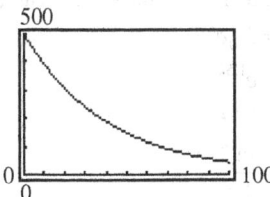

c. $A(10) = 500e^{(-0.0244)(10)} \approx 391.7$ grams

d. Find t when $A = 400$:

$400 = 500e^{-0.0244t}$

$0.8 = e^{-0.0244t}$

$\ln 0.8 = -0.0244t$

$t = \dfrac{\ln 0.8}{-0.0244} \approx 9.1$ years

e. Find t when $A = 250$:

$250 = 500e^{-0.0244t}$

$0.5 = e^{-0.0244t}$

$\ln 0.5 = -0.0244t$

$t = \dfrac{\ln 0.5}{-0.0244} \approx 28.4$ years

5. a. $N(t) = N_0 e^{kt}$

b. If $N(t) = 1800$, $N_0 = 1000$, and $t = 1$, then

$1800 = 1000e^{k(1)}$

$1.8 = e^k$

$k = \ln 1.8$

If $t = 3$, then $N(3) = 1000e^{(\ln 1.8)(3)} = 5832$ mosquitoes.

c. Find t when $N(t) = 10,000$:

$$10,000 = 1000e^{(\ln 1.8)t}$$

$$10 = e^{(\ln 1.8)t}$$

$$\ln 10 = (\ln 1.8)t$$

$$t = \frac{\ln 10}{\ln 1.8} \approx 3.9 \text{ days}$$

7. a. $N(t) = N_0 e^{kt}$

b. Note that 18 months = 1.5 years, so $t = 1.5$.

$$2N_0 = N_0 e^{k(1.5)}$$

$$2 = e^{1.5k}$$

$$\ln 2 = 1.5k$$

$$k = \frac{\ln 2}{1.5}$$

If $N_0 = 10,000$ and $t = 2$, then

$$P(2) = 10,000e^{\left(\frac{\ln 2}{1.5}\right)(2)} \approx 25,198$$

The population 2 years from now will be 25,198.

9. Use $A = A_0 e^{kt}$ and solve for k :

$$0.5A_0 = A_0 e^{k(1690)}$$

$$0.5 = e^{1690k}$$

$$\ln 0.5 = 1690k$$

$$k = \frac{\ln 0.5}{1690}$$

When $A_0 = 10$ and $t = 50$:

$$A = 10e^{\left(\frac{\ln 0.5}{1690}\right)(50)} \approx 9.797 \text{ grams}$$

11. a. Use $A = A_0 e^{kt}$ and solve for k :

half-life = 5730 years

$$0.5A_0 = A_0 e^{k(5730)}$$

$$0.5 = e^{5730k}$$

$$\ln 0.5 = 5730k$$

$$k = \frac{\ln 0.5}{5730}$$

Solve for t when $A = 0.3A_0$:

$$0.3A_0 = A_0 e^{\frac{\ln 0.5}{5730}t}$$

$$0.3 = e^{\frac{\ln 0.5}{5730}t}$$

$$\ln 0.3 = \frac{\ln 0.5}{5730}t$$

$$t = \frac{\ln 0.3}{\frac{\ln 0.5}{5730}} \approx 9953$$

The tree died approximately 9953 years ago.

b. $Y_1 = e^{\frac{\ln 0.5}{5730}t}$

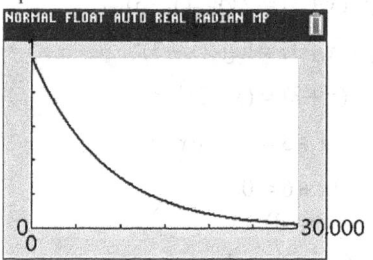

c. $Y_1 = e^{\frac{\ln 0.5}{5730}t}$; $Y_2 = 0.5$

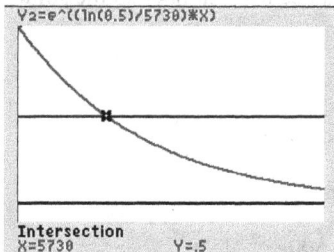

Thus, 5730 years will elapse until half of the carbon-14 remains.

d. $Y_1 = e^{\frac{\ln 0.5}{5730}t}$; $Y_2 = 0.3$

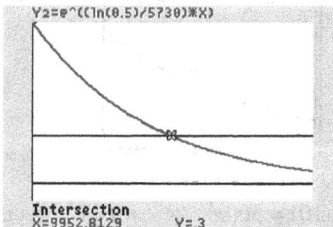

This verifies that the tree died approximately 9953 years ago.

13. a. Using $u = T + (u_0 - T)e^{kt}$ with $t = 5$,
$T = 70$, $u_0 = 450$, and $u = 300$:

$$300 = 70 + (450 - 70)e^{k(5)}$$

$$230 = 380e^{5k}$$

$$\frac{230}{380} = e^{5k}$$

$$\ln\left(\frac{23}{38}\right) = 5k$$

$$k = \frac{\ln\left(\frac{23}{38}\right)}{5} \approx -0.1004$$

$T = 70$, $u_0 = 450$, $u = 135$:

$$135 = 70 + (450 - 70)e^{\frac{\ln(23/38)}{5}t}$$

$$65 = 380e^{\frac{\ln(23/38)}{5}t}$$

$$\frac{65}{380} = e^{\frac{\ln(23/38)}{5}t}$$

$$\ln\frac{65}{380} = \frac{\ln(23/38)}{5}t$$

$$t = \frac{\ln(65/380)}{\left(\frac{\ln(23/38)}{5}\right)} \approx 18 \text{ minutes}$$

The pizza will be cool enough to eat at about 5:18 PM.

b. $Y_1 = 70 + (450 - 70)e^{\frac{\ln(23/38)}{5}x}$

c. $Y_1 = 70 + (450 - 70)e^{\frac{\ln(23/38)}{5}x}$; $Y_2 = 160$

The pizza will be 160°F after about 14.3 minutes.

d. As time passes, the temperature gets closer to 70°F.

15. a. Using $u = T + (u_0 - T)e^{kt}$ with $t = 3$,
$T = 35$, $u_0 = 8$, and $u = 15$:

$$15 = 35 + (8 - 35)e^{k(3)}$$

$$-20 = -27e^{3k}$$

$$\frac{20}{27} = e^{3k}$$

$$\ln\left(\frac{20}{27}\right) = 3k$$

$$k = \frac{\ln(20/27)}{3}$$

At $t = 5$: $u = 35 + (8 - 35)e^{\left(\frac{\ln(20/27)}{3}\right)(5)} \approx 18.63°C$

After 5 minutes, the thermometer will read approximately 18.63°C.

At $t = 10$:

$$u = 35 + (8 - 35)e^{\left(\frac{\ln(20/27)}{3}\right)(10)} \approx 25.07°C$$

After 10 minutes, the thermometer will read approximately 25.07°C

b. $Y_1 = 35 + (8 - 35)e^{\left(\frac{\ln(20/27)}{3}\right)x}$

17. Use $A = A_0 e^{kt}$ and solve for k:

$$2.2 = 2.5e^{k(24)}$$

$$0.88 = e^{24k}$$

$$\ln 0.88 = 24k$$

$$k = \frac{\ln 0.88}{24}$$

When $A_0 = 2.5$ and $t = 72$:

$$A = 2.5e^{\left(\frac{\ln 0.88}{24}\right)(72)} \approx 1.70$$

After 3 days (72 hours), the amount of free chlorine will be 1.70 parts per million.

Find t when $A = 1$:

$$1 = 2.5e^{\left(\frac{\ln 0.88}{24}\right)t}$$

$$0.4 = e^{\left(\frac{\ln 0.88}{24}\right)t}$$

$$\ln 0.4 = \left(\frac{\ln 0.88}{24}\right)t$$

$$t = \frac{24}{\ln 0.88} \cdot \ln 0.4 \approx 172$$

Ben will have to shock his pool again after 172 hours (or 7.17 days) when the level of free chlorine reaches 1.0 parts per million.

19. Use $A = A_0 e^{kt}$ and solve for k:

$$0.36 = 0.40e^{k(30)}$$

$$0.9 = e^{30k}$$

$$\ln 0.9 = 30k$$

$$k = \frac{\ln 0.9}{30}$$

Note that 2 hours = 120 minutes.
When $A_0 = 0.40$ and $t = 120$:

$$A = 0.40e^{\left(\frac{\ln 0.9}{30}\right)(120)} \approx 0.26$$

After 2 hours, approximately 0.26 M of sucrose will remain.

Find t when $A = 0.10$:

$$0.10 = 0.40e^{\left(\frac{\ln 0.9}{30}\right)t}$$

$$0.25 = e^{\left(\frac{\ln 0.9}{30}\right)t}$$

$$\ln 0.25 = \left(\frac{\ln 0.9}{30}\right)t$$

$$t = \frac{30}{\ln 0.9} \cdot \ln 0.25 \approx 395$$

It will take approximately 395 minutes (or 6.58 hours) until 0.10 M of sucrose remains.

21. Use $A = A_0 e^{kt}$ and solve for k:

$$0.5A_0 = A_0 e^{k(8)}$$

$$0.5 = e^{8k}$$

$$\ln 0.5 = 8k$$

$$k = \frac{\ln 0.5}{8}$$

Find t when $A = 0.1A_0$:

$$0.1A_0 = A_0 e^{\left(\frac{\ln 0.5}{8}\right)t}$$

$$0.1 = e^{\left(\frac{\ln 0.5}{8}\right)t}$$

$$\ln 0.1 = \left(\frac{\ln 0.5}{8}\right)t$$

$$t = \frac{8}{\ln 0.5} \cdot \ln 0.1 \approx 26.6$$

The farmers need to wait about 26.6 days before using the hay.

23. a. As $t \to \infty$, $e^{-0.439t} \to 0$. Thus, $P(t) \to 1000$. The carrying capacity is 1000 grams of bacteria.

b. Growth rate $= 0.439 = 43.9\%$.

c. $P(0) = \dfrac{1000}{1 + 32.33e^{-0.439(0)}} = \dfrac{1000}{33.33} = 30$
The initial population was 30 grams of bacteria.

d. $Y_1 = \dfrac{1000}{1 + 32.33e^{-0.439x}}$

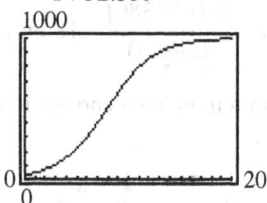

e. $P(9) = \dfrac{1000}{1 + 32.33e^{-0.439(9)}} \approx 616.6$
After 9 hours, the population of bacteria will be about 616.8 grams.

f. We need to find t such that $P = 700$:

$$Y_1 = \frac{1000}{1 + 32.33e^{-0.439x}}; \; Y_2 = 700$$

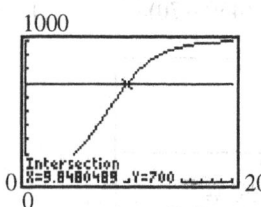

Thus, $t \approx 9.85$. The population of bacteria will be 700 grams after about 9.85 hours.

g. We need to find t such that

$$P = \frac{1}{2}(1000) = 500:$$

$$Y_1 = \frac{1000}{1 + 32.33e^{-0.439x}}; \; Y_2 = 500$$

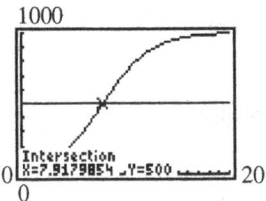

Thus, $t \approx 7.9$. The population of bacteria will reach one-half of is carrying capacity after about 7.9 hours.

25. a. $P(0) = \dfrac{431}{1 + 7.91e^{-0.017(0)}} \approx 48$

In 1900 the number of invasive species present in the Great Lakes was approximately 48.

b. The growth rate of invasive species is 1.7%.

c. $Y_1 = \dfrac{431}{1 + 7.91e^{-0.017x}}$

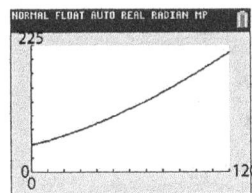

d. $y = \dfrac{431}{1 + 7.91e^{-0.017(100)}} \approx 176$

The number of invasive species present in 2000 was approximately 176.

e. We need to find t such that $P = 175$.

$$Y_1 = \frac{431}{1 + 7.91e^{-0.071x}}; \; Y_2 = 175$$

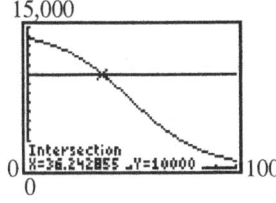

Thus, $t \approx 99$. Now, $1900 + 99 = 1999$. There were 175 invasive species in 1999.

27. a. $P(0) = \dfrac{95.4993}{1 + 0.0405e^{0.1968(0)}} = \dfrac{95.4993}{1.0405} \approx 91.8$

In 1984, about 91.8% of households did not own a PC.

b. $Y_1 = \dfrac{95.4993}{1 + 0.0405e^{0.1968x}}$

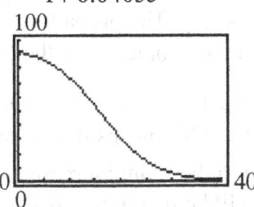

c. $t = 1995 - 1984 = 11$

$$P(11) = \frac{95.4993}{1 + 0.0405e^{0.1968(11)}} \approx 70.6$$

In 1995, about 70.6% of households did not own a PC.

d. We need to find t such that $P = 10$

$$Y_1 = \frac{95.4993}{1 + 0.0405e^{0.1968x}}; \; Y_2 = 10$$

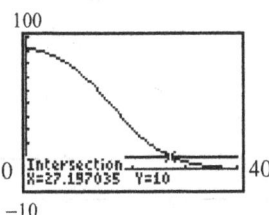

Thus, $t \approx 27.2$. Now, $1984 + 27.2 = 2011.2$. The percentage of households that do not own a PC will reach 10% during 2011.

29. a. $Y_1 = \dfrac{113.3198}{1 + 0.115e^{0.0912x}}$

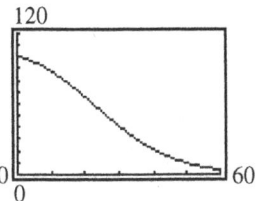

b. $P(15) = \dfrac{113.3198}{1 + 0.115e^{0.0912(15)}} \approx 78$

In a room of 15 people, the probability that no two people share the same birthday is about 78% or 0.78.

c. We need to find n such that $P = 10$.

$$Y_1 = \frac{113.3198}{1 + 0.115e^{0.0912x}}; \; Y_2 = 10$$

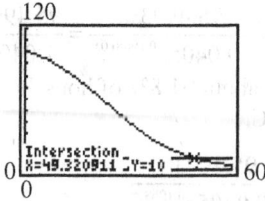

Thus, $t \approx 49.3$. The probability falls below 10% when 50 people are in the room.

d. As $n \to \infty$, $1+0.115e^{0.0912n} \to \infty$. Thus, $P(n) \to 0$. This means that as the number of people in the room increases, the more likely it will be that two will share the same birthday.

31. a. $n = 8$; $P_0 = 25,000$

$$P(t) = 25,000(2)^{t/8}$$

b.

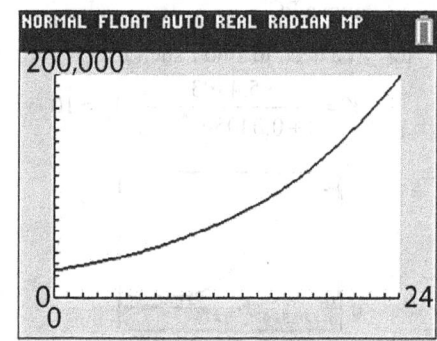

c. $t = 3$, then $P(3) = 25,000(2)^{\frac{3}{8}} \approx 32,421$

The population 3 years from now will be 32,421.

d.

$$80,000 = 25,000(2)^{\frac{t}{8}}$$

$$\frac{80,000}{25,000} = (2)^{\frac{t}{8}}$$

$$\ln(3.2) = \ln(2)^{\frac{t}{8}}$$

$$\ln(3.2) = \frac{t}{8}\ln(2)$$

$$\frac{\ln(3.2)}{\ln 2} = \frac{t}{8}$$

$$t = 8\frac{\ln(3.2)}{\ln 2} \approx 13.42$$

The population will reach 80,000 in 13.42 years.

e. $y = 25,000(2)^{t/8}$

$$= 25,000\left(e^{\ln 2}\right)^{t/8}$$

$$= 25,000e^{\ln 2(t/8)}$$

$$= 25,000e^{0.087t}$$

$$P(t) = 25,000e^{0.087t}$$

33. $m = \dfrac{-5-1}{8-4} = \dfrac{-6}{4} = -\dfrac{3}{2}$

$$(y-y_1) = m(x-x_1)$$

$$y-1 = -\frac{3}{2}(x-4)$$

$$y-1 = -\frac{3}{2}x+6$$

$$y = -\frac{3}{2}x+7$$

35. $\ln \dfrac{x^2\sqrt{y}}{z} = \ln \dfrac{x^2 y^{\frac{1}{2}}}{z}$

$= \ln\left(x^2 y^{\frac{1}{2}}\right) - \ln(z)$

$= \ln(x^2) + \ln\left(y^{\frac{1}{2}}\right) - \ln(z)$

$= 2\ln x + \dfrac{1}{2}\ln y - \ln z$

Section 6.9

1. a.

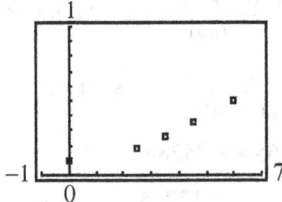

b. Using EXPonential REGression on the data yields: $y = 0.0903(1.3384)^x$

c. $y = 0.0903(1.3384)^x$

$= 0.0903\left(e^{\ln(1.3384)}\right)^x$

$= 0.0903 e^{\ln(1.3384)x}$

$N(t) = 0.0903 e^{0.2915t}$

d. $Y_1 = 0.0903 e^{0.2915x}$

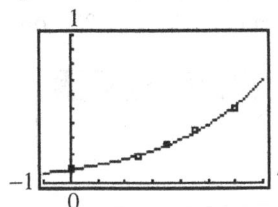

e. $N(7) = 0.0903 e^{(0.2915)\cdot 7} \approx 0.69$ bacteria

f. We need to find t when $N = 0.75$:

$0.0903 e^{(0.2915)\cdot t} = 0.75$

$e^{(0.2915)\cdot t} = \dfrac{0.75}{0.0903}$

$0.2915t = \ln\left(\dfrac{0.75}{0.0903}\right)$

$t \approx \dfrac{\ln\left(\dfrac{0.75}{0.0903}\right)}{0.2915} \approx 7.26$ hours

3. a.

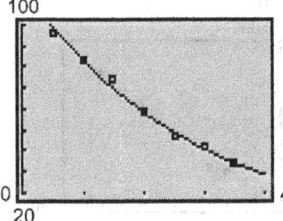

b. Using EXPonential REGression on the data yields: $y = 118.7226(0.7013)^x$

c. $y = 118.7226(0.7013)^x$

$= 118.7226\left(e^{\ln(0.7013)}\right)^x$

$= 118.7226 e^{\ln(0.7013)x}$

$A(t) = 118.7226 e^{(-0.3548)t}$

d. $Y_1 = 118.7226 e^{(-0.3548)x}$

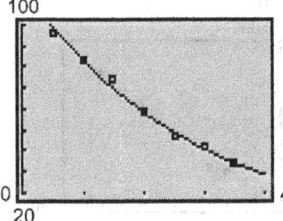

e. $A(4) = 118.7226 e^{(-0.3548)\cdot 4} \approx 28.7\%$

f. $k = -0.3548 = -35.48\%$ is the exponential decay rate. It represents the rate at which the percentage of patients surviving advanced-stage breast cancer is decreasing.

5. a.

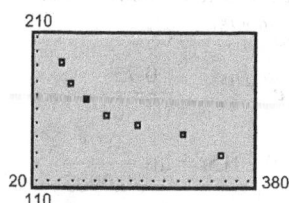

b. Using lnREGression on the data
yields: $y = 330.0549 - 34.5008 \ln x$

c. $Y_1 = 330.0549 - 34.5008 \ln x$

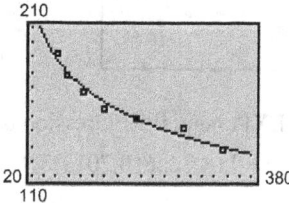

d. Note that 2008 is represented by $t = 28$.
$y = 300.0549 - 34.5008 \ln(28) \approx 185$ billion pounds.

e. It is under by 5 billion pounds.

7. a. Let $x = 0$ correspond to 1900, $x = 10$ correspond to 1910, $x = 20$ correspond to 1920, etc.

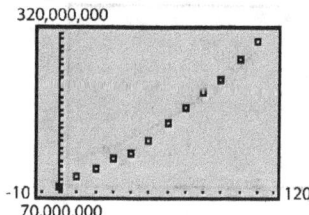

b. Using LOGISTIC REGression on the data
yields: $y = \dfrac{762,176,844.4}{1 + 8.7428e^{-0.0162x}}$

c. $Y_1 = \dfrac{762,176,844.4}{1 + 8.7428e^{-0.0162x}}$

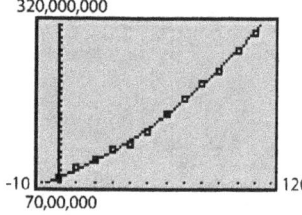

d. As $x \to \infty$, $8.7428e^{-0.0162x} \to 0$, which means $1 + 8.7428e^{-0.0162x} \to 1$, so
$$y = \frac{762,176,844.4}{1 + 8.7428e^{-0.0162x}} \to 762,176,844.4$$
Therefore, the carrying capacity of the United States is approximately 762,176,844 people.

e. The year 2012 corresponds to $x = 112$, so
$$y = \frac{762,176,844.4}{1 + 8.7428e^{-0.0162(112)}}$$
$$\approx 314,362,768 \text{ people}$$

f. Find x when $y = 350,000,000$
$$\frac{762,176,844.4}{1 + 8.7428e^{-0.0162x}} = 350,000,000$$
$$762,176,844.4 = 350,000,000\left(1 + 8.7428e^{-0.0162x}\right)$$
$$\frac{762,176,844.4}{350,000,000} = 1 + 8.7428e^{-0.0162x}$$
$$\frac{762,176,844.4}{350,000,000} - 1 = 8.7428e^{-0.0162x}$$
$$1.17765 \approx 8.7428e^{-0.0162x}$$
$$\frac{1.17765}{8.7428} \approx e^{-0.0162x}$$
$$\ln\left(\frac{1.17765}{8.7428}\right) \approx -0.0162x$$
$$\frac{\ln\left(\dfrac{1.17765}{8.7428}\right)}{-0.0162} \approx x$$
$$x \approx 123.75$$
Therefore, the United States population will be 350,000,000 in the year 2023.

9. a. Let $x = 1$ correspond to 1985, $x = 10$ correspond to 1995, $x = 28$ correspond to 2012, etc.

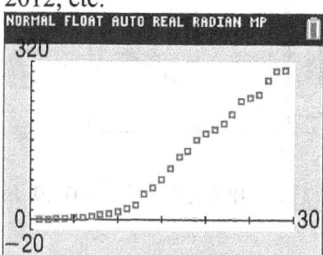

b. Using LOGISTIC REGression on the data
yields: $y = \dfrac{321.0384}{1 + 135.3081e^{-0.2516x}}$

c. $Y_1 = \dfrac{321.0384}{1+135.3081e^{-0.2516x}}$

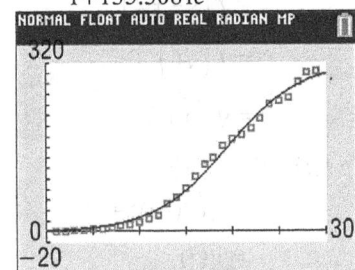

d. As $x \to \infty$, $135.3081e^{-0.2516x} \to 0$, which means $1+135.3081e^{-0.2516x} \to 1$, so
$$y = \dfrac{321.0384}{1+135.3081e^{-0.2516x}} \to 321.0384$$

Therefore, the predicted carrying capacity for cell sites in the United States is 321.0 thousand.

e. The year 2019 corresponds to $x = 35$, so
$$y = \dfrac{321.0384}{1+135.3081e^{-0.2516(35)}} = 314.7.$$
In 2019, there will be approximately 314.7 thousand cell sites in the U.S.

11. a.

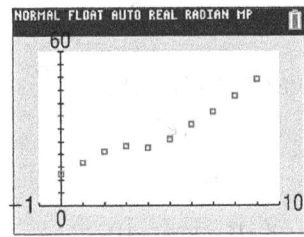

b. Based on the shape of the graph, a cubic model might best describe the data.

c. Using CubicReg, the cubic model is
$y = 0.0691x^3 - 0.6538x^2 + 4.4323x + 13.0352$

d.

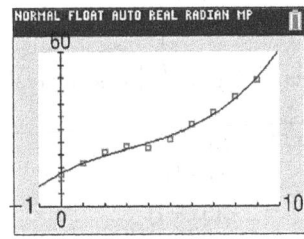

e. $y(11) = 0.0691(11)^3 - 0.6538(11)^2$
$+ 4.4323(11) + 13.0352$
≈ 74.6
The model predicts an online advertising revenue of \$74.6 billion in 2016.

13. a.

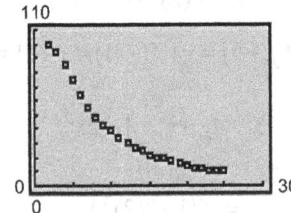

b. Based on the graph, an exponential model would best describe the data.

c. Using EXPonential REGression, the model is $y = 115.5779(0.9012)^x$.
$= 115.5449e^{-0.1040x}$

d.

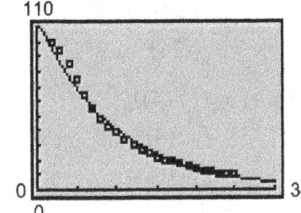

e. $y = 115.5779(0.9012)^{30}$
≈ 5.1
The model predicts the expected percentage of a 30-foot putt to be made is 5.1%.

15. $\dfrac{3}{\sqrt{2}} = \dfrac{3}{\sqrt{2}} \dfrac{\sqrt{2}}{\sqrt{2}} = \dfrac{3\sqrt{2}}{2}$

17. The equation represents a circle with center $(3,0)$ and radius 5.

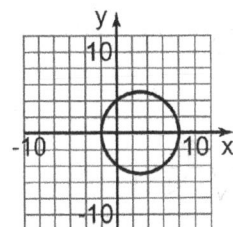

Chapter 6 Review Exercises

1. a. $(g \circ f)(-8) = g(f(-8)) = g(2) = -4$

b. $(f \circ g)(-8) = f(g(-8)) = f(-2) = 1$

c. $(g \circ g)(7) = g(g(7)) = g(0) = -6$

d. $(g \circ f)(-5) = g(f(-5)) = g(4) = -6$

2. $f(x) = 3x - 5 \qquad g(x) = 1 - 2x^2$

a. $(f \circ g)(2) = f(g(2))$
$= f\left(1 - 2(2)^2\right)$
$= f(-7)$
$= 3(-7) - 5$
$= -26$

b. $(g \circ f)(-2) = g(f(-2))$
$= g(3(-2) - 5)$
$= g(-11)$
$= 1 - 2(-11)^2$
$= -241$

c. $(f \circ f)(4) = f(f(4))$
$= f(3(4) - 5)$
$= f(7)$
$= 3(7) - 5$
$= 16$

d. $(g \circ g)(-1) = g(g(-1))$
$= g\left(1 - 2(-1)^2\right)$
$= g(-1)$
$= 1 - 2(-1)^2$
$= -1$

3. $f(x) = \sqrt{x+2} \qquad g(x) = 2x^2 + 1$

a. $(f \circ g)(2) = f(g(2))$
$= f\left(2(2)^2 + 1\right)$
$= f(9)$
$= \sqrt{9+2}$
$= \sqrt{11}$

b. $(g \circ f)(-2) = g(f(-2))$
$= g\left(\sqrt{-2+2}\right)$
$= g(0)$
$= 2(0)^2 + 1$
$= 1$

c. $(f \circ f)(4) = f(f(4))$
$= f\left(\sqrt{4+2}\right)$
$= f\left(\sqrt{6}\right)$
$= \sqrt{\sqrt{6}+2}$

d. $(g \circ g)(-1) = g(g(-1))$
$= g\left(2(-1)^2 + 1\right)$
$= g(3)$
$= 2(3)^2 + 1$
$= 19$

4. $f(x) = e^x \qquad g(x) = 3x - 2$

a. $(f \circ g)(2) = f(g(2))$
$= f(3(2) - 2)$
$= f(4)$
$= e^4$

b. $(g \circ f)(-2) = g(f(-2))$
$= g\left(e^{-2}\right)$
$= 3e^{-2} - 2$
$= \dfrac{3}{e^2} - 2$

c. $(f \circ f)(4) = f(f(4))$
$= f\left(e^4\right)$
$= e^{e^4}$

d. $(g \circ g)(-1) = g(g(-1))$
$= g(3(-1) - 2)$
$= g(-5)$
$= 3(-5) - 2$
$= -17$

5. $f(x) = 2 - x \qquad g(x) = 3x + 1$

The domain of f is $\{x \mid x \text{ is any real number}\}$.

The domain of g is $\{x \mid x \text{ is any real number}\}$.

$(f \circ g)(x) = f(g(x))$
$= f(3x + 1)$
$= 2 - (3x + 1)$
$= 2 - 3x - 1$
$= 1 - 3x$

Domain: $\{x \mid x \text{ is any real number}\}$.

$$(g \circ f)(x) = g(f(x))$$
$$= g(2-x)$$
$$= 3(2-x)+1$$
$$= 6-3x+1$$
$$= 7-3x$$

Domain: $\{x \mid x \text{ is any real number}\}$.

$$(f \circ f)(x) = f(f(x))$$
$$= f(2-x)$$
$$= 2-(2-x)$$
$$= 2-2+x$$
$$= x$$

Domain: $\{x \mid x \text{ is any real number}\}$.

$$(g \circ g)(x) = g(g(x))$$
$$= g(3x+1)$$
$$= 3(3x+1)+1$$
$$= 9x+3+1$$
$$= 9x+4$$

Domain: $\{x \mid x \text{ is any real number}\}$.

6. $f(x) = \sqrt{3x} \qquad g(x) = 1+x+x^2$

The domain of f is $\{x \mid x \geq 0\}$.

The domain of g is $\{x \mid x \text{ is any real number}\}$.

$$(f \circ g)(x) = f(g(x))$$
$$= f(1+x+x^2)$$
$$= \sqrt{3(1+x+x^2)}$$
$$= \sqrt{3+3x+3x^2}$$

Domain: $\{x \mid x \text{ is any real number}\}$.

$$(g \circ f)(x) = g(f(x))$$
$$= g(\sqrt{3x})$$
$$= 1+\sqrt{3x}+(\sqrt{3x})^2$$
$$= 1+\sqrt{3x}+3x$$

Domain: $\{x \mid x \geq 0\}$.

$$(f \circ f)(x) = f(f(x)) = f(\sqrt{3x}) = \sqrt{3\sqrt{3x}}$$

Domain: $\{x \mid x \geq 0\}$.

$$(g \circ g)(x) = g(g(x))$$
$$= g(1+x+x^2)$$
$$= 1+(1+x+x^2)+(1+x+x^2)^2$$
$$= 1+1+x+x^2+1+2x+3x^2+2x^3+x^4$$
$$= 3+3x+4x^2+2x^3+x^4$$

Domain: $\{x \mid x \text{ is any real number}\}$.

7. $f(x) = \dfrac{x+1}{x-1} \qquad g(x) = \dfrac{1}{x}$

The domain of f is $\{x \mid x \neq 1\}$.

The domain of g is $\{x \mid x \neq 0\}$.

$$(f \circ g)(x) = f(g(x))$$
$$= f\left(\frac{1}{x}\right) = \frac{\frac{1}{x}+1}{\frac{1}{x}-1}$$
$$= \frac{\left(\frac{1}{x}+1\right)x}{\left(\frac{1}{x}-1\right)x} = \frac{1+x}{1-x}$$

Domain $\{x \mid x \neq 0, x \neq 1\}$.

$$(g \circ f)(x) = g(f(x))$$
$$= g\left(\frac{x+1}{x-1}\right) = \frac{1}{\left(\frac{x+1}{x-1}\right)} = \frac{x-1}{x+1}$$

Domain $\{x \mid x \neq -1, x \neq 1\}$.

$$(f \circ f)(x) = f(f(x))$$
$$= f\left(\frac{x+1}{x-1}\right) = \frac{\frac{x+1}{x-1}+1}{\frac{x+1}{x-1}-1}$$
$$= \frac{\left(\frac{x+1}{x-1}+1\right)(x-1)}{\left(\frac{x+1}{x-1}-1\right)(x-1)}$$
$$= \frac{x+1+x-1}{x+1-(x-1)} = \frac{2x}{2} = x$$

Domain $\{x \mid x \neq 1\}$.

$$(g \circ g)(x) = g(g(x)) = g\left(\frac{1}{x}\right) = \frac{1}{\left(\frac{1}{x}\right)} = x$$

Domain $\{x \mid x \neq 0\}$.

8. a. The function is one-to-one because there are no two distinct inputs that correspond to the same output.

   b. The inverse is $\{(2,1),(5,3),(8,5),(10,6)\}$.

9. The function f is one-to-one because every horizontal line intersects the graph at exactly one point.

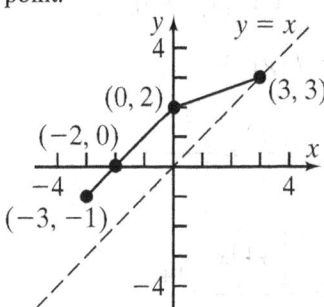

10. $f(x) = \dfrac{2x+3}{5x-2}$

$y = \dfrac{2x+3}{5x-2}$

$x = \dfrac{2y+3}{5y-2}$ Inverse

$x(5y-2) = 2y+3$

$5xy - 2x = 2y + 3$

$5xy - 2y = 2x + 3$

$y(5x-2) = 2x + 3$

$y = \dfrac{2x+3}{5x-2}$

$f^{-1}(x) = \dfrac{2x+3}{5x-2}$

Domain of f = Range of f^{-1}

$\qquad$ = All real numbers except $\dfrac{2}{5}$.

Range of f = Domain of f^{-1}

$\qquad$ = All real numbers except $\dfrac{2}{5}$.

11. $f(x) = \dfrac{1}{x-1}$

$y = \dfrac{1}{x-1}$

$x = \dfrac{1}{y-1}$ Inverse

$x(y-1) = 1$

$xy - x = 1$

$xy = x + 1$

$y = \dfrac{x+1}{x}$

$f^{-1}(x) = \dfrac{x+1}{x}$

Domain of f = Range of f^{-1}

$\qquad$ = All real numbers except 1

Range of f = Domain of f^{-1}

$\qquad$ = All real numbers except 0

12. $f(x) = \sqrt{x-2}$

$y = \sqrt{x-2}$

$x = \sqrt{y-2}$ Inverse

$x^2 = y - 2 \quad x \geq 0$

$y = x^2 + 2 \quad x \geq 0$

$f^{-1}(x) = x^2 + 2 \quad x \geq 0$

Domain of f = Range of f^{-1} = $\{x \mid x \geq 2\}$ or $[2, \infty)$

Range of f = Domain of f^{-1} = $\{x \mid x \geq 0\}$ or $[0, \infty)$

13. $f(x) = x^{1/3} + 1$

$y = x^{1/3} + 1$

$x = y^{1/3} + 1$ Inverse

$y^{1/3} = x - 1$

$y = (x-1)^3$

$f^{-1}(x) = (x-1)^3$

Domain of f = Range of f^{-1}

$\qquad$ = All real numbers or $(-\infty, \infty)$

Range of f = Domain of f^{-1}

$\qquad$ = All real numbers or $(-\infty, \infty)$

14. a. $f(4) = 3^4 = 81$

b. $g(9) = \log_3(9) = \log_3(3^2) = 2$

c. $f(-2) = 3^{-2} = \dfrac{1}{9}$

d. $g\left(\dfrac{1}{27}\right) = \log_3\left(\dfrac{1}{27}\right) = \log_3(3^{-3}) = -3$

15. $5^2 = z$ is equivalent to $2 = \log_5 z$

16. $\log_5 u = 13$ is equivalent to $5^{13} = u$

17. $f(x) = \log(3x - 2)$ requires:

$3x - 2 > 0$

$x > \dfrac{2}{3}$

Domain: $\left\{x \mid x > \dfrac{2}{3}\right\}$ or $\left(\dfrac{2}{3}, \infty\right)$

18. $H(x) = \log_2\left(x^2 - 3x + 2\right)$ requires

$p(x) = x^2 - 3x + 2 > 0$

$(x - 2)(x - 1) > 0$

$x = 2$ and $x = 1$ are the zeros of p.

Interval	$(-\infty, 1)$	$(1, 2)$	$(2, \infty)$
Test Value	0	$\dfrac{3}{2}$	3
Value of p	2	$-\dfrac{1}{4}$	2
Conclusion	positive	negative	positive

Thus, the domain of $H(x) = \log_2\left(x^2 - 3x + 2\right)$

is $\{x \mid x < 1 \text{ or } x > 2\}$ or $(-\infty, 1) \cup (2, \infty)$.

19. $\log_2\left(\dfrac{1}{8}\right) = \log_2 2^{-3} = -3 \log_2 2 = -3$

20. $\ln e^{\sqrt{2}} = \sqrt{2}$

21. $2^{\log_2 0.4} = 0.4$

22. $\log_3\left(\dfrac{uv^2}{w}\right) = \log_3 uv^2 - \log_3 w$

$= \log_3 u + \log_3 v^2 - \log_3 w$

$= \log_3 u + 2\log_3 v - \log_3 w$

23. $\log_2\left(a^2\sqrt{b}\right)^4 = 4\log_2\left(a^2\sqrt{b}\right)$

$= 4\left(\log_2 a^2 + \log_2 b^{1/2}\right)$

$= 4\left(2\log_2 a + \dfrac{1}{2}\log_2 b\right)$

$= 8\log_2 a + 2\log_2 b$

24. $\log\left(x^2\sqrt{x^3 + 1}\right) = \log x^2 + \log\left(x^3 + 1\right)^{1/2}$

$= 2\log x + \dfrac{1}{2}\log\left(x^3 + 1\right)$

25. $\ln\left(\dfrac{2x + 3}{x^2 - 3x + 2}\right)^2$

$= 2\ln\left(\dfrac{2x + 3}{x^2 - 3x + 2}\right)$

$= 2\left(\ln(2x + 3) - \ln[(x - 1)(x - 2)]\right)$

$= 2\left(\ln(2x + 3) - \ln(x - 1) - \ln(x - 2)\right)$

$= 2\ln(2x + 3) - 2\ln(x - 1) - 2\ln(x - 2)$

26. $3\log_4 x^2 + \dfrac{1}{2}\log_4 \sqrt{x} = \log_4\left(x^2\right)^3 + \log_4\left(x^{1/2}\right)^{1/2}$

$= \log_4 x^6 + \log_4 x^{1/4}$

$= \log_4\left(x^6 \cdot x^{1/4}\right)$

$= \log_4 x^{25/4}$

$= \dfrac{25}{4}\log_4 x$

27. $\ln\left(\dfrac{x - 1}{x}\right) + \ln\left(\dfrac{x}{x + 1}\right) - \ln\left(x^2 - 1\right)$

$= \ln\left(\dfrac{x - 1}{x} \cdot \dfrac{x}{x + 1}\right) - \ln\left(x^2 - 1\right)$

$= \ln\left[\dfrac{\dfrac{x - 1}{x + 1}}{x^2 - 1}\right]$

$= \ln\left(\dfrac{x - 1}{x + 1} \cdot \dfrac{1}{(x - 1)(x + 1)}\right)$

$= \ln\dfrac{1}{(x + 1)^2}$

$= \ln(x + 1)^{-2}$

$= -2\ln(x + 1)$

28. $\frac{1}{2}\ln\left(x^2+1\right)-4\ln\frac{1}{2}-\frac{1}{2}\left[\ln(x-4)+\ln x\right]$

$=\ln\left(x^2+1\right)^{1/2}-\ln\left(\frac{1}{2}\right)^4-\ln\left(x(x-4)\right)^{1/2}$

$=\ln\left(\dfrac{\left(x^2+1\right)^{1/2}}{\dfrac{1}{16}\left[x(x-4)\right]^{1/2}}\right)$

$=\ln\left(\dfrac{16\sqrt{x^2+1}}{\sqrt{x(x-4)}}\right)$

29. $\log_4 19 = \dfrac{\ln 19}{\ln 4} \approx 2.124$

30. $Y_1 = \log_3 x = \dfrac{\ln x}{\ln 3}$

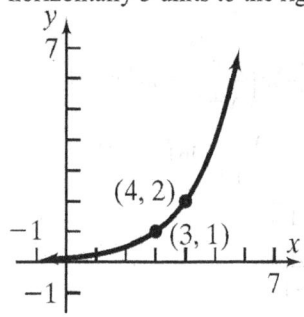

31. $f(x) = 2^{x-3}$

  a. Domain: $(-\infty, \infty)$

  b. Using the graph of $y = 2^x$, shift the graph horizontally 3 units to the right.

         (graph showing points $(4, 2)$ and $(3, 1)$)

  c. Range: $(0, \infty)$

     Horizontal Asymptote: $y = 0$

  d. $\quad f(x) = 2^{x-3}$

        $y = 2^{x-3}$

        $x = 2^{y-3}\quad$ Inverse

        $y-3 = \log_2 x$

        $y = 3 + \log_2 x$

        $f^{-1}(x) = 3 + \log_2 x$

  e. Range of f = Domain f^{-1}: $(0, \infty)$

     Domain of f = Range of f^{-1}: $(-\infty, \infty)$

  f. Using the graph of $y = \log_2 x$, shift the graph vertically 3 units up.

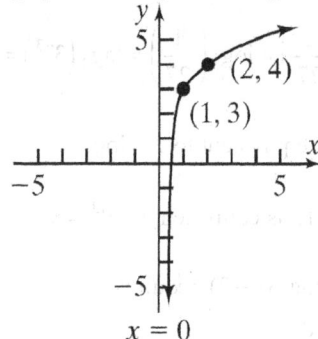

         (graph showing points $(2, 4)$ and $(1, 3)$; $x = 0$)

32. $f(x) = 1 + 3^{-x}$

  a. Domain: $(-\infty, \infty)$

  b. Using the graph of $y = 3^x$, reflect the graph about the y-axis, and shift vertically 1 unit up.

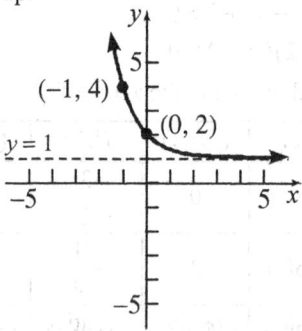

         (graph showing points $(-1, 4)$ and $(0, 2)$; $y = 1$)

  c. Range: $(1, \infty)$

     Horizontal Asymptote: $y = 1$

  d. $\quad f(x) = 1 + 3^{-x}$

        $y = 1 + 3^{-x}$

        $x = 1 + 3^{-y}\quad$ Inverse

        $x - 1 = 3^{-y}$

        $-y = \log_3 (x-1)$

        $y = -\log_3 (x-1)$

        $f^{-1}(x) = -\log_3 (x-1)$

  e. $\quad x - 1 > 0$

        $x > 1$

     Range of f = Domain f^{-1}: $(1, \infty)$

Domain of f = Range of f^{-1}: $(-\infty, \infty)$

f. Using the graph of $y = \log_3 x$, shift the graph horizontally to the right 1 unit, and reflect vertically about the x-axis.

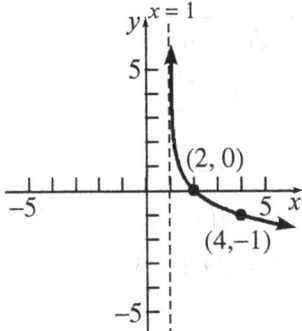

33. $f(x) = 3e^{x-2}$

a. Domain: $(-\infty, \infty)$

b. Using the graph of $y = e^x$, shift the graph two units horizontally to the right, and stretch vertically by a factor of 3.

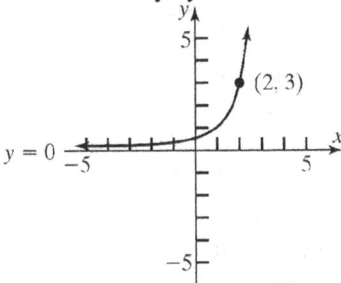

c. Range: $(0, \infty)$
Horizontal Asymptote: $y = 0$

d. $f(x) = 3e^{x-2}$
$y = 3e^{x-2}$
$x = 3e^{y-2}$ Inverse
$\dfrac{x}{3} = e^{y-2}$
$y - 2 = \ln\left(\dfrac{x}{3}\right)$
$y = 2 + \ln\left(\dfrac{x}{3}\right)$
$f^{-1}(x) = 2 + \ln\left(\dfrac{x}{3}\right)$

e. Range of f = Domain f^{-1}: $(0, \infty)$
Domain of f = Range of f^{-1}: $(-\infty, \infty)$

f. Using the graph of $y = \ln x$, stretch horizontally by a factor of 3, and shift vertically up 2 units.

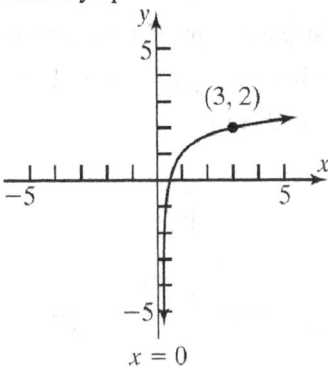

34. $f(x) = \dfrac{1}{2}\ln(x+3)$

a. Domain: $(-3, \infty)$

b. Using the graph of $y = \ln x$, shift the graph to the left 3 units and compress vertically by a factor of $\dfrac{1}{2}$.

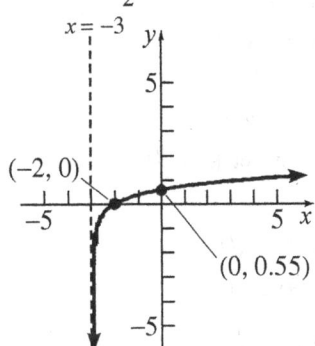

c. Range: $(-\infty, \infty)$
Vertical Asymptote: $x = -3$

d. $f(x) = \dfrac{1}{2}\ln(x+3)$
$y = \dfrac{1}{2}\ln(x+3)$
$x = \dfrac{1}{2}\ln(y+3)$ Inverse
$2x = \ln(y+3)$
$y + 3 = e^{2x}$
$y = e^{2x} - 3$
$f^{-1}(x) = e^{2x} - 3$

e. Range of f = Domain f^{-1}: $(-\infty, \infty)$

Domain of f = Range of f^{-1}: $(-3, \infty)$

f. Using the graph of $y = e^x$, compress horizontally by a factor of $\frac{1}{2}$, and shift down 3 units.

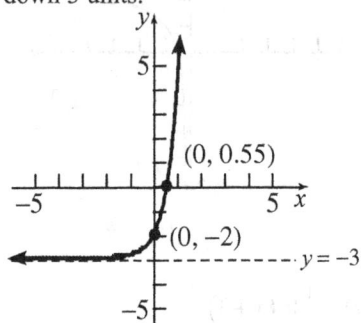

35. $8^{6+3x} = 4$

$\left(2^3\right)^{6+3x} = 2^2$

$2^{18+9x} = 2^2$

$18 + 9x = 2$

$9x = -16$

$x = -\frac{16}{9}$

The solution set is $\left\{-\frac{16}{9}\right\}$.

36. $3^{x^2+x} = \sqrt{3}$

$3^{x^2+x} = 3^{1/2}$

$x^2 + x = \frac{1}{2}$

$2x^2 + 2x - 1 = 0$

$x = \dfrac{-2 \pm \sqrt{2^2 - 4(2)(-1)}}{2(2)}$

$= \dfrac{-2 \pm \sqrt{12}}{4} = \dfrac{-2 \pm 2\sqrt{3}}{4} = \dfrac{-1 \pm \sqrt{3}}{2}$

The solution is $\left\{\dfrac{-1-\sqrt{3}}{2}, \dfrac{-1+\sqrt{3}}{2}\right\} \approx \{-1.366, 0.366\}$.

37. $\log_x 64 = -3$

$x^{-3} = 64$

$\left(x^{-3}\right)^{-1/3} = 64^{-1/3}$

$x = \dfrac{1}{\sqrt[3]{64}} = \dfrac{1}{4}$

The solution set is $\left\{\dfrac{1}{4}\right\}$.

38. $5^x = 3^{x+2}$

$\ln\left(5^x\right) = \ln\left(3^{x+2}\right)$

$x \ln 5 = (x+2)\ln 3$

$x \ln 5 = x \ln 3 + 2 \ln 3$

$x \ln 5 - x \ln 3 = 2 \ln 3$

$x(\ln 5 - \ln 3) = 2 \ln 3$

$x = \dfrac{2 \ln 3}{\ln 5 - \ln 3} \approx 4.301$

The solution set is $\left\{\dfrac{2 \ln 3}{\ln 5 - \ln 3}\right\} \approx \{4.301\}$.

39. $25^{2x} = 5^{x^2-12}$

$\left(5^2\right)^{2x} = 5^{x^2-12}$

$5^{4x} = 5^{x^2-12}$

$4x = x^2 - 12$

$x^2 - 4x - 12 = 0$

$(x-6)(x+2) = 0$

$x = 6 \text{ or } x = -2$

The solution set is $\{-2, 6\}$.

40. $\log_3 \sqrt{x-2} = 2$

$\sqrt{x-2} = 3^2$

$\sqrt{x-2} = 9$

$x - 2 = 9^2$

$x - 2 = 81$

$x = 83$

Check: $\log_3 \sqrt{83-2} = \log_3 \sqrt{81}$

$= \log_3 9$

$= 2$

The solution set is $\{83\}$.

41. $8 = 4^{x^2} \cdot 2^{5x}$

$2^3 = \left(2^2\right)^{x^2} \cdot 2^{5x}$

$2^3 = 2^{2x^2+5x}$

$3 = 2x^2 + 5x$

$0 = 2x^2 + 5x - 3$

$0 = (2x-1)(x+3)$

$x = \dfrac{1}{2}$ or $x = -3$

The solution set is $\left\{-3, \dfrac{1}{2}\right\}$.

42. $2^x \cdot 5 = 10^x$

$\ln\left(2^x \cdot 5\right) = \ln 10^x$

$\ln 2^x + \ln 5 = \ln 10^x$

$x \ln 2 + \ln 5 = x \ln 10$

$\ln 5 = x \ln 10 - x \ln 2$

$\ln 5 = x(\ln 10 - \ln 2)$

$\dfrac{\ln 5}{\ln 10 - \ln 2} = x$

$x = \dfrac{\ln 5}{\ln \dfrac{10}{2}} = \dfrac{\ln 5}{\ln 5} = 1$

The solution set is $\{1\}$.

43. $\log_6(x+3) + \log_6(x+4) = 1$

$\log_6\left((x+3)(x+4)\right) = 1$

$(x+3)(x+4) = 6^1$

$x^2 + 7x + 12 = 6$

$x^2 + 7x + 6 = 0$

$(x+6)(x+1) = 0$

$x = -6$ or $x = -1$

Since $\log_6(-6+3) = \log_6(-3)$ is undefined, the solution set is $\{-1\}$.

44. $e^{1-x} = 5$

$1 - x = \ln 5$

$-x = -1 + \ln 5$

$x = 1 - \ln 5 \approx -0.609$

The solution set is $\{1 - \ln 5\} \approx \{-0.609\}$.

45. $9^x + 4 \cdot 3^x - 3 = 0$

$\left(3^2\right)^x + 4 \cdot 3^x - 3 = 0$

$\left(3^x\right)^2 + 4 \cdot 3^x - 3 = 0$

Let $u = 3^x$.

$u^2 + 4u - 3 = 0$

$a = 1, b = 4, c = -3$

$u = \dfrac{-(4) \pm \sqrt{(4)^2 - 4(1)(-3)}}{2(1)}$

$= \dfrac{-4 \pm \sqrt{28}}{2} = \dfrac{-4 \pm 2\sqrt{7}}{2} = -2 \pm \sqrt{7}$

$\cancel{3^x = -2 - \sqrt{7}}$ or $3^x = -2 + \sqrt{7}$

3^x can't be negative $x = \log_3\left(-2 + \sqrt{7}\right)$

The solution set is $\left\{\log_3\left(-2+\sqrt{7}\right)\right\} \approx \{-0.398\}$.

46. a. $f(x) = \log_2(x-2) + 1$

Using the graph of $y = \log_2 x$, shift the graph right 2 units and up 1 unit.

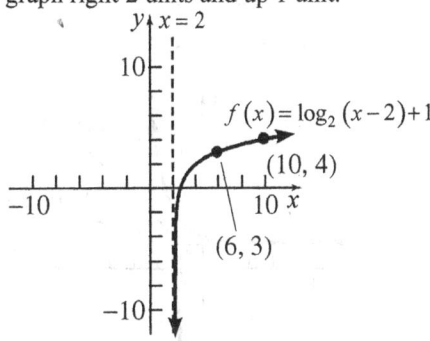

b. $f(6) = \log_2(6-2) + 1$

$= \log_2(4) + 1 = 2 + 1 = 3$

The point $(6, 3)$ is on the graph of f.

c. $f(x) = 4$

$\log_2(x-2) + 1 = 4$

$\log_2(x-2) = 3$

$x - 2 = 2^3$

$x - 2 = 8$

$x = 10$

The solution set is $\{10\}$. The point $(10, 4)$ is on the graph of f.

d.
$$f(x) = 0$$
$$\log_2(x-2)+1 = 0$$
$$\log_2(x-2) = -1$$
$$x-2 = 2^{-1}$$
$$x-2 = \frac{1}{2}$$
$$x = \frac{5}{2}$$

Based on the graph drawn in part (a),

$f(x) > 0$ when $x > \dfrac{5}{2}$. The solution set is

$\left\{ x \mid x > \dfrac{5}{2} \right\}$ or $\left(\dfrac{5}{2}, \infty \right)$.

e.
$$f(x) = \log_2(x-2)+1$$
$$y = \log_2(x-2)+1$$
$$x = \log_2(y-2)+1 \quad \text{Inverse}$$
$$x-1 = \log_2(y-2)$$
$$y-2 = 2^{x-1}$$
$$y = 2^{x-1}+2$$
$$f^{-1}(x) = 2^{x-1}+2$$

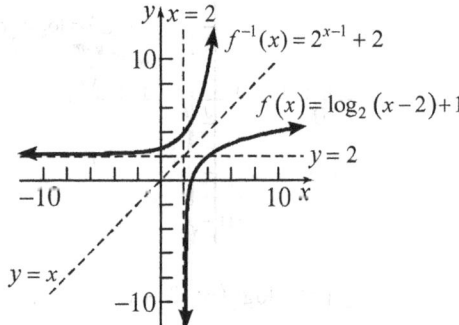

47. $P = 25e^{0.1d}$

a. $P = 25e^{0.1(4)} = 25e^{0.4} \approx 37.3$ watts

b.
$$50 = 25e^{0.1d}$$
$$2 = e^{0.1d}$$
$$\ln 2 = 0.1d$$
$$d = \frac{\ln 2}{0.1} \approx 6.9 \text{ decibels}$$

48. $L = 9+(5.1)\log d$

a. $L = 9+(5.1)\log 3.5 \approx 11.77$

b.
$$14 = 9+(5.1)\log d$$
$$5 = (5.1)\log d$$
$$\log d = \frac{5}{5.1} \approx 0.9804$$
$$d \approx 10^{0.9804} \approx 9.56 \text{ inches}$$

49. a. $n = \dfrac{\log 10{,}000 - \log 90{,}000}{\log(1-0.20)} \approx 9.85$ years

b. $n = \dfrac{\log(0.5i) - \log(i)}{\log(1-0.15)}$

$$= \frac{\log\left(\dfrac{0.5i}{i}\right)}{\log 0.85} = \frac{\log 0.5}{\log 0.85} \approx 4.27 \text{ years}$$

50. In 18 years, $A = 10{,}000\left(1+\dfrac{0.04}{2}\right)^{(2)(18)}$

$$= 10{,}000(1.02)^{36}$$
$$\approx \$20{,}398.87$$

The effective interest rate is computed as follows:

When $t = 1$, $A = 10{,}000\left(1+\dfrac{0.04}{2}\right)^{(2)(1)}$

$$= 10{,}000(1.02)^2$$
$$= \$10{,}404$$

Note, $\dfrac{10{,}404-10{,}000}{10{,}000} = \dfrac{404}{10{,}000} = 0.0404$, so the effective interest rate is 4.04%.

In order for the bond to double in value, we have the equation: $A = 2P$.

$$10{,}000\left(1+\frac{0.04}{2}\right)^{2t} = 20{,}000$$
$$(1.02)^{2t} = 2$$
$$2t\ln 1.02 = \ln 2$$
$$t = \frac{\ln 2}{2\ln 1.02} \approx 17.5 \text{ years}$$

51. $P = A\left(1+\dfrac{r}{n}\right)^{-nt} = 85,000\left(1+\dfrac{0.04}{2}\right)^{-2(18)}$

$\approx \$41,668.97$

52. $\quad A = A_0 e^{kt}$

$0.5A_0 = A_0 e^{k(5700)}$

$0.5 = e^{5700k}$

$\ln 0.5 = 5700k$

$k = \dfrac{\ln 0.5}{5700}$

$0.05A_0 = A_0 e^{\left(\frac{\ln 0.5}{5700}\right)t}$

$0.05 = e^{\left(\frac{\ln 0.5}{5700}\right)t}$

$\ln 0.05 = \left(\dfrac{\ln 0.5}{5700}\right)t$

$t = \dfrac{\ln 0.05}{\left(\dfrac{\ln 0.5}{5700}\right)} \approx 24,635$

The man died approximately 24,635 years ago.

53. Using $u = T + (u_0 - T)e^{kt}$, with $t = 5$, $T = 70$, $u_0 = 450$, and $u = 400$.

$400 = 70 + (450 - 70)e^{k(5)}$

$330 = 380e^{5k}$

$\dfrac{330}{380} = e^{5k}$

$\ln\left(\dfrac{330}{380}\right) = 5k$

$k = \dfrac{\ln(330/380)}{5}$

Find time for temperature of 150°F:

$150 = 70 + (450-70)e^{\left(\frac{\ln(330/380)}{5}\right)t}$

$80 = 380e^{\left(\frac{\ln(330/380)}{5}\right)t}$

$\dfrac{80}{380} = e^{\left(\frac{\ln(330/380)}{5}\right)t}$

$\ln\left(\dfrac{80}{380}\right) = \left(\dfrac{\ln(330/380)}{5}\right)t$

$t = \dfrac{\ln\left(\dfrac{80}{380}\right)}{\dfrac{\ln(330/380)}{5}} \approx 55.22$

The temperature of the skillet will be 150°F after approximately 55.22 minutes (or 55 minutes, 13 seconds).

54. $P_0 = 7,214,958,996$, $k = 0.0108$, and

$t = 2020 - 2015 = 5$

$P = P_0 e^{kt} = 7,214,958,996 e^{0.0108(5)}$

$\approx 7,615,278,125$ people

55. Use $A = A_0 e^{kt}$ and solve for k:

$0.5A_0 = A_0 e^{k(5.27)}$

$0.5 = e^{(5.27)k}$

$\ln 0.5 = 5.27k$

$k = \dfrac{\ln 0.5}{5.27}$

When $A_0 = 100$ and $t = 20$:

$A = 100e^{\frac{\ln 0.5}{5.27}(20)} \approx 7.204$ grams

When $A_0 = 100$ and $t = 40$:

$A = 100e^{\frac{\ln 0.5}{5.27}(40)} \approx 0.519$ grams

56. a. $P(0) = \dfrac{0.8}{1+1.67e^{-0.16(0)}} = \dfrac{0.8}{1+1.67} \approx 0.3$

In 2006, about 30% of cars had a GPS.

b. The maximum proportion is the carrying capacity, $c = 0.8 = 80\%$.

c. $Y_1 = \dfrac{0.8}{1+1.67e^{-0.16x}}$

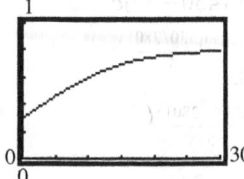

d. Find t such that $P(t) = 0.75$.

$$\frac{0.8}{1+1.67e^{-0.16t}} = 0.75$$

$$0.8 = 0.75\left(1+1.67e^{-0.16t}\right)$$

$$\frac{0.8}{0.75} = 1+1.67e^{-0.16t}$$

$$\frac{0.8}{0.75}-1 = 1.67e^{-0.16t}$$

$$\frac{\frac{0.8}{0.75}-1}{1.67} = e^{-0.16t}$$

$$\ln\left(\frac{\frac{0.8}{0.75}-1}{1.67}\right) = -0.16t$$

$$t = \frac{\ln\left(\frac{\frac{0.8}{0.75}-1}{1.67}\right)}{-0.16} \approx 20.13$$

Note that $2006 + 20.13 = 2026.13$, so 75% of new cars will have GPS in 2026.

57. a.

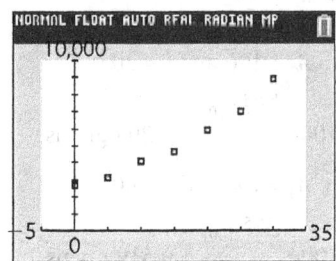

b. Using EXPonential REGression on the data yields: $y = 2638.26(1.0407)^x$

c. $y = 2638.26(1.0407)^x$

$$= 2638.26\left(e^{\ln(1.0407)}\right)^x$$

$$= 2638.26e^{\ln(1.0407)x}$$

$$A(t) = 2638.26e^{0.0399x}$$

d. $Y_1 = 2638.26(1.0407)^x$

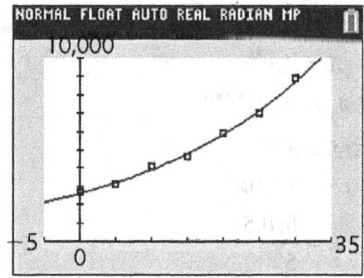

e. Find x when $y = 12000$.

$$2638.26(1.0407)^x = 12000$$

$$(1.0407)^x = \frac{12000}{2638.26}$$

$$x\ln 1.0407 = \ln\frac{12000}{2638.26}$$

$$x = \frac{\ln\dfrac{12000}{2638.26}}{\ln 1.0407} \approx 38$$

Therefore, it will take approximately 38 years for the tuition to reach $12,000 or in 2021-22.

58. a.

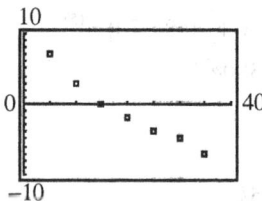

b. Using LnREGression on the data yields: $y = 18.9028 - 7.0963\ln x$ where $y =$ wind chill and $x =$ wind speed.

c. $Y_1 = 18.9028 - 7.0963\ln x$

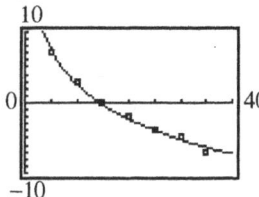

d. If $x = 23$, then
$y = 18.9028 - 7.0963\ln 23 \approx -3°\,\text{F}$.

59. a.

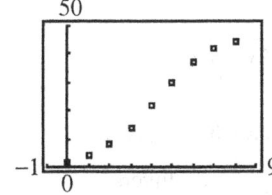

The data appear to have a logistic relation.

b. Using LOGISTIC REGression on the data yields:

$$C = \frac{46.93}{1+21.273e^{-0.7306t}}$$

c. $Y_1 = \frac{46.93}{1+21.273e^{-0.7306t}}$

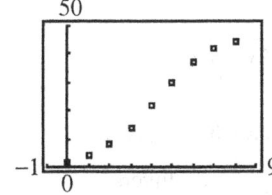

d. As $t \to \infty$, $21.2733e^{-0.7306t} \to 0$, which means $1+21.2733e^{-0.7306t} \to 1$, so

$$C = \frac{46.9292}{1+21.2733e^{-0.7306t}} \to 46.9292$$

Therefore, according to the function, a

maximum of about 47 people can catch the cold.

In reality, all 50 people living in the town might catch the cold.

e. Find t when $C = 10$.

$$\frac{46.9292}{1+21.2733e^{-0.7306t}} = 10$$

$$46.9292 = 10\left(1+21.2733e^{-0.7306t}\right)$$

$$\frac{46.9292}{10} = 1+21.2733e^{-0.7306t}$$

$$\frac{46.9292}{10} - 1 = 21.2733e^{-0.7306t}$$

$$3.69292 = 21.2733e^{-0.7306t}$$

$$\frac{3.69292}{21.2733} = e^{-0.7306t}$$

$$\ln\left(\frac{3.69292}{21.2733}\right) = -0.7306t$$

$$\frac{\ln\left(\frac{3.69292}{21.2733}\right)}{-0.7306} = t$$

$$t \approx 2.4$$

Therefore, after approximately 2.4 days (during the 10th hour on the 3rd day), 10 people had caught the cold.

f. Find t when $C = 46$.

$$\frac{46.9292}{1+21.2733e^{-0.7306t}} = 46$$

$$46.9292 = 46\left(1+21.2733e^{-0.7306t}\right)$$

$$\frac{46.9292}{46} = 1+21.2733e^{-0.7306t}$$

$$\frac{46.9292}{46} - 1 = 21.2733e^{-0.7306t}$$

$$0.0202 = 21.2733e^{-0.7306t}$$

$$\frac{0.0202}{21.2733} = e^{-0.7306t}$$

$$\frac{0.0202}{21.2733} = e^{-0.7306t}$$

$$\ln\left(\frac{0.0202}{21.2733}\right) = -0.7306t$$

$$\frac{\ln\left(\frac{0.0202}{21.2733}\right)}{-0.7306} = t \approx 9.5$$

Therefore, after approximately 9.5 days (during the 12th hour on the 10th day), 46 people had caught the cold.

Chapter 6 Test

1. $f(x) = \frac{x+2}{x-2}$ $g(x) = 2x+5$

The domain of f is $\{x \mid x \neq 2\}$.

The domain of g is all real numbers.

a. $(f \circ g)(x) = f(g(x))$

$$= f(2x+5)$$

$$= \frac{(2x+5)+2}{(2x+5)-2}$$

$$= \frac{2x+7}{2x+3}$$

Domain $\left\{x \mid x \neq -\frac{3}{2}\right\}$.

409

b. $(g \circ f)(-2) = g(f(-2))$

$= g\left(\dfrac{-2+2}{-2-2}\right)$

$= g(0)$

$= 2(0)+5$

$= 5$

c. $(f \circ g)(-2) = f(g(-2)) = f(2(-2)+5)$

$= f(1) = \dfrac{1+2}{1-2} = \dfrac{3}{-1} = -3$

2. a. Graph $y = 4x^2 + 3$:

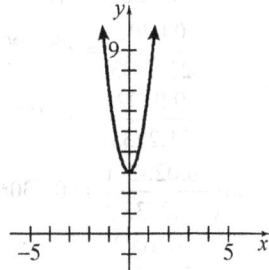

The function is not one-to-one because it fails the horizontal line test. A horizontal line (for example, $y = 4$) intersects the graph twice.

b. Graph $y = \sqrt{x+3} - 5$:

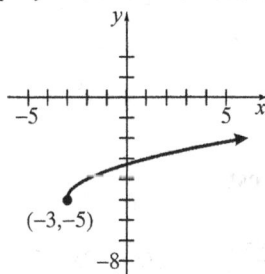

$(-3,-5)$

The function is one-to-one because it passes the horizontal line test. Every horizontal line intersects the graph at most once.

3. $f(x) = \dfrac{2}{3x-5}$

$y = \dfrac{2}{3x-5}$

$x = \dfrac{2}{3y-5}$ Inverse

$x(3y-5) = 2$

$3xy - 5x = 2$

$3xy = 5x + 2$

$y = \dfrac{5x+2}{3x}$

$f^{-1}(x) = \dfrac{5x+2}{3x}$

Domain of f = Range of f^{-1}

$= \left\{x \mid x \neq \tfrac{5}{3}\right\}.$

Range of f = Domain of f^{-1}

$= \{x \mid x \neq 0\}$

4. If the point $(3, -5)$ is on the graph of f, then the point $(-5, 3)$ must be on the graph of f^{-1}.

5. $3^x = 243$

$3^x = 3^5$

$x = 5$

6. $\log_b 16 = 2$

$b^2 = 16$

$b = \pm\sqrt{16} = \pm 4$

Since the base of a logarithm must be positive, the only viable solution is $b = 4$.

7. $\log_5 x = 4$

$x = 5^4$

$x = 625$

8. $e^3 + 2 \approx 22.086$

9. $\log 20 \approx 1.301$

10. $\log_3 21 = \dfrac{\ln 21}{\ln 3} \approx 2.771$

11. $\ln 133 \approx 4.890$

12. $f(x) = 4^{x+1} - 2$

   a. Domain: $(-\infty, \infty)$

   b. Using the graph of $y = 4^x$, shift the graph 1 unit to the left, and shift 2 units down.

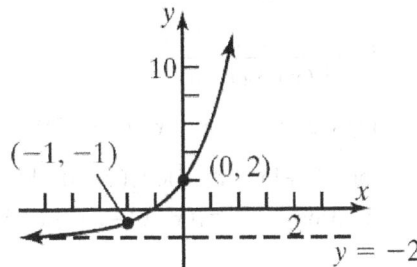

   c. Range: $(-2, \infty)$

      Horizontal Asymptote: $y = -2$

   d.
$$f(x) = 4^{x+1} - 2$$
$$y = 4^{x+1} - 2$$
$$x = 4^{y+1} - 2 \quad \text{Inverse}$$
$$x + 2 = 4^{y+1}$$
$$y + 1 = \log_4(x+2)$$
$$y = \log_4(x+2) - 1$$
$$f^{-1}(x) = \log_4(x+2) - 1$$

   e. Range of f = Domain f^{-1}: $(-2, \infty)$

      Domain of f = Range of f^{-1}: $(-\infty, \infty)$

   f. Using the graph of $y = \log_4 x$, shift the graph 2 units to the left, and shift down 1 unit.

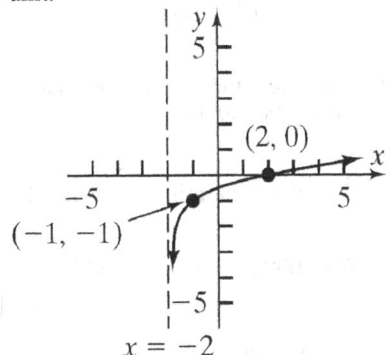

13. $f(x) = 1 - \log_5(x-2)$

   a. Domain: $(2, \infty)$

   b. Using the graph of $y = \log_5 x$, shift the graph to the right 2 units, reflect vertically about the y-axis, and shift up 1 unit.

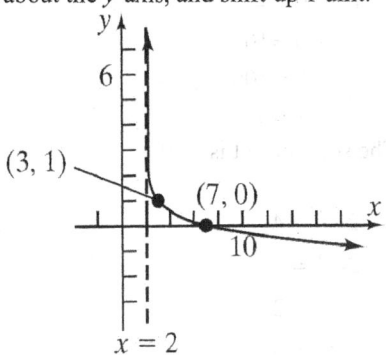

   c. Range: $(-\infty, \infty)$

      Vertical Asymptote: $x = 2$

   d.
$$f(x) = 1 - \log_5(x-2)$$
$$y = 1 - \log_5(x-2)$$
$$x = 1 - \log_5(y-2) \quad \text{Inverse}$$
$$x - 1 = -\log_5(y-2)$$
$$1 - x = \log_5(y-2)$$
$$y - 2 = 5^{1-x}$$
$$y = 5^{1-x} + 2$$
$$f^{-1}(x) = 5^{1-x} + 2$$

   e. Range of f = Domain f^{-1}: $(-\infty, \infty)$

      Domain of f = Range of f^{-1}: $(2, \infty)$

   f. Using the graph of $y = 5^x$, reflect the graph horizontally about the y-axis, shift to the right 1 unit, and shift up 2 units.

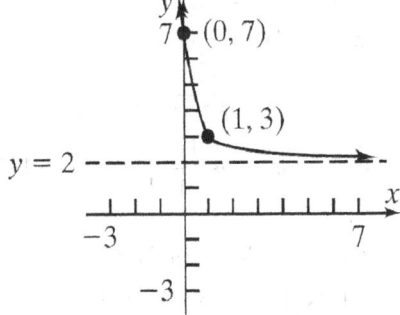

14. $5^{x+2} = 125$
$$5^{x+2} = 5^3$$
$$x + 2 = 3$$
$$x = 1$$
The solution set is $\{1\}$.

15. $\log(x+9) = 2$

$x+9 = 10^2$

$x+9 = 100$

$x = 91$

The solution set is {91}.

16. $8 - 2e^{-x} = 4$

$-2e^{-x} = -4$

$e^{-x} = 2$

$-x = \ln 2$

$x = -\ln 2 \approx -0.693$

The solution set is $\{-\ln 2\} \approx \{-0.693\}$.

17. $\log(x^2 + 3) = \log(x+6)$

$x^2 + 3 = x + 6$

$x^2 - x - 3 = 0$

$x = \dfrac{-(-1) \pm \sqrt{(-1)^2 - 4(1)(-3)}}{2(1)} = \dfrac{1 \pm \sqrt{13}}{2}$

The solution set is $\left\{\dfrac{1-\sqrt{13}}{2}, \dfrac{1+\sqrt{13}}{2}\right\}$

$\approx \{-1.303, \ 2.303\}$.

18. $7^{x+3} = e^x$

$\ln 7^{x+3} = \ln e^x$

$(x+3)\ln 7 = x$

$x \ln 7 + 3\ln 7 = x$

$x \ln 7 - x = -3\ln 7$

$x(\ln 7 - 1) = -3\ln 7$

$x = \dfrac{-3\ln 7}{\ln 7 - 1} = \dfrac{3\ln 7}{1 - \ln 7} \approx -6.172$

The solution set is $\left\{\dfrac{3\ln 7}{1-\ln 7}\right\} \approx \{-6.172\}$.

19. $\log_2(x-4) + \log_2(x+4) = 3$

$\log_2[(x-4)(x+4)] = 3$

$\log_2(x^2 - 16) = 3$

$x^2 - 16 = 2^3$

$x^2 - 16 = 8$

$x^2 = 24$

$x = \pm\sqrt{24} = \pm 2\sqrt{6}$

Because $x = -2\sqrt{6}$ results in a negative arguments for the original logarithms, the only

viable solution is $x = 2\sqrt{6}$. That is, the solution set is $\{2\sqrt{6}\} \approx \{4.899\}$.

20. $\log_2\left(\dfrac{4x^3}{x^2 - 3x - 18}\right)$

$= \log_2\left(\dfrac{2^2 x^3}{(x+3)(x-6)}\right)$

$= \log_2(2^2 x^3) - \log_2[(x-6)(x+3)]$

$= \log_2 2^2 + \log_2 x^3 - [\log_2(x-6) + \log_2(x+3)]$

$= 2 + 3\log_2 x - \log_2(x-6) - \log_2(x+3)$

21. $A = A_0 e^{kt}$

$34 = 50e^{k(30)}$

$0.68 = e^{30k}$

$\ln 0.68 = 30k$

$k = \dfrac{\ln 0.68}{30}$

Thus, the decay model is $A = 50e^{\left(\frac{\ln 0.68}{30}\right)t}$.

We need to find t when $A = 2$:

$2 = 50e^{\left(\frac{\ln 0.68}{30}\right)t}$

$0.04 = e^{\left(\frac{\ln 0.68}{30}\right)t}$

$\ln 0.04 = \left(\dfrac{\ln 0.68}{30}\right)t$

$t = \dfrac{\ln 0.04}{\left(\dfrac{\ln 0.68}{30}\right)} \approx 250.39$

There will be 2 mg of the substance remaining after about 250.39 days.

22. a. Note that 8 months $= \dfrac{2}{3}$ year. Thus,

$P = 1000$, $r = 0.05$, $n = 12$, and $t = \dfrac{2}{3}$.

So, $A = 1000\left(1 + \dfrac{0.05}{12}\right)^{(12)(2/3)}$

$= 1000\left(1 + \dfrac{0.05}{12}\right)^8$

$\approx \$1033.82$

b. Note that 9 months $= \dfrac{3}{4}$ year. Thus,

$A = 1000$, $r = 0.05$, $n = 4$, and $t = \dfrac{3}{4}$. So,

$$1000 = A_0 \left(1 + \frac{0.05}{4}\right)^{(4)(3/4)}$$

$$1000 = A_0 (1.0125)^3$$

$$A_0 = \frac{1000}{(1.0125)^3} \approx \$963.42$$

c. $r = 0.06$ and $n = 1$. So,

$$2A_0 = A_0 \left(1 + \frac{0.06}{1}\right)^{(1)t}$$

$$2A_0 = A_0 (1.06)^t$$

$$2 = (1.06)^t$$

$$t = \log_{1.06} 2 = \frac{\ln 2}{\ln 1.06} \approx 11.9$$

It will take about 11.9 years to double your money under these conditions.

23. a. $\quad 80 = 10 \log \left(\dfrac{I}{10^{-12}}\right)$

$$8 = \log \left(\frac{I}{10^{-12}}\right)$$

$$8 = \log I - \log 10^{-12}$$

$$8 = \log I - (-12)$$

$$8 = \log I + 12$$

$$-4 = \log I$$

$$I = 10^{-4} = 0.0001$$

If one person shouts, the intensity is 10^{-4} watts per square meter. Thus, if two people shout at the same time, the intensity will be 2×10^{-4} watts per square meter. Thus, the loudness will be

$$D = 10 \log \left(\frac{2 \times 10^{-4}}{10^{-12}}\right) = 10 \log \left(2 \times 10^8\right) \approx 83$$

decibels

b. Let n represent the number of people who must shout. Then the intensity will be $n \times 10^{-4}$. If $D = 125$, then

$$125 = 10 \log \left(\frac{n \times 10^{-4}}{10^{-12}}\right)$$

$$125 = 10 \log \left(n \times 10^8\right)$$

$$12.5 = \log \left(n \times 10^8\right)$$

$$n \times 10^8 = 10^{12.5}$$

$$n = 10^{4.5} \approx 31,623$$

About 31,623 people would have to shout at the same time in order for the resulting sound level to meet the pain threshold.

Chapter 6 Cumulative Review

1. a. The graph represents a function since it passes the Vertical Line Test.

The function is not a one-to-one function since the graph fails the Horizontal Line Test.

b. The graph appears to be the graph of a polynomial function. It has no gaps, holes or cusps. It has two turning points and three real zeroes. The degree must be at least three.

2. $f(x) = 2x^2 - 3x + 1$

a. $f(3) = 2(3)^2 - 3(3) + 1 = 18 - 9 + 1 = 10$

b. $f(-x) = 2(-x)^2 - 3(-x) + 1 = 2x^2 + 3x + 1$

c. $f(x+h) = 2(x+h)^2 - 3(x+h) + 1$
$$= 2\left(x^2 + 2xh + h^2\right) - 3x - 3h + 1$$
$$= 2x^2 + 4xh + 2h^2 - 3x - 3h + 1$$

3. $x^2 + y^2 = 1$

a. $\left(\dfrac{1}{2}\right)^2 + \left(\dfrac{1}{2}\right)^2 = \dfrac{1}{4} + \dfrac{1}{4} = \dfrac{1}{2} \neq 1$; $\left(\dfrac{1}{2}, \dfrac{1}{2}\right)$ is not on the graph.

b. $\left(\dfrac{1}{2}\right)^2 + \left(\dfrac{\sqrt{3}}{2}\right)^2 = \dfrac{1}{4} + \dfrac{3}{4} = 1$; $\left(\dfrac{1}{2}, \dfrac{\sqrt{3}}{2}\right)$ is on the graph.

4. $3(x-2)=4(x+5)$

$3x-6=4x+20$

$-26=x$

The solution set is $\{-26\}$.

5. $2x-4y=16$

x-intercept: $\qquad$ y-intercept:

$2x-4(0)=16$ $\qquad$ $2(0)-4y=16$

$2x=16$ $\qquad$ $-4y=16$

$x=8$ $\qquad$ $y=-4$

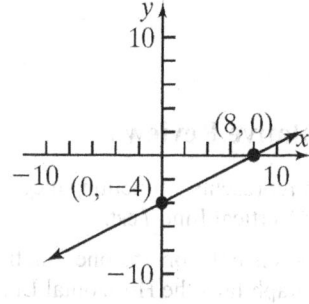

6. a. $f(x)=-x^2+2x-3$; $a=-1, b=2, c=-3$.

Since $a=-1<0$, the graph opens down.

The x-coordinate of the vertex is

$$x=-\frac{b}{2a}=-\frac{2}{2(-1)}=-\frac{2}{-2}=1.$$

The y-coordinate of the vertex is

$$f\left(-\frac{b}{2a}\right)=f(1)$$

$$=-1^2+2(1)-3$$

$$=-1+2-3$$

$$=-2$$

Thus, the vertex is $(1,-2)$.

The axis of symmetry is the line $x=1$.

The discriminant is:

$b^2-4ac=2^2-4(-1)(-3)=4-12=-8<0$.

The graph has no x-intercepts.

The y-intercept is $f(0)=-0^2+2(0)-3=-3$.

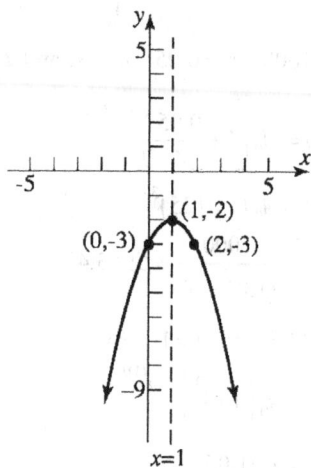

b. The graph of $f(x)=-x^2+2x-3$ indicates that $f(x)\le 0$ for all values of x. Thus, the solution to $f(x)\le 0$ is $(-\infty,\infty)$.

7. Given that the graph of $f(x)=ax^2+bx+c$ has vertex $(4,-8)$ and passes through the point $(0,24)$, we can conclude $-\frac{b}{2a}=4$, $f(4)=-8$, and $f(0)=24$. Notice that

$$f(0)=24$$

$$a(0)^2+b(0)+c=24$$

$$c=24$$

Therefore, $f(x)=ax^2+bx+c=ax^2+bx+24$.

Furthermore, $-\frac{b}{2a}=4$, so that $b=-8a$, and

$$f(4)=-8$$

$$a(4)^2+b(4)+24=-8$$

$$16a+4b+24=-8$$

$$16a+4b=-32$$

$$4a+b=-8$$

Replacing b with $-8a$ in this equation yields

$$4a-8a=-8$$

$$-4a=-8$$

$$a=2$$

So $b=-8a=-8(2)=-16$.

Therefore, we have the function

$$f(x)=2x^2-16x+24.$$

8.

x	$y = f(x)$	$\dfrac{f(x+1)}{f(x)}$
-1	$\dfrac{2}{3}$	$\dfrac{2}{(2/3)} = 3$
0	2	$\dfrac{6}{2} = 3$
1	6	

The ratio of consecutive outputs is a constant, 3.
This is an exponential function with growth
factor $a = 3$. The initial value of the exponential
function is $C = 2$. Therefore, the exponential
function that models the data is
$$f(x) = Ca^x = 2 \cdot (3)^x = 2 \cdot 3^x.$$

9. $f(x) = 3(x+1)^3 - 2$

Using the graph of $y = x^3$, shift the graph 1 unit
to the left, stretch vertically by a factor of 3, and
shift 2 units down.

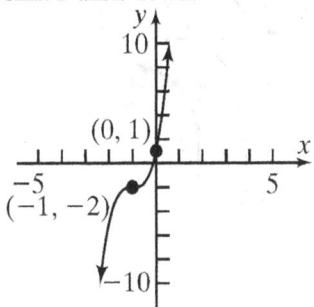

10. a. $f(x) = x^2 + 2 \qquad g(x) = \dfrac{2}{x-3}$

$$f(g(x)) = f\left(\dfrac{2}{x-3}\right)$$
$$= \left(\dfrac{2}{x-3}\right)^2 + 2$$
$$= \dfrac{4}{(x-3)^2} + 2$$

The domain of f is $\{x \mid x \text{ is any real number}\}$.

The domain of g is $\{x \mid x \neq 3\}$.

So, the domain of $f(g(x))$ is $\{x \mid x \neq 3\}$.

$$f(g(5)) = \dfrac{4}{(5-3)^2} + 2 = \dfrac{4}{2^2} + 2 = \dfrac{4}{4} + 2 = 3$$

b. $f(x) = x + 2; \ g(x) = \log_2 x$

$$f(g(x)) = f(\log_2 x) = \log_2 x + 2$$

Domain: $\{x \mid x > 0\}$ or $(0, \infty)$

$$f(g(14)) = f(\log_2 14) = \log_2 14 + 2$$

11. $f(x) = 4x^3 + 9x^2 - 30x - 8$

a. The graph of $Y_1 = 4x^3 + 9x^2 - 30x - 8$
appears to indicate zeros at $x = -4$ and
$x = 2$.

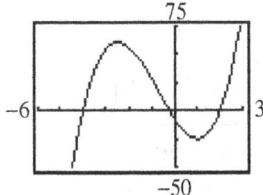

$$f(-4) = 4(-4)^3 + 9(-4)^2 - 30(-4) - 8$$
$$= -256 + 144 + 120 - 8$$
$$= 0$$
$$f(2) = 4(2)^3 + 9(2)^2 - 30(2) - 8$$
$$= 32 + 36 - 60 - 8$$
$$= 0$$

Therefore, $x = -4$ and $x = 2$ are real zeros
for f.

Using synthetic division:

$$
\begin{array}{r|rrrr}
2 & 4 & 9 & -30 & -8 \\
  &   & 8 & 34 & 8 \\
\hline
  & 4 & 17 & 4 & 0
\end{array}
$$

$$f(x) = 4x^3 + 9x^2 - 30x - 8$$
$$= (x-2)(4x^2 + 17x + 4)$$
$$= (x-2)(x+4)(4x+1)$$

Therefore, $x = 2$, $x = -\dfrac{1}{4}$ and $x = -4$ are
real zeros of f.

b. f has x-intercepts at $x = 2$, $x = -\dfrac{1}{4}$ and
$x = -4$.

f has y-intercept at
$$f(0) = 4 \cdot 0^3 + 9 \cdot 0^2 - 30 \cdot 0 - 8 = -8$$

c. Use MAXIMUM to determine that f has a
local maximum at the point $(-2.5, 60.75)$.

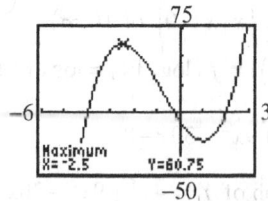

Use MINIMUM to determine that f has a local minimum at the point $(1,-25)$.

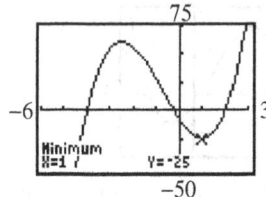

Thus, f has a local maximum of 60.75 that occurs at $x = -2.5$, and f has a local minimum of -25 that occurs at $x = 1$.

d. Graphing by hand:

The graph of f is above the x-axis for $\left(-4, -\dfrac{1}{4}\right)$ and $(2, \infty)$.

The graph of f is below the x-axis for $(-\infty, -4)$.

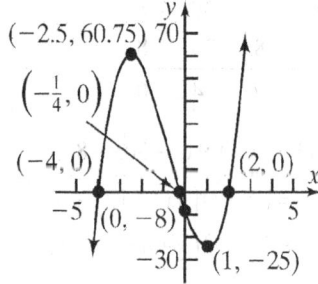

12. a. $g(x) = 3^x + 2$

Using the graph of $y = 3^x$, shift up 2 units.

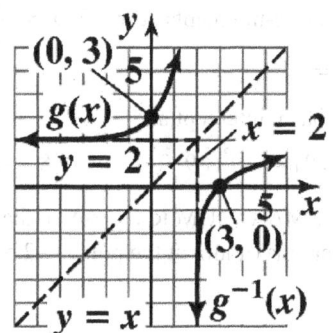

Domain of g: $(-\infty, \infty)$

Range of g: $(2, \infty)$

Horizontal Asymptote for g: $y = 2$

b. $\qquad g(x) = 3^x + 2$

$\qquad\qquad y = 3^x + 2$

$\qquad\qquad x = 3^y + 2 \qquad$ Inverse

$\qquad x - 2 = 3^y$

$\qquad\qquad y = \log_3(x - 2)$

$\qquad g^{-1}(x) = \log_3(x - 2)$

Domain of g^{-1}: $(2, \infty)$

Range of g^{-1}: $(-\infty, \infty)$

Vertical Asymptote for g^{-1}: $x = 2$

c.

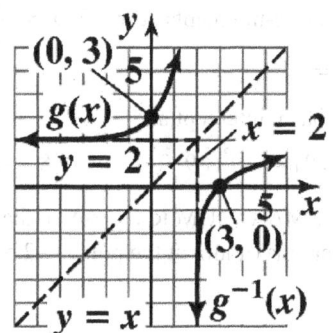

13. $\qquad 4^{x-3} = 8^{2x}$

$\qquad \left(2^2\right)^{x-3} = \left(2^3\right)^{2x}$

$\qquad\quad 2^{2x-6} = 2^{6x}$

$\qquad\quad 2x - 6 = 6x$

$\qquad\qquad -6 = 4x$

$\qquad\qquad x = -\dfrac{6}{4} = -\dfrac{3}{2}$

The solution set is $\left\{-\dfrac{3}{2}\right\}$.

14. $\log_3(x+1) + \log_3(2x-3) = \log_9 9$

$\qquad \log_3\big((x+1)(2x-3)\big) = 1$

$\qquad\qquad (x+1)(2x-3) = 3^1$

$\qquad\qquad\qquad 2x^2 - x - 3 = 3$

$\qquad\qquad\qquad 2x^2 - x - 6 = 0$

$\qquad\qquad (2x+3)(x-2) = 0$

$\qquad\qquad x = -\dfrac{3}{2} \ \text{ or } \ x = 2$

Since $\log_3\left(-\frac{3}{2}+1\right) = \log_3\left(-\frac{1}{2}\right)$ is undefined

the solution set is $\{2\}$.

15. **a.** $\log_3(x+2) = 0$

$$x+2 = 3^0$$
$$x+2 = 1$$
$$x = -1$$

The solution set is $\{-1\}$.

b. $\log_3(x+2) > 0$

$$x+2 > 3^0$$
$$x+2 > 1$$
$$x > -1$$

The solution set is $\{x \mid x > -1\}$ or $(-1, \infty)$.

c. $\log_3(x+2) = 3$

$$x+2 = 3^3$$
$$x+2 = 27$$
$$x = 25$$

The solution set is $\{25\}$.

16. **a.**

b. Logarithmic: $y = 49.293 - 10.563 \ln x$

c. Answers will vary.

Chapter 6 Projects

Project I – Internet-based Project - Answers will vary

Project II

a. Newton's Law of Cooling:

$$u(t) = T + (u_0 - T)e^{kt}, \, k < 0$$

Container 1: $u_0 = 200°F$, T = 70°F, u(30)=100°F, t = 30 mins.

$$100 = 70 + (200 - 70)e^{30k}$$
$$30 = 130e^{30k}$$
$$\frac{30}{130} = e^{30k}$$
$$30k = \ln\left(\frac{30}{130}\right)$$
$$k = \frac{1}{30}\ln\left(\frac{30}{130}\right) \approx -0.04888$$
$$u_1(t) = 70 + 130e^{-0.04888t}$$

Container 2: $u_0 = 200°F$, T = 60°F, u(25)=110°F, t = 25 mins.

$$100 = 60 + (200 - 60)e^{25k}$$
$$50 = 140e^{25k}$$
$$\frac{50}{140} = e^{25k}$$
$$25k = \ln\left(\frac{50}{140}\right)$$
$$k = \frac{\ln\left(\frac{50}{140}\right)}{25} \approx -0.04118$$
$$u_2(t) = 60 + 140e^{-0.04118t}$$

Container 3: $u_0 = 200°F$, T = 65°F, u(20)=120°F, t = 20 mins.

$$100 = 65 + (200 - 65)e^{20k}$$
$$55 = 135e^{20k}$$
$$\frac{55}{135} = e^{20k}$$
$$20k = \ln\left(\frac{55}{135}\right)$$
$$k = \frac{\ln\left(\frac{55}{135}\right)}{20} \approx -0.04490$$
$$u_3(t) = 65 + 135e^{-0.04490t}$$

b. We need time for each of the problems, so solve for t first then substitute the specific values for each container:

$$u = T + (u_0 - T)e^{kt}$$

$$u - T = (u_0 - T)e^{kt} \Rightarrow \frac{u-T}{u_0-T} = e^{kt}$$

$$kt = \ln\left(\frac{u-T}{u_0-T}\right) \Rightarrow t = \frac{\ln\left(\frac{u-T}{u_0-T}\right)}{k}$$

Container 1:

$$t = \frac{\ln\left(\frac{130-70}{200-70}\right)}{-0.04888} \approx 15.82 \text{ minutes}$$

Container 2:

$$t = \frac{\ln\left(\frac{130-60}{200-60}\right)}{-0.04118} \approx 16.83 \text{ minutes}$$

Container 3:

$$t = \frac{\ln\left(\frac{130-65}{200-65}\right)}{-0.04490} \approx 16.28 \text{ minutes}$$

c. Container 1:

$$t = \frac{\ln\left(\frac{110-70}{130-70}\right)}{-0.04888} \approx 8.295$$

It will remain between 110° and 130° for about 8.3 minutes.

Container 2:

$$t = \frac{\ln\left(\frac{110-60}{130-60}\right)}{-0.04118} \approx 8.171$$

It will remain between 110° and 130° for about 8.17 minutes

Container 3:

$$t = \frac{\ln\left(\frac{110-65}{130-65}\right)}{-0.04490} \approx 8.190$$

It will remain between 110° and 130° for about 8.19 minutes.

d. All three graphs basically lie on top of each other.

e. Container 1 would be the best. It cools off the quickest but it stays in a warm beverage range the longest.

f. Since all three containers are within seconds of each other in cooling and staying warm, the cost would have an effect. The cheaper one would be the best recommendation.

Project III

Solder Joint Strain, εp	$X = \ln(\varepsilon p)$	Fatigue Cycles, Nf	$Y = \ln(Nf)$
0.01	−4.605	10,000	9.210
0.035	−3.352	1000	6.908
0.1	−2.303	100	4.605
0.4	−0.916	10	2.303
1.5	0.405	1	0

1.

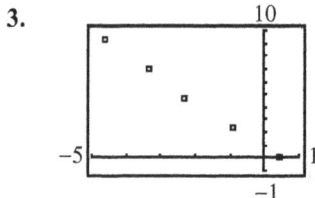

2.

The shape becomes exponential.

3.

The shape became linear.

4.

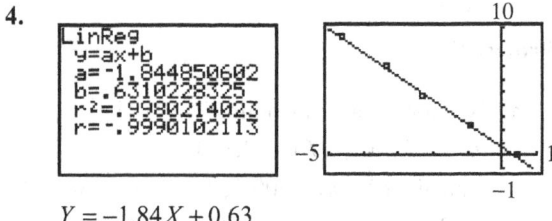

$$Y = -1.84X + 0.63$$

5. $Y = -1.84X + 0.63$

   $\ln(Nf) = -1.84 \ln(\varepsilon p) + 0.63$

   $\ln(Nf) = \ln\left((\varepsilon p)^{-1.84}\right) + \ln(e^{0.63})$

   $\ln(Nf) = \ln\left(\left((\varepsilon p)^{-1.84}\right)(e^{0.63})\right)$

   $Nf = \left((\varepsilon p)^{-1.84}\right)(e^{0.63})$

   $Nf = e^{0.63}(\varepsilon p)^{-1.84}$

6. $Nf = e^{0.63}(0.02)^{-1.84}$

   $Nf = 2510.21 \; cycles$

   $Nf = e^{0.63}(\varepsilon p)^{-1.84}$

   $3000 = e^{0.63}(\varepsilon p)^{-1.84}$

   $\dfrac{3000}{e^{0.63}} = (\varepsilon p)^{-1.84}$

   $\varepsilon p = \left(\dfrac{3000}{e^{0.63}}\right)^{-1/1.84}$

   $\varepsilon p = 0.018$

7. $Nf = e^{0.63}(\varepsilon p)^{-1.84}$ $\varepsilon p = 1.41(Nf)^{-.543}$

   $Nf = 1.88(\varepsilon p)^{-1.84}$ $\varepsilon p = 1.41(3000)^{-.543}$

   $\dfrac{Nf}{1.88} = (\varepsilon p)^{-1.84}$ $\varepsilon p = 0.018$

   $\varepsilon p = (0.53 Nf)^{-1/1.84}$

   $\varepsilon p = (0.53 Nf)^{-.543}$

   $\varepsilon p = 1.41(Nf)^{-.543}$

Chapter 7

Analytic Geometry

Section 7.1

Not applicable

Section 7.2

1. $\sqrt{(x_2 - x_1)^2 + (y_2 - y_1)^2}$

3. $(x+4)^2 = 9$

$x + 4 = \pm 3$

$x + 4 = 3$ or $x + 4 = -3$

$x = -1$ or $x = -7$

The solution set is $\{-7, -1\}$.

5. 3, up

7. axis of symmetry

9. c

11. d

13. (b); the graph has a vertex $(h, k) = (0, 0)$ and opens up. Therefore, the equation of the graph has the form $x^2 = 4ay$. The graph passes through the point $(2, 1)$ so we have

$(2)^2 = 4a(1)$

$4 = 4a$

$1 = a$

Thus, the equation of the graph is $x^2 = 4y$.

15. (e); the graph has vertex $(h, k) = (1, 1)$ and opens to the right. Therefore, the equation of the graph has the form $(y-1)^2 = 4a(x-1)$.

17. (h); the graph has vertex $(h, k) = (-1, -1)$ and opens down. Therefore, the equation of the graph

19. (c); the graph has vertex $(h, k) = (0, 0)$ and opens to the left. Therefore, the equation of the graph has the form $y^2 = -4ax$. The graph passes through the point $(-1, -2)$ so we have

$(-2)^2 = -4a(-1)$

$4 = 4a$

$1 = a$

Thus, the equation of the graph is $y^2 = -4x$.

21. The focus is (4, 0) and the vertex is (0, 0). Both lie on the horizontal line $y = 0$. $a = 4$ and since (4, 0) is to the right of (0, 0), the parabola opens to the right. The equation of the parabola is:

$y^2 = 4ax$

$y^2 = 4 \cdot 4 \cdot x$

$y^2 = 16x$

Letting $x = 4$, we find $y^2 = 64$ or $y = \pm 8$.

The points (4, 8) and (4, –8) define the latus rectum.

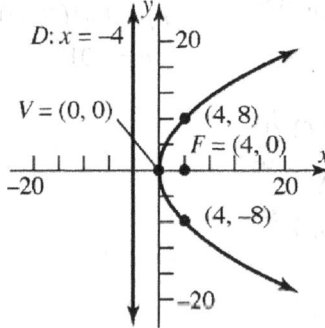

23. The focus is (0, –3) and the vertex is (0, 0). Both lie on the vertical line $x = 0$. $a = 3$ and since (0, –3) is below (0, 0), the parabola opens down. The equation of the parabola is:

$x^2 = -4ay$

$x^2 = -4 \cdot 3 \cdot y$

$x^2 = -12y$

Letting $y = -3$, we find $x^2 = 36$ or $x = \pm 6$.

The points $(-6, -3)$ and $(6, -3)$ define the latus rectum.

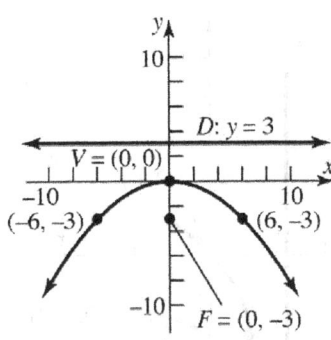

25. The focus is $(-2, 0)$ and the directrix is $x = 2$. The vertex is $(0, 0)$. $a = 2$ and since $(-2, 0)$ is to the left of $(0, 0)$, the parabola opens to the left. The equation of the parabola is:

$$y^2 = -4ax$$
$$y^2 = -4 \cdot 2 \cdot x$$
$$y^2 = -8x$$

Letting $x = -2$, we find $y^2 = 16$ or $y = \pm 4$. The points $(-2, 4)$ and $(-2, -4)$ define the latus rectum.

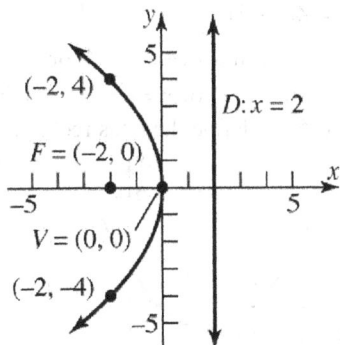

27. The directrix is $y = -\frac{1}{2}$ and the vertex is $(0, 0)$.

The focus is $\left(0, \frac{1}{2}\right)$. $a = \frac{1}{2}$ and since $\left(0, \frac{1}{2}\right)$ is above $(0, 0)$, the parabola opens up. The equation of the parabola is:

$$x^2 = 4ay$$
$$x^2 = 4 \cdot \frac{1}{2} \cdot y$$
$$x^2 = 2y$$

Letting $y = \frac{1}{2}$, we find $x^2 = 1$ or $x = \pm 1$.

The points $\left(1, \frac{1}{2}\right)$ and $\left(-1, \frac{1}{2}\right)$ define the latus rectum.

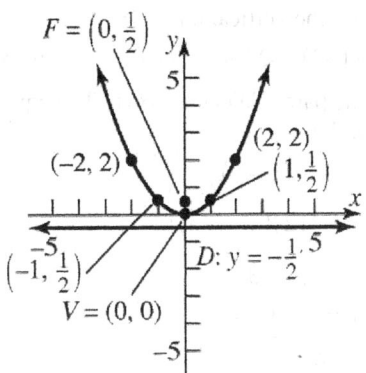

29. Vertex: $(0, 0)$. Since the axis of symmetry is vertical, the parabola opens up or down. Since $(2, 3)$ is above $(0, 0)$, the parabola opens up. The equation has the form $x^2 = 4ay$. Substitute the coordinates of $(2, 3)$ into the equation to find a:

$$2^2 = 4a \cdot 3$$
$$4 = 12a$$
$$a = \frac{1}{3}$$

The equation of the parabola is: $x^2 = \frac{4}{3}y$. The focus is $\left(0, \frac{1}{3}\right)$. Letting $y = \frac{1}{3}$, we find $x^2 = \frac{4}{9}$ or $x = \pm\frac{2}{3}$. The points $\left(\frac{2}{3}, \frac{1}{3}\right)$ and $\left(-\frac{2}{3}, \frac{1}{3}\right)$ define the latus rectum.

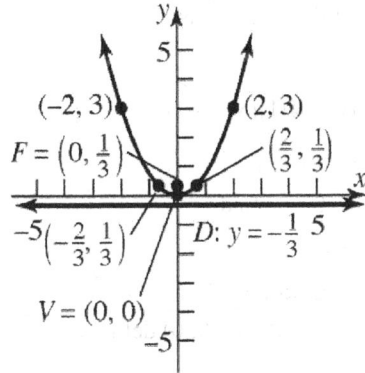

31. The vertex is (2, –3) and the focus is (2, –5).
Both lie on the vertical line $x = 2$.

$a = |-5 - (-3)| = 2$ and since (2, –5) is below
(2, –3), the parabola opens down. The equation
of the parabola is:

$$(x-h)^2 = -4a(y-k)$$

$$(x-2)^2 = -4(2)(y-(-3))$$

$$(x-2)^2 = -8(y+3)$$

Letting $y = -5$, we find

$$(x-2)^2 = 16$$

$$x - 2 = \pm 4 \Rightarrow x = -2 \text{ or } x = 6$$

The points (–2, –5) and (6, –5) define the latus
rectum.

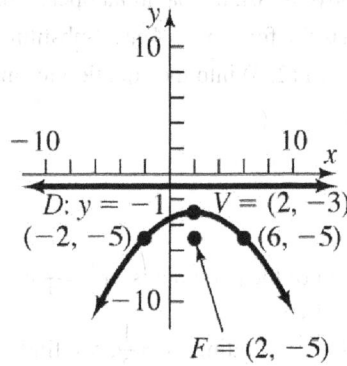

33. The vertex is (–1, –2) and the focus is (0, –2).
Both lie on the horizontal line $y = -2$.

$a = |-1 - 0| = 1$ and since (0, –2) is to the right of
(–1, –2), the parabola opens to the right. The
equation of the parabola is:

$$(y-k)^2 = 4a(x-h)$$

$$(y-(-2))^2 = 4(1)(x-(-1))$$

$$(y+2)^2 = 4(x+1)$$

Letting $x = 0$, we find

$$(y+2)^2 = 4$$

$$y + 2 = \pm 2 \Rightarrow y = -4 \text{ or } y = 0$$

The points (0, –4) and (0, 0) define the latus
rectum.

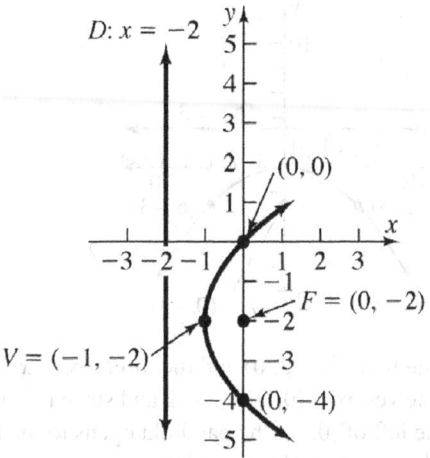

35. The directrix is $y = 2$ and the focus is (–3, 4).
This is a vertical case, so the vertex is (–3, 3).
$a = 1$ and since (–3, 4) is above $y = 2$, the
parabola opens up. The equation of the parabola
is: $(x-h)^2 = 4a(y-k)$

$$(x-(-3))^2 = 4 \cdot 1 \cdot (y-3)$$

$$(x+3)^2 = 4(y-3)$$

Letting $y = 4$, we find $(x+3)^2 = 4$ or
$x + 3 = \pm 2$. So, $x = -1$ or $x = -5$. The points
(–1, 4) and (–5, 4) define the latus rectum.

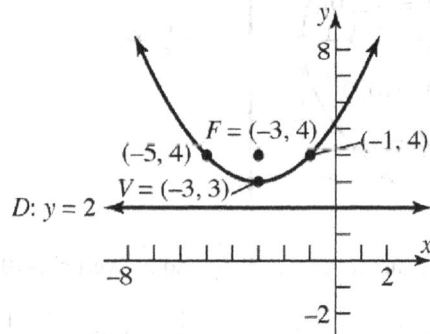

37. The directrix is $x = 1$ and the focus is $(-3, -2)$. This is a horizontal case, so the vertex is $(-1, -2)$. $a = 2$ and since $(-3, -2)$ is to the left of $x = 1$, the parabola opens to the left. The equation of the parabola is:

$$(y - k)^2 = -4a(x - h)$$

$$(y - (-2))^2 = -4 \cdot 2 \cdot (x - (-1))$$

$$(y + 2)^2 = -8(x + 1)$$

Letting $x = -3$, we find $(y + 2)^2 = 16$ or $y + 2 = \pm 4$. So, $y = 2$ or $y = -6$. The points $(-3, 2)$ and $(-3, -6)$ define the latus rectum.

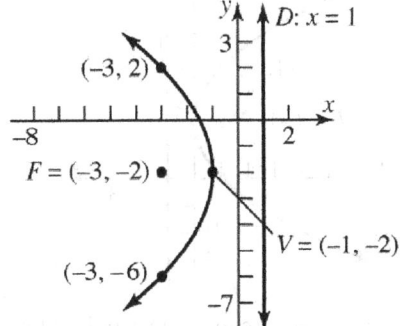

39. The equation $x^2 = 4y$ is in the form $x^2 = 4ay$ where $4a = 4$ or $a = 1$.
Thus, we have:
Vertex: $(0, 0)$
Focus: $(0, 1)$
Directrix: $y = -1$

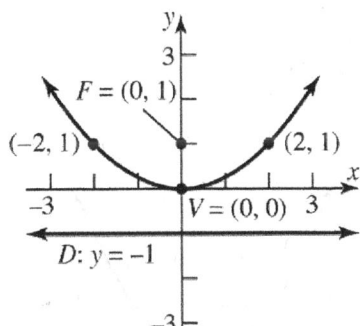

41. The equation
$y^2 = -16x$ is in the form $y^2 = -4ax$ where
$-4a = -16$ or $a = 4$. Thus, we have:
Vertex: $(0, 0)$
Focus: $(-4, 0)$
Directrix: $x = 4$

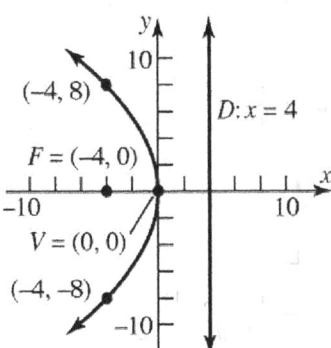

43. The equation $(y - 2)^2 = 8(x + 1)$ is in the form
$(y - k)^2 = 4a(x - h)$ where $4a = 8$ or $a = 2$,
$h = -1$, and $k = 2$. Thus, we have:
Vertex: $(-1, 2)$;
Focus: $(1, 2)$;
Directrix: $x = -3$

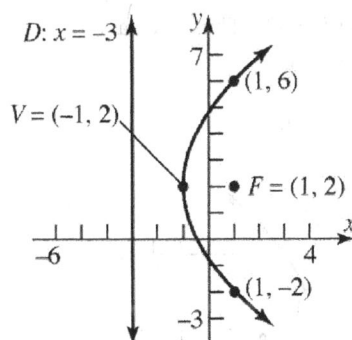

45. a. The equation $(x-3)^2 = -(y+1)$ is in the form $(x-h)^2 = -4a(y-k)$ where

$-4a = -1$ or $a = \dfrac{1}{4}$, $h = 3$, and $k = -1$. Thus, we have:

Vertex: $(3, -1)$;

Focus: $\left(3, -\dfrac{5}{4}\right)$;

Directrix: $y = -\dfrac{3}{4}$

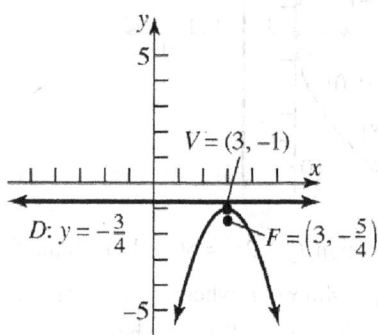

47. The equation $(y+3)^2 = 8(x-2)$ is in the form $(y-k)^2 = 4a(x-h)$ where $4a = 8$ or $a = 2$, $h = 2$, and $k = -3$. Thus, we have:

Vertex: $(2, -3)$;
Focus: $(4, -3)$
Directrix: $x = 0$

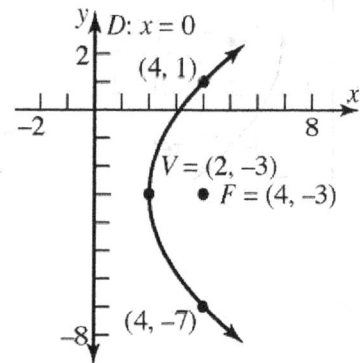

49. Complete the square to put in standard form:
$$y^2 - 4y + 4x + 4 = 0$$
$$y^2 - 4y + 4 = -4x$$
$$(y-2)^2 = -4x$$

The equation is in the form
$(y-k)^2 = -4a(x-h)$ where
$-4a = -4$ or $a = 1$, $h = 0$, and $k = 2$. Thus, we have:

Vertex: $(0, 2)$; Focus: $(-1, 2)$; Directrix: $x = 1$

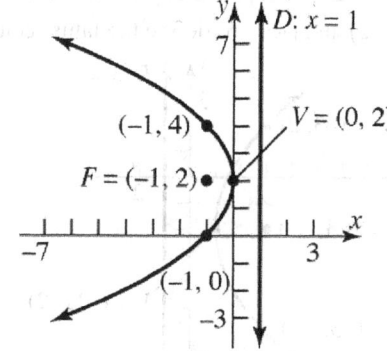

51. Complete the square to put in standard form:
$$x^2 + 8x = 4y - 8$$
$$x^2 + 8x + 16 = 4y - 8 + 16$$
$$(x+4)^2 = 4(y+2)$$

The equation is in the form $(x-h)^2 = 4a(y-k)$ where $4a = 4$ or $a = 1$, $h = -4$, and $k = -2$.

Thus, we have:
Vertex: $(-4, -2)$;
Focus: $(-4, -1)$
Directrix: $y = -3$

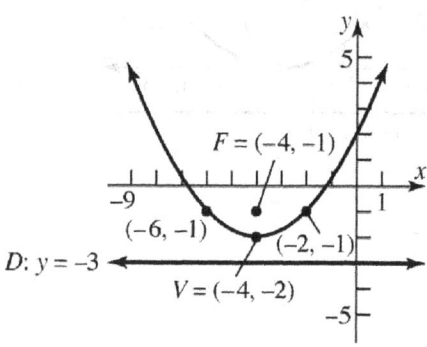

53. Complete the square to put in standard form:

$y^2 + 2y - x = 0$

$y^2 + 2y + 1 = x + 1$

$(y+1)^2 = x + 1$

The equation is in the form $(y-k)^2 = 4a(x-h)$

where $4a = 1$ or $a = \dfrac{1}{4}$, $h = -1$, and $k = -1$.

Thus, we have:

Vertex: $(-1, -1)$; Focus: $\left(-\dfrac{3}{4}, -1\right)$

Directrix: $x = -\dfrac{5}{4}$

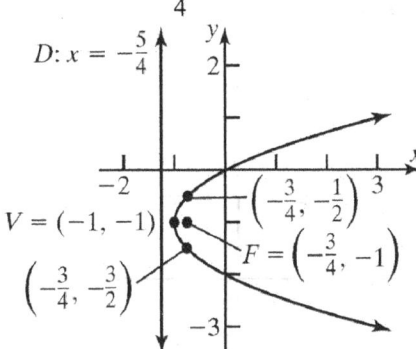

55. Complete the square to put in standard form:

$x^2 - 4x = y + 4$

$x^2 - 4x + 4 = y + 4 + 4$

$(x-2)^2 = y + 8$

The equation is in the form $(x-h)^2 = 4a(y-k)$

where $4a = 1$ or $a = \dfrac{1}{4}$, $h = 2$, and $k = -8$. Thus, we have:

Vertex: $(2, -8)$; Focus: $\left(2, -\dfrac{31}{4}\right)$

Directrix: $y = -\dfrac{33}{4}$

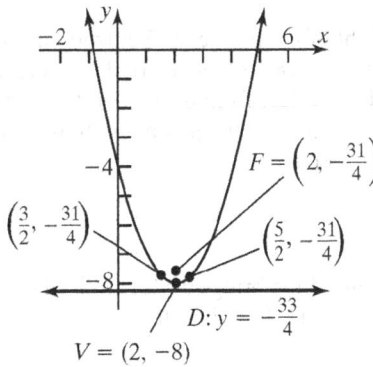

57. $(y-1)^2 = c(x-0)$

$(y-1)^2 = cx$

$(2-1)^2 = c(1) \Rightarrow 1 = c$

$(y-1)^2 = x$

59. $(y-1)^2 = c(x-2)$

$(0-1)^2 = c(1-2)$

$1 = -c \Rightarrow c = -1$

$(y-1)^2 = -(x-2)$

61. $(x-0)^2 = c(y-1)$

$x^2 = c(y-1)$

$2^2 = c(2-1)$

$4 = c$

$x^2 = 4(y-1)$

63. $(y-0)^2 = c(x-(-2))$

$y^2 = c(x+2)$

$1^2 = c(0+2) \Rightarrow 1 = 2c \Rightarrow c = \dfrac{1}{2}$

$y^2 = \dfrac{1}{2}(x+2)$

65. Set up the problem so that the vertex of the parabola is at $(0, 0)$ and it opens up. Then the equation of the parabola has the form:

$x^2 = 4ay$. Since the parabola is 10 feet across and 4 feet deep, the points $(5, 4)$ and $(-5, 4)$ are on the parabola. Substitute and solve for a:

$5^2 = 4a(4) \Rightarrow 25 = 16a \Rightarrow a = \dfrac{25}{16}$

a is the distance from the vertex to the focus. Thus, the receiver (located at the focus) is

$\dfrac{25}{16} = 1.5625$ feet, or 18.75 inches from the base of the dish, along the axis of the parabola.

67. Set up the problem so that the vertex of the parabola is at $(0, 0)$ and it opens up. Then the equation of the parabola has the form: $x^2 = 4ay$.

Since the parabola is 4 inches across and 1 inch deep, the points $(2, 1)$ and $(-2, 1)$ are on the parabola. Substitute and solve for a:

$2^2 = 4a(1) \Rightarrow 4 = 4a \Rightarrow a = 1$

a is the distance from the vertex to the focus.

425

Thus, the bulb (located at the focus) should be 1 inch from the vertex.

69. Set up the problem so that the vertex of the parabola is at (0, 0) and it opens up. Then the equation of the parabola has the form: $x^2 = cy$.

The point (300, 80) is a point on the parabola. Solve for c and find the equation:

$300^2 = c(80) \Rightarrow c = 1125$

$x^2 = 1125y$

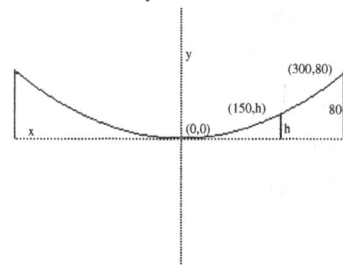

Since the height of the cable 150 feet from the center is to be found, the point $(150, h)$ is a point on the parabola. Solve for h:

$150^2 = 1125h$

$22,500 = 1125h$

$20 = h$

The height of the cable 150 feet from the center is 20 feet.

71. Set up the problem so that the vertex of the parabola is at (0, 0) and it opens up. Then the equation of the parabola has the form: $x^2 = 4ay$.

a is the distance from the vertex to the focus (where the source is located), so $a = 2$. Since the opening is 5 feet across, there is a point $(2.5, y)$ on the parabola.

Solve for y: $x^2 = 8y$

$2.5^2 = 8y$

$6.25 = 8y$

$y = 0.78125$ feet

The depth of the searchlight should be 0.78125 feet.

73. Set up the problem so that the vertex of the parabola is at (0, 0) and it opens up. Then the equation of the parabola has the form: $x^2 = 4ay$.

Since the parabola is 20 feet across and 6 feet deep, the points (10, 6) and (−10, 6) are on the parabola. Substitute and solve for a:

$10^2 = 4a(6)$

$100 = 24a$

$a \approx 4.17$ feet

The heat will be concentrated about 4.17 feet from the base, along the axis of symmetry.

75. Set up the problem so that the vertex of the parabola is at (0, 0) and it opens down. Then the equation of the parabola has the form: $x^2 = cy$.

The point (60, −25) is a point on the parabola. Solve for c and find the equation:

$60^2 = c(-25) \Rightarrow c = -144$

$x^2 = -144y$

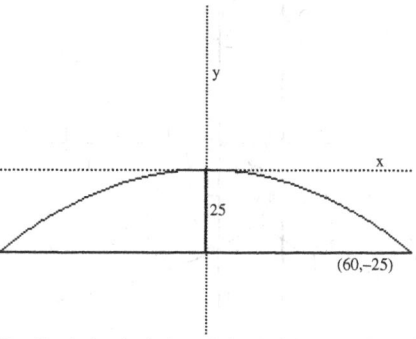

To find the height of the bridge 10 feet from the center the point (10, y) is a point on the parabola. Solve for y:

$10^2 = -144y$

$100 = -144y$

$-0.69 \approx y$

The height of the bridge 10 feet from the center is about $25 - 0.69 = 24.31$ feet. To find the height of the bridge 30 feet from the center the point (30, y) is a point on the parabola. Solve for y:

$30^2 = -144y$

$900 = -144y$

$-6.25 = y$

The height of the bridge 30 feet from the center is $25 - 6.25 = 18.75$ feet. To find the height of the bridge, 50 feet from the center, the point (50, y) is a point on the parabola. Solve for y:

$50^2 = -144y$

$2500 = -144y$

$y = -17.36$

The height of the bridge 50 feet from the center is about $25 - 17.36 = 7.64$ feet.

77. a. Imagine placing the Arch along the x-axis with the peak along the y-axis. Since the Arch is 630 feet high and is 630 feet wide at its base, we would have the points $(-315, 0)$, $(0, 630)$, and $(315, 0)$. The equation of the parabola would have the form $y = ax^2 + c$. Using the point $(0, 630)$ we have

$$630 = a(0)^2 + c$$
$$630 = c$$

The model then becomes $y = ax^2 + 630$.

Next, using the point $(315, 0)$ we get

$$0 = a(315)^2 + 630$$
$$-630 = (315)^2 a$$
$$a = -\frac{630}{(315)^2}$$

Thus, the equation of the parabola with the same given dimensions is

$$y = -\frac{630}{(315)^2} x^2 + 630.$$

b. Using $y = -\frac{630}{(315)^2} x^2 + 630$, we get

Width (ft)	x	Height (ft), model
567	283.5	119.7
478	239	267.3
308	154	479.4

c. No; the heights computed by using the model do not fit the actual heights.

79. $Cy^2 + Dx = 0 \quad C \neq 0, \ D \neq 0$

$$Cy^2 = -Dx$$
$$y^2 = -\frac{D}{C} x$$

This is the equation of a parabola with vertex at $(0, 0)$ and whose axis of symmetry is the x-axis.

The focus is $\left(-\frac{D}{4C}, 0 \right)$. The directrix is

$x = \frac{D}{4C}$. The parabola opens to the right if

$-\frac{D}{C} > 0$ and to the left if $-\frac{D}{C} < 0$.

81. $Cy^2 + Dx + Ey + F = 0 \quad C \neq 0$

a. If $D \neq 0$, then:

$$Cy^2 + Ey = -Dx - F$$
$$C\left(y^2 + \frac{E}{C} y + \frac{E^2}{4C^2} \right) = -Dx - F + \frac{E^2}{4C}$$
$$\left(y + \frac{E}{2C} \right)^2 = \frac{1}{C}\left(-Dx - F + \frac{E^2}{4C} \right)$$
$$\left(y + \frac{E}{2C} \right)^2 = \frac{-D}{C}\left(x + \frac{F}{D} - \frac{E^2}{4CD} \right)$$
$$\left(y + \frac{E}{2C} \right)^2 = \frac{-D}{C}\left(x - \frac{E^2 - 4CF}{4CD} \right)$$

This is the equation of a parabola whose

vertex is $\left(\dfrac{E^2 - 4CF}{4CD}, -\dfrac{E}{2C} \right)$, and whose

axis of symmetry is parallel to the x-axis.

b. If $D = 0$, then

$$Cy^2 + Ey + F = 0 \Rightarrow y = \frac{-E \pm \sqrt{E^2 - 4CF}}{2C}$$

If $E^2 - 4CF = 0$, then $y = -\dfrac{E}{2C}$ is a single horizontal line.

c. If $D = 0$, then

$$Cy^2 + Ey + F = 0 \Rightarrow y = \frac{-E \pm \sqrt{E^2 - 4CF}}{2C}$$

If $E^2 - 4CF > 0$, then

$$y = \frac{-E + \sqrt{E^2 - 4CF}}{2C} \text{ and}$$
$$y = \frac{-E - \sqrt{E^2 - 4CF}}{2C} \text{ are two horizontal}$$

lines.

d. If $D = 0$, then

$$Cy^2 + Ey + F = 0 \Rightarrow y = \frac{-E \pm \sqrt{E^2 - 4CF}}{2C}$$

If $E^2 - 4CF < 0$, then there is no real solution. The graph contains no points.

83. $4^{x+1} = 8^{x-1}$

$2^{2(x+1)} = 2^{3(x-1)}$

$2(x+1) = 3(x-1)$

$2x + 2 = 3x - 3$

$-x = -5$

$x = 5$

The solution set is $\{5\}$.

85. $f(x) = x^2 + 2x - 3$

$\dfrac{f(x+h) - f(x)}{h}$

$= \dfrac{(x+h)^2 + 2(x+h) - 3 - (x^2 + 2x - 3)}{h}$

$\dfrac{(x+h)^2 + 2(x+h) - 3 - (x^2 + 2x - 3)}{h}$

$= \dfrac{x^2 + 2xh + h^2 + 2x + 2h - 3 - x^2 - 2x + 3}{h}$

$= \dfrac{2xh + h^2 + 2h}{h}$

$= \dfrac{h(2x + h + 2)}{h}$

$= 2x + h + 2$

Section 7.3

1. $d = \sqrt{(4-2)^2 + (-2-(-5))^2} = \sqrt{2^2 + 3^2} = \sqrt{13}$

3. x-intercepts: $\quad 0^2 = 16 - 4x^2$

$4x^2 = 16$

$x^2 = 4$

$x = \pm 2 \rightarrow (-2, 0), (2, 0)$

y-intercepts: $\quad y^2 = 16 - 4(0)^2$

$y^2 = 16$

$y = \pm 4 \rightarrow (0, -4), (0, 4)$

The intercepts are $(-2, 0)$, $(2, 0)$, $(0, -4)$, and $(0, 4)$.

5. left 1; down 4

7. ellipse

9. $(0, -5); (0, 5)$

11. $(-2, -3) \ (6, -3)$

13. (c); the major axis is along the x-axis and the vertices are at $(-4, 0)$ and $(4, 0)$.

15. (b); the major axis is along the y-axis and the vertices are at $(0, -2)$ and $(0, 2)$.

17. $\dfrac{x^2}{25} + \dfrac{y^2}{4} = 1$

The center of the ellipse is at the origin.

$a = 5, \ b = 2$. The vertices are (5, 0) and (–5, 0).

Find the value of c:

$c^2 = a^2 - b^2 = 25 - 4 = 21 \rightarrow c = \sqrt{21}$

The foci are $\left(\sqrt{21}, 0\right)$ and $\left(-\sqrt{21}, 0\right)$.

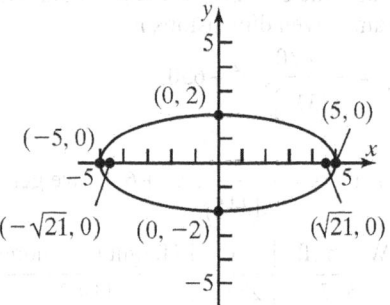

19. $\dfrac{x^2}{9} + \dfrac{y^2}{25} = 1$

The center of the ellipse is at the origin.

$a = 5, \ b = 3$. The vertices are (0, 5) and (0, –5).

Find the value of c:

$c^2 = a^2 - b^2 = 25 - 9 = 16$

$c = 4$

The foci are (0, 4) and (0, –4).

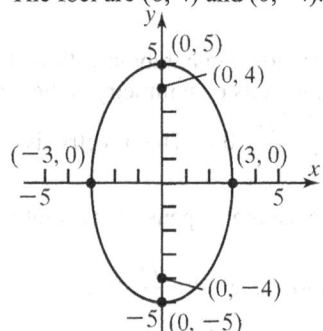

21. $4x^2 + y^2 = 16$

Divide by 16 to put in standard form:

$$\frac{4x^2}{16} + \frac{y^2}{16} = \frac{16}{16}$$

$$\frac{x^2}{4} + \frac{y^2}{16} = 1$$

The center of the ellipse is at the origin.
$a = 4$, $b = 2$.

The vertices are $(0, 4)$ and $(0, -4)$. Find the value of c:

$$c^2 = a^2 - b^2 = 16 - 4 = 12$$

$$c = \sqrt{12} = 2\sqrt{3}$$

The foci are $\left(0, 2\sqrt{3}\right)$ and $\left(0, -2\sqrt{3}\right)$.

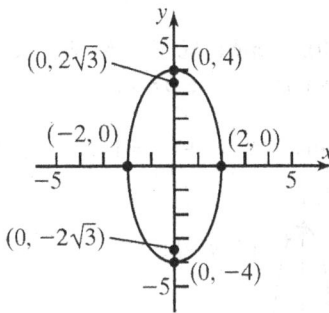

23. $4y^2 + x^2 = 8$

Divide by 8 to put in standard form:

$$\frac{4y^2}{8} + \frac{x^2}{8} = \frac{8}{8}$$

$$\frac{x^2}{8} + \frac{y^2}{2} = 1$$

The center of the ellipse is at the origin.
$a = \sqrt{8} = 2\sqrt{2}$, $b = \sqrt{2}$.

The vertices are $\left(2\sqrt{2}, 0\right)$ and $\left(-2\sqrt{2}, 0\right)$. Find the value of c:

$$c^2 = a^2 - b^2 = 8 - 2 = 6$$

$$c = \sqrt{6}$$

The foci are $\left(\sqrt{6}, 0\right)$ and $\left(-\sqrt{6}, 0\right)$.

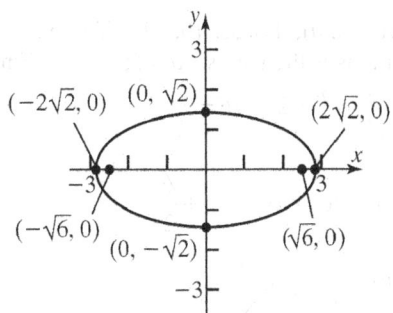

25. $x^2 + y^2 = 16$

$$\frac{x^2}{16} + \frac{y^2}{16} = 1$$

This is a circle whose center is at $(0, 0)$ and radius = 4. The focus of the ellipse is $(0,0)$ and the vertices are $(-4,0)$, $(4,0)$, $(0,-4)$, $(0,4)$.

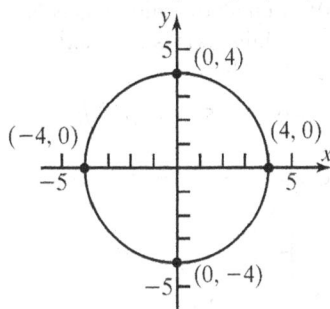

27. Center: $(0, 0)$; Focus: $(3, 0)$; Vertex: $(5, 0)$;
Major axis is the x-axis; $a = 5$; $c = 3$. Find b:

$$b^2 = a^2 - c^2 = 25 - 9 = 16$$

$$b = 4$$

Write the equation: $\dfrac{x^2}{25} + \dfrac{y^2}{16} = 1$

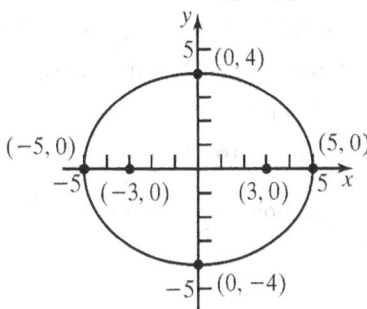

29. Center: $(0,0)$; Focus: $(0,-4)$; Vertex: $(0,5)$; Major axis is the y-axis; $a = 5$; $c = 4$. Find b:

$b^2 = a^2 - c^2 = 25 - 16 = 9$

$b = 3$

Write the equation: $\dfrac{x^2}{9} + \dfrac{y^2}{25} = 1$

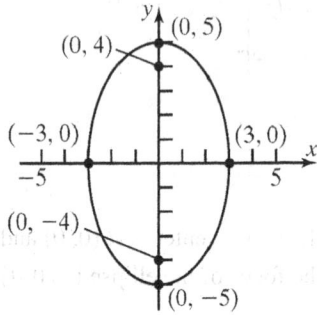

31. Foci: $(\pm 2, 0)$; Length of major axis is 6. Center: $(0,0)$; Major axis is the x-axis; $a = 3$; $c = 2$. Find b:

$b^2 = a^2 - c^2 = 9 - 4 = 5 \;\rightarrow\; b = \sqrt{5}$

Write the equation: $\dfrac{x^2}{9} + \dfrac{y^2}{5} = 1$

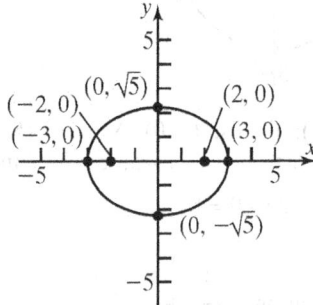

33. Focus: $(-4, 0)$; Vertices: $(-5, 0)$ and $(5, 0)$; Center: $(0, 0)$; Major axis is the x-axis.

$a = 5$; $c = 4$. Find b:

$b^2 = a^2 - c^2 = 25 - 16 = 9 \;\rightarrow\; b = 3$

Write the equation: $\dfrac{x^2}{25} + \dfrac{y^2}{9} = 1$

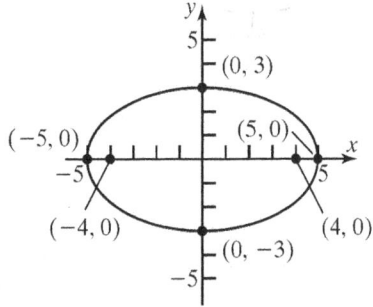

35. Foci: $(0, \pm 3)$; x-intercepts are ± 2. Center: $(0, 0)$; Major axis is the y-axis; $c = 3$; $b = 2$. Find a:

$a^2 = b^2 + c^2 = 4 + 9 = 13 \;\rightarrow\; a = \sqrt{13}$

Write the equation: $\dfrac{x^2}{4} + \dfrac{y^2}{13} = 1$

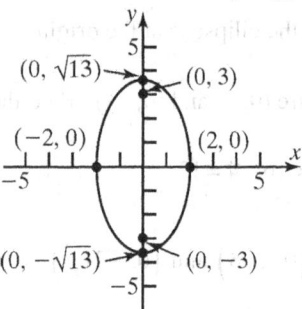

37. Center: $(0, 0)$; Vertex: $(0, 4)$; $b = 1$; Major axis is the y-axis; $a = 4$; $b = 1$.

Write the equation: $x^2 + \dfrac{y^2}{16} = 1$

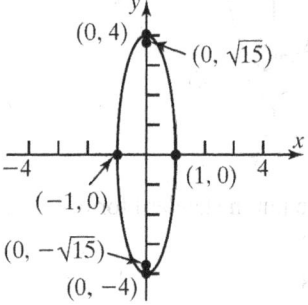

39. Center: $(-1, 1)$

Major axis: parallel to x-axis

Length of major axis: $4 = 2a \rightarrow a = 2$

Length of minor axis: $2 = 2b \rightarrow b = 1$

$\dfrac{(x+1)^2}{4} + (y-1)^2 = 1$

41. Center: $(1, 0)$

Major axis: parallel to y-axis

Length of major axis: $4 = 2a \rightarrow a = 2$

Length of minor axis: $2 = 2b \rightarrow b = 1$

$(x-1)^2 + \dfrac{y^2}{4} = 1$

43. The equation $\dfrac{(x-3)^2}{4}+\dfrac{(y+1)^2}{9}=1$ is in the

form $\dfrac{(x-h)^2}{b^2}+\dfrac{(y-k)^2}{a^2}=1$ (major axis parallel

to the y-axis) where $a=3$, $b=2$,

$h=3$, and $k=-1$. Solving for c:

$c^2=a^2-b^2=9-4=5\rightarrow c=\sqrt{5}$

Thus, we have:

Center: $(3,-1)$

Foci: $\left(3,-1+\sqrt{5}\right),\ \left(3,-1-\sqrt{5}\right)$

Vertices: $(3,2),(3,-4)$

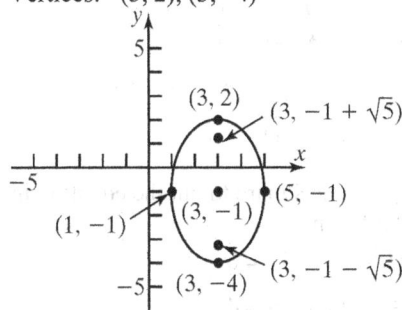

45. Divide by 16 to put the equation in standard
form:

$(x+5)^2+4(y-4)^2=16$

$\dfrac{(x+5)^2}{16}+\dfrac{4(y-4)^2}{16}=\dfrac{16}{16}$

$\dfrac{(x+5)^2}{16}+\dfrac{(y-4)^2}{4}=1$

The equation is in the form

$\dfrac{(x-h)^2}{a^2}+\dfrac{(y-k)^2}{b^2}=1$ (major axis parallel to the

x-axis) where $a=4$, $b=2$, $h=-5$, and $k=4$.

Solving for c:

$c^2=a^2-b^2=16-4=12\rightarrow c=\sqrt{12}=2\sqrt{3}$

Thus, we have:

Center: $(-5,4)$

Foci: $\left(-5-2\sqrt{3},4\right),\ \left(-5+2\sqrt{3},4\right)$

Vertices: $(-9,4),(-1,4)$

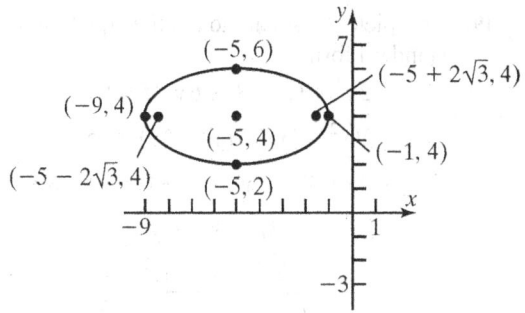

47. Complete the square to put the equation in
standard form:

$x^2+4x+4y^2-8y+4=0$

$(x^2+4x+4)+4(y^2-2y+1)=-4+4+4$

$(x+2)^2+4(y-1)^2=4$

$\dfrac{(x+2)^2}{4}+\dfrac{4(y-1)^2}{4}=\dfrac{4}{4}$

$\dfrac{(x+2)^2}{4}+(y-1)^2=1$

The equation is in the form

$\dfrac{(x-h)^2}{a^2}+\dfrac{(y-k)^2}{b^2}=1$ (major axis parallel to the

x-axis) where $a=2$, $b=1$, $h=-2$, and $k=1$.

Solving for c: $c^2=a^2-b^2=4-1=3\rightarrow c=\sqrt{3}$

Thus, we have:

Center: $(-2,1)$

Foci: $\left(-2-\sqrt{3},1\right),\ \left(-2+\sqrt{3},1\right)$

Vertices: $(-4,1),\ (0,1)$

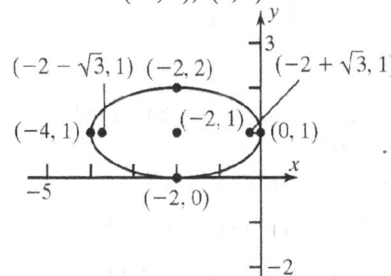

49. Complete the square to put the equation in standard form:

$$2x^2 + 3y^2 - 8x + 6y + 5 = 0$$

$$2(x^2 - 4x) + 3(y^2 + 2y) = -5$$

$$2(x^2 - 4x + 4) + 3(y^2 + 2y + 1) = -5 + 8 + 3$$

$$2(x-2)^2 + 3(y+1)^2 = 6$$

$$\frac{2(x-2)^2}{6} + \frac{3(y+1)^2}{6} = \frac{6}{6}$$

$$\frac{(x-2)^2}{3} + \frac{(y+1)^2}{2} = 1$$

The equation is in the form

$$\frac{(x-h)^2}{a^2} + \frac{(y-k)^2}{b^2} = 1 \text{ (major axis parallel to the}$$

x-axis) where

$a = \sqrt{3}$, $b = \sqrt{2}$, $h = 2$, and $k = -1$.

Solving for c: $c^2 = a^2 - b^2 = 3 - 2 = 1 \rightarrow c = 1$

Thus, we have:

Center: $(2, -1)$

Foci: $(1, -1)$, $(3, -1)$

Vertices: $(2 - \sqrt{3}, -1)$, $(2 + \sqrt{3}, -1)$

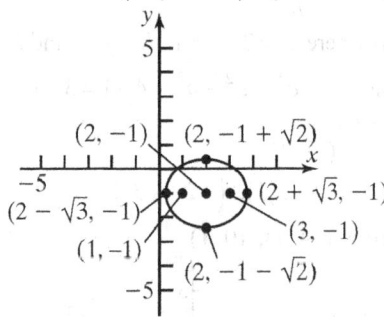

51. Complete the square to put the equation in standard form:

$$9x^2 + 4y^2 - 18x + 16y - 11 = 0$$

$$9(x^2 - 2x) + 4(y^2 + 4y) = 11$$

$$9(x^2 - 2x + 1) + 4(y^2 + 4y + 4) = 11 + 9 + 16$$

$$9(x-1)^2 + 4(y+2)^2 = 36$$

$$\frac{9(x-1)^2}{36} + \frac{4(y+2)^2}{36} = \frac{36}{36}$$

$$\frac{(x-1)^2}{4} + \frac{(y+2)^2}{9} = 1$$

The equation is in the form

$$\frac{(x-h)^2}{b^2} + \frac{(y-k)^2}{a^2} = 1 \text{ (major axis parallel to the}$$

y-axis) where $a = 3$, $b = 2$, $h = 1$, and $k = -2$.

Solving for c:

$$c^2 = a^2 - b^2 = 9 - 4 = 5 \rightarrow c = \sqrt{5}$$

Thus, we have:

Center: $(1, -2)$

Foci: $(1, -2 + \sqrt{5})$, $(1, -2 - \sqrt{5})$

Vertices: $(1, 1)$, $(1, -5)$

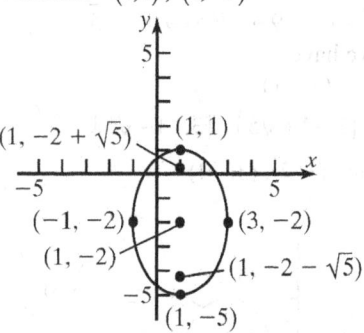

53. Complete the square to put the equation in standard form:

$$4x^2 + y^2 + 4y = 0$$

$$4x^2 + y^2 + 4y + 4 = 4$$

$$4x^2 + (y+2)^2 = 4$$

$$\frac{4x^2}{4} + \frac{(y+2)^2}{4} = \frac{4}{4}$$

$$x^2 + \frac{(y+2)^2}{4} = 1$$

The equation is in the form

$$\frac{(x-h)^2}{b^2} + \frac{(y-k)^2}{a^2} = 1 \text{ (major axis parallel to the}$$

y-axis) where $a = 2$, $b = 1$, $h = 0$, and $k = -2$.

Solving for c:

$$c^2 = a^2 - b^2 = 4 - 1 = 3 \rightarrow c = \sqrt{3}$$

Thus, we have:

Center: $(0, -2)$

Foci: $(0, -2 + \sqrt{3})$, $(0, -2 - \sqrt{3})$

Vertices: $(0, 0)$, $(0, -4)$

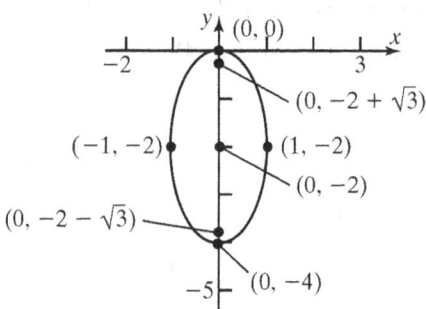

55. Center: $(2, -2)$; Vertex: $(7, -2)$; Focus: $(4, -2)$;
Major axis parallel to the x-axis; $a = 5$; $c = 2$.
Find b:
$$b^2 = a^2 - c^2 = 25 - 4 = 21 \rightarrow b = \sqrt{21}$$
Write the equation: $\dfrac{(x-2)^2}{25} + \dfrac{(y+2)^2}{21} = 1$

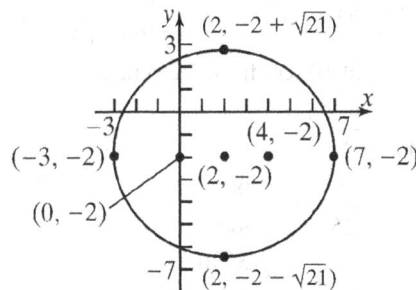

57. Vertices: $(4, 3)$, $(4, 9)$; Focus: $(4, 8)$;
Center: $(4, 6)$; Major axis parallel to the y-axis;
$a = 3$; $c = 2$. Find b:
$$b^2 = a^2 - c^2 = 9 - 4 = 5 \rightarrow b = \sqrt{5}$$
Write the equation: $\dfrac{(x-4)^2}{5} + \dfrac{(y-6)^2}{9} = 1$

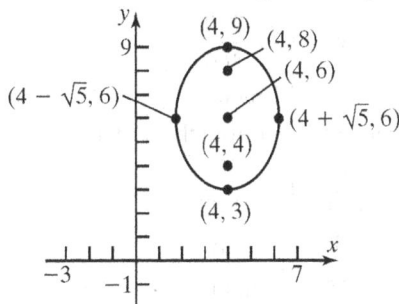

59. Foci: $(5, 1)$, $(-1, 1)$;
Length of the major axis $= 8$; Center: $(2, 1)$;
Major axis parallel to the x-axis; $a = 4$; $c = 3$.
Find b:
$$b^2 = a^2 - c^2 = 16 - 9 = 7 \rightarrow b = \sqrt{7}$$
Write the equation: $\dfrac{(x-2)^2}{16} + \dfrac{(y-1)^2}{7} = 1$

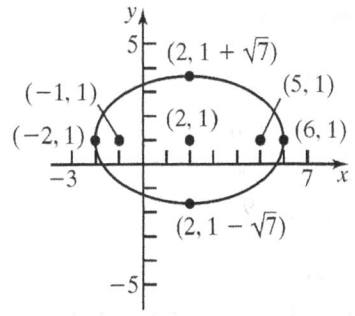

61. Center: $(1, 2)$; Focus: $(4, 2)$; contains the point
$(1, 3)$; Major axis parallel to the x-axis; $c = 3$.
The equation has the form:
$$\frac{(x-1)^2}{a^2} + \frac{(y-2)^2}{b^2} = 1$$
Since the point $(1, 3)$ is on the curve:
$$\frac{0}{a^2} + \frac{1}{b^2} = 1$$
$$\frac{1}{b^2} = 1 \rightarrow b^2 = 1 \rightarrow b = 1$$
Find a:
$$a^2 = b^2 + c^2 = 1 + 9 = 10 \rightarrow a = \sqrt{10}$$
Write the equation: $\dfrac{(x-1)^2}{10} + (y-2)^2 = 1$

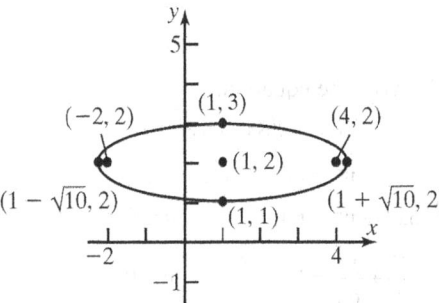

63. Center: $(1, 2)$; Vertex: $(4, 2)$; contains the point
$(1, 5)$; Major axis parallel to the x-axis; $a = 3$.
The equation has the form:
$$\frac{(x-1)^2}{a^2} + \frac{(y-2)^2}{b^2} = 1$$
Since the point $(1, 5)$ is on the curve:
$$\frac{0}{9} + \frac{3^2}{b^2} = 1$$
$$\frac{9}{b^2} = 1 \rightarrow b^2 = 9 \rightarrow b = 3$$
Solve for c:
$$c^2 = a^2 - b^2 = 9 - 9 = 0. \text{ Thus, } c = 0.$$
Write the equation: $\dfrac{(x-1)^2}{9} + \dfrac{(y-2)^2}{9} = 1$

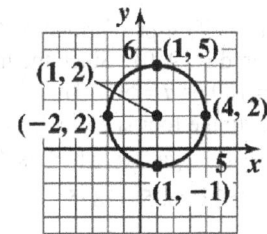

65. Rewrite the equation:

$$y = \sqrt{16 - 4x^2}$$
$$y^2 = 16 - 4x^2, \quad y \geq 0$$
$$4x^2 + y^2 = 16, \quad y \geq 0$$
$$\frac{x^2}{4} + \frac{y^2}{16} = 1, \quad y \geq 0$$

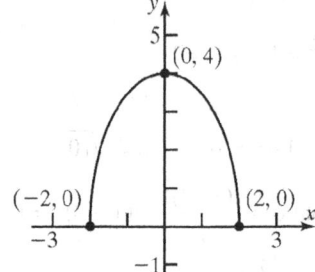

67. Rewrite the equation:

$$y = -\sqrt{64 - 16x^2}$$
$$y^2 = 64 - 16x^2, \quad y \leq 0$$
$$16x^2 + y^2 = 64, \quad y \leq 0$$
$$\frac{x^2}{4} + \frac{y^2}{64} = 1, \quad y \leq 0$$

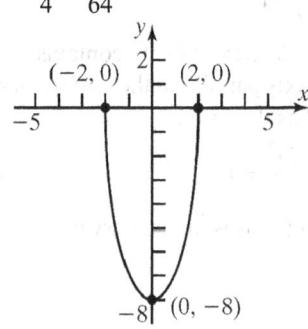

69. The center of the ellipse is (0, 0). The length of the major axis is 20, so $a = 10$. The length of half the minor axis is 6, so $b = 6$. The ellipse is situated with its major axis on the x-axis. The equation is: $\dfrac{x^2}{100} + \dfrac{y^2}{36} = 1$.

71. Assume that the half ellipse formed by the gallery is centered at (0, 0). Since the hall is 100 feet long, $2a = 100$ or $a = 50$. The distance from the center to the foci is 25 feet, so $c = 25$. Find the height of the gallery which is b:

$$b^2 = a^2 - c^2 = 2500 - 625 = 1875$$
$$b = \sqrt{1875} \approx 43.3$$

The ceiling will be 43.3 feet high in the center.

73. Place the semi-elliptical arch so that the x-axis coincides with the water and the y-axis passes through the center of the arch. Since the bridge has a span of 120 feet, the length of the major axis is 120, or $2a = 120$ or $a = 60$. The maximum height of the bridge is 25 feet, so $b = 25$. The equation is: $\dfrac{x^2}{3600} + \dfrac{y^2}{625} = 1$.

The height 10 feet from the center:

$$\frac{10^2}{3600} + \frac{y^2}{625} = 1$$
$$\frac{y^2}{625} = 1 - \frac{100}{3600}$$
$$y^2 = 625 \cdot \frac{3500}{3600}$$
$$y \approx 24.65 \text{ feet}$$

The height 30 feet from the center:

$$\frac{30^2}{3600} + \frac{y^2}{625} = 1$$
$$\frac{y^2}{625} = 1 - \frac{900}{3600}$$
$$y^2 = 625 \cdot \frac{2700}{3600}$$
$$y \approx 21.65 \text{ feet}$$

The height 50 feet from the center:

$$\frac{50^2}{3600} + \frac{y^2}{625} = 1$$
$$\frac{y^2}{625} = 1 - \frac{2500}{3600}$$
$$y^2 = 625 \cdot \frac{1100}{3600}$$
$$y \approx 13.82 \text{ feet}$$

75. If the x-axis is placed along the 100 foot length and the y-axis is placed along the 50 foot length, the equation for the ellipse is: $\dfrac{x^2}{50^2} + \dfrac{y^2}{25^2} = 1$.

Find y when $x = 40$:

$$\frac{40^2}{50^2} + \frac{y^2}{25^2} = 1$$
$$\frac{y^2}{625} = 1 - \frac{1600}{2500}$$
$$y^2 = 625 \cdot \frac{9}{25}$$
$$y = 15 \text{ feet}$$

To get the width of the ellipse at $x = 40$, we

need to double the y value. Thus, the width 10 feet from a vertex is 30 feet.

77. Because of the pitch of the roof, the major axis will run parallel to the direction of the pitch and the minor axis will run perpendicular to the direction of the pitch. The length of the major axis can be determined from the pitch by using the Pythagorean Theorem. The length of the minor axis is 8 inches (the diameter of the pipe).

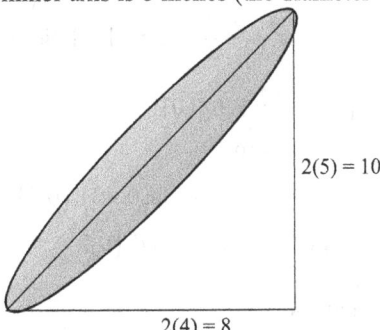

The length of the major axis is

$\sqrt{(8)^2 + (10)^2} = \sqrt{164} = 2\sqrt{41}$ inches.

79. Since the mean distance is 93 million miles, $a = 93$ million. The length of the major axis is 186 million. The perihelion is 186 million $-$ 94.5 million $=$ 91.5 million miles.

The distance from the center of the ellipse to the Sun (focus) is 93 million $-$ 91.5 million $=$ 1.5 million miles. Therefore, $c = 1.5$ million. Find b:

$b^2 = a^2 - c^2$

$= (93 \times 10^6)^2 - (1.5 \times 10^6)^2$

$= 8.64675 \times 10^{15}$

$= 8646.75 \times 10^{12}$

$b = 92.99 \times 10^6$

The equation of the orbit is:

$$\frac{x^2}{(93 \times 10^6)^2} + \frac{y^2}{(92.99 \times 10^6)^2} = 1$$

We can simplify the equation by letting our units for x and y be millions of miles. The equation then becomes:

$$\frac{x^2}{8649} + \frac{y^2}{8646.75} = 1$$

81. The mean distance is 507 million $-$ 23.2 million $=$ 483.8 million miles.

The perihelion is 483.8 million $-$ 23.2 million $=$ 460.6 million miles.

Since $a = 483.8 \times 10^6$ and $c = 23.2 \times 10^6$, we can find b:

$b^2 = a^2 - c^2$

$= (483.8 \times 10^6)^2 - (23.2 \times 10^6)^2$

$= 2.335242 \times 10^{17}$

$b = 483.2 \times 10^6$

The equation of the orbit of Jupiter is:

$$\frac{x^2}{(483.8 \times 10^6)^2} + \frac{y^2}{(483.2 \times 10^6)^2} = 1$$

We can simplify the equation by letting our units for x and y be millions of miles. The equation then becomes:

$$\frac{x^2}{234,062.44} + \frac{y^2}{233,524.2} = 1$$

83. $e = \dfrac{c}{a} = 0.75$. Perihelion: $a - c = 5$

$c = 0.75a$

$a - 0.75a = 5$

$0.25a = 5$

$a = 20$

$a - c = 5$

$20 - c = 5$

$c = 15$

So the aphelion is $a + c = 35$ million mi.

85. $\dfrac{x^2}{20} + \dfrac{y^2}{4} = 1$

So the vertices are at $(-2\sqrt{5}, 0)$ and $(2\sqrt{5}, 0)$.

The endpoints of the minor axis are at $(0, -2)$ and $(0, 2)$.

$C = (-2\sqrt{5}, 0)$ and $B = (0, 2)$ and $M = (-5, y)$

$d_1 = d_2$

$\sqrt{(-2\sqrt{5} - (-5))^2 + (0 - y)^2} = \sqrt{(0 - (-5))^2 + (2 - y)^2}$

$(-2\sqrt{5} - (-5))^2 + (0 - y)^2 = (0 - (-5))^2 + (2 - y)^2$

$20 - 20\sqrt{5} + 25 + y^2 = 25 + 4 - 4y + y^2$

$16 - 20\sqrt{5} = -4y$

$5\sqrt{5} - 4 = y$

87. Complete the square on the given equation:

$$Ax^2 + Cy^2 + Dx + Ey + F = 0, \quad A \neq 0, C \neq 0$$

$$Ax^2 + Cy^2 + Dx + Ey = -F$$

$$A\left(x^2 + \frac{D}{A}x\right) + C\left(y^2 + \frac{E}{C}y\right) = -F$$

$$A\left(x + \frac{D}{2A}\right)^2 + C\left(y + \frac{E}{2C}\right)^2 = \frac{D^2}{4A} + \frac{E^2}{4C} - F$$

where $A \cdot C > 0$.

Let $U = \frac{D^2}{4A} + \frac{E^2}{4C} - F$.

a. If U is of the same sign as A (and C), then

$$\frac{\left(x + \frac{D}{2A}\right)^2}{\frac{U}{A}} + \frac{\left(y + \frac{E}{2C}\right)^2}{\frac{U}{C}} = 1$$

This is the equation of an ellipse whose center is $\left(\frac{-D}{2A}, \frac{-E}{2C}\right)$.

b. If $U = 0$, the graph is the single point $\left(\frac{-D}{2A}, \frac{-E}{2C}\right)$.

c. If U is of the opposite sign as A (and C), this graph contains no points since the left side always has the opposite sign of the right side.

89. $f(x) = (x-5)^2 - 12$

$$0 = (x-5)^2 - 12$$

$$12 = (x-5)^2$$

$$\pm\sqrt{12} = x - 5$$

$$x = 5 \pm \sqrt{12} = 5 \pm 2\sqrt{3}$$

The zeros are $x = 5 + 2\sqrt{3}$ and $5 - 2\sqrt{3}$.
The x-intercepts are $5 + 2\sqrt{3}$ and $5 - 2\sqrt{3}$.

91. $|4 - 5x| \leq 11$

$$-11 \leq 4 - 5x \leq 11$$

$$-15 \leq -5x \leq 7$$

$$3 \geq -5x \geq -\frac{7}{5}$$

$$-\frac{7}{5} \leq -5x \leq 3 \quad \text{or} \quad -\frac{7}{5}, 3$$

Section 7.4

1. $d = \sqrt{(-2-3)^2 + (1-(-4))^2}$

$$= \sqrt{(-5)^2 + (5)^2} = \sqrt{25 + 25} = \sqrt{50} = 5\sqrt{2}$$

3. x-intercepts: $0^2 = 9 + 4x^2$

$$4x^2 = -9$$

$$x^2 = -\frac{9}{4} \quad \text{(no real solution)}$$

y-intercepts: $y^2 = 9 + 4(0)^2$

$$y^2 = 9$$

$$y = \pm 3 \quad \rightarrow \quad (0, -3), (0, 3)$$

The intercepts are $(0, -3)$ and $(0, 3)$.

5. right 5; down 4

7. hyperbola

9. b

11. $(2, 6); (2, -4)$

13. 2; 3; x

15. (b); the hyperbola opens to the left and right, and has vertices at $(\pm 1, 0)$. Thus, the graph has an equation of the form $x^2 - \frac{y^2}{b^2} = 1$.

17. (a); the hyperbola opens to the left and right, and has vertices at $(\pm 2, 0)$. Thus, the graph has an equation of the form $\frac{x^2}{4} - \frac{y^2}{b^2} = 1$.

19. Center: (0, 0); Focus: (3, 0); Vertex: (1, 0); Transverse axis is the x-axis; $a = 1$; $c = 3$. Find the value of b:

$$b^2 = c^2 - a^2 = 9 - 1 = 8$$

$$b = \sqrt{8} = 2\sqrt{2}$$

Write the equation: $x^2 - \dfrac{y^2}{8} = 1$.

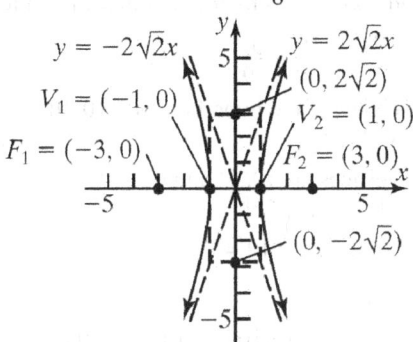

21. Center: $(0, 0)$; Focus: $(0, -6)$; Vertex: $(0, 4)$
Transverse axis is the y-axis; $a = 4$; $c = 6$.
Find the value of b:
$$b^2 = c^2 - a^2 = 36 - 16 = 20$$
$$b = \sqrt{20} = 2\sqrt{5}$$
Write the equation: $\dfrac{y^2}{16} - \dfrac{x^2}{20} = 1$.

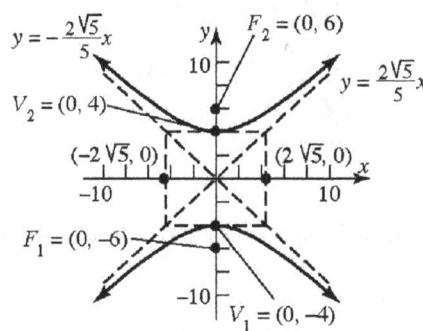

23. Foci: $(-5, 0)$, $(5, 0)$; Vertex: $(3, 0)$
Center: $(0, 0)$; Transverse axis is the x-axis;
$a = 3$; $c = 5$.
Find the value of b:
$$b^2 = c^2 - a^2 = 25 - 9 = 16 \Rightarrow b = 4$$
Write the equation: $\dfrac{x^2}{9} - \dfrac{y^2}{16} = 1$.

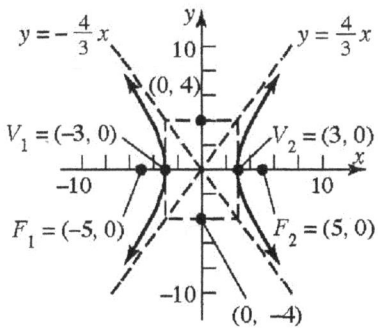

25. Vertices: $(0, -6)$, $(0, 6)$; asymptote: $y = 2x$;
Center: $(0, 0)$; Transverse axis is the y-axis;
$a = 6$. Find the value of b using the slope of the

asymptote: $\dfrac{a}{b} = \dfrac{6}{b} = 2 \Rightarrow 2b = 6 \Rightarrow b = 3$

Find the value of c:
$$c^2 = a^2 + b^2 = 36 + 9 = 45$$
$$c = 3\sqrt{5}$$

Write the equation: $\dfrac{y^2}{36} - \dfrac{x^2}{9} = 1$.

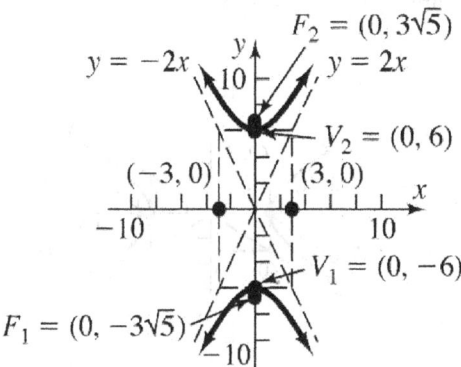

27. Foci: $(-4, 0)$, $(4, 0)$; asymptote: $y = -x$;
Center: $(0, 0)$; Transverse axis is the x-axis;
$c = 4$. Using the slope of the asymptote:
$$-\dfrac{b}{a} = -1 \Rightarrow -b = -a \Rightarrow b = a.$$
Find the value of b:
$$b^2 = c^2 - a^2 \Rightarrow a^2 + b^2 = c^2 \quad (c = 4)$$
$$b^2 + b^2 = 16 \Rightarrow 2b^2 = 16 \Rightarrow b^2 = 8$$
$$b = \sqrt{8} = 2\sqrt{2}$$
$$a = \sqrt{8} = 2\sqrt{2} \quad (a = b)$$
Write the equation: $\dfrac{x^2}{8} - \dfrac{y^2}{8} = 1$.

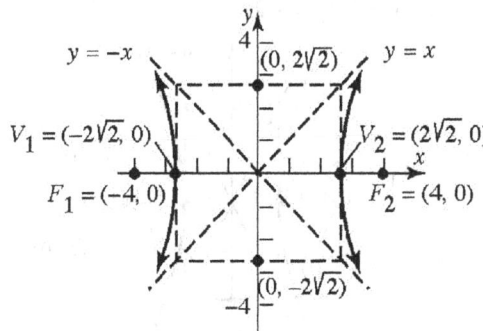

29. $\dfrac{x^2}{25} - \dfrac{y^2}{9} = 1$

The center of the hyperbola is at (0, 0).
$a = 5$, $b = 3$. The vertices are $(5,0)$ and

$(-5,0)$. Find the value of c:

$c^2 = a^2 + b^2 = 25 + 9 = 34 \Rightarrow c = \sqrt{34}$

The foci are $\left(\sqrt{34},0\right)$ and $\left(-\sqrt{34},0\right)$.

The transverse axis is the x-axis. The asymptotes

are $y = \dfrac{3}{5}x;\ y = -\dfrac{3}{5}x$.

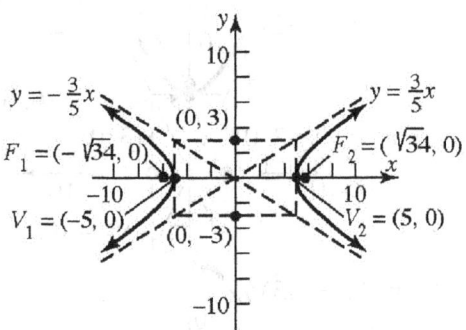

31. $4x^2 - y^2 = 16$

Divide both sides by 16 to put in standard form:

$\dfrac{4x^2}{16} - \dfrac{y^2}{16} = \dfrac{16}{16} \Rightarrow \dfrac{x^2}{4} - \dfrac{y^2}{16} = 1$

The center of the hyperbola is at (0, 0).
$a = 2$, $b = 4$.
The vertices are (2, 0) and (–2, 0).
Find the value of c:

$c^2 = a^2 + b^2 = 4 + 16 = 20$

$c = \sqrt{20} = 2\sqrt{5}$

The foci are $\left(2\sqrt{5},0\right)$ and $\left(-2\sqrt{5},0\right)$.

The transverse axis is the x-axis. The asymptotes
are $y = 2x;\ y = -2x$.

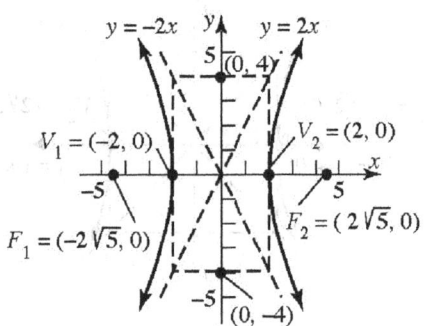

33. $y^2 - 9x^2 = 9$

Divide both sides by 9 to put in standard form:

$\dfrac{y^2}{9} - \dfrac{9x^2}{9} = \dfrac{9}{9} \Rightarrow \dfrac{y^2}{9} - x^2 = 1$

The center of the hyperbola is at (0, 0).
$a = 3$, $b = 1$.

The vertices are (0, 3) and (0, –3).
Find the value of c:

$c^2 = a^2 + b^2 = 9 + 1 = 10$

$c = \sqrt{10}$

The foci are $\left(0,\sqrt{10}\right)$ and $\left(0,-\sqrt{10}\right)$.

The transverse axis is the y-axis.
The asymptotes are $y = 3x;\ y = -3x$.

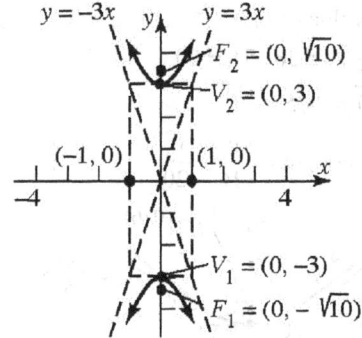

35. $y^2 - x^2 = 25$

Divide both sides by 25 to put in standard form:

$\dfrac{y^2}{25} - \dfrac{x^2}{25} = 1$.

The center of the hyperbola is at (0, 0).
$a = 5$, $b = 5$. The vertices are $(0,5)$ and

$(0,-5)$. Find the value of c:

$c^2 = a^2 + b^2 = 25 + 25 = 50$

$c = \sqrt{50} = 5\sqrt{2}$

The foci are $\left(0,5\sqrt{2}\right)$ and $\left(0,-5\sqrt{2}\right)$.

The transverse axis is the y-axis.

The asymptotes are $y = x; y = -x$.

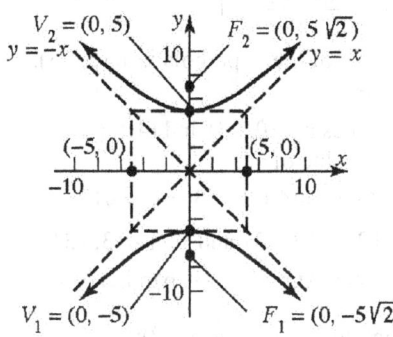

37. The center of the hyperbola is at $(0, 0)$.
$a = 1, \ b = 1$. The vertices are $(1, 0)$ and $(-1, 0)$.

Find the value of c:

$c^2 = a^2 + b^2 = 1 + 1 = 2$

$c = \sqrt{2}$

The foci are $\left(\sqrt{2}, 0\right)$ and $\left(-\sqrt{2}, 0\right)$.

The transverse axis is the x-axis.
The asymptotes are $y = x; y = -x$.

The equation is: $x^2 - y^2 = 1$.

39. The center of the hyperbola is at $(0, 0)$.
$a = 6, \ b = 3$.

The vertices are $(0, -6)$ and $(0, 6)$. Find the value of c:

$c^2 = a^2 + b^2 = 36 + 9 = 45$

$c = \sqrt{45} = 3\sqrt{5}$

The foci are $\left(0, -3\sqrt{5}\right)$ and $\left(0, 3\sqrt{5}\right)$.

The transverse axis is the y-axis.
The asymptotes are $y = 2x; y = -2x$. The

41. Center: $(4, -1)$; Focus: $(7, -1)$; Vertex: $(6, -1)$;
Transverse axis is parallel to the x-axis;
$a = 2; \ c = 3$.

Find the value of b:

$b^2 = c^2 - a^2 = 9 - 4 = 5 \Rightarrow b = \sqrt{5}$

Write the equation: $\dfrac{(x-4)^2}{4} - \dfrac{(y+1)^2}{5} = 1$.

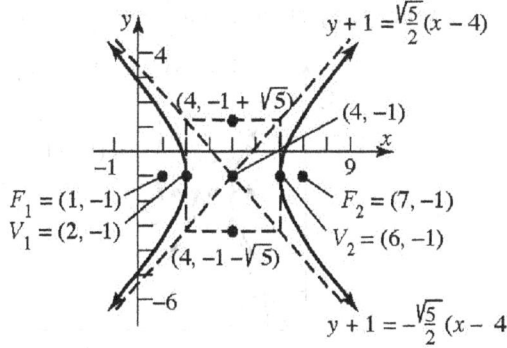

43. Center: $(-3, -4)$; Focus: $(-3, -8)$;
Vertex: $(-3, -2)$;
Transverse axis is parallel to the y-axis;
$a = 2; \ c = 4$.

Find the value of b:

$b^2 = c^2 - a^2 = 16 - 4 = 12$

$b = \sqrt{12} = 2\sqrt{3}$

Write the equation: $\dfrac{(y+4)^2}{4} - \dfrac{(x+3)^2}{12} = 1$.

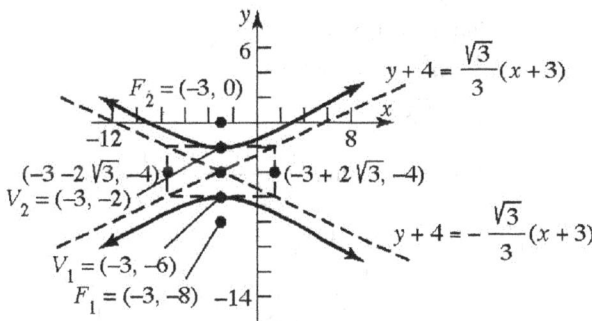

45. Foci: $(3, 7), (7, 7)$; Vertex: $(6, 7)$;
Center: $(5, 7)$; Transverse axis is parallel to the
x-axis; $a = 1$; $c = 2$.
Find the value of b:
$$b^2 = c^2 - a^2 = 4 - 1 = 3$$
$$b = \sqrt{3}$$

Write the equation: $(x-5)^2 - \dfrac{(y-7)^2}{3} = 1$.

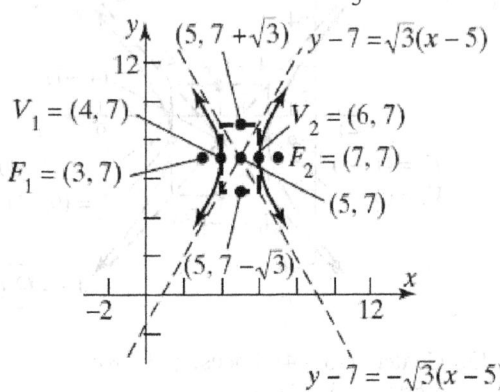

47. Vertices: $(-1, -1), (3, -1)$; Center: $(1, -1)$;
Transverse axis is parallel to the x-axis; $a = 2$.

Asymptote: $y + 1 = \dfrac{3}{2}(x-1)$

Using the slope of the asymptote, find the value
of b:
$$\frac{b}{a} = \frac{b}{2} = \frac{3}{2} \Rightarrow b = 3$$
Find the value of c:
$$c^2 = a^2 + b^2 = 4 + 9 = 13$$
$$c = \sqrt{13}$$

Write the equation: $\dfrac{(x-1)^2}{4} - \dfrac{(y+1)^2}{9} = 1$.

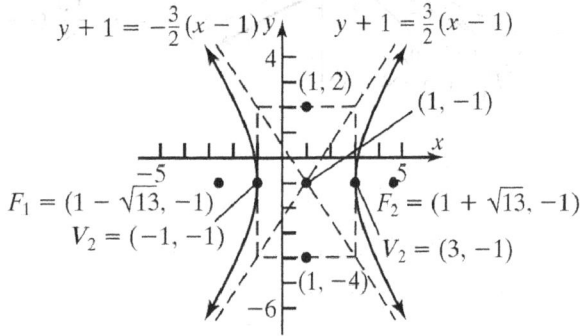

49. $\dfrac{(x-2)^2}{4} - \dfrac{(y+3)^2}{9} = 1$

The center of the hyperbola is at $(2, -3)$.
$a = 2$, $b = 3$.
The vertices are $(0, -3)$ and $(4, -3)$.
Find the value of c:
$$c^2 = a^2 + b^2 = 4 + 9 = 13 \Rightarrow c = \sqrt{13}$$
Foci: $\left(2 - \sqrt{13}, -3\right)$ and $\left(2 + \sqrt{13}, -3\right)$.
Transverse axis: $y = -3$, parallel to x-axis.

Asymptotes: $y + 3 = \dfrac{3}{2}(x-2)$;
$$y + 3 = -\frac{3}{2}(x-2)$$

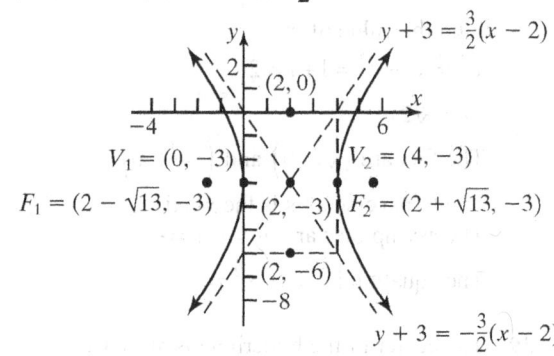

51. $(y-2)^2 - 4(x+2)^2 = 4$
Divide both sides by 4 to put in standard form:
$$\frac{(y-2)^2}{4} - (x+2)^2 = 1.$$
The center of the hyperbola is at $(-2, 2)$.
$a = 2$, $b = 1$.
The vertices are $(-2, 4)$ and $(-2, 0)$. Find the
value of c:
$$c^2 = a^2 + b^2 = 4 + 1 = 5 \Rightarrow c = \sqrt{5}$$
Foci: $\left(-2, 2 - \sqrt{5}\right)$ and $\left(-2, 2 + \sqrt{5}\right)$.
Transverse axis: $x = -2$, parallel to the y-axis.
Asymptotes: $y - 2 = 2(x+2)$; $y - 2 = -2(x+2)$.

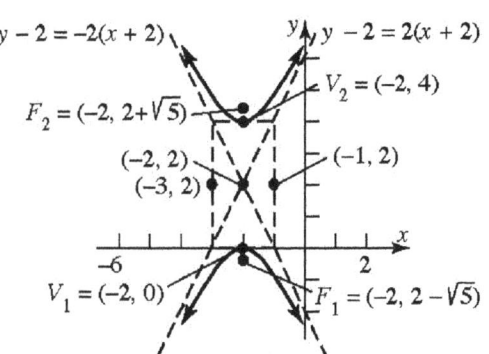

53. $(x+1)^2 - (y+2)^2 = 4$

Divide both sides by 4 to put in standard form:

$\dfrac{(x+1)^2}{4} - \dfrac{(y+2)^2}{4} = 1$.

The center of the hyperbola is $(-1, -2)$.

$a = 2, \; b = 2$.

The vertices are $(-3, -2)$ and $(1, -2)$.

Find the value of c:

$c^2 = a^2 + b^2 = 4 + 4 = 8 \Rightarrow c = \sqrt{8} = 2\sqrt{2}$

Foci: $\left(-1 - 2\sqrt{2}, -2\right)$ and $\left(-1 + 2\sqrt{2}, -2\right)$

Transverse axis: $y = -2$, parallel to the x-axis.

Asymptotes: $y + 2 = x + 1; \; y + 2 = -(x+1)$

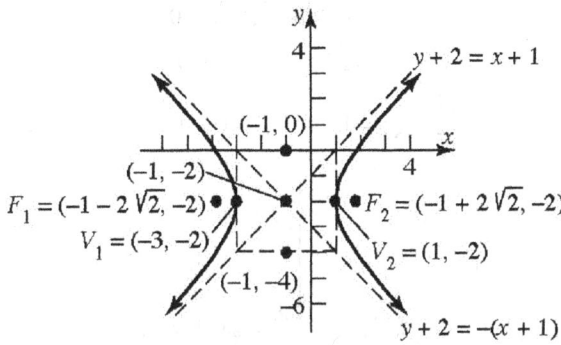

55. Complete the squares to put in standard form:

$$x^2 - y^2 - 2x - 2y - 1 = 0$$
$$(x^2 - 2x + 1) - (y^2 + 2y + 1) = 1 + 1 - 1$$
$$(x-1)^2 - (y+1)^2 = 1$$

The center of the hyperbola is $(1, -1)$.

$a = 1, \; b = 1$. The vertices are $(0, -1)$ and $(2, -1)$. Find the value of c:

$c^2 = a^2 + b^2 = 1 + 1 = 2 \Rightarrow c = \sqrt{2}$

Foci: $\left(1 - \sqrt{2}, -1\right)$ and $\left(1 + \sqrt{2}, -1\right)$.

Transverse axis: $y = -1$, parallel to x-axis.

Asymptotes: $y + 1 = x - 1; \; y + 1 = -(x-1)$.

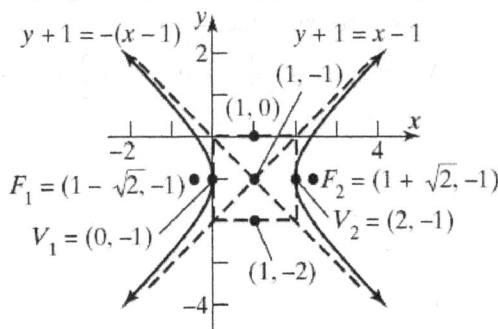

57. Complete the squares to put in standard form:

$$y^2 - 4x^2 - 4y - 8x - 4 = 0$$
$$(y^2 - 4y + 4) - 4(x^2 + 2x + 1) = 4 + 4 - 4$$
$$(y-2)^2 - 4(x+1)^2 = 4$$
$$\dfrac{(y-2)^2}{4} - (x+1)^2 = 1$$

The center of the hyperbola is $(-1, 2)$.

$a = 2, \; b = 1$.

The vertices are $(-1, 4)$ and $(-1, 0)$.

Find the value of c:

$c^2 = a^2 + b^2 = 4 + 1 = 5 \Rightarrow c = \sqrt{5}$

Foci: $\left(-1, 2 - \sqrt{5}\right)$ and $\left(-1, 2 + \sqrt{5}\right)$.

Transverse axis: $x = -1$, parallel to the y-axis.

Asymptotes: $y - 2 = 2(x+1); \; y - 2 = -2(x+1)$.

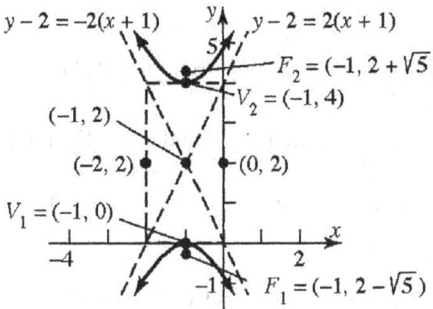

59. Complete the squares to put in standard form:

$$4x^2 - y^2 - 24x - 4y + 16 = 0$$
$$4(x^2 - 6x + 9) - (y^2 + 4y + 4) = -16 + 36 - 4$$
$$4(x-3)^2 - (y+2)^2 = 16$$
$$\dfrac{(x-3)^2}{4} - \dfrac{(y+2)^2}{16} = 1$$

The center of the hyperbola is $(3, -2)$.

$a = 2, \; b = 4$.

The vertices are $(1, -2)$ and $(5, -2)$. Find the value of c:

$c^2 = a^2 + b^2 = 4 + 16 = 20$

$c = \sqrt{20} = 2\sqrt{5}$

Foci: $\left(3 - 2\sqrt{5}, -2\right)$ and $\left(3 + 2\sqrt{5}, -2\right)$.

Transverse axis: $y = -2$, parallel to x-axis.

Asymptotes: $y + 2 = 2(x-3)$;

$$y + 2 = -2(x-3)$$

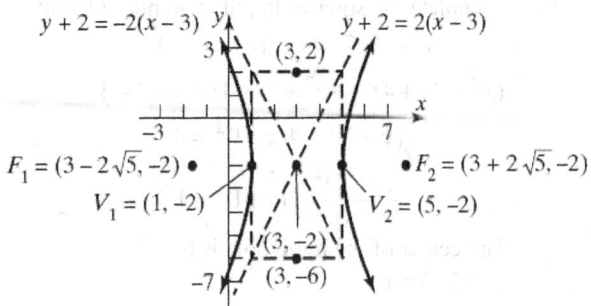

61. Complete the squares to put in standard form:

$$y^2 - 4x^2 - 16x - 2y - 19 = 0$$
$$(y^2 - 2y + 1) - 4(x^2 + 4x + 4) = 19 + 1 - 16$$
$$(y-1)^2 - 4(x+2)^2 = 4$$
$$\frac{(y-1)^2}{4} - (x+2)^2 = 1$$

The center of the hyperbola is (–2, 1).
$a = 2,\ b = 1$.
The vertices are (–2, 3) and (–2, –1). Find the value of c:

$$c^2 = a^2 + b^2 = 4 + 1 = 5$$
$$c = \sqrt{5}$$

Foci: $\left(-2, 1-\sqrt{5}\right)$ and $\left(-2, 1+\sqrt{5}\right)$.

Transverse axis: $x = -2$, parallel to the y-axis.

Asymptotes: $y - 1 = 2(x+2);\ y - 1 = -2(x+2)$.

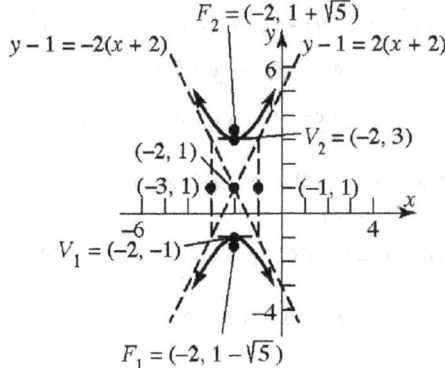

63. Rewrite the equation:

$$y = \sqrt{16 + 4x^2}$$
$$y^2 = 16 + 4x^2, \quad y \geq 0$$
$$y^2 - 4x^2 = 16, \quad\quad y \geq 0$$
$$\frac{y^2}{16} - \frac{x^2}{4} = 1, \quad\quad y \geq 0$$

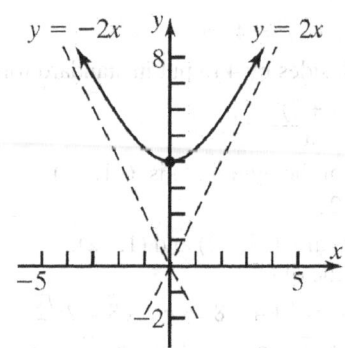

65. Rewrite the equation:

$$y = -\sqrt{-25 + x^2}$$
$$y^2 = -25 + x^2, \quad y \leq 0$$
$$x^2 - y^2 = 25, \quad\quad y \leq 0$$
$$\frac{x^2}{25} - \frac{y^2}{25} = 1, \quad\quad y \leq 0$$

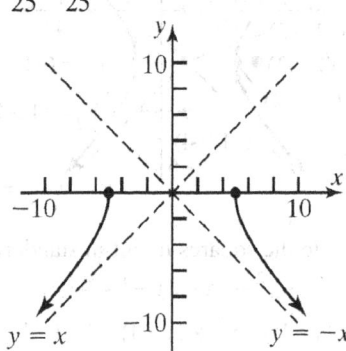

67. $\dfrac{(x-3)^2}{4} - \dfrac{y^2}{25} = 1$

The graph will be a hyperbola. The center of the hyperbola is at (3, 0). $a = 2,\ b = 5$.
The vertices are (5, 0) and (1, 0).
Find the value of c:

$$c^2 = a^2 + b^2 = 4 + 25 = 29 \Rightarrow c = \sqrt{29}$$

Foci: $\left(3 - \sqrt{29}, 0\right)$ and $\left(3 + \sqrt{29}, 0\right)$

Transverse axis is the x-axis.

Asymptotes: $y = \dfrac{5}{2}(x-3);\ y = -\dfrac{5}{2}(x-3)$

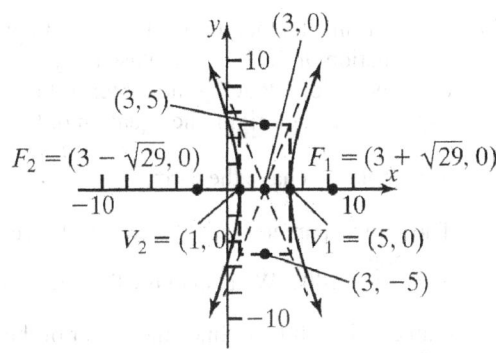

Foci: (5, 4), (5, –4)
Vertices: (5, 5), (5, –5)

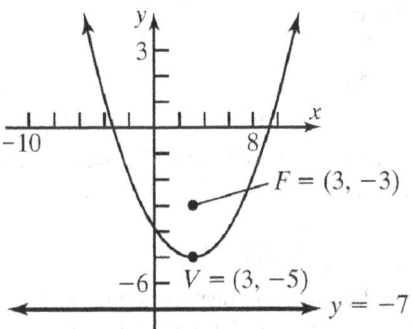

69. $x^2 = 16(y-3)$

The graph will be a parabola. The equation is in the form $(x-h)^2 = 4a(y-k)$ where $4a = 16$ or $a = 4$, $h = 0$, and $k = 3$. Thus, we have:

Vertex: (0, 3); Focus: (0, 7); Directrix: $y = -1$

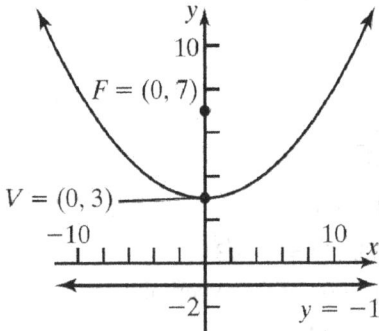

71. The graph will be an ellipse. Complete the square to put the equation in standard form:

$$25x^2 + 9y^2 - 250x + 400 = 0$$
$$(25x^2 - 250x) + 9y^2 = -400$$
$$25(x^2 - 10x) + 9y^2 = -400$$
$$25(x^2 - 10x + 25) + 9y^2 = -400 + 625$$
$$25(x-5)^2 + 9y^2 = 225$$
$$\frac{25(x-5)^2}{225} + \frac{9y^2}{225} = \frac{225}{225}$$
$$\frac{(x-5)^2}{9} + \frac{y^2}{25} = 1$$

The equation is in the form

$\dfrac{(x-h)^2}{b^2} + \dfrac{(y-k)^2}{a^2} = 1$ (major axis parallel to the

y-axis) where $a = 5$, $b = 3$, $h = 5$, and $k = 0$.

Solving for c: $c^2 = a^2 - b^2 = 25 - 9 = 16 \rightarrow c = 4$
Thus, we have:
Center: (5, 0)

73. The graph will be a parabola. Complete the square to put the equation in standard form:

$$x^2 - 6x - 8y - 31 = 0$$
$$x^2 - 6x = 8y + 31$$
$$x^2 - 6x + 9 = 8y + 31 + 9$$
$$(x-3)^2 = 8y + 40$$
$$(x-3)^2 = 8(y+5)$$

The equation is in the form
$(x-h)^2 = 4a(y-k)$ where $4a = 8$ or $a = 2$,
$h = 3$, and $k = -5$. Thus, we have:

Vertex: $(3,-5)$;

Focus: $(3,-3)$;

Directrix: $y = -7$

443

75. First note that all points where a burst could take place, such that the time difference would be the same as that for the first burst, would form a hyperbola with A and B as the foci. Start with a diagram:

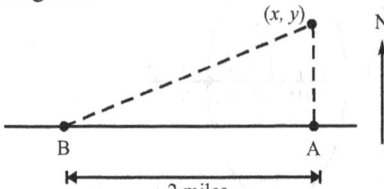

Assume a coordinate system with the x-axis containing $\overline{BA}$ and the origin at the midpoint of $\overline{BA}$.

The ordered pair (x, y) represents the location of the fireworks. We know that sound travels at 1100 feet per second, so the person at point A is 1100 feet closer to the fireworks display than the person at point B. Since the difference of the distance from (x, y) to A and from (x, y) to B is the constant 1100, the point (x, y) lies on a hyperbola whose foci are at A and B. The hyperbola has the equation

$$\frac{x^2}{a^2} - \frac{y^2}{b^2} = 1$$

where $2a = 1100$, so $a = 550$. Because the distance between the two people is 2 miles (10,560 feet) and each person is at a focus of the hyperbola, we have

$$2c = 10,560$$
$$c = 5280$$
$$b^2 = c^2 - a^2 = 5280^2 - 550^2 = 27,575,900$$

The equation of the hyperbola that describes the location of the fireworks display is

$$\frac{x^2}{550^2} - \frac{y^2}{27,575,900} = 1$$

Since the fireworks display is due north of the individual at A, we let $x = 5280$ and solve the equation for y.

$$\frac{5280^2}{550^2} - \frac{y^2}{27,575,900} = 1$$

$$-\frac{y^2}{27,575,900} = -91.16$$

$$y^2 = 2,513,819,044$$

$$y = 50,138$$

Therefore, the fireworks display was 50,138 feet (approximately 9.5 miles) due north of the person at point A.

77. To determine the height, we first need to obtain the equation of the hyperbola used to generate the hyperboloid. Placing the center of the hyperbola at the origin, the equation of the hyperbola will have the form $\frac{x^2}{a^2} - \frac{y^2}{b^2} = 1$.

The center diameter is 200 feet so we have $a = \frac{200}{2} = 100$. We also know that the base diameter is 400 feet. Since the center of the hyperbola is at the origin, the points $(200, -360)$ and $(-200, -360)$ must be on the graph of our hyperbola (recall the center is 360 feet above ground). Therefore,

$$\frac{(200)^2}{(100)^2} - \frac{(-360)^2}{b^2} = 1$$

$$4 - \frac{360^2}{b^2} = 1$$

$$3 = \frac{360^2}{b^2}$$

$$b^2 = 43,200$$

$$b = \sqrt{43,200} = 120\sqrt{3}$$

The equation of the hyperbola is

$$\frac{x^2}{10,000} - \frac{y^2}{43,200} = 1$$

At the top of the tower we have $x = \frac{300}{2} = 150$.

$$\frac{150^2}{10,000} - \frac{y^2}{43,200} = 1$$

$$\frac{y^2}{43,200} = 1.25$$

$$y^2 = 54000$$

$$y \approx 232.4$$

The height of the tower is approximately $232.4 + 360 = 592.4$ feet.

79. a. Since the particles are deflected at a $45°$ angle, the asymptotes will be $y = \pm x$.

b. Since the vertex is 10 cm from the center of the hyperbola, we know that $a = 10$. The slope of the asymptotes is given by $\pm\dfrac{b}{a}$.

Therefore, we have
$$\frac{b}{a} = 1 \implies \frac{b}{10} = 1 \implies b = 10$$

Using the origin as the center of the hyperbola, the equation of the particle path would be
$$\frac{x^2}{100} - \frac{y^2}{100} = 1, \quad x \ge 0$$

81. Assume $\dfrac{x^2}{a^2} - \dfrac{y^2}{b^2} = 1$.

If the eccentricity is close to 1, then $c \approx a$ and $b \approx 0$. When b is close to 0, the hyperbola is very narrow, because the slopes of the asymptotes are close to 0.

If the eccentricity is very large, then c is much larger than a and b is very large. The result is a hyperbola that is very wide, because the slopes of the asymptotes are very large.

For $\dfrac{y^2}{a^2} - \dfrac{x^2}{b^2} = 1$, the opposite is true. When the eccentricity is close to 1, the hyperbola is very wide because the slopes of the asymptotes are close to 0. When the eccentricity is very large, the hyperbola is very narrow because the slopes of the asymptotes are very large.

83. $\dfrac{x^2}{4} - y^2 = 1 \quad (a = 2, \ b = 1)$

This is a hyperbola with horizontal transverse axis, centered at $(0, 0)$ and has asymptotes:
$$y = \pm\frac{1}{2}x$$

$y^2 - \dfrac{x^2}{4} = 1 \quad (a = 1, \ b = 2)$

This is a hyperbola with vertical transverse axis, centered at $(0, 0)$ and has asymptotes: $y = \pm\dfrac{1}{2}x$.

Since the two hyperbolas have the same asymptotes, they are conjugates.

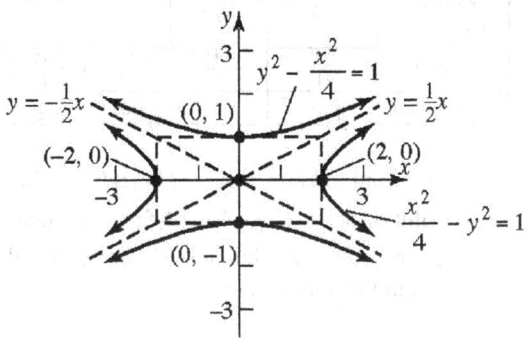

85. Put the equation in standard hyperbola form:
$$Ax^2 + Cy^2 + F = 0 \qquad A \cdot C < 0, \ F \ne 0$$
$$Ax^2 + Cy^2 = -F$$
$$\frac{Ax^2}{-F} + \frac{Cy^2}{-F} = 1$$
$$\frac{x^2}{\left(-\dfrac{F}{A}\right)} + \frac{y^2}{\left(-\dfrac{F}{C}\right)} = 1$$

Since $-F/A$ and $-F/C$ have opposite signs, this is a hyperbola with center $(0, 0)$.
The transverse axis is the x-axis if $-F/A > 0$.
The transverse axis is the y-axis if $-F/A < 0$.

87. $(8x^3 - 12x^2) - (50x - 75) = 0$
$$4x^2(2x - 3) - 25(2x - 3) = 0$$
$$(4x^2 - 25)(2x - 3) = 0$$
$$(2x - 5)(2x + 5)(2x - 3) = 0$$

$$x = \frac{5}{2} \text{ or } x = -\frac{5}{2} \text{ or } x = \frac{3}{2}$$

The solution set is: $\left\{-\dfrac{5}{2}, \dfrac{3}{2}, \dfrac{5}{2}\right\}$

89.
$$\frac{x^2 - 16}{x^2 - 25} \le 0$$
$$\frac{(x-4)(x+4)}{(x-5)(x+5)} \le 0$$

The zeros and values where f is undefined are $x = -5, -4, 4$ and 5

Interval	$(-\infty, -5)$	$(-5, -4)$	$(-4, 4)$	$(4, 5)$	$(5, \infty)$
Number Chosen	-6	-4.5	0	4.5	6
Value of f	$\frac{20}{11}$	$-\frac{17}{19}$	$\frac{16}{25}$	$-\frac{17}{19}$	$\frac{20}{11}$
Conclusion	pos	neg	pos	neg	pos

We want to know where $f(x) \le 0$, so the solution set is $(-5, -4] \cup [4, 5)$. Note that -5 and 5 are not in the solution set because -5 and 5 are not in the domain of f.

Chapter 7 Review Exercises

1. $y^2 = -16x$

This is a parabola.

$a = 4$

Vertex: $(0, 0)$

Focus: $(-4, 0)$

Directrix: $x = 4$

2. $\dfrac{x^2}{25} - y^2 = 1$

This is a hyperbola.

$a = 5,\ b = 1$.

Find the value of c:

$c^2 = a^2 + b^2 = 25 + 1 = 26$

$c = \sqrt{26}$

Center: $(0, 0)$

Vertices: $(5, 0), (-5, 0)$

Foci: $\left(\sqrt{26}, 0\right), \left(-\sqrt{26}, 0\right)$

Asymptotes: $y = \dfrac{1}{5}x;\ \ y = -\dfrac{1}{5}x$

3. $\dfrac{y^2}{25} + \dfrac{x^2}{16} = 1$

This is an ellipse.

$a = 5,\ b = 4$.

Find the value of c:

$c^2 = a^2 - b^2 = 25 - 16 = 9$

$c = 3$

Center: $(0, 0)$

Vertices: $(0, 5), (0, -5)$

Foci: $(0, 3), (0, -3)$

4. $x^2 + 4y = 4$

This is a parabola.

Write in standard form:

$x^2 = -4y + 4$

$x^2 = -4(y - 1)$

$a = 1$

Vertex: $(0, 1)$

Focus: $(0, 0)$ Directrix: $y = 2$

5. $4x^2 - y^2 = 8$

This is a hyperbola.

Write in standard form:

$\dfrac{x^2}{2} - \dfrac{y^2}{8} = 1$

$a = \sqrt{2},\ b = \sqrt{8} = 2\sqrt{2}$.

Find the value of c:

$c^2 = a^2 + b^2 = 2 + 8 = 10$

$c = \sqrt{10}$

Center: $(0, 0)$

Vertices: $\left(-\sqrt{2}, 0\right), \left(\sqrt{2}, 0\right)$

Foci: $\left(-\sqrt{10}, 0\right), \left(\sqrt{10}, 0\right)$

Asymptotes: $y = 2x;\ \ y = -2x$

6. $x^2 - 4x = 2y$

This is a parabola.

Write in standard form:

$x^2 - 4x + 4 = 2y + 4$

$(x - 2)^2 = 2(y + 2)$

$a = \dfrac{1}{2}$

Vertex: $(2, -2)$

Focus: $\left(2, -\dfrac{3}{2}\right)$

Directrix: $y = -\dfrac{5}{2}$

7. $y^2 - 4y - 4x^2 + 8x = 4$

This is a hyperbola.
Write in standard form:
$(y^2 - 4y + 4) - 4(x^2 - 2x + 1) = 4 + 4 - 4$
$$(y-2)^2 - 4(x-1)^2 = 4$$
$$\frac{(y-2)^2}{4} - \frac{(x-1)^2}{1} = 1$$
$a = 2, \ b = 1$.
Find the value of c:
$c^2 = a^2 + b^2 = 4 + 1 = 5$
$c = \sqrt{5}$
Center: $(1, 2)$
Vertices: $(1, 0), (1, 4)$
Foci: $\left(1, 2 - \sqrt{5}\right), \left(1, 2 + \sqrt{5}\right)$
Asymptotes: $y - 2 = 2(x-1); \ y - 2 = -2(x-1)$

8. $4x^2 + 9y^2 - 16x - 18y = 11$

This is an ellipse.
Write in standard form:
$$4x^2 + 9y^2 - 16x - 18y = 11$$
$$4(x^2 - 4x + 4) + 9(y^2 - 2y + 1) = 11 + 16 + 9$$
$$4(x-2)^2 + 9(y-1)^2 = 36$$
$$\frac{(x-2)^2}{9} + \frac{(y-1)^2}{4} = 1$$
$a = 3, \ b = 2$. Find the value of c:
$c^2 = a^2 - b^2 = 9 - 4 = 5 \ \rightarrow \ c = \sqrt{5}$
Center: $(2, 1)$; Vertices: $(-1, 1), (5, 1)$
Foci: $\left(2 - \sqrt{5}, 1\right), \left(2 + \sqrt{5}, 1\right)$

9. $4x^2 - 16x + 16y + 32 = 0$

This is a parabola.
Write in standard form:
$$4(x^2 - 4x + 4) = -16y - 32 + 16$$
$$4(x-2)^2 = -16(y+1)$$
$$(x-2)^2 = -4(y+1)$$
$a = 1$
Vertex: $(2, -1)$; Focus: $(2, -2)$;
Directrix: $y = 0$

10. $9x^2 + 4y^2 - 18x + 8y = 23$

This is an ellipse.
Write in standard form:
$$9(x^2 - 2x + 1) + 4(y^2 + 2y + 1) = 23 + 9 + 4$$
$$9(x-1)^2 + 4(y+1)^2 = 36$$
$$\frac{(x-1)^2}{4} + \frac{(y+1)^2}{9} = 1$$
$a = 3, \ b = 2$.
Find the value of c:
$c^2 = a^2 - b^2 = 9 - 4 = 5$
$c = \sqrt{5}$
Center: $(1, -1)$
Vertices: $(1, -4), (1, 2)$
Foci: $\left(1, -1 - \sqrt{5}\right), \left(1, -1 + \sqrt{5}\right)$

11. Parabola: The focus is $(-2, 0)$ and the directrix is $x = 2$. The vertex is $(0, 0)$. $a = 2$ and since $(-2, 0)$ is to the left of $(0, 0)$, the parabola opens to the left. The equation of the parabola is:
$y^2 = -4ax$
$y^2 = -4 \cdot 2 \cdot x$
$y^2 = -8x$

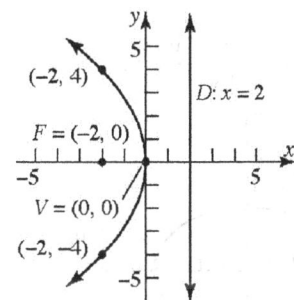

12. Hyperbola: Center: $(0,0)$;
Focus: $(0,4)$; Vertex: $(0,-2)$;
Transverse axis is the y-axis; $a=2$; $c=4$.
Find b:
$$b^2 = c^2 - a^2 = 16 - 4 = 12$$
$$b = \sqrt{12} = 2\sqrt{3}$$

Write the equation: $\dfrac{y^2}{4} - \dfrac{x^2}{12} = 1$

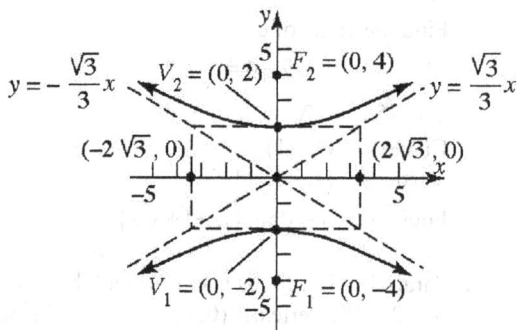

13. Ellipse: Foci: $(-3,0)$, $(3,0)$; Vertex: $(4,0)$;
Center: $(0,0)$; Major axis is the x-axis;
$a=4$; $c=3$. Find b:
$$b^2 = a^2 - c^2 = 16 - 9 = 7$$
$$b = \sqrt{7}$$

Write the equation: $\dfrac{x^2}{16} + \dfrac{y^2}{7} = 1$

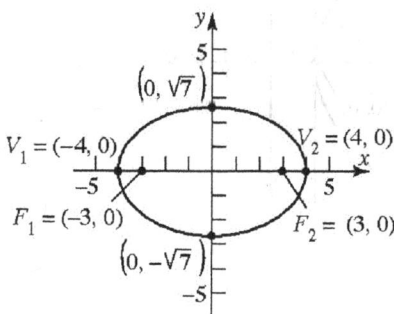

14. Parabola: The focus is $(2,-4)$ and the vertex is
$(2,-3)$. Both lie on the vertical line $x=2$.
$a=1$ and since $(2,-4)$ is below $(2,-3)$, the
parabola opens down. The equation of the
parabola is:

$$(x-h)^2 = -4a(y-k)$$
$$(x-2)^2 = -4\cdot 1\cdot(y-(-3))$$
$$(x-2)^2 = -4(y+3)$$

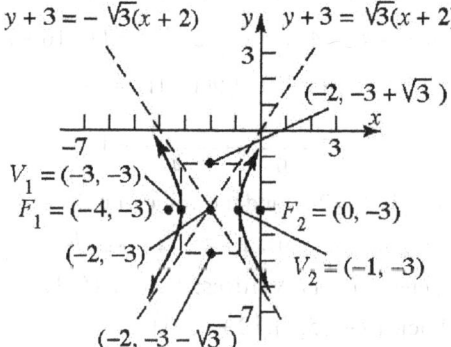

15. Hyperbola: Center: $(-2,-3)$; Focus: $(-4,-3)$;
Vertex: $(-3,-3)$; Transverse axis is parallel to
the x-axis; $a=1$; $c=2$. Find b:
$$b^2 = c^2 - a^2 = 4 - 1 = 3$$
$$b = \sqrt{3}$$

Write the equation: $\dfrac{(x+2)^2}{1} - \dfrac{(y+3)^2}{3} = 1$

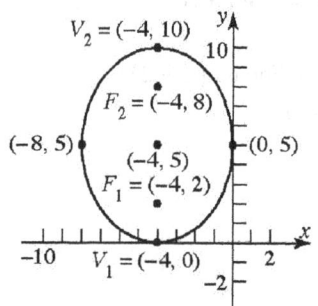

16. Ellipse: Foci: $(-4,2)$, $(-4,8)$; Vertex: $(-4,10)$;
Center: $(-4,5)$; Major axis is parallel to the y-
axis; $a=5$; $c=3$. Find b:
$$b^2 = a^2 - c^2 = 25 - 9 = 16 \quad \rightarrow \quad b = 4$$

Write the equation: $\dfrac{(x+4)^2}{16} + \dfrac{(y-5)^2}{25} = 1$

Copyright © 2017 Pearson Education, Inc.

17. Hyperbola: Center: $(-1, 2)$; $a = 3$; $c = 4$;
Transverse axis parallel to the x-axis;
Find b: $b^2 = c^2 - a^2 = 16 - 9 = 7 \rightarrow b = \sqrt{7}$

Write the equation: $\dfrac{(x+1)^2}{9} - \dfrac{(y-2)^2}{7} = 1$

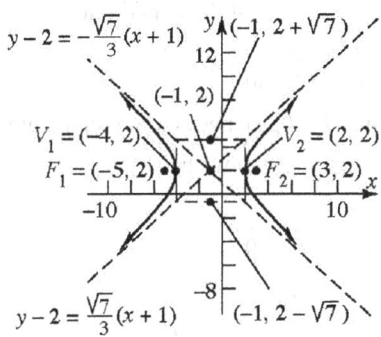

18. Hyperbola: Vertices: $(0, 1), (6, 1)$; Asymptote: $3y + 2x - 9 = 0$; Center: $(3, 1)$; Transverse axis is parallel to the x-axis; $a = 3$; The slope of the asymptote is $-\dfrac{2}{3}$; Find b:

$\dfrac{-b}{a} = \dfrac{-b}{3} = \dfrac{-2}{3} \rightarrow -3b = -6 \rightarrow b = 2$

Write the equation: $\dfrac{(x-3)^2}{9} - \dfrac{(y-1)^2}{4} = 1$

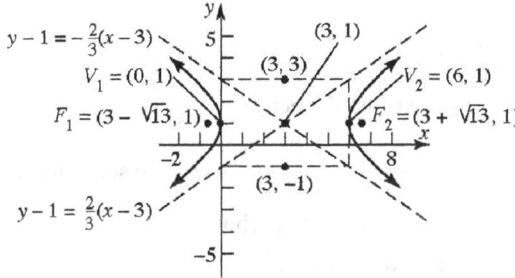

19. Write the equation in standard form:

$4x^2 + 9y^2 = 36 \rightarrow \dfrac{x^2}{9} + \dfrac{y^2}{4} = 1$

The center of the ellipse is $(0, 0)$. The major axis is the x-axis.
$a = 3$; $b = 2$;
$c^2 = a^2 - b^2 = 9 - 4 = 5 \rightarrow c = \sqrt{5}$.
For the ellipse:
Vertices: $(-3, 0), (3, 0)$;
Foci: $\left(-\sqrt{5}, 0\right), \left(\sqrt{5}, 0\right)$

For the hyperbola:
Foci: $(-3, 0), (3, 0)$;

Vertices: $\left(-\sqrt{5}, 0\right), \left(\sqrt{5}, 0\right)$;
Center: $(0, 0)$
$a = \sqrt{5}$; $c = 3$;
$b^2 = c^2 - a^2 = 9 - 5 = 4 \rightarrow b = 2$

The equation of the hyperbola is: $\dfrac{x^2}{5} - \dfrac{y^2}{4} = 1$

20. Let (x, y) be any point in the collection of points.
The distance from
(x, y) to $(3, 0) = \sqrt{(x-3)^2 + y^2}$.
The distance from
(x, y) to the line $x = \dfrac{16}{3}$ is $\left| x - \dfrac{16}{3} \right|$.
Relating the distances, we have:
$\sqrt{(x-3)^2 + y^2} = \dfrac{3}{4} \left| x - \dfrac{16}{3} \right|$

$(x-3)^2 + y^2 = \dfrac{9}{16} \left(x - \dfrac{16}{3} \right)^2$

$x^2 - 6x + 9 + y^2 = \dfrac{9}{16} \left(x^2 - \dfrac{32}{3}x + \dfrac{256}{9} \right)$

$16x^2 - 96x + 144 + 16y^2 = 9x^2 - 96x + 256$

$7x^2 + 16y^2 = 112$

$\dfrac{7x^2}{112} + \dfrac{16y^2}{112} = 1$

$\dfrac{x^2}{16} + \dfrac{y^2}{7} = 1$

The set of points is an ellipse.

21. Locate the parabola so that the vertex is at $(0, 0)$ and opens up. It then has the equation:
$x^2 = 4ay$. Since the light source is located at the focus and is 1 foot from the base, $a = 1$. Thus, $x^2 = 4y$. The width of the opening is 2, so the point $(1, y)$ is located on the parabola. Solve for y:
$1^2 = 4y \rightarrow 1 = 4y \rightarrow y = 0.25$ feet
The mirror is 0.25 feet, or 3 inches, deep.

22. Place the semi-elliptical arch so that the x-axis coincides with the water and the y-axis passes through the center of the arch. Since the bridge has a span of 60 feet, the length of the major axis is 60, or $2a = 60$ or $a = 30$. The maximum height of the bridge is 20 feet, so $b = 20$. The

equation is: $\dfrac{x^2}{900} + \dfrac{y^2}{400} = 1$.

The height 5 feet from the center:

$$\frac{5^2}{900} + \frac{y^2}{400} = 1$$

$$\frac{y^2}{400} = 1 - \frac{25}{900}$$

$$y^2 = 400 \cdot \frac{875}{900} \rightarrow y \approx 19.72 \text{ feet}$$

The height 10 feet from the center:

$$\frac{10^2}{900} + \frac{y^2}{400} = 1$$

$$\frac{y^2}{400} = 1 - \frac{100}{900}$$

$$y^2 = 400 \cdot \frac{800}{900} \rightarrow y \approx 18.86 \text{ feet}$$

The height 20 feet from the center:

$$\frac{20^2}{900} + \frac{y^2}{400} = 1$$

$$\frac{y^2}{400} = 1 - \frac{400}{900}$$

$$y^2 = 400 \cdot \frac{500}{900} \rightarrow y \approx 14.91 \text{ feet}$$

23. First note that all points where an explosion could take place, such that the time difference would be the same as that for the first detonation, would form a hyperbola with A and B as the foci.
Start with a diagram:

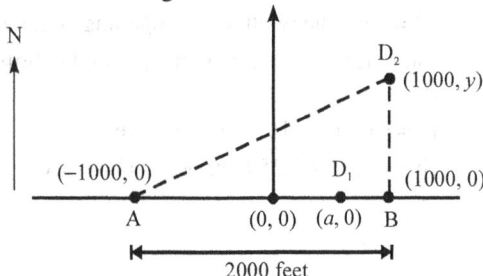

Since A and B are the foci, we have
$2c = 2000$

$c = 1000$

Since D_1 is on the transverse axis and is on the hyperbola, then it must be a vertex of the hyperbola. Since it is 200 feet from B, we have $a = 800$. Finally,

$b^2 = c^2 - a^2 = 1000^2 - 800^2 = 360,000$

Thus, the equation of the hyperbola is

$$\frac{x^2}{640,000} - \frac{y^2}{360,000} = 1$$

The point $(1000, y)$ needs to lie on the graph of the hyperbola. Thus, we have

$$\frac{(1000)^2}{640,000} - \frac{y^2}{360,000} = 1$$

$$-\frac{y^2}{360,000} = -\frac{9}{16}$$

$$y^2 = 202,500$$

$$y = 450$$

The second explosion should be set off 450 feet due north of point B.

Chapter 7 Test

1. $\dfrac{(x+1)^2}{4} - \dfrac{y^2}{9} = 1$

Rewriting the equation as

$\dfrac{(x-(-1))^2}{2^2} - \dfrac{(y-0)^2}{3^2} = 1$, we see that this is the

equation of a hyperbola in the form

$\dfrac{(x-h)^2}{a^2} - \dfrac{(y-k)^2}{b^2} = 1$. Therefore, we have

$h = -1$, $k = 0$, $a = 2$, and $b = 3$. Since $a^2 = 4$ and $b^2 = 9$, we get $c^2 = a^2 + b^2 = 4 + 9 = 13$, or $c = \sqrt{13}$. The center is at $(-1, 0)$ and the transverse axis is the x-axis. The vertices are at $(h \pm a, k) = (-1 \pm 2, 0)$, or $(-3, 0)$ and $(1, 0)$.

The foci are at $(h \pm c, k) = \left(-1 \pm \sqrt{13}, 0\right)$, or

$\left(-1 - \sqrt{13}, 0\right)$ and $\left(-1 + \sqrt{13}, 0\right)$. The asymptotes

are $y - 0 = \pm\dfrac{3}{2}(x - (-1))$, or $y = -\dfrac{3}{2}(x+1)$ and

$y = \dfrac{3}{2}(x+1)$.

2. $8y = (x-1)^2 - 4$

Rewriting gives

$(x-1)^2 = 8y + 4$

$(x-1)^2 = 8\left(y - \left(-\dfrac{1}{2}\right)\right)$

$(x-1)^2 = 4(2)\left(y - \left(-\dfrac{1}{2}\right)\right)$

This is the equation of a parabola in the form $(x-h)^2 = 4a(y-k)$. Therefore, the axis of symmetry is parallel to the y-axis and we have $(h,k) = \left(1, -\dfrac{1}{2}\right)$ and $a = 2$. The vertex is at $(h,k) = \left(1, -\dfrac{1}{2}\right)$, the axis of symmetry is $x = 1$,

the focus is at $(h, k+a) = \left(1, -\dfrac{1}{2} + 2\right) = \left(1, \dfrac{3}{2}\right)$,

and the directrix is given by the line $y = k - a$,

or $y = -\dfrac{5}{2}$.

3. $2x^2 + 3y^2 + 4x - 6y = 13$

Rewrite the equation by completing the square in x and y.

$$2x^2 + 3y^2 + 4x - 6y = 13$$
$$2x^2 + 4x + 3y^2 - 6y = 13$$
$$2(x^2 + 2x) + 3(y^2 - 2y) = 13$$
$$2(x^2 + 2x + 1) + 3(y^2 - 2y + 1) = 13 + 2 + 3$$
$$2(x+1)^2 + 3(y-1)^2 = 18$$
$$\dfrac{(x-(-1))^2}{9} + \dfrac{(y-1)^2}{6} = 1$$

This is the equation of an ellipse with center at $(-1,1)$ and major axis parallel to the x-axis.

Since $a^2 = 9$ and $b^2 = 6$, we have $a = 3$, $b = \sqrt{6}$, and $c^2 = a^2 - b^2 = 9 - 6 = 3$, or $c = \sqrt{3}$. The foci are $(h \pm c, k) = \left(-1 \pm \sqrt{3}, 1\right)$ or $\left(-1 - \sqrt{3}, 1\right)$ and $\left(-1 + \sqrt{3}, 1\right)$. The vertices are at $(h \pm a, k) = (-1 \pm 3, 1)$, or $(-4, 1)$ and $(2, 1)$.

4. The vertex $(-1,3)$ and the focus $(-1, 4.5)$ both lie on the vertical line $x = -1$ (the axis of symmetry). The distance a from the vertex to the focus is $a = 1.5$. Because the focus lies above the vertex, we know the parabola opens upward. As a result, the form of the equation is

$(x-h)^2 = 4a(y-k)$

where $(h,k) = (-1,3)$ and $a = 1.5$. Therefore, the equation is

$(x+1)^2 = 4(1.5)(y-3)$

$(x+1)^2 = 6(y-3)$

The points $(h \pm 2a, k)$, that is $(-4, 4.5)$ and $(2, 4.5)$, define the lattice rectum; the line $y = 1.5$ is the directrix.

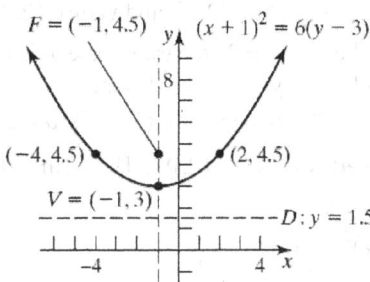

5. The center is $(h,k) = (0,0)$ so $h = 0$ and $k = 0$. Since the center, focus, and vertex all lie on the line $x = 0$, the major axis is the y-axis. The distance from the center $(0,0)$ to a focus $(0,3)$ is $c = 3$. The distance from the center $(0,0)$ to a vertex $(0,-4)$ is $a = 4$. Then,

$b^2 = a^2 - c^2 = 4^2 - 3^2 = 16 - 9 = 7$

The form of the equation is

$\dfrac{(x-h)^2}{b^2} + \dfrac{(y-k)^2}{a^2} = 1$

where $h = 0$, $k = 0$, $a = 4$, and $b = \sqrt{7}$. Thus, we get

$\dfrac{x^2}{7} + \dfrac{y^2}{16} = 1$

To graph the equation, we use the center $(h,k) = (0,0)$ to locate the vertices. The major axis is the y-axis, so the vertices are $a = 4$ units above and below the center. Therefore, the vertices are $V_1 = (0,4)$ and $V_2 = (0,-4)$. Since $c = 3$ and the major axis is the y-axis, the foci are 3 units above and below the center. Therefore, the foci are $F_1 = (0,3)$ and $F_2 = (0,-3)$. Finally, we use the value $b = \sqrt{7}$ to find the two points left and right of the center: $\left(-\sqrt{7},0\right)$ and $\left(\sqrt{7},0\right)$.

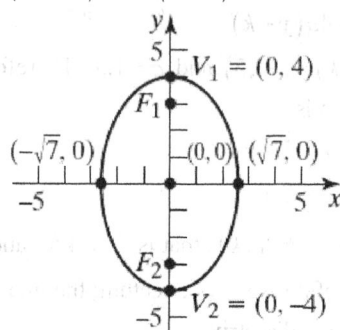

6. The center $(h,k) = (2,2)$ and vertex $(2,4)$ both lie on the line $x = 2$, the transverse axis is parallel to the y-axis. The distance from the center $(2,2)$ to the vertex $(2,4)$ is $a = 2$, so the other vertex must be $(2,0)$. The form of the equation is

$$\frac{(y-k)^2}{a^2} - \frac{(x-h)^2}{b^2} = 1$$

where $h = 2$, $k = 2$, and $a = 2$. This gives us

$$\frac{(y-2)^2}{4} - \frac{(x-2)^2}{b^2} = 1$$

Since the graph contains the point $(x,y) = \left(2 + \sqrt{10}, 5\right)$, we can use this point to determine the value for b.

$$\frac{(5-2)^2}{4} - \frac{\left(2+\sqrt{10}-2\right)^2}{b^2} = 1$$

$$\frac{9}{4} - \frac{10}{b^2} = 1$$

$$\frac{5}{4} = \frac{10}{b^2}$$

$$b^2 = 8$$

$$b = 2\sqrt{2}$$

Therefore, the equation becomes

$$\frac{(y-2)^2}{4} - \frac{(x-2)^2}{8} = 1$$

Since $c^2 = a^2 + b^2 = 4 + 8 = 12$, the distance from the center to either focus is $c = 2\sqrt{3}$.

Therefore, the foci are $c = 2\sqrt{3}$ units above and below the center. The foci are $F_1 = \left(2, 2+2\sqrt{3}\right)$ and $F_2 = \left(2, 2-2\sqrt{3}\right)$. The asymptotes are given by the lines

$y - k = \pm\dfrac{a}{b}(x-h)$. Therefore, the asymptotes are

$$y - 2 = \pm\frac{2}{2\sqrt{2}}(x-2)$$

$$y = \pm\frac{\sqrt{2}}{2}(x-2) + 2$$

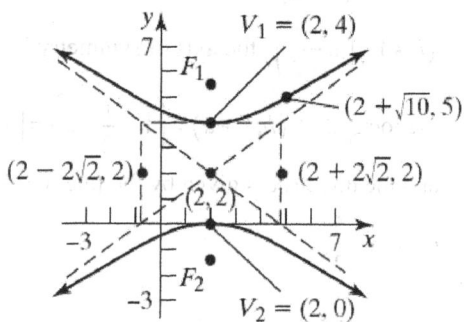

7. We can draw the parabola used to form the reflector on a rectangular coordinate system so that the vertex of the parabola is at the origin and its focus is on the positive y-axis.

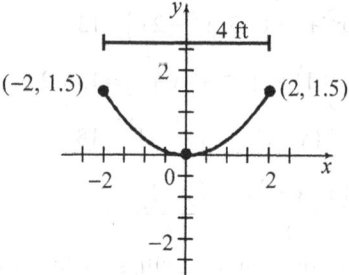

The form of the equation of the parabola is $x^2 = 4ay$ and its focus is at $(0,a)$. Since the point $(2, 1.5)$ is on the graph, we have

$$2^2 = 4a(1.5)$$

$$4 = 6a$$

$$a = \frac{2}{3}$$

The microphone should be located $\frac{2}{3}$ feet (or 8 inches) from the base of the reflector, along its axis of symmetry.

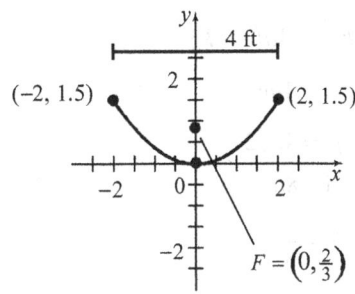

Chapter 7 Cumulative Review

1. $\dfrac{f(x+h)-f(x)}{h}$

$= \dfrac{-3(x+h)^2 +5(x+h)-2-\left(-3x^2+5x-2\right)}{h}$

$= \dfrac{-3\left(x^2+2xh+h^2\right)+5x+5h-2+3x^2-5x+2}{h}$

$= \dfrac{-3x^2-6xh-3h^2+5x+5h-2+3x^2-5x+2}{h}$

$= \dfrac{-6xh-3h^2+5h}{h} = -6x-3h+5$

2. $9x^4+33x^3-71x^2-57x-10=0$
There are at most 4 real zeros.
Possible rational zeros:
$p = \pm 1, \pm 2, \pm 5, \pm 10; \quad q = \pm 1, \pm 3, \pm 9;$

$\dfrac{p}{q} = \pm 1, \pm \dfrac{1}{3}, \pm \dfrac{1}{9}, \pm 2, \pm \dfrac{2}{3}, \pm \dfrac{2}{9}, \pm 5,$

$\pm \dfrac{5}{3}, \pm \dfrac{5}{9}, \pm 10, \pm \dfrac{10}{3}, \pm \dfrac{10}{9}$

Graphing $y_1 = 9x^4+33x^3-71x^2-57x-10$
indicates that there appear to be zeros at $x = -5$
and at $x = 2$.
Using synthetic division with $x = -5$:

```
-5) 9   33   -71   -57   -10
        -45   60    55    10
    9   -12  -11    -2     0
```

Since the remainder is 0, –5 is a zero for f. So
$x-(-5)=x+5$ is a factor.
The other factor is the quotient:

$9x^3-12x^2-11x-2$.
Thus, $f(x)=(x+5)\left(9x^3-12x^2-11x-2\right)$.
Using synthetic division on the quotient and
$x = 2$:

```
2) 9   -12   -11   -2
        18    12    2
   9     6     1    0
```

Since the remainder is 0, 2 is a zero for f. So
$x-2$ is a factor; thus,

$f(x)=(x+5)(x-2)\left(9x^2+6x+1\right)$.

$= (x+5)(x-2)(3x+1)(3x+1)$

Therefore, $x = -\dfrac{1}{3}$ is also a zero for f (with

multiplicity 2). Solution set: $\left\{-5, -\dfrac{1}{3}, 2\right\}$.

3.
$$6-x \geq x^2$$
$$0 \geq x^2 +x-6$$
$$x^2 +x-6 \leq 0$$
$$(x+3)(x-2) \leq 0$$
$$f(x) = x^2 +x-6$$

$x = -3,\ x = 2$ are the zeros of f.

Interval	$(-\infty,-3)$	$(-3,2)$	$(2,\infty)$
Test Value	–4	0	3
Value of f	6	–6	6
Conclusion	Positive	Negative	Positive

The solution set is $\left\{ x \mid -3 \leq x \leq 2 \right\}$, or $[-3,2]$.

4. $f(x) = 3^x +2$

a. Domain: $(-\infty,\infty)$; Range: $(2,\infty)$.

b. $f(x) = 3^x +2$

$y = 3^x +2$

$x = 3^y +2$ Inverse

$x-2 = 3^y$

$\log_3 (x-2) = y \Rightarrow f^{-1}(x) = \log_3 (x-2)$

Domain of f^{-1} = range of f = $(2,\infty)$.

Range of f^{-1} = domain of f = $(-\infty,\infty)$.

5. $f(x) = \log_4(x-2)$

    a. $f(x) = \log_4(x-2) = 2$

        $x - 2 = 4^2$

        $x - 2 = 16$

          $x = 18$

    The solution set is $\{18\}$.

    b. $f(x) = \log_4(x-2) \le 2$

        $x - 2 \le 4^2$ and $x - 2 > 0$

        $x - 2 \le 16$ and $x > 2$

          $x \le 18$ and $x > 2$

          $2 < x \le 18$

            $(2, 18]$

6. a. This graph is a line containing points $(0, -2)$ and $(1, 0)$.

$$\text{slope} = \frac{\Delta y}{\Delta x} = \frac{0 - (-2)}{1 - 0} = \frac{2}{1} = 2$$

using $y - y_1 = m(x - x_1)$

$y - 0 = 2(x - 1)$

    $y = 2x - 2$ or $2x - y - 2 = 0$

    b. This graph is a circle with center point $(2, 0)$ and radius 2.

$$(x-h)^2 + (y-k)^2 = r^2$$

$$(x-2)^2 + (y-0)^2 = 2^2$$

$$(x-2)^2 + y^2 = 4$$

    c. This graph is an ellipse with center point $(0, 0)$; vertices $(\pm 3, 0)$ and y-intercepts $(0, \pm 2)$.

$$\frac{(x-h)^2}{a^2} + \frac{(y-k)^2}{b^2} = 1$$

$$\frac{(x-0)^2}{3^2} + \frac{(y-0)^2}{2^2} = 1 \implies \frac{x^2}{9} + \frac{y^2}{4} = 1$$

    d. This graph is a parabola with vertex $(1, 0)$ and y-intercept $(0, 2)$.

$$(x-h)^2 = 4a(y-k)$$

$$(x-1)^2 = 4ay$$

$$(0-1)^2 = 4a(2)$$

$$1 = 8a$$

$$a = \frac{1}{8}$$

$$(x-1)^2 = \frac{1}{2}y \text{ or } y = 2(x-1)^2$$

    e. This graph is a hyperbola with center $(0, 0)$ and vertices $(0, \pm 1)$, containing the point $(3, 2)$.

$$\frac{(y-k)^2}{a^2} - \frac{(x-h)^2}{b^2} = 1$$

$$\frac{y^2}{1} - \frac{x^2}{b^2} = 1$$

$$\frac{(2)^2}{1} - \frac{(3)^2}{b^2} = 1$$

$$\frac{4}{1} - \frac{9}{b^2} = 1$$

$$4 - \frac{9}{b^2} = 1$$

$$3 = \frac{9}{b^2}$$

$$3b^2 = 9$$

$$b^2 = 3$$

The equation of the hyperbola is:

$$\frac{y^2}{1} - \frac{x^2}{3} = 1$$

    f. This is the graph of an exponential function with y-intercept $(0, 1)$, containing the point $(1, 4)$.

$y = A \cdot b^x$

y-intercept $(0, 1) \implies 1 = A \cdot b^0 = A \cdot 1 \implies A = 1$

point $(1, 4) \implies 4 = b^1 = b$

Therefore, $y = 4^x$.

Chapter 7 Projects

Project I – Internet-based Project

Project II

1. Figure:

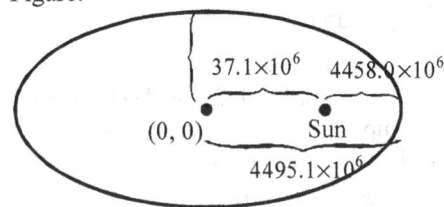

$c = 37.2 \times 10^6$

$b^2 = a^2 - c^2$

$b^2 = \left(4495.1 \times 10^6\right)^2 - \left(37.1 \times 10^6\right)^2$

$b = 4494.9 \times 10^6$

$\dfrac{x^2}{(4495.1\,x10^6)^2} + \dfrac{y^2}{(4494.9\,x10^6)^2} = 1$

2. Figure:

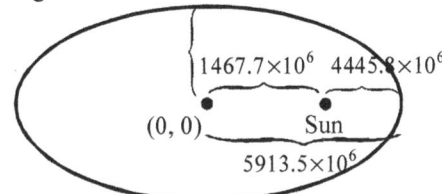

$77381.2 \times 10^6 + 4445.8 \times 10^6 = 11827 \times 10^6$

$a = 0.5\left(11827 \times 10^6\right) = 5913.5 \times 10^6$

$c = 1467.7 \times 10^6$

$b^2 = \left(5913.5 \times 10^6\right)^2 - \left(1467.7 \times 10^6\right)^2$

$b = 5728.5 \times 10^6$

$\dfrac{x^2}{(5913.5\,x10^6)^2} + \dfrac{y^2}{(5728.5\,x10^6)^2} = 1$

3. The two graphs are being graphed with the same center. Actually, the Sun should remain in the same place for each graph. This means that the graph of Pluto needs to be adjusted.

4. Shift = Pluto's distance – Neptune's distance

$= 1467.7 \times 10^6 - 37.1 \times 10^6$

$= 1430.6 \times 10^6$

$\dfrac{(x + 1430.6\,x10^6)^2}{(5913.5\,x10^6)^2} + \dfrac{y^2}{(5728.5\,x10^6)^2} = 1$

5. Yes. One must adjust the scale accordingly to see it.

6. $\left(4431.6 \times 10^6, 752.6 \times 10^6\right)$,

$\left(4431.6 \times 10^6, 752.6 \times 10^6\right)$

7. No, The timing is different. They do not both pass through those points at the same time.

Project III

1. As an example, T_1 will be used. (Note that any of the targets will yield the same result.)

$z = 4ax^2 + 4ay^2$

$0.5 = 4a(0)^2 + 4a(-2)^2$

$0.5 = 16a$

$a = \dfrac{1}{32}$

The focal length is 0.03125 m.

$z = \dfrac{1}{8}x^2 + \dfrac{1}{8}y^2$

2.

Target	x	R	θ	Z	y
T_1	0	9.551	−11.78	0.5	−1.950
T_2	0	9.948	−5.65	0.125	−0.979
T_3	0	9.928	5.90	0.125	1.021
T_4	0	9.708	11.89	0.5	2.000
T_5	9.510	9.722	−11.99	11.31	0
T_6	9.865	9.917	−5.85	12.165	0
T_7	9.875	9.925	5.78	12.189	0
T_8	9.350	9.551	11.78	10.928	0

$z = \dfrac{1}{8}x^2 + \dfrac{1}{8}y^2$, y = Rsinθ, x = Rcosθ

4. T_1 through T_4 do not need to be adjusted. T_5 must move 11.510 m toward the y-axis and the z coordinate must move down 10.81 m. T_6 must move 10.865 m toward the y-axis and the z coordinate must move down 12.04 m. T_7 must move 8.875 toward the y-axis and z must move down 12.064m. T_8 must move 7.35 m toward the y-axis and z must move down 10.425 m.

Project IV

Figure 1

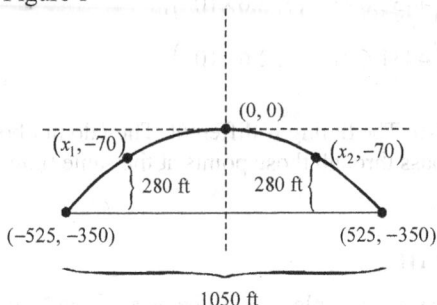

1. $x^2 = -4ay$

$(525)^2 = -4a(-350)$

$272625 = 1400a$

$a = 196.875$

$x^2 = -787.5y$

2. Let $y = -70$. (The arch needs to be 280 ft high. Remember the vertex is at $(0, 0)$, so we must measure down to the arch from the x-axis at the point where the arch's height is 280 ft.)

$x^2 = -787.5(-70)$

$x^2 = 55125 \quad \rightarrow \quad x = \pm234.8$

The channel will be 469.6 ft wide.

3. Figure 2

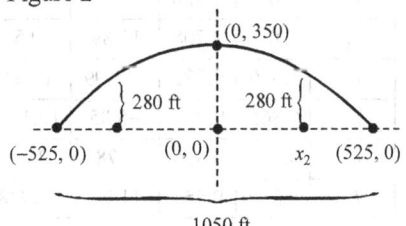

$\dfrac{x^2}{a^2} + \dfrac{y^2}{b^2} = 1$

$\dfrac{x^2}{275625} + \dfrac{y^2}{122500} = 1$

4. $\dfrac{x^2}{275625} + \dfrac{(280)^2}{122500} = 1$

$x^2 = \left(1 - \dfrac{(280)^2}{122500}\right)275625$

$x^2 = 99225$

$x = \pm315$

The channel will be 630 feet wide.

5. If the river rises 10 feet, then we need to look for how wide the channel is when the height is 290 ft.

For the parabolic shape:

$x^2 = -787.5(-60)$

$x^2 = 47250$

$x = \pm217.4$

There is still a 435 ft wide channel for the ship.

For the semi-ellipse:

$\dfrac{x^2}{275625} + \dfrac{(290)^2}{122500} = 1$

$x^2 = \left(1 - \dfrac{(290)^2}{122500}\right)275625$

$x^2 = 86400$

$x = \pm293.9$

The ship has a 588-ft channel. A semi-ellipse would be more practical since the channel doesn't shrink in width in a flood as fast as a parabola.

Chapter 8
Systems of Equations and Inequalities

Section 8.1

1. $3x + 4 = 8 - x$
$4x = 4$
$x = 1$
The solution set is $\{1\}$.

3. inconsistent

5. $(3, -2)$

7. b

9. $\begin{cases} 2x - y = 5 \\ 5x + 2y = 8 \end{cases}$

Substituting the values of the variables:
$\begin{cases} 2(2) - (-1) = 4 + 1 = 5 \\ 5(2) + 2(-1) = 10 - 2 = 8 \end{cases}$
Each equation is satisfied, so $x = 2$, $y = -1$, or
$(2, -1)$, is a solution of the system of equations.

11. $\begin{cases} 3x - 4y = 4 \\ \frac{1}{2}x - 3y = -\frac{1}{2} \end{cases}$

Substituting the values of the variables:
$\begin{cases} 3(2) - 4\left(\frac{1}{2}\right) = 6 - 2 = 4 \\ \frac{1}{2}(2) - 3\left(\frac{1}{2}\right) = 1 - \frac{3}{2} = -\frac{1}{2} \end{cases}$

Each equation is satisfied, so $x = 2$, $y = \frac{1}{2}$, or

$\left(2, \frac{1}{2}\right)$, is a solution of the system of equations.

13. $\begin{cases} x - y = 3 \\ \frac{1}{2}x + y = 3 \end{cases}$

Substituting the values of the variables, we obtain:
$\begin{cases} 4 - 1 = 3 \\ \frac{1}{2}(4) + 1 = 2 + 1 = 3 \end{cases}$
Each equation is satisfied, so $x = 4$, $y = 1$, or
$(4, 1)$, is a solution of the system of equations.

15. $\begin{cases} 3x + 3y + 2z = 4 \\ x - y - z = 0 \\ 2y - 3z = -8 \end{cases}$

Substituting the values of the variables:
$\begin{cases} 3(1) + 3(-1) + 2(2) = 3 - 3 + 4 = 4 \\ 1 - (-1) - 2 = 1 + 1 - 2 = 0 \\ 2(-1) - 3(2) = -2 - 6 = -8 \end{cases}$
Each equation is satisfied, so $x = 1$, $y = -1$, $z = 2$,
or $(1, -1, 2)$, is a solution of the system of
equations.

17. $\begin{cases} 3x + 3y + 2z = 4 \\ x - 3y + z = 10 \\ 5x - 2y - 3z = 8 \end{cases}$

Substituting the values of the variables:
$\begin{cases} 3(2) + 3(-2) + 2(2) = 6 - 6 + 4 = 4 \\ 2 - 3(-2) + 2 = 2 + 6 + 2 = 10 \\ 5(2) - 2(-2) - 3(2) = 10 + 4 - 6 = 8 \end{cases}$
Each equation is satisfied, so $x = 2$, $y = -2$,
$z = 2$, or $(2, -2, 2)$ is a solution of the system of
equations.

19. $\begin{cases} x + y = 8 \\ x - y = 4 \end{cases}$

Solve the first equation for y, substitute into the
second equation and solve:
$\begin{cases} y = 8 - x \\ x - y = 4 \end{cases}$
$x - (8 - x) = 4$
$x - 8 + x = 4$
$2x = 12$
$x = 6$
Since $x = 6$, $y = 8 - 6 = 2$. The solution of the
system is $x = 6$, $y = 2$ or using ordered pairs
$(6, 2)$.

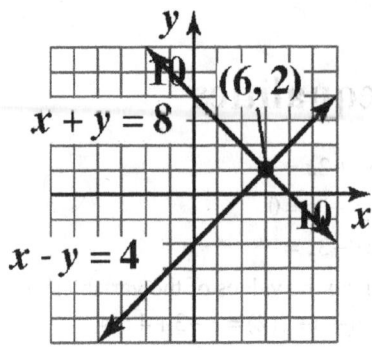

21. $\begin{cases} 5x - y = 21 \\ 2x + 3y = -12 \end{cases}$

Multiply each side of the first equation by 3 and add the equations to eliminate y:

$\begin{cases} 15x - 3y = 63 \\ 2x + 3y = -12 \end{cases}$

$\qquad 17x \quad\;\; = 51$

$\qquad\qquad x = 3$

Substitute and solve for y:

$5(3) - y = 21$

$\quad 15 - y = 21$

$\qquad -y = 6$

$\qquad\;\; y = -6$

The solution of the system is $x = 3, \; y = -6$ or using ordered an pair $(3, -6)$.

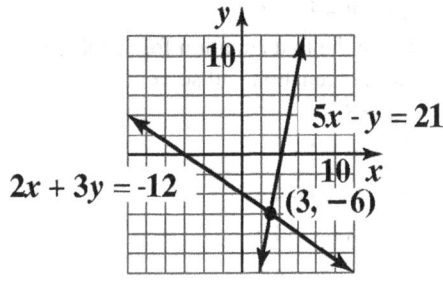

23. $\begin{cases} 3x \qquad = 24 \\ x + 2y = \;\; 0 \end{cases}$

Solve the first equation for x and substitute into the second equation:

$\begin{cases} x = 8 \\ x + 2y = \;\; 0 \end{cases}$

$8 + 2y = 0$

$\quad 2y = -8$

$\quad\; y = -4$

The solution of the system is $x = 8, \; y = -4$ or using ordered pairs $(8, -4)$

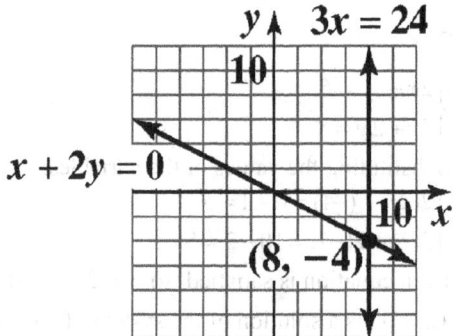

25. $\begin{cases} 3x - 6y = 2 \\ 5x + 4y = 1 \end{cases}$

Multiply each side of the first equation by 2 and each side of the second equation by 3, then add to eliminate y:

$\begin{cases} 6x - 12y = 4 \\ 15x + 12y = 3 \end{cases}$

$\quad 21x \qquad = 7$

$\qquad\qquad x = \dfrac{1}{3}$

Substitute and solve for y:

$3(1/3) - 6y = 2$

$\qquad 1 - 6y = 2$

$\qquad\; -6y = 1$

$\qquad\qquad y = -\dfrac{1}{6}$

The solution of the system is $x = \dfrac{1}{3}, \; y = -\dfrac{1}{6}$ or using ordered pairs $\left(\dfrac{1}{3}, -\dfrac{1}{6} \right)$.

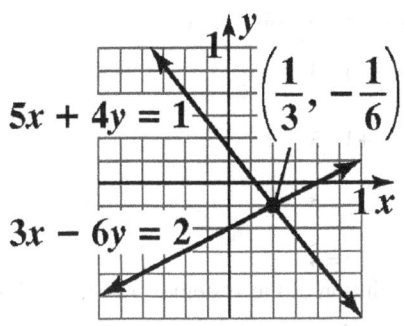

$5x + 4y = 1$

$\left(\dfrac{1}{3}, -\dfrac{1}{6}\right)$

$3x - 6y = 2$

27. $\begin{cases} 2x + y = 1 \\ 4x + 2y = 3 \end{cases}$

Solve the first equation for y, substitute into the second equation and solve:

$\begin{cases} y = 1 - 2x \\ 4x + 2y = 3 \end{cases}$

$4x + 2(1 - 2x) = 3$
$4x + 2 - 4x = 3$
$0 = 1$

This equation is false, so the system is inconsistent.

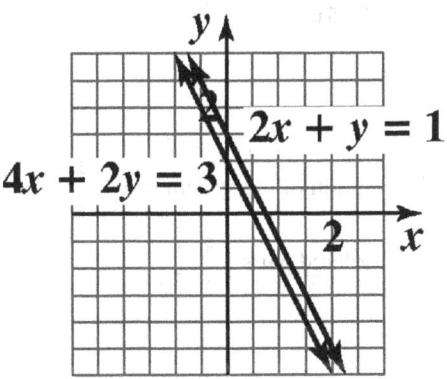

$2x + y = 1$

$4x + 2y = 3$

29. $\begin{cases} 2x - y = 0 \\ 4x + 2y = 12 \end{cases}$

Solve the first equation for y, substitute into the second equation and solve:

$\begin{cases} y = 2x \\ 4x + 2y = 12 \end{cases}$

$4x + 2(2x) = 12$
$4x + 4x = 12$
$8x = 12$
$x = \dfrac{3}{2}$

Since $x = \dfrac{3}{2}$, $y = 2\left(\dfrac{3}{2}\right) = 3$

The solution of the system is $x = \dfrac{3}{2}$, $y = 3$ or

using ordered pairs $\left(\dfrac{3}{2}, 3\right)$.

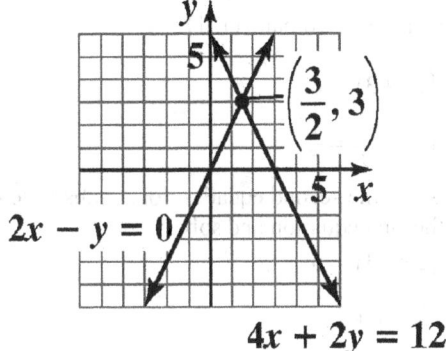

$2x - y = 0$

$\left(\dfrac{3}{2}, 3\right)$

$4x + 2y = 12$

31. $\begin{cases} x + 2y = 4 \\ 2x + 4y = 8 \end{cases}$

Solve the first equation for x, substitute into the second equation and solve:

$\begin{cases} x = 4 - 2y \\ 2x + 4y = 8 \end{cases}$

$2(4 - 2y) + 4y = 8$
$8 - 4y + 4y = 8$
$0 = 0$

These equations are dependent. The solution of the system is either $x = 4 - 2y$, where y is any real

number or $y = \dfrac{4 - x}{2}$, where x is any real number.

Using ordered pairs, we write the solution as
$\{(x, y) \mid x = 4 - 2y, y \text{ is any real number}\}$ or as

$\left\{(x, y) \mid y = \dfrac{4 - x}{2}, x \text{ is any real number}\right\}$.

33. $\begin{cases} 2x - 3y = -1 \\ 10x + y = 11 \end{cases}$

Multiply each side of the first equation by -5, and add the equations to eliminate x:

$\begin{cases} -10x + 15y = 5 \\ 10x + y = 11 \end{cases}$

$16y = 16$
$y = 1$

Substitute and solve for x:

$$2x - 3(1) = -1$$
$$2x - 3 = -1$$
$$2x = 2$$
$$x = 1$$

The solution of the system is $x = 1$, $y = 1$ or using ordered pairs (1, 1).

35. $\begin{cases} 2x + 3y = 6 \\ x - y = \dfrac{1}{2} \end{cases}$

Solve the second equation for x, substitute into the first equation and solve:

$$\begin{cases} 2x + 3y = 6 \\ x = y + \dfrac{1}{2} \end{cases}$$

$$2\left(y + \frac{1}{2}\right) + 3y = 6$$
$$2y + 1 + 3y = 6$$
$$5y = 5$$
$$y = 1$$

Since $y = 1$, $x = 1 + \dfrac{1}{2} = \dfrac{3}{2}$. The solution of the

system is $x = \dfrac{3}{2}$, $y = 1$ or using ordered pairs

$\left(\dfrac{3}{2}, 1\right)$.

37. $\begin{cases} \dfrac{1}{2}x + \dfrac{1}{3}y - 3 \\ \dfrac{1}{4}x - \dfrac{2}{3}y = -1 \end{cases}$

Multiply each side of the first equation by –6 and each side of the second equation by 12, then add to eliminate x:

$$\begin{cases} -3x - 2y = -18 \\ 3x - 8y = -12 \end{cases}$$
$$\underline{}$$
$$-10y = -30$$
$$y = 3$$

Substitute and solve for x:

$$\frac{1}{2}x + \frac{1}{3}(3) = 3$$
$$\frac{1}{2}x + 1 = 3$$
$$\frac{1}{2}x = 2$$
$$x = 4$$

The solution of the system is $x = 4$, $y = 3$ or using ordered pairs (4, 3).

39. $\begin{cases} 3x - 5y = 3 \\ 15x + 5y = 21 \end{cases}$

Add the equations to eliminate y:

$$\begin{cases} 3x - 5y = 3 \\ 15x + 5y = 21 \end{cases}$$
$$\overline{18x \qquad = 24}$$
$$x = \frac{4}{3}$$

Substitute and solve for y:

$$3(4/3) - 5y = 3$$
$$4 - 5y = 3$$
$$-5y = -1$$
$$y = \frac{1}{5}$$

The solution of the system is $x = \dfrac{4}{3}$, $y = \dfrac{1}{5}$ or

using ordered pairs $\left(\dfrac{4}{3}, \dfrac{1}{5}\right)$.

41. $\begin{cases} \dfrac{1}{x} + \dfrac{1}{y} = 8 \\ \dfrac{3}{x} - \dfrac{5}{y} = 0 \end{cases}$

Rewrite letting $u = \dfrac{1}{x}$, $v = \dfrac{1}{y}$:

$$\begin{cases} u + v = 8 \\ 3u - 5v = 0 \end{cases}$$

Solve the first equation for u, substitute into the second equation and solve:

$$\begin{cases} u = 8 - v \\ 3u - 5v = 0 \end{cases}$$

$$3(8-v)-5v=0$$
$$24-3v-5v=0$$
$$-8v=-24$$
$$v=3$$

Since $v=3$, $u=8-3=5$. Thus, $x=\dfrac{1}{u}=\dfrac{1}{5}$,

$y=\dfrac{1}{v}=\dfrac{1}{3}$. The solution of the system is

$x=\dfrac{1}{5}$, $y=\dfrac{1}{3}$ or using ordered pairs $\left(\dfrac{1}{5},\dfrac{1}{3}\right)$.

43. $\begin{cases} x-\ y=\ 6 \\ 2x-3z=16 \\ 2y+\ z=\ 4 \end{cases}$

Multiply each side of the first equation by –2 and add to the second equation to eliminate x:

$$-2x+2y\qquad =-12$$
$$\underline{\ 2x\qquad -3z=\ 16\ }$$
$$2y-3z=\ \ 4$$

Multiply each side of the result by –1 and add to the original third equation to eliminate y:

$$-2y+3z=-4$$
$$\underline{\ 2y+\ z=\ \ 4\ }$$
$$4z=\ \ 0$$
$$z=0$$

Substituting and solving for the other variables:

$$2y+0=4 \qquad\qquad 2x-3(0)=16$$
$$2y=4 \qquad\qquad\quad 2x=16$$
$$y=2 \qquad\qquad\quad\ x=8$$

The solution is $x=8$, $y=2$, $z=0$ or using ordered triples (8, 2, 0).

45. $\begin{cases} x-2y+3z=\ \ \ 7 \\ 2x+\ y+\ z=\ \ \ 4 \\ -3x+2y-2z=-10 \end{cases}$

Multiply each side of the first equation by –2 and add to the second equation to eliminate x; and multiply each side of the first equation by 3 and add to the third equation to eliminate x:

$$-2x+4y-6z=-14$$
$$\underline{\ 2x+\ y+\ z=\ \ \ 4\ }$$
$$5y-5z=-10$$

$$3x-6y+9z=\ \ 21$$
$$\underline{-3x+2y-2z=-10\ }$$
$$-4y+7z=\ \ 11$$

Multiply each side of the first result by $\dfrac{4}{5}$ and add to the second result to eliminate y:

$$4y-4z=-8$$
$$\underline{-4y+7z=11\ }$$
$$3z=3$$
$$z=1$$

Substituting and solving for the other variables:

$$y-1=-2 \qquad\qquad x-2(-1)+3(1)=7$$
$$y=-1 \qquad\qquad\qquad x+2+3=7$$
$$\qquad\qquad\qquad\qquad\qquad x=2$$

The solution is $x=2$, $y=-1$, $z=1$ or using ordered triples $(2,-1,1)$.

47. $\begin{cases} x-\ y-z=1 \\ 2x+3y+z=2 \\ 3x+2y\ \ \ \ =0 \end{cases}$

Add the first and second equations to eliminate z:

$$x-\ y-z=1$$
$$\underline{2x+3y+z=2\ }$$
$$3x+2y\ \ \ \ =3$$

Multiply each side of the result by –1 and add to the original third equation to eliminate y:

$$-3x-2y=-3$$
$$\underline{\ 3x+2y=\ 0\ }$$
$$0=-3$$

This equation is false, so the system is inconsistent.

49. $\begin{cases} x-\ y-\ z=\ \ 1 \\ -x+2y-3z=-4 \\ 3x-2y-7z=\ \ 0 \end{cases}$

Add the first and second equations to eliminate x; multiply the first equation by –3 and add to the third equation to eliminate x:

$$x-\ y-\ z=\ \ 1$$
$$\underline{-x+2y-3z=-4\ }$$
$$y-4z=-3$$

$$-3x+3y+3z=-3$$
$$\underline{\ 3x-2y-7z=\ \ 0\ }$$
$$y-4z=-3$$

Multiply each side of the first result by –1 and add to the second result to eliminate y:

$$-y+4z = 3$$
$$\underline{y-4z = -3}$$
$$0 = 0$$

The system is dependent. If z is any real number, then $y = 4z-3$.

Solving for x in terms of z in the first equation:
$$x-(4z-3)-z = 1$$
$$x-4z+3-z = 1$$
$$x-5z+3 = 1$$
$$x = 5z-2$$

The solution is $\{(x,y,z)|\ x = 5z-2,\ y = 4z-3,$ z is any real number$\}$.

51. $\begin{cases} 2x-2y+3z = 6 \\ 4x-3y+2z = 0 \\ -2x+3y-7z = 1 \end{cases}$

Multiply the first equation by –2 and add to the second equation to eliminate x; add the first and third equations to eliminate x:
$$-4x+4y-6z = -12$$
$$\underline{4x-3y+2z = \ \ 0}$$
$$y-4z = -12$$

$$2x-2y+3z = 6$$
$$\underline{-2x+3y-7z = 1}$$
$$y-4z = 7$$

Multiply each side of the first result by –1 and add to the second result to eliminate y:
$$-y+4z - 12$$
$$\underline{y-4z = 7}$$
$$0 = 19$$

This result is false, so the system is inconsistent.

53. $\begin{cases} x+\ y-\ z = \ 6 \\ 3x-2y+\ z = -5 \\ x+3y-2z = 14 \end{cases}$

Add the first and second equations to eliminate z; multiply the second equation by 2 and add to the third equation to eliminate z:
$$x+\ y-\ z = \ 6$$
$$\underline{3x-2y+\ z = -5}$$
$$4x-\ y\ \ \ \ = 1$$

$$6x-4y+2z = -10$$
$$\underline{x+3y-2z = \ 14}$$
$$7x-\ y\ \ \ \ = \ 4$$

Multiply each side of the first result by –1 and add to the second result to eliminate y:
$$-4x+y = -1$$
$$\underline{7x-y = \ 4}$$
$$3x = 3$$
$$x = 1$$

Substituting and solving for the other variables:
$$\begin{array}{ll} 4(1)-y = 1 & 3(1)-2(3)+z = -5 \\ -y = -3 & 3-6+z = -5 \\ y = 3 & z = -2 \end{array}$$

The solution is $x = 1,\ y = 3,\ z = -2$ or using ordered triplets $(1,3,-2)$.

55. $\begin{cases} x+2y-\ z = -3 \\ 2x-4y+\ z = -7 \\ -2x+2y-3z = \ 4 \end{cases}$

Add the first and second equations to eliminate z; multiply the second equation by 3 and add to the third equation to eliminate z:
$$x+2y-\ z = -3$$
$$\underline{2x-4y+\ z = -7}$$
$$3x-2y\ \ \ \ = -10$$

$$6x-12y+3z = -21$$
$$\underline{-2x+\ 2y-3z = \ \ 4}$$
$$4x-10y\ \ \ \ = -17$$

Multiply each side of the first result by –5 and add to the second result to eliminate y:
$$-15x+10y = \ 50$$
$$\underline{4x-10y = -17}$$
$$-11x\ \ \ \ = \ 33$$
$$x = -3$$

Substituting and solving for the other variables:
$$3(-3)-2y = -10$$
$$-9-2y = -10$$
$$-2y = -1$$
$$y = \frac{1}{2}$$

$$-3 + 2\left(\frac{1}{2}\right) - z = -3$$
$$-3 + 1 - z = -3$$
$$-z = -1$$
$$z = 1$$

The solution is $x = -3$, $y = \frac{1}{2}$, $z = 1$ or using ordered triplets $\left(-3, \frac{1}{2}, 1\right)$.

57. Let l be the length of the rectangle and w be the width of the rectangle. Then:
$l = 2w$ and $2l + 2w = 90$

Solve by substitution:
$$2(2w) + 2w = 90$$
$$4w + 2w = 90$$
$$6w = 90$$
$$w = 15 \text{ feet}$$
$$l = 2(15) = 30 \text{ feet}$$
The floor is 15 feet by 30 feet.

59. Let x = the number of commercial launches and y = the number of noncommercial launches.
Then: $x + y = 92$ and $y = 3x$

Solve by substitution:
$$x + (3x) = 92 \qquad y = 3(23)$$
$$4x = 92 \qquad y = 69$$
$$x = 23$$
In 2014 there were 23 commercial launches and 69 noncommercial launches.

61. Let x = the number of pounds of cashews.
Let y = is the number of pounds in the mixture.
The value of the cashews is $5x$.
The value of the peanuts is $1.50(30) = 45$.
The value of the mixture is $3y$.

Then $x + 30 = y$ represents the amount of mixture.
$5x + 45 = 3y$ represents the value of the mixture.

Solve by substitution:
$$5x + 45 = 3(x + 30)$$
$$2x = 45$$
$$x = 22.5$$
So, 22.5 pounds of cashews should be used in the mixture.

63. Let s = the price of a smartphone and t = the price of a tablet. Then:
$$s + t = 965$$
$$340s + 250t = 270500$$
Solve the first equation for t: $t = 965 - s$
Solve by substitution:
$$340s + 250(965 - s) = 270500$$
$$340s + 241250 - 250s = 270500$$
$$90s = 29250$$
$$s = 325$$
$$t = 965 - 325 = 640$$
The price of the smartphone is $325.00 and the price of the tablet is $640.00.

65. Let x = the plane's average airspeed and y = the average wind speed.

	Rate	Time	Distance
With Wind	$x + y$	3	600
Against	$x - y$	4	600

$$\begin{cases} (x + y)(3) = 600 \\ (x - y)(4) = 600 \end{cases}$$

Multiply each side of the first equation by $\frac{1}{3}$, multiply each side of the second equation by $\frac{1}{4}$, and add the result to eliminate y
$$x + y = 200$$
$$\underline{x - y = 150}$$
$$2x = 350$$
$$x = 175$$
$$175 + y = 200$$
$$y = 25$$
The average airspeed of the plane is 175 mph, and the average wind speed is 25 mph.

67. Let x = the number of $25-design.
Let y = the number of $45-design.
Then $x + y$ = the total number of sets of dishes.
$25x + 45y$ = the cost of the dishes.

Setting up the equations and solving by substitution:
$$\begin{cases} x + y = 200 \\ 25x + 45y = 7400 \end{cases}$$
Solve the first equation for y, the solve by substitution: $y = 200 - x$

$$25x + 45(200 - x) = 7400$$
$$25x + 9000 - 45x = 7400$$
$$-20x = -1600$$
$$x = 80$$
$$y = 200 - 80 = 120$$

Thus, 80 sets of the $25 dishes and 120 sets of the $45 dishes should be ordered.

69. Let x = the cost per package of bacon.
Let y = the cost of a carton of eggs.

Set up a system of equations for the problem:
$$\begin{cases} 3x + 2y = 13.45 \\ 2x + 3y = 11.45 \end{cases}$$

Multiply each side of the first equation by 3 and each side of the second equation by –2 and solve by elimination:
$$9x + 6y = 40.35$$
$$-4x - 6y = -22.90$$
$$\overline{5x = 17.45}$$
$$x = 3.49$$

Substitute and solve for y:
$$3(3.49) + 2y = 13.45$$
$$10.47 + 2y = 13.45$$
$$2y = 2.98$$
$$y = 1.49$$

A package of bacon costs $3.49 and a carton of eggs cost $1.49. The refund for 2 packages of bacon and 2 cartons of eggs will be $2($3.49) + 2($1.49) = $9.96.

71. Let x = the # of mg of compound 1.
Let y = the # of mg of compound 2.

Setting up the equations and solving by substitution:
$$\begin{cases} 0.2x + 0.4y = 40 & \text{vitamin C} \\ 0.3x + 0.2y = 30 & \text{vitamin D} \end{cases}$$

Multiplying each equation by 10 yields
$$\begin{cases} 2x + 4y = 400 \\ 6x + 4y = 600 \end{cases}$$

Subtracting the bottom equation from the top equation yields
$$2x + 4y - (6x + 4y) = 400 - 600$$
$$2x - 6x = -200$$
$$-4x = -200$$
$$x = 50$$

$$2(50) + 4y = 400$$
$$100 + 4y = 400$$
$$4y = 300$$
$$y = \frac{300}{4} = 75$$

So 50 mg of compound 1 should be mixed with 75 mg of compound 2.

73. $y = ax^2 + bx + c$

At (–1, 4) the equation becomes:
$$4 = a(-1)^2 + b(-1) + c$$
$$4 = a - b + c$$

At (2, 3) the equation becomes:
$$3 = a(2)^2 + b(2) + c$$
$$3 = 4a + 2b + c$$

At (0, 1) the equation becomes:
$$1 = a(0)^2 + b(0) + c$$
$$1 = c$$

The system of equations is:
$$\begin{cases} a - b + c = 4 \\ 4a + 2b + c = 3 \\ \phantom{4a + 2b + {}} c = 1 \end{cases}$$

Substitute $c = 1$ into the first and second equations and simplify:
$$a - b + 1 = 4 \qquad 4a + 2b + 1 = 3$$
$$a - b = 3 \qquad\quad 4a + 2b = 2$$
$$a = b + 3$$

Solve the first result for a, substitute into the second result and solve:
$$4(b + 3) + 2b = 2$$
$$4b + 12 + 2b = 2$$
$$6b = -10$$
$$b = -\frac{5}{3}$$
$$a = -\frac{5}{3} + 3 = \frac{4}{3}$$

The solution is $a = \frac{4}{3}$, $b = -\frac{5}{3}$, $c = 1$. The equation is $y = \frac{4}{3}x^2 - \frac{5}{3}x + 1$.

75. $\begin{cases} 0.06Y - 5000r = 240 \\ 0.06Y + 6000r = 900 \end{cases}$

Multiply the first equation by -1, the add the result to the second equation to eliminate Y.

$-0.06Y + 5000r = -240$

$\underline{0.06Y + 6000r = 900}$

$\qquad\quad 11000r = 660$

$\qquad\qquad\quad r = 0.06$

Substitute this result into the first equation to find Y.

$0.06Y - 5000(0.06) = 240$

$0.06Y - 300 = 240$

$0.06Y = 540$

$Y = 9000$

The equilibrium level of income and interest rates is $9000 million and 6%.

77. $\begin{cases} I_2 = I_1 + I_3 \\ 5 - 3I_1 - 5I_2 = 0 \\ 10 - 5I_2 - 7I_3 = 0 \end{cases}$

Substitute the expression for I_2 into the second and third equations and simplify:

$5 - 3I_1 - 5(I_1 + I_3) = 0$

$-8I_1 - 5I_3 = -5$

$10 - 5(I_1 + I_3) - 7I_3 = 0$

$-5I_1 - 12I_3 = -10$

Multiply both sides of the first result by 5 and multiply both sides of the second result by -8 to eliminate I_1:

$-40I_1 - 25I_3 = -25$

$\underline{40I_1 + 96I_3 = 80}$

$\qquad\quad 71I_3 = 55$

$\qquad\qquad I_3 = \dfrac{55}{71}$

Substituting and solving for the other variables:

$-8I_1 - 5\left(\dfrac{55}{71}\right) = -5$

$-8I_1 - \dfrac{275}{71} = -5$

$-8I_1 = -\dfrac{80}{71}$

$I_1 = \dfrac{10}{71}$

$I_2 = \left(\dfrac{10}{71}\right) + \dfrac{55}{71} = \dfrac{65}{71}$

The solution is $I_1 = \dfrac{10}{71}$, $I_2 = \dfrac{65}{71}$, $I_3 = \dfrac{55}{71}$.

79. Let x = the number of orchestra seats.

Let y = the number of main seats.

Let z = the number of balcony seats.

Since the total number of seats is 500, $x + y + z = 500$.

Since the total revenue is $64,250 if all seats are sold, $150x + 135y + 110z = 64,250$.

If only half of the orchestra seats are sold, the revenue is $56,750.

So, $150\,\dfrac{1}{2}x + 135y + 110z = 56,750$.

Thus, we have the following system:

$\begin{aligned} x + y + z &= 500 \\ 150x + 135y + 110z &= 64,250 \\ 75x + 135y + 110z &= 56,750 \end{aligned}$

Multiply each side of the first equation by -110 and add to the second equation to eliminate z; multiply each side of the third equation by -1 and add to the second equation to eliminate z:

$-110x - 110y - 110z = -55,000$

$\underline{150x + 135y + 110z = 64,250}$

$40x + 25y \qquad = 9250$

$150x + 135y + 110z = 64,250$

$\underline{-75x - 135y - 110z = -56,750}$

$75x \qquad\qquad = 7500$

$x = 100$

Substituting and solving for the other variables:

$40(100) + 25y = 9250 \qquad 100 + 210 + z = 500$

$4000 + 25y = 9250 \qquad\qquad 310 + z = 500$

$25y = 5250 \qquad\qquad\qquad z = 190$

$y = 210$

There are 100 orchestra seats, 210 main seats, and 190 balcony seats.

465

81. Let x = the number of servings of chicken.
Let y = the number of servings of corn.
Let z = the number of servings of 2% milk.

Protein equation: $30x + 3y + 9z = 66$
Carbohydrate equation: $35x + 16y + 13z = 94.5$
Calcium equation: $200x + 10y + 300z = 910$

Multiply each side of the first equation by -16 and multiply each side of the second equation by 3 and add them to eliminate y; multiply each side of the second equation by -5 and multiply each side of the third equation by 8 and add to eliminate y:

$$-480x - 48y - 144z = -1056$$
$$\underline{105x + 48y + 39z = 283.5}$$
$$-375x \qquad -105z = -772.5$$

$$-175x - 80y - 65z = -472.5$$
$$\underline{1600x + 80y + 2400z = 7280}$$
$$1425x \qquad +2335z = 6807.5$$

Multiply each side of the first result by 19 and multiply each side of the second result by 5 to eliminate x:

$$-7125x - 1995z = -14,677.5$$
$$\underline{7125x + 11,675z = 34,037.5}$$
$$9680z = 19,360$$
$$z = 2$$

Substituting and solving for the other variables:
$$-375x - 105(2) = -772.5$$
$$-375x - 210 = -772.5$$
$$-375x = -562.5$$
$$x = 1.5$$

$$30(1.5) + 3y + 9(2) = 66$$
$$45 + 3y + 18 = 66$$
$$3y = 3$$
$$y = 1$$

The dietitian should serve 1.5 servings of chicken, 1 serving of corn, and 2 servings of 2% milk.

83. Let x = the price of 1 hamburger.
Let y = the price of 1 order of fries.
Let z = the price of 1 drink.

We can construct the system
$$8x + 6y + 6z = 52.20$$
$$10x + 6y + 8z = 63.20$$

A system involving only 2 equations that contain 3 or more unknowns cannot be solved uniquely.

Multiply the first equation by $-\dfrac{1}{2}$ and the second equation by $\dfrac{1}{2}$, then add to eliminate y:

$$-4x - 3y - 3z = -26.10$$
$$\underline{5x + 3y + 4z = 31.60}$$
$$x \qquad + z = 5.50$$
$$x = 5.50 - z$$

Substitute and solve for y in terms of z:
$$5(5.50 - z) + 3y + 4z = 31.60$$
$$27.50 + 3y - z = 31.60$$
$$3y = z + 4.10$$
$$y = \frac{1}{3}z + \frac{4.10}{3}$$

Solutions of the system are: $x = 5.50 - z$,
$y = \dfrac{1}{3}z + \dfrac{4.10}{3}$.

Since we are given that $1.20 \le z \le 1.80$, we choose values of z that give two-decimal-place values of x and y with $3.50 \le x \le 4.50$ and $1.50 \le y \le 2.00$.

The possible values of x, y, and z are shown in the table.

x	y	z
4.23	1.78	1.24
4.02	1.86	1.48
3.72	1.96	1.78

85. Let x = Beth's time working alone.
Let y = Bill's time working alone.
Let z = Edie's time working alone.

We can use the following tables to organize our work:

	Beth	Bill	Edie
Hours to do job	x	y	z
Part of job done in 1 hour	$\dfrac{1}{x}$	$\dfrac{1}{y}$	$\dfrac{1}{z}$

In 10 hours they complete 1 entire job, so

$$10\left(\frac{1}{x}+\frac{1}{y}+\frac{1}{z}\right)=1$$

$$\frac{1}{x}+\frac{1}{y}+\frac{1}{z}=\frac{1}{10}$$

	Bill	Edie
Hours to do job	y	z
Part of job done in 1 hour	$\dfrac{1}{y}$	$\dfrac{1}{z}$

In 15 hours they complete 1 entire job, so

$$15\left(\frac{1}{y}+\frac{1}{z}\right)=1.$$

$$\frac{1}{y}+\frac{1}{z}=\frac{1}{15}$$

	Beth	Bill	Edie
Hours to do job	x	y	z
Part of job done in 1 hour	$\dfrac{1}{x}$	$\dfrac{1}{y}$	$\dfrac{1}{z}$

With all 3 working for 4 hours and Beth and Bill working for an additional 8 hours, they complete

1 entire job, so $4\left(\dfrac{1}{x}+\dfrac{1}{y}+\dfrac{1}{z}\right)+8\left(\dfrac{1}{x}+\dfrac{1}{y}\right)=1$

$$\frac{12}{x}+\frac{12}{y}+\frac{4}{z}=1$$

We have the system

$$\begin{cases} \dfrac{1}{x}+\dfrac{1}{y}+\dfrac{1}{z}=\dfrac{1}{10} \\ \dfrac{1}{y}+\dfrac{1}{z}=\dfrac{1}{15} \\ \dfrac{12}{x}+\dfrac{12}{y}+\dfrac{4}{z}=1 \end{cases}$$

Subtract the second equation from the first equation:

$$\frac{1}{x}+\frac{1}{y}+\frac{1}{z}=\frac{1}{10}$$
$$\frac{1}{y}+\frac{1}{z}=\frac{1}{15}$$
$$\overline{\hspace{0.5cm}}$$
$$\frac{1}{x}\hspace{1.2cm}=\frac{1}{30}$$
$$x=30$$

Substitute $x = 30$ into the third equation:

$$\frac{12}{30}+\frac{12}{y}+\frac{4}{z}=1$$

$$\frac{12}{y}+\frac{4}{z}=\frac{3}{5}$$

Now consider the system consisting of the last result and the second original equation. Multiply the second original equation by -12 and add it to the last result to eliminate y:

$$\frac{-12}{y}+\frac{-12}{z}=\frac{-12}{15}$$
$$\frac{12}{y}+\frac{4}{z}=\frac{3}{5}$$
$$\overline{\hspace{0.5cm}}$$
$$-\frac{8}{z}=-\frac{3}{15}$$
$$z=40$$

Plugging $z = 40$ to find y:

$$\frac{12}{y}+\frac{4}{z}=\frac{3}{5}$$

$$\frac{12}{y}+\frac{4}{40}=\frac{3}{5}$$

$$\frac{12}{y}=\frac{1}{2}$$

$$y=24$$

Working alone, it would take Beth 30 hours, Bill 24 hours, and Edie 40 hours to complete the job.

87. Answers will vary.

89.

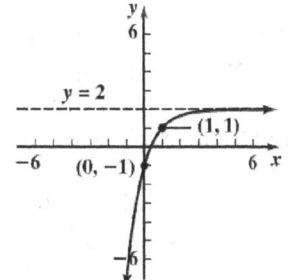

Copyright © 2017 Pearson Education, Inc.

91. $|3x-2|+5 \le 9$

$|3x-2| \le 4$

$-\dfrac{2}{3} \le 3x \le 2$

So the solution is: $-\dfrac{2}{3}, 2$

Section 8.2

1. matrix

3. third; fifth

5. b

7. Writing the augmented matrix for the system of equations:

$$\begin{cases} x-5y=5 \\ 4x+3y=6 \end{cases} \rightarrow \begin{bmatrix} 1 & -5 & | & 5 \\ 4 & 3 & | & 6 \end{bmatrix}$$

9. $\begin{cases} 2x+3y-6=0 \\ 4x-6y+2=0 \end{cases}$

Write the system in standard form and then write the augmented matrix for the system of equations:

$$\begin{cases} 2x+3y=6 \\ 4x-6y=-2 \end{cases} \rightarrow \begin{bmatrix} 2 & 3 & | & 6 \\ 4 & -6 & | & -2 \end{bmatrix}$$

11. Writing the augmented matrix for the system of equations:

$$\begin{cases} 0.01x-0.03y=0.06 \\ 0.13x+0.10y=0.20 \end{cases} \rightarrow \begin{bmatrix} 0.01 & -0.03 & | & 0.06 \\ 0.13 & 0.10 & | & 0.20 \end{bmatrix}$$

13. Writing the augmented matrix for the system of equations:

$$\begin{cases} x-y+z=10 \\ 3x+3y =5 \\ x+y+2z=2 \end{cases} \rightarrow \begin{bmatrix} 1 & -1 & 1 & | & 10 \\ 3 & 3 & 0 & | & 5 \\ 1 & 1 & 2 & | & 2 \end{bmatrix}$$

15. Writing the augmented matrix for the system of equations:

$$\begin{cases} x+y-z=2 \\ 3x-2y =2 \\ 5x+3y-z=1 \end{cases} \rightarrow \begin{bmatrix} 1 & 1 & -1 & | & 2 \\ 3 & -2 & 0 & | & 2 \\ 5 & 3 & -1 & | & 1 \end{bmatrix}$$

17. Writing the augmented matrix for the system of equations:

$$\begin{cases} x-y-z=10 \\ 2x+y+2z=-1 \\ -3x+4y=5 \\ 4x-5y+z=0 \end{cases} \rightarrow \begin{bmatrix} 1 & -1 & -1 & | & 10 \\ 2 & 1 & 2 & | & -1 \\ -3 & 4 & 0 & | & 5 \\ 4 & -5 & 1 & | & 0 \end{bmatrix}$$

19. $\begin{bmatrix} 1 & -3 & | & -2 \\ 2 & -5 & | & 5 \end{bmatrix} \rightarrow \begin{cases} x-3y=-2 \\ 2x-5y=5 \end{cases}$

$R_2 = -2r_1 + r_2$

$$\begin{bmatrix} 1 & -3 & | & -2 \\ 2 & -5 & | & 5 \end{bmatrix} \rightarrow \begin{bmatrix} 1 & -3 & | & -2 \\ -2(1)+2 & 2(-3)-5 & | & -2(-2)+5 \end{bmatrix}$$

$$\rightarrow \begin{bmatrix} 1 & -3 & | & -2 \\ 0 & 1 & | & 9 \end{bmatrix}$$

21. $\begin{bmatrix} 1 & -3 & 4 & | & 3 \\ 3 & -5 & 6 & | & 6 \\ -5 & 3 & 4 & | & 6 \end{bmatrix} \rightarrow \begin{cases} x-3y+4z=3 \\ 3x-5y+6z=6 \\ -5x+3y+4z=6 \end{cases}$

$R_2 = -3r_1 + r_2$

$$\begin{bmatrix} 1 & -3 & 4 & | & 3 \\ 3 & -5 & 6 & | & 6 \\ -5 & 3 & 4 & | & 6 \end{bmatrix}$$

$$\rightarrow \begin{bmatrix} 1 & -3 & 4 & | & 3 \\ -3(1)+3 & -3(-3)-5 & -3(4)+6 & | & -3(3)+6 \\ -5 & 3 & 4 & | & 6 \end{bmatrix}$$

$$\rightarrow \begin{bmatrix} 1 & -3 & 4 & | & 3 \\ 0 & 4 & -6 & | & -3 \\ -5 & 3 & 4 & | & 6 \end{bmatrix}$$

$R_3 = 5r_1 + r_3$

$$\begin{bmatrix} 1 & -3 & 4 & | & 3 \\ 0 & 4 & -6 & | & -3 \\ -5 & 3 & 4 & | & 6 \end{bmatrix}$$

$$\rightarrow \begin{bmatrix} 1 & -3 & 4 & | & 3 \\ 0 & 4 & -6 & | & -3 \\ 5(1)-5 & 5(-3)+3 & 5(4)+4 & | & 5(3)+6 \end{bmatrix}$$

$$\rightarrow \begin{bmatrix} 1 & -3 & 4 & | & 3 \\ 0 & 4 & -6 & | & -3 \\ 0 & -12 & 24 & | & 21 \end{bmatrix}$$

23. $\begin{bmatrix} 1 & -3 & 2 & | & -6 \\ 2 & -5 & 3 & | & -4 \\ -3 & -6 & 4 & | & 6 \end{bmatrix} \rightarrow \begin{cases} x - 3y + 2z = -6 \\ 2x - 5y + 3z = -4 \\ -3x - 6y + 4z = 6 \end{cases}$

$R_2 = -2r_1 + r_2$

$\begin{bmatrix} 1 & -3 & 2 & | & -6 \\ 2 & -5 & 3 & | & -4 \\ -3 & -6 & 4 & | & 6 \end{bmatrix}$

$\rightarrow \begin{bmatrix} 1 & -3 & 2 & | & -6 \\ -2(1)+2 & -2(-3)-5 & -2(2)+3 & | & -2(-6)-4 \\ -3 & -6 & 4 & | & 6 \end{bmatrix}$

$\rightarrow \begin{bmatrix} 1 & -3 & 2 & | & -6 \\ 0 & 1 & -1 & | & 8 \\ -3 & -6 & 4 & | & 6 \end{bmatrix}$

$R_3 = 3r_1 + r_3$

$\begin{bmatrix} 1 & -3 & 2 & | & -6 \\ 0 & 1 & -1 & | & 8 \\ -3 & -6 & 4 & | & 6 \end{bmatrix}$

$\rightarrow \begin{bmatrix} 1 & -3 & 2 & | & -6 \\ 0 & 1 & -1 & | & 8 \\ 3(1)-3 & 3(-3)-6 & 3(2)+4 & | & 3(-6)+6 \end{bmatrix}$

$\rightarrow \begin{bmatrix} 1 & -3 & 2 & | & -6 \\ 0 & 1 & -1 & | & 8 \\ 0 & -15 & 10 & | & -12 \end{bmatrix}$

25. $\begin{bmatrix} 5 & -3 & 1 & | & -2 \\ 2 & -5 & 6 & | & -2 \\ -4 & 1 & 4 & | & 6 \end{bmatrix} \rightarrow \begin{cases} 5x - 3y + z = -2 \\ 2x - 5y + 6z = -2 \\ -4x + y + 4z = 6 \end{cases}$

$R_1 = -2r_2 + r_1$

$\begin{bmatrix} 5 & -3 & 1 & | & -2 \\ 2 & -5 & 6 & | & -2 \\ -4 & 1 & 4 & | & 6 \end{bmatrix}$

$\rightarrow \begin{bmatrix} -2(2)+5 & -2(-5)-3 & -2(6)+1 & | & -2(-2)-2 \\ 2 & -5 & 6 & | & -2 \\ -4 & 1 & 4 & | & 6 \end{bmatrix}$

$\rightarrow \begin{bmatrix} 1 & 7 & -11 & | & 2 \\ 2 & -5 & 6 & | & -2 \\ -4 & 1 & 4 & | & 6 \end{bmatrix}$

$R_3 = 2r_2 + r_3$

$\begin{bmatrix} 1 & 7 & -11 & | & 2 \\ 2 & -5 & 6 & | & -2 \\ -4 & 1 & 4 & | & 6 \end{bmatrix}$

$\rightarrow \begin{bmatrix} 1 & 7 & -11 & | & 2 \\ 2 & -5 & 6 & | & -2 \\ 2(2)+(-4) & 2(-5)+1 & 2(6)+4 & | & 2(-2)+6 \end{bmatrix}$

$\rightarrow \begin{bmatrix} 1 & 7 & -11 & | & 2 \\ 2 & -5 & 6 & | & -2 \\ 0 & -9 & 16 & | & 2 \end{bmatrix}$

27. $\begin{cases} x = 5 \\ y = -1 \end{cases}$

Consistent; $x = 5$, $y = -1$, or using ordered pairs $(5, -1)$.

29. $\begin{cases} x = 1 \\ y = 2 \\ 0 = 3 \end{cases}$

Inconsistent

31. $\begin{cases} x + 2z = -1 \\ y - 4z = -2 \\ \quad\ 0 = 0 \end{cases}$

Consistent;

$\begin{cases} x = -1 - 2z \\ y = -2 + 4z \\ z \text{ is any real number} \end{cases}$

or $\{(x, y, z) \,|\, x = -1 - 2z,\ y = -2 + 4z,\ z \text{ is any real number}\}$

33. $\begin{cases} x_1 = 1 \\ x_2 + x_4 = 2 \\ x_3 + 2x_4 = 3 \end{cases}$

Consistent;

$\begin{cases} x_1 = 1 \\ x_2 = 2 - x_4 \\ x_3 = 3 - 2x_4 \\ x_4 \text{ is any real number} \end{cases}$

or $\{(x_1, x_2, x_3, x_4) \,|\, x_1 = 1,\ x_2 = 2 - x_4,\ x_3 = 3 - 2x_4, x_4 \text{ is any real number}\}$

35. $\begin{cases} x_1 + 4x_4 = 2 \\ x_2 + x_3 + 3x_4 = 3 \\ 0 = 0 \end{cases}$

Consistent;

$\begin{cases} x_1 = 2 - 4x_4 \\ x_2 = 3 - x_3 - 3x_4 \\ x_3, \ x_4 \text{ are any real numbers} \end{cases}$

or $\{(x_1, x_2, x_3, x_4) \mid x_1 = 2 - 4x_4, \ x_2 = 3 - x_3 - 3x_4,$

$x_3 \text{ and } x_4 \text{ are any real numbers}\}$

37. $\begin{cases} x_1 + x_4 = -2 \\ x_2 + 2x_4 = 2 \\ x_3 - x_4 = 0 \\ 0 = 0 \end{cases}$

Consistent;

$\begin{cases} x_1 = -2 - x_4 \\ x_2 = 2 - 2x_4 \\ x_3 = x_4 \\ x_4 \text{ is any real number} \end{cases}$

or $\{(x_1, x_2, x_3, x_4) \mid x_1 = -2 - x_4, x_2 = 2 - 2x_4,$

$x_3 = x_4, \ x_4 \text{ is any real number}\}$

39. $\begin{cases} x + y = 8 \\ x - y = 4 \end{cases}$

Write the augmented matrix:

$\begin{bmatrix} 1 & 1 & | & 8 \\ 1 & -1 & | & 4 \end{bmatrix} \rightarrow \begin{bmatrix} 1 & 1 & | & 8 \\ 0 & -2 & | & -4 \end{bmatrix} \ (R_2 = -r_1 + r_2)$

$\rightarrow \begin{bmatrix} 1 & 1 & | & 8 \\ 0 & 1 & | & 2 \end{bmatrix} \quad (R_2 = -\tfrac{1}{2}r_2)$

$\rightarrow \begin{bmatrix} 1 & 0 & | & 6 \\ 0 & 1 & | & 2 \end{bmatrix} \quad (R_1 = -r_2 + r_1)$

The solution is $x = 6$, $y = 2$ or using ordered pairs (6, 2).

41. $\begin{cases} 2x - 4y = -2 \\ 3x + 2y = 3 \end{cases}$

Write the augmented matrix:

$\begin{bmatrix} 2 & -4 & | & -2 \\ 3 & 2 & | & 3 \end{bmatrix} \rightarrow \begin{bmatrix} 1 & -2 & | & -1 \\ 3 & 2 & | & 3 \end{bmatrix} \ \left(R_1 = \tfrac{1}{2}r_1\right)$

$\rightarrow \begin{bmatrix} 1 & -2 & | & -1 \\ 0 & 8 & | & 6 \end{bmatrix} \quad (R_2 = -3r_1 + r_2)$

$\rightarrow \begin{bmatrix} 1 & -2 & | & -1 \\ 0 & 1 & | & \frac{3}{4} \end{bmatrix} \quad \left(R_2 = \tfrac{1}{8}r_2\right)$

$\rightarrow \begin{bmatrix} 1 & 0 & | & \frac{1}{2} \\ 0 & 1 & | & \frac{3}{4} \end{bmatrix} \quad (R_1 = 2r_2 + r_1)$

The solution is $x = \dfrac{1}{2}, \ y = \dfrac{3}{4}$ or using ordered pairs $\left(\dfrac{1}{2}, \dfrac{3}{4}\right)$.

43. $\begin{cases} x + 2y = 4 \\ 2x + 4y = 8 \end{cases}$

Write the augmented matrix:

$\begin{bmatrix} 1 & 2 & | & 4 \\ 2 & 4 & | & 8 \end{bmatrix} \rightarrow \begin{bmatrix} 1 & 2 & | & 4 \\ 0 & 0 & | & 0 \end{bmatrix} \ (R_2 = -2r_1 + r_2)$

This is a dependent system.

$x + 2y = 4$

$x = 4 - 2y$

The solution is $x = 4 - 2y$, y is any real number or $\{(x, y) \mid x = 4 - 2y, y \text{ is any real number}\}$

45. $\begin{cases} 2x + 3y = 6 \\ x - y = \dfrac{1}{2} \end{cases}$

Write the augmented matrix:

$\begin{bmatrix} 2 & 3 & | & 6 \\ 1 & -1 & | & \frac{1}{2} \end{bmatrix} \rightarrow \begin{bmatrix} 1 & \frac{3}{2} & | & 3 \\ 1 & -1 & | & \frac{1}{2} \end{bmatrix} \ \left(R_1 = \tfrac{1}{2}r_1\right)$

$\rightarrow \begin{bmatrix} 1 & \frac{3}{2} & | & 3 \\ 0 & -\frac{5}{2} & | & -\frac{5}{2} \end{bmatrix} \quad (R_2 = -r_1 + r_2)$

$\rightarrow \begin{bmatrix} 1 & \frac{3}{2} & | & 3 \\ 0 & 1 & | & 1 \end{bmatrix} \quad \left(R_2 = -\tfrac{2}{5}r_2\right)$

$\rightarrow \begin{bmatrix} 1 & 0 & | & \frac{3}{2} \\ 0 & 1 & | & 1 \end{bmatrix} \quad \left(R_1 = -\tfrac{3}{2}r_2 + r_1\right)$

The solution is $x = \dfrac{3}{2}, \ y = 1$ or $\left(\dfrac{3}{2}, 1\right)$.

47. $\begin{cases} 3x - 5y = 3 \\ 15x + 5y = 21 \end{cases}$

Write the augmented matrix:

$\begin{bmatrix} 3 & -5 & | & 3 \\ 15 & 5 & | & 21 \end{bmatrix} \rightarrow \begin{bmatrix} 1 & -\frac{5}{3} & | & 1 \\ 15 & 5 & | & 21 \end{bmatrix} \quad \left(R_1 = \frac{1}{3} r_1 \right)$

$\rightarrow \begin{bmatrix} 1 & -\frac{5}{3} & | & 1 \\ 0 & 30 & | & 6 \end{bmatrix} \quad \left(R_2 = -15 r_1 + r_2 \right)$

$\rightarrow \begin{bmatrix} 1 & -\frac{5}{3} & | & 1 \\ 0 & 1 & | & \frac{1}{5} \end{bmatrix} \quad \left(R_2 = \frac{1}{30} r_2 \right)$

$\rightarrow \begin{bmatrix} 1 & 0 & | & \frac{4}{3} \\ 0 & 1 & | & \frac{1}{5} \end{bmatrix} \quad \left(R_1 = \frac{5}{3} r_2 + r_1 \right)$

The solution is $x = \dfrac{4}{3}, \; y = \dfrac{1}{5}$ or $\left(\dfrac{4}{3}, \dfrac{1}{5} \right)$.

49. $\begin{cases} x - y = 6 \\ 2x - 3z = 16 \\ 2y + z = 4 \end{cases}$

Write the augmented matrix:

$\begin{bmatrix} 1 & -1 & 0 & | & 6 \\ 2 & 0 & -3 & | & 16 \\ 0 & 2 & 1 & | & 4 \end{bmatrix}$

$\rightarrow \begin{bmatrix} 1 & -1 & 0 & | & 6 \\ 0 & 2 & -3 & | & 4 \\ 0 & 2 & 1 & | & 4 \end{bmatrix} \quad \left(R_2 = -2 r_1 + r_2 \right)$

$\rightarrow \begin{bmatrix} 1 & -1 & 0 & | & 6 \\ 0 & 1 & -\frac{3}{2} & | & 2 \\ 0 & 2 & 1 & | & 4 \end{bmatrix} \quad \left(R_2 = \frac{1}{2} r_2 \right)$

$\rightarrow \begin{bmatrix} 1 & 0 & -\frac{3}{2} & | & 8 \\ 0 & 1 & -\frac{3}{2} & | & 2 \\ 0 & 0 & 4 & | & 0 \end{bmatrix} \quad \begin{pmatrix} R_1 = r_2 + r_1 \\ R_3 = -2 r_2 + r_3 \end{pmatrix}$

$\rightarrow \begin{bmatrix} 1 & 0 & -\frac{3}{2} & | & 8 \\ 0 & 1 & -\frac{3}{2} & | & 2 \\ 0 & 0 & 1 & | & 0 \end{bmatrix} \quad \left(R_3 = \frac{1}{4} r_3 \right)$

$\rightarrow \begin{bmatrix} 1 & 0 & 0 & | & 8 \\ 0 & 1 & 0 & | & 2 \\ 0 & 0 & 1 & | & 0 \end{bmatrix} \quad \begin{pmatrix} R_1 = \frac{3}{2} r_3 + r_1 \\ R_2 = \frac{3}{2} r_3 + r_2 \end{pmatrix}$

The solution is $x = 8, \; y = 2, \; z = 0$ or $(8, 2, 0)$.

51. $\begin{cases} x - 2y + 3z = 7 \\ 2x + y + z = 4 \\ -3x + 2y - 2z = -10 \end{cases}$

Write the augmented matrix:

$\begin{bmatrix} 1 & -2 & 3 & | & 7 \\ 2 & 1 & 1 & | & 4 \\ -3 & 2 & -2 & | & -10 \end{bmatrix}$

$\rightarrow \begin{bmatrix} 1 & -2 & 3 & | & 7 \\ 0 & 5 & -5 & | & -10 \\ 0 & -4 & 7 & | & 11 \end{bmatrix} \quad \begin{pmatrix} R_2 = -2 r_1 + r_2 \\ R_3 = 3 r_1 + r_3 \end{pmatrix}$

$\rightarrow \begin{bmatrix} 1 & -2 & 3 & | & 7 \\ 0 & 1 & -1 & | & -2 \\ 0 & -4 & 7 & | & 11 \end{bmatrix} \quad \left(R_2 = \frac{1}{5} r_2 \right)$

$\rightarrow \begin{bmatrix} 1 & 0 & 1 & | & 3 \\ 0 & 1 & -1 & | & -2 \\ 0 & 0 & 3 & | & 3 \end{bmatrix} \quad \begin{pmatrix} R_1 = 2 r_2 + r_1 \\ R_3 = 4 r_2 + r_3 \end{pmatrix}$

$\rightarrow \begin{bmatrix} 1 & 0 & 1 & | & 3 \\ 0 & 1 & -1 & | & -2 \\ 0 & 0 & 1 & | & 1 \end{bmatrix} \quad \left(R_3 = \frac{1}{3} r_3 \right)$

$\rightarrow \begin{bmatrix} 1 & 0 & 0 & | & 2 \\ 0 & 1 & 0 & | & -1 \\ 0 & 0 & 1 & | & 1 \end{bmatrix} \quad \begin{pmatrix} R_1 = -r_3 + r_1 \\ R_2 = r_3 + r_2 \end{pmatrix}$

The solution is $x = 2, \; y = -1, \; z = 1$ or $(2, -1, 1)$.

53. $\begin{cases} 2x - 2y - 2z = 2 \\ 2x + 3y + z = 2 \\ 3x + 2y = 0 \end{cases}$

Write the augmented matrix:

$$\begin{bmatrix} 2 & -2 & -2 & | & 2 \\ 2 & 3 & 1 & | & 2 \\ 3 & 2 & 0 & | & 0 \end{bmatrix}$$

$$\rightarrow \begin{bmatrix} 1 & -1 & -1 & | & 1 \\ 2 & 3 & 1 & | & 2 \\ 3 & 2 & 0 & | & 0 \end{bmatrix} \quad \left(R_1 = \tfrac{1}{2} r_1 \right)$$

$$\rightarrow \begin{bmatrix} 1 & -1 & -1 & | & 1 \\ 0 & 5 & 3 & | & 0 \\ 0 & 5 & 3 & | & -3 \end{bmatrix} \quad \begin{pmatrix} R_2 = -2r_1 + r_2 \\ R_3 = -3r_1 + r_3 \end{pmatrix}$$

$$\rightarrow \begin{bmatrix} 1 & -1 & -1 & | & 1 \\ 0 & 5 & 3 & | & 0 \\ 0 & 0 & 0 & | & -3 \end{bmatrix} \quad \left(R_3 = -r_2 + r_3 \right)$$

There is no solution. The system is inconsistent.

55. $\begin{cases} -x + y + z = -1 \\ -x + 2y - 3z = -4 \\ 3x - 2y - 7z = 0 \end{cases}$

Write the augmented matrix:

$$\begin{bmatrix} -1 & 1 & 1 & | & -1 \\ -1 & 2 & -3 & | & -4 \\ 3 & -2 & -7 & | & 0 \end{bmatrix}$$

$$\rightarrow \begin{bmatrix} 1 & -1 & -1 & | & 1 \\ -1 & 2 & -3 & | & -4 \\ 3 & -2 & -7 & | & 0 \end{bmatrix} \quad \left(R_1 = -r_1 \right)$$

$$\rightarrow \begin{bmatrix} 1 & -1 & -1 & | & 1 \\ 0 & 1 & -4 & | & -3 \\ 0 & 1 & -4 & | & -3 \end{bmatrix} \quad \begin{pmatrix} R_2 = r_1 + r_2 \\ R_3 = -3r_1 + r_3 \end{pmatrix}$$

$$\rightarrow \begin{bmatrix} 1 & 0 & -5 & | & -2 \\ 0 & 1 & -4 & | & -3 \\ 0 & 0 & 0 & | & 0 \end{bmatrix} \quad \begin{pmatrix} R_1 = r_2 + r_1 \\ R_3 = -r_2 + r_3 \end{pmatrix}$$

The matrix in the last step represents the system
$\begin{cases} x - 5z = -2 \\ y - 4z = -3 \\ 0 = 0 \end{cases}$ or, equivalently, $\begin{cases} x = 5z - 2 \\ y = 4z - 3 \\ 0 = 0 \end{cases}$

The solution is $x = 5z - 2$, $y = 4z - 3$, z is any real number or $\{(x, y, z) \mid x = 5z - 2, y = 4z - 3, z$ is any real number$\}$.

57. $\begin{cases} 2x - 2y + 3z = 6 \\ 4x - 3y + 2z = 0 \\ -2x + 3y - 7z = 1 \end{cases}$

Write the augmented matrix:

$$\begin{bmatrix} 2 & -2 & 3 & | & 6 \\ 4 & -3 & 2 & | & 0 \\ -2 & 3 & -7 & | & 1 \end{bmatrix}$$

$$\rightarrow \begin{bmatrix} 1 & -1 & \frac{3}{2} & 3 \\ 4 & -3 & 2 & 0 \\ -2 & 3 & -7 & 1 \end{bmatrix} \quad \left(R_1 = \tfrac{1}{2} r_1 \right)$$

$$\rightarrow \begin{bmatrix} 1 & -1 & \frac{3}{2} & 3 \\ 0 & 1 & -4 & -12 \\ 0 & 1 & -4 & 7 \end{bmatrix} \quad \begin{pmatrix} R_2 = -4r_1 + r_2 \\ R_3 = 2r_1 + r_3 \end{pmatrix}$$

$$\rightarrow \begin{bmatrix} 1 & 0 & -\frac{5}{2} & -9 \\ 0 & 1 & -4 & -12 \\ 0 & 0 & 0 & 19 \end{bmatrix} \quad \begin{pmatrix} R_1 = r_2 + r_1 \\ R_3 = -r_2 + r_3 \end{pmatrix}$$

There is no solution. The system is inconsistent.

59. $\begin{cases} x + y - z = 6 \\ 3x - 2y + z = -5 \\ x + 3y - 2z = 14 \end{cases}$

Write the augmented matrix:

$$\begin{bmatrix} 1 & 1 & -1 & 6 \\ 3 & -2 & 1 & -5 \\ 1 & 3 & -2 & 14 \end{bmatrix}$$

$$\rightarrow \begin{bmatrix} 1 & 1 & -1 & 6 \\ 0 & -5 & 4 & -23 \\ 0 & 2 & -1 & 8 \end{bmatrix} \quad \begin{pmatrix} R_2 = -3r_1 + r_2 \\ R_3 = -r_1 + r_3 \end{pmatrix}$$

$$\rightarrow \begin{bmatrix} 1 & 1 & -1 & 6 \\ 0 & 1 & -\frac{4}{5} & \frac{23}{5} \\ 0 & 2 & -1 & 8 \end{bmatrix} \quad \left(R_2 = -\tfrac{1}{5} r_2 \right)$$

$$\rightarrow \begin{bmatrix} 1 & 0 & -\frac{1}{5} & \frac{7}{5} \\ 0 & 1 & -\frac{4}{5} & \frac{23}{5} \\ 0 & 0 & \frac{3}{5} & -\frac{6}{5} \end{bmatrix} \quad \begin{pmatrix} R_1 = -r_2 + r_1 \\ R_3 = -2r_2 + r_3 \end{pmatrix}$$

$$\rightarrow \begin{bmatrix} 1 & 0 & -\frac{1}{5} & \frac{7}{5} \\ 0 & 1 & -\frac{4}{5} & \frac{23}{5} \\ 0 & 0 & 1 & -2 \end{bmatrix} \quad \left(R_3 = \tfrac{5}{3} r_3 \right)$$

$$\rightarrow \begin{bmatrix} 1 & 0 & 0 & 1 \\ 0 & 1 & 0 & 3 \\ 0 & 0 & 1 & -2 \end{bmatrix} \quad \begin{pmatrix} R_1 = \tfrac{1}{5} r_3 + r_1 \\ R_2 = \tfrac{4}{5} r_3 + r_2 \end{pmatrix}$$

The solution is $x = 1, y = 3, z = -2$, or $(1, 3, -2)$.

61. $\begin{cases} x + 2y - z = -3 \\ 2x - 4y + z = -7 \\ -2x + 2y - 3z = 4 \end{cases}$

Write the augmented matrix:

$$\begin{bmatrix} 1 & 2 & -1 & -3 \\ 2 & -4 & 1 & -7 \\ -2 & 2 & -3 & 4 \end{bmatrix}$$

$$\rightarrow \begin{bmatrix} 1 & 2 & -1 & -3 \\ 0 & -8 & 3 & -1 \\ 0 & 6 & -5 & -2 \end{bmatrix} \quad \begin{pmatrix} R_2 = -2r_1 + r_2 \\ R_3 = 2r_1 + r_3 \end{pmatrix}$$

$$\rightarrow \begin{bmatrix} 1 & 2 & -1 & -3 \\ 0 & 1 & -\frac{3}{8} & \frac{1}{8} \\ 0 & 6 & -5 & -2 \end{bmatrix} \quad \left(R_2 = -\tfrac{1}{8} r_2 \right)$$

$$\rightarrow \begin{bmatrix} 1 & 0 & -\frac{1}{4} & -\frac{13}{4} \\ 0 & 1 & -\frac{3}{8} & \frac{1}{8} \\ 0 & 0 & -\frac{11}{4} & -\frac{11}{4} \end{bmatrix} \quad \begin{pmatrix} R_1 = -2r_2 + r_1 \\ R_3 = -6r_2 + r_3 \end{pmatrix}$$

$$\rightarrow \begin{bmatrix} 1 & 0 & -\frac{1}{4} & -\frac{13}{4} \\ 0 & 1 & -\frac{3}{8} & \frac{1}{8} \\ 0 & 0 & 1 & 1 \end{bmatrix} \quad \left(R_3 = -\tfrac{4}{11} r_3 \right)$$

$$\rightarrow \begin{bmatrix} 1 & 0 & 0 & -3 \\ 0 & 1 & 0 & \frac{1}{2} \\ 0 & 0 & 1 & 1 \end{bmatrix} \quad \begin{pmatrix} R_1 = \tfrac{1}{4} r_3 + r_1 \\ R_2 = \tfrac{3}{8} r_3 + r_2 \end{pmatrix}$$

The solution is $x = -3, y = \dfrac{1}{2}, z = 1$ or $\left(-3, \dfrac{1}{2}, 1 \right)$.

63. $\begin{cases} 3x + y - z = \dfrac{2}{3} \\ 2x - y + z = 1 \\ 4x + 2y = \dfrac{8}{3} \end{cases}$

Write the augmented matrix:

$$\begin{bmatrix} 3 & 1 & -1 & \frac{2}{3} \\ 2 & -1 & 1 & 1 \\ 4 & 2 & 0 & \frac{8}{3} \end{bmatrix}$$

$$\rightarrow \begin{bmatrix} 1 & \frac{1}{3} & -\frac{1}{3} & \frac{2}{9} \\ 2 & -1 & 1 & 1 \\ 4 & 2 & 0 & \frac{8}{3} \end{bmatrix} \quad \left(R_1 = \tfrac{1}{3} r_1 \right)$$

473

Copyright © 2017 Pearson Education, Inc.

$$\rightarrow \begin{bmatrix} 1 & -\frac{1}{3} & -\frac{1}{3} & \frac{2}{9} \\ 0 & -\frac{5}{3} & \frac{5}{3} & \frac{5}{9} \\ 0 & \frac{2}{3} & \frac{4}{3} & \frac{16}{9} \end{bmatrix} \quad \begin{pmatrix} R_2 = -2r_1 + r_2 \\ R_3 = -4r_1 + r_3 \end{pmatrix}$$

$$\rightarrow \begin{bmatrix} 1 & \frac{1}{3} & -\frac{1}{3} & \frac{2}{9} \\ 0 & 1 & -1 & -\frac{1}{3} \\ 0 & \frac{2}{3} & \frac{4}{3} & \frac{16}{9} \end{bmatrix} \quad \left(R_2 = -\frac{3}{5}r_2 \right)$$

$$\rightarrow \begin{bmatrix} 1 & 0 & 0 & \frac{1}{3} \\ 0 & 1 & -1 & -\frac{1}{3} \\ 0 & 0 & 2 & 2 \end{bmatrix} \quad \begin{pmatrix} R_1 = -\frac{1}{3}r_2 + r_1 \\ R_3 = -\frac{2}{3}r_2 + r_3 \end{pmatrix}$$

$$\rightarrow \begin{bmatrix} 1 & 0 & 0 & \frac{1}{3} \\ 0 & 1 & -1 & -\frac{1}{3} \\ 0 & 0 & 1 & 1 \end{bmatrix} \quad \left(R_3 = \frac{1}{2}r_3 \right)$$

$$\rightarrow \begin{bmatrix} 1 & 0 & 0 & \frac{1}{3} \\ 0 & 1 & 0 & \frac{2}{3} \\ 0 & 0 & 1 & 1 \end{bmatrix} \quad \left(R_2 = r_3 + r_2 \right)$$

The solution is $x = \frac{1}{3}, y = \frac{2}{3}, z = 1$ or $\left(\frac{1}{3}, \frac{2}{3}, 1 \right)$.

65. $\begin{cases} x + y + z + w = 4 \\ 2x - y + z = 0 \\ 3x + 2y + z - w = 6 \\ x - 2y - 2z + 2w = -1 \end{cases}$

Write the augmented matrix:

$$\begin{bmatrix} 1 & 1 & 1 & 1 & 4 \\ 2 & -1 & 1 & 0 & 0 \\ 3 & 2 & 1 & -1 & 6 \\ 1 & -2 & -2 & 2 & -1 \end{bmatrix}$$

$$\rightarrow \begin{bmatrix} 1 & 1 & 1 & 1 & 4 \\ 0 & -3 & -1 & -2 & -8 \\ 0 & -1 & -2 & -4 & -6 \\ 0 & -3 & -3 & 1 & -5 \end{bmatrix} \quad \begin{pmatrix} R_2 = -2r_1 + r_2 \\ R_3 = -3r_1 + r_3 \\ R_4 = -r_1 + r_4 \end{pmatrix}$$

$$\rightarrow \begin{bmatrix} 1 & 1 & 1 & 1 & 4 \\ 0 & -1 & -2 & -4 & -6 \\ 0 & -3 & -1 & -2 & -8 \\ 0 & -3 & -3 & 1 & -5 \end{bmatrix} \quad \begin{pmatrix} \text{Interchange} \\ r_2 \text{ and } r_3 \end{pmatrix}$$

$$\rightarrow \begin{bmatrix} 1 & 1 & 1 & 1 & 4 \\ 0 & 1 & 2 & 4 & 6 \\ 0 & -3 & -1 & -2 & -8 \\ 0 & -3 & -3 & 1 & -5 \end{bmatrix} \quad \left(R_2 = -r_2 \right)$$

$$\rightarrow \begin{bmatrix} 1 & 0 & -1 & -3 & -2 \\ 0 & 1 & 2 & 4 & 6 \\ 0 & 0 & 5 & 10 & 10 \\ 0 & 0 & 3 & 13 & 13 \end{bmatrix} \quad \begin{pmatrix} R_1 = -r_2 + r_1 \\ R_3 = 3r_2 + r_3 \\ R_4 = 3r_2 + r_4 \end{pmatrix}$$

$$\rightarrow \begin{bmatrix} 1 & 0 & -1 & -3 & -2 \\ 0 & 1 & 2 & 4 & 6 \\ 0 & 0 & 1 & 2 & 2 \\ 0 & 0 & 3 & 13 & 13 \end{bmatrix} \quad \left(R_3 = \frac{1}{5}r_3 \right)$$

$$\rightarrow \begin{bmatrix} 1 & 0 & 0 & -1 & 0 \\ 0 & 1 & 0 & 0 & 2 \\ 0 & 0 & 1 & 2 & 2 \\ 0 & 0 & 0 & 7 & 7 \end{bmatrix} \quad \begin{pmatrix} R_1 = r_3 + r_1 \\ R_2 = -2r_3 + r_2 \\ R_4 = -3r_3 + r_4 \end{pmatrix}$$

$$\rightarrow \begin{bmatrix} 1 & 0 & 0 & -1 & 0 \\ 0 & 1 & 0 & 0 & 2 \\ 0 & 0 & 1 & 2 & 2 \\ 0 & 0 & 0 & 1 & 1 \end{bmatrix} \quad \left(R_4 = \frac{1}{7}r_4 \right)$$

$$\rightarrow \begin{bmatrix} 1 & 0 & 0 & 0 & 1 \\ 0 & 1 & 0 & 0 & 2 \\ 0 & 0 & 1 & 0 & 0 \\ 0 & 0 & 0 & 1 & 1 \end{bmatrix} \quad \begin{pmatrix} R_1 = r_4 + r_1 \\ R_3 = -2r_4 + r_3 \end{pmatrix}$$

The solution is $x = 1, y = 2, z = 0, w = 1$ or $(1, 2, 0, 1)$.

67. $\begin{cases} x + 2y + z = 1 \\ 2x - y + 2z = 2 \\ 3x + y + 3z = 3 \end{cases}$

Write the augmented matrix:

$$\begin{bmatrix} 1 & 2 & 1 & 1 \\ 2 & -1 & 2 & 2 \\ 3 & 1 & 3 & 3 \end{bmatrix}$$

$$\rightarrow \begin{bmatrix} 1 & 2 & 1 & | & 1 \\ 0 & -5 & 0 & | & 0 \\ 0 & -5 & 0 & | & 0 \end{bmatrix} \quad \begin{pmatrix} R_2 = -2r_1 + r_2 \\ R_3 = -3r_1 + r_3 \end{pmatrix}$$

$$\rightarrow \begin{bmatrix} 1 & 2 & 1 & | & 1 \\ 0 & -5 & 0 & | & 0 \\ 0 & 0 & 0 & | & 0 \end{bmatrix} \quad (R_3 = -r_2 + r_3)$$

The matrix in the last step represents the system
$$\begin{cases} x + 2y + z = 1 \\ \quad -5y = 0 \\ \qquad\qquad 0 = 0 \end{cases}$$

Substitute and solve:
$$-5y = 0 \qquad\qquad x + 2(0) + z = 1$$
$$y = 0 \qquad\qquad\qquad z = 1 - x$$

The solution is $y = 0$, $z = 1 - x$, x is any real number or $\{(x, y, z) \mid y = 0, z = 1 - x, x \text{ is any real number}\}$.

69. $\begin{cases} x - y + z = 5 \\ 3x + 2y - 2z = 0 \end{cases}$

Write the augmented matrix:
$$\begin{bmatrix} 1 & -1 & 1 & | & 5 \\ 3 & 2 & -2 & | & 0 \end{bmatrix}$$

$$\rightarrow \begin{bmatrix} 1 & -1 & 1 & | & 5 \\ 0 & 5 & -5 & | & -15 \end{bmatrix} \quad (R_2 = -3r_1 + r_2)$$

$$\rightarrow \begin{bmatrix} 1 & -1 & 1 & | & 5 \\ 0 & 1 & -1 & | & -3 \end{bmatrix} \quad (R_2 = \tfrac{1}{5}r_2)$$

$$\rightarrow \begin{bmatrix} 1 & 0 & 0 & | & 2 \\ 0 & 1 & -1 & | & -3 \end{bmatrix} \quad (R_1 = r_2 + r_1)$$

The matrix in the last step represents the system
$$\begin{cases} x = 2 \\ y - z = -3 \end{cases} \text{ or, equivalently, } \begin{cases} x = 2 \\ y = z - 3 \end{cases}$$

Thus, the solution is $x = 2$, $y = z - 3$, z is any real number or $\{(x, y, z) \mid x = 2, y = z - 3, z \text{ is any real number}\}$.

71. $\begin{cases} 2x + 3y - z = 3 \\ x - y - z = 0 \\ -x + y + z = 0 \\ x + y + 3z = 5 \end{cases}$

Write the augmented matrix:

$$\begin{bmatrix} 2 & 3 & -1 & | & 3 \\ 1 & -1 & -1 & | & 0 \\ -1 & 1 & 1 & | & 0 \\ 1 & 1 & 3 & | & 5 \end{bmatrix}$$

$$\rightarrow \begin{bmatrix} 1 & -1 & -1 & | & 0 \\ 2 & 3 & -1 & | & 3 \\ -1 & 1 & 1 & | & 0 \\ 1 & 1 & 3 & | & 5 \end{bmatrix} \quad \begin{pmatrix} \text{interchange} \\ r_1 \text{ and } r_2 \end{pmatrix}$$

$$\rightarrow \begin{bmatrix} 1 & -1 & -1 & | & 0 \\ 0 & 5 & 1 & | & 3 \\ 0 & 0 & 0 & | & 0 \\ 0 & 2 & 4 & | & 5 \end{bmatrix} \quad \begin{pmatrix} R_2 = -2r_1 + r_2 \\ R_3 = r_1 + r_3 \\ R_4 = -r_1 + r_4 \end{pmatrix}$$

$$\rightarrow \begin{bmatrix} 1 & -1 & -1 & | & 0 \\ 0 & 5 & 1 & | & 3 \\ 0 & 2 & 4 & | & 5 \\ 0 & 0 & 0 & | & 0 \end{bmatrix} \quad \begin{pmatrix} \text{interchange} \\ r_3 \text{ and } r_4 \end{pmatrix}$$

$$\rightarrow \begin{bmatrix} 1 & -1 & -1 & | & 0 \\ 0 & 1 & -7 & | & -7 \\ 0 & 2 & 4 & | & 5 \\ 0 & 0 & 0 & | & 0 \end{bmatrix} \quad (R_2 = -2r_3 + r_2)$$

$$\rightarrow \begin{bmatrix} 1 & 0 & -8 & | & -7 \\ 0 & 1 & -7 & | & -7 \\ 0 & 0 & 18 & | & 19 \\ 0 & 0 & 0 & | & 0 \end{bmatrix} \quad \begin{pmatrix} R_1 = r_2 + r_1 \\ R_3 = -2r_2 + r_3 \end{pmatrix}$$

$$\rightarrow \begin{bmatrix} 1 & 0 & -8 & | & -7 \\ 0 & 1 & -7 & | & -7 \\ 0 & 0 & 1 & | & \frac{19}{18} \\ 0 & 0 & 0 & | & 0 \end{bmatrix} \quad (R_3 = \tfrac{1}{18}r_3)$$

The matrix in the last step represents the system
$$\begin{cases} x - 8z = -7 \\ y - 7z = -7 \\ z = \dfrac{19}{18} \end{cases}$$

Substitute and solve:
$$y - 7\left(\frac{19}{18}\right) = 7 \qquad\qquad x - 8\left(\frac{19}{18}\right) = -7$$
$$y = \frac{7}{18} \qquad\qquad\qquad x = \frac{13}{9}$$

Thus, the solution is $x = \dfrac{13}{9}$, $y = \dfrac{7}{18}$, $z = \dfrac{19}{18}$ or $\left(\dfrac{13}{9}, \dfrac{7}{18}, \dfrac{19}{18}\right)$.

475

73. $\begin{cases} 4x+y+z-w=4 \\ x-y+2z+3w=3 \end{cases}$

Write the augmented matrix:

$$\begin{bmatrix} 4 & 1 & 1 & -1 & | & 4 \\ 1 & -1 & 2 & 3 & | & 3 \end{bmatrix}$$

$$\rightarrow \begin{bmatrix} 1 & -1 & 2 & 3 & | & 3 \\ 4 & 1 & 1 & -1 & | & 4 \end{bmatrix} \begin{pmatrix} \text{interchange} \\ r_1 \text{ and } r_2 \end{pmatrix}$$

$$\rightarrow \begin{bmatrix} 1 & -1 & 2 & 3 & | & 3 \\ 0 & 5 & -7 & -13 & | & -8 \end{bmatrix} \quad (R_2 = -4r_1 + r_2)$$

The matrix in the last step represents the system

$$\begin{cases} x-y+2z+3w=3 \\ 5y-7z-13w=-8 \end{cases}$$

The second equation yields

$5y-7z-13w=-8$

$5y = 7z+13w-8$

$y = \dfrac{7}{5}z+\dfrac{13}{5}w-\dfrac{8}{5}$

The first equation yields

$x-y+2z+3w=3$

$x = 3+y-2z-3w$

Substituting for y:

$x = 3+\left(-\dfrac{8}{5}+\dfrac{7}{5}z+\dfrac{13}{5}w\right)-2z-3w$

$x = -\dfrac{3}{5}z-\dfrac{2}{5}w+\dfrac{7}{5}$

Thus, the solution is $x = -\dfrac{3}{5}z-\dfrac{2}{5}w+\dfrac{7}{5}$,

$y = \dfrac{7}{5}z+\dfrac{13}{5}w-\dfrac{8}{5}$, z and w are any real numbers or

$\left\{(x,y,z,w) \left| \; x = -\dfrac{3}{5}z-\dfrac{2}{5}w+\dfrac{7}{5}, \; y = \dfrac{7}{5}z+\dfrac{13}{5}w-\dfrac{8}{5}, \right. \right.$

z and w are any real numbers$\Big\}$.

75. Each of the points must satisfy the equation
$y = ax^2 + bx + c$.

$\quad (1,2): \qquad 2 = a+b+c$

$\quad (-2,\ 7): \quad -7 = 4a-2b+c$

$\quad (2,-3): \qquad -3 = 4a+2b+c$

Set up a matrix and solve:

$$\begin{bmatrix} 1 & 1 & 1 & | & 2 \\ 4 & -2 & 1 & | & -7 \\ 4 & 2 & 1 & | & -3 \end{bmatrix}$$

$$\rightarrow \begin{bmatrix} 1 & 1 & 1 & | & 2 \\ 0 & -6 & -3 & | & -15 \\ 0 & -2 & -3 & | & -11 \end{bmatrix} \begin{pmatrix} R_2 = -4r_1 + r_2 \\ R_3 = -4r_1 + r_3 \end{pmatrix}$$

$$\rightarrow \begin{bmatrix} 1 & 1 & 1 & | & 2 \\ 0 & 1 & \frac{1}{2} & | & \frac{5}{2} \\ 0 & -2 & -3 & | & -11 \end{bmatrix} \left(R_2 = -\tfrac{1}{6}r_2\right)$$

$$\rightarrow \begin{bmatrix} 1 & 0 & \frac{1}{2} & | & -\frac{1}{2} \\ 0 & 1 & \frac{1}{2} & | & \frac{5}{2} \\ 0 & 0 & -2 & | & -6 \end{bmatrix} \begin{pmatrix} R_1 = -r_2 + r_1 \\ R_3 = 2r_2 + r_3 \end{pmatrix}$$

$$\rightarrow \begin{bmatrix} 1 & 0 & \frac{1}{2} & | & -\frac{1}{2} \\ 0 & 1 & \frac{1}{2} & | & \frac{5}{2} \\ 0 & 0 & 1 & | & 3 \end{bmatrix} \rightarrow \left(R_3 = -\tfrac{1}{2}r_3\right)$$

$$\rightarrow \begin{bmatrix} 1 & 0 & 0 & | & -2 \\ 0 & 1 & 0 & | & 1 \\ 0 & 0 & 1 & | & 3 \end{bmatrix} \begin{pmatrix} R_1 = -\tfrac{1}{2}r_3 + r_1 \\ R_2 = -\tfrac{1}{2}r_3 + r_2 \end{pmatrix}$$

The solution is $a = -2, b = 1, c = 3$; so the equation is $y = -2x^2 + x + 3$.

77. Each of the points must satisfy the equation
$f(x) = ax^3 + bx^2 + cx + d$.

$\quad f(-3) = -112: \quad -27a+9b-3c+d = -112$

$\quad f(-1) = -2: \qquad\quad -a+b-c+d = -2$

$\quad f(1) = 4: \qquad\qquad\quad a+b+c+d = 4$

$\quad f(2) = 13: \qquad\quad 8a+4b+2c+d = 13$

Set up a matrix and solve:

$$\begin{bmatrix} -27 & 9 & -3 & 1 & | & -112 \\ -1 & 1 & -1 & 1 & | & -2 \\ 1 & 1 & 1 & 1 & | & 4 \\ 8 & 4 & 2 & 1 & | & 13 \end{bmatrix}$$

$$\rightarrow \begin{bmatrix} 1 & 1 & 1 & 1 & | & 4 \\ -1 & 1 & -1 & 1 & | & -2 \\ -27 & 9 & -3 & 1 & | & -112 \\ 8 & 4 & 2 & 1 & | & 13 \end{bmatrix} \begin{pmatrix} \text{Interchange} \\ r_3 \text{ and } r_1 \end{pmatrix}$$

$$\rightarrow \begin{bmatrix} 1 & 1 & 1 & 1 & | & 4 \\ 0 & 2 & 0 & 2 & | & 2 \\ 0 & 36 & 24 & 28 & | & -4 \\ 0 & -4 & -6 & -7 & | & -19 \end{bmatrix} \begin{pmatrix} R_2 = r_1 + r_2 \\ R_3 = 27 r_1 + r_3 \\ R_4 = -8 r_1 + r_4 \end{pmatrix}$$

$$\rightarrow \begin{bmatrix} 1 & 1 & 1 & 1 & | & 4 \\ 0 & 1 & 0 & 1 & | & 1 \\ 0 & 36 & 24 & 28 & | & -4 \\ 0 & -4 & -6 & -7 & | & -19 \end{bmatrix} \left(R_2 = \tfrac{1}{2} r_2 \right)$$

$$\rightarrow \begin{bmatrix} 1 & 0 & 1 & 0 & | & 3 \\ 0 & 1 & 0 & 1 & | & 1 \\ 0 & 0 & 24 & -8 & | & -40 \\ 0 & 0 & -6 & -3 & | & -15 \end{bmatrix} \begin{pmatrix} R_1 = -r_2 + r_1 \\ R_3 = -36 r_2 + r_3 \\ R_4 = 4 r_2 + r_4 \end{pmatrix}$$

$$\rightarrow \begin{bmatrix} 1 & 0 & 1 & 0 & | & 3 \\ 0 & 1 & 0 & 1 & | & 1 \\ 0 & 0 & 1 & -\tfrac{5}{3} & | & -\tfrac{5}{3} \\ 0 & 0 & -6 & -3 & | & -15 \end{bmatrix} \left(R_3 = \tfrac{1}{24} r_3 \right)$$

$$\rightarrow \begin{bmatrix} 1 & 0 & 0 & \tfrac{1}{3} & | & \tfrac{14}{3} \\ 0 & 1 & 0 & 1 & | & 1 \\ 0 & 0 & 1 & -\tfrac{1}{3} & | & -\tfrac{5}{3} \\ 0 & 0 & 0 & -5 & | & -25 \end{bmatrix} \begin{pmatrix} R_1 = -r_3 + r_1 \\ R_4 = 6 r_3 + r_4 \end{pmatrix}$$

$$\rightarrow \begin{bmatrix} 1 & 0 & 0 & \tfrac{1}{3} & | & \tfrac{14}{3} \\ 0 & 1 & 0 & 1 & | & 1 \\ 0 & 0 & 1 & -\tfrac{1}{3} & | & -\tfrac{5}{3} \\ 0 & 0 & 0 & 1 & | & 5 \end{bmatrix} \left(R_4 = -\tfrac{1}{5} r_4 \right)$$

$$\rightarrow \begin{bmatrix} 1 & 0 & 0 & 0 & | & 3 \\ 0 & 1 & 0 & 0 & | & -4 \\ 0 & 0 & 1 & 0 & | & 0 \\ 0 & 0 & 0 & 1 & | & 5 \end{bmatrix} \begin{pmatrix} R_1 = -\tfrac{1}{3} r_4 + r_1 \\ R_2 = -r_4 + r_2 \\ R_3 = \tfrac{1}{3} r_4 + r_3 \end{pmatrix}$$

The solution is $a = 3, b = -4, c = 0, d = 5$; so the equation is $f(x) = 3x^3 - 4x^2 + 5$.

79. Let x = the number of servings of salmon steak.
Let y = the number of servings of baked eggs.
Let z = the number of servings of acorn squash.
Protein equation: $30x + 15y + 3z = 78$
Carbohydrate equation: $20x + 2y + 25z = 59$
Vitamin A equation: $2x + 20y + 32z = 75$
Set up a matrix and solve:

$$\begin{bmatrix} 30 & 15 & 3 & | & 78 \\ 20 & 2 & 25 & | & 59 \\ 2 & 20 & 32 & | & 75 \end{bmatrix}$$

$$\rightarrow \begin{bmatrix} 2 & 20 & 32 & | & 75 \\ 20 & 2 & 25 & | & 59 \\ 30 & 15 & 3 & | & 78 \end{bmatrix} \begin{pmatrix} \text{Interchange} \\ r_3 \text{ and } r_1 \end{pmatrix}$$

$$\rightarrow \begin{bmatrix} 1 & 10 & 16 & | & 37.5 \\ 20 & 2 & 25 & | & 59 \\ 30 & 15 & 3 & | & 78 \end{bmatrix} \left(R_1 = \tfrac{1}{2} r_1 \right)$$

$$\rightarrow \begin{bmatrix} 1 & 10 & 16 & | & 37.5 \\ 0 & -198 & -295 & | & -691 \\ 0 & -285 & -477 & | & -1047 \end{bmatrix} \begin{pmatrix} R_2 = -20 r_1 + r_2 \\ R_3 = -30 r_1 + r_3 \end{pmatrix}$$

$$\rightarrow \begin{bmatrix} 1 & 10 & 16 & | & 37.5 \\ 0 & -198 & -295 & | & -691 \\ 0 & 0 & -\tfrac{3457}{66} & | & -\tfrac{3457}{66} \end{bmatrix} \left(R_3 = -\tfrac{95}{66} r_2 + r_3 \right)$$

$$\rightarrow \begin{bmatrix} 1 & 10 & 16 & | & 37.5 \\ 0 & -198 & -295 & | & -691 \\ 0 & 0 & 1 & | & 1 \end{bmatrix} \left(R_3 = -\tfrac{66}{3457} r_3 \right)$$

Substitute $z = 1$ and solve:
$-198y - 295(1) = -691$
$-198y = -396$
$y = 2$
$x + 10(2) + 16(1) = 37.5$
$x + 36 = 37.5$
$x = 1.5$

The dietitian should serve 1.5 servings of salmon steak, 2 servings of baked eggs, and 1 serving of acorn squash.

81. Let x = the amount invested in Treasury bills.
Let y = the amount invested in Treasury bonds.
Let z = the amount invested in corporate bonds.
Total investment equation:
$$x + y + z = 10,000$$
Annual income equation:
$$0.06x + 0.07y + 0.08z = 680$$
Condition on investment equation:
$$z = 0.5x$$
$$x - 2z = 0$$
Set up a matrix and solve:

$$\begin{bmatrix} 1 & 1 & 1 & | & 10,000 \\ 0.06 & 0.07 & 0.08 & | & 680 \\ 1 & 0 & -2 & | & 0 \end{bmatrix}$$

$$\rightarrow \begin{bmatrix} 1 & 1 & 1 & | & 10,000 \\ 0 & 0.01 & 0.02 & | & 80 \\ 0 & -1 & -3 & | & -10,000 \end{bmatrix} \begin{pmatrix} R_2 = -0.06r_1 + r_2 \\ R_3 = -r_1 + r_3 \end{pmatrix}$$

$$\rightarrow \begin{bmatrix} 1 & 1 & 1 & | & 10,000 \\ 0 & 1 & 2 & | & 8000 \\ 0 & -1 & -3 & | & -10,000 \end{bmatrix} (R_2 = 100r_2)$$

$$\rightarrow \begin{bmatrix} 1 & 0 & -1 & | & 2000 \\ 0 & 1 & 2 & | & 8000 \\ 0 & 0 & -1 & | & -2000 \end{bmatrix} \begin{pmatrix} R_1 = -r_2 + r_1 \\ R_3 = r_2 + r_3 \end{pmatrix}$$

$$\rightarrow \begin{bmatrix} 1 & 0 & -1 & | & 2000 \\ 0 & 1 & 2 & | & 8000 \\ 0 & 0 & 1 & | & 2000 \end{bmatrix} (R_3 = -r_3)$$

$$\rightarrow \begin{bmatrix} 1 & 0 & 0 & | & 4000 \\ 0 & 1 & 0 & | & 4000 \\ 0 & 0 & 1 & | & 2000 \end{bmatrix} \begin{pmatrix} R_1 = r_3 + r_1 \\ R_2 = -2r_3 + r_2 \end{pmatrix}$$

Carletta should invest $4000 in Treasury bills, $4000 in Treasury bonds, and $2000 in corporate bonds.

83. Let x = the number of Deltas produced.
Let y = the number of Betas produced.
Let z = the number of Sigmas produced.
Painting equation: $10x + 16y + 8z = 240$
Drying equation: $3x + 5y + 2z = 69$
Polishing equation: $2x + 3y + z = 41$
Set up a matrix and solve:

$$\begin{bmatrix} 10 & 16 & 8 & | & 240 \\ 3 & 5 & 2 & | & 69 \\ 2 & 3 & 1 & | & 41 \end{bmatrix}$$

$$\rightarrow \begin{bmatrix} 1 & 1 & 2 & | & 33 \\ 3 & 5 & 2 & | & 69 \\ 2 & 3 & 1 & | & 41 \end{bmatrix} (R_1 = -3r_2 + r_1)$$

$$\rightarrow \begin{bmatrix} 1 & 1 & 2 & | & 33 \\ 0 & 2 & -4 & | & -30 \\ 0 & 1 & -3 & | & -25 \end{bmatrix} \begin{pmatrix} R_2 = -3r_1 + r_2 \\ R_3 = -2r_1 + r_3 \end{pmatrix}$$

$$\rightarrow \begin{bmatrix} 1 & 1 & 2 & | & 33 \\ 0 & 1 & -2 & | & -15 \\ 0 & 1 & -3 & | & -25 \end{bmatrix} (R_2 = \tfrac{1}{2}r_2)$$

$$\rightarrow \begin{bmatrix} 1 & 0 & 4 & | & 48 \\ 0 & 1 & -2 & | & -15 \\ 0 & 0 & -1 & | & -10 \end{bmatrix} \begin{pmatrix} R_1 = r_1 - r_2 \\ R_3 = r_3 - r_2 \end{pmatrix}$$

$$\rightarrow \begin{bmatrix} 1 & 0 & 4 & | & 48 \\ 0 & 1 & -2 & | & -15 \\ 0 & 0 & 1 & | & 10 \end{bmatrix} (R_3 = -r_3)$$

$$\rightarrow \begin{bmatrix} 1 & 0 & 0 & | & 8 \\ 0 & 1 & 0 & | & 5 \\ 0 & 0 & 1 & | & 10 \end{bmatrix} \begin{pmatrix} R_1 = -4r_3 + r_1 \\ R_2 = 2r_3 + r_2 \end{pmatrix}$$

The company should produce 8 Deltas, 5 Betas, and 10 Sigmas.

85. Rewrite the system to set up the matrix and solve:

$$\begin{cases} -4+8-2I_2 = 0 \\ 8 = 5I_4 + I_1 \\ 4 = 3I_3 + I_1 \\ I_3 + I_4 = I_1 \end{cases} \rightarrow \begin{cases} 2I_2 = 4 \\ I_1 + 5I_4 = 8 \\ I_1 + 3I_3 = 4 \\ I_1 - I_3 - I_4 = 0 \end{cases}$$

$$\begin{bmatrix} 0 & 2 & 0 & 0 & | & 4 \\ 1 & 0 & 0 & 5 & | & 8 \\ 1 & 0 & 3 & 0 & | & 4 \\ 1 & 0 & -1 & -1 & | & 0 \end{bmatrix}$$

$$\rightarrow \begin{bmatrix} 1 & 0 & 0 & 5 & | & 8 \\ 0 & 2 & 0 & 0 & | & 4 \\ 1 & 0 & 3 & 0 & | & 4 \\ 1 & 0 & -1 & -1 & | & 0 \end{bmatrix} \begin{pmatrix} \text{Interchange} \\ r_2 \text{ and } r_1 \end{pmatrix}$$

$$\rightarrow \begin{bmatrix} 1 & 0 & 0 & 5 & | & 8 \\ 0 & 1 & 0 & 0 & | & 2 \\ 0 & 0 & 3 & -5 & | & -4 \\ 0 & 0 & -1 & -6 & | & -8 \end{bmatrix} \begin{pmatrix} R_2 = \frac{1}{2}r_2 \\ R_3 = -r_1 + r_3 \\ R_4 = -r_1 + r_4 \end{pmatrix}$$

$$\rightarrow \begin{bmatrix} 1 & 0 & 0 & 5 & | & 8 \\ 0 & 1 & 0 & 0 & | & 2 \\ 0 & 0 & -1 & -6 & | & -8 \\ 0 & 0 & 3 & -5 & | & -4 \end{bmatrix} \begin{pmatrix} \text{Interchange} \\ r_3 \text{ and } r_4 \end{pmatrix}$$

$$\rightarrow \begin{bmatrix} 1 & 0 & 0 & 5 & | & 8 \\ 0 & 1 & 0 & 0 & | & 2 \\ 0 & 0 & 1 & 6 & | & 8 \\ 0 & 0 & 0 & -23 & | & -28 \end{bmatrix} \begin{pmatrix} R_3 = -r_3 \\ R_4 = -3r_3 + r_4 \end{pmatrix}$$

$$\rightarrow \begin{bmatrix} 1 & 0 & 0 & 5 & | & 8 \\ 0 & 1 & 0 & 0 & | & 2 \\ 0 & 0 & 1 & 6 & | & 8 \\ 0 & 0 & 0 & 1 & | & \frac{28}{23} \end{bmatrix} \left(R_4 = -\frac{1}{23}r_4 \right)$$

$$\rightarrow \begin{bmatrix} 1 & 0 & 0 & 0 & | & \frac{44}{23} \\ 0 & 1 & 0 & 0 & | & 2 \\ 0 & 0 & 1 & 0 & | & \frac{16}{23} \\ 0 & 0 & 0 & 1 & | & \frac{28}{23} \end{bmatrix} \begin{pmatrix} R_1 = -5r_4 + r_1 \\ R_3 = -6r_4 + r_3 \end{pmatrix}$$

The solution is $I_1 = \dfrac{44}{23}$, $I_2 = 2$, $I_3 = \dfrac{16}{23}$,

$I_4 = \dfrac{28}{23}$.

87. Let x = the amount invested in Treasury bills.
Let y = the amount invested in corporate bonds.
Let z = the amount invested in junk bonds.

a. Total investment equation:
$$x + y + z = 20,000$$
Annual income equation:
$$0.07x + 0.09y + 0.11z = 2000$$
Set up a matrix and solve:

$$\begin{bmatrix} 1 & 1 & 1 & | & 20,000 \\ 0.07 & 0.09 & 0.11 & | & 2000 \end{bmatrix}$$

$$\rightarrow \begin{bmatrix} 1 & 1 & 1 & | & 20,000 \\ 7 & 9 & 11 & | & 200,000 \end{bmatrix} \left(R_2 = 100r_2 \right)$$

$$\rightarrow \begin{bmatrix} 1 & 1 & 1 & | & 20,000 \\ 0 & 2 & 4 & | & 60,000 \end{bmatrix} \left(R_2 = r_2 - 7r_1 \right)$$

$$\rightarrow \begin{bmatrix} 1 & 1 & 1 & | & 20,000 \\ 0 & 1 & 2 & | & 30,000 \end{bmatrix} \left(R_2 = \frac{1}{2}r_2 \right)$$

$$\rightarrow \begin{bmatrix} 1 & 0 & -1 & | & -10,000 \\ 0 & 1 & 2 & | & 30,000 \end{bmatrix} \left(R_1 = r_1 - r_2 \right)$$

The matrix in the last step represents the
system $\begin{cases} x - z = -10,000 \\ y + 2z = 30,000 \end{cases}$

Therefore the solution is $x = -10,000 + z$,
$y = 30,000 - 2z$, z is any real number.

Possible investment strategies:

Amount Invested At		
7%	9%	11%
0	10,000	10,000
1000	8000	11,000
2000	6000	12,000
3000	4000	13,000
4000	2000	14,000
5000	0	15,000

b. Total investment equation:
$$x + y + z = 25,000$$
Annual income equation:
$$0.07x + 0.09y + 0.11z = 2000$$
Set up a matrix and solve:

479

Copyright © 2017 Pearson Education, Inc.

$$\begin{bmatrix} 1 & 1 & 1 & | & 25,000 \\ 0.07 & 0.09 & 0.11 & | & 2000 \end{bmatrix}$$

$$\rightarrow \begin{bmatrix} 1 & 1 & 1 & | & 25,000 \\ 7 & 9 & 11 & | & 200,000 \end{bmatrix} \quad (R_2 = 100r_2)$$

$$\rightarrow \begin{bmatrix} 1 & 1 & 1 & | & 25,000 \\ 0 & 2 & 4 & | & 25,000 \end{bmatrix} \quad (R_2 = r_2 - 7r_1)$$

$$\rightarrow \begin{bmatrix} 1 & 1 & 1 & | & 25,000 \\ 0 & 1 & 2 & | & 12,500 \end{bmatrix} \quad \left(R_2 = \tfrac{1}{2}r_2\right)$$

$$\rightarrow \begin{bmatrix} 1 & 0 & -1 & | & 12,500 \\ 0 & 1 & 2 & | & 12,500 \end{bmatrix} \quad (R_1 = r_1 - r_2)$$

The matrix in the last step represents the
system $\begin{cases} x - z = 12,500 \\ y + 2z = 12,500 \end{cases}$

Thus, the solution is $x = z + 12,500$,
$y = -2z + 12,500$, z is any real number.

Possible investment strategies:

	Amount Invested At	
7%	9%	11%
12,500	12,500	0
14,500	8500	2000
16,500	4500	4000
18,750	0	6250

c. Total investment equation:
$x + y + z = 30,000$
Annual income equation:
$0.07x + 0.09y + 0.11z = 2000$
Set up a matrix and solve:

$$\begin{bmatrix} 1 & 1 & 1 & | & 30,000 \\ 0.07 & 0.09 & 0.11 & | & 2000 \end{bmatrix}$$

$$\rightarrow \begin{bmatrix} 1 & 1 & 1 & | & 30,000 \\ 7 & 9 & 11 & | & 200,000 \end{bmatrix} \quad (R_2 = 100r_2)$$

$$\rightarrow \begin{bmatrix} 1 & 1 & 1 & | & 30,000 \\ 0 & 2 & 4 & | & -10,000 \end{bmatrix} \quad (R_1 = r_2 - 7r_1)$$

$$\rightarrow \begin{bmatrix} 1 & 1 & 1 & | & 30,000 \\ 0 & 1 & 2 & | & -5000 \end{bmatrix} \quad \left(R_2 = \tfrac{1}{2}r_2\right)$$

$$\rightarrow \begin{bmatrix} 1 & 0 & -1 & | & 35,000 \\ 0 & 1 & 2 & | & -5000 \end{bmatrix} \quad (R_1 = r_1 - r_2)$$

The matrix in the last step represents the

system $\begin{cases} x - z = 35,000 \\ y + 2z = -5000 \end{cases}$

Thus, the solution is $x = z + 35,000$,
$y = -2z - 5000$, z is any real number.
However, y and z cannot be negative. From
$y = -2z - 5000$, we must have $y = z = 0$.

One possible investment strategy

	Amount Invested At	
7%	9%	11%
30,000	0	0

This will yield ($30,000)(0.07) = $2100,
which is more than the required income.

d. Answers will vary.

89. Let x = the amount of supplement 1.
Let y = the amount of supplement 2.
Let z = the amount of supplement 3.
$\begin{cases} 0.20x + 0.40y + 0.30z = 40 & \text{Vitamin C} \\ 0.30x + 0.20y + 0.50z = 30 & \text{Vitamin D} \end{cases}$
Multiplying each equation by 10 yields
$\begin{cases} 2x + 4y + 3z = 400 \\ 3x + 2y + 5z = 300 \end{cases}$

Set up a matrix and solve:

$$\begin{bmatrix} 2 & 4 & 3 & | & 400 \\ 3 & 2 & 5 & | & 300 \end{bmatrix}$$

$$\rightarrow \begin{bmatrix} 1 & 2 & \tfrac{3}{2} & | & 200 \\ 3 & 2 & 5 & | & 300 \end{bmatrix} \quad \left(R_1 = \tfrac{1}{2}r_1\right)$$

$$\rightarrow \begin{bmatrix} 1 & 2 & \tfrac{3}{2} & | & 200 \\ 0 & -4 & \tfrac{1}{2} & | & -300 \end{bmatrix} \quad (R_2 = r_2 - 3r_1)$$

$$\rightarrow \begin{bmatrix} 1 & 2 & \tfrac{3}{2} & | & 200 \\ 0 & 1 & -\tfrac{1}{8} & | & 75 \end{bmatrix} \quad \left(R_2 = -\tfrac{1}{4}r_2\right)$$

$$\rightarrow \begin{bmatrix} 1 & 0 & \tfrac{7}{4} & | & 50 \\ 0 & 1 & -\tfrac{1}{8} & | & 75 \end{bmatrix} \quad (R_1 = r_1 - 2r_2)$$

The matrix in the last step represents the system
$\begin{cases} x + \tfrac{7}{4}z = 50 \\ y - \tfrac{1}{8}z = 75 \end{cases}$

Therefore the solution is $x = 50 - \dfrac{7}{4}z$,

$y = 75 + \dfrac{1}{8}z$, z is any real number.

Possible combinations:

Supplement 1	Supplement 2	Supplement 3
50mg	75mg	0mg
36mg	76mg	8mg
22mg	77mg	16mg
8mg	78mg	24mg

91 – 93. Answers will vary.

95. $R(x) = \dfrac{2x^2 - x - 1}{x^2 + 2x + 1} = \dfrac{(2x+1)(x-1)}{(x+1)(x+1)}$ $\quad p(x) = 2x^2 - x - 1; \quad q(x) = x^2 + 2x + 1;$

Domain: $\{x \mid x \neq -1\}$. $R(x) = \dfrac{2x^2 - x - 1}{x^2 + 2x + 1}$ is in lowest terms.

The y-intercept is $f(0) = \dfrac{2 \cdot 0^2 - 0 - 1}{0^2 + 2 \cdot 0 + 1} = \dfrac{-1}{1} = -1$. Plot the point $(0, -1)$.

The x-intercepts are the zeros of $p(x)$: 1 and $-\dfrac{1}{2}$.

$R(x) = \dfrac{2x^2 - x - 1}{x^2 + 2x + 1}$ is in lowest terms. The vertical asymptotes are the zeros of $q(x)$:

$\quad x = -1$. Graph this asymptote with dashed lines.

Since $n = m$, the line $y = \dfrac{2}{1} = 2$ is the horizontal asymptote. Solve to find intersection points:

$\quad$ Plot the line $y = 2$ using dashes.

Graph:

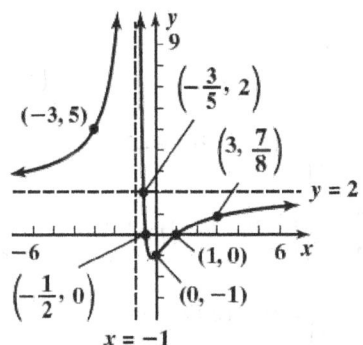

97. $f(x) = 4x^4 + 21x^2 - 100$

$4x^4 + 21x^2 - 100 = 0$

$(x^2 - 4)(x^2 + 25) = 0$

$x^2 - 4 = 0 \quad$ or $\quad x^2 + 25 = 0$

$\quad x = \pm 2 \quad$ or $\quad\quad x = \pm 5i$

So the solution set is: $\{-2, 2, -5i, 5i\}$

Section 8.3

1. $ad - bc$

3. False; If ad=bc, the the det = 0.

5. False; See Thm (14)

7. $\begin{vmatrix} 6 & 4 \\ -1 & 3 \end{vmatrix} = 6(3)-(-1)(4) = 18+4 = 22$

9. $\begin{vmatrix} -3 & -1 \\ 4 & 2 \end{vmatrix} = -3(2)-4(-1) = -6+4 = -2$

11. $\begin{vmatrix} 3 & 4 & 2 \\ 1 & -1 & 5 \\ 1 & 2 & -2 \end{vmatrix} = 3\begin{vmatrix} -1 & 5 \\ 2 & -2 \end{vmatrix} - 4\begin{vmatrix} 1 & 5 \\ 1 & -2 \end{vmatrix} + 2\begin{vmatrix} 1 & -1 \\ 1 & 2 \end{vmatrix}$

$= 3\big[(-1)(-2)-2(5)\big] - 4\big[1(-2)-1(5)\big]$
$\qquad\qquad + 2\big[1(2)-1(-1)\big]$

$= 3(-8)-4(-7)+2(3)$

$= -24+28+6$

$= 10$

13. $\begin{vmatrix} 4 & -1 & 2 \\ 6 & -1 & 0 \\ 1 & -3 & 4 \end{vmatrix} = 4\begin{vmatrix} -1 & 0 \\ -3 & 4 \end{vmatrix} - (-1)\begin{vmatrix} 6 & 0 \\ 1 & 4 \end{vmatrix} + 2\begin{vmatrix} 6 & -1 \\ 1 & -3 \end{vmatrix}$

$= 4\big[-1(4)-0(-3)\big] + 1\big[6(4)-1(0)\big]$
$\qquad\qquad + 2\big[6(-3)-1(-1)\big]$

$= 4(-4)+1(24)+2(-17)$

$= -16+24-34$

$= -26$

15. $\begin{cases} x+y=8 \\ x-y=4 \end{cases}$

$D = \begin{vmatrix} 1 & 1 \\ 1 & -1 \end{vmatrix} = -1-1 = -2$

$D_x = \begin{vmatrix} 8 & 1 \\ 4 & -1 \end{vmatrix} = -8-4 = -12$

$D_y = \begin{vmatrix} 1 & 8 \\ 1 & 4 \end{vmatrix} = 4-8 = -4$

Find the solutions by Cramer's Rule:

$x = \dfrac{D_x}{D} = \dfrac{-12}{-2} = 6 \qquad y = \dfrac{D_y}{D} = \dfrac{-4}{-2} = 2$

The solution is (6, 2).

17. $\begin{cases} 5x-y=13 \\ 2x+3y=12 \end{cases}$

$D = \begin{vmatrix} 5 & -1 \\ 2 & 3 \end{vmatrix} = 15+2 = 17$

$D_x = \begin{vmatrix} 13 & -1 \\ 12 & 3 \end{vmatrix} = 39+12 = 51$

$D_y = \begin{vmatrix} 5 & 13 \\ 2 & 12 \end{vmatrix} = 60-26 = 34$

Find the solutions by Cramer's Rule:

$x = \dfrac{D_x}{D} = \dfrac{51}{17} = 3 \qquad y = \dfrac{D_y}{D} = \dfrac{34}{17} = 2$

The solution is (3, 2).

19. $\begin{cases} 3x = 24 \\ x+2y = 0 \end{cases}$

$D = \begin{vmatrix} 3 & 0 \\ 1 & 2 \end{vmatrix} = 6-0 = 6$

$D_x = \begin{vmatrix} 24 & 0 \\ 0 & 2 \end{vmatrix} = 48-0 = 48$

$D_y = \begin{vmatrix} 3 & 24 \\ 1 & 0 \end{vmatrix} = 0-24 = -24$

Find the solutions by Cramer's Rule:

$x = \dfrac{D_x}{D} = \dfrac{48}{6} = 8 \qquad y = \dfrac{D_y}{D} = \dfrac{-24}{6} = -4$

The solution is $(8,-4)$.

21. $\begin{cases} 3x-6y=24 \\ 5x+4y = 12 \end{cases}$

$D = \begin{vmatrix} 3 & -6 \\ 5 & 4 \end{vmatrix} = 12-(-30) = 42$

$D_x = \begin{vmatrix} 24 & -6 \\ 12 & 4 \end{vmatrix} = 96-(-72) = 168$

$D_y = \begin{vmatrix} 3 & 24 \\ 5 & 12 \end{vmatrix} = 36-120 = -84$

Find the solutions by Cramer's Rule:

$x = \dfrac{D_x}{D} = \dfrac{168}{42} = 4 \qquad y = \dfrac{D_y}{D} = \dfrac{-84}{42} = -2$

The solution is $(4,-2)$.

23. $\begin{cases} 3x - 2y = 4 \\ 6x - 4y = 0 \end{cases}$

$D = \begin{vmatrix} 3 & -2 \\ 6 & -4 \end{vmatrix} = -12 - (-12) = 0$

Since $D = 0$, Cramer's Rule does not apply.

25. $\begin{cases} 2x - 4y = -2 \\ 3x + 2y = 3 \end{cases}$

$D = \begin{vmatrix} 2 & -4 \\ 3 & 2 \end{vmatrix} = 4 + 12 = 16$

$D_x = \begin{vmatrix} -2 & -4 \\ 3 & 2 \end{vmatrix} = -4 + 12 = 8$

$D_y = \begin{vmatrix} 2 & -2 \\ 3 & 3 \end{vmatrix} = 6 + 6 = 12$

Find the solutions by Cramer's Rule:

$x = \dfrac{D_x}{D} = \dfrac{8}{16} = \dfrac{1}{2}$ $\qquad y = \dfrac{D_y}{D} = \dfrac{12}{16} = \dfrac{3}{4}$

The solution is $\left(\dfrac{1}{2}, \dfrac{3}{4} \right)$.

27. $\begin{cases} 2x - 3y = -1 \\ 10x + 10y = 5 \end{cases}$

$D = \begin{vmatrix} 2 & -3 \\ 10 & 10 \end{vmatrix} = 20 - (-30) = 50$

$D_x = \begin{vmatrix} -1 & -3 \\ 5 & 10 \end{vmatrix} = -10 - (-15) = 5$

$D_y = \begin{vmatrix} 2 & -1 \\ 10 & 5 \end{vmatrix} = 10 - (-10) = 20$

Find the solutions by Cramer's Rule:

$x = \dfrac{D_x}{D} = \dfrac{5}{50} = \dfrac{1}{10}$ $\qquad y = \dfrac{D_y}{D} = \dfrac{20}{50} = \dfrac{2}{5}$

The solution is $\left(\dfrac{1}{10}, \dfrac{2}{5} \right)$.

29. $\begin{cases} 2x + 3y = 6 \\ x - y = \dfrac{1}{2} \end{cases}$

$D = \begin{vmatrix} 2 & 3 \\ 1 & -1 \end{vmatrix} = -2 - 3 = -5$

$D_x = \begin{vmatrix} 6 & 3 \\ \frac{1}{2} & -1 \end{vmatrix} = -6 - \dfrac{3}{2} = -\dfrac{15}{2}$

$D_y = \begin{vmatrix} 2 & 6 \\ 1 & \frac{1}{2} \end{vmatrix} = 1 - 6 = -5$

Find the solutions by Cramer's Rule:

$x = \dfrac{D_x}{D} = \dfrac{-\frac{15}{2}}{-5} = \dfrac{3}{2}$ $\qquad y = \dfrac{D_y}{D} = \dfrac{-5}{-5} = 1$

The solution is $\left(\dfrac{3}{2}, 1 \right)$.

31. $\begin{cases} 3x - 5y = 3 \\ 15x + 5y = 21 \end{cases}$

$D = \begin{vmatrix} 3 & -5 \\ 15 & 5 \end{vmatrix} = 15 - (-75) = 90$

$D_x = \begin{vmatrix} 3 & -5 \\ 21 & 5 \end{vmatrix} = 15 - (-105) = 120$

$D_y = \begin{vmatrix} 3 & 3 \\ 15 & 21 \end{vmatrix} = 63 - 45 = 18$

Find the solutions by Cramer's Rule:

$x = \dfrac{D_x}{D} = \dfrac{120}{90} = \dfrac{4}{3}$ $\qquad y = \dfrac{D_y}{D} = \dfrac{18}{90} = \dfrac{1}{5}$

The solution is $\left(\dfrac{4}{3}, \dfrac{1}{5} \right)$.

33. $\begin{cases} x + y - z = 6 \\ 3x - 2y + z = -5 \\ x + 3y - 2z = 14 \end{cases}$

$D = \begin{vmatrix} 1 & 1 & -1 \\ 3 & -2 & 1 \\ 1 & 3 & -2 \end{vmatrix}$

$= 1 \begin{vmatrix} -2 & 1 \\ 3 & -2 \end{vmatrix} - 1 \begin{vmatrix} 3 & 1 \\ 1 & -2 \end{vmatrix} + (-1) \begin{vmatrix} 3 & -2 \\ 1 & 3 \end{vmatrix}$

$= 1(4 - 3) - 1(-6 - 1) - 1(9 + 2)$

$= 1 + 7 - 11$

$= -3$

483

$$D_x = \begin{vmatrix} 6 & 1 & -1 \\ -5 & -2 & 1 \\ 14 & 3 & -2 \end{vmatrix}$$

$$= 6\begin{vmatrix} -2 & 1 \\ 3 & -2 \end{vmatrix} - 1\begin{vmatrix} -5 & 1 \\ 14 & -2 \end{vmatrix} + (-1)\begin{vmatrix} -5 & -2 \\ 14 & 3 \end{vmatrix}$$

$$= 6(4-3) - 1(10-14) - 1(-15+28)$$

$$= 6 + 4 - 13$$

$$= -3$$

$$D_y = \begin{vmatrix} 1 & 6 & -1 \\ 3 & -5 & 1 \\ 1 & 14 & -2 \end{vmatrix}$$

$$= 1\begin{vmatrix} -5 & 1 \\ 14 & -2 \end{vmatrix} - 6\begin{vmatrix} 3 & 1 \\ 1 & -2 \end{vmatrix} + (-1)\begin{vmatrix} 3 & -5 \\ 1 & 14 \end{vmatrix}$$

$$= 1(10-14) - 6(-6-1) - 1(42+5)$$

$$= -4 + 42 - 47$$

$$= -9$$

$$D_z = \begin{vmatrix} 1 & 1 & 6 \\ 3 & -2 & -5 \\ 1 & 3 & 14 \end{vmatrix}$$

$$= 1\begin{vmatrix} -2 & -5 \\ 3 & 14 \end{vmatrix} - 1\begin{vmatrix} 3 & -5 \\ 1 & 14 \end{vmatrix} + 6\begin{vmatrix} 3 & -2 \\ 1 & 3 \end{vmatrix}$$

$$= 1(-28+15) - 1(42+5) + 6(9+2)$$

$$= -13 - 47 + 66$$

$$= 6$$

Find the solutions by Cramer's Rule:

$$x = \frac{D_x}{D} = \frac{-3}{-3} = 1 \qquad y = \frac{D_y}{D} = \frac{-9}{-3} = 3$$

$$z = \frac{D_z}{D} = \frac{6}{-3} = -2$$

The solution is $(1, 3, -2)$.

35. $\begin{cases} x+2y-z = -3 \\ 2x-4y+z = -7 \\ -2x+2y-3z = 4 \end{cases}$

$$D = \begin{vmatrix} 1 & 2 & -1 \\ 2 & -4 & 1 \\ -2 & 2 & -3 \end{vmatrix}$$

$$= 1\begin{vmatrix} -4 & 1 \\ 2 & -3 \end{vmatrix} - 2\begin{vmatrix} 2 & 1 \\ -2 & -3 \end{vmatrix} + (-1)\begin{vmatrix} 2 & -4 \\ -2 & 2 \end{vmatrix}$$

$$= 1(12-2) - 2(-6+2) - 1(4-8)$$

$$= 10 + 8 + 4$$

$$= 22$$

$$D_x = \begin{vmatrix} -3 & 2 & -1 \\ -7 & -4 & 1 \\ 4 & 2 & -3 \end{vmatrix}$$

$$= -3\begin{vmatrix} -4 & 1 \\ 2 & -3 \end{vmatrix} - 2\begin{vmatrix} -7 & 1 \\ 4 & -3 \end{vmatrix} + (-1)\begin{vmatrix} -7 & -4 \\ 4 & 2 \end{vmatrix}$$

$$= -3(12-2) - 2(21-4) - 1(-14+16)$$

$$= -30 - 34 - 2$$

$$= -66$$

$$D_y = \begin{vmatrix} 1 & -3 & -1 \\ 2 & -7 & 1 \\ -2 & 4 & -3 \end{vmatrix}$$

$$= 1\begin{vmatrix} -7 & 1 \\ 4 & -3 \end{vmatrix} - (-3)\begin{vmatrix} 2 & 1 \\ -2 & -3 \end{vmatrix} + (-1)\begin{vmatrix} 2 & -7 \\ -2 & 4 \end{vmatrix}$$

$$= 1(21-4) + 3(-6+2) - 1(8-14)$$

$$= 17 - 12 + 6$$

$$= 11$$

$$D_z = \begin{vmatrix} 1 & 2 & -3 \\ 2 & -4 & -7 \\ -2 & 2 & 4 \end{vmatrix}$$

$$= 1\begin{vmatrix} -4 & -7 \\ 2 & 4 \end{vmatrix} - 2\begin{vmatrix} 2 & -7 \\ -2 & 4 \end{vmatrix} + (-3)\begin{vmatrix} 2 & -4 \\ -2 & 2 \end{vmatrix}$$

$$= 1(-16+14) - 2(8-14) - 3(4-8)$$

$$= -2 + 12 + 12$$

$$= 22$$

Find the solutions by Cramer's Rule:

$$x = \frac{D_x}{D} = \frac{-66}{22} = -3 \qquad y = \frac{D_y}{D} = \frac{11}{22} = \frac{1}{2}$$

$$z = \frac{D_z}{D} = \frac{22}{22} = 1$$

The solution is $\left(-3, \frac{1}{2}, 1\right)$.

37. $\begin{cases} x-2y+3z=1 \\ 3x+y-2z=0 \\ 2x-4y+6z=2 \end{cases}$

$D = \begin{vmatrix} 1 & -2 & 3 \\ 3 & 1 & -2 \\ 2 & -4 & 6 \end{vmatrix}$

$= 1\begin{vmatrix} 1 & -2 \\ -4 & 6 \end{vmatrix} - (-2)\begin{vmatrix} 3 & -2 \\ 2 & 6 \end{vmatrix} + 3\begin{vmatrix} 3 & 1 \\ 2 & -4 \end{vmatrix}$

$= 1(6-8) + 2(18+4) + 3(-12-2)$

$= -2 + 44 - 42$

$= 0$

Since $D = 0$, Cramer's Rule does not apply.

39. $\begin{cases} x+2y-z=0 \\ 2x-4y+z=0 \\ -2x+2y-3z=0 \end{cases}$

$D = \begin{vmatrix} 1 & 2 & -1 \\ 2 & -4 & 1 \\ -2 & 2 & -3 \end{vmatrix}$

$= 1\begin{vmatrix} -4 & 1 \\ 2 & -3 \end{vmatrix} - 2\begin{vmatrix} 2 & 1 \\ -2 & -3 \end{vmatrix} + (-1)\begin{vmatrix} 2 & -4 \\ -2 & 2 \end{vmatrix}$

$= 1(12-2) - 2(-6+2) - 1(4-8)$

$= 10 + 8 + 4$

$= 22$

$D_x = \begin{vmatrix} 0 & 2 & -1 \\ 0 & -4 & 1 \\ 0 & 2 & -3 \end{vmatrix} = 0$ [By Theorem (12)]

$D_y = \begin{vmatrix} 1 & 0 & -1 \\ 2 & 0 & 1 \\ -2 & 0 & -3 \end{vmatrix} = 0$ [By Theorem (12)]

$D_z = \begin{vmatrix} 1 & 2 & 0 \\ 2 & -4 & 0 \\ -2 & 2 & 0 \end{vmatrix} = 0$ [By Theorem (12)]

Find the solutions by Cramer's Rule:

$x = \dfrac{D_x}{D} = \dfrac{0}{22} = 0 \qquad y = \dfrac{D_y}{D} = \dfrac{0}{22} = 0$

$z = \dfrac{D_z}{D} = \dfrac{0}{22} = 0$

The solution is (0, 0, 0).

41. $\begin{cases} x-2y+3z=0 \\ 3x+y-2z=0 \\ 2x-4y+6z=0 \end{cases}$

$D = \begin{vmatrix} 1 & -2 & 3 \\ 3 & 1 & -2 \\ 2 & -4 & 6 \end{vmatrix}$

$= 1\begin{vmatrix} 1 & -2 \\ -4 & 6 \end{vmatrix} - (-2)\begin{vmatrix} 3 & -2 \\ 2 & 6 \end{vmatrix} + 3\begin{vmatrix} 3 & 1 \\ 2 & -4 \end{vmatrix}$

$= 1(6-8) + 2(18+4) + 3(-12-2)$

$= -2 + 44 - 42$

$= 0$

Since $D = 0$, Cramer's Rule does not apply.

43. $\begin{vmatrix} x & y & z \\ u & v & w \\ 1 & 2 & 3 \end{vmatrix} = 4$

By Theorem (11), the value of a determinant changes sign if any two rows are interchanged.

Thus, $\begin{vmatrix} 1 & 2 & 3 \\ u & v & w \\ x & y & z \end{vmatrix} = -4$.

45. Let $\begin{vmatrix} x & y & z \\ u & v & w \\ 1 & 2 & 3 \end{vmatrix} = 4$.

$\begin{vmatrix} x & y & z \\ -3 & -6 & -9 \\ u & v & w \end{vmatrix} = -3\begin{vmatrix} x & y & z \\ 1 & 2 & 3 \\ u & v & w \end{vmatrix}$ [Theorem (14)]

$\qquad = -3(-1)\begin{vmatrix} x & y & z \\ u & v & w \\ 1 & 2 & 3 \end{vmatrix}$ [Theorem (11)]

$\qquad = 3(4) = 12$

47. Let $\begin{vmatrix} x & y & z \\ u & v & w \\ 1 & 2 & 3 \end{vmatrix} = 4$

$$\begin{vmatrix} 1 & 2 & 3 \\ x-3 & y-6 & z-9 \\ 2u & 2v & 2w \end{vmatrix}$$

$$= 2 \begin{vmatrix} 1 & 2 & 3 \\ x-3 & y-6 & z-9 \\ u & v & w \end{vmatrix} \quad \text{[Theorem (14)]}$$

$$= 2(-1) \begin{vmatrix} x-3 & y-6 & z-9 \\ 1 & 2 & 3 \\ u & v & w \end{vmatrix} \quad \text{[Theorem (11)]}$$

$$= 2(-1)(-1) \begin{vmatrix} x-3 & y-6 & z-9 \\ u & v & w \\ 1 & 2 & 3 \end{vmatrix} \quad \text{[Theorem (11)]}$$

$$= 2(-1)(-1) \begin{vmatrix} x & y & z \\ u & v & w \\ 1 & 2 & 3 \end{vmatrix} \quad \begin{array}{l} \text{[Theorem (15)]} \\ (R_1 = -3r_3 + r_1) \end{array}$$

$$= 2(-1)(-1)(4) = 8$$

49. Let $\begin{vmatrix} x & y & z \\ u & v & w \\ 1 & 2 & 3 \end{vmatrix} = 4$

$$\begin{vmatrix} 1 & 2 & 3 \\ 2x & 2y & 2z \\ u-1 & v-2 & w-3 \end{vmatrix}$$

$$= 2 \begin{vmatrix} 1 & 2 & 3 \\ x & y & z \\ u-1 & v-2 & w-3 \end{vmatrix} \quad \text{[Theorem (14)]}$$

$$= 2(-1) \begin{vmatrix} x & y & z \\ 1 & 2 & 3 \\ u-1 & v-2 & w-3 \end{vmatrix} \quad \text{[Theorem (11)]}$$

$$= 2(-1)(-1) \begin{vmatrix} x & y & z \\ u-1 & v-2 & w-3 \\ 1 & 2 & 3 \end{vmatrix} \quad \text{[Theorem (11)]}$$

$$= 2(-1)(-1) \begin{vmatrix} x & y & z \\ u & v & w \\ 1 & 2 & 3 \end{vmatrix} \quad \begin{array}{l} \text{[Theorem (15)]} \\ (R_2 = -r_3 + r_2) \end{array}$$

$$= 2(-1)(-1)(4)$$

$$= 8$$

51. Solve for x:

$$\begin{vmatrix} x & x \\ 4 & 3 \end{vmatrix} = 5$$

$$3x - 4x = 5$$

$$-x = 5$$

$$x = -5$$

53. Solve for x:

$$\begin{vmatrix} x & 1 & 1 \\ 4 & 3 & 2 \\ -1 & 2 & 5 \end{vmatrix} = 2$$

$$x \begin{vmatrix} 3 & 2 \\ 2 & 5 \end{vmatrix} - 1 \begin{vmatrix} 4 & 2 \\ -1 & 5 \end{vmatrix} + 1 \begin{vmatrix} 4 & 3 \\ -1 & 2 \end{vmatrix} = 2$$

$$x(15-4) - (20+2) + (8+3) = 2$$

$$11x - 22 + 11 = 2$$

$$11x = 13$$

$$x = \frac{13}{11}$$

55. Solve for x:

$$\begin{vmatrix} x & 2 & 3 \\ 1 & x & 0 \\ 6 & 1 & -2 \end{vmatrix} = 7$$

$$x \begin{vmatrix} x & 0 \\ 1 & -2 \end{vmatrix} - 2 \begin{vmatrix} 1 & 0 \\ 6 & -2 \end{vmatrix} + 3 \begin{vmatrix} 1 & x \\ 6 & 1 \end{vmatrix} = 7$$

$$x(-2x) - 2(-2) + 3(1-6x) = 7$$

$$-2x^2 + 4 + 3 - 18x = 7$$

$$-2x^2 - 18x = 0$$

$$-2x(x+9) = 0$$

$$x = 0 \ \text{ or } \ x = -9$$

57. Expanding the determinant:

$$\begin{vmatrix} x & y & 1 \\ x_1 & y_1 & 1 \\ x_2 & y_2 & 1 \end{vmatrix} = 0$$

$$x \begin{vmatrix} y_1 & 1 \\ y_2 & 1 \end{vmatrix} - y \begin{vmatrix} x_1 & 1 \\ x_2 & 1 \end{vmatrix} + 1 \begin{vmatrix} x_1 & y_1 \\ x_2 & y_2 \end{vmatrix} = 0$$

$$x(y_1 - y_2) - y(x_1 - x_2) + (x_1 y_2 - x_2 y_1) = 0$$

$$x(y_1 - y_2) + y(x_2 - x_1) = x_2 y_1 - x_1 y_2$$

$$y(x_2 - x_1) = x_2 y_1 - x_1 y_2 + x(y_2 - y_1)$$

$$y(x_2 - x_1) - y_1(x_2 - x_1)$$
$$= x_2 y_1 - x_1 y_2 + x(y_2 - y_1) - y_1(x_2 - x_1)$$
$$(x_2 - x_1)(y - y_1)$$
$$= x(y_2 - y_1) + x_2 y_1 - x_1 y_2 - y_1 x_2 + y_1 x_1$$
$$(x_2 - x_1)(y - y_1) = (y_2 - y_1)x - (y_2 - y_1)x_1$$
$$(x_2 - x_1)(y - y_1) = (y_2 - y_1)(x - x_1)$$
$$(y - y_1) = \frac{(y_2 - y_1)}{(x_2 - x_1)}(x - x_1)$$

This is the 2-point form of the equation for a line.

59. If the vertices of a triangle are (2, 3), (5, 2), and (6, 5), then:

$$D = \frac{1}{2}\begin{vmatrix} 2 & 5 & 6 \\ 3 & 2 & 5 \\ 1 & 1 & 1 \end{vmatrix}$$

$$= \frac{1}{2}\left(2\begin{vmatrix} 2 & 5 \\ 1 & 1 \end{vmatrix} - 5\begin{vmatrix} 3 & 5 \\ 1 & 1 \end{vmatrix} + 6\begin{vmatrix} 3 & 2 \\ 1 & 1 \end{vmatrix}\right)$$

$$= \frac{1}{2}[2(2-5) - 5(3-5) + 6(3-2)]$$

$$= \frac{1}{2}[2(-3) - 5(-2) + 6(1)]$$

$$= \frac{1}{2}[-6 + 10 + 6]$$

$$= 5$$

The area of the triangle is $|5| = 5$ square units.

61. $A = \frac{1}{2}\left(\begin{vmatrix} 6 & 8 \\ 1 & 6 \end{vmatrix} + \begin{vmatrix} 1 & 6 \\ -1 & 3 \end{vmatrix} + \begin{vmatrix} -1 & 3 \\ 6 & -2 \end{vmatrix} + \begin{vmatrix} 6 & -2 \\ 8 & 4 \end{vmatrix} + \begin{vmatrix} 8 & 4 \\ 6 & 8 \end{vmatrix}\right)$

$$= \frac{1}{2}[(36-8) + (3+6) + (2-18) + (24+16) + (64-24)]$$

$$= \frac{1}{2}[28 + 9 - 16 + 40 + 40] = 50.5 \text{ square units}$$

63.

$$\begin{vmatrix} 1 & 2 & 2 & 2 \\ x & 7 & 3 & 6 \\ y & -5 & 3 & 2 \\ x^2 + y^2 & 74 & 18 & 40 \end{vmatrix}$$

$$= \begin{vmatrix} 7 & 3 & 6 \\ -5 & 3 & 2 \\ 74 & 18 & 40 \end{vmatrix} - x\begin{vmatrix} 1 & 1 & 1 \\ -5 & 3 & 2 \\ 74 & 18 & 40 \end{vmatrix}$$

$$+ y\begin{vmatrix} 1 & 1 & 1 \\ 7 & 3 & 6 \\ 74 & 18 & 40 \end{vmatrix} - (x^2 + y^2)\begin{vmatrix} 1 & 1 & 1 \\ 7 & 3 & 6 \\ -5 & 3 & 2 \end{vmatrix}$$

$$= (-240) - x(120) + y(80) - (x^2 + y^2)(-20)$$
$$= -240 - 120x + 80y + 20x^2 + 20y^2$$

Now set this expression equal to 0. Then complete the square to obtain the standard form.

$$-240 - 120x + 80y + 20x^2 + 20y^2 = 0$$
$$20x^2 - 120x + 20y^2 + 80y = 240$$
$$x^2 - 6x + y^2 + 4y = 12$$
$$x^2 - 6x + 9 + y^2 + 4y + 4 = 12 + 9 + 4$$
$$(x - 3)^2 + (y + 2)^2 = 25$$

65. If $a = 0$, then $b \neq 0$ and $c \neq 0$ since $ad - bc \neq 0$, and the system is $\begin{cases} by = s \\ cx + dy = t \end{cases}$.

The solution of the system is $y = \dfrac{s}{b}$,

$$x = \frac{t - dy}{c} = \frac{t - d\left(\frac{s}{b}\right)}{c} = \frac{tb - sd}{bc}. \text{ Using Cramer's}$$

Rule, we get $D = \begin{vmatrix} 0 & b \\ c & d \end{vmatrix} = -bc$,

$$D_x = \begin{vmatrix} s & b \\ t & d \end{vmatrix} = sd - tb,$$

$$D_y = \begin{vmatrix} 0 & s \\ c & t \end{vmatrix} = 0 - sc = -sc, \text{ so}$$

$$x = \frac{D_x}{D} = \frac{ds - tb}{-bc} = \frac{td - sd}{bc} \text{ and}$$

$$y = \frac{D_y}{D} = \frac{-sc}{-bc} = \frac{s}{b}, \text{ which is the solution. Note}$$

that these solutions agree if $d = 0$.

If $b = 0$, then $a \neq 0$ and $d \neq 0$ since $ad - bc \neq 0$, and the system is $\begin{cases} ax = s \\ cx + dy = t \end{cases}$.

The solution of the system is $x = \dfrac{s}{a}$,

$$y = \frac{t - cx}{d} = \frac{at - cs}{ad}. \text{ Using Cramer's Rule, we}$$

get $D = \begin{vmatrix} a & 0 \\ c & d \end{vmatrix} = ad$, $D_x = \begin{vmatrix} s & 0 \\ t & d \end{vmatrix} = sd$, and

$$D_y = \begin{vmatrix} a & s \\ c & t \end{vmatrix} = at - cs, \text{ so } x = \frac{D_x}{D} = \frac{sd}{ad} = \frac{s}{a}$$

and $y = \dfrac{D_y}{D} = \dfrac{at - cs}{ad}$, which is the solution.

Note that these solutions agree if $c = 0$.

If $c = 0$, then $a \neq 0$ and $d \neq 0$ since

$ad - bc \neq 0$, and the system is $\begin{cases} ax + by = s \\ \qquad dy = t \end{cases}$.

The solution of the system is $y = \dfrac{t}{d}$,

$x = \dfrac{s - by}{a} = \dfrac{sd - tb}{ad}$. Using Cramer's Rule, we

get $D = \begin{vmatrix} a & b \\ 0 & d \end{vmatrix} = ad$, $D_x = \begin{vmatrix} s & b \\ t & d \end{vmatrix} = sd - tb$,

and $D_y = \begin{vmatrix} a & s \\ 0 & t \end{vmatrix} = at$, so $x = \dfrac{D_x}{D} = \dfrac{sd - tb}{ad}$ and

$y = \dfrac{D_y}{D} = \dfrac{at}{ad} = \dfrac{t}{d}$, which is the solution. Note

that these solutions agree if $b = 0$.

If $d = 0$, then $b \neq 0$ and $c \neq 0$ since

$ad - bc \neq 0$, and the system is $\begin{cases} ax + by = s \\ cx \qquad = t \end{cases}$.

The solution of the system is $x = \dfrac{t}{c}$,

$y = \dfrac{s - ax}{b} = \dfrac{cs - at}{bc}$. Using Cramer's Rule, we

get $D = \begin{vmatrix} a & b \\ c & 0 \end{vmatrix} = 0 - bc = -bc$,

$D_x = \begin{vmatrix} s & b \\ t & 0 \end{vmatrix} = 0 - tb = -tb$, and

$D_y = \begin{vmatrix} a & s \\ c & t \end{vmatrix} = at - cs$, so $x = \dfrac{D_x}{D} = \dfrac{-tb}{-bc} = \dfrac{t}{c}$ and

$y = \dfrac{D_y}{D} = \dfrac{at - cs}{-bc} = \dfrac{cs - at}{bc}$, which is the

solution. Note that these solutions agree if $a = 0$.

67. Evaluating the determinant to show the relationship:

$\begin{vmatrix} a_{11} & a_{12} & a_{13} \\ ka_{21} & ka_{22} & ka_{23} \\ a_{31} & a_{32} & a_{33} \end{vmatrix}$

$= a_{11} \begin{vmatrix} ka_{22} & ka_{23} \\ a_{32} & a_{33} \end{vmatrix} - a_{12} \begin{vmatrix} ka_{21} & ka_{23} \\ a_{31} & a_{33} \end{vmatrix} + a_{13} \begin{vmatrix} ka_{21} & ka_{22} \\ a_{31} & a_{32} \end{vmatrix}$

$= a_{11}(ka_{22}a_{33} - ka_{23}a_{32}) - a_{12}(ka_{21}a_{33} - ka_{23}a_{31})$
$\qquad + a_{13}(ka_{21}a_{32} - ka_{22}a_{31})$

$= ka_{11}(a_{22}a_{33} - a_{23}a_{32}) - ka_{12}(a_{21}a_{33} - a_{23}a_{31})$
$\qquad + ka_{13}(a_{21}a_{32} - a_{22}a_{31})$

$= k \big(a_{11}(a_{22}a_{33} - a_{23}a_{32}) - a_{12}(a_{21}a_{33} - a_{23}a_{31})$
$\qquad + a_{13}(a_{21}a_{32} - a_{22}a_{31}) \big)$

$= k \left(a_{11} \begin{vmatrix} a_{22} & a_{23} \\ a_{32} & a_{33} \end{vmatrix} - a_{12} \begin{vmatrix} a_{21} & a_{23} \\ a_{31} & a_{33} \end{vmatrix} + a_{13} \begin{vmatrix} a_{21} & a_{22} \\ a_{31} & a_{32} \end{vmatrix} \right)$

$= k \begin{vmatrix} a_{11} & a_{12} & a_{13} \\ a_{21} & a_{22} & a_{23} \\ a_{31} & a_{32} & a_{33} \end{vmatrix}$

69. Evaluating the determinant to show the relationship:

$$\begin{vmatrix} a_{11}+ka_{21} & a_{12}+ka_{22} & a_{13}+ka_{23} \\ a_{21} & a_{22} & a_{23} \\ a_{31} & a_{32} & a_{33} \end{vmatrix}$$

$$= (a_{11}+ka_{21})\begin{vmatrix} a_{22} & a_{23} \\ a_{32} & a_{33} \end{vmatrix} - (a_{12}+ka_{22})\begin{vmatrix} a_{21} & a_{23} \\ a_{31} & a_{33} \end{vmatrix}$$

$$\quad + (a_{13}+ka_{23})\begin{vmatrix} a_{21} & a_{22} \\ a_{31} & a_{32} \end{vmatrix}$$

$$= (a_{11}+ka_{21})(a_{22}a_{33}-a_{23}a_{32})$$
$$\quad - (a_{12}+ka_{22})(a_{21}a_{33}-a_{23}a_{31})$$
$$\quad + (a_{13}+ka_{23})(a_{21}a_{32}-a_{22}a_{31})$$

$$= a_{11}(a_{22}a_{33}-a_{23}a_{32}) + ka_{21}(a_{22}a_{33}-a_{23}a_{32})$$
$$\quad - a_{12}(a_{21}a_{33}-a_{23}a_{31}) - ka_{22}(a_{21}a_{33}-a_{23}a_{31})$$
$$\quad + a_{13}(a_{21}a_{32}-a_{22}a_{31}) + ka_{23}(a_{21}a_{32}-a_{22}a_{31})$$

$$= a_{11}(a_{22}a_{33}-a_{23}a_{32}) + ka_{21}a_{22}a_{33}$$
$$\quad - ka_{21}a_{23}a_{32} - a_{12}(a_{21}a_{33}-a_{23}a_{31})$$
$$\quad - ka_{22}a_{21}a_{33} + ka_{22}a_{23}a_{31}$$
$$\quad + a_{13}(a_{21}a_{32}-a_{22}a_{31}) + ka_{23}a_{21}a_{32}$$
$$\quad - ka_{23}a_{22}a_{31}$$

$$= a_{11}(a_{22}a_{33}-a_{23}a_{32}) - a_{12}(a_{21}a_{33}-a_{23}a_{31})$$
$$\quad + a_{13}(a_{21}a_{32}-a_{22}a_{31})$$

$$= a_{11}\begin{vmatrix} a_{22} & a_{23} \\ a_{32} & a_{33} \end{vmatrix} - a_{12}\begin{vmatrix} a_{21} & a_{23} \\ a_{31} & a_{33} \end{vmatrix} + a_{13}\begin{vmatrix} a_{21} & a_{22} \\ a_{31} & a_{32} \end{vmatrix}$$

$$= \begin{vmatrix} a_{11} & a_{12} & a_{13} \\ a_{21} & a_{22} & a_{23} \\ a_{31} & a_{32} & a_{33} \end{vmatrix}$$

71. $f(x) = 2x^3 - 5x^2 + x - 10$

p must be a factor of 10: $p = \pm 1, \pm 2, \pm 5, \pm 10$

q must be a factor of 2: $q = \pm 1, \pm 2$

The possible rational zeros are:

$$\frac{p}{q} = \pm\frac{1}{2}, \pm\frac{5}{2}, \pm 1, \pm 2, \pm 5, \pm 10$$

73. $5^x = y \leftrightarrow x = \log_5 y$

Section 8.4

1. square

3. false

5. True

7. a

9. $A + B = \begin{bmatrix} 0 & 3 & -5 \\ 1 & 2 & 6 \end{bmatrix} + \begin{bmatrix} 4 & 1 & 0 \\ -2 & 3 & -2 \end{bmatrix}$

$$= \begin{bmatrix} 0+4 & 3+1 & -5+0 \\ 1+(-2) & 2+3 & 6+(-2) \end{bmatrix}$$

$$= \begin{bmatrix} 4 & 4 & -5 \\ -1 & 5 & 4 \end{bmatrix}$$

11. $4A = 4\begin{bmatrix} 0 & 3 & -5 \\ 1 & 2 & 6 \end{bmatrix}$

$$= \begin{bmatrix} 4\cdot 0 & 4\cdot 3 & 4(-5) \\ 4\cdot 1 & 4\cdot 2 & 4\cdot 6 \end{bmatrix}$$

$$= \begin{bmatrix} 0 & 12 & -20 \\ 4 & 8 & 24 \end{bmatrix}$$

13. $3A - 2B = 3\begin{bmatrix} 0 & 3 & -5 \\ 1 & 2 & 6 \end{bmatrix} - 2\begin{bmatrix} 4 & 1 & 0 \\ -2 & 3 & -2 \end{bmatrix}$

$$= \begin{bmatrix} 0 & 9 & -15 \\ 3 & 6 & 18 \end{bmatrix} - \begin{bmatrix} 8 & 2 & 0 \\ -4 & 6 & -4 \end{bmatrix}$$

$$= \begin{bmatrix} -8 & 7 & -15 \\ 7 & 0 & 22 \end{bmatrix}$$

15. $AC = \begin{bmatrix} 0 & 3 & -5 \\ 1 & 2 & 6 \end{bmatrix} \cdot \begin{bmatrix} 4 & 1 \\ 6 & 2 \\ -2 & 3 \end{bmatrix}$

$$= \begin{bmatrix} 0(4)+3(6)+(-5)(-2) & 0(1)+3(2)+(-5)(3) \\ 1(4)+2(6)+6(-2) & 1(1)+2(2)+6(3) \end{bmatrix}$$

$$= \begin{bmatrix} 28 & -9 \\ 4 & 23 \end{bmatrix}$$

17. $AB = \begin{bmatrix} 0 & 3 & -5 \\ 1 & 2 & 6 \end{bmatrix}\begin{bmatrix} 4 & 1 & 0 \\ -2 & 3 & -2 \end{bmatrix}$

Not defined since the number of columns in A is not equal to the number of rows in B.

19. $CA = \begin{bmatrix} 4 & 1 \\ 6 & 2 \\ -2 & 3 \end{bmatrix} \cdot \begin{bmatrix} 0 & 3 & -5 \\ 1 & 2 & 6 \end{bmatrix}$

$= \begin{bmatrix} 4(0)+1(1) & 4(3)+1(2) & 4(-5)+1(6) \\ 6(0)+2(1) & 6(3)+2(2) & 6(-5)+2(6) \\ -2(0)+3(1) & -2(3)+3(2) & -2(-5)+3(6) \end{bmatrix}$

$= \begin{bmatrix} 1 & 14 & -14 \\ 2 & 22 & -18 \\ 3 & 0 & 28 \end{bmatrix}$

21. $C(A+B) = \begin{bmatrix} 4 & 1 \\ 6 & 2 \\ -2 & 3 \end{bmatrix} \left(\begin{bmatrix} 0 & 3 & -5 \\ 1 & 2 & 6 \end{bmatrix} + \begin{bmatrix} 4 & 1 & 0 \\ -2 & 3 & -2 \end{bmatrix} \right)$

$= \begin{bmatrix} 4 & 1 \\ 6 & 2 \\ -2 & 3 \end{bmatrix} \cdot \begin{bmatrix} 4 & 4 & -5 \\ -1 & 5 & 4 \end{bmatrix}$

$= \begin{bmatrix} 15 & 21 & -16 \\ 22 & 34 & -22 \\ -11 & 7 & 22 \end{bmatrix}$

23. $AC - 3I_2 = \begin{bmatrix} 0 & 3 & -5 \\ 1 & 2 & 6 \end{bmatrix} \cdot \begin{bmatrix} 4 & 1 \\ 6 & 2 \\ -2 & 3 \end{bmatrix} - 3\begin{bmatrix} 1 & 0 \\ 0 & 1 \end{bmatrix}$

$= \begin{bmatrix} 28 & -9 \\ 4 & 23 \end{bmatrix} - \begin{bmatrix} 3 & 0 \\ 0 & 3 \end{bmatrix}$

$= \begin{bmatrix} 25 & -9 \\ 4 & 20 \end{bmatrix}$

25. $CA - CB$

$= \begin{bmatrix} 4 & 1 \\ 6 & 2 \\ -2 & 3 \end{bmatrix} \cdot \begin{bmatrix} 0 & 3 & -5 \\ 1 & 2 & 6 \end{bmatrix} - \begin{bmatrix} 4 & 1 \\ 6 & 2 \\ -2 & 3 \end{bmatrix} \cdot \begin{bmatrix} 4 & 1 & 0 \\ -2 & 3 & -2 \end{bmatrix}$

$= \begin{bmatrix} 1 & 14 & -14 \\ 2 & 22 & -18 \\ 3 & 0 & 28 \end{bmatrix} - \begin{bmatrix} 14 & 7 & -2 \\ 20 & 12 & -4 \\ -14 & 7 & -6 \end{bmatrix}$

$= \begin{bmatrix} -13 & 7 & -12 \\ -18 & 10 & -14 \\ 17 & -7 & 34 \end{bmatrix}$

27. $a_{11} = 2(2)+(-2)(3) = -2$

$a_{12} = 2(1)+(-2)(-1) = 4$

$a_{13} = 2(4)+(-2)(3) = 2$

$a_{14} = 2(6)+(-2)(2) = 8$

$a_{21} = 1(2)+0(3) = 2$

$a_{22} = 1(1)+0(-1) = 1$

$a_{23} = 1(4)+0(3) = 4$

$a_{24} = 1(6)+0(2) = 6$

$\begin{bmatrix} 2 & -2 \\ 1 & 0 \end{bmatrix} \begin{bmatrix} 2 & 1 & 4 & 6 \\ 3 & -1 & 3 & 2 \end{bmatrix} = \begin{bmatrix} -2 & 4 & 2 & 8 \\ 2 & 1 & 4 & 6 \end{bmatrix}$

29. $\begin{bmatrix} 1 & 2 & 3 \\ 0 & -1 & 4 \end{bmatrix} \begin{bmatrix} 1 & 2 \\ -1 & 0 \\ 2 & 4 \end{bmatrix}$

$= \begin{bmatrix} 1(1)+2(-1)+3(2) & 1(2)+2(0)+3(4) \\ 0(1)+(-1)(-1)+4(2) & 0(2)+(-1)(0)+4(4) \end{bmatrix}$

$= \begin{bmatrix} 5 & 14 \\ 9 & 16 \end{bmatrix}$

31. $\begin{array}{ccc} -4 & 1 & 0 \\ 2 & 3 & -1 \end{array}$

Not defined since the number of columns in the first matrix is not equal to the number of rows in the second matrix.

33. $\begin{bmatrix} 1 & 0 & 1 \\ 2 & 4 & 1 \\ 3 & 6 & 1 \end{bmatrix} \begin{bmatrix} 1 & 3 \\ 6 & 2 \\ 8 & -1 \end{bmatrix}$

$= \begin{bmatrix} 1(1)+0(6)+1(8) & 1(3)+0(2)+1(-1) \\ 2(1)+4(6)+1(8) & 2(3)+4(2)+1(-1) \\ 3(1)+6(6)+1(8) & 3(3)+6(2)+1(-1) \end{bmatrix}$

$= \begin{bmatrix} 9 & 2 \\ 34 & 13 \\ 47 & 20 \end{bmatrix}$

35. $A = \begin{bmatrix} 2 & 1 \\ 1 & 1 \end{bmatrix}$

Augment the matrix with the identity and use row operations to find the inverse:

$$\begin{bmatrix} 2 & 1 & | & 1 & 0 \\ 1 & 1 & | & 0 & 1 \end{bmatrix}$$

$$\rightarrow \begin{bmatrix} 1 & 1 & | & 0 & 1 \\ 2 & 1 & | & 1 & 0 \end{bmatrix} \begin{pmatrix} \text{Interchange} \\ r_1 \text{ and } r_2 \end{pmatrix}$$

$$\rightarrow \begin{bmatrix} 1 & 1 & | & 0 & 1 \\ 0 & -1 & | & 1 & -2 \end{bmatrix} (R_2 = -2r_1 + r_2)$$

$$\rightarrow \begin{bmatrix} 1 & 1 & | & 0 & 1 \\ 0 & 1 & | & -1 & 2 \end{bmatrix} (R_2 = -r_2)$$

$$\rightarrow \begin{bmatrix} 1 & 0 & | & 1 & -1 \\ 0 & 1 & | & -1 & 2 \end{bmatrix} (R_1 = -r_2 + r_1)$$

Thus, $A^{-1} = \begin{bmatrix} 1 & -1 \\ -1 & 2 \end{bmatrix}$.

37. $A = \begin{bmatrix} 6 & 5 \\ 2 & 2 \end{bmatrix}$

Augment the matrix with the identity and use row operations to find the inverse:

$$\begin{bmatrix} 6 & 5 & | & 1 & 0 \\ 2 & 2 & | & 0 & 1 \end{bmatrix}$$

$$\rightarrow \begin{bmatrix} 2 & 2 & | & 0 & 1 \\ 6 & 5 & | & 1 & 0 \end{bmatrix} \begin{pmatrix} \text{Interchange} \\ r_1 \text{ and } r_2 \end{pmatrix}$$

$$\rightarrow \begin{bmatrix} 2 & 2 & | & 0 & 1 \\ 0 & -1 & | & 1 & -3 \end{bmatrix} (R_2 = -3r_1 + r_2)$$

$$\rightarrow \begin{bmatrix} 1 & 1 & | & 0 & \frac{1}{2} \\ 0 & 1 & | & -1 & 3 \end{bmatrix} \begin{pmatrix} R_1 = \frac{1}{2}r_1 \\ R_2 = -r_2 \end{pmatrix}$$

$$\rightarrow \begin{bmatrix} 1 & 0 & | & 1 & -\frac{5}{2} \\ 0 & 1 & | & -1 & 3 \end{bmatrix} (R_1 = -r_2 + r_1)$$

Thus, $A^{-1} = \begin{bmatrix} 1 & -\frac{5}{2} \\ -1 & 3 \end{bmatrix}$.

39. $A = \begin{bmatrix} 2 & 1 \\ a & a \end{bmatrix}$ where $a \neq 0$.

Augment the matrix with the identity and use row operations to find the inverse:

$$\begin{bmatrix} 2 & 1 & | & 1 & 0 \\ a & a & | & 0 & 1 \end{bmatrix}$$

$$\rightarrow \begin{bmatrix} 1 & \frac{1}{2} & | & \frac{1}{2} & 0 \\ a & a & | & 0 & 1 \end{bmatrix} \left(R_1 = \frac{1}{2}r_1 \right)$$

$$\rightarrow \begin{bmatrix} 1 & \frac{1}{2} & | & \frac{1}{2} & 0 \\ 0 & \frac{1}{2}a & | & -\frac{1}{2}a & 1 \end{bmatrix} (R_2 = -ar_1 + r_2)$$

$$\rightarrow \begin{bmatrix} 1 & \frac{1}{2} & | & \frac{1}{2} & 0 \\ 0 & 1 & | & -1 & \frac{2}{a} \end{bmatrix} \left(R_2 = \frac{2}{a}r_2 \right)$$

$$\rightarrow \begin{bmatrix} 1 & 0 & | & 1 & -\frac{1}{a} \\ 0 & 1 & | & -1 & \frac{2}{a} \end{bmatrix} \left(R_1 = -\frac{1}{2}r_2 + r_1 \right)$$

Thus, $A^{-1} = \begin{bmatrix} 1 & -\frac{1}{a} \\ -1 & \frac{2}{a} \end{bmatrix}$.

41. $A = \begin{bmatrix} 1 & -1 & 1 \\ 0 & -2 & 1 \\ -2 & -3 & 0 \end{bmatrix}$

Augment the matrix with the identity and use row operations to find the inverse:

$$\begin{bmatrix} 1 & -1 & 1 & | & 1 & 0 & 0 \\ 0 & -2 & 1 & | & 0 & 1 & 0 \\ -2 & -3 & 0 & | & 0 & 0 & 1 \end{bmatrix}$$

$$\rightarrow \begin{bmatrix} 1 & -1 & 1 & | & 1 & 0 & 0 \\ 0 & -2 & 1 & | & 0 & 1 & 0 \\ 0 & -5 & 2 & | & 2 & 0 & 1 \end{bmatrix} \quad (R_3 = 2r_1 + r_3)$$

$$\rightarrow \begin{bmatrix} 1 & -1 & 1 & | & 1 & 0 & 0 \\ 0 & 1 & -\frac{1}{2} & | & 0 & -\frac{1}{2} & 0 \\ 0 & -5 & 2 & | & 2 & 0 & 1 \end{bmatrix} \quad \left(R_2 = -\frac{1}{2}r_2\right)$$

$$\rightarrow \begin{bmatrix} 1 & 0 & \frac{1}{2} & | & 1 & -\frac{1}{2} & 0 \\ 0 & 1 & -\frac{1}{2} & | & 0 & -\frac{1}{2} & 0 \\ 0 & 0 & -\frac{1}{2} & | & 2 & -\frac{5}{2} & 1 \end{bmatrix} \quad \begin{pmatrix} R_1 = r_2 + r_1 \\ R_3 = 5r_2 + r_3 \end{pmatrix}$$

$$\rightarrow \begin{bmatrix} 1 & 0 & \frac{1}{2} & | & 1 & -\frac{1}{2} & 0 \\ 0 & 1 & -\frac{1}{2} & | & 0 & -\frac{1}{2} & 0 \\ 0 & 0 & 1 & | & -4 & 5 & -2 \end{bmatrix} \quad (R_3 = -2r_3)$$

$$\rightarrow \begin{bmatrix} 1 & 0 & 0 & | & 3 & -3 & 1 \\ 0 & 1 & 0 & | & -2 & 2 & -1 \\ 0 & 0 & 1 & | & -4 & 5 & -2 \end{bmatrix} \quad \begin{pmatrix} R_1 = -\frac{1}{2}r_3 + r_1 \\ R_2 = \frac{1}{2}r_3 + r_2 \end{pmatrix}$$

Thus, $A^{-1} = \begin{bmatrix} 3 & -3 & 1 \\ -2 & 2 & -1 \\ -4 & 5 & -2 \end{bmatrix}$.

43. $A = \begin{bmatrix} 1 & 1 & 1 \\ 3 & 2 & -1 \\ 3 & 1 & 2 \end{bmatrix}$

Augment the matrix with the identity and use row operations to find the inverse:

$$\begin{bmatrix} 1 & 1 & 1 & | & 1 & 0 & 0 \\ 3 & 2 & -1 & | & 0 & 1 & 0 \\ 3 & 1 & 2 & | & 0 & 0 & 1 \end{bmatrix}$$

$$\rightarrow \begin{bmatrix} 1 & 1 & 1 & | & 1 & 0 & 0 \\ 0 & -1 & -4 & | & -3 & 1 & 0 \\ 0 & -2 & -1 & | & -3 & 0 & 1 \end{bmatrix} \quad \begin{pmatrix} R_2 = -3r_1 + r_2 \\ R_3 = -3r_1 + r_3 \end{pmatrix}$$

$$\rightarrow \begin{bmatrix} 1 & 1 & 1 & | & 1 & 0 & 0 \\ 0 & 1 & 4 & | & 3 & -1 & 0 \\ 0 & -2 & -1 & | & -3 & 0 & 1 \end{bmatrix} \quad (R_2 = -r_2)$$

$$\rightarrow \begin{bmatrix} 1 & 0 & -3 & | & -2 & 1 & 0 \\ 0 & 1 & 4 & | & 3 & -1 & 0 \\ 0 & 0 & 1 & | & \frac{3}{7} & -\frac{2}{7} & \frac{1}{7} \end{bmatrix} \quad \left(R_3 = \frac{1}{7}r_3\right)$$

$$\rightarrow \begin{bmatrix} 1 & 0 & 0 & | & -\frac{5}{7} & \frac{1}{7} & \frac{3}{7} \\ 0 & 1 & 0 & | & \frac{9}{7} & \frac{1}{7} & -\frac{4}{7} \\ 0 & 0 & 1 & | & \frac{3}{7} & -\frac{2}{7} & \frac{1}{7} \end{bmatrix} \quad \begin{pmatrix} R_1 = 3r_3 + r_1 \\ R_2 = -4r_3 + r_2 \end{pmatrix}$$

Thus, $A^{-1} = \begin{bmatrix} -\frac{5}{7} & \frac{1}{7} & \frac{3}{7} \\ \frac{9}{7} & \frac{1}{7} & -\frac{4}{7} \\ \frac{3}{7} & -\frac{2}{7} & \frac{1}{7} \end{bmatrix}$.

45. $\begin{cases} 2x + y = 8 \\ x + y = 5 \end{cases}$

Rewrite the system of equations in matrix form:

$A = \begin{bmatrix} 2 & 1 \\ 1 & 1 \end{bmatrix}, \quad X = \begin{bmatrix} x \\ y \end{bmatrix}, \quad B = \begin{bmatrix} 8 \\ 5 \end{bmatrix}$

Find the inverse of A and solve $X = A^{-1}B$:

From Problem 29, $A^{-1} = \begin{bmatrix} 1 & -1 \\ -1 & 2 \end{bmatrix}$, so

$X = A^{-1}B = \begin{bmatrix} 1 & -1 \\ -1 & 2 \end{bmatrix}\begin{bmatrix} 8 \\ 5 \end{bmatrix} = \begin{bmatrix} 3 \\ 2 \end{bmatrix}$.

The solution is $x = 3$, $y = 2$ or $(3, 2)$.

47. $\begin{cases} 2x + y = 0 \\ x + y = 5 \end{cases}$

Rewrite the system of equations in matrix form:

$A = \begin{bmatrix} 2 & 1 \\ 1 & 1 \end{bmatrix}, \quad X = \begin{bmatrix} x \\ y \end{bmatrix}, \quad B = \begin{bmatrix} 0 \\ 5 \end{bmatrix}$

Find the inverse of A and solve $X = A^{-1}B$:

From Problem 29, $A^{-1} = \begin{bmatrix} 1 & -1 \\ -1 & 2 \end{bmatrix}$, so

$X = A^{-1}B = \begin{bmatrix} 1 & -1 \\ -1 & 2 \end{bmatrix}\begin{bmatrix} 0 \\ 5 \end{bmatrix} = \begin{bmatrix} -5 \\ 10 \end{bmatrix}$.

The solution is $x = -5$, $y = 10$ or $(-5, 10)$.

49. $\begin{cases} 6x+5y=7 \\ 2x+2y=2 \end{cases}$

Rewrite the system of equations in matrix form:

$A = \begin{bmatrix} 6 & 5 \\ 2 & 2 \end{bmatrix}$, $X = \begin{bmatrix} x \\ y \end{bmatrix}$, $B = \begin{bmatrix} 7 \\ 2 \end{bmatrix}$

Find the inverse of A and solve $X = A^{-1}B$:

From Problem 31, $A^{-1} = \begin{bmatrix} 1 & -\frac{5}{2} \\ -1 & 3 \end{bmatrix}$, so

$X = A^{-1}B = \begin{bmatrix} 1 & -\frac{5}{2} \\ -1 & 3 \end{bmatrix}\begin{bmatrix} 7 \\ 2 \end{bmatrix} = \begin{bmatrix} 2 \\ -1 \end{bmatrix}$.

The solution is $x = 2$, $y = -1$ or $(2, -1)$.

51. $\begin{cases} 6x+5y=13 \\ 2x+2y=5 \end{cases}$

Rewrite the system of equations in matrix form:

$A = \begin{bmatrix} 6 & 5 \\ 2 & 2 \end{bmatrix}$, $X = \begin{bmatrix} x \\ y \end{bmatrix}$, $B = \begin{bmatrix} 13 \\ 5 \end{bmatrix}$

Find the inverse of A and solve $X = A^{-1}B$:

From Problem 31, $A^{-1} = \begin{bmatrix} 1 & -\frac{5}{2} \\ -1 & 3 \end{bmatrix}$, so

$X = A^{-1}B = \begin{bmatrix} 1 & -\frac{5}{2} \\ -1 & 3 \end{bmatrix}\begin{bmatrix} 13 \\ 5 \end{bmatrix} = \begin{bmatrix} \frac{1}{2} \\ 2 \end{bmatrix}$.

The solution is $x = \frac{1}{2}$, $y = 2$ or $\left(\frac{1}{2}, 2\right)$.

53. $\begin{cases} 2x+y=-3 \\ ax+ay=-a \end{cases} \quad a \neq 0$

Rewrite the system of equations in matrix form:

$A = \begin{bmatrix} 2 & 1 \\ a & a \end{bmatrix}$, $X = \begin{bmatrix} x \\ y \end{bmatrix}$, $B = \begin{bmatrix} -3 \\ -a \end{bmatrix}$

Find the inverse of A and solve $X = A^{-1}B$:

From Problem 33, $A^{-1} = \begin{bmatrix} 1 & -\frac{1}{a} \\ -1 & \frac{2}{a} \end{bmatrix}$, so

$X = A^{-1}B = \begin{bmatrix} 1 & -\frac{1}{a} \\ -1 & \frac{2}{a} \end{bmatrix}\begin{bmatrix} -3 \\ -a \end{bmatrix} = \begin{bmatrix} -2 \\ 1 \end{bmatrix}$.

The solution is $x = -2$, $y = 1$ or $(-2, 1)$.

55. $\begin{cases} 2x+y=\dfrac{7}{a} \\ ax+ay=5 \end{cases} \quad a \neq 0$

Rewrite the system of equations in matrix form:

$A = \begin{bmatrix} 2 & 1 \\ a & a \end{bmatrix}$, $X = \begin{bmatrix} x \\ y \end{bmatrix}$, $B = \begin{bmatrix} \frac{7}{a} \\ 5 \end{bmatrix}$

Find the inverse of A and solve $X = A^{-1}B$:

From Problem 33, $A^{-1} = \begin{bmatrix} 1 & -\frac{1}{a} \\ -1 & \frac{2}{a} \end{bmatrix}$, so

$X = A^{-1}B = \begin{bmatrix} 1 & -\frac{1}{a} \\ -1 & \frac{2}{a} \end{bmatrix}\begin{bmatrix} \frac{7}{a} \\ 5 \end{bmatrix} = \begin{bmatrix} \frac{2}{a} \\ \frac{3}{a} \end{bmatrix}$.

The solution is $x = \dfrac{2}{a}$, $y = \dfrac{3}{a}$ or $\left(\dfrac{2}{a}, \dfrac{3}{a}\right)$.

57. $\begin{cases} x-y+z=0 \\ -2y+z=-1 \\ -2x-3y=-5 \end{cases}$

Rewrite the system of equations in matrix form:

$A = \begin{bmatrix} 1 & -1 & 1 \\ 0 & -2 & 1 \\ -2 & -3 & 0 \end{bmatrix}$, $X = \begin{bmatrix} x \\ y \\ z \end{bmatrix}$, $B = \begin{bmatrix} 0 \\ -1 \\ -5 \end{bmatrix}$

Find the inverse of A and solve $X = A^{-1}B$:

From Problem 35, $A^{-1} = \begin{bmatrix} 3 & -3 & 1 \\ -2 & 2 & -1 \\ -4 & 5 & -2 \end{bmatrix}$, so

$X = A^{-1}B = \begin{bmatrix} 3 & -3 & 1 \\ -2 & 2 & -1 \\ -4 & 5 & -2 \end{bmatrix}\begin{bmatrix} 0 \\ -1 \\ -5 \end{bmatrix} = \begin{bmatrix} -2 \\ 3 \\ 5 \end{bmatrix}$.

The solution is $x = -2$, $y = 3$, $z = 5$ or $(-2, 3, 5)$.

59. $\begin{cases} x-y+z=2 \\ -2y+z=2 \\ -2x-3y=\dfrac{1}{2} \end{cases}$

Rewrite the system of equations in matrix form:

$A = \begin{bmatrix} 1 & -1 & 1 \\ 0 & -2 & 1 \\ -2 & -3 & 0 \end{bmatrix}$, $X = \begin{bmatrix} x \\ y \\ z \end{bmatrix}$, $B = \begin{bmatrix} 2 \\ 2 \\ \frac{1}{2} \end{bmatrix}$

Find the inverse of A and solve $X = A^{-1}B$:

From Problem 35, $A^{-1} = \begin{bmatrix} 3 & -3 & 1 \\ -2 & 2 & -1 \\ -4 & 5 & -2 \end{bmatrix}$, so

$X = A^{-1}B = \begin{bmatrix} 3 & -3 & 1 \\ -2 & 2 & -1 \\ -4 & 5 & -2 \end{bmatrix} \begin{bmatrix} 2 \\ 2 \\ \frac{1}{2} \end{bmatrix} = \begin{bmatrix} \frac{1}{2} \\ -\frac{1}{2} \\ 1 \end{bmatrix}$

The solution is $x = \frac{1}{2}$, $y = -\frac{1}{2}$, $z = 1$ or $\left(\frac{1}{2}, -\frac{1}{2}, 1 \right)$.

61. $\begin{cases} x + y + z = 9 \\ 3x + 2y - z = 8 \\ 3x + y + 2z = 1 \end{cases}$

Rewrite the system of equations in matrix form:

$A = \begin{bmatrix} 1 & 1 & 1 \\ 3 & 2 & -1 \\ 3 & 1 & 2 \end{bmatrix}$, $X = \begin{bmatrix} x \\ y \\ z \end{bmatrix}$, $B = \begin{bmatrix} 9 \\ 8 \\ 1 \end{bmatrix}$

Find the inverse of A and solve $X = A^{-1}B$:

From Problem 37, $A^{-1} = \begin{bmatrix} -\frac{5}{7} & \frac{1}{7} & \frac{3}{7} \\ \frac{9}{7} & \frac{1}{7} & -\frac{4}{7} \\ \frac{3}{7} & -\frac{2}{7} & \frac{1}{7} \end{bmatrix}$, so

$X = A^{-1}B = \begin{bmatrix} -\frac{5}{7} & \frac{1}{7} & \frac{3}{7} \\ \frac{9}{7} & \frac{1}{7} & -\frac{4}{7} \\ \frac{3}{7} & -\frac{2}{7} & \frac{1}{7} \end{bmatrix} \begin{bmatrix} 9 \\ 8 \\ 1 \end{bmatrix} = \begin{bmatrix} -\frac{34}{7} \\ \frac{85}{7} \\ \frac{12}{7} \end{bmatrix}$.

The solution is $x = -\frac{34}{7}$, $y = \frac{85}{7}$, $z = \frac{12}{7}$ or $\left(-\frac{34}{7}, \frac{85}{7}, \frac{12}{7} \right)$.

63. $\begin{cases} x + y + z = 2 \\ 3x + 2y - z = \frac{7}{3} \\ 3x + y + 2z = \frac{10}{3} \end{cases}$

Rewrite the system of equations in matrix form:

$A = \begin{bmatrix} 1 & 1 & 1 \\ 3 & 2 & -1 \\ 3 & 1 & 2 \end{bmatrix}$, $X = \begin{bmatrix} x \\ y \\ z \end{bmatrix}$, $B = \begin{bmatrix} 2 \\ \frac{7}{3} \\ \frac{10}{3} \end{bmatrix}$

Find the inverse of A and solve $X = A^{-1}B$:

From Problem 37, $A^{-1} = \begin{bmatrix} -\frac{5}{7} & \frac{1}{7} & \frac{3}{7} \\ \frac{9}{7} & \frac{1}{7} & -\frac{4}{7} \\ \frac{3}{7} & -\frac{2}{7} & \frac{1}{7} \end{bmatrix}$, so

$X = A^{-1}B = \begin{bmatrix} -\frac{5}{7} & \frac{1}{7} & \frac{3}{7} \\ \frac{9}{7} & \frac{1}{7} & -\frac{4}{7} \\ \frac{3}{7} & -\frac{2}{7} & \frac{1}{7} \end{bmatrix} \begin{bmatrix} 2 \\ \frac{7}{3} \\ \frac{10}{3} \end{bmatrix} = \begin{bmatrix} \frac{1}{3} \\ 1 \\ \frac{2}{3} \end{bmatrix}$.

The solution is $x = \frac{1}{3}$, $y = 1$, $z = \frac{2}{3}$ or $\left(\frac{1}{3}, 1, \frac{2}{3} \right)$.

65. $A = \begin{bmatrix} 4 & 2 \\ 2 & 1 \end{bmatrix}$

Augment the matrix with the identity and use row operations to find the inverse:

$\begin{bmatrix} 4 & 2 & | & 1 & 0 \\ 2 & 1 & | & 0 & 1 \end{bmatrix}$

$\rightarrow \begin{bmatrix} 4 & 2 & | & 1 & 0 \\ 0 & 0 & | & -\frac{1}{2} & 1 \end{bmatrix}$ $\left(R_2 = -\frac{1}{2}r_1 + r_2 \right)$

$\rightarrow \begin{bmatrix} 1 & \frac{1}{2} & | & \frac{1}{4} & 0 \\ 0 & 0 & | & -\frac{1}{2} & 1 \end{bmatrix}$ $\left(R_1 = \frac{1}{4}r_1 \right)$

There is no way to obtain the identity matrix on the left. Thus, this matrix has no inverse.

67. $A = \begin{bmatrix} 15 & 3 \\ 10 & 2 \end{bmatrix}$

Augment the matrix with the identity and use row operations to find the inverse:

$\begin{bmatrix} 15 & 3 & | & 1 & 0 \\ 10 & 2 & | & 0 & 1 \end{bmatrix}$

$\rightarrow \begin{bmatrix} 15 & 3 & | & 1 & 0 \\ 0 & 0 & | & -\frac{2}{3} & 1 \end{bmatrix}$ $\left(R_2 = -\frac{2}{3}r_1 + r_2 \right)$

$\rightarrow \begin{bmatrix} 1 & \frac{1}{5} & | & \frac{1}{15} & 0 \\ 0 & 0 & | & -\frac{2}{3} & 1 \end{bmatrix}$ $\left(R_1 = \frac{1}{15}r_1 \right)$

There is no way to obtain the identity matrix on the left; thus, there is no inverse.

69. $A = \begin{bmatrix} -3 & 1 & -1 \\ 1 & -4 & -7 \\ 1 & 2 & 5 \end{bmatrix}$

Augment the matrix with the identity and use row operations to find the inverse:

$\begin{bmatrix} -3 & 1 & -1 & | & 1 & 0 & 0 \\ 1 & -4 & -7 & | & 0 & 1 & 0 \\ 1 & 2 & 5 & | & 0 & 0 & 1 \end{bmatrix}$

$\rightarrow \begin{bmatrix} 1 & 2 & 5 & | & 0 & 0 & 1 \\ 1 & -4 & -7 & | & 0 & 1 & 0 \\ -3 & 1 & -1 & | & 1 & 0 & 0 \end{bmatrix}$ $\begin{pmatrix} \text{Interchange} \\ r_1 \text{ and } r_3 \end{pmatrix}$

$\rightarrow \begin{bmatrix} 1 & 2 & 5 & | & 0 & 0 & 1 \\ 0 & -6 & -12 & | & 0 & 1 & -1 \\ 0 & 7 & 14 & | & 1 & 0 & 3 \end{bmatrix}$ $\begin{pmatrix} R_2 = -r_1 + r_2 \\ R_3 = 3r_1 + r_3 \end{pmatrix}$

$\rightarrow \begin{bmatrix} 1 & 2 & 5 & | & 0 & 0 & 1 \\ 0 & 1 & 2 & | & 0 & -\frac{1}{6} & \frac{1}{6} \\ 0 & 7 & 14 & | & 1 & 0 & 3 \end{bmatrix}$ $\left(R_2 = -\frac{1}{6} r_2 \right)$

$\rightarrow \begin{bmatrix} 1 & 0 & 1 & | & 0 & \frac{1}{3} & \frac{2}{3} \\ 0 & 1 & 2 & | & 0 & -\frac{1}{6} & \frac{1}{6} \\ 0 & 0 & 0 & | & 1 & \frac{7}{6} & \frac{11}{6} \end{bmatrix}$ $\begin{pmatrix} R_1 = -2r_2 + r_1 \\ R_3 = -7r_2 + r_3 \end{pmatrix}$

There is no way to obtain the identity matrix on the left; thus, there is no inverse.

71. $A = \begin{bmatrix} 25 & 61 & -12 \\ 18 & -2 & 4 \\ 8 & 35 & 21 \end{bmatrix}$

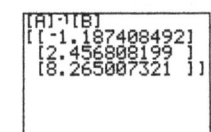

Thus, $A^{-1} \approx \begin{bmatrix} 0.01 & 0.05 & -0.01 \\ 0.01 & -0.02 & 0.01 \\ -0.02 & 0.01 & 0.03 \end{bmatrix}$

73. $A = \begin{bmatrix} 44 & 21 & 18 & 6 \\ -2 & 10 & 15 & 5 \\ 21 & 12 & -12 & 4 \\ -8 & -16 & 4 & 9 \end{bmatrix}$

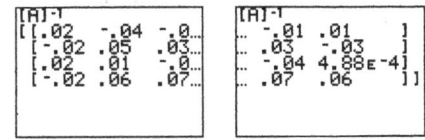

Thus, $A^{-1} \approx \begin{bmatrix} 0.02 & -0.04 & -0.01 & 0.01 \\ -0.02 & 0.05 & 0.03 & -0.03 \\ 0.02 & 0.01 & -0.04 & 0.00 \\ -0.02 & 0.06 & 0.07 & 0.06 \end{bmatrix}$.

75. $A = \begin{bmatrix} 25 & 61 & -12 \\ 18 & -12 & 7 \\ 3 & 4 & -1 \end{bmatrix}$; $B = \begin{bmatrix} 10 \\ -9 \\ 12 \end{bmatrix}$

Enter the matrices into a graphing utility and use $A^{-1}B$ to solve the system. The result is shown below:

```
[A]⁻¹[B]
[[4.566617862 ]
 [-6.44363104  ]
 [-24.07467057]]
```

Thus, the solution to the system is $x \approx 4.57$, $y \approx -6.44$, $z \approx -24.07$ or $(4.57, -6.44, -24.07)$.

77. $A = \begin{bmatrix} 25 & 61 & -12 \\ 18 & -12 & 7 \\ 3 & 4 & -1 \end{bmatrix}$; $B = \begin{bmatrix} 21 \\ 7 \\ -2 \end{bmatrix}$

```
[A]⁻¹[B]
[[-1.187408492]
 [2.456808199 ]
 [8.265007321 ]]
```

Thus, the solution to the system is $x \approx -1.19$, $y \approx 2.46$, $z \approx 8.27$ or $(-1.19, 2.46, 8.27)$.

79. $\begin{cases} 2x + 3y = 11 \\ 5x + 7y = 24 \end{cases}$

Multiply each side of the first equation by 5, and each side of the second equation by -2. Then add the equations to eliminate x:

$\begin{cases} 10x + 15y = 55 \\ \underline{-10x - 14y = -48} \\ \quad y = 7 \end{cases}$

Substitute and solve for x:

$2x + 3(7) = 11$

$2x + 21 = 11$

$2x = -10$

$x = -5$

The solution of the system is $x = -5$, $y = 7$ or using ordered pairs $(-5, 7)$.

81. $\begin{cases} x - 2y + 4z = 2 \\ -3x + 5y - 2z = 17 \\ 4x - 3y = -22 \end{cases}$

Write the augmented matrix:

$\begin{bmatrix} 1 & -2 & 4 & | & 2 \\ -3 & 5 & -2 & | & 17 \\ 4 & -3 & 0 & | & -22 \end{bmatrix}$

$\begin{bmatrix} 1 & -2 & 4 & | & 2 \\ 0 & -1 & 10 & | & 23 \\ 0 & 5 & -16 & | & -30 \end{bmatrix} \begin{matrix} R_2 = 3r_1 + r_2 \\ R_3 = -4r_1 + r_3 \end{matrix}$

$\begin{bmatrix} 1 & -2 & 4 & | & 2 \\ 0 & 1 & -10 & | & -23 \\ 0 & 5 & -16 & | & -30 \end{bmatrix} R_2 = -r_2$

$\begin{bmatrix} 1 & -2 & 4 & | & 2 \\ 0 & 1 & -10 & | & -23 \\ 0 & 0 & 34 & | & 85 \end{bmatrix} R_3 = -5r_2 + r_3$

$\begin{bmatrix} 1 & -2 & 4 & | & 2 \\ 0 & 1 & -10 & | & -23 \\ 0 & 0 & 1 & | & \frac{5}{2} \end{bmatrix} R_3 = r_3 / 34$

$\begin{bmatrix} 1 & -2 & 0 & | & -8 \\ 0 & 1 & 0 & | & 2 \\ 0 & 0 & 1 & | & \frac{5}{2} \end{bmatrix} \begin{matrix} R_1 = -4r_3 + r_1 \\ R_2 = 10r_3 + r_2 \end{matrix}$

$\begin{bmatrix} 1 & 0 & 0 & | & -4 \\ 0 & 1 & 0 & | & 2 \\ 0 & 0 & 1 & | & \frac{5}{2} \end{bmatrix} R_1 = 2r_2 + r_1$

The solution is $x = -4$, $y = 2$, $z = \frac{5}{2}$ or $\left(-4, 2, \frac{5}{2}\right)$.

83. $\begin{cases} 5x - y + 4z = 2 \\ -x + 5y - 4z = 3 \\ 7x + 13y - 4z = 17 \end{cases}$

Write the augmented matrix:

$\begin{bmatrix} 5 & -1 & 4 & | & 2 \\ -1 & 5 & -4 & | & 3 \\ 7 & 13 & -4 & | & 17 \end{bmatrix}$

$\begin{bmatrix} 1 & -5 & 4 & | & -3 \\ 5 & -1 & 4 & | & 2 \\ 7 & 13 & -4 & | & 17 \end{bmatrix} \begin{matrix} R_1 = -r_2 \\ R_2 = r_1 \end{matrix}$

$\begin{bmatrix} 1 & -5 & 4 & | & -3 \\ 0 & 24 & -16 & | & 17 \\ 0 & 48 & -32 & | & 38 \end{bmatrix} \begin{matrix} R_2 = -5r_1 + r_2 \\ R_3 = -7r_1 + r_3 \end{matrix}$

$\begin{bmatrix} 1 & -5 & 4 & | & -3 \\ 0 & 1 & -\frac{2}{3} & | & \frac{17}{24} \\ 0 & 48 & -32 & | & 38 \end{bmatrix} R_2 = r_2 / 24$

$\begin{bmatrix} 1 & -5 & 4 & | & -3 \\ 0 & 1 & -\frac{2}{3} & | & \frac{17}{24} \\ 0 & 0 & 0 & | & 4 \end{bmatrix} R_3 = -48r_2 + r_3$

The last row of our matrix is a contradiction. Therefore, the system is inconsistent. The solution set is $\{ \ \}$, or $\varnothing$.

85. $\begin{cases} 2x - 3y + z = 4 \\ -3x + 2y - z = -3 \\ -5y + z = 6 \end{cases}$

Write the augmented matrix:

$\begin{bmatrix} 2 & -3 & 1 & | & 4 \\ -3 & 2 & -1 & | & -3 \\ 0 & -5 & 1 & | & 6 \end{bmatrix}$

$\begin{bmatrix} 1 & -\frac{3}{2} & \frac{1}{2} & | & 2 \\ -3 & 2 & -1 & | & -3 \\ 0 & -5 & 1 & | & 6 \end{bmatrix} R_1 = r_1 / 2$

$\begin{bmatrix} 1 & -\frac{3}{2} & \frac{1}{2} & | & 2 \\ 0 & -\frac{5}{2} & \frac{1}{2} & | & 3 \\ 0 & -5 & 1 & | & 6 \end{bmatrix} R_2 = 3r_1 + r_2$

$\begin{bmatrix} 1 & -\frac{3}{2} & \frac{1}{2} & | & 2 \\ 0 & 1 & -\frac{1}{5} & | & -\frac{6}{5} \\ 0 & -5 & 1 & | & 6 \end{bmatrix} R_2 = -\frac{2}{5}r_2$

$\begin{bmatrix} 1 & -\frac{3}{2} & \frac{1}{2} & | & 2 \\ 0 & 1 & -\frac{1}{5} & | & -\frac{6}{5} \\ 0 & 0 & 0 & | & 0 \end{bmatrix} R_3 = 5r_2 + r_3$

$\begin{bmatrix} 1 & 0 & \frac{1}{5} & | & \frac{1}{5} \\ 0 & 1 & -\frac{1}{5} & | & -\frac{6}{5} \\ 0 & 0 & 0 & | & 0 \end{bmatrix} R_1 = \frac{3}{2}r_2 + r_1$

Since the last row yields an identity, and no contradictions exist in the other rows, there are an infinite number of solutions. The solution is $x = -\frac{1}{5}z + \frac{1}{5}$, $y = \frac{1}{5}z - \frac{6}{5}$, and z is any real number. That is,

$$\left\{(x,y,z) \mid x = -\frac{1}{5}z + \frac{1}{5}, \ y = \frac{1}{5}z - \frac{6}{5}, \right.$$

z is any real number$\}$

87. a. $A = \begin{matrix} 6 & 9 \\ 3 & 12 \end{matrix}$; $B = \begin{matrix} 128.00 \\ 341.60 \end{matrix}$

b. $AB = \begin{matrix} 6 & 9 \\ 3 & 12 \end{matrix} \begin{matrix} 128.00 \\ 341.60 \end{matrix}$

$= \begin{matrix} 6(128.00) + 9(341.60) \\ 3(128.00) + 12(341.60) \end{matrix} = \begin{matrix} 3842.40 \\ 4483.20 \end{matrix}$

Nikki's total tuition is $3842.40, and Joe's total tuition is $4483.20.

89. a. The rows of the 2 by 3 matrix represent stainless steel and aluminum. The columns represent 10-gallon, 5-gallon, and 1-gallon.

The 2 by 3 matrix is: $\begin{bmatrix} 500 & 350 & 400 \\ 700 & 500 & 850 \end{bmatrix}$.

The 3 by 2 matrix is: $\begin{bmatrix} 500 & 700 \\ 350 & 500 \\ 400 & 850 \end{bmatrix}$.

b. The 3 by 1 matrix representing the amount of

material is: $\begin{bmatrix} 15 \\ 8 \\ 3 \end{bmatrix}$.

c. The days usage of materials is:

$\begin{bmatrix} 500 & 350 & 400 \\ 700 & 500 & 850 \end{bmatrix} \cdot \begin{bmatrix} 15 \\ 8 \\ 3 \end{bmatrix} = \begin{bmatrix} 11,500 \\ 17,050 \end{bmatrix}$

Thus, 11,500 pounds of stainless steel and 17,050 pounds of aluminum were used that day.

d. The 1 by 2 matrix representing cost is: $\begin{bmatrix} 0.10 & 0.05 \end{bmatrix}$.

e. The total cost of the day's production was:

$\begin{bmatrix} 0.10 & 0.05 \end{bmatrix} \cdot \begin{bmatrix} 11,500 \\ 17,050 \end{bmatrix} = \begin{bmatrix} 2002.50 \end{bmatrix}$.

The total cost of the day's production was $2002.50.

91. a. $K = \begin{bmatrix} 2 & 1 & 1 \\ 1 & 1 & 0 \\ 1 & 1 & 1 \end{bmatrix}$

Augment the matrix with the identity and use row operations to find the inverse:

$\begin{bmatrix} 2 & 1 & 1 & | & 1 & 0 & 0 \\ 1 & 1 & 0 & | & 0 & 1 & 0 \\ 1 & 1 & 1 & | & 0 & 0 & 1 \end{bmatrix}$

$\rightarrow \begin{bmatrix} 1 & 1 & 0 & | & 0 & 1 & 0 \\ 2 & 1 & 1 & | & 1 & 0 & 0 \\ 1 & 1 & 1 & | & 0 & 0 & 1 \end{bmatrix} \begin{pmatrix} \text{Interchange} \\ r_1 \text{ and } r_2 \end{pmatrix}$

$\rightarrow \begin{bmatrix} 1 & 1 & 0 & | & 0 & 1 & 0 \\ 0 & -1 & 1 & | & 1 & -2 & 0 \\ 0 & 0 & 1 & | & 0 & -1 & 1 \end{bmatrix} \begin{pmatrix} R_2 = -2r_1 + r_2 \\ R_3 = -r_1 + r_3 \end{pmatrix}$

$\rightarrow \begin{bmatrix} 1 & 1 & 0 & | & 0 & 1 & 0 \\ 0 & 1 & -1 & | & -1 & 2 & 0 \\ 0 & 0 & 1 & | & 0 & -1 & 1 \end{bmatrix} (R_2 = -r_2)$

$\rightarrow \begin{bmatrix} 1 & 1 & 0 & | & 0 & 1 & 0 \\ 0 & 1 & 0 & | & -1 & 1 & 1 \\ 0 & 0 & 1 & | & 0 & -1 & 1 \end{bmatrix} (R_2 = r_2 + r_3)$

$\rightarrow \begin{bmatrix} 1 & 0 & 0 & | & 1 & 0 & -1 \\ 0 & 1 & 0 & | & -1 & 1 & 1 \\ 0 & 0 & 1 & | & 0 & -1 & 1 \end{bmatrix} (R_1 = r_1 - r_2)$

Thus, $K^{-1} = \begin{bmatrix} 1 & 0 & -1 \\ -1 & 1 & 1 \\ 0 & -1 & 1 \end{bmatrix}$.

b. $M = E \cdot K^{-1} = \begin{bmatrix} 47 & 34 & 33 \\ 44 & 36 & 27 \\ 47 & 41 & 20 \end{bmatrix} \begin{bmatrix} 1 & 0 & -1 \\ -1 & 1 & 1 \\ 0 & -1 & 1 \end{bmatrix}$

$= \begin{bmatrix} 13 & 1 & 20 \\ 8 & 9 & 19 \\ 6 & 21 & 14 \end{bmatrix}$

because
$$a_{11} = 47(1) + 34(-1) + 33(0) = 13$$
$$a_{12} = 47(0) + 34(1) + 33(-1) = 1$$
$$a_{13} = 47(-1) + 34(1) + 33(1) = 20$$
$$a_{21} = 44(1) + 36(-1) + 27(0) = 8$$
$$a_{22} = 44(0) + 36(1) + 27(-1) = 9$$
$$a_{23} = 44(-1) + 36(1) + 27(1) = 19$$
$$a_{31} = 47(1) + 41(-1) + 20(0) = 6$$
$$a_{32} = 47(0) + 41(1) + 20(-1) = 21$$
$$a_{33} = 47(-1) + 41(1) + 20(1) = 14$$

c. $13 \to M; 1 \to A; 20 \to T; 8 \to H; 9 \to I;$
$19 \to S; 6 \to F; 21 \to U; 14 \to N$
The message: Math is fun.

93. $A = \begin{bmatrix} a & b \\ c & d \end{bmatrix}$

If $D = ad - bc \neq 0$, then $a \neq 0$ and $d \neq 0$, or $b \neq 0$ and $c \neq 0$. Assuming the former, then

$$\begin{bmatrix} a & b & | & 1 & 0 \\ c & d & | & 0 & 1 \end{bmatrix}$$

$$\to \begin{bmatrix} 1 & \frac{b}{a} & | & \frac{1}{a} & 0 \\ c & d & | & 0 & 1 \end{bmatrix} \qquad \left(R_1 = \frac{1}{a} \cdot r_1 \right)$$

$$\to \begin{bmatrix} 1 & \frac{b}{a} & | & \frac{1}{a} & 0 \\ 0 & d - \frac{bc}{a} & | & -\frac{c}{a} & 1 \end{bmatrix} \qquad \left(R_2 = -c \cdot r_1 + r_2 \right)$$

$$\to \begin{bmatrix} 1 & \frac{b}{a} & | & \frac{1}{a} & 0 \\ 0 & \frac{ad-bc}{a} & | & -\frac{c}{a} & 1 \end{bmatrix}$$

$$\to \begin{bmatrix} 1 & \frac{b}{a} & | & \frac{1}{a} & 0 \\ 0 & 1 & | & \frac{-c}{ad-bc} & \frac{a}{ad-bc} \end{bmatrix} \qquad \left(R_2 = \frac{a}{ad-bc} \cdot r_2 \right)$$

$$\to \begin{bmatrix} 1 & 0 & | & \frac{1}{a} + \frac{bc}{a(ad-bc)} & \frac{-b}{ad-bc} \\ 0 & 1 & | & \frac{-c}{ad-bc} & \frac{a}{ad-bc} \end{bmatrix} \qquad \left(R_1 = -\frac{b}{a} \cdot r_2 + r_1 \right)$$

$$\to \begin{bmatrix} 1 & 0 & | & \frac{d}{ad-bc} & \frac{-b}{ad-bc} \\ 0 & 1 & | & \frac{-c}{ad-bc} & \frac{a}{ad-bc} \end{bmatrix}$$

$$\to \begin{bmatrix} 1 & 0 & | & \frac{d}{D} & -\frac{b}{D} \\ 0 & 1 & | & -\frac{c}{D} & \frac{a}{D} \end{bmatrix}$$

Thus, $A^{-1} = \begin{bmatrix} \frac{d}{D} & -\frac{b}{D} \\ -\frac{c}{D} & \frac{a}{D} \end{bmatrix} = \frac{1}{D} \begin{bmatrix} d & -b \\ -c & a \end{bmatrix}$

where $D = ad - bc$.

95. $A = \begin{bmatrix} 0 & 1 & 1 & 0 & 0 \\ 1 & 0 & 0 & 1 & 1 \\ 1 & 0 & 0 & 1 & 0 \\ 0 & 0 & 1 & 0 & 1 \\ 0 & 1 & 0 & 0 & 0 \end{bmatrix}$

a.

$$A^2 = \begin{bmatrix} 2 & 0 & 0 & 2 & 1 \\ 0 & 2 & 2 & 0 & 1 \\ 0 & 1 & 2 & 0 & 1 \\ 1 & 1 & 0 & 1 & 0 \\ 1 & 0 & 0 & 1 & 1 \end{bmatrix} \quad A^3 = \begin{bmatrix} 0 & 3 & 4 & 0 & 2 \\ 4 & 1 & 0 & 4 & 2 \\ 3 & 1 & 0 & 3 & 1 \\ 1 & 1 & 2 & 1 & 2 \\ 0 & 2 & 2 & 0 & 1 \end{bmatrix}$$

$$A + A^2 + A^3 = \begin{bmatrix} 2 & 4 & 5 & 2 & 3 \\ 5 & 3 & 2 & 5 & 4 \\ 4 & 2 & 2 & 4 & 2 \\ 2 & 2 & 3 & 2 & 3 \\ 1 & 3 & 2 & 1 & 2 \end{bmatrix}$$

Yes, all pages can reach every other page within 3 clicks.

b. The largest number in row 1 (page 1) is 5 which corresponds to page 3.

97. a. $XR = \begin{bmatrix} \frac{1}{2} & -\frac{\sqrt{3}}{2} & 0 \\ \frac{\sqrt{3}}{2} & \frac{1}{2} & 0 \\ 0 & 0 & 1 \end{bmatrix} \begin{bmatrix} 6 \\ 4 \\ 1 \end{bmatrix} = \begin{bmatrix} 3 - 2\sqrt{3} \\ 3\sqrt{3} + 2 \\ 1 \end{bmatrix}$

The coordinates would be
$\left(3 - 2\sqrt{3}, 3\sqrt{3} + 2 \right)$

b. $R^{-1} = \begin{bmatrix} \frac{1}{2} & \frac{\sqrt{3}}{2} & 0 \\ -\frac{\sqrt{3}}{2} & \frac{1}{2} & 0 \\ 0 & 0 & 1 \end{bmatrix}$; This is the rotation

matrix needed to get the translated coordinates back to the original coordinates.

99. Since the product is found by multiplying the components from the columns of the first matrix by the components in the rows of the second matrix and then adding those products, then the number of columns in the first must equal the number of rows in the second.

101. If the inverse of A exists:

$$AX = 0$$

$$A^{-1}AX = A^{-1}0$$

$$IX = 0$$

$$X = 0$$

If the inverse of A does not exist, then A is singular and you would not be able to multiply A by its inverse to give the identity inverse and X would have no solution.

103.
$$\frac{3}{8} = \frac{f(12) - f(0)}{12 - 0}$$

$$\frac{3}{8} = \frac{f(12) - \frac{1}{2}}{12}$$

$$12 \cdot \frac{3}{8} = f(12) - \frac{1}{2}$$

$$\frac{9}{2} = f(12) - \frac{1}{2}$$

$$f(12) = \frac{10}{2} = 5$$

105. $D(x) = 3500 - x^2$

$$D(32) = 3500 - (32)^2$$

$$= 3500 - 1024$$

$$= 2476$$

The numbers of units sold is 2,476,000.

Section 8.5

1. True

3. $3x^4 + 6x^3 + 3x^2 = 3x^2 \left(x^2 + 2x + 1\right)$
$$= 3x^2 (x+1)^2$$

5. The rational expression $\dfrac{x}{x^2 - 1}$ is proper, since the degree of the numerator is less than the degree of the denominator.

7. The rational expression $\dfrac{x^2 + 5}{x^2 - 4}$ is improper, so perform the division:

$$x^2 - 4 \overline{\smash{\big)}\, \begin{matrix} 1 \\ x^2 + 5 \end{matrix}}$$
$$\underline{x^2 - 4}$$
$$9$$

The proper rational expression is:
$$\frac{x^2 + 5}{x^2 - 4} = 1 + \frac{9}{x^2 - 4}$$

9. The rational expression $\dfrac{5x^3 + 2x - 1}{x^2 - 4}$ is improper, so perform the division:

$$x^2 - 4 \overline{\smash{\big)}\, \begin{matrix} 5x \\ 5x^3 \quad + 2x - 1 \end{matrix}}$$
$$\underline{5x^3 \quad - 20x}$$
$$22x - 1$$

The proper rational expression is:
$$\frac{5x^3 + 2x - 1}{x^2 - 4} = 5x + \frac{22x - 1}{x^2 - 4}$$

11. The rational expression
$$\frac{x(x-1)}{(x+4)(x-3)} = \frac{x^2 - x}{x^2 + x - 12}$$ is improper, so perform the division:

$$x^2 + x - 12 \overline{\smash{\big)}\, \begin{matrix} 1 \\ x^2 - x + 0 \end{matrix}}$$
$$\underline{x^2 + x - 12}$$
$$-2x + 12$$

The proper rational expression is:
$$\frac{x(x-1)}{(x+4)(x-3)} = 1 + \frac{-2x+12}{x^2 + x - 12} = 1 + \frac{-2(x-6)}{(x+4)(x-3)}$$

13. Find the partial fraction decomposition:
$$\frac{4}{x(x-1)} = \frac{A}{x} + \frac{B}{x-1}$$

$$x(x-1)\left(\frac{4}{x(x-1)}\right) = x(x-1)\left(\frac{A}{x} + \frac{B}{x-1}\right)$$
$$4 = A(x-1) + Bx$$

Let $x = 1$, then $4 = A(0) + B$

$$B = 4$$

Let $x = 0$, then $4 = A(-1) + B(0)$

$$A = -4$$

$$\frac{4}{x(x-1)} = \frac{-4}{x} + \frac{4}{x-1}$$

15. Find the partial fraction decomposition:

$$\frac{1}{x(x^2+1)} = \frac{A}{x} + \frac{Bx+C}{x^2+1}$$

$$x(x^2+1)\left(\frac{1}{x(x^2+1)}\right) = x(x^2+1)\left(\frac{A}{x} + \frac{Bx+C}{x^2+1}\right)$$

$$1 = A(x^2+1) + (Bx+C)x$$

Let $x = 0$, then $1 = A(0^2+1) + (B(0)+C)(0)$

$$A = 1$$

Let $x = 1$, then $1 = A(1^2+1) + (B(1)+C)(1)$

$$1 = 2A + B + C$$
$$1 = 2(1) + B + C$$
$$B + C = -1$$

Let $x = -1$, then

$$1 = A((-1)^2+1) + (B(-1)+C)(-1)$$
$$1 = A(1+1) + (-B+C)(-1)$$
$$1 = 2A + B - C$$
$$1 = 2(1) + B - C$$
$$B - C = -1$$

Solve the system of equations:

$$\begin{array}{rcl} B+C & = & -1 \\ B-C & = & -1 \\ \hline 2B & = & -2 \\ B & = & -1 \end{array}$$

$$-1 + C = -1$$
$$C = 0$$

$$\frac{1}{x(x^2+1)} = \frac{1}{x} + \frac{-x}{x^2+1}$$

17. Find the partial fraction decomposition:

$$\frac{x}{(x-1)(x-2)} = \frac{A}{x-1} + \frac{B}{x-2}$$

Multiplying both sides by $(x-1)(x-2)$, we

obtain: $x = A(x-2) + B(x-1)$

Let $x = 1$, then $1 = A(1-2) + B(1-1)$

$$1 = -A$$
$$A = -1$$

Let $x = 2$, then $2 = A(2-2) + B(2-1)$

$$2 = B$$

$$\frac{x}{(x-1)(x-2)} = \frac{-1}{x-1} + \frac{2}{x-2}$$

19. Find the partial fraction decomposition:

$$\frac{x^2}{(x-1)^2(x+1)} = \frac{A}{x-1} + \frac{B}{(x-1)^2} + \frac{C}{x+1}$$

Multiplying both sides by $(x-1)^2(x+1)$, we

obtain: $x^2 = A(x-1)(x+1) + B(x+1) + C(x-1)^2$

Let $x = 1$, then

$$1^2 = A(1-1)(1+1) + B(1+1) + C(1-1)^2$$
$$1 = A(0)(2) + B(2) + C(0)^2$$
$$1 = 2B$$
$$B = \frac{1}{2}$$

Let $x = -1$, then

$$(-1)^2 = A(-1-1)(-1+1) + B(-1+1) + C(-1-1)^2$$
$$1 = A(-2)(0) + B(0) + C(-2)^2$$
$$1 = 4C$$
$$C = \frac{1}{4}$$

Let $x = 0$, then

$$0^2 = A(0-1)(0+1) + B(0+1) + C(0-1)^2$$
$$0 = -A + B + C$$
$$A = B + C$$
$$A = \frac{1}{2} + \frac{1}{4} = \frac{3}{4}$$

$$\frac{x^2}{(x-1)^2(x+1)} = \frac{\frac{3}{4}}{x-1} + \frac{\frac{1}{2}}{(x-1)^2} + \frac{\frac{1}{4}}{x+1}$$

21. Find the partial fraction decomposition:

$$\frac{1}{x^3-8} = \frac{1}{(x-2)(x^2+2x+4)}$$

$$\frac{1}{(x-2)(x^2+2x+4)} = \frac{A}{x-2} + \frac{Bx+C}{x^2+2x+4}$$

Multiplying both sides by $(x-2)(x^2+2x+4)$,

we obtain: $1 = A(x^2+2x+4)+(Bx+C)(x-2)$

Let $x=2$, then

$$1 = A\left(2^2+2(2)+4\right)+(B(2)+C)(2-2)$$

$$1 = 12A$$

$$A = \frac{1}{12}$$

Let $x=0$, then

$$1 = A\left(0^2+2(0)+4\right)+(B(0)+C)(0-2)$$

$$1 = 4A-2C$$

$$1 = 4(1/12)-2C$$

$$-2C = \frac{2}{3}$$

$$C = -\frac{1}{3}$$

Let $x=1$, then

$$1 = A\left(1^2+2(1)+4\right)+(B(1)+C)(1-2)$$

$$1 = 7A-B-C$$

$$1 = 7(1/12)-B+\frac{1}{3}$$

$$B = -\frac{1}{12}$$

23. Find the partial fraction decomposition:

$$\frac{x^2}{(x-1)^2(x+1)^2} = \frac{A}{x-1}+\frac{B}{(x-1)^2}+\frac{C}{x+1}+\frac{D}{(x+1)^2}$$

Multiplying both sides by $(x-1)^2(x+1)^2$, we obtain:

$$x^2 = A(x-1)(x+1)^2+B(x+1)^2$$
$$+C(x-1)^2(x+1)+D(x-1)^2$$

Let $x=1$, then

$$1^2 = A(1-1)(1+1)^2+B(1+1)^2$$
$$+C(1-1)^2(1+1)+D(1-1)^2$$

$$1 = 4B$$

$$B = \frac{1}{4}$$

Let $x=-1$, then

$$(-1)^2 = A(-1-1)(-1+1)^2+B(-1+1)^2$$
$$+C(-1-1)^2(-1+1)+D(-1-1)^2$$

$$1 = 4D$$

$$D = \frac{1}{4}$$

Let $x=0$, then

$$0^2 = A(0-1)(0+1)^2+B(0+1)^2$$
$$+C(0-1)^2(0+1)+D(0-1)^2$$

$$0 = -A+B+C+D$$

$$A-C = B+D$$

$$A-C = \frac{1}{4}+\frac{1}{4} = \frac{1}{2}$$

Let $x=2$, then

$$2^2 = A(2-1)(2+1)^2+B(2+1)^2$$
$$+C(2-1)^2(2+1)+D(2-1)^2$$

$$4 = 9A+9B+3C+D$$

$$9A+3C = 4-9B-D$$

$$9A+3C = 4-9\left(\frac{1}{4}\right)-\frac{1}{4} = \frac{3}{2}$$

$$3A+C = \frac{1}{2}$$

Solve the system of equations:

$$A-C = \tfrac{1}{2}$$

$$\underline{3A+C = \tfrac{1}{2}}$$

$$4A \quad = 1$$

$$A = \tfrac{1}{4}$$

$$\tfrac{3}{4}+C = \tfrac{1}{2}$$

$$C = -\tfrac{1}{4}$$

$$\frac{x^2}{(x-1)^2(x+1)^2} = \frac{\frac{1}{4}}{x-1}+\frac{\frac{1}{4}}{(x-1)^2}+\frac{-\frac{1}{4}}{x+1}+\frac{\frac{1}{4}}{(x+1)^2}$$

25. Find the partial fraction decomposition:

$$\frac{x-3}{(x+2)(x+1)^2} = \frac{A}{x+2}+\frac{B}{x+1}+\frac{C}{(x+1)^2}$$

Multiplying both sides by $(x+2)(x+1)^2$, we

obtain: $x-3 = A(x+1)^2+B(x+2)(x+1)+C(x+2)$

Let $x=-2$, then

$$-2-3 = A(-2+1)^2+B(-2+2)(-2+1)+C(-2+2)$$

$$-5 = A$$

$$A = -5$$

501

Copyright © 2017 Pearson Education, Inc.

Let $x = -1$, then
$$-1 - 3 = A(-1+1)^2 + B(-1+2)(-1+1) + C(-1+2)$$
$$-4 = C$$
$$C = -4$$
Let $x = 0$, then
$$0 - 3 = A(0+1)^2 + B(0+2)(0+1) + C(0+2)$$
$$-3 = A + 2B + 2C$$
$$-3 = -5 + 2B + 2(-4)$$
$$2B = 10$$
$$B = 5$$
$$\frac{x-3}{(x+2)(x+1)^2} = \frac{-5}{x+2} + \frac{5}{x+1} + \frac{-4}{(x+1)^2}$$

27. Find the partial fraction decomposition:
$$\frac{x+4}{x^2(x^2+4)} = \frac{A}{x} + \frac{B}{x^2} + \frac{Cx+D}{x^2+4}$$
Multiplying both sides by $x^2(x^2+4)$, we obtain:
$$x + 4 = Ax(x^2+4) + B(x^2+4) + (Cx+D)x^2$$
Let $x = 0$, then
$$0 + 4 = A(0)(0^2+4) + B(0^2+4) + (C(0)+D)(0)^2$$
$$4 = 4B$$
$$B = 1$$
Let $x = 1$, then
$$1 + 4 = A(1)(1^2+4) + B(1^2+4) + (C(1)+D)(1)^2$$
$$5 = 5A + 5B + C + D$$
$$5 = 5A + 5 + C + D$$
$$5A + C + D = 0$$
Let $x = -1$, then
$$-1 + 4 = A(-1)((-1)^2+4) + B((-1)^2+4)$$
$$+ (C(-1)+D)(-1)^2$$
$$3 = -5A + 5B - C + D$$
$$3 = -5A + 5 - C + D$$
$$-5A - C + D = -2$$
Let $x = 2$, then
$$2 + 4 = A(2)(2^2+4) + B(2^2+4) + (C(2)+D)(2)^2$$
$$6 = 16A + 8B + 8C + 4D$$
$$6 = 16A + 8 + 8C + 4D$$
$$16A + 8C + 4D = -2$$
Solve the system of equations:
$$5A + C + D = 0$$
$$\underline{-5A - C + D = -2}$$
$$2D = -2$$
$$D = -1$$

$$5A + C - 1 = 0$$
$$C = 1 - 5A$$
$$16A + 8(1-5A) + 4(-1) = -2$$
$$16A + 8 - 40A - 4 = -2$$
$$-24A = -6$$
$$A = \frac{1}{4}$$
$$C = 1 - 5\left(\frac{1}{4}\right) = 1 - \frac{5}{4} = -\frac{1}{4}$$
$$\frac{x+4}{x^2(x^2+4)} = \frac{\frac{1}{4}}{x} + \frac{1}{x^2} + \frac{-\frac{1}{4}x-1}{x^2+4}$$
$$= \frac{\frac{1}{4}}{x} + \frac{1}{x^2} + \frac{-\frac{1}{4}(x+4)}{x^2+4}$$

29. Find the partial fraction decomposition:
$$\frac{x^2+2x+3}{(x+1)(x^2+2x+4)} = \frac{A}{x+1} + \frac{Bx+C}{x^2+2x+4}$$
Multiplying both sides by $(x+1)(x^2+2x+4)$, we obtain:
$$x^2 + 2x + 3 = A(x^2+2x+4) + (Bx+C)(x+1)$$
Let $x = -1$, then
$$(-1)^2 + 2(-1) + 3 = A((-1)^2 + 2(-1) + 4)$$
$$+ (B(-1)+C)(-1+1)$$
$$2 = 3A$$
$$A = \frac{2}{3}$$
Let $x = 0$, then
$$0^2 + 2(0) + 3 = A(0^2 + 2(0) + 4) + (B(0)+C)(0+1)$$
$$3 = 4A + C$$
$$3 = 4(2/3) + C$$
$$C = \frac{1}{3}$$
Let $x = 1$, then
$$1^2 + 2(1) + 3 = A(1^2 + 2(1) + 4) + (B(1)+C)(1+1)$$
$$6 = 7A + 2B + 2C$$
$$6 = 7(2/3) + 2B + 2(1/3)$$
$$2B = 6 - \frac{14}{3} - \frac{2}{3} = \frac{2}{3}$$
$$B = \frac{1}{3}$$
$$\frac{x^2+2x+3}{(x+1)(x^2+2x+4)} = \frac{\frac{2}{3}}{x+1} + \frac{\frac{1}{3}x+\frac{1}{3}}{x^2+2x+4}$$
$$= \frac{\frac{2}{3}}{x+1} + \frac{\frac{1}{3}(x+1)}{x^2+2x+4}$$

31. Find the partial fraction decomposition:

$$\frac{x}{(3x-2)(2x+1)} = \frac{A}{3x-2} + \frac{B}{2x+1}$$

Multiplying both sides by $(3x-2)(2x+1)$, we obtain: $x = A(2x+1) + B(3x-2)$

Let $x = -\frac{1}{2}$, then

$$-\frac{1}{2} = A\big(2(-1/2)+1\big) + B\big(3(-1/2)-2\big)$$

$$-\frac{1}{2} = -\frac{7}{2}B$$

$$B = \frac{1}{7}$$

Let $x = \frac{2}{3}$, then

$$\frac{2}{3} = A\big(2(2/3)+1\big) + B\big(3(2/3)-2\big)$$

$$\frac{2}{3} = \frac{7}{3}A$$

$$A = \frac{2}{7}$$

$$\frac{x}{(3x-2)(2x+1)} = \frac{\frac{2}{7}}{3x-2} + \frac{\frac{1}{7}}{2x+1}$$

33. Find the partial fraction decomposition:

$$\frac{x}{x^2+2x-3} = \frac{x}{(x+3)(x-1)} = \frac{A}{x+3} + \frac{B}{x-1}$$

Multiplying both sides by $(x+3)(x-1)$, we obtain: $x = A(x-1) + B(x+3)$

Let $x = 1$, then $\quad 1 = A(1-1) + B(1+3)$

$$1 = 4B$$

$$B = \frac{1}{4}$$

Let $x = -3$, then $-3 = A(-3-1) + B(-3+3)$

$$-3 = -4A$$

$$A = \frac{3}{4}$$

$$\frac{x}{x^2+2x-3} = \frac{\frac{3}{4}}{x+3} + \frac{\frac{1}{4}}{x-1}$$

35. Find the partial fraction decomposition:

$$\frac{x^2+2x+3}{(x^2+4)^2} = \frac{Ax+B}{x^2+4} + \frac{Cx+D}{(x^2+4)^2}$$

Multiplying both sides by $(x^2+4)^2$, we obtain:

$$x^2+2x+3 = (Ax+B)(x^2+4) + Cx+D$$

$$x^2+2x+3 = Ax^3 + Bx^2 + 4Ax + 4B + Cx + D$$

$$x^2+2x+3 = Ax^3 + Bx^2 + (4A+C)x + 4B + D$$

$$A = 0; \quad B = 1;$$

$$4A+C = 2 \qquad\qquad 4B+D = 3$$
$$4(0)+C = 2 \qquad\qquad 4(1)+D = 3$$
$$C = 2 \qquad\qquad\qquad D = -1$$

$$\frac{x^2+2x+3}{(x^2+4)^2} = \frac{1}{x^2+4} + \frac{2x-1}{(x^2+4)^2}$$

37. Find the partial fraction decomposition:

$$\frac{7x+3}{x^3-2x^2-3x} = \frac{7x+3}{x(x-3)(x+1)}$$

$$= \frac{A}{x} + \frac{B}{x-3} + \frac{C}{x+1}$$

Multiplying both sides by $x(x-3)(x+1)$, we obtain:

$$7x+3 = A(x-3)(x+1) + Bx(x+1) + Cx(x-3)$$

Let $x = 0$, then

$$7(0)+3 = A(0-3)(0+1) + B(0)(0+1) + C(0)(0-3)$$

$$3 = -3A$$

$$A = -1$$

Let $x = 3$, then

$$7(3)+3 = A(3-3)(3+1) + B(3)(3+1) + C(3)(3-3)$$

$$24 = 12B$$

$$B = 2$$

Let $x = -1$, then

$$7(-1)+3 = A(-1-3)(-1+1) + B(-1)(-1+1)$$
$$+ C(-1)(-1-3)$$

$$-4 = 4C$$

$$C = -1$$

$$\frac{7x+3}{x^3-2x^2-3x} = \frac{-1}{x} + \frac{2}{x-3} + \frac{-1}{x+1}$$

39. Perform synthetic division to find a factor:

$$2\overline{)1 \quad -4 \quad 5 \quad -2}$$
$$\underline{\quad\quad 2 \quad -4 \quad 2}$$
$$1 \quad -2 \quad 1 \quad 0$$

$$x^3 - 4x^2 + 5x - 2 = (x-2)(x^2 - 2x + 1)$$
$$= (x-2)(x-1)^2$$

Find the partial fraction decomposition:

$$\frac{x^2}{x^3 - 4x^2 + 5x - 2} = \frac{x^2}{(x-2)(x-1)^2}$$
$$= \frac{A}{x-2} + \frac{B}{x-1} + \frac{C}{(x-1)^2}$$

Multiplying both sides by $(x-2)(x-1)^2$, we obtain:

$$x^2 = A(x-1)^2 + B(x-2)(x-1) + C(x-2)$$

Let $x = 2$, then

$$2^2 = A(2-1)^2 + B(2-2)(2-1) + C(2-2)$$
$$4 = A$$

Let $x = 1$, then

$$1^2 = A(1-1)^2 + B(1-2)(1-1) + C(1-2)$$
$$1 = -C$$
$$C = -1$$

Let $x = 0$, then

$$0^2 = A(0-1)^2 + B(0-2)(0-1) + C(0-2)$$
$$0 = A + 2B - 2C$$
$$0 = 4 + 2B - 2(-1)$$
$$-2B = 6$$
$$B = -3$$

$$\frac{x^2}{x^3 - 4x^2 + 5x - 2} = \frac{4}{x-2} + \frac{-3}{x-1} + \frac{-1}{(x-1)^2}$$

41. Find the partial fraction decomposition:

$$\frac{x^3}{(x^2+16)^3} = \frac{Ax+B}{x^2+16} + \frac{Cx+D}{(x^2+16)^2} + \frac{Ex+F}{(x^2+16)^3}$$

Multiplying both sides by $(x^2+16)^3$, we obtain:

$$x^3 = (Ax+B)(x^2+16)^2 + (Cx+D)(x^2+16)$$
$$+ Ex + F$$
$$x^3 = (Ax+B)(x^4 + 32x^2 + 256) + Cx^3 + Dx^2$$
$$+ 16Cx + 16D + Ex + F$$
$$x^3 = Ax^5 + Bx^4 + 32Ax^3 + 32Bx^2 + 256Ax$$
$$+ 256B + Cx^3 + Dx^2$$
$$+ 16Cx + 16D + Ex + F$$

$$x^3 = Ax^5 + Bx^4 + (32A+C)x^3 + (32B+D)x^2$$
$$+ (256A + 16C + E)x$$
$$+ (256B + 16D + F)$$

$$A = 0; \quad B = 0; \qquad 32A + C = 1$$
$$32(0) + C = 1$$
$$C = 1$$

$$32B + D = 0 \qquad\qquad 256A + 16C + E = 0$$
$$32(0) + D = 0 \qquad\qquad 256(0) + 16(1) + E = 0$$
$$D = 0 \qquad\qquad\qquad E = -16$$

$$256B + 16D + F = 0$$
$$256(0) + 16(0) + F = 0$$
$$F = 0$$

$$\frac{x^3}{(x^2+16)^3} = \frac{x}{(x^2+16)^2} + \frac{-16x}{(x^2+16)^3}$$

43. Find the partial fraction decomposition:

$$\frac{4}{2x^2 - 5x - 3} = \frac{4}{(x-3)(2x+1)} = \frac{A}{x-3} + \frac{B}{2x+1}$$

Multiplying both sides by $(x-3)(2x+1)$, we obtain: $4 = A(2x+1) + B(x-3)$

Let $x = -\frac{1}{2}$, then

$$4 = A\left(2\left(-\frac{1}{2}\right)+1\right) + B\left(-\frac{1}{2}-3\right)$$
$$4 = -\frac{7}{2}B$$
$$B = -\frac{8}{7}$$

Let $x = 3$, then $4 = A(2(3)+1) + B(3-3)$
$$4 = 7A$$
$$A = \frac{4}{7}$$

$$\frac{4}{2x^2 - 5x - 3} = \frac{\frac{4}{7}}{x-3} + \frac{-\frac{8}{7}}{2x+1}$$

45. Find the partial fraction decomposition:

$$\frac{2x+3}{x^4-9x^2} = \frac{2x+3}{x^2(x-3)(x+3)}$$

$$= \frac{A}{x} + \frac{B}{x^2} + \frac{C}{x-3} + \frac{D}{x+3}$$

Multiplying both sides by $x^2(x-3)(x+3)$, we obtain:

$$2x+3 = Ax(x-3)(x+3) + B(x-3)(x+3)$$
$$+ Cx^2(x+3) + Dx^2(x-3)$$

Let $x=0$, then

$$2\cdot0+3 = A\cdot0(0-3)(0+3) + B(0-3)(0+3)$$
$$+ C\cdot0^2(0+3) + D\cdot0^2(0-3)$$

$$3 = -9B$$

$$B = -\frac{1}{3}$$

Let $x=3$, then

$$2\cdot3+3 = A\cdot3(3-3)(3+3) + B(3-3)(3+3)$$
$$+ C\cdot3^2(3+3) + D\cdot3^2(3-3)$$

$$9 = 54C$$

$$C = \frac{1}{6}$$

Let $x=-3$, then

$$2(-3)+3 = A(-3)(-3-3)(-3+3)$$
$$+ B(-3-3)(-3+3)$$
$$+ C(-3)^2(-3+3)$$
$$+ D(-3)^2(-3-3)$$

$$-3 = -54D$$

$$D = \frac{1}{18}$$

Let $x=1$, then

$$2\cdot1+3 = A\cdot1(1-3)(1+3) + B(1-3)(1+3)$$
$$+ C\cdot1^2(1+3) + D\cdot1^2(1-3)$$

$$5 = -8A - 8B + 4C - 2D$$

$$5 = -8A - 8(-1/3) + 4(1/6) - 2(1/18)$$

$$5 = -8A + \frac{8}{3} + \frac{2}{3} - \frac{1}{9}$$

$$-8A = \frac{16}{9}$$

$$A = -\frac{2}{9}$$

$$\frac{2x+3}{x^4-9x^2} = \frac{-\frac{2}{9}}{x} + \frac{-\frac{1}{3}}{x^2} + \frac{\frac{1}{6}}{x-3} + \frac{\frac{1}{18}}{x+3}$$

47. $\dfrac{x^3+x^2-3}{x^2+3x-4}$

Dividing:

$$\begin{array}{r} x-2 \\ x^2+3x-4 \overline{)\ x^3+x^2+0x-3} \\ -(x^3+3x^2-4x) \\ \hline -2x^2+4x-3 \\ -(-2x^2-6x+8) \\ \hline 10x-11 \end{array}$$

$$\frac{x^3+x^2-3}{x^2+3x-4} = x-2 + \frac{10x-11}{x^2+3x-4}, \quad x \neq -4, 1$$

Find the partial fraction decomposition:

$$\frac{10x-11}{x^2+3x-4} = \frac{10x-11}{(x+4)(x-1)} = \frac{A}{x+4} + \frac{B}{x-1}$$

Multiplying both sides by $(x+4)(x-1)$, we obtain:

$$10x-11 = A(x-1) + B(x+4)$$

Let $x=1$, then

$$10(1)-11 = A(1-1) + B(1+4)$$

$$-1 = 5B$$

$$B = -\frac{1}{5}$$

Let $x=-4$, then

$$10(-4)-11 = A((-4)-1) + B(-4+4)$$

$$-51 = -5A$$

$$A = \frac{51}{5}$$

$$\frac{10x-11}{x^2+3x-4} = \frac{\frac{51}{5}}{x+4} + \frac{-\frac{1}{5}}{x-1}$$

Thus, $\dfrac{x^3+x^2-3}{x^2+3x-4} = x-2 + \dfrac{\frac{51}{5}}{x+4} + \dfrac{-\frac{1}{5}}{x-1}$

49. $\dfrac{x^3}{x^2+1}$

Dividing:

$$\begin{array}{r} x \\ x^2+0x+1 \overline{)\ x^3+0x^2+0x} \\ -(x^3+0x^2+x) \\ \hline -x \end{array}$$

$$\frac{x^3}{x^2+1} = x + \frac{-x}{x^2+1}$$

Since $x^2 + 1$ is irreducible then we cannot go any further.

51. $\dfrac{x^4 - 5x^3 + x - 4}{x^2 + 4x + 4}$

Dividing:

$$\begin{array}{r} x^2 - 4x + 7 \\ x^2 + 4x + 4 \overline{) \; x^4 + 0x^3 - 5x^2 + x - 4} \\ -\left(x^4 + 4x^3 + 4x^2\right) \\ \hline -4x^3 - 9x^2 \;\; + x \\ -\left(-4x^3 - 16x^2 - 16x\right) \\ \hline 7x^2 + 17x - 4 \\ -\left(7x^2 + 28x + 28\right) \\ \hline -11x - 32 \end{array}$$

$\dfrac{x^4 - 5x^3 + x - 4}{x^2 + 4x + 4} = x^2 - 4x + 7 + \dfrac{-11x - 32}{x^2 + 4x + 4}$,

$x \neq -2$

Find the partial fraction decomposition:

$\dfrac{-11x - 32}{x^2 + 4x + 4} = \dfrac{-11x - 32}{(x+2)(x+2)} = \dfrac{A}{x+2} + \dfrac{B}{(x+2)^2}$

Multiplying both sides by $(x+2)^2$, we obtain:

$-11x - 32 = A(x+2) + B$

$-11x - 32 = Ax + 2A + B$

$-11x - 32 = Ax + (2A + B)$

Since the coefficient of x is A then $A = -11$.

Let $A = -11$, then

$-32 = 2A + B$

$-32 = 2(-11) + B$

$-10 = B$

$\dfrac{-11x - 32}{x^2 + 4x + 4} = \dfrac{-11}{x+2} + \dfrac{-10}{(x+2)^2}$

Thus,

$\dfrac{x^4 - 5x^3 + x - 4}{x^2 + 4x + 4} = x^2 - 4x + 7 + \dfrac{-11}{x+2} + \dfrac{-10}{(x+2)^2}$

53. $\dfrac{x^5 + x^4 - x^2 + 2}{x^4 - 2x^2 + 1}$

Dividing:

$$\begin{array}{r} x + 1 \\ x^4 + 0x^3 - 2x^2 + 0x + 1 \overline{) \; x^5 + x^4 + 0x^3 - x^2 + 0x + 2} \\ -\left(x^5 + 0x^4 - 2x^3 + 0x^2 \;\; + x\right) \\ \hline x^4 + 2x^3 - x^2 \;\; - x + 2 \\ -\left(x^4 + 0x^3 - 2x^2 + 0x + 1\right) \\ \hline 2x^3 - x^2 \;\; - x + 1 \end{array}$$

$\dfrac{x^5 + x^4 - x^2 + 2}{x^4 - 2x^2 + 1} = x + 1 + \dfrac{2x^3 + x^2 - x + 1}{x^4 - 2x^2 + 1}$,

$x \neq 1, -1$

Find the partial fraction decomposition:

$\dfrac{2x^3 + x^2 - x + 1}{(x+1)^2 (x-1)^2} = \dfrac{A}{x+1} + \dfrac{B}{(x+1)^2} + \dfrac{C}{x-1} + \dfrac{D}{(x-1)^2}$

Multiplying both sides by $(x+1)^2 (x-1)^2$, we obtain:

$2x^3 + x^2 - x + 1 = A(x+1)(x-1)^2 + B(x-1)^2$

$\qquad + C(x+1)^2 (x-1) + D(x+1)^2$

Let $x = -1$, then

$2(-1)^3 + (-1)^2 - (-1) + 1 = A(-1+1)(-1-1)^2$

$\qquad + B(-1-1)^2 + C(-1+1)^2(-1-1) + D(-1+1)^2$

$1 = 4B$

$\dfrac{1}{4} = B$

Let $x = 1$, then

$2(1)^3 + (1)^2 - (1) + 1 = A(1+1)(1-1)^2$

$\qquad + B(1-1)^2 + C(1+1)^2(1-1) + D(1+1)^2$

$3 = 4D$

$\dfrac{3}{4} = D$

Let $x = 0$, then

$2(0)^3 + (0)^2 - (0) + 1 = A(0+1)(0-1)^2$

$\qquad + \dfrac{1}{4}(0-1)^2 + C(0+1)^2(0-1) + \dfrac{3}{4}(0+1)^2$

$1 = A + \dfrac{1}{4} - C + \dfrac{3}{4}$

$0 = A - C$

Let $x = 2$, then

$2(2)^3 + (2)^2 - (2) + 1 = A(2+1)(2-1)^2$

$\qquad + \dfrac{1}{4}(2-1)^2 + C(2+1)^2(2-1) + \dfrac{3}{4}(2+1)^2$

$19 = 3A + \dfrac{1}{4} + 9C + \dfrac{27}{4}$

$12 = 3A + 9C$

$\quad 0 = A - C$

$12 = 3A + 9C$

$\quad A = C$

$12 = 3A + 9A$

$12 = 12A$

$\quad 1 = A$ and $1 = C$

$\dfrac{2x^3 + x^2 - x + 1}{x^4 - 2x^2 + 1} = \dfrac{1}{x+1} + \dfrac{\frac{1}{4}}{(x+1)^2} + \dfrac{1}{x-1} + \dfrac{\frac{3}{4}}{(x-1)^2}$

Thus, $\dfrac{x^5 + x^4 - x^2 + 2}{x^4 - 2x^2 + 1} = x + 1$

$\qquad + \dfrac{\frac{1}{4}}{(x+1)^2} + \dfrac{1}{x-1} + \dfrac{\frac{3}{4}}{(x-1)^2}$

55. $A = P\left(1 + \dfrac{r}{n}\right)^{nt}$

$8400 = 4200\left(1 + \dfrac{0.18}{365}\right)^{365t}$

$2 = (1.000493)^{365t}$

$\ln 2 = \ln(1.000493)^{365t}$

$\ln 2 = 365t \ln(1.000493)$

$t = \dfrac{\ln 2}{365 \ln(1.000493)}$

$\quad = 3.85$ years

57. Since a is negative then the graph opens up so the function has a maximum value. To find the maximum value we can find the vertex.

$x = -\dfrac{b}{2a} = -\dfrac{-120}{2(-3)} = 20$

$f(20) = -3(20)^2 + 120(20) + 50 = 1250$

The vertex is $(20, 1250)$ so the maximum value of the function is 1250.

Section 8.6

1. $y = 3x + 2$

The graph is a line.

x-intercept:

$\quad 0 = 3x + 2$

$\quad 3x = -2$

$\quad x = -\dfrac{2}{3}$

y-intercept: $y = 3(0) + 2 = 2$

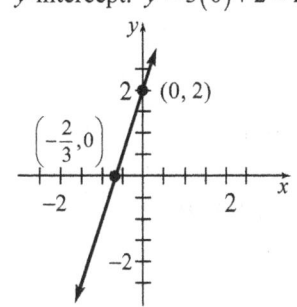

3. $\quad y^2 = x^2 - 1$

$\quad x^2 - y^2 = 1$

$\quad \dfrac{x^2}{1^2} - \dfrac{y^2}{1^2} = 1$

The graph is a hyperbola with center $(0, 0)$, transverse axis along the x-axis, and vertices at $(-1, 0)$ and $(1, 0)$. The asymptotes are $y = -x$ and $y = x$.

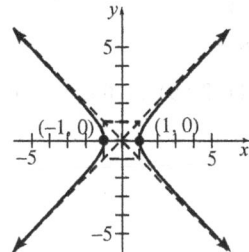

5. $\begin{cases} y = x^2 + 1 \\ y = x + 1 \end{cases}$

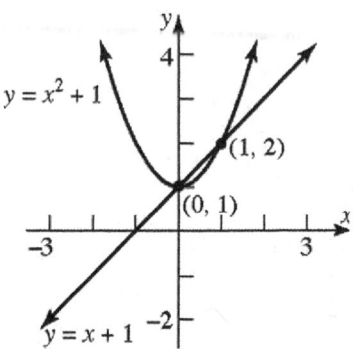

$(0, 1)$ and $(1, 2)$ are the intersection points.

Solve by substitution:

$x^2 + 1 = x + 1$

$x^2 - x = 0$

$x(x - 1) = 0$

$\quad x = 0 \text{ or } x = 1$

$\quad y = 1 \qquad y = 2$

Solutions: $(0, 1)$ and $(1, 2)$

7. $\begin{cases} y = \sqrt{36 - x^2} \\ y = 8 - x \end{cases}$

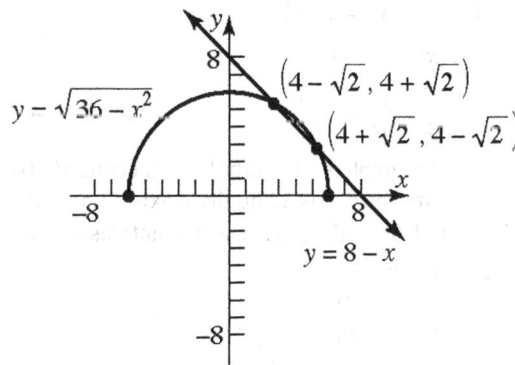

$(2.59, 5.41)$ and $(5.41, 2.59)$ are the intersection points.

Solve by substitution:

$$\sqrt{36 - x^2} = 8 - x$$

$$36 - x^2 = 64 - 16x + x^2$$

$$2x^2 - 16x + 28 = 0$$

$$x^2 - 8x + 14 = 0$$

$$x = \frac{8 \pm \sqrt{64 - 56}}{2}$$

$$= \frac{8 \pm 2\sqrt{2}}{2}$$

$$= 4 \pm \sqrt{2}$$

If $x = 4 + \sqrt{2}$, $y = 8 - \left(4 + \sqrt{2}\right) = 4 - \sqrt{2}$

If $x = 4 - \sqrt{2}$, $y = 8 - \left(4 - \sqrt{2}\right) = 4 + \sqrt{2}$

Solutions: $\left(4 + \sqrt{2}, 4 - \sqrt{2}\right)$ and $\left(4 - \sqrt{2}, 4 + \sqrt{2}\right)$

9. $\begin{cases} y = \sqrt{x} \\ y = 2 - x \end{cases}$

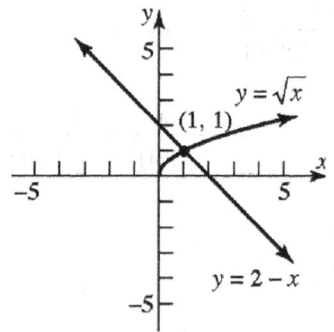

$(1, 1)$ is the intersection point.

Solve by substitution:

$$\sqrt{x} = 2 - x$$

$$x = 4 - 4x + x^2$$

$$x^2 - 5x + 4 = 0$$

$$(x - 4)(x - 1) = 0$$

$$x = 4 \quad \text{ or } x = 1$$

$$y = -2 \text{ or } y = 1$$

Eliminate $(4, -2)$; we must have $y \geq 0$.

Solution: $(1, 1)$

11. $\begin{cases} x = 2y \\ x = y^2 - 2y \end{cases}$

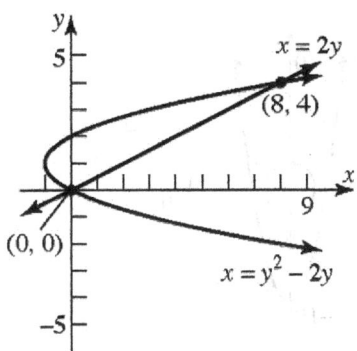

$(0, 0)$ and $(8, 4)$ are the intersection points.

Solve by substitution:

$$2y = y^2 - 2y$$
$$y^2 - 4y = 0$$
$$y(y - 4) = 0$$
$$y = 0 \text{ or } y = 4$$
$$x = 0 \text{ or } x = 8$$

Solutions: $(0, 0)$ and $(8, 4)$

13. $\begin{cases} x^2 + y^2 = 4 \\ x^2 + 2x + y^2 = 0 \end{cases}$

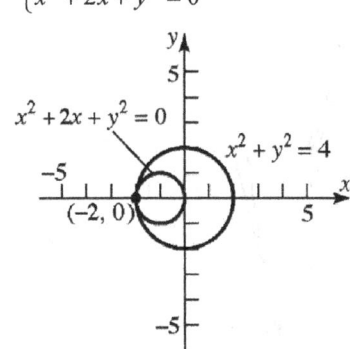

$(-2, 0)$ is the intersection point.

Substitute 4 for $x^2 + y^2$ in the second equation.

$$2x + 4 = 0$$
$$2x = -4$$
$$x = -2$$
$$y = \sqrt{4 - (-2)^2} = 0$$

Solution: $(-2, 0)$

15. $\begin{cases} y = 3x - 5 \\ x^2 + y^2 = 5 \end{cases}$

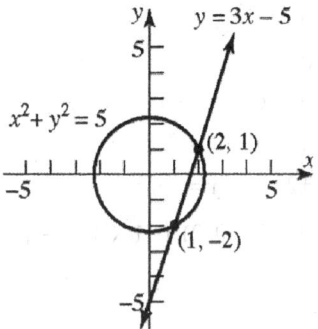

$(1, -2)$ and $(2, 1)$ are the intersection points.

Solve by substitution:

$$x^2 + (3x - 5)^2 = 5$$
$$x^2 + 9x^2 - 30x + 25 = 5$$
$$10x^2 - 30x + 20 = 0$$
$$x^2 - 3x + 2 = 0$$
$$(x - 1)(x - 2) = 0$$
$$x = 1 \quad \text{or} \quad x = 2$$
$$y = -2 \qquad y = 1$$

Solutions: $(1, -2)$ and $(2, 1)$

17. $\begin{cases} x^2 + y^2 = 4 \\ y^2 - x = 4 \end{cases}$

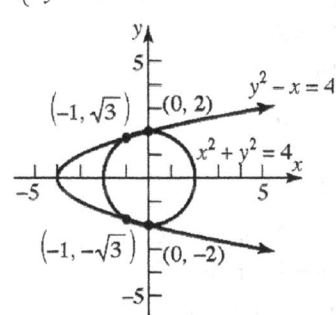

$(-1, 1.73)$, $(-1, -1.73)$, $(0, 2)$, and $(0, -2)$ are the intersection points.

Substitute $x + 4$ for y^2 in the first equation:

$$x^2 + x + 4 = 4$$
$$x^2 + x = 0$$
$$x(x+1) = 0$$
$$x = 0 \quad \text{or} \quad x = -1$$
$$y^2 = 4 \qquad y^2 = 3$$
$$y = \pm 2 \qquad y = \pm\sqrt{3}$$

Solutions: $(0, -2), (0, 2), \left(-1, \sqrt{3}\right), \left(-1, -\sqrt{3}\right)$

19. $\begin{cases} xy = 4 \\ x^2 + y^2 = 8 \end{cases}$

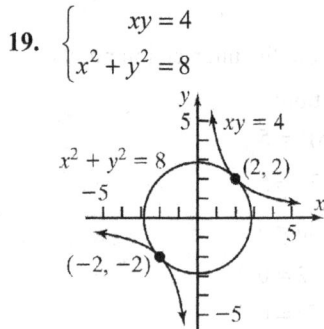

$(-2, -2)$ and $(2, 2)$ are the intersection points.

Solve by substitution:

$$x^2 + \left(\frac{4}{x}\right)^2 = 8$$
$$x^2 + \frac{16}{x^2} = 8$$
$$x^4 + 16 = 8x^2$$
$$x^4 - 8x^2 + 16 = 0$$
$$(x^2 - 4)^2 = 0$$
$$x^2 - 4 = 0$$
$$x^2 = 4$$
$$x = 2 \quad \text{or} \quad x = -2$$
$$y = 2 \quad \text{or} \quad y = -2$$

Solutions: $(-2, -2)$ and $(2, 2)$

21. $\begin{cases} x^2 + y^2 = 4 \\ y = x^2 - 9 \end{cases}$

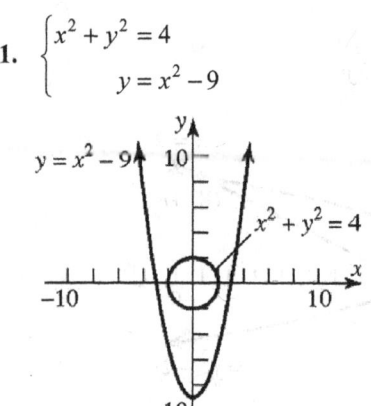

No solution; Inconsistent.

Solve by substitution:

$$x^2 + (x^2 - 9)^2 = 4$$
$$x^2 + x^4 - 18x^2 + 81 = 4$$
$$x^4 - 17x^2 + 77 = 0$$
$$x^2 = \frac{17 \pm \sqrt{289 - 4(77)}}{2}$$
$$= \frac{17 \pm \sqrt{-19}}{2}$$

There are no real solutions to this expression. Inconsistent.

23. $\begin{cases} y = x^2 - 4 \\ y = 6x - 13 \end{cases}$

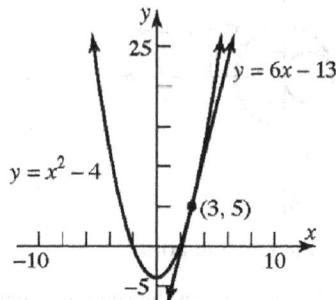

$(3, 5)$ is the intersection point.

Solve by substitution:

$$x^2 - 4 = 6x - 13$$
$$x^2 - 6x + 9 = 0$$
$$(x - 3)^2 = 0$$
$$x - 3 = 0$$
$$x = 3$$
$$y = (3)^2 - 4 = 5$$

Solution: $(3, 5)$

25. Solve the second equation for y, substitute into the first equation and solve:

$$\begin{cases} 2x^2 + y^2 = 18 \\ xy = 4 \Rightarrow y = \dfrac{4}{x} \end{cases}$$

$$2x^2 + \left(\frac{4}{x}\right)^2 = 18$$

$$2x^2 + \frac{16}{x^2} = 18$$

$$2x^4 + 16 = 18x^2$$

$$2x^4 - 18x^2 + 16 = 0$$

$$x^4 - 9x^2 + 8 = 0$$

$$\left(x^2 - 8\right)\left(x^2 - 1\right) = 0$$

$$x^2 = 8 \qquad \text{or} \quad x^2 = 1$$

$$x = \pm\sqrt{8} = \pm 2\sqrt{2} \quad \text{or} \qquad x = \pm 1$$

If $x = 2\sqrt{2}$: $\quad y = \dfrac{4}{2\sqrt{2}} = \sqrt{2}$

If $x = -2\sqrt{2}$: $\quad y = \dfrac{4}{-2\sqrt{2}} = -\sqrt{2}$

If $x = 1$: $\quad y = \dfrac{4}{1} = 4$

If $x = -1$: $\quad y = \dfrac{4}{-1} = -4$

Solutions:
$\left(2\sqrt{2}, \sqrt{2}\right), \left(-2\sqrt{2}, -\sqrt{2}\right), (1, 4), (-1, -4)$

27. Substitute the first equation into the second equation and solve:

$$\begin{cases} y = 2x + 1 \\ 2x^2 + y^2 = 1 \end{cases}$$

$$2x^2 + (2x + 1)^2 = 1$$

$$2x^2 + 4x^2 + 4x + 1 = 1$$

$$6x^2 + 4x = 0$$

$$2x(3x + 2) = 0$$

$$2x = 0 \quad \text{or} \quad 3x + 2 = 0$$

$$x = 0 \quad \text{or} \quad x = -\frac{2}{3}$$

If $x = 0$: $\quad y = 2(0) + 1 = 1$

If $x = -\dfrac{2}{3}$: $\quad y = 2\left(-\dfrac{2}{3}\right) + 1 = -\dfrac{4}{3} + 1 = -\dfrac{1}{3}$

Solutions: $(0, 1), \left(-\dfrac{2}{3}, -\dfrac{1}{3}\right)$

29. Solve the first equation for y, substitute into the second equation and solve:

$$\begin{cases} x + y + 1 = 0 \Rightarrow y = -x - 1 \\ x^2 + y^2 + 6y - x = -5 \end{cases}$$

$$x^2 + (-x - 1)^2 + 6(-x - 1) - x = -5$$

$$x^2 + x^2 + 2x + 1 - 6x - 6 - x = -5$$

$$2x^2 - 5x = 0$$

$$x(2x - 5) = 0$$

$$x = 0 \quad \text{or} \quad x = \frac{5}{2}$$

If $x = 0$: $\quad y = -(0) - 1 = -1$

If $x = \dfrac{5}{2}$: $\quad y = -\dfrac{5}{2} - 1 = -\dfrac{7}{2}$

Solutions: $(0, -1), \left(\frac{5}{2}, -\frac{7}{2}\right)$

31. Solve the second equation for y, substitute into the first equation and solve:

$$\begin{cases} 4x^2 - 3xy + 9y^2 = 15 \\ 2x + 3y = 5 \Rightarrow y = -\dfrac{2}{3}x + \dfrac{5}{3} \end{cases}$$

$$4x^2 - 3x\left(-\frac{2}{3}x + \frac{5}{3}\right) + 9\left(-\frac{2}{3}x + \frac{5}{3}\right)^2 = 15$$

$$4x^2 + 2x^2 - 5x + 4x^2 - 20x + 25 = 15$$

$$10x^2 - 25x + 10 = 0$$

$$2x^2 - 5x + 2 = 0$$

$$(2x - 1)(x - 2) = 0$$

$$x = \frac{1}{2} \quad \text{or} \quad x = 2$$

If $x = \dfrac{1}{2}$: $\quad y = -\dfrac{2}{3}\left(\dfrac{1}{2}\right) + \dfrac{5}{3} = \dfrac{4}{3}$

If $x = 2$: $\quad y = -\dfrac{2}{3}(2) + \dfrac{5}{3} = \dfrac{1}{3}$

Solutions: $\left(\dfrac{1}{2}, \dfrac{4}{3}\right), \left(2, \dfrac{1}{3}\right)$

33. Multiply each side of the second equation by 4 and add the equations to eliminate y:

$$\begin{cases} x^2 - 4y^2 = -7 \longrightarrow \quad x^2 - 4y^2 = -7 \\ 3x^2 + y^2 = 31 \overset{4}{\longrightarrow} \underline{12x^2 + 4y^2 = 124} \end{cases}$$

$$13x^2 \qquad = 117$$

$$x^2 = 9$$

$$x = \pm 3$$

If $x = 3$: $3(3)^2 + y^2 = 31 \Rightarrow y^2 = 4 \Rightarrow y = \pm 2$

If $x = -3$: $3(-3)^2 + y^2 = 31 \Rightarrow y^2 = 4 \Rightarrow y = \pm 2$

Solutions: $(3, 2), (3, -2), (-3, 2), (-3, -2)$

35. $\begin{cases} 7x^2 - 3y^2 + 5 = 0 \\ 3x^2 + 5y^2 = 12 \end{cases}$

$\begin{cases} 7x^2 - 3y^2 = -5 \\ 3x^2 + 5y^2 = 12 \end{cases}$

Multiply each side of the first equation by 5 and each side of the second equation by 3 and add the equations to eliminate y:

$35x^2 - 15y^2 = -25$

$\dfrac{9x^2 + 15y^2 = 36}{44x^2 = 11}$

$x^2 = \dfrac{1}{4}$

$x = \pm \dfrac{1}{2}$

If $x = \dfrac{1}{2}$:

$3\left(\dfrac{1}{2}\right)^2 + 5y^2 = 12 \Rightarrow y^2 = \dfrac{9}{4} \Rightarrow y = \pm \dfrac{3}{2}$

If $x = -\dfrac{1}{2}$:

$3\left(-\dfrac{1}{2}\right)^2 + 5y^2 = 12 \Rightarrow y^2 = \dfrac{9}{4} \Rightarrow y = \pm \dfrac{3}{2}$

Solutions:

$\left(\dfrac{1}{2}, \dfrac{3}{2}\right), \left(\dfrac{1}{2}, -\dfrac{3}{2}\right), \left(-\dfrac{1}{2}, \dfrac{3}{2}\right), \left(-\dfrac{1}{2}, -\dfrac{3}{2}\right)$

37. Multiply each side of the second equation by 2 and add the equations to eliminate xy:

$\begin{cases} x^2 + 2xy = 10 \longrightarrow \quad x^2 + 2xy = 10 \\ 3x^2 - xy = 2 \xrightarrow{\ 2\ } 6x^2 - 2xy = 4 \end{cases}$

$\qquad \qquad 7x^2 \qquad = 14$

$\qquad \qquad x^2 = 2$

$\qquad \qquad x = \pm \sqrt{2}$

If $x = \sqrt{2}$:

$3\left(\sqrt{2}\right)^2 - \sqrt{2} \cdot y = 2$

$\Rightarrow -\sqrt{2} \cdot y = -4 \Rightarrow y = \dfrac{4}{\sqrt{2}} \Rightarrow y = 2\sqrt{2}$

If $x = -\sqrt{2}$:

$3\left(-\sqrt{2}\right)^2 - \left(-\sqrt{2}\right)y = 2$

$\Rightarrow \sqrt{2} \cdot y = -4 \Rightarrow y = \dfrac{-4}{\sqrt{2}} \Rightarrow y = -2\sqrt{2}$

Solutions: $\left(\sqrt{2}, 2\sqrt{2}\right), \left(-\sqrt{2}, -2\sqrt{2}\right)$

39. $\begin{cases} 2x^2 + y^2 = 2 \\ x^2 - 2y^2 + 8 = 0 \end{cases}$

$\begin{cases} 2x^2 + y^2 = 2 \\ x^2 - 2y^2 = -8 \end{cases}$

Multiply each side of the first equation by 2 and add the equations to eliminate y:

$4x^2 + 2y^2 = 4$

$\dfrac{x^2 - 2y^2 = -8}{5x^2 = -4}$

$x^2 = -\dfrac{4}{5}$

No real solution exists. The system is inconsistent.

41. $\begin{cases} x^2 + 2y^2 = 16 \\ 4x^2 - y^2 = 24 \end{cases}$

Multiply each side of the second equation by 2 and add the equations to eliminate y:

$x^2 + 2y^2 = 16$

$\dfrac{8x^2 - 2y^2 = 48}{9x^2 = 64}$

$x^2 = \dfrac{64}{9}$

$x = \pm \dfrac{8}{3}$

If $x = \dfrac{8}{3}$:

$\left(\dfrac{8}{3}\right)^2 + 2y^2 = 16 \Rightarrow 2y^2 = \dfrac{80}{9}$

$\Rightarrow y^2 = \dfrac{40}{9} \Rightarrow y = \pm \dfrac{2\sqrt{10}}{3}$

If $x = -\dfrac{8}{3}$:

$\left(-\dfrac{8}{3}\right)^2 + 2y^2 = 16 \Rightarrow 2y^2 = \dfrac{80}{9}$

$\Rightarrow y^2 = \dfrac{40}{9} \Rightarrow y = \pm\dfrac{2\sqrt{10}}{3}$

Solutions:

$\left(\dfrac{8}{3}, \dfrac{2\sqrt{10}}{3}\right), \left(\dfrac{8}{3}, -\dfrac{2\sqrt{10}}{3}\right), \left(-\dfrac{8}{3}, \dfrac{2\sqrt{10}}{3}\right),$

$\left(-\dfrac{8}{3}, -\dfrac{2\sqrt{10}}{3}\right)$

43. $\begin{cases} \dfrac{5}{x^2} - \dfrac{2}{y^2} + 3 = 0 \\ \dfrac{3}{x^2} + \dfrac{1}{y^2} = 7 \end{cases}$

$\begin{cases} \dfrac{5}{x^2} - \dfrac{2}{y^2} = -3 \\ \dfrac{3}{x^2} + \dfrac{1}{y^2} = 7 \end{cases}$

Multiply each side of the second equation by 2 and add the equations to eliminate y:

$\dfrac{5}{x^2} - \dfrac{2}{y^2} = -3$

$\dfrac{6}{x^2} + \dfrac{2}{y^2} = 14$

$\rule{3cm}{0.4pt}$

$\dfrac{11}{x^2} = 11$

$x^2 = 1$

$x = \pm 1$

If $x = 1$:

$\dfrac{3}{(1)^2} + \dfrac{1}{y^2} = 7 \Rightarrow \dfrac{1}{y^2} = 4 \Rightarrow y^2 = \dfrac{1}{4}$

$\Rightarrow y = \pm\dfrac{1}{2}$

If $x = -1$:

$\dfrac{3}{(-1)^2} + \dfrac{1}{y^2} = 7 \Rightarrow \dfrac{1}{y^2} = 4 \Rightarrow y^2 = \dfrac{1}{4}$

$\Rightarrow y = \pm\dfrac{1}{2}$

Solutions: $\left(1, \dfrac{1}{2}\right), \left(1, -\dfrac{1}{2}\right), \left(-1, \dfrac{1}{2}\right), \left(-1, -\dfrac{1}{2}\right)$

45. $\begin{cases} \dfrac{1}{x^4} + \dfrac{6}{y^4} = 6 \\ \dfrac{2}{x^4} - \dfrac{2}{y^4} = 19 \end{cases}$

Multiply each side of the first equation by -2 and add the equations to eliminate x:

$\dfrac{-2}{x^4} - \dfrac{12}{y^4} = -12$

$\dfrac{2}{x^4} - \dfrac{2}{y^4} = 19$

$\rule{3cm}{0.4pt}$

$-\dfrac{14}{y^4} = 7$

$y^4 = -2$

There are no real solutions. The system is inconsistent.

47. $\begin{cases} x^2 - 3xy + 2y^2 = 0 \\ x^2 + xy = 6 \end{cases}$

Subtract the second equation from the first to eliminate the x^2 term.

$-4xy + 2y^2 = -6$

$2xy - y^2 = 3$

Since $y \neq 0$, we can solve for x in this equation to get

$x = \dfrac{y^2 + 3}{2y}, \quad y \neq 0$

Now substitute for x in the second equation and solve for y.

$x^2 + xy = 6$

$\left(\dfrac{y^2 + 3}{2y}\right)^2 + \left(\dfrac{y^2 + 3}{2y}\right)y = 6$

$\dfrac{y^4 + 6y^2 + 9}{4y^2} + \dfrac{y^2 + 3}{2} = 6$

$y^4 + 6y^2 + 9 + 2y^4 + 6y^2 = 24y^2$

$3y^4 - 12y^2 + 9 = 0$

$y^4 - 4y^2 + 3 = 0$

$(y^2 - 3)(y^2 - 1) = 0$

Thus, $y = \pm\sqrt{3}$ or $y = \pm 1$.

If $y = 1$: $\quad x = 2 \cdot 1 = 2$

If $y = -1$: $\quad x = 2(-1) = -2$

If $y = \sqrt{3}$: $x = \sqrt{3}$

If $y = -\sqrt{3}$: $x = -\sqrt{3}$

Solutions: $(2, 1), (2, -1), \left(\sqrt{3}, \sqrt{3}\right), \left(-\sqrt{3}, -\sqrt{3}\right)$

49. $\begin{cases} y^2 + y + x^2 - x - 2 = 0 \\ y + 1 + \dfrac{x-2}{y} = 0 \end{cases}$

Multiply each side of the second equation by $-y$ and add the equations to eliminate y:

$$y^2 + y + x^2 - x - 2 = 0$$
$$\underline{-y^2 - y \qquad -x + 2 = 0}$$
$$x^2 - 2x = 0$$
$$x(x-2) = 0$$
$$x = 0 \text{ or } x = 2$$

If $x = 0$:

$y^2 + y + 0^2 - 0 - 2 = 0 \Rightarrow y^2 + y - 2 = 0$

$\Rightarrow (y+2)(y-1) = 0 \Rightarrow y = -2$ or $y = 1$

If $x = 2$:

$y^2 + y + 2^2 - 2 - 2 = 0 \Rightarrow y^2 + y = 0$

$\Rightarrow y(y+1) = 0 \Rightarrow y = 0$ or $y = -1$

Note: $y \neq 0$ because of division by zero.

Solutions: $(0, -2), (0, 1), (2, -1)$

51. Rewrite each equation in exponential form:

$\begin{cases} \log_x y = 3 \;\rightarrow\; y = x^3 \\ \log_x (4y) = 5 \;\rightarrow\; 4y = x^5 \end{cases}$

Substitute the first equation into the second and solve:

$$4x^3 = x^5$$
$$x^5 - 4x^3 = 0$$
$$x^3(x^2 - 4) = 0$$
$$x^3 = 0 \text{ or } x^2 = 4 \Rightarrow x = 0 \text{ or } x = \pm 2$$

The base of a logarithm must be positive, thus $x \neq 0$ and $x \neq -2$.

If $x = 2$: $y = 2^3 = 8$

Solution: $(2, 8)$

53. Rewrite each equation in exponential form:

$\ln x = 4 \ln y \Rightarrow x = e^{4\ln y} = e^{\ln y^4} = y^4$

$\log_3 x = 2 + 2\log_3 y$

$x = 3^{2 + 2\log_3 y} = 3^2 \cdot 3^{2\log_3 y} = 3^2 \cdot 3^{\log_3 y^2} = 9y^2$

So we have the system $\begin{cases} x = y^4 \\ x = 9y^2 \end{cases}$

Therefore we have :

$9y^2 = y^4 \Rightarrow 9y^2 - y^4 = 0 \Rightarrow y^2(9 - y^2) = 0$

$y^2(3+y)(3-y) = 0$

$y = 0$ or $y = -3$ or $y = 3$

Since $\ln y$ is undefined when $y \leq 0$, the only solution is $y = 3$.

If $y = 3$: $x = y^4 \Rightarrow x = 3^4 = 81$

Solution: $(81, 3)$

55. $\begin{cases} x^2 + x + y^2 - 3y + 2 = 0 \\ x + 1 + \dfrac{y^2 - y}{x} = 0 \end{cases}$

$\begin{cases} \left(x + \frac{1}{2}\right)^2 + \left(y - \frac{3}{2}\right)^2 = \frac{1}{2} \\ \left(x + \frac{1}{2}\right)^2 + \left(y - \frac{1}{2}\right)^2 = \frac{1}{2} \end{cases}$

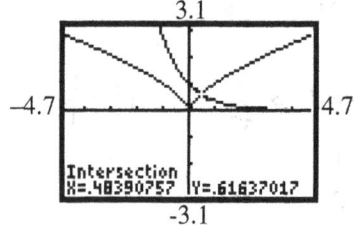

57. Graph: $y_1 = x \wedge (2/3)$; $y_2 = e \wedge (-x)$

Use INTERSECT to solve:

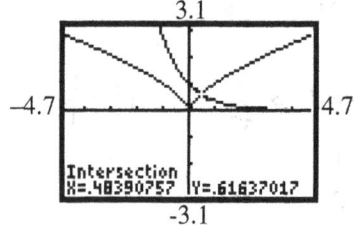

Solution: $x = 0.48, y = 0.62$ or $(0.48, 0.62)$

59. Graph: $y_1 = \sqrt[3]{2 - x^2}$; $y_2 = 4/x^3$

Use INTERSECT to solve:

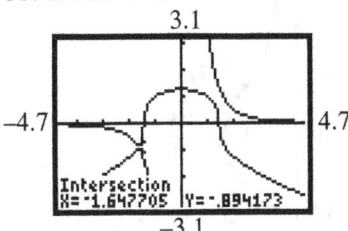

Solution: $x = -1.65, y = -0.89$ or $(-1.65, -0.89)$

61. Graph: $y_1 = \sqrt[4]{12 - x^4}$; $y_2 = -\sqrt[4]{12 - x^4}$;

$y_3 = \sqrt{2/x}$; $y_4 = -\sqrt{2/x}$

Use INTERSECT to solve:

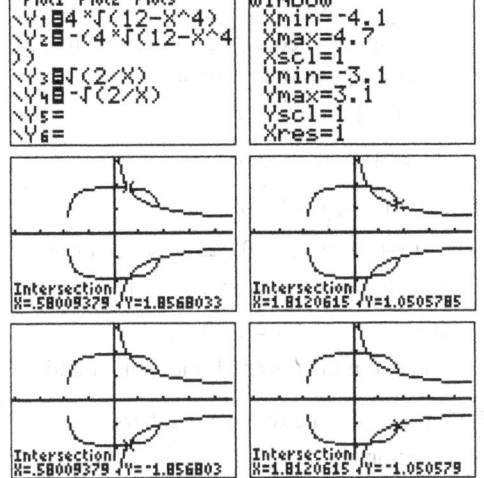

Solutions: $x = 0.58, y = 1.86$; $x = 1.81, y = 1.05$;

$x = 1.81, y = -1.05$; ; $x = 0.58, y = -1.86$ or

$(0.58, 1.86), (1.81, 1.05), (1.81, -1.05),$

$(0.58, -1.86)$

63. Graph: $y_1 = 2/x$; $y_2 = \ln x$

Use INTERSECT to solve:

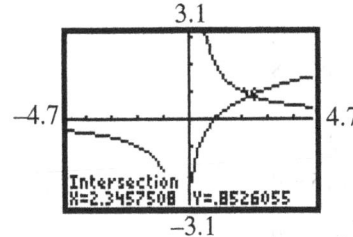

Solution: $x = 2.35, y = 0.85$ or $(2.35, 0.85)$

65. Solve the first equation for x, substitute into the second equation and solve:

$$\begin{cases} x + 2y = 0 \Rightarrow x = -2y \\ (x-1)^2 + (y-1)^2 = 5 \end{cases}$$

$$(-2y-1)^2 + (y-1)^2 = 5$$

$$4y^2 + 4y + 1 + y^2 - 2y + 1 = 5 \Rightarrow 5y^2 + 2y - 3 = 0$$

$$(5y-3)(y+1) = 0$$

$$y = \frac{3}{5} = 0.6 \quad \text{or} \quad y = -1$$

$$x = -\frac{6}{5} = -1.2 \quad \text{or} \quad x = 2$$

The points of intersection are $\left(-\frac{6}{5}, \frac{3}{5} \right), (2, -1)$.

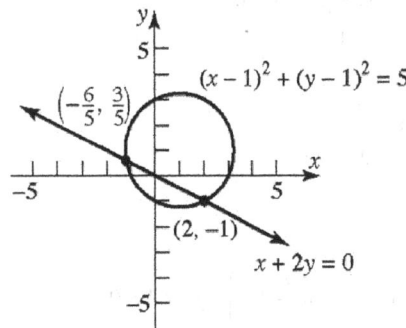

67. Complete the square on the second equation.

$$y^2 + 4y + 4 = x - 1 + 4$$

$$(y+2)^2 = x + 3$$

Substitute this result into the first equation.

$$(x-1)^2 + x + 3 = 4$$

$$x^2 - 2x + 1 + x + 3 = 4$$

$$x^2 - x = 0$$

$$x(x-1) = 0$$

$$x = 0 \quad \text{or} \quad x = 1$$

If $x = 0$: $(y+2)^2 = 0 + 3$

$y + 2 = \pm\sqrt{3} \Rightarrow y = -2 \pm \sqrt{3}$

If $x = 1$: $(y+2)^2 = 1 + 3$

$y + 2 = \pm 2 \Rightarrow y = -2 \pm 2$

The points of intersection are:

$\left(0, -2 - \sqrt{3} \right), \left(0, -2 + \sqrt{3} \right), (1, -4), (1, 0)$.

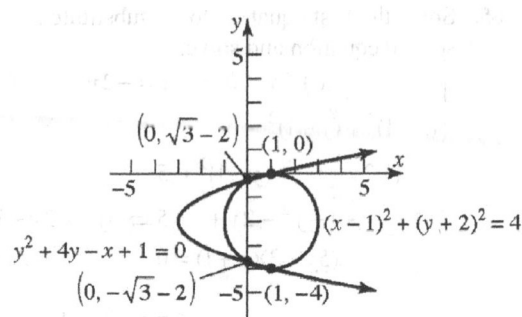

$y^2 + 4y - x + 1 = 0$

$(x-1)^2 + (y+2)^2 = 4$

points labeled: $(0, \sqrt{3}-2)$, $(1, 0)$, $(0, -\sqrt{3}-2)$, $(1, -4)$

69. Solve the first equation for x, substitute into the second equation and solve:

$$\begin{cases} y = \dfrac{4}{x-3} \\ x^2 - 6x + y^2 + 1 = 0 \end{cases}$$

$$y = \frac{4}{x-3}$$

$$x - 3 = \frac{4}{y}$$

$$x = \frac{4}{y} + 3$$

$$\left(\frac{4}{y}+3\right)^2 - 6\left(\frac{4}{y}+3\right) + y^2 + 1 = 0$$

$$\frac{16}{y^2} + \frac{24}{y} + 9 - \frac{24}{y} - 18 + y^2 + 1 = 0$$

$$\frac{16}{y^2} + y^2 - 8 = 0$$

$$16 + y^4 - 8y^2 = 0$$

$$y^4 - 8y^2 + 16 = 0$$

$$(y^2 - 4)^2 = 0$$

$$y^2 - 4 = 0$$

$$y^2 = 4$$

$$y = \pm 2$$

If $y = 2$: $\quad x = \dfrac{4}{2} + 3 = 5$

If $y = -2$: $\quad x = \dfrac{4}{-2} + 3 = 1$

The points of intersection are: $(1, -2), (5, 2)$.

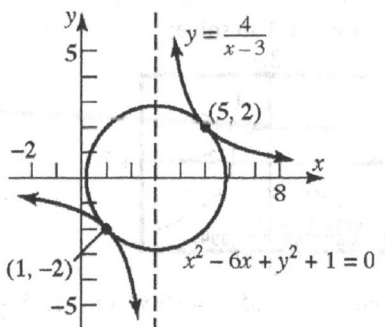

$y = \dfrac{4}{x-3}$

points labeled: $(5, 2)$, $(1, -2)$

$x^2 - 6x + y^2 + 1 = 0$

71. Let x and y be the two numbers. The system of equations is:

$$\begin{cases} x - y = 2 & \Rightarrow x = y + 2 \\ x^2 + y^2 = 10 \end{cases}$$

Solve the first equation for x, substitute into the second equation and solve:

$$(y+2)^2 + y^2 = 10$$

$$y^2 + 4y + 4 + y^2 = 10$$

$$y^2 + 2y - 3 = 0$$

$$(y+3)(y-1) = 0 \quad \Rightarrow \quad y = -3 \text{ or } y = 1$$

If $y = -3$: $\quad x = -3 + 2 = -1$

If $y = 1$: $\quad x = 1 + 2 = 3$

The two numbers are 1 and 3 or -1 and -3.

73. Let x and y be the two numbers. The system of equations is:

$$\begin{cases} xy = 4 \Rightarrow x = \dfrac{4}{y} \\ x^2 + y^2 = 8 \end{cases}$$

Solve the first equation for x, substitute into the second equation and solve:

$$\left(\frac{4}{y}\right)^2 + y^2 = 8$$

$$\frac{16}{y^2} + y^2 = 8$$

$$16 + y^4 = 8y^2$$

$$y^4 - 8y^2 + 16 = 0$$

$$(y^2 - 4)^2 = 0$$

$$y^2 = 4$$

$$y = \pm 2$$

If $y = 2$: $x = \dfrac{4}{2} = 2$; If $y = -2$: $x = \dfrac{4}{-2} = -2$

The two numbers are 2 and 2 or –2 and –2.

75. Let x and y be the two numbers. The system of equations is:
$$\begin{cases} x - y = xy \\ \dfrac{1}{x} + \dfrac{1}{y} = 5 \end{cases}$$

Solve the first equation for x, substitute into the second equation and solve:

$x - xy = y$

$x(1 - y) = y \Rightarrow x = \dfrac{y}{1-y}$

$\dfrac{1}{\frac{y}{1-y}} + \dfrac{1}{y} = 5$

$\dfrac{1-y}{y} + \dfrac{1}{y} = 5$

$\dfrac{2-y}{y} = 5$

$2 - y = 5y$

$6y = 2$

$y = \dfrac{1}{3} \Rightarrow x = \dfrac{\frac{1}{3}}{1 - \frac{1}{3}} = \dfrac{\frac{1}{3}}{\frac{2}{3}} = \dfrac{1}{2}$

The two numbers are $\dfrac{1}{2}$ and $\dfrac{1}{3}$.

77. $\begin{cases} \dfrac{a}{b} = \dfrac{2}{3} \\ a + b = 10 \Rightarrow a = 10 - b \end{cases}$

Solve the second equation for a, substitute into the first equation and solve:

$\dfrac{10-b}{b} = \dfrac{2}{3}$

$3(10 - b) = 2b$

$30 - 3b = 2b$

$30 = 5b$

$b = 6 \Rightarrow a = 4$

$a + b = 10$; $b - a = 2$

The ratio of $a + b$ to $b - a$ is $\dfrac{10}{2} = 5$.

79. Let x = the width of the rectangle.
Let y = the length of the rectangle.
$$\begin{cases} 2x + 2y = 16 \\ xy = 15 \end{cases}$$
Solve the first equation for y, substitute into the second equation and solve.
$2x + 2y = 16$

$2y = 16 - 2x$

$y = 8 - x$

$x(8 - x) = 15$

$8x - x^2 = 15$

$x^2 - 8x + 15 = 0$

$(x - 5)(x - 3) = 0$

$x = 5$ or $x = 3$

The dimensions of the rectangle are 3 inches by 5 inches.

81. Let x = the radius of the first circle.
Let y = the radius of the second circle.
$$\begin{cases} 2\pi x + 2\pi y = 12\pi \\ \pi x^2 + \pi y^2 = 20\pi \end{cases}$$
Solve the first equation for y, substitute into the second equation and solve:
$2\pi x + 2\pi y = 12\pi$

$x + y = 6$

$y = 6 - x$

$\pi x^2 + \pi y^2 = 20\pi$

$x^2 + y^2 = 20$

$x^2 + (6 - x)^2 = 20$

$x^2 + 36 - 12x + x^2 = 20 \Rightarrow 2x^2 - 12x + 16 = 0$

$x^2 - 6x + 8 = 0 \Rightarrow (x - 4)(x - 2) = 0$

$x = 4$ or $x = 2$

$y = 2 \qquad y = 4$

The radii of the circles are 2 centimeters and 4 centimeters.

83. The tortoise takes $9 + 3 = 12$ minutes or 0.2 hour longer to complete the race than the hare.
Let r = the rate of the hare.
Let t = the time for the hare to complete the race. Then $t + 0.2$ = the time for the tortoise and $r - 0.5$ = the rate for the tortoise. Since the length of the race is 21 meters, the distance equations are:

$$\begin{cases} rt = 21 \Rightarrow r = \dfrac{21}{t} \\ (r-0.5)(t+0.2) = 21 \end{cases}$$

Solve the first equation for r, substitute into the second equation and solve:

$$\left(\frac{21}{t} - 0.5\right)(t+0.2) = 21$$

$$21 + \frac{4.2}{t} - 0.5t - 0.1 = 21$$

$$10t\left(21 + \frac{4.2}{t} - 0.5t - 0.1\right) = 10t \cdot (21)$$

$$210t + 42 - 5t^2 - t = 210t$$

$$5t^2 + t - 42 = 0$$

$$(5t-14)(t+3) = 0$$

$$5t - 14 = 0 \qquad \text{or} \quad t + 3 = 0$$

$$5t = 14 \qquad\qquad t = -3$$

$$t = \frac{14}{5} = 2.8$$

$t = -3$ makes no sense, since time cannot be negative.

Solve for r:

$$r = \frac{21}{2.8} = 7.5$$

The average speed of the hare is 7.5 meters per hour, and the average speed for the tortoise is 7 meters per hour.

85. Let x = the width of the cardboard. Let y = the length of the cardboard. The width of the box will be $x-4$, the length of the box will be $y-4$, and the height is 2. The volume is $V = (x-4)(y-4)(2)$.

Solve the system of equations:

$$\begin{cases} xy = 216 \qquad \Rightarrow y = \dfrac{216}{x} \\ 2(x-4)(y-4) = 224 \end{cases}$$

Solve the first equation for y, substitute into the second equation and solve.

$$(2x-8)\left(\frac{216}{x} - 4\right) = 224$$

$$432 - 8x - \frac{1728}{x} + 32 = 224$$

$$432x - 8x^2 - 1728 + 32x = 224x$$

$$8x^2 - 240x + 1728 = 0$$

$$x^2 - 30x + 216 = 0$$

$$(x-12)(x-18) = 0$$

$$x - 12 = 0 \quad \text{or} \quad x - 18 = 0$$

$$x = 12 \qquad\qquad x = 18$$

The cardboard should be 12 centimeters by 18 centimeters.

87. Find equations relating area and perimeter:

$$\begin{cases} x^2 + y^2 = 4500 \\ 3x + 3y + (x-y) = 300 \end{cases}$$

Solve the second equation for y, substitute into the first equation and solve:

$$4x + 2y = 300$$

$$2y = 300 - 4x$$

$$y = 150 - 2x$$

$$x^2 + (150 - 2x)^2 = 4500$$

$$x^2 + 22{,}500 - 600x + 4x^2 = 4500$$

$$5x^2 - 600x + 18{,}000 = 0$$

$$x^2 - 120x + 3600 = 0$$

$$(x-60)^2 = 0$$

$$x - 60 = 0$$

$$x = 60$$

$$y = 150 - 2(60) = 30$$

The sides of the squares are 30 feet and 60 feet.

89. Solve the system for l and w:
$$\begin{cases} 2l+2w = P \\ \quad lw = A \end{cases}$$

Solve the first equation for l, substitute into the second equation and solve.
$$2l = P-2w$$
$$l = \frac{P}{2}-w$$
$$\left(\frac{P}{2}-w\right)w = A$$
$$\frac{P}{2}w-w^2 = A$$
$$w^2 - \frac{P}{2}w + A = 0$$
$$w = \frac{\frac{P}{2}\pm\sqrt{\frac{P^2}{4}-4A}}{2} = \frac{\frac{P}{2}\pm\sqrt{\frac{P^2}{4}-\frac{16A}{4}}}{2}$$
$$= \frac{\frac{P}{2}\pm\frac{\sqrt{P^2-16A}}{2}}{2} = \frac{P\pm\sqrt{P^2-16A}}{4}$$

If $w = \dfrac{P+\sqrt{P^2-16A}}{4}$ then
$$l = \frac{P}{2}-\frac{P+\sqrt{P^2-16A}}{4} = \frac{P-\sqrt{P^2-16A}}{4}$$

If $w = \dfrac{P-\sqrt{P^2-16A}}{4}$ then
$$l = \frac{P}{2}-\frac{P-\sqrt{P^2-16A}}{4} = \frac{P+\sqrt{P^2-16A}}{4}$$

If it is required that length be greater than width, then the solution is:
$$w = \frac{P-\sqrt{P^2-16A}}{4} \text{ and } l = \frac{P+\sqrt{P^2-16A}}{4}$$

91. Solve the equation: $m^2 - 4(2m-4) = 0$
$$m^2 - 8m + 16 = 0$$
$$(m-4)^2 = 0$$
$$m = 4$$

Use the point-slope equation with slope 4 and the point (2, 4) to obtain the equation of the tangent line:
$$y-4 = 4(x-2) \Rightarrow y-4 = 4x-8 \Rightarrow y = 4x-4$$

93. Solve the system:
$$\begin{cases} y = x^2+2 \\ y = mx+b \end{cases}$$

Solve the system by substitution:
$$x^2+2 = mx+b \Rightarrow x^2-mx+2-b = 0$$
Note that the tangent line passes through (1, 3).
Find the relation between m and b:
$$3 = m(1)+b \Rightarrow b = 3-m$$
Substitute into the quadratic to eliminate b:
$$x^2-mx+2-(3-m) = 0 \Rightarrow x^2-mx+(m-1) = 0$$
Find when the discriminant equals 0:
$$(-m)^2 - 4(1)(m-1) = 0$$
$$m^2 - 4m + 4 = 0$$
$$(m-2)^2 = 0$$
$$m-2 = 0$$
$$m = 2$$
$$b = 3-m = 3-2 = 1$$
The equation of the tangent line is $y = 2x+1$.

95. Solve the system:
$$\begin{cases} 2x^2+3y^2 = 14 \\ \quad\quad y = mx+b \end{cases}$$

Solve the system by substitution:
$$2x^2+3(mx+b)^2 = 14$$
$$2x^2+3m^2x^2+6mbx+3b^2 = 14$$
$$(3m^2+2)x^2+6mbx+3b^2-14 = 0$$

Note that the tangent line passes through (1, 2).
Find the relation between m and b:
$$2 = m(1)+b \Rightarrow b = 2-m$$
Substitute into the quadratic to eliminate b:
$$(3m^2+2)x^2+6m(2-m)x+3(2-m)^2-14 = 0$$
$$(3m^2+2)x^2+(12m-6m^2)x+(3m^2-12m-2) = 0$$
Find when the discriminant equals 0:
$$(12m-6m^2)^2 - 4(3m^2+2)(3m^2-12m-2) = 0$$
$$144m^2+96m+16 = 0$$
$$9m^2+6m+1 = 0$$
$$(3m+1)^2 = 0$$
$$3m+1 = 0$$
$$m = -\frac{1}{3}$$

$$b = 2 - m = 2 - \left(-\frac{1}{3}\right) = \frac{7}{3}$$

The equation of the tangent line is $y = -\frac{1}{3}x + \frac{7}{3}$.

97. Solve the system:
$$\begin{cases} x^2 - y^2 = 3 \\ y = mx + b \end{cases}$$
Solve the system by substitution:
$$x^2 - (mx + b)^2 = 3$$
$$x^2 - m^2 x^2 - 2mbx - b^2 = 3$$
$$(1 - m^2)x^2 - 2mbx - b^2 - 3 = 0$$
Note that the tangent line passes through (2, 1).
Find the relation between m and b:
$$1 = m(2) + b \Rightarrow b = 1 - 2m$$
Substitute into the quadratic to eliminate b:
$$(1 - m^2)x^2 - 2m(1 - 2m)x - (1 - 2m)^2 - 3 = 0$$
$$(1 - m^2)x^2 + (-2m + 4m^2)x - 1 + 4m - 4m^2 - 3 = 0$$
$$(1 - m^2)x^2 + (-2m + 4m^2)x + (-4m^2 + 4m - 4) = 0$$
Find when the discriminant equals 0:
$$\left(-2m + 4m^2\right)^2 - 4\left(1 - m^2\right)\left(-4m^2 + 4m - 4\right) = 0$$
$$4m^2 - 16m^3 + 16m^4 - 16m^4 + 16m^3 - 16m + 16 = 0$$
$$4m^2 - 16m + 16 = 0$$
$$m^2 - 4m + 4 = 0$$
$$(m - 2)^2 = 0$$
$$m = 2$$
The equation of the tangent line is $y = 2x - 3$.

99. Solve for r_1 and r_2:
$$\begin{cases} r_1 + r_2 = -\dfrac{b}{a} \\ r_1 r_2 = \dfrac{c}{a} \end{cases}$$
Substitute and solve:
$$r_1 = -r_2 - \frac{b}{a}$$
$$\left(-r_2 - \frac{b}{a}\right)r_2 = \frac{c}{a}$$
$$-r_2^2 - \frac{b}{a}r_2 - \frac{c}{a} = 0$$
$$ar_2^2 + br_2 + c = 0$$

$$r_2 = \frac{-b \pm \sqrt{b^2 - 4ac}}{2a}$$
$$r_1 = -r_2 - \frac{b}{a} =$$
$$= -\left(\frac{-b \pm \sqrt{b^2 - 4ac}}{2a}\right) - \frac{b}{a}$$
$$= \frac{b \mp \sqrt{b^2 - 4ac}}{2a} - \frac{2b}{2a}$$
$$= \frac{-b \mp \sqrt{b^2 - 4ac}}{2a}$$
The solutions are:
$$r_1 = \frac{-b + \sqrt{b^2 - 4ac}}{2a} \text{ and } r_2 = \frac{-b - \sqrt{b^2 - 4ac}}{2a}.$$

101. Since the area of the square piece of sheet metal is 100 square feet, the sheet's dimensions are 10 feet by 10 feet. Let $x =$ the length of the cut.

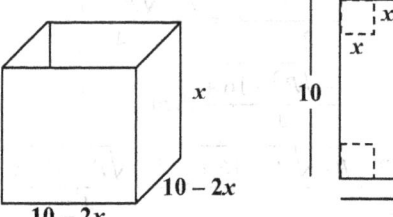

The dimensions of the box are: length $= 10 - 2x$; width $= 10 - 2x$; height $= x$. Note that each of these expressions must be positive. So we must have $x > 0$ and $10 - 2x > 0 \Rightarrow x < 5$, that is, $0 < x < 5$. So the volume of the box is given by
$$V = (\text{length}) \cdot (\text{width}) \cdot (\text{height})$$
$$= (10 - 2x)(10 - 2x)(x)$$
$$= (10 - 2x)^2 (x)$$

a. In order to get a volume equal to 9 cubic feet, we solve $(10 - 2x)^2 (x) = 9$.
$$(10 - 2x)^2 (x) = 9$$
$$\left(100 - 40x + 4x^2\right)x = 9$$
$$100x - 40x^2 + 4x^3 = 9$$
So we need to solve the equation
$$4x^3 - 40x^2 + 100x - 9 = 0.$$

Graphing $y_1 = 4x^3 - 40x^2 + 100x - 9$ on a calculator yields the graph

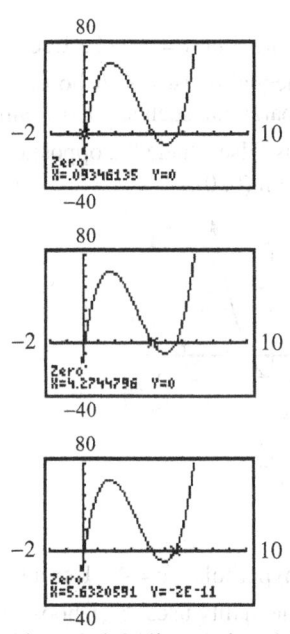

The graph indicates that there are three real zeros on the interval [0, 6]. Using the ZERO feature of a graphing calculator, we find that the three roots shown occur at $x \approx 0.093$, $x \approx 4.274$ and $x \approx 5.632$. But we've already noted that we must have $0<x<5$, so the only practical values for the sides of the square base are $x \approx 0.093$ feet and $x \approx 4.274$ feet.

b. Answers will vary.

103. $y - y_1 = m(x - x_1)$

$y - (-7) = -\frac{2}{5}(x - 10)$

$y + 7 = -\frac{2}{5}x + 4 \rightarrow y = -\frac{2}{5}x - 3$

105. $P = A\left(1 + \frac{r}{n}\right)^{nt}$

$1800 = 1500\left(1 + \frac{r}{4}\right)^{4 \cdot 3}$

$\frac{1800}{1500} = \left(1 + \frac{r}{4}\right)^{12}$

$\sqrt[12]{\frac{6}{5}} = 1 + \frac{r}{4}$

$\sqrt[12]{\frac{6}{5}} - 1 = \frac{r}{4}$

$4\left(\sqrt[12]{\frac{6}{5}} - 1\right) = r$

$r \approx .0612$ or 6.12%

Section 8.7

1. $3x + 4 < 8 - x$

$4x < 4$

$x < 1$

$\{x \mid x < 1\}$ or $(-\infty, 1)$

3. $x^2 + y^2 = 9$

The graph is a circle. Center: $(0, 0)$; Radius: 3

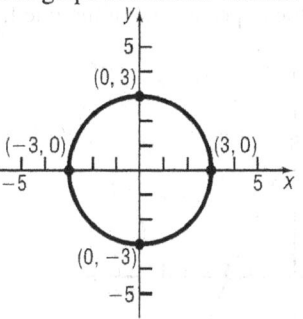

5. True

7. dashes; solid

9. False, see example 7b.

11. $x \geq 0$

Graph the line $x = 0$. Use a solid line since the inequality uses $\geq$. Choose a test point not on the line, such as $(2, 0)$. Since $2 \geq 0$ is true, shade the side of the line containing $(2, 0)$.

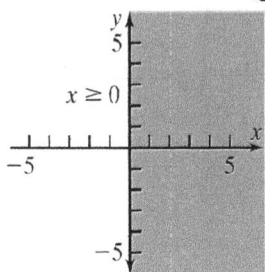

521

13. $x \geq 4$

Graph the line $x = 4$. Use a solid line since the inequality uses $\geq$. Choose a test point not on the line, such as $(5, 0)$. Since $5 \geq 0$ is true, shade the side of the line containing $(5, 0)$.

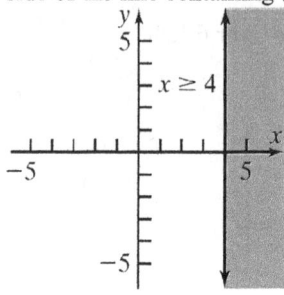

15. $2x + y \geq 6$

Graph the line $2x + y = 6$. Use a solid line since the inequality uses $\geq$. Choose a test point not on the line, such as $(0, 0)$. Since $2(0) + 0 \geq 6$ is false, shade the opposite side of the line from $(0, 0)$.

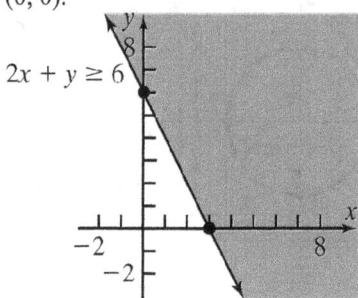

17. $x^2 + y^2 > 1$

Graph the circle $x^2 + y^2 = 1$. Use a dashed line since the inequality uses $>$. Choose a test point not on the circle, such as $(0, 0)$. Since $0^2 + 0^2 > 1$ is false, shade the opposite side of the circle from $(0, 0)$.

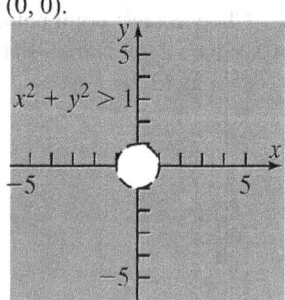

19. $y \leq x^2 - 1$

Graph the parabola $y = x^2 - 1$. Use a solid line since the inequality uses $\leq$. Choose a test point not on the parabola, such as $(0, 0)$. Since $0 \leq 0^2 - 1$ is false, shade the opposite side of the parabola from $(0, 0)$.

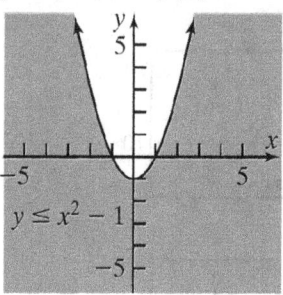

21. $xy \geq 4$

Graph the hyperbola $xy = 4$. Use a solid line since the inequality uses $\geq$. Choose a test point not on the hyperbola, such as $(0, 0)$. Since $0 \cdot 0 \geq 4$ is false, shade the opposite side of the hyperbola from $(0, 0)$.

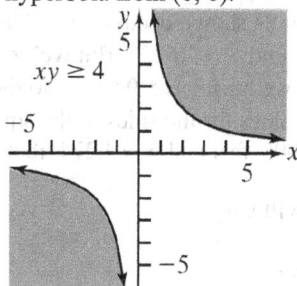

23. $\begin{cases} x + y \leq 2 \\ 2x + y \geq 4 \end{cases}$

Graph the line $x + y = 2$. Use a solid line since the inequality uses $\leq$. Choose a test point not on the line, such as $(0, 0)$. Since $0 + 0 \leq 2$ is true, shade the side of the line containing $(0, 0)$. Graph the line $2x + y = 4$. Use a solid line since the inequality uses $\geq$. Choose a test point not on the line, such as $(0, 0)$. Since $2(0) + 0 \geq 4$ is false, shade the opposite side of the line from $(0, 0)$. The overlapping region is the solution.

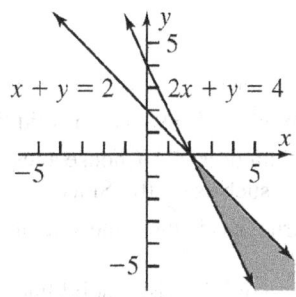

25. $\begin{cases} 2x - y \le 4 \\ 3x + 2y \ge -6 \end{cases}$

Graph the line $2x - y = 4$. Use a solid line since the inequality uses $\le$. Choose a test point not on the line, such as $(0, 0)$. Since $2(0) - 0 \le 4$ is true, shade the side of the line containing $(0, 0)$. Graph the line $3x + 2y = -6$. Use a solid line since the inequality uses $\ge$. Choose a test point not on the line, such as $(0, 0)$. Since $3(0) + 2(0) \ge -6$ is true, shade the side of the line containing $(0, 0)$. The overlapping region is the solution.

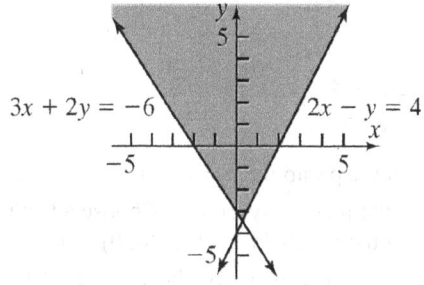

27. $\begin{cases} 2x - 3y \le 0 \\ 3x + 2y \le 6 \end{cases}$

Graph the line $2x - 3y = 0$. Use a solid line since the inequality uses $\le$. Choose a test point not on the line, such as $(0, 3)$. Since $2(0) - 3(3) \le 0$ is true, shade the side of the line containing $(0, 3)$. Graph the line $3x + 2y = 6$. Use a solid line since the inequality uses $\le$. Choose a test point not on the line, such as $(0, 0)$. Since $3(0) + 2(0) \le 6$ is true, shade the side of the line containing $(0, 0)$.

The overlapping region is the solution.

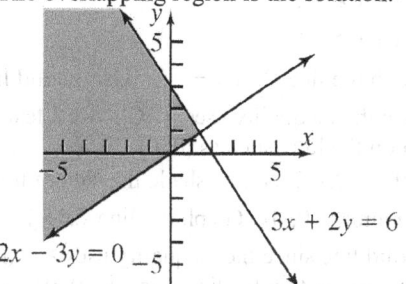

29. $\begin{cases} x - 2y \le 6 \\ 2x - 4y \ge 0 \end{cases}$

Graph the line $x - 2y = 6$. Use a solid line since the inequality uses $\le$. Choose a test point not on the line, such as $(0, 0)$. Since $0 - 2(0) \le 6$ is true, shade the side of the line containing $(0, 0)$. Graph the line $2x - 4y = 0$. Use a solid line since the inequality uses $\ge$. Choose a test point not on the line, such as $(0, 2)$. Since $2(0) - 4(2) \ge 0$ is false, shade the opposite side of the line from $(0, 2)$. The overlapping region is the solution.

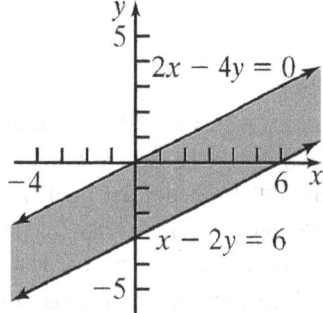

31. $\begin{cases} 2x+y \geq -2 \\ 2x+y \geq 2 \end{cases}$

Graph the line $2x+y=-2$. Use a solid line since the inequality uses $\geq$. Choose a test point not on the line, such as (0, 0). Since $2(0)+0 \geq -2$ is true, shade the side of the line containing (0, 0). Graph the line $2x+y=2$. Use a solid line since the inequality uses $\geq$. Choose a test point not on the line, such as (0, 0). Since $2(0)+0 \geq 2$ is false, shade the opposite side of the line from (0, 0). The overlapping region is the solution.

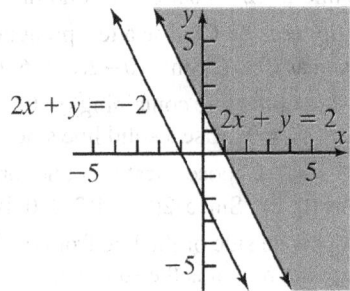

33. $\begin{cases} 2x+3y \geq 6 \\ 2x+3y \leq 0 \end{cases}$

Graph the line $2x+3y=6$. Use a solid line since the inequality uses $\geq$. Choose a test point not on the line, such as (0, 0). Since $2(0)+3(0) \geq 6$ is false, shade the opposite side of the line from (0, 0). Graph the line $2x+3y=0$. Use a solid line since the inequality uses $\leq$. Choose a test point not on the line, such as (0, 2). Since $2(0)+3(2) \leq 0$ is false, shade the opposite side of the line from (0, 2). Since the regions do not overlap, the solution is an empty set.

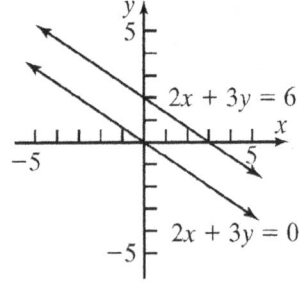

35. $\begin{cases} x^2+y^2 \leq 9 \\ x+y \geq 3 \end{cases}$

Graph the circle $x^2+y^2=9$. Use a solid line since the inequality uses $\leq$. Choose a test point not on the circle, such as (0, 0). Since $0^2+0^2 \leq 9$ is true, shade the same side of the circle as (0, 0).
Graph the line $x+y=3$. Use a solid line since the inequality uses $\geq$. Choose a test point not on the line, such as (0, 0). Since $0+0 \geq 3$ is false, shade the opposite side of the line from (0, 0). The overlapping region is the solution.

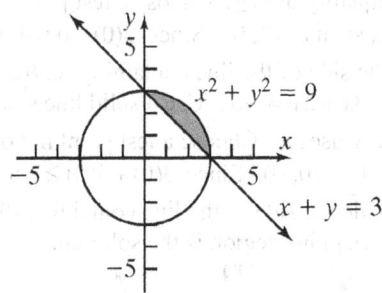

37. $\begin{cases} y \geq x^2-4 \\ y \leq x-2 \end{cases}$

Graph the parabola $y=x^2-4$. Use a solid line since the inequality uses $\geq$. Choose a test point not on the parabola, such as (0, 0). Since $0 \geq 0^2-4$ is true, shade the same side of the parabola as (0, 0). Graph the line $y=x-2$. Use a solid line since the inequality uses $\leq$. Choose a test point not on the line, such as (0, 0). Since $0 \leq 0-2$ is false, shade the opposite side of the line from (0, 0). The overlapping region is the solution.

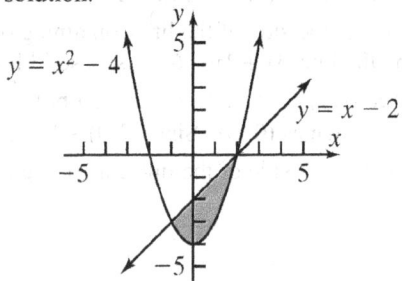

39. $\begin{cases} x^2 + y^2 \le 16 \\ y \ge x^2 - 4 \end{cases}$

Graph the circle $x^2 + y^2 = 16$. Use a sold line since the inequality is not strict. Choose a test point not on the circle, such as $(0,0)$. Since $0^2 + 0^2 \le 16$ is true, shade the side of the circle containing $(0,0)$. Graph the parabola $y = x^2 - 4$. Use a solid line since the inequality is not strict. Choose a test point not on the parabola, such as $(0,0)$. Since $0 \ge 0^2 - 4$ is true, shade the side of the parabola that contains $(0,0)$. The overlapping region is the solution.

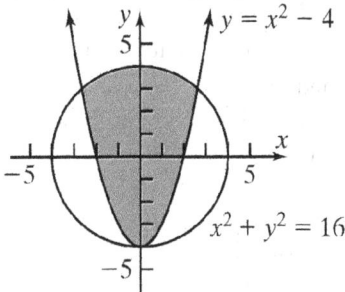

41. $\begin{cases} xy \ge 4 \\ y \ge x^2 + 1 \end{cases}$

Graph the hyperbola $xy = 4$. Use a solid line since the inequality uses $\ge$. Choose a test point not on the parabola, such as $(0, 0)$. Since $0 \cdot 0 \ge 4$ is false, shade the opposite side of the hyperbola from $(0, 0)$. Graph the parabola $y = x^2 + 1$. Use a solid line since the inequality uses $\ge$. Choose a test point not on the parabola, such as $(0, 0)$. Since $0 \ge 0^2 + 1$ is false, shade the opposite side of the parabola from $(0, 0)$. The overlapping region is the solution.

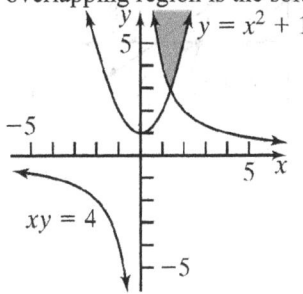

43. $\begin{cases} x \ge 0 \\ y \ge 0 \\ 2x + y \le 6 \\ x + 2y \le 6 \end{cases}$

Graph $x \ge 0$; $y \ge 0$. Shaded region is the first quadrant. Graph the line $2x + y = 6$. Use a solid line since the inequality uses $\le$. Choose a test point not on the line, such as $(0, 0)$. Since $2(0) + 0 \le 6$ is true, shade the side of the line containing $(0, 0)$. Graph the line $x + 2y = 6$. Use a solid line since the inequality uses $\le$. Choose a test point not on the line, such as $(0, 0)$. Since $0 + 2(0) \le 6$ is true, shade the side of the line containing $(0, 0)$. The overlapping region is the solution. The graph is bounded. Find the vertices:

The x-axis and y-axis intersect at $(0, 0)$. The intersection of $x + 2y = 6$ and the y-axis is $(0, 3)$. The intersection of $2x + y = 6$ and the x-axis is $(3, 0)$. To find the intersection of $x + 2y = 6$ and $2x + y = 6$, solve the system:

$\begin{cases} x + 2y = 6 \\ 2x + y = 6 \end{cases}$

Solve the first equation for x: $x = 6 - 2y$.
Substitute and solve:
$2(6 - 2y) + y = 6$
$12 - 4y + y = 6$
$12 - 3y = 6$
$-3y = -6$
$y = 2$
$x = 6 - 2(2) = 2$
The point of intersection is $(2, 2)$.
The four corner points are $(0, 0)$, $(0, 3)$, $(3, 0)$, and $(2, 2)$.

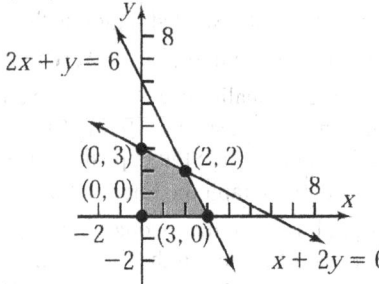

45. $\begin{cases} x \geq 0 \\ y \geq 0 \\ x+y \geq 2 \\ 2x+y \geq 4 \end{cases}$

Graph $x \geq 0$; $y \geq 0$. Shaded region is the first quadrant. Graph the line $x+y=2$. Use a solid line since the inequality uses $\geq$. Choose a test point not on the line, such as (0, 0). Since $0+0 \geq 2$ is false, shade the opposite side of the line from (0, 0). Graph the line $2x+y=4$. Use a solid line since the inequality uses $\geq$. Choose a test point not on the line, such as (0, 0). Since $2(0)+0 \geq 4$ is false, shade the opposite side of the line from (0, 0). The overlapping region is the solution. The graph is unbounded.

Find the vertices:
The intersection of $x+y=2$ and the x-axis is (2, 0). The intersection of $2x+y=4$ and the y-axis is (0, 4). The two corner points are (2, 0), and (0, 4).

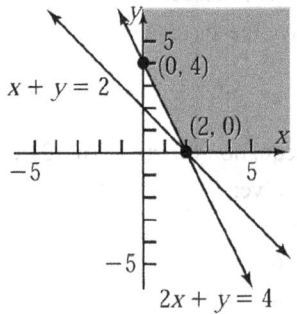

47. $\begin{cases} x \geq 0 \\ y \geq 0 \\ x+y \geq 2 \\ 2x+3y \leq 12 \\ 3x+y \leq 12 \end{cases}$

Graph $x \geq 0$; $y \geq 0$. Shaded region is the first quadrant. Graph the line $x+y=2$. Use a solid line since the inequality uses $\geq$. Choose a test point not on the line, such as (0, 0). Since $0+0 \geq 2$ is false, shade the opposite side of the line from (0, 0). Graph the line $2x+3y=12$.

Use a solid line since the inequality uses $\leq$. Choose a test point not on the line, such as (0, 0). Since $2(0)+3(0) \leq 12$ is true, shade the side of the line containing (0, 0). Graph the line $3x+y=12$. Use a solid line since the inequality uses $\leq$. Choose a test point not on the line, such

as (0, 0). Since $3(0)+0 \leq 12$ is true, shade the side of the line containing (0, 0). The overlapping region is the solution. The graph is bounded.

Find the vertices:
The intersection of $x+y=2$ and the y-axis is (0, 2). The intersection of $x+y=2$ and the x-axis is (2, 0). The intersection of $2x+3y=12$ and the y-axis is (0, 4). The intersection of $3x+y=12$ and the x-axis is (4, 0).

To find the intersection of $2x+3y=12$ and $3x+y=12$, solve the system:
$\begin{cases} 2x+3y=12 \\ 3x+y=12 \end{cases}$
Solve the second equation for y: $y=12-3x$.
Substitute and solve:
$$2x+3(12-3x)=12$$
$$2x+36-9x=12$$
$$-7x=-24$$
$$x=\frac{24}{7}$$
$$y=12-3\left(\frac{24}{7}\right)=12-\frac{72}{7}=\frac{12}{7}$$
The point of intersection is $\left(\frac{24}{7}, \frac{12}{7}\right)$.

The five corner points are (0, 2), (0, 4), (2, 0), (4, 0), and $\left(\frac{24}{7}, \frac{12}{7}\right)$.

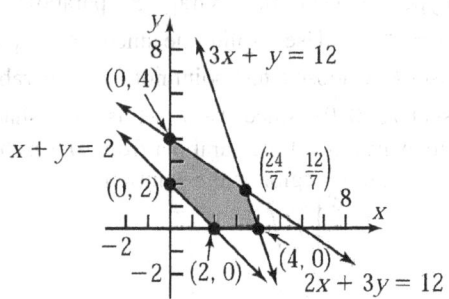

49. $\begin{cases} x \geq 0 \\ y \geq 0 \\ x+y \geq 2 \\ x+y \leq 8 \\ 2x+y \leq 10 \end{cases}$

Graph $x \geq 0$; $y \geq 0$. Shaded region is the first quadrant. Graph the line $x+y=2$. Use a solid line since the inequality uses $\geq$. Choose a test point not on the line, such as (0, 0). Since $0+0 \geq 2$ is false, shade the opposite side of the line from (0, 0). Graph the line $x+y=8$. Use a solid line since the inequality uses $\leq$. Choose a test point not on the line, such as (0, 0). Since $0+0 \leq 8$ is true, shade the side of the line containing (0, 0). Graph the line $2x+y=10$. Use a solid line since the inequality uses $\leq$. Choose a test point not on the line, such as (0, 0). Since $2(0)+0 \leq 10$ is true, shade the side of the line containing (0, 0). The overlapping region is the solution. The graph is bounded.

Find the vertices:
The intersection of $x+y=2$ and the y-axis is (0, 2). The intersection of $x+y=2$ and the x-axis is (2, 0). The intersection of $x+y=8$ and the y-axis is (0, 8). The intersection of $2x+y=10$ and the x-axis is (5, 0). To find the intersection of $x+y=8$ and $2x+y=10$, solve the system:

$\begin{cases} x+y=8 \\ 2x+y=10 \end{cases}$

Solve the first equation for y: $y=8-x$.
Substitute and solve:
$2x+8-x=10$
$\qquad x=2$
$y=8-2=6$
The point of intersection is (2, 6).
The five corner points are (0, 2), (0, 8), (2, 0), (5, 0), and (2, 6).

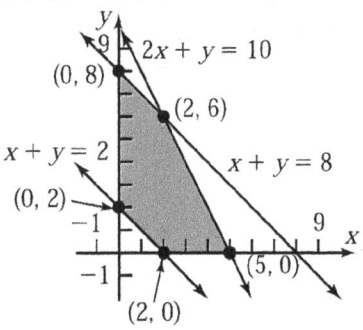

51. $\begin{cases} x \geq 0 \\ y \geq 0 \\ x+2y \geq 1 \\ x+2y \leq 10 \end{cases}$

Graph $x \geq 0$; $y \geq 0$. Shaded region is the first quadrant. Graph the line $x+2y=1$. Use a solid line since the inequality uses $\geq$. Choose a test point not on the line, such as (0, 0). Since $0+2(0) \geq 1$ is false, shade the opposite side of the line from (0, 0). Graph the line $x+2y=10$. Use a solid line since the inequality uses $\leq$. Choose a test point not on the line, such as (0, 0). Since $0+2(0) \leq 10$ is true, shade the side of the line containing (0, 0). The overlapping region is the solution. The graph is bounded.

Find the vertices:
The intersection of $x+2y=1$ and the y-axis is (0, 0.5). The intersection of $x+2y=1$ and the x-axis is (1, 0). The intersection of $x+2y=10$ and the y-axis is (0, 5). The intersection of $x+2y=10$ and the x-axis is (10, 0). The four corner points are (0, 0.5), (0, 5), (1, 0), and (10, 0).

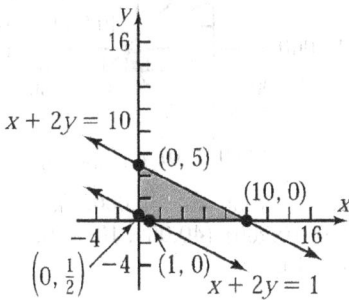

53. The system of linear inequalities is:

$\begin{cases} x \leq 4 \\ x+y \leq 6 \\ x \geq 0 \\ y \geq 0 \end{cases}$

55. The system of linear inequalities is:

$\begin{cases} x \leq 20 \\ y \geq 15 \\ x+y \leq 50 \\ x-y \leq 0 \\ x \geq 0 \end{cases}$

57. a. Let x = the amount invested in Treasury bills, and let y = the amount invested in corporate bonds.
The constraints are:
$x + y \le 50,000$ because the total investment cannot exceed $50,000.
$x \ge 35,000$ because the amount invested in Treasury bills must be at least $35,000.
$y \le 10,000$ because the amount invested in corporate bonds must not exceed $10,000.
$x \ge 0$, $y \ge 0$ because a non-negative amount must be invested.
The system is
$$\begin{cases} x + y \le 50,000 \\ x \ge 35,000 \\ y \le 10,000 \\ x \ge 0 \\ y \ge 0 \end{cases}$$

b. Graph the system.

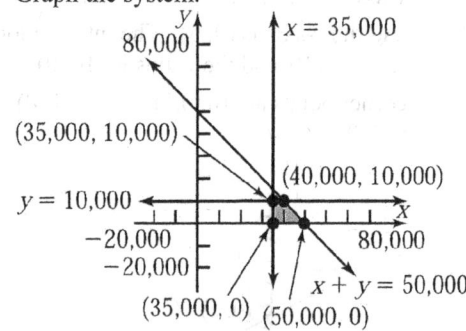

The corner points are (35,000, 0), (35,000, 10,000), (40,000, 10,000), (50,000, 0).

59. a. Let x = the # of packages of the economy blend, and let y = the # of packages of the superior blend.
The constraints are:
$x \ge 0$, $y \ge 0$ because a non-negative # of packages must be produced.
$4x + 8y \le 75 \cdot 16$ because the total amount of "A grade" coffee cannot exceed 75 pounds.
(Note: 75 pounds = (75)(16) ounces.)
$12x + 8y \le 120 \cdot 16$ because the total amount of "B grade" coffee cannot exceed 120 pounds.

(Note: 120 pounds = (120)(16) ounces.)
Simplifying the inequalities, we obtain:

$4x + 8y \le 75 \cdot 16$	$12x + 8y \le 120 \cdot 16$
$x + 2y \le 75 \cdot 4$	$3x + 2y \le 120 \cdot 4$
$x + 2y \le 300$	$3x + 2y \le 480$

The system is:
$$\begin{cases} x \ge 0 \\ y \ge 0 \\ x + 2y \le 300 \\ 3x + 2y \le 480 \end{cases}$$

b. Graph the system.

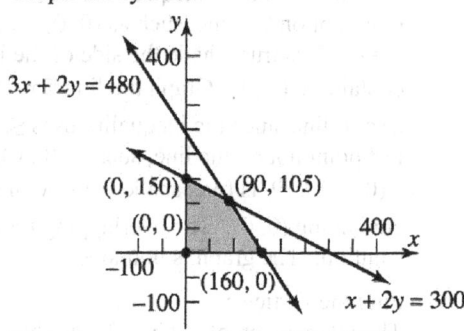

The corner points are (0, 0), (0, 150), (90, 105), (160, 0).

61. a. Let x = the # of microwaves, and let y = the # of printers.
The constraints are:
$x \ge 0$, $y \ge 0$ because a non-negative # of items must be shipped.
$30x + 20y \le 1600$ because a total cargo weight cannot exceed 1600 pounds.
$2x + 3y \le 150$ because the total cargo volume cannot exceed 150 cubic feet. Note that the inequality $30x + 20y \le 1600$ can be simplified: $3x + 2y \le 160$.
The system is:
$$\begin{cases} 3x + 2y \le 160 \\ 2x + 3y \le 150 \\ x \ge 0; \ y \ge 0 \end{cases}$$

b. Graph the system.

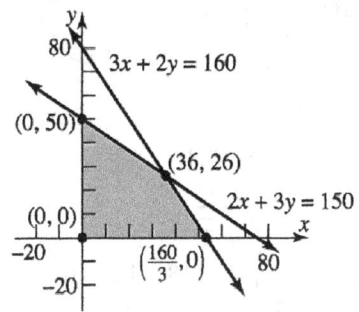

The corner points are $(0, 0)$, $(0, 50)$, $(36, 26)$, $\left(\frac{160}{3}, 0\right)$.

63. Distance:

$$d = \sqrt{(0-7)^2 + (-3-(-8))^2}$$
$$= \sqrt{(-7)^2 + (5)^2}$$
$$= \sqrt{49+25} = \sqrt{74}$$

Midpoint:

$$\frac{7+0}{2}, \frac{-8+(-3)}{2}$$
$$\frac{7}{2}, -\frac{11}{2}$$

65. $f(x) = \dfrac{5x-2}{x+3}$

The vertical asymptote is where the denominator is undefined: $x = -3$

The horizontal asymptote is $y = \dfrac{5}{1} = 5$.

Section 8.8

1. objective function

3. $z = x + y$

Vertex	Value of $z = x + y$
$(0, 3)$	$z = 0+3 = 3$
$(0, 6)$	$z = 0+6 = 6$
$(5, 6)$	$z = 5+6 = 11$
$(5, 2)$	$z = 5+2 = 7$
$(4, 0)$	$z = 4+0 = 4$

The maximum value is 11 at $(5, 6)$, and the minimum value is 3 at $(0, 3)$.

5. $z = x + 10y$

Vertex	Value of $z = x + 10y$
$(0, 3)$	$z = 0+10(3) = 30$
$(0, 6)$	$z = 0+10(6) = 60$
$(5, 6)$	$z = 5+10(6) = 65$
$(5, 2)$	$z = 5+10(2) = 25$
$(4, 0)$	$z = 4+10(0) = 4$

The maximum value is 65 at $(5, 6)$, and the minimum value is 4 at $(4, 0)$.

7. $z = 5x + 7y$

Vertex	Value of $z = 5x + 7y$
$(0, 3)$	$z = 5(0)+7(3) = 21$
$(0, 6)$	$z = 5(0)+7(6) = 42$
$(5, 6)$	$z = 5(5)+7(6) = 67$
$(5, 2)$	$z = 5(5)+7(2) = 39$
$(4, 0)$	$z = 5(4)+7(0) = 20$

The maximum value is 67 at $(5, 6)$, and the minimum value is 20 at $(4, 0)$.

9. Maximize $z = 2x + y$ subject to $x \geq 0$, $y \geq 0$, $x+y \leq 6$, $x+y \geq 1$. Graph the constraints.

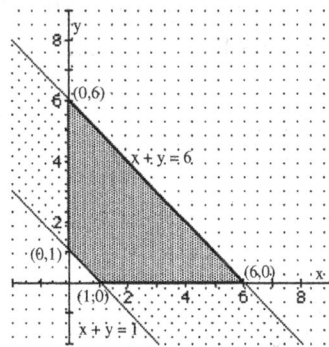

The corner points are $(0, 1)$, $(1, 0)$, $(0, 6)$, $(6, 0)$.
Evaluate the objective function:

Vertex	Value of $z = 2x + y$
$(0, 1)$	$z = 2(0)+1 = 1$
$(0, 6)$	$z = 2(0)+6 = 6$
$(1, 0)$	$z = 2(1)+0 = 2$
$(6, 0)$	$z = 2(6)+0 = 12$

The maximum value is 12 at $(6, 0)$.

11. Minimize $z = 2x + 5y$ subject to $x \geq 0$, $y \geq 0$, $x + y \geq 2$, $x \leq 5$, $y \leq 3$. Graph the constraints.

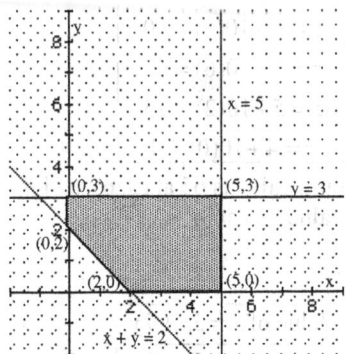

The corner points are (0, 2), (2, 0), (0, 3), (5, 0), (5, 3). Evaluate the objective function:

Vertex	Value of $z = 2x + 5y$
(0, 2)	$z = 2(0) + 5(2) = 10$
(0, 3)	$z = 2(0) + 5(3) = 15$
(2, 0)	$z = 2(2) + 5(0) = 4$
(5, 0)	$z = 2(5) + 5(0) = 10$
(5, 3)	$z = 2(5) + 5(3) = 25$

The minimum value is 4 at (2, 0).

13. Maximize $z = 3x + 5y$ subject to $x \geq 0$, $y \geq 0$, $x + y \geq 2$, $2x + 3y \leq 12$, $3x + 2y \leq 12$. Graph the constraints.

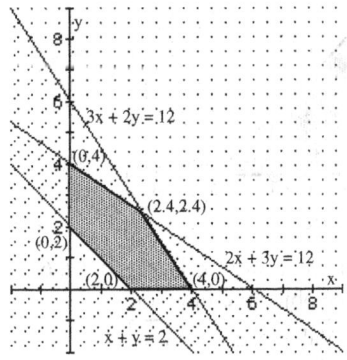

To find the intersection of $2x + 3y = 12$ and $3x + 2y = 12$, solve the system:

$$\begin{cases} 2x + 3y = 12 \\ 3x + 2y = 12 \end{cases}$$

Solve the second equation for y: $y = 6 - \dfrac{3}{2}x$

Substitute and solve:

$$2x + 3\left(6 - \frac{3}{2}x\right) = 12$$

$$2x + 18 - \frac{9}{2}x = 12$$

$$-\frac{5}{2}x = -6$$

$$x = \frac{12}{5}$$

$$y = 6 - \frac{3}{2}\left(\frac{12}{5}\right) = 6 - \frac{18}{5} = \frac{12}{5}$$

The point of intersection is $(2.4, 2.4)$.

The corner points are (0, 2), (2, 0), (0, 4), (4, 0), (2.4, 2.4). Evaluate the objective function:

Vertex	Value of $z = 3x + 5y$
(0, 2)	$z = 3(0) + 5(2) = 10$
(0, 4)	$z = 3(0) + 5(4) = 20$
(2, 0)	$z = 3(2) + 5(0) = 6$
(4, 0)	$z = 3(4) + 5(0) = 12$
(2.4, 2.4)	$z = 3(2.4) + 5(2.4) = 19.2$

The maximum value is 20 at (0, 4).

15. Minimize $z = 5x + 4y$ subject to $x \geq 0$, $y \geq 0$, $x + y \geq 2$, $2x + 3y \leq 12$, $3x + y \leq 12$. Graph the constraints.

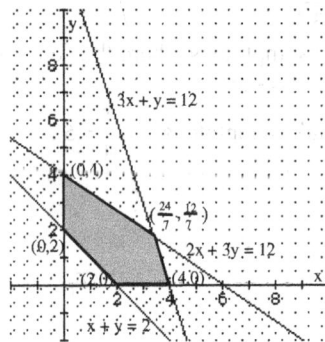

To find the intersection of $2x + 3y = 12$ and $3x + y = 12$, solve the system:

$$\begin{cases} 2x + 3y = 12 \\ 3x + y = 12 \end{cases}$$

Solve the second equation for y: $y = 12 - 3x$

Substitute and solve:

$$2x + 3(12 - 3x) = 12$$

$$2x + 36 - 9x = 12$$

$$-7x = -24$$

$$x = \frac{24}{7}$$

$$y = 12 - 3\left(\frac{24}{7}\right) = 12 - \frac{72}{7} = \frac{12}{7}$$

The point of intersection is $\left(\frac{24}{7}, \frac{12}{7}\right)$.

The corner points are (0, 2), (2, 0), (0, 4), (4, 0), $\left(\frac{24}{7}, \frac{12}{7}\right)$. Evaluate the objective function:

Vertex	Value of $z = 5x + 4y$
(0, 2)	$z = 5(0) + 4(2) = 8$
(0, 4)	$z = 5(0) + 4(4) = 16$
(2, 0)	$z = 5(2) + 4(0) = 10$
(4, 0)	$z = 5(4) + 4(0) = 20$
$\left(\frac{24}{7}, \frac{12}{7}\right)$	$z = 5\left(\frac{24}{7}\right) + 4\left(\frac{12}{7}\right) = 24$

The minimum value is 8 at (0, 2).

17. Maximize $z = 5x + 2y$ subject to $x \geq 0$, $y \geq 0$, $x + y \leq 10$, $2x + y \geq 10$, $x + 2y \geq 10$.
Graph the constraints.

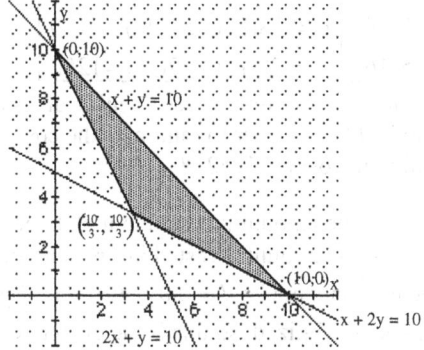

To find the intersection of $2x + y = 10$ and $x + 2y = 10$, solve the system:
$$\begin{cases} 2x + y = 10 \\ x + 2y = 10 \end{cases}$$
Solve the first equation for y: $y = 10 - 2x$.
Substitute and solve:
$$x + 2(10 - 2x) = 10$$
$$x + 20 - 4x = 10$$
$$-3x = -10$$
$$x = \frac{10}{3}$$
$$y = 10 - 2\left(\frac{10}{3}\right) = 10 - \frac{20}{3} = \frac{10}{3}$$
The point of intersection is (10/3 10/3).
The corner points are (0, 10), (10, 0), (10/3, 10/3). Evaluate the objective function:

Vertex	Value of $z = 5x + 2y$
(0, 10)	$z = 5(0) + 2(10) = 20$
(10, 0)	$z = 5(10) + 2(0) = 50$
$\left(\frac{10}{3}, \frac{10}{3}\right)$	$z = 5\left(\frac{10}{3}\right) + 2\left(\frac{10}{3}\right) = \frac{70}{3} = 23\frac{1}{3}$

The maximum value is 50 at (10, 0).

19. Let x = the number of downhill skis produced, and let y = the number of cross-country skis produced. The total profit is: $P = 70x + 50y$.
Profit is to be maximized, so this is the objective function. The constraints are:
$x \geq 0$, $y \geq 0$ A positive number of skis must be produced.
$2x + y \leq 40$ Manufacturing time available.
$x + y \leq 32$ Finishing time available.
Graph the constraints.

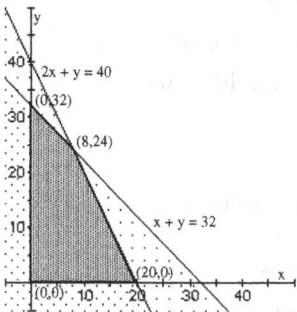

To find the intersection of $x + y = 32$ and $2x + y = 40$, solve the system:
$$\begin{cases} x + y = 32 \\ 2x + y = 40 \end{cases}$$
Solve the first equation for y: $y = 32 - x$.
Substitute and solve:
$$2x + (32 - x) = 40$$
$$x = 8$$
$$y = 32 - 8 = 24$$
The point of intersection is (8, 24).
The corner points are (0, 0), (0, 32), (20, 0), (8, 24). Evaluate the objective function:

Vertex	Value of $P = 70x + 50y$
(0, 0)	$P = 70(0) + 50(0) = 0$
(0, 32)	$P = 70(0) + 50(32) = 1600$
(20, 0)	$P = 70(20) + 50(0) = 1400$
(8, 24)	$P = 70(8) + 50(24) = 1760$

The maximum profit is $1760, when 8 downhill skis and 24 cross-country skis are produced.

With the increase of the manufacturing time to 48 hours, we do the following:

The constraints are:

$x \geq 0, \ y \geq 0$ — A positive number of skis must be produced.

$2x + y \leq 48$ — Manufacturing time available.

$x + y \leq 32$ — Finishing time available.

Graph the constraints.

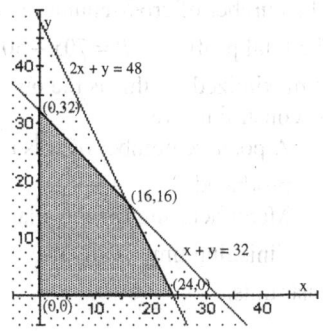

To find the intersection of $x + y = 32$ and $2x + y = 48$, solve the system:

$$\begin{cases} x + y = 32 \\ 2x + y = 48 \end{cases}$$

Solve the first equation for y: $y = 32 - x$.

Substitute and solve:

$$2x + (32 - x) = 48$$
$$x = 16$$
$$y = 32 - 16 = 16$$

The point of intersection is (16, 16).

The corner points are (0, 0), (0, 32), (24, 0), (16, 16). Evaluate the objective function:

Vertex	Value of $P = 70x + 50y$
(0, 0)	$P = 70(0) + 50(0) = 0$
(0, 32)	$P = 70(0) + 50(32) = 1600$
(24, 0)	$P = 70(24) + 50(0) = 1680$
(16, 16)	$P = 70(16) + 50(16) = 1920$

The maximum profit is $1920, when 16 downhill skis and 16 cross-country skis are produced.

21. Let x = the number of rectangular tables rented, and let y = the number of round tables rented.

The cost for the tables is: $C = 28x + 52y$. Cost is to be minimized, so this is the objective function. The constraints are:

$x \geq 0, \ y \geq 0$ — A non-negative number of tables must be used.

$x + y \leq 35$ — Maximum number of tables.

$6x + 10y \geq 250$ — Number of guests.

$x \leq 15$ — Rectangular tables available.

Graph the constraints.

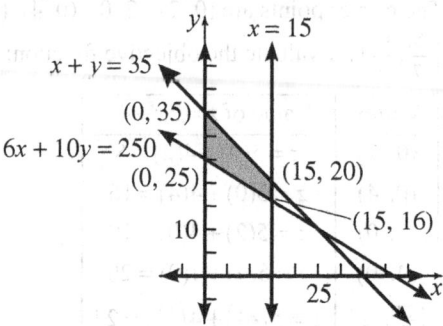

The corner points are (0, 25), (0, 35), (15, 20), (15, 16). Evaluate the objective function:

Vertex	Value of $C = 28x + 52y$
(0, 25)	$C = 28(0) + 52(25) = 1300$
(0, 35)	$C = 28(0) + 52(35) = 1820$
(15, 20)	$C = 28(15) + 52(20) = 1460$
(15, 16)	$C = 28(15) + 52(16) = 1252$

Kathleen should rent 15 rectangular tables and 16 round tables in order to minimize the cost. The minimum cost is $1252.00.

23. Let x = the amount invested in junk bonds, and let y = the amount invested in Treasury bills.

The total income is: $I = 0.09x + 0.07y$. Income is to be maximized, so this is the objective function. The constraints are:

$x \geq 0, \ y \geq 0$ — A non-negative amount must be invested.

$x + y \leq 20,000$ — Total investment cannot exceed $20,000.

$x \leq 12,000$ — Amount invested in junk bonds must not exceed $12,000.

$y \geq 8000$ — Amount invested in Treasury bills must be at least $8000.

a. $y \geq x$ — Amount invested in Treasury bills must be equal to or greater than the amount invested in junk bonds.

Graph the constraints.

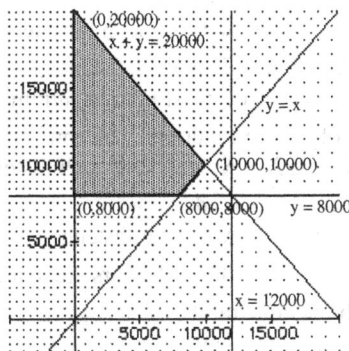

The corner points are (0, 20,000), (0, 8000), (8000, 8000), (10,000, 10,000).
Evaluate the objective function:

Vertex	Value of $I = 0.09x + 0.07y$
(0, 20000)	$I = 0.09(0) + 0.07(20000)$
	$= 1400$
(0, 8000)	$I = 0.09(0) + 0.07(8000)$
	$= 560$
(8000, 8000)	$I = 0.09(8000) + 0.07(8000)$
	$= 1280$
(10000, 10000)	$I = 0.09(10000) + 0.07(10000)$
	$= 1600$

The maximum income is $1600, when $10,000 is invested in junk bonds and $10,000 is invested in Treasury bills.

b. $y \leq x$ Amount invested in Treasury bills must not exceed the amount invested in junk bonds.
Graph the constraints.

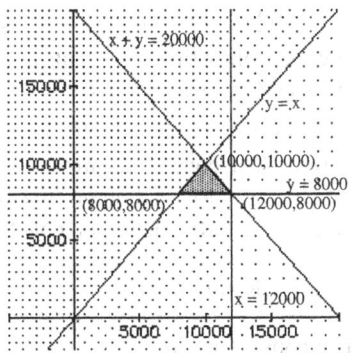

The corner points are (12,000, 8000), (8000, 8000), (10,000, 10,000).
Evaluate the objective function:

Vertex	Value of $I = 0.09x + 0.07y$
(12000, 8000)	$I = 0.09(12000) + 0.07(8000)$
	$= 1640$
(8000, 8000)	$I = 0.09(8000) + 0.07(8000)$
	$= 1280$
(10000, 10000)	$I = 0.09(10000) + 0.07(10000)$
	$= 1600$

The maximum income is $1640, when $12,000 is invested in junk bonds and $8000 is invested in Treasury bills.

25. Let x = the number of pounds of ground beef, and let y = the number of pounds of ground pork. The total cost is: $C = 0.75x + 0.45y$. Cost is to be minimized, so this is the objective function. The constraints are:

$x \geq 0$, $y \geq 0$ A positive number of pounds must be used.

$x \leq 200$ Only 200 pounds of ground beef are available.

$y \geq 50$ At least 50 pounds of ground pork must be used.

$0.75x + 0.60y \geq 0.70(x + y)$ Leanness condition

(Note that the last equation will simplify to $y \leq \frac{1}{2}x$.) Graph the constraints.

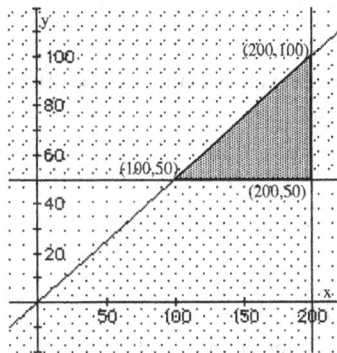

The corner points are (100, 50), (200, 50), (200, 100). Evaluate the objective function:

Vertex	Value of $C = 0.75x + 0.45y$
(100, 50)	$C = 0.75(100) + 0.45(50) = 97.50$
(200, 50)	$C = 0.75(200) + 0.45(50) = 172.50$
(200, 100)	$C = 0.75(200) + 0.45(100) = 195$

The minimum cost is $97.50, when 100 pounds of ground beef and 50 pounds of ground pork are used.

27. Let x = the number of racing skates manufactured, and let y = the number of figure skates manufactured. The total profit is: $P = 10x + 12y$. Profit is to be maximized, so this is the objective function. The constraints are:

$x \geq 0,\ y \geq 0$ A positive number of skates must be manufactured.

$6x + 4y \leq 120$ Only 120 hours are available for fabrication.

$x + 2y \leq 40$ Only 40 hours are available for finishing.

Graph the constraints.

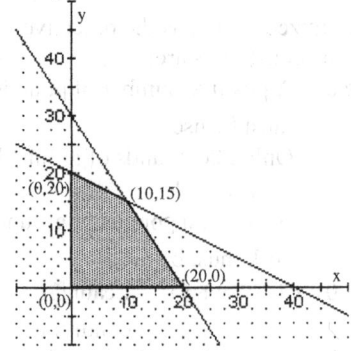

To find the intersection of $6x + 4y = 120$ and $x + 2y = 40$, solve the system:

$$\begin{cases} 6x + 4y = 120 \\ x + 2y = 40 \end{cases}$$

Solve the second equation for x: $x = 40 - 2y$

Substitute and solve.

$6(40 - 2y) + 4y = 120$

$240 - 12y + 4y = 120$

$-8y = -120$

$y = 15$

$x = 40 - 2(15) = 10$

The point of intersection is (10, 15).
The corner points are (0, 0), (0, 20), (20, 0), (10, 15). Evaluate the objective function:

Vertex	Value of $P = 10x + 12y$
(0, 0)	$P = 10(0) + 12(0) = 0$
(0, 20)	$P = 10(0) + 12(20) = 240$
(20, 0)	$P = 10(20) + 12(0) = 200$
(10, 15)	$P = 10(10) + 12(15) = 280$

The maximum profit is $280, when 10 racing skates and 15 figure skates are produced.

29. Let x = the number of metal fasteners, and let y = the number of plastic fasteners. The total cost is: $C = 9x + 4y$. Cost is to be minimized, so this is the objective function. The constraints are:

$x \geq 2,\ y \geq 2$ At least 2 of each fastener must be made.

$x + y \geq 6$ At least 6 fasteners are needed.

$4x + 2y \leq 24$ Only 24 hours are available.

Graph the constraints.

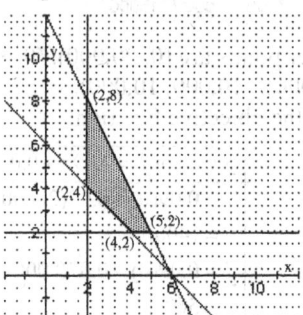

The corner points are (2, 4), (2, 8), (4, 2), (5, 2).
Evaluate the objective function:

Vertex	Value of $C = 9x + 4y$
(2, 4)	$C = 9(2) + 4(4) = 34$
(2, 8)	$C = 9(2) + 4(8) = 50$
(4, 2)	$C = 9(4) + 4(2) = 44$
(5, 2)	$C = 9(5) + 4(2) = 53$

The minimum cost is $34, when 2 metal fasteners and 4 plastic fasteners are ordered.

31. Let x = the number of first class seats, and let y = the number of coach seats. Using the hint, the revenue from x first class seats and y coach seats is $Fx + Cy$, where $F > C > 0$. Thus, $R = Fx + Cy$ is the objective function to be maximized. The constraints are:

$8 \leq x \leq 16$ Restriction on first class seats.

$80 \leq y \leq 120$ Restriction on coach seats.

a. $\dfrac{x}{y} \leq \dfrac{1}{12}$ Ratio of seats.

The constraints are:

$8 \leq x \leq 16$

$80 \leq y \leq 120$

$12x \leq y$

Graph the constraints.

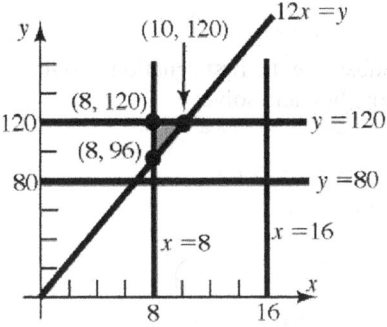

The corner points are (8, 96), (8, 120), and (10, 120). Evaluate the objective function:

Vertex	Value of $R = Fx + Cy$
(8, 96)	$R = 8F + 96C$
(8, 120)	$R = 8F + 120C$
(10, 120)	$R = 10F + 120C$

Since $C > 0$, $120C > 96C$, so $8F + 120C > 8F + 96C$.
Since $F > 0$, $10F > 8F$, so $10F + 120C > 8F + 120C$.
Thus, the maximum revenue occurs when the aircraft is configured with 10 first class seats and 120 coach seats.

b. $\dfrac{x}{y} \le \dfrac{1}{8}$

The constraints are:
$8 \le x \le 16$
$80 \le y \le 120$
$8x \le y$

Graph the constraints.

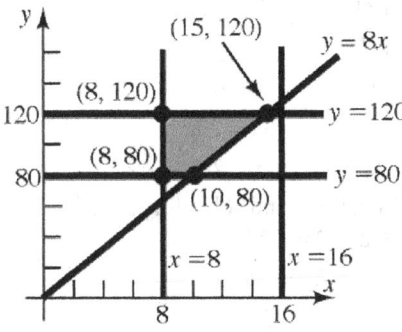

The corner points are (8, 80), (8, 120), (15, 120), and (10, 80).
Evaluate the objective function:

Vertex	Value of $R = Fx + Cy$
(8, 80)	$R = 8F + 80C$
(8, 120)	$R = 8F + 120C$
(15, 120)	$R = 15F + 120C$
(10, 80)	$R = 10F + 80C$

Since $F > 0$ and $C > 0$, $120C > 96C$, the maximum value of R occurs at (15, 120). The maximum revenue occurs when the aircraft is configured with 15 first class seats and 120 coach seats.

c. Answers will vary.

33.
$$2m^{2/5} - m^{1/5} = 1$$
$$2m^{2/5} - m^{1/5} - 1 = 0$$
$$(2m^{1/5} + 1)(m^{1/5} - 1) = 0$$

$$(2m^{1/5} + 1) = 0 \text{ or } (m^{1/5} - 1) = 0$$
$$2m^{1/5} = -1 \quad \text{ or } \quad m^{1/5} = 1$$
$$m^{1/5} = -\frac{1}{2} \quad \text{ or } \quad m^{1/5} = 1$$
$$m = \left(-\frac{1}{2}\right)^5 \quad \text{ or } \quad m = 1^5$$
$$m = -\frac{1}{32} \quad \text{ or } \quad m = 1$$

35.
$$A(t) = A_0 e^{rt}$$
$$1 = 2e^{63t}$$
$$\ln 0.5 = 63t$$
$$r = -0.011$$
Find t when $A = 75$ and $A_0 = 200$:
$$75 = 200e^{-0.011t}$$
$$0.375 = e^{-0.011t}$$
$$\ln 0.375 = -0.011t$$
$$t = \frac{\ln 0.375}{-0.011} \approx 89.1 \text{ years}$$

Chapter 8 Review Exercises

1. $\begin{cases} 2x - y = 5 \\ 5x + 2y = 8 \end{cases}$

Solve the first equation for y: $y = 2x - 5$.

Substitute and solve:
$5x + 2(2x - 5) = 8$
$5x + 4x - 10 = 8$
$9x = 18$
$x = 2$
$y = 2(2) - 5 = 4 - 5 = -1$

The solution is $x = 2$, $y = -1$ or $(2, -1)$.

2. $\begin{cases} 3x - 4y = 4 \\ x - 3y = \dfrac{1}{2} \end{cases}$

Solve the second equation for x: $x = 3y + \dfrac{1}{2}$

Substitute into the first equation and solve:
$3\left(3y + \dfrac{1}{2} \right) - 4y = 4$
$9y + \dfrac{3}{2} - 4y = 4$
$5y = \dfrac{5}{2}$
$y = \dfrac{1}{2}$
$x = 3\left(\dfrac{1}{2} \right) + \dfrac{1}{2} = 2$

The solution is $x = 2$, $y = \dfrac{1}{2}$ or $\left(2, \dfrac{1}{2} \right)$.

3. $\begin{cases} x - 2y - 4 = 0 \\ 3x + 2y - 4 = 0 \end{cases}$

Solve the first equation for x: $x = 2y + 4$

Substitute into the second equation and solve:
$3(2y + 4) + 2y - 4 = 0$
$6y + 12 + 2y - 4 = 0$
$8y = -8$
$y = -1$
$x = 2(-1) + 4 = 2$

The solution is $x = 2$, $y = -1$ or $(2, -1)$.

4. $\begin{cases} y = 2x - 5 \\ x = 3y + 4 \end{cases}$

Substitute the first equation into the second equation and solve:
$x = 3(2x - 5) + 4$
$x = 6x - 15 + 4$
$-5x = -11$
$x = \dfrac{11}{5}$
$y = 2\left(\dfrac{11}{5} \right) - 5 = -\dfrac{3}{5}$

The solution is $x = \dfrac{11}{5}$, $y = -\dfrac{3}{5}$ or $\left(\dfrac{11}{5}, -\dfrac{3}{5} \right)$.

5. $\begin{cases} x - 3y + 4 = 0 \\ \dfrac{1}{2}x - \dfrac{3}{2}y + \dfrac{4}{3} = 0 \end{cases}$

Multiply each side of the first equation by 3 and each side of the second equation by -6 and add:
$\begin{cases} 3x - 9y + 12 = 0 \\ -3x + 9y - 8 = 0 \end{cases}$
$ 4 = 0$

There is no solution to the system. The system is inconsistent.

6. $\begin{cases} 2x + 3y - 13 = 0 \\ 3x - 2y = 0 \end{cases}$

Multiply each side of the first equation by 2 and each side of the second equation by 3, and add to eliminate y:
$\begin{cases} 4x + 6y - 26 = 0 \\ 9x - 6y = 0 \end{cases}$
$13x - 26 = 0$
$ 13x = 26$
$ x = 2$

Substitute and solve for y:
$3(2) - 2y = 0$
$-2y = -6$
$y = 3$

The solution is $x = 2$, $y = 3$ or $(2, 3)$.

7. $\begin{cases} 2x+5y=10 \\ 4x+10y=20 \end{cases}$

Multiply each side of the first equation by –2 and add to eliminate x:

$\begin{cases} -4x-10y=-20 \\ 4x+10y=20 \end{cases}$

$0=0$

The system is dependent.

$2x+5y=10$

$5y=-2x+10$

$y=-\dfrac{2}{5}x+2$

The solution is $y=-\dfrac{2}{5}x+2$, x is any real number

or $\left\{(x,y)\Big| y=-\dfrac{2}{5}x+2, x \text{ is any real number}\right\}$.

8. $\begin{cases} x+2y-z=6 \\ 2x-y+3z=-13 \\ 3x-2y+3z=-16 \end{cases}$

Multiply each side of the first equation by –2 and add to the second equation to eliminate x;

$\begin{cases} -2x-4y+2z=-12 \\ 2x-y+3z=-13 \end{cases}$

$-5y+5z=-25$

$y-z=5$

Multiply each side of the first equation by –3 and add to the third equation to eliminate x:

$-3x-6y+3z=-18$

$\underline{3x-2y+3z=-16}$

$-8y+6z=-34$

Multiply each side of the first result by 8 and add to the second result to eliminate y:

$8y-8z=40$

$\underline{-8y+6z=-34}$

$-2z=6$

$z=-3$

Substituting and solving for the other variables:

$y-(-3)=5 \quad x+2(2)-(-3)=6$

$y=2 x+4+3=6$

$x=-1$

The solution is $x=-1, y=2, z=-3$ or $(-1,2,-3)$.

9. $\begin{cases} 2x-4y+z=-15 \\ x+2y-4z=27 \\ 5x-6y-2z=-3 \end{cases}$

Multiply the first equation by -1 and the second equation by 2, and then add to eliminate x:

$\begin{cases} -2x+4y-z=15 \\ 2x+4y-8z=54 \end{cases}$

$8y-9z=69$

Multiply the second equation by -5 and add to the third equation to eliminate x:

$-5x-10y+20z=-135$

$\underline{5x-6y-2z=-3}$

$-16y+18z=-138$

Multiply both sides of the first result by 2 and add to the second result to eliminate y:

$16y-18z=138$

$\underline{-16y+18z=-138}$

$0=0$

The system is dependent.

$-16y+18z=-138$

$18z+138=16y$

$y=\dfrac{9}{8}z+\dfrac{69}{8}$

Substituting into the second equation and solving for x:

$x+2\left(\dfrac{9}{8}z+\dfrac{69}{8}\right)-4z=27$

$x+\dfrac{9}{4}z+\dfrac{69}{4}-4z=27$

$x=\dfrac{7}{4}z+\dfrac{39}{4}$

The solution is $x=\dfrac{7}{4}z+\dfrac{39}{4}$, $y=\dfrac{9}{8}z+\dfrac{69}{8}$, z is

any real number or $\left\{(x,y,z)\Big| x=\dfrac{7}{4}z+\dfrac{39}{4},\right.$

$\left. y=\dfrac{9}{8}z+\dfrac{69}{8}, z \text{ is any real number}\right\}$.

10. $\begin{cases} x-4y+3z=15 \\ -3x+y-5z=-5 \\ -7x-5y-9z=10 \end{cases}$

Multiply the first equation by 3 and then add the second equation to eliminate x:

$$\begin{cases} 3x - 12y + 9z = 45 \\ -3x + y - 5z = -5 \end{cases}$$
$$\overline{\qquad -11y + 4z = 40}$$

Multiply the first equation by 7 and add to the third equation to eliminate x:
$$7x - 28y + 21z = 105$$
$$\underline{-7x - 5y - 9z = 10}$$
$$-33y + 12z = 115$$
$$-11y + 4z = \frac{115}{3}$$

Multiply the first result by -1 and adding it to the second result:
$$\begin{cases} 11y - 4z = -40 \\ -11y + 4z = \dfrac{115}{3} \end{cases}$$
$$\overline{\qquad 0 = -\dfrac{5}{3}}$$

The system has no solution. The system is inconsistent.

11. $\begin{cases} 3x + 2y = 8 \\ x + 4y = -1 \end{cases}$

12. $\begin{cases} x + 2y + 5z = -2 \\ 5x \qquad - 3z = 8 \\ 2x - y \qquad = 0 \end{cases}$

13. $A + C = \begin{bmatrix} 1 & 0 \\ 2 & 4 \\ -1 & 2 \end{bmatrix} + \begin{bmatrix} 3 & -4 \\ 1 & 5 \\ 5 & 2 \end{bmatrix}$

$$= \begin{bmatrix} 1+3 & 0+(-4) \\ 2+1 & 4+5 \\ -1+5 & 2+2 \end{bmatrix} = \begin{bmatrix} 4 & -4 \\ 3 & 9 \\ 4 & 4 \end{bmatrix}$$

14. $6A = 6 \cdot \begin{bmatrix} 1 & 0 \\ 2 & 4 \\ -1 & 2 \end{bmatrix} = \begin{bmatrix} 6 \cdot 1 & 6 \cdot 0 \\ 6 \cdot 2 & 6 \cdot 4 \\ 6(-1) & 6 \cdot 2 \end{bmatrix} = \begin{bmatrix} 6 & 0 \\ 12 & 24 \\ -6 & 12 \end{bmatrix}$

15. $AB = \begin{bmatrix} 1 & 0 \\ 2 & 4 \\ -1 & 2 \end{bmatrix} \cdot \begin{bmatrix} 4 & -3 & 0 \\ 1 & 1 & -2 \end{bmatrix}$

$$= \begin{bmatrix} 1(4)+0(1) & 1(-3)+0(1) & 1(0)+0(-2) \\ 2(4)+4(1) & 2(-3)+4(1) & 2(0)+4(-2) \\ -1(4)+2(1) & -1(-3)+2(1) & -1(0)+2(-2) \end{bmatrix}$$

$$= \begin{bmatrix} 4 & -3 & 0 \\ 12 & -2 & -8 \\ -2 & 5 & -4 \end{bmatrix}$$

16. $BC = \begin{bmatrix} 4 & -3 & 0 \\ 1 & 1 & -2 \end{bmatrix} \cdot \begin{bmatrix} 3 & -4 \\ 1 & 5 \\ 5 & 2 \end{bmatrix}$

$$= \begin{bmatrix} 4(3)-3(1)+0(5) & 4(-4)-3(5)+0(2) \\ 1(3)+1(1)-2(5) & 1(-4)+1(5)-2(2) \end{bmatrix}$$

$$= \begin{bmatrix} 9 & -31 \\ -6 & -3 \end{bmatrix}$$

17. $A = \begin{bmatrix} 4 & 6 \\ 1 & 3 \end{bmatrix}$

Augment the matrix with the identity and use row operations to find the inverse:

$$\begin{bmatrix} 4 & 6 & | & 1 & 0 \\ 1 & 3 & | & 0 & 1 \end{bmatrix}$$

$$\rightarrow \begin{bmatrix} 1 & 3 & | & 0 & 1 \\ 4 & 6 & | & 1 & 0 \end{bmatrix} \begin{pmatrix} \text{Interchange} \\ r_1 \text{ and } r_2 \end{pmatrix}$$

$$\rightarrow \begin{bmatrix} 1 & 3 & | & 0 & 1 \\ 0 & -6 & | & 1 & -4 \end{bmatrix} \left(R_2 = -4r_1 + r_2 \right)$$

$$\rightarrow \begin{bmatrix} 1 & 3 & | & 0 & 1 \\ 0 & 1 & | & -\frac{1}{6} & \frac{2}{3} \end{bmatrix} \left(R_2 = -\frac{1}{6}r_2 \right)$$

$$\rightarrow \begin{bmatrix} 1 & 0 & | & \frac{1}{2} & -1 \\ 0 & 1 & | & -\frac{1}{6} & \frac{2}{3} \end{bmatrix} \left(R_1 = -3r_2 + r_1 \right)$$

Thus, $A^{-1} = \begin{bmatrix} \frac{1}{2} & -1 \\ -\frac{1}{6} & \frac{2}{3} \end{bmatrix}$.

18. $A = \begin{bmatrix} 1 & 3 & 3 \\ 1 & 2 & 1 \\ 1 & -1 & 2 \end{bmatrix}$

Augment the matrix with the identity and use row operations to find the inverse:

$\begin{bmatrix} 1 & 3 & 3 & | & 1 & 0 & 0 \\ 1 & 2 & 1 & | & 0 & 1 & 0 \\ 1 & -1 & 2 & | & 0 & 0 & 1 \end{bmatrix}$

$\rightarrow \begin{bmatrix} 1 & 3 & 3 & | & 1 & 0 & 0 \\ 0 & -1 & -2 & | & -1 & 1 & 0 \\ 0 & -4 & -1 & | & -1 & 0 & 1 \end{bmatrix} \begin{pmatrix} R_2 = -r_1 + r_2 \\ R_3 = -r_1 + r_3 \end{pmatrix}$

$\rightarrow \begin{bmatrix} 1 & 3 & 3 & | & 1 & 0 & 0 \\ 0 & 1 & 2 & | & 1 & -1 & 0 \\ 0 & -4 & -1 & | & -1 & 0 & 1 \end{bmatrix} (R_2 = -r_2)$

$\rightarrow \begin{bmatrix} 1 & 0 & -3 & | & -2 & 3 & 0 \\ 0 & 1 & 2 & | & 1 & -1 & 0 \\ 0 & 0 & 7 & | & 3 & -4 & 1 \end{bmatrix} \begin{pmatrix} R_1 = -3r_2 + r_1 \\ R_3 = 4r_2 + r_3 \end{pmatrix}$

$\rightarrow \begin{bmatrix} 1 & 0 & -3 & | & -2 & 3 & 0 \\ 0 & 1 & 2 & | & 1 & -1 & 0 \\ 0 & 0 & 1 & | & \frac{3}{7} & -\frac{4}{7} & \frac{1}{7} \end{bmatrix} (R_3 = \frac{1}{7} r_3)$

$\rightarrow \begin{bmatrix} 1 & 0 & 0 & | & -\frac{5}{7} & \frac{9}{7} & \frac{3}{7} \\ 0 & 1 & 0 & | & \frac{1}{7} & \frac{1}{7} & -\frac{2}{7} \\ 0 & 0 & 1 & | & \frac{3}{7} & -\frac{4}{7} & \frac{1}{7} \end{bmatrix} \begin{pmatrix} R_1 = 3r_3 + r_1 \\ R_2 = -2r_3 + r_2 \end{pmatrix}$

Thus, $A^{-1} = \begin{bmatrix} -\frac{5}{7} & \frac{9}{7} & \frac{3}{7} \\ \frac{1}{7} & \frac{1}{7} & -\frac{2}{7} \\ \frac{3}{7} & -\frac{4}{7} & \frac{1}{7} \end{bmatrix}$.

19. $A = \begin{bmatrix} 4 & -8 \\ -1 & 2 \end{bmatrix}$

Augment the matrix with the identity and use row operations to find the inverse:

$\begin{bmatrix} 4 & -8 & | & 1 & 0 \\ -1 & 2 & | & 0 & 1 \end{bmatrix}$

$\rightarrow \begin{bmatrix} -1 & 2 & | & 0 & 1 \\ 4 & -8 & | & 1 & 0 \end{bmatrix} \begin{pmatrix} \text{Interchange} \\ r_1 \text{ and } r_2 \end{pmatrix}$

$\rightarrow \begin{bmatrix} -1 & 2 & | & 0 & 1 \\ 0 & 0 & | & 1 & 4 \end{bmatrix} (R_2 = 4r_1 + r_2)$

$\rightarrow \begin{bmatrix} 1 & -2 & | & 0 & -1 \\ 0 & 0 & | & 1 & 4 \end{bmatrix} (R_1 = -r_1)$

There is no inverse because there is no way to

obtain the identity on the left side. The matrix is singular.

20. $\begin{cases} 3x - 2y = 1 \\ 10x + 10y = 5 \end{cases}$

Write the augmented matrix:

$\begin{bmatrix} 3 & -2 & | & 1 \\ 10 & 10 & | & 5 \end{bmatrix}$

$\rightarrow \begin{bmatrix} 3 & -2 & | & 1 \\ 1 & 16 & | & 2 \end{bmatrix} (R_2 = -3r_1 + r_2)$

$\rightarrow \begin{bmatrix} 1 & 16 & | & 2 \\ 3 & -2 & | & 1 \end{bmatrix} \begin{pmatrix} \text{Interchange} \\ r_1 \text{ and } r_2 \end{pmatrix}$

$\rightarrow \begin{bmatrix} 1 & 16 & | & 2 \\ 0 & -50 & | & -5 \end{bmatrix} (R_2 = -3r_1 + r_2)$

$\rightarrow \begin{bmatrix} 1 & 16 & | & 2 \\ 0 & 1 & | & \frac{1}{10} \end{bmatrix} (R_2 = -\frac{1}{50} r_2)$

$\rightarrow \begin{bmatrix} 1 & 0 & | & \frac{2}{5} \\ 0 & 1 & | & \frac{1}{10} \end{bmatrix} (R_1 = -16r_2 + r_1)$

The solution is $x = \frac{2}{5}, y = \frac{1}{10}$ or $\left(\frac{2}{5}, \frac{1}{10} \right)$.

21. $\begin{cases} 5x - 6y - 3z = 6 \\ 4x - 7y - 2z = -3 \\ 3x + y - 7z = 1 \end{cases}$

Write the augmented matrix:

$\begin{bmatrix} 5 & -6 & -3 & | & 6 \\ 4 & -7 & -2 & | & -3 \\ 3 & 1 & -7 & | & 1 \end{bmatrix}$

$\rightarrow \begin{bmatrix} 1 & 1 & -1 & | & 9 \\ 4 & -7 & -2 & | & -3 \\ 3 & 1 & -7 & | & 1 \end{bmatrix} (R_1 = -r_2 + r_1)$

$\rightarrow \begin{bmatrix} 1 & 1 & -1 & | & 9 \\ 0 & -11 & 2 & | & -39 \\ 0 & -2 & -4 & | & -26 \end{bmatrix} \begin{pmatrix} R_2 = -4r_1 + r_2 \\ R_3 = -3r_1 + r_3 \end{pmatrix}$

$\rightarrow \begin{bmatrix} 1 & 1 & -1 & | & 9 \\ 0 & 1 & -\frac{2}{11} & | & \frac{39}{11} \\ 0 & 1 & 2 & | & 13 \end{bmatrix} \begin{pmatrix} R_2 = -\frac{1}{11} r_2 \\ R_3 = -\frac{1}{2} r_3 \end{pmatrix}$

$$\rightarrow \begin{bmatrix} 1 & 0 & -\frac{9}{11} & \frac{60}{11} \\ 0 & 1 & -\frac{2}{11} & \frac{39}{11} \\ 0 & 0 & \frac{24}{11} & \frac{104}{11} \end{bmatrix} \quad \begin{pmatrix} R_1 = -r_2 + r_1 \\ R_3 = -r_2 + r_3 \end{pmatrix}$$

$$\rightarrow \begin{bmatrix} 1 & 0 & -\frac{9}{11} & \frac{60}{11} \\ 0 & 1 & -\frac{2}{11} & \frac{39}{11} \\ 0 & 0 & 1 & \frac{13}{3} \end{bmatrix} \quad \left(R_3 = \frac{11}{24} r_3 \right)$$

$$\rightarrow \begin{bmatrix} 1 & 0 & 0 & 9 \\ 0 & 1 & 0 & \frac{13}{3} \\ 0 & 0 & 1 & \frac{13}{3} \end{bmatrix} \quad \begin{pmatrix} R_1 = \frac{9}{11} r_3 + r_1 \\ R_2 = \frac{2}{11} r_3 + r_2 \end{pmatrix}$$

The solution is $x = 9$, $y = \dfrac{13}{3}$, $z = \dfrac{13}{3}$ or

$\left(9, \dfrac{13}{3}, \dfrac{13}{3} \right)$.

22. $\begin{cases} 2x + y + z = 5 \\ 4x - y - 3z = 1 \\ 8x + y - z = 5 \end{cases}$

Write the augmented matrix:

$$\begin{bmatrix} 2 & 1 & 1 & 5 \\ 4 & -1 & -3 & 1 \\ 8 & 1 & -1 & 5 \end{bmatrix}$$

$$\rightarrow \begin{bmatrix} 2 & 1 & 1 & 5 \\ 0 & -3 & -5 & -9 \\ 0 & -3 & -5 & -15 \end{bmatrix} \quad \begin{pmatrix} R_2 = -2r_1 + r_2 \\ R_3 = -4r_1 + r_3 \end{pmatrix}$$

$$\rightarrow \begin{bmatrix} 1 & \frac{1}{2} & \frac{1}{2} & \frac{5}{2} \\ 0 & 1 & \frac{5}{3} & 3 \\ 0 & -3 & -5 & -15 \end{bmatrix} \quad \begin{pmatrix} R_1 = \frac{1}{2} r_1 \\ R_2 = -\frac{1}{3} r_2 \end{pmatrix}$$

$$\rightarrow \begin{bmatrix} 1 & 0 & -\frac{1}{3} & 1 \\ 0 & 1 & \frac{5}{3} & 3 \\ 0 & 0 & 0 & -6 \end{bmatrix} \quad \begin{pmatrix} R_1 = -\frac{1}{2} r_2 + r_1 \\ R_3 = 3r_2 + r_3 \end{pmatrix}$$

There is no solution; the system is inconsistent.

23. $\begin{cases} x \quad\;\;\; - 2z = 1 \\ 2x + 3y \quad\;\;\; = -3 \\ 4x - 3y - 4z = 3 \end{cases}$

Write the augmented matrix:

$$\begin{bmatrix} 1 & 0 & -2 & 1 \\ 2 & 3 & 0 & -3 \\ 4 & -3 & -4 & 3 \end{bmatrix}$$

$$\rightarrow \begin{bmatrix} 1 & 0 & -2 & 1 \\ 0 & 3 & 4 & -5 \\ 0 & -3 & 4 & -1 \end{bmatrix} \quad \begin{pmatrix} R_2 = -2r_1 + r_2 \\ R_3 = -4r_1 + r_3 \end{pmatrix}$$

$$\rightarrow \begin{bmatrix} 1 & 0 & -2 & 1 \\ 0 & 1 & \frac{4}{3} & -\frac{5}{3} \\ 0 & -3 & 4 & -1 \end{bmatrix} \quad \left(R_2 = \frac{1}{3} r_2 \right)$$

$$\rightarrow \begin{bmatrix} 1 & 0 & -2 & 1 \\ 0 & 1 & \frac{4}{3} & -\frac{5}{3} \\ 0 & 0 & 8 & -6 \end{bmatrix} \quad \left(R_3 = 3r_2 + r_3 \right)$$

$$\rightarrow \begin{bmatrix} 1 & 0 & -2 & 1 \\ 0 & 1 & \frac{4}{3} & -\frac{5}{3} \\ 0 & 0 & 1 & -\frac{3}{4} \end{bmatrix} \quad \left(R_3 = \frac{1}{8} r_3 \right)$$

$$\rightarrow \begin{bmatrix} 1 & 0 & 0 & -\frac{1}{2} \\ 0 & 1 & 0 & -\frac{2}{3} \\ 0 & 0 & 1 & -\frac{3}{4} \end{bmatrix} \quad \begin{pmatrix} R_1 = 2r_3 + r_1 \\ R_2 = -\frac{4}{3} r_3 + r_2 \end{pmatrix}$$

The solution is $x = -\dfrac{1}{2}$, $y = -\dfrac{2}{3}$, $z = -\dfrac{3}{4}$ or

$\left(-\dfrac{1}{2}, -\dfrac{2}{3}, -\dfrac{3}{4} \right)$.

24. $\begin{cases} x - y + z = 0 \\ x - y - 5z = 6 \\ 2x - 2y + z = 1 \end{cases}$

Write the augmented matrix:

$$\begin{bmatrix} 1 & -1 & 1 & 0 \\ 1 & -1 & -5 & 6 \\ 2 & -2 & 1 & 1 \end{bmatrix}$$

$$\rightarrow \begin{bmatrix} 1 & -1 & 1 & 0 \\ 0 & 0 & -6 & 6 \\ 0 & 0 & -1 & 1 \end{bmatrix} \quad \begin{pmatrix} R_2 = -r_1 + r_2 \\ R_3 = -2r_1 + r_3 \end{pmatrix}$$

$$\rightarrow \begin{bmatrix} 1 & -1 & 1 & 0 \\ 0 & 0 & 1 & -1 \\ 0 & 0 & -1 & 1 \end{bmatrix} \quad \left(R_2 = -\frac{1}{6} r_2 \right)$$

$$\rightarrow \begin{bmatrix} 1 & -1 & 0 & 1 \\ 0 & 0 & 1 & -1 \\ 0 & 0 & 0 & 0 \end{bmatrix} \quad \begin{pmatrix} R_1 = -r_2 + r_1 \\ R_3 = r_2 + r_3 \end{pmatrix}$$

The system is dependent.

$\begin{cases} x = y+1 \\ z = -1 \end{cases}$

The solution is $x = y+1$, $z = -1$, y is any real number or $\{(x, y, z) | x = y+1, z = -1, y$ is any real number$\}$.

25. $\begin{cases} x - y - z - t = 1 \\ 2x + y - z + 2t = 3 \\ x - 2y - 2z - 3t = 0 \\ 3x - 4y + z + 5t = -3 \end{cases}$

Write the augmented matrix:

$\begin{bmatrix} 1 & -1 & -1 & -1 & | & 1 \\ 2 & 1 & -1 & 2 & | & 3 \\ 1 & -2 & -2 & -3 & | & 0 \\ 3 & -4 & 1 & 5 & | & -3 \end{bmatrix}$

$\rightarrow \begin{bmatrix} 1 & -1 & -1 & -1 & | & 1 \\ 0 & 3 & 1 & 4 & | & 1 \\ 0 & -1 & -1 & -2 & | & -1 \\ 0 & -1 & 4 & 8 & | & -6 \end{bmatrix} \begin{pmatrix} R_2 = -2r_1 + r_2 \\ R_3 = -r_1 + r_3 \\ R_4 = -3r_1 + r_4 \end{pmatrix}$

$\rightarrow \begin{bmatrix} 1 & -1 & -1 & -1 & | & 1 \\ 0 & -1 & -1 & -2 & | & -1 \\ 0 & 3 & 1 & 4 & | & 1 \\ 0 & -1 & 4 & 8 & | & -6 \end{bmatrix} \begin{pmatrix} \text{Interchange} \\ r_2 \text{ and } r_3 \end{pmatrix}$

$\rightarrow \begin{bmatrix} 1 & -1 & -1 & -1 & | & 1 \\ 0 & 1 & 1 & 2 & | & 1 \\ 0 & 3 & 1 & 4 & | & 1 \\ 0 & -1 & 4 & 8 & | & -6 \end{bmatrix} (R_2 = -r_2)$

$\rightarrow \begin{bmatrix} 1 & 0 & 0 & 1 & | & 2 \\ 0 & 1 & 1 & 2 & | & 1 \\ 0 & 0 & -2 & -2 & | & -2 \\ 0 & 0 & 5 & 10 & | & -5 \end{bmatrix} \begin{pmatrix} R_1 = r_2 + r_1 \\ R_3 = -3r_2 + r_3 \\ R_4 = r_2 + r_4 \end{pmatrix}$

$\rightarrow \begin{bmatrix} 1 & 0 & 0 & 1 & | & 2 \\ 0 & 1 & 1 & 2 & | & 1 \\ 0 & 0 & 1 & 1 & | & 1 \\ 0 & 0 & 1 & 2 & | & -1 \end{bmatrix} \begin{pmatrix} R_3 = -\frac{1}{2}r_3 \\ R_4 = \frac{1}{5}r_4 \end{pmatrix}$

$\rightarrow \begin{bmatrix} 1 & 0 & 0 & 1 & | & 2 \\ 0 & 1 & 0 & 1 & | & 0 \\ 0 & 0 & 1 & 1 & | & 1 \\ 0 & 0 & 0 & 1 & | & -2 \end{bmatrix} \begin{pmatrix} R_2 = -r_3 + r_2 \\ R_4 = -r_3 + r_4 \end{pmatrix}$

$\rightarrow \begin{bmatrix} 1 & 0 & 0 & 0 & | & 4 \\ 0 & 1 & 0 & 0 & | & 2 \\ 0 & 0 & 1 & 0 & | & 3 \\ 0 & 0 & 0 & 1 & | & -2 \end{bmatrix} \begin{pmatrix} R_1 = -r_4 + r_1 \\ R_2 = -r_4 + r_2 \\ R_3 = -r_4 + r_3 \end{pmatrix}$

The solution is $x = 4, y = 2, z = 3, t = -2$ or $(4, 2, 3, -2)$.

26. $\begin{vmatrix} 3 & 4 \\ 1 & 3 \end{vmatrix} = 3(3) - 4(1) = 9 - 4 = 5$

27. $\begin{vmatrix} 1 & 4 & 0 \\ -1 & 2 & 6 \\ 4 & 1 & 3 \end{vmatrix} = 1 \begin{vmatrix} 2 & 6 \\ 1 & 3 \end{vmatrix} - 4 \begin{vmatrix} -1 & 6 \\ 4 & 3 \end{vmatrix} + 0 \begin{vmatrix} -1 & 2 \\ 4 & 1 \end{vmatrix}$

$= 1(6-6) - 4(-3-24) + 0(-1-8)$
$= 1(0) - 4(-27) + 0(-9) = 0 + 108 + 0$
$= 108$

28. $\begin{vmatrix} 2 & 1 & -3 \\ 5 & 0 & 1 \\ 2 & 6 & 0 \end{vmatrix} = 2 \begin{vmatrix} 0 & 1 \\ 6 & 0 \end{vmatrix} - 1 \begin{vmatrix} 5 & 1 \\ 2 & 0 \end{vmatrix} + (-3) \begin{vmatrix} 5 & 0 \\ 2 & 6 \end{vmatrix}$

$= 2(0-6) - 1(0-2) - 3(30-0)$
$= 2(-6) - 1(-2) - 3(30)$
$= -12 + 2 - 90$
$= -100$

29. $\begin{cases} x - 2y = 4 \\ 3x + 2y = 4 \end{cases}$

Set up and evaluate the determinants to use Cramer's Rule:

$D = \begin{vmatrix} 1 & -2 \\ 3 & 2 \end{vmatrix} = 1(2) - 3(-2) = 2 + 6 = 8$

$D_x = \begin{vmatrix} 4 & -2 \\ 4 & 2 \end{vmatrix} = 4(2) - 4(-2) = 8 + 8 = 16$

$D_y = \begin{vmatrix} 1 & 4 \\ 3 & 4 \end{vmatrix} = 1(4) - 3(4) = 4 - 12 = -8$

The solution is $x = \dfrac{D_x}{D} = \dfrac{16}{8} = 2$, $y = \dfrac{D_y}{D} = \dfrac{-8}{8} = -1$ or $(2, -1)$.

30. $\begin{cases} 2x+3y-13=0 \\ 3x-2y=0 \end{cases}$

Write the system is standard form:

$\begin{cases} 2x+3y=13 \\ 3x-2y=0 \end{cases}$

Set up and evaluate the determinants to use Cramer's Rule:

$D = \begin{vmatrix} 2 & 3 \\ 3 & -2 \end{vmatrix} = -4-9 = -13$

$D_x = \begin{vmatrix} 13 & 3 \\ 0 & -2 \end{vmatrix} = -26-0 = -26$

$D_y = \begin{vmatrix} 2 & 13 \\ 3 & 0 \end{vmatrix} = 0-39 = -39$

The solution is $x = \dfrac{D_x}{D} = \dfrac{-26}{-13} = 2$,

$y = \dfrac{D_y}{D} = \dfrac{-39}{-13} = 3$ or (2, 3).

31. $\begin{cases} x+2y-z=6 \\ 2x-y+3z=-13 \\ 3x-2y+3z=-16 \end{cases}$

Set up and evaluate the determinants to use Cramer's Rule:

$D = \begin{vmatrix} 1 & 2 & -1 \\ 2 & -1 & 3 \\ 3 & -2 & 3 \end{vmatrix}$

$= 1\begin{vmatrix} -1 & 3 \\ -2 & 3 \end{vmatrix} - 2\begin{vmatrix} -1 & 3 \\ -2 & 3 \end{vmatrix} + (-1)\begin{vmatrix} 2 & -1 \\ 3 & -2 \end{vmatrix}$

$= 1(-3+6) - 2(-3+6) + (-1)(-4+3)$

$= 3+6+1 = 10$

$D_x = \begin{vmatrix} 6 & 2 & -1 \\ -13 & -1 & 3 \\ -16 & -2 & 3 \end{vmatrix}$

$= 6\begin{vmatrix} -1 & 3 \\ -2 & 3 \end{vmatrix} - 2\begin{vmatrix} -13 & 3 \\ -16 & 3 \end{vmatrix} + (-1)\begin{vmatrix} -13 & -1 \\ -16 & -2 \end{vmatrix}$

$= 6(-3+6) - 2(-39+48) + (-1)(26-16)$

$= 18-18-10 = -10$

$D_y = \begin{vmatrix} 1 & 6 & -1 \\ 2 & -13 & 3 \\ 3 & -16 & 3 \end{vmatrix}$

$= 1\begin{vmatrix} -13 & 3 \\ -16 & 3 \end{vmatrix} - 6\begin{vmatrix} 2 & 3 \\ 3 & 3 \end{vmatrix} + (-1)\begin{vmatrix} 2 & -13 \\ 3 & -16 \end{vmatrix}$

$= 1(-39+48) - 6(6-9) + (-1)(-32+39)$

$= 9+18-7 = 20$

$D_z = \begin{vmatrix} 1 & 2 & 6 \\ 2 & -1 & -13 \\ 3 & -2 & -16 \end{vmatrix}$

$= 1\begin{vmatrix} -1 & -13 \\ -2 & -16 \end{vmatrix} - 2\begin{vmatrix} 2 & -13 \\ 3 & -16 \end{vmatrix} + 6\begin{vmatrix} 2 & -1 \\ 3 & -2 \end{vmatrix}$

$= 1(16-26) - 2(-32+39) + 6(-4+3)$

$= -10-14-6 = -30$

The solution is $x = \dfrac{D_x}{D} = \dfrac{-10}{10} = -1$,

$y = \dfrac{D_y}{D} = \dfrac{20}{10} = 2$, $z = \dfrac{D_z}{D} = \dfrac{-30}{10} = -3$ or

$(-1, 2, -3)$.

32. Let $\begin{vmatrix} x & y \\ a & b \end{vmatrix} = 8$.

Then $\begin{vmatrix} 2x & y \\ 2a & b \end{vmatrix} = 2(8) = 16$ by Theorem (14).

The value of the determinant is multiplied by k when the elements of a column are multiplied by k.

33. Let $\begin{vmatrix} x & y \\ a & b \end{vmatrix} = 8$.

Then $\begin{vmatrix} y & x \\ b & a \end{vmatrix} = -8$ by Theorem (11). The

value of the determinant changes sign when any 2 columns are interchanged.

34. Find the partial fraction decomposition:

$x(x-4)\left(\dfrac{6}{x(x-4)}\right) = x(x-4)\left(\dfrac{A}{x} + \dfrac{B}{x-4}\right)$

$6 = A(x-4) + Bx$

Let $x = 4$, then $6 = A(4-4) + B(4)$

$4B = 6$

$B = \dfrac{3}{2}$

Let $x = 0$, then $6 = A(0-4) + B(0)$

$-4A = 6$

$A = -\dfrac{3}{2}$

$\dfrac{6}{x(x-4)} = \dfrac{-\dfrac{3}{2}}{x} + \dfrac{\dfrac{3}{2}}{x-4}$

35. Find the partial fraction decomposition:

$$\frac{x-4}{x^2(x-1)} = \frac{A}{x} + \frac{B}{x^2} + \frac{C}{x-1}$$

Multiply both sides by $x^2(x-1)$

$$x-4 = Ax(x-1) + B(x-1) + Cx^2$$

Let $x = 1$, then

$$1-4 = A(1)(1-1) + B(1-1) + C(1)^2$$
$$-3 = C$$
$$C = -3$$

Let $x = 0$, then

$$0-4 = A(0)(0-1) + B(0-1) + C(0)^2$$
$$-4 = -B$$
$$B = 4$$

Let $x = 2$, then

$$2-4 = A(2)(2-1) + B(2-1) + C(2)^2$$
$$-2 = 2A + B + 4C$$
$$2A = -2 - 4 - 4(-3)$$
$$2A = 6$$
$$A = 3$$

$$\frac{x-4}{x^2(x-1)} = \frac{3}{x} + \frac{4}{x^2} + \frac{-3}{x-1}$$

36. Find the partial fraction decomposition:

$$\frac{x}{(x^2+9)(x+1)} = \frac{A}{x+1} + \frac{Bx+C}{x^2+9}$$

Multiply both sides by $(x+1)(x^2+9)$.

$$x = A(x^2+9) + (Bx+C)(x+1)$$

Let $x = -1$, then

$$-1 = A\left((-1)^2+9\right) + \left(B(-1)+C\right)(-1+1)$$
$$-1 = A(10) + (-B+C)(0)$$
$$-1 = 10A$$
$$A = -\frac{1}{10}$$

Let $x = 0$, then

$$0 = A\left(0^2+9\right) + \left(B(0)+C\right)(0+1)$$
$$0 = 9A + C$$
$$0 = 9\left(-\frac{1}{10}\right) + C$$
$$C = \frac{9}{10}$$

Let $x = 1$, then

$$1 = A\left(1^2+9\right) + \left(B(1)+C\right)(1+1)$$
$$1 = A(10) + (B+C)(2)$$
$$1 = 10A + 2B + 2C$$
$$1 = 10\left(-\frac{1}{10}\right) + 2B + 2\left(\frac{9}{10}\right)$$
$$1 = -1 + 2B + \frac{9}{5}$$
$$2B = \frac{1}{5}$$
$$B = \frac{1}{10}$$

$$\frac{x}{(x^2+9)(x+1)} = \frac{-\frac{1}{10}}{x+1} + \frac{\frac{1}{10}x + \frac{9}{10}}{x^2+9}$$

37. Find the partial fraction decomposition:

$$\frac{x^3}{(x^2+4)^2} = \frac{Ax+B}{x^2+4} + \frac{Cx+D}{(x^2+4)^2}$$

Multiply both sides by $(x^2+4)^2$.

$$x^3 = (Ax+B)(x^2+4) + Cx + D$$
$$x^3 = Ax^3 + Bx^2 + 4Ax + 4B + Cx + D$$
$$x^3 = Ax^3 + Bx^2 + (4A+C)x + 4B + D$$
$$A = 1; \quad B = 0$$
$$4A + C = 0$$
$$4(1) + C = 0$$
$$C = -4$$
$$4B + D = 0$$
$$4(0) + D = 0$$
$$D = 0$$

$$\frac{x^3}{(x^2+4)^2} = \frac{x}{x^2+4} + \frac{-4x}{(x^2+4)^2}$$

38. Find the partial fraction decomposition:

$$\frac{x^2}{(x^2+1)(x^2-1)} = \frac{x^2}{(x^2+1)(x-1)(x+1)}$$
$$= \frac{A}{x-1} + \frac{B}{x+1} + \frac{Cx+D}{x^2+1}$$

Multiply both sides by $(x-1)(x+1)(x^2+1)$.

$$x^2 = A(x+1)(x^2+1) + B(x-1)(x^2+1)$$
$$+ (Cx+D)(x-1)(x+1)$$

Let $x = 1$, then

$1^2 = A(1+1)(1^2+1) + B(1-1)(1^2+1)$
$\qquad\qquad + (C(1)+D)(1-1)(1+1)$

$1 = 4A$

$A = \dfrac{1}{4}$

Let $x = -1$, then

$(-1)^2 = A(-1+1)((-1)^2+1)$
$\qquad\qquad + B(-1-1)((-1)^2+1)$
$\qquad\qquad + (C(-1)+D)(-1-1)(-1+1)$

$1 = -4B$

$B = -\dfrac{1}{4}$

Let $x = 0$, then

$0^2 = A(0+1)(0^2+1) + B(0-1)(0^2+1)$
$\qquad\qquad + (C(0)+D)(0-1)(0+1)$

$0 = A - B - D$

$0 = \dfrac{1}{4} - \left(-\dfrac{1}{4}\right) - D$

$D = \dfrac{1}{2}$

Let $x = 2$, then

$2^2 = A(2+1)(2^2+1) + B(2-1)(2^2+1)$
$\qquad\qquad + (C(2)+D)(2-1)(2+1)$

$4 = 15A + 5B + 6C + 3D$

$4 = 15\left(\dfrac{1}{4}\right) + 5\left(-\dfrac{1}{4}\right) + 6C + 3\left(\dfrac{1}{2}\right)$

$6C = 4 - \dfrac{15}{4} + \dfrac{5}{4} - \dfrac{3}{2}$

$6C = 0$

$C = 0$

$\dfrac{x^2}{\left(x^2+1\right)\left(x^2-1\right)} = \dfrac{\frac{1}{4}}{x-1} + \dfrac{-\frac{1}{4}}{x+1} + \dfrac{\frac{1}{2}}{x^2+1}$

39. Solve the first equation for y, substitute into the second equation and solve:

$\begin{cases} 2x + y + 3 = 0 \rightarrow y = -2x - 3 \\ x^2 + y^2 = 5 \end{cases}$

$x^2 + (-2x-3)^2 = 5 \rightarrow x^2 + 4x^2 + 12x + 9 = 5$

$5x^2 + 12x + 4 = 0 \rightarrow (5x+2)(x+2) = 0$

$\Rightarrow x = -\dfrac{2}{5} \qquad \text{or} \qquad x = -2$

$y = -\dfrac{11}{5} \qquad\qquad y = 1$

Solutions: $\left(-\dfrac{2}{5}, -\dfrac{11}{5}\right), (-2, 1)$.

40. Multiply each side of the second equation by 2 and add the equations to eliminate xy:

$\begin{cases} 2xy + y^2 = 10 \xrightarrow{} \quad 2xy + y^2 = 10 \\ -xy + 3y^2 = 2 \xrightarrow{2} \quad -2xy + 6y^2 = 4 \end{cases}$

$\qquad\qquad\qquad\qquad\qquad\qquad 7y^2 = 14$

$\qquad\qquad\qquad\qquad\qquad\qquad\quad y^2 = 2$

$\qquad\qquad\qquad\qquad\qquad\qquad\quad y = \pm\sqrt{2}$

If $y = \sqrt{2}$:

$2x\left(\sqrt{2}\right) + \left(\sqrt{2}\right)^2 = 10 \rightarrow 2\sqrt{2}x = 8 \rightarrow x = 2\sqrt{2}$

If $y = -\sqrt{2}$:

$2x\left(-\sqrt{2}\right) + \left(-\sqrt{2}\right)^2 = 10 \rightarrow -2\sqrt{2}x = 8$

$\qquad\qquad\qquad\qquad\qquad\qquad \rightarrow x = -2\sqrt{2}$

Solutions: $\left(2\sqrt{2}, \sqrt{2}\right), \left(-2\sqrt{2}, -\sqrt{2}\right)$

41. Substitute into the second equation into the first equation and solve:

$\begin{cases} x^2 + y^2 = 6y \\ \quad x^2 = 3y \end{cases}$

$3y + y^2 = 6y$

$y^2 - 3y = 0$

$y(y-3) = 0 \rightarrow y = 0 \text{ or } y = 3$

If $y = 0$: $\quad x^2 = 3(0) \rightarrow x^2 = 0 \rightarrow x = 0$

If $y = 3$: $\quad x^2 = 3(3) \rightarrow x^2 = 9 \rightarrow x = \pm 3$

Solutions: $(0, 0), (-3, 3), (3, 3)$.

42. Factor the second equation, solve for x, substitute into the first equation and solve:

$\begin{cases} 3x^2 + 4xy + 5y^2 = 8 \\ \quad x^2 + 3xy + 2y^2 = 0 \end{cases}$

$x^2 + 3xy + 2y^2 = 0$

$(x+2y)(x+y) = 0 \rightarrow x = -2y \text{ or } x = -y$

Substitute $x = -2y$ and solve:

$$3x^2 + 4xy + 5y^2 = 8$$
$$3(-2y)^2 + 4(-2y)y + 5y^2 = 8$$
$$12y^2 - 8y^2 + 5y^2 = 8$$
$$9y^2 = 8$$
$$y^2 = \frac{8}{9} \Rightarrow y = \pm\frac{2\sqrt{2}}{3}$$

Substitute $x = -y$ and solve:

$$3x^2 + 4xy + 5y^2 = 8$$
$$3(-y)^2 + 4(-y)y + 5y^2 = 8$$
$$3y^2 - 4y^2 + 5y^2 = 8$$
$$4y^2 = 8$$
$$y^2 = 2 \Rightarrow y = \pm\sqrt{2}$$

If $y = \frac{2\sqrt{2}}{3}$: $\quad x = -2\left(\frac{2\sqrt{2}}{3}\right) = \frac{-4\sqrt{2}}{3}$

If $y = \frac{-2\sqrt{2}}{3}$: $\quad x = -2\left(\frac{-2\sqrt{2}}{3}\right) = \frac{4\sqrt{2}}{3}$

If $y = \sqrt{2}$: $\quad x = -\sqrt{2}$

If $y = -\sqrt{2}$: $\quad x = \sqrt{2}$

Solutions:
$$\left(\frac{-4\sqrt{2}}{3}, \frac{2\sqrt{2}}{3}\right), \left(\frac{4\sqrt{2}}{3}, \frac{-2\sqrt{2}}{3}\right), \left(-\sqrt{2}, \sqrt{2}\right),$$
$$\left(\sqrt{2}, -\sqrt{2}\right)$$

43. $\begin{cases} x^2 - 3x + y^2 + y = -2 \\ \dfrac{x^2 - x}{y} + y + 1 = 0 \end{cases}$

Multiply each side of the second equation by $-y$ and add the equations to eliminate y:

$$x^2 - 3x + y^2 + y = -2$$
$$\underline{-x^2 + x - y^2 - y = 0}$$
$$-2x = -2$$
$$x = 1$$

If $x = 1$: $\quad 1^2 - 3(1) + y^2 + y = -2$
$$y^2 + y = 0$$
$$y(y+1) = 0$$
$$y = 0 \text{ or } y = -1$$

Note that $y \neq 0$ because that would cause

division by zero in the original system.
Solution: $(1, -1)$

44. $3x + 4y \leq 12$

Graph the line $3x + 4y = 12$. Use a solid line since the inequality uses $\leq$. Choose a test point not on the line, such as (0, 0). Since $3(0) + 4(0) \leq 12$ is true, shade the side of the line containing (0, 0).

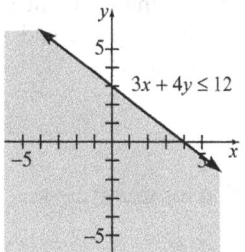

45. $y \leq x^2$

Graph the parabola $y = x^2$. Use a solid curve since the inequality uses $\leq$. Choose a test point not on the parabola, such as (0, 1). Since $0 \leq 1^2$ is false, shade the opposite side of the parabola from (0, 1).

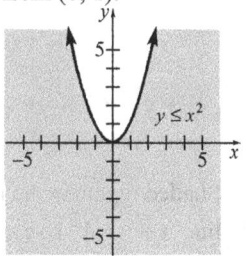

46. $\begin{cases} -2x + y \leq 2 \\ x + y \geq 2 \end{cases}$

Graph the line $-2x + y = 2$. Use a solid line since the inequality uses $\leq$. Choose a test point not on the line, such as (0, 0). Since $-2(0) + 0 \leq 2$ is true, shade the side of the line containing (0, 0). Graph the line $x + y = 2$. Use a solid line since the inequality uses $\geq$. Choose a test point not on the line, such as (0, 0). Since $0 + 0 \geq 2$ is false, shade the opposite side of the line from (0, 0). The overlapping region is the solution.

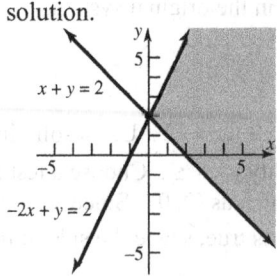

The graph is unbounded. Find the vertices:
To find the intersection of $x + y = 2$ and
$-2x + y = 2$, solve the system:

$$\begin{cases} x + y = 2 \\ -2x + y = 2 \end{cases}$$

Solve the first equation for x: $x = 2 - y$.
Substitute and solve:

$$-2(2 - y) + y = 2$$
$$-4 + 2y + y = 2$$
$$3y = 6$$
$$y = 2$$
$$x = 2 - 2 = 0$$

The point of intersection is $(0, 2)$.
The corner point is $(0, 2)$.

47. $\begin{cases} x \geq 0 \\ y \geq 0 \\ x + y \leq 4 \\ 2x + 3y \leq 6 \end{cases}$

Graph $x \geq 0$; $y \geq 0$. Shaded region is the first
quadrant. Graph the line $x + y = 4$. Use a solid
line since the inequality uses $\leq$. Choose a test
point not on the line, such as $(0, 0)$. Since
$0 + 0 \leq 4$ is true, shade the side of the line
containing $(0, 0)$. Graph the line $2x + 3y = 6$.
Use a solid line since the inequality uses $\leq$.
Choose a test point not on the line, such as $(0, 0)$.
Since $2(0) + 3(0) \leq 6$ is true, shade the side of the
line containing $(0, 0)$.

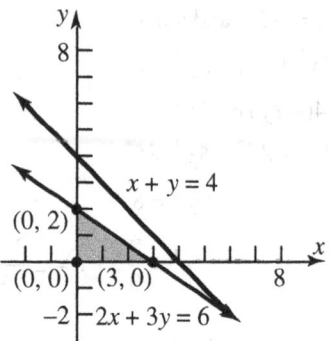

The overlapping region is the solution. The graph
is bounded. Find the vertices: The x-axis and y-
axis intersect at $(0, 0)$. The intersection of
$2x + 3y = 6$ and the y-axis is $(0, 2)$. The
intersection of $2x + 3y = 6$ and the x-axis is
$(3, 0)$. The three corner points are $(0, 0)$, $(0, 2)$,
and $(3, 0)$.

48. $\begin{cases} x \geq 0 \\ y \geq 0 \\ 2x + y \leq 8 \\ x + 2y \geq 2 \end{cases}$

Graph $x \geq 0$; $y \geq 0$. Shaded region is the first
quadrant. Graph the line $2x + y = 8$. Use a solid
line since the inequality uses $\leq$. Choose a test
point not on the line, such as $(0, 0)$. Since
$2(0) + 0 \leq 8$ is true, shade the side of the line
containing $(0, 0)$. Graph the line $x + 2y = 2$. Use
a solid line since the inequality uses $\geq$. Choose a
test point not on the line, such as $(0, 0)$. Since
$0 + 2(0) \geq 2$ is false, shade the opposite side of the
line from $(0, 0)$.

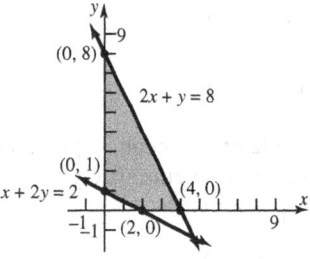

The overlapping region is the solution. The graph
is bounded. Find the vertices: The intersection of
$x + 2y = 2$ and the y-axis is $(0, 1)$. The
intersection of $x + 2y = 2$ and the x-axis is $(2, 0)$.
The intersection of $2x + y = 8$ and the y-axis is
$(0, 8)$. The intersection of $2x + y = 8$ and the x-
axis is $(4, 0)$. The four corner points are $(0, 1)$,
$(0, 8)$, $(2, 0)$, and $(4, 0)$.

49. Graph the system of inequalities:
$$\begin{cases} x^2 + y^2 \le 16 \\ x + y \ge 2 \end{cases}$$

Graph the circle $x^2 + y^2 = 16$. Use a solid line since the inequality uses $\le$. Choose a test point not on the circle, such as (0, 0). Since $0^2 + 0^2 \le 16$ is true, shade the side of the circle containing (0, 0).

Graph the line $x + y = 2$. Use a solid line since the inequality uses $\ge$. Choose a test point not on the line, such as (0, 0). Since $0 + 0 \ge 2$ is false, shade the opposite side of the line from (0, 0). The overlapping region is the solution.

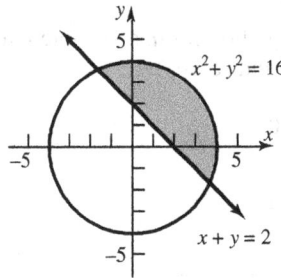

50. Graph the system of inequalities:
$$\begin{cases} x^2 + y^2 \le 25 \\ xy \le 4 \end{cases}$$

Graph the circle $x^2 + y^2 = 25$. Use a solid line since the inequality uses $\le$. Choose a test point not on the circle, such as (0, 0). Since $0^2 + 0^2 \le 25$ is true, shade the side of the circle containing (0, 0).

Graph the hyperbola $xy = 4$. Use a solid line since the inequality uses $\le$. Choose a test point not on the hyperbola, such as (1, 2). Since $1 \cdot 2 \le 4$ is true, shade the same side of the hyperbola as (1, 2). The overlapping region is the solution.

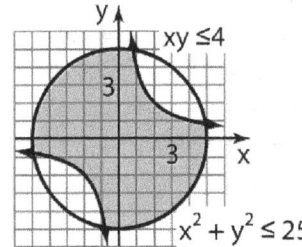

51. Maximize $z = 3x + 4y$ subject to $x \ge 0$, $y \ge 0$, $3x + 2y \ge 6$, $x + y \le 8$. Graph the constraints.

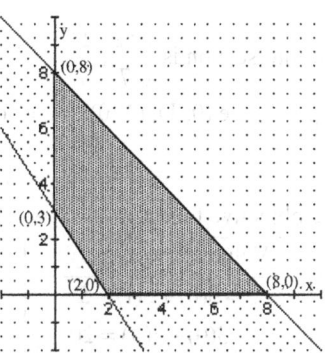

The corner points are (0, 3), (2, 0), (0, 8), (8, 0). Evaluate the objective function:

Vertex	Value of $z = 3x + 4y$
(0, 3)	$z = 3(0) + 4(3) = 12$
(0, 8)	$z = 3(0) + 4(8) = 32$
(2, 0)	$z = 3(2) + 4(0) = 6$
(8, 0)	$z = 3(8) + 4(0) = 24$

The maximum value is 32 at (0, 8).

52. Minimize $z = 3x + 5y$ subject to $x \ge 0$, $y \ge 0$, $x + y \ge 1$, $3x + 2y \le 12$, $x + 3y \le 12$.

Graph the constraints.

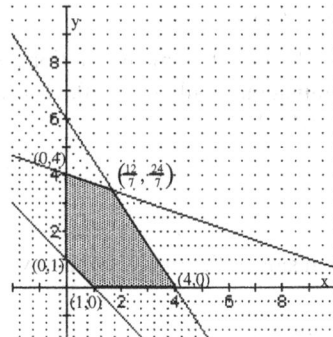

To find the intersection of $3x + 2y = 12$ and $x + 3y = 12$, solve the system:
$$\begin{cases} 3x + 2y = 12 \\ x + 3y = 12 \end{cases}$$

Solve the second equation for x: $x = 12 - 3y$

Substitute and solve:
$$3(12 - 3y) + 2y = 12$$
$$36 - 9y + 2y = 12$$
$$-7y = -24$$
$$y = \frac{24}{7}$$

$$x = 12 - 3\left(\frac{24}{7}\right) = 12 - \frac{72}{7} = \frac{12}{7}$$

The point of intersection is $\left(\dfrac{12}{7}, \dfrac{24}{7}\right)$.

The corner points are (0, 1), (1, 0), (0, 4), (4, 0), $\left(\dfrac{12}{7}, \dfrac{24}{7}\right)$.

Evaluate the objective function:

Vertex	Value of $z = 3x + 5y$
(0, 1)	$z = 3(0) + 5(1) = 5$
(0, 4)	$z = 3(0) + 5(4) = 20$
(1, 0)	$z = 3(1) + 5(0) = 3$
(4, 0)	$z = 3(4) + 5(0) = 12$
$\left(\dfrac{12}{7}, \dfrac{24}{7}\right)$	$z = 3\left(\dfrac{12}{7}\right) + 5\left(\dfrac{24}{7}\right) = \dfrac{156}{7}$

The minimum value is 3 at (1, 0).

53. $\begin{cases} 2x + 5y = 5 \\ 4x + 10y = A \end{cases}$

Multiply each side of the first equation by -2 and eliminate x:

$\begin{cases} -4x - 10y = -10 \\ 4x + 10y = A \end{cases}$

$0 = A - 10$

If there are to be infinitely many solutions, the result of elimination should be $0 = 0$. Therefore, $A - 10 = 0$ or $A = 10$.

54. $\begin{cases} 2x + 5y = 5 \\ 4x + 10y = A \end{cases}$

Multiply each side of the first equation by -2 and eliminate x:

$\begin{cases} -4x - 10y = -10 \\ 4x + 10y = A \end{cases}$

$0 = A - 10$

If the system is to be inconsistent, the result of elimination should be $0 =$ any number except 0. Therefore, $A - 10 \neq 0$ or $A \neq 10$.

55. $y = ax^2 + bx + c$

At (0, 1) the equation becomes:
$1 = a(0)^2 + b(0) + c$
$c = 1$

At (1, 0) the equation becomes:
$0 = a(1)^2 + b(1) + c$
$0 = a + b + c$
$a + b + c = 0$

At (−2, 1) the equation becomes:
$1 = a(-2)^2 + b(-2) + c$
$1 = 4a - 2b + c$
$4a - 2b + c = 1$

The system of equations is:
$\begin{cases} a + b + c = 0 \\ 4a - 2b + c = 1 \\ c = 1 \end{cases}$

Substitute $c = 1$ into the first and second equations and simplify:

$a + b + 1 = 0 \qquad 4a - 2b + 1 = 1$
$a + b = -1 \qquad\quad 4a - 2b = 0$
$a = -b - 1$

Solve the first equation for a, substitute into the second equation and solve:

$4(-b - 1) - 2b = 0$
$-4b - 4 - 2b = 0$
$-6b = 4$
$b = -\dfrac{2}{3}$

$a = \dfrac{2}{3} - 1 = -\dfrac{1}{3}$

The quadratic function is $y = -\dfrac{1}{3}x^2 - \dfrac{2}{3}x + 1$.

56. Let $x =$ the number of pounds of coffee that costs $6.00 per pound, and let $y =$ the number of pounds of coffee that costs $9.00 per pound. Then $x + y = 100$ represents the total amount of coffee in the blend. The value of the blend will be represented by the equation: $6x + 9y = 6.90(100)$. Solve the system of equations:

$\begin{cases} x + y = 100 \\ 6x + 9y = 690 \end{cases}$

Solve the first equation for y: $y = 100 - x$.

Solve by substitution:
$6x + 9(100 - x) = 690$
$6x + 900 - 9x = 690$
$-3x = -210$
$x = 70$
$y = 100 - 70 = 30$

The blend is made up of 70 pounds of the $6.00-per-pound coffee and 30 pounds of the $9.00-per-pound coffee.

57. Let x = the number of small boxes, let y = the number of medium boxes, and let z = the number of large boxes.
Oatmeal raisin equation: $x+2y+2z=15$
Chocolate chip equation: $x+y+2z=10$
Shortbread equation: $y+3z=11$

$$\begin{cases} x+2y+2z=15 \\ x+y+2z=10 \\ y+3z=11 \end{cases}$$

Multiply each side of the second equation by -1 and add to the first equation to eliminate x:

$$\begin{cases} x+2y+2z=15 \\ -x-y-2z=-10 \end{cases}$$
$$y+3z=11$$
$$y=5$$

Substituting and solving for the other variables:

$$5+3z=11 \qquad x+5+2(2)=10$$
$$3z=6 \qquad\qquad x+9=10$$
$$z=2 \qquad\qquad x=1$$

Thus, 1 small box, 5 medium boxes, and 2 large boxes of cookies should be purchased.

58. a. Let x = the number of lower-priced packages, and let y = the number of quality packages.
Peanut inequality: $\quad 8x+6y \le 120(16)$
$$4x+3y \le 960$$
Cashew inequality: $\quad 4x+6y \le 72(16)$
$$2x+3y \le 576$$

The system of inequalities is:

$$\begin{cases} x \ge 0 \\ y \ge 0 \\ 4x+3y \le 960 \\ 2x+3y \le 576 \end{cases}$$

b. Graphing:

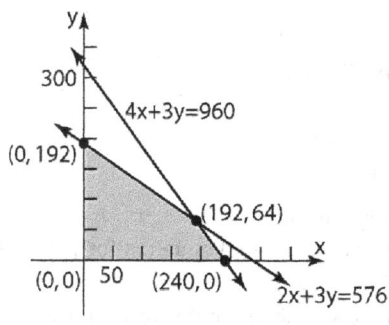

To find the intersection of $2x+3y=576$ and $4x+3y=960$, solve the system:

$$\begin{cases} 4x+3y=960 \\ 2x+3y=576 \end{cases}$$

Subtract the second equation from the first:

$$4x+3y=960$$
$$\underline{-2x-3y=576}$$
$$2x=384$$
$$x=192$$

Substitute and solve:
$$2(192)+3y=576$$
$$3y=192$$
$$y=64$$

The corner points are $(0, 0)$, $(0, 192)$, $(240, 0)$, and $(192, 64)$.

59. Let x = the speed of the boat in still water, and let y = the speed of the river current. The distance from Chiritza to the Flotel Orellana is 100 kilometers.

	Rate	Time	Distance
trip downstream	$x+y$	$5/2$	100
trip downstream	$x-y$	3	100

The system of equations is:

$$\begin{cases} \dfrac{5}{2}(x+y)=100 \\ 3(x-y)=100 \end{cases}$$

Multiply both sides of the first equation by 6, multiply both sides of the second equation by 5, and add the results.

$$15x+15y=600$$
$$\underline{15x-15y=500}$$
$$30x=1100$$
$$x=\frac{1100}{30}=\frac{110}{3}$$

$$3\left(\frac{110}{3}\right)-3y=100$$
$$110-3y=100$$
$$10=3y$$
$$y=\frac{10}{3}$$

The speed of the boat is $110/3 \approx 36.67$ km/hr; the speed of the current is $10/3 \approx 3.33$ km/hr.

60. Let x = the number of hours for Bruce to do the job alone, let y = the number of hours for Bryce to do the job alone, and let z = the number of hours for Marty to do the job alone. Then $1/x$ represents the fraction of the job that Bruce does in one hour.

$1/y$ represents the fraction of the job that Bryce does in one hour.

$1/z$ represents the fraction of the job that Marty does in one hour.

The equation representing Bruce and Bryce working together is:

$$\frac{1}{x}+\frac{1}{y}=\frac{1}{(4/3)}=\frac{3}{4}=0.75$$

The equation representing Bryce and Marty working together is:

$$\frac{1}{y}+\frac{1}{z}=\frac{1}{(8/5)}=\frac{5}{6}=0.625$$

The equation representing Bruce and Marty working together is:

$$\frac{1}{x}+\frac{1}{z}=\frac{1}{(8/3)}=\frac{3}{8}=0.375$$

Solve the system of equations:

$$\begin{cases} x^{-1}+y^{-1}=0.75 \\ y^{-1}+z^{-1}=0.625 \\ x^{-1}+z^{-1}=0.375 \end{cases}$$

Let $u=x^{-1}$, $v=y^{-1}$, $w=z^{-1}$

$$\begin{cases} u+v=0.75 \\ v+w=0.625 \\ u+w=0.375 \end{cases}$$

Solve the first equation for u: $u=0.75-v$.
Solve the second equation for w: $w=0.625-v$.
Substitute into the third equation and solve:
$$(0.75-v)+(0.625-v)=0.375$$
$$-2v=-1$$
$$v=0.5$$
$u=0.75-0.5=0.25$
$w=0.625-0.5=0.125$
Solve for
$x, y,$ and $z: x=4,\ y=2,\ z=8$ (reciprocals)
Bruce can do the job in 4 hours, Bryce in 2 hours, and Marty in 8 hours.

61. Let x = the number of gasoline engines produced each week, and let y = the number of diesel engines produced each week. The total cost is:

$C=450x+550y$. Cost is to be minimized; thus, this is the objective function. The constraints are:
$20 \le x \le 60$ number of gasoline engines needed and capacity each week.
$15 \le y \le 40$ number of diesel engines needed and capacity each week.
$x+y \ge 50$ number of engines produced to prevent layoffs.
Graph the constraints.

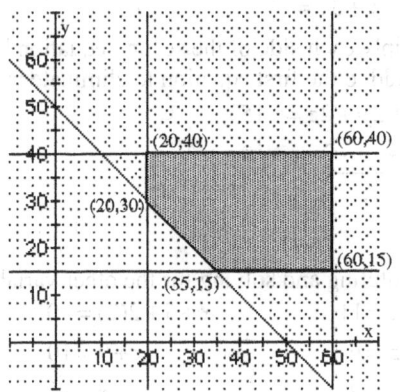

The corner points are (20, 30), (20, 40), (35, 15), (60, 15), (60, 40)
Evaluate the objective function:

Vertex	Value of $C=450x+550y$
(20, 30)	$C=450(20)+550(30)=25,500$
(20, 40)	$C=450(35)+550(40)=31,000$
(35, 15)	$C=450(35)+550(15)=24,000$
(60, 15)	$C=450(60)+550(15)=35,250$
(60, 40)	$C=450(60)+550(40)=49,000$

The minimum cost is $24,000, when 35 gasoline engines and 15 diesel engines are produced. The excess capacity is 15 gasoline engines, since only 20 gasoline engines had to be delivered.

62. Answers will vary.

Chapter 8 Test

1. $\begin{cases} -2x+y=-7 \\ 4x+3y=9 \end{cases}$

Substitution:
We solve the first equation for y, obtaining
$y=2x-7$
Next we substitute this result for y in the second

equation and solve for x.
$$4x+3y=9$$
$$4x+3(2x-7)=9$$
$$4x+6x-21=9$$
$$10x=30$$
$$x=\frac{30}{10}=3$$

We can now obtain the value for y by letting $x=3$ in our substitution for y.
$$y=2x-7$$
$$y=2(3)-7=6-7=-1$$

The solution of the system is $x=3$, $y=-1$ or $(3,-1)$.

Elimination:
Multiply each side of the first equation by 2 so that the coefficients of x in the two equations are negatives of each other. The result is the equivalent system
$$\begin{cases} -4x+2y=-14 \\ 4x+3y=9 \end{cases}$$

We can replace the second equation of this system by the sum of the two equations. The result is the equivalent system
$$\begin{cases} -4x+2y=-14 \\ 5y=-5 \end{cases}$$

Now we solve the second equation for y.
$$5y=-5$$
$$y=\frac{-5}{5}=-1$$

We back-substitute this value for y into the original first equation and solve for x.
$$-2x+y=-7$$
$$-2x+(-1)=-7$$
$$-2x=-6$$
$$x=\frac{-6}{-2}=3$$

The solution of the system is $x=3$, $y=-1$ or $(3,-1)$.

2. $\begin{cases} \dfrac{1}{3}x-2y=1 \\ 5x-30y=18 \end{cases}$

We choose to use the method of elimination and multiply the first equation by -15 to obtain the equivalent system

$$\begin{cases} -5x+30y=-15 \\ 5x-30y=18 \end{cases}$$

We replace the second equation by the sum of the two equations to obtain the equivalent system
$$\begin{cases} -5x+30y=-15 \\ 0=3 \end{cases}$$

The second equation is a contradiction and has no solution. This means that the system itself has no solution and is therefore inconsistent.

3. $\begin{cases} x-y+2z=5 & (1) \\ 3x+4y-z=-2 & (2) \\ 5x+2y+3z=8 & (3) \end{cases}$

We use the method of elimination and begin by eliminating the variable y from equation (2). Multiply each side of equation (1) by 4 and add the result to equation (2). This result becomes our new equation (2).

$$\begin{array}{ll} x-y+2z=5 & 4x-4y+8z=20 \\ 3x+4y-z=-2 & \underline{3x+4y-z=-2} \\ & 7x \quad +7z=18 \;\; (2) \end{array}$$

We now eliminate the variable y from equation (3) by multiplying each side of equation (1) by 2 and adding the result to equation (3). The result becomes our new equation (3).

$$\begin{array}{ll} x-y+2z=5 & 2x-2y+4z=10 \\ 5x+2y+3z=8 & \underline{5x+2y+3z=8} \\ & 7x \quad +7z=18 \;\; (3) \end{array}$$

Our (equivalent) system now looks like
$$\begin{cases} x-y+2z=5 & (1) \\ 7x \quad +7z=18 & (2) \\ 7x \quad +7z=18 & (3) \end{cases}$$

Treat equations (2) and (3) as a system of two equations containing two variables, and eliminate the x variable by multiplying each side of equation (2) by -1 and adding the result to equation (3). The result becomes our new equation (3).

$$\begin{array}{ll} 7x+7z=18 & -7x-7z=-18 \\ 7x+7z=18 & \underline{7x+7z=18} \\ & 0=0 \;\; (3) \end{array}$$

We now have the equivalent system
$$\begin{cases} x - y + 2z = 5 & (1) \\ 7x \quad\;\; + 7z = 18 & (2) \\ \qquad\qquad 0 = 0 & (3) \end{cases}$$

This is equivalent to a system of two equations with three variables. Since one of the equations contains three variables and one contains only two variables, the system will be dependent. There are infinitely many solutions.
We solve equation (2) for x and determine that
$x = -z + \dfrac{18}{7}$. Substitute this expression into equation (1) to obtain y in terms of z.
$$x - y + 2z = 5$$
$$\left(-z + \frac{18}{7}\right) - y + 2z = 5$$
$$-z + \frac{18}{7} - y + 2z = 5$$
$$-y + z = \frac{17}{7}$$
$$y = z - \frac{17}{7}$$

The solution is $x = -z + \dfrac{18}{7}$, $y = z - \dfrac{17}{7}$,
z is any real number or $\left\{ (x, y, z) \;\middle|\; x = -z + \dfrac{18}{7}, \right.$
$\left. y = z - \dfrac{17}{7}, z \text{ is any real number} \right\}$.

4. $\begin{cases} 3x + 2y - 8z = -3 & (1) \\ -x - \frac{2}{3}y + z = 1 & (2) \\ 6x - 3y + 15z = 8 & (3) \end{cases}$

We start by clearing the fraction in equation (2) by multiplying both sides of the equation by 3.
$$\begin{cases} 3x + 2y - 8z = -3 & (1) \\ -3x - 2y + 3z = 3 & (2) \\ 6x - 3y + 15z = 8 & (3) \end{cases}$$
We use the method of elimination and begin by eliminating the variable x from equation (2). The coefficients on x in equations (1) and (2) are negatives of each other so we simply add the two equations together. This result becomes our new equation (2).
$$3x + 2y - 8z = -3$$
$$\underline{-3x - 2y + 3z = 3}$$
$$\qquad\qquad\;\; -5z = 0 \quad (2)$$

We now eliminate the variable x from equation (3) by multiplying each side of equation (1) by -2 and adding the result to equation (3). The result becomes our new equation (3).
$$3x + 2y - 8z = -3 \qquad -6x - 4y + 16z = 6$$
$$6x - 3y + 15z = 8 \qquad \underline{6x - 3y + 15z = 8}$$
$$\qquad\qquad\qquad\qquad\qquad -7y + 31z = 14 \quad (3)$$

Our (equivalent) system now looks like
$$\begin{cases} 3x + 2y - 8z = -3 & (1) \\ \qquad\qquad -5z = 0 & (2) \\ -7y + 31z = 14 & (3) \end{cases}$$
We solve equation (2) for z by dividing both sides of the equation by -5.
$$-5z = 0$$
$$z = 0$$
Back-substitute $z = 0$ into equation (3) and solve for y.
$$-7y + 31z = 14$$
$$-7y + 31(0) = 14$$
$$-7y = 14$$
$$y = -2$$
Finally, back-substitute $y = -2$ and $z = 0$ into equation (1) and solve for x.
$$3x + 2y - 8z = -3$$
$$3x + 2(-2) - 8(0) = -3$$
$$3x - 4 = -3$$
$$3x = 1$$
$$x = \frac{1}{3}$$
The solution of the original system is
$$x = \frac{1}{3}, \; y = -2, \; z = 0 \text{ or } \left(\frac{1}{3}, -2, 0\right).$$

5. $\begin{cases} 4x - 5y + z = 0 \\ -2x - y + 6 = -19 \\ x + 5y - 5z = 10 \end{cases}$

We first check the equations to make sure that all variable terms are on the left side of the equation and the constants are on the right side. If a variable is missing, we put it in with a coefficient of 0. Our system can be rewritten as
$$\begin{cases} 4x - 5y + z = 0 \\ -2x - y + 0z = -25 \\ x + 5y - 5z = 10 \end{cases}$$

The augmented matrix is

$$\begin{bmatrix} 4 & -5 & 1 & | & 0 \\ -2 & -1 & 0 & | & -25 \\ 1 & 5 & -5 & | & 10 \end{bmatrix}$$

6. The matrix has three rows and represents a system with three equations. The three columns to the left of the vertical bar indicate that the system has three variables. We can let x, y, and z denote these variables. The column to the right of the vertical bar represents the constants on the right side of the equations. The system is

$$\begin{cases} 3x + 2y + 4z = -6 \\ 1x + 0y + 8z = 2 \\ -2x + 1y + 3z = -11 \end{cases} \quad \text{or} \quad \begin{cases} 3x + 2y + 4z = -6 \\ x + 8z = 2 \\ -2x + y + 3z = -11 \end{cases}$$

7. $2A + C = 2\begin{bmatrix} 1 & -1 \\ 0 & -4 \\ 3 & 2 \end{bmatrix} + \begin{bmatrix} 4 & 6 \\ 1 & -3 \\ -1 & 8 \end{bmatrix}$

$= \begin{bmatrix} 2 & -2 \\ 0 & -8 \\ 6 & 4 \end{bmatrix} + \begin{bmatrix} 4 & 6 \\ 1 & -3 \\ -1 & 8 \end{bmatrix} = \begin{bmatrix} 6 & 4 \\ 1 & -11 \\ 5 & 12 \end{bmatrix}$

8. $A - 3C = \begin{bmatrix} 1 & -1 \\ 0 & -4 \\ 3 & 2 \end{bmatrix} - 3\begin{bmatrix} 4 & 6 \\ 1 & -3 \\ -1 & 8 \end{bmatrix}$

$= \begin{bmatrix} 1 & -1 \\ 0 & -4 \\ 3 & 2 \end{bmatrix} - \begin{bmatrix} 12 & 18 \\ 3 & -9 \\ -3 & 24 \end{bmatrix} = \begin{bmatrix} -11 & -19 \\ -3 & 5 \\ 6 & -22 \end{bmatrix}$

9. Here we are taking the product of a 3×2 matrix and a 2×3 matrix. Since the number of columns in the first matrix is the same as the number of rows in the second matrix (2 in both cases), the operation can be performed and will result in a 3×3 matrix.

$CB = \begin{bmatrix} 4 & 6 \\ 1 & -3 \\ -1 & 8 \end{bmatrix}\begin{bmatrix} 1 & -2 & 5 \\ 0 & 3 & 1 \end{bmatrix}$

$= \begin{bmatrix} 4 \cdot 1 + 6 \cdot 0 & 4(-2) + 6 \cdot 3 & 4 \cdot 5 + 6 \cdot 1 \\ 1 \cdot 1 + (-3)0 & 1(-2) + (-3)3 & 1 \cdot 5 + (-3)1 \\ (-1)1 + 8 \cdot 0 & (-1)(-2) + 8 \cdot 3 & (-1)5 + 8 \cdot 1 \end{bmatrix}$

$= \begin{bmatrix} 4 & 10 & 26 \\ 1 & -11 & 2 \\ -1 & 26 & 3 \end{bmatrix}$

10. Here we are taking the product of a 2×3 matrix and a 3×2 matrix. Since the number of columns in the first matrix is the same as the number of rows in the second matrix (3 in both cases), the operation can be performed and will result in a 2×2 matrix.

$BA = \begin{bmatrix} 1 & -2 & 5 \\ 0 & 3 & 1 \end{bmatrix}\begin{bmatrix} 1 & -1 \\ 0 & -4 \\ 3 & 2 \end{bmatrix}$

$= \begin{bmatrix} 1 \cdot 1 + (-2) \cdot 0 + 5 \cdot 3 & 1 \cdot (-1) + (-2) \cdot (-4) + 5 \cdot 2 \\ 0 \cdot 1 + 3 \cdot 0 + 1 \cdot 3 & 0 \cdot (-1) + 3(-4) + 1 \cdot 2 \end{bmatrix}$

$= \begin{bmatrix} 16 & 17 \\ 3 & -10 \end{bmatrix}$

11. We first form the matrix

$$[A \mid I_2] = \begin{bmatrix} 3 & 2 & | & 1 & 0 \\ 5 & 4 & | & 0 & 1 \end{bmatrix}$$

Next we use row operations to transform $[A \mid I_2]$ into reduced row echelon form.

$\begin{bmatrix} 3 & 2 & | & 1 & 0 \\ 5 & 4 & | & 0 & 1 \end{bmatrix} \rightarrow \begin{bmatrix} 1 & \frac{2}{3} & | & \frac{1}{3} & 0 \\ 5 & 4 & | & 0 & 1 \end{bmatrix} \quad \left(R_1 = \frac{1}{3}r_1\right)$

$\rightarrow \begin{bmatrix} 1 & \frac{2}{3} & | & \frac{1}{3} & 0 \\ 0 & \frac{2}{3} & | & -\frac{5}{3} & 1 \end{bmatrix} \quad \left(R_2 = -5r_1 + r_2\right)$

$\rightarrow \begin{bmatrix} 1 & \frac{2}{3} & | & \frac{1}{3} & 0 \\ 0 & 1 & | & -\frac{5}{2} & \frac{3}{2} \end{bmatrix} \quad \left(R_2 = \frac{3}{2}r_2\right)$

$\rightarrow \begin{bmatrix} 1 & 0 & | & 2 & -1 \\ 0 & 1 & | & -\frac{5}{2} & \frac{3}{2} \end{bmatrix} \quad \left(R_1 = -\frac{2}{3}r_2 + r_1\right)$

Therefore, $A^{-1} = \begin{bmatrix} 2 & -1 \\ -\frac{5}{2} & \frac{3}{2} \end{bmatrix}$.

12. We first form the matrix

$$[B \mid I_3] = \begin{bmatrix} 1 & -1 & 1 & | & 1 & 0 & 0 \\ 2 & 5 & -1 & | & 0 & 1 & 0 \\ 2 & 3 & 0 & | & 0 & 0 & 1 \end{bmatrix}$$

Next we use row operations to transform $[B \mid I_3]$ into reduced row echelon form.

$$\begin{bmatrix} 1 & -1 & 1 & | & 1 & 0 & 0 \\ 2 & 5 & -1 & | & 0 & 1 & 0 \\ 2 & 3 & 0 & | & 0 & 0 & 1 \end{bmatrix}$$

$$\rightarrow \begin{bmatrix} 1 & -1 & 1 & | & 1 & 0 & 0 \\ 0 & 7 & -3 & | & -2 & 1 & 0 \\ 0 & 5 & -2 & | & -2 & 0 & 1 \end{bmatrix} \begin{pmatrix} R_2 = -2r_1 + r_2 \\ R_3 = -2r_1 + r_3 \end{pmatrix}$$

$$\rightarrow \begin{bmatrix} 1 & -1 & 1 & | & 1 & 0 & 0 \\ 0 & 1 & -\frac{3}{7} & | & -\frac{2}{7} & \frac{1}{7} & 0 \\ 0 & 5 & -2 & | & -2 & 0 & 1 \end{bmatrix} \left(R_2 = \frac{1}{7}r_2 \right)$$

$$\rightarrow \begin{bmatrix} 1 & 0 & \frac{4}{7} & | & \frac{5}{7} & \frac{1}{7} & 0 \\ 0 & 1 & -\frac{3}{7} & | & -\frac{2}{7} & \frac{1}{7} & 0 \\ 0 & 0 & \frac{1}{7} & | & -\frac{4}{7} & -\frac{5}{7} & 1 \end{bmatrix} \begin{pmatrix} R_1 = r_2 + r_1 \\ R_3 = -5r_2 + r_3 \end{pmatrix}$$

$$\rightarrow \begin{bmatrix} 1 & 0 & \frac{4}{7} & | & \frac{5}{7} & \frac{1}{7} & 0 \\ 0 & 1 & -\frac{3}{7} & | & -\frac{2}{7} & \frac{1}{7} & 0 \\ 0 & 0 & 1 & | & -4 & -5 & 7 \end{bmatrix} (R_3 = 7r_3)$$

$$\rightarrow \begin{bmatrix} 1 & 0 & 0 & | & 3 & 3 & -4 \\ 0 & 1 & 0 & | & -2 & -2 & 3 \\ 0 & 0 & 1 & | & -4 & -5 & 7 \end{bmatrix} \begin{pmatrix} R_1 = -\frac{4}{7}r_3 + r_1 \\ R_2 = \frac{3}{7}r_3 + r_2 \end{pmatrix}$$

Thus, $B^{-1} = \begin{bmatrix} 3 & 3 & -4 \\ -2 & -2 & 3 \\ -4 & -5 & 7 \end{bmatrix}$

13. $\begin{cases} 6x + 3y = 12 \\ 2x - y = -2 \end{cases}$

We start by writing the augmented matrix for the system.

$$\begin{bmatrix} 6 & 3 & | & 12 \\ 2 & -1 & | & -2 \end{bmatrix}$$

Next we use row operations to transform the augmented matrix into row echelon form.

$$\begin{bmatrix} 6 & 3 & | & 12 \\ 2 & -1 & | & -2 \end{bmatrix} \rightarrow \begin{bmatrix} 2 & -1 & | & -2 \\ 6 & 3 & | & 12 \end{bmatrix} \begin{pmatrix} R_1 = r_2 \\ R_2 = r_1 \end{pmatrix}$$

$$\rightarrow \begin{bmatrix} 1 & -\frac{1}{2} & | & -1 \\ 6 & 3 & | & 12 \end{bmatrix} \left(R_1 = \frac{1}{2}r_1 \right)$$

$$\rightarrow \begin{bmatrix} 1 & -\frac{1}{2} & | & -1 \\ 0 & 6 & | & 18 \end{bmatrix} (R_2 = -6r_1 + r_2)$$

$$\rightarrow \begin{bmatrix} 1 & -\frac{1}{2} & | & -1 \\ 0 & 1 & | & 3 \end{bmatrix} \left(R_2 = \frac{1}{6}r_2 \right)$$

$$\rightarrow \begin{bmatrix} 1 & 0 & | & \frac{1}{2} \\ 0 & 1 & | & 3 \end{bmatrix} \left(R_2 = \frac{1}{2}r_2 + r_1 \right)$$

The solution of the system is $x = \frac{1}{2}$, $y = 3$ or $\left(\frac{1}{2}, 3 \right)$

14. $\begin{cases} x + \dfrac{1}{4}y = 7 \\ 8x + 2y = 56 \end{cases}$

We start by writing the augmented matrix for the system.

$$\begin{bmatrix} 1 & \frac{1}{4} & | & 7 \\ 8 & 2 & | & 56 \end{bmatrix}$$

Next we use row operations to transform the augmented matrix into row echelon form.

$$\begin{bmatrix} 1 & \frac{1}{4} & | & 7 \\ 8 & 2 & | & 56 \end{bmatrix} R_2 = -8R_1 + r_2$$

$$\begin{bmatrix} 1 & \frac{1}{4} & | & 7 \\ 0 & 0 & | & 0 \end{bmatrix}$$

The augmented matrix is now in row echelon form. Because the bottom row consists entirely of 0's, the system actually consists of one equation in two variables. The system is dependent and therefore has an infinite number of solutions. Any ordered pair satisfying the equation $x + \dfrac{1}{4}y = 7$, or $y = -4x + 28$, is a solution to the system.

15. $\begin{cases} x+2y+4z = -3 \\ 2x+7y+15z = -12 \\ 4x+7y+13z = -10 \end{cases}$

We start by writing the augmented matrix for the system.

$$\begin{bmatrix} 1 & 2 & 4 & | & -3 \\ 2 & 7 & 15 & | & -12 \\ 4 & 7 & 13 & | & -10 \end{bmatrix}$$

Next we use row operations to transform the augmented matrix into row echelon form.

$$\begin{bmatrix} 1 & 2 & 4 & | & -3 \\ 2 & 7 & 15 & | & -12 \\ 4 & 7 & 13 & | & -10 \end{bmatrix}$$

$$\rightarrow \begin{bmatrix} 1 & 2 & 4 & | & -3 \\ 0 & 3 & 7 & | & -6 \\ 0 & -1 & -3 & | & 2 \end{bmatrix} \begin{pmatrix} R_2 = -2r_1 + r_2 \\ R_3 = -4r_1 + r_3 \end{pmatrix}$$

$$\rightarrow \begin{bmatrix} 1 & 2 & 4 & | & -3 \\ 0 & 1 & 3 & | & -2 \\ 0 & 3 & 7 & | & -6 \end{bmatrix} \begin{pmatrix} R_2 = -r_3 \\ R_3 = r_2 \end{pmatrix}$$

$$\rightarrow \begin{bmatrix} 1 & 2 & 4 & | & -3 \\ 0 & 1 & 3 & | & -2 \\ 0 & 0 & -2 & | & 0 \end{bmatrix} \left(R_3 = -3r_2 + r_3 \right)$$

$$\rightarrow \begin{bmatrix} 1 & 2 & 4 & | & -3 \\ 0 & 1 & 3 & | & -2 \\ 0 & 0 & 1 & | & 0 \end{bmatrix} \left(R_3 = -\frac{1}{2}r_3 \right)$$

The matrix is now in row echelon form. The last row represents the equation $z = 0$. Using $z = 0$ we back-substitute into the equation $y + 3z = -2$ (from the second row) and obtain

$y + 3z = -2$

$y + 3(0) = -2$

$y = -2$

Using $y = -2$ and $z = 0$, we back-substitute into the equation $x + 2y + 4z = -3$ (from the first row) and obtain

$x + 2y + 4z = -3$

$x + 2(-2) + 4(0) = -3$

$x = 1$

The solution is $x = 1$, $y = -2$, $z = 0$ or $(1, -2, 0)$.

16. $\begin{cases} 2x+2y-3z = 5 \\ x-y+2z = 8 \\ 3x+5y-8z = -2 \end{cases}$

We start by writing the augmented matrix for the system.

$$\begin{bmatrix} 2 & 2 & -3 & | & 5 \\ 1 & -1 & 2 & | & 8 \\ 3 & 5 & -8 & | & -2 \end{bmatrix}$$

Next we use row operations to transform the augmented matrix into row echelon form.

$$\begin{bmatrix} 2 & 2 & -3 & | & 5 \\ 1 & -1 & 2 & | & 8 \\ 3 & 5 & -8 & | & -2 \end{bmatrix}$$

$$= \begin{bmatrix} 1 & -1 & 2 & | & 8 \\ 2 & 2 & -3 & | & 5 \\ 3 & 5 & -8 & | & -2 \end{bmatrix} \begin{pmatrix} R_1 = r_2 \\ R_2 = r_1 \end{pmatrix}$$

$$= \begin{bmatrix} 1 & -1 & 2 & | & 8 \\ 0 & 4 & -7 & | & -11 \\ 0 & 8 & -14 & | & -26 \end{bmatrix} \begin{pmatrix} R_2 = -2r_1 + r_2 \\ R_3 = -3r_1 + r_3 \end{pmatrix}$$

$$= \begin{bmatrix} 1 & -1 & 2 & | & 8 \\ 0 & 1 & -\frac{7}{4} & | & -\frac{11}{4} \\ 0 & 8 & -14 & | & -26 \end{bmatrix} \left(R_2 = \frac{1}{4}r_2 \right)$$

$$= \begin{bmatrix} 1 & -1 & 2 & | & 8 \\ 0 & 1 & -\frac{7}{4} & | & -\frac{11}{4} \\ 0 & 0 & 0 & | & -4 \end{bmatrix} \left(R_3 = -8r_2 + r_3 \right)$$

The last row represents the equation $0 = -4$ which is a contradiction. Therefore, the system has no solution and is be inconsistent.

17. $\begin{vmatrix} -2 & 5 \\ 3 & 7 \end{vmatrix} = (-2)(7) - (5)(3) = -14 - 15 = -29$

18. $\begin{vmatrix} 2 & -4 & 6 \\ 1 & 4 & 0 \\ -1 & 2 & -4 \end{vmatrix}$

$= 2\begin{vmatrix} 4 & 0 \\ 2 & -4 \end{vmatrix} - (-4)\begin{vmatrix} 1 & 0 \\ -1 & -4 \end{vmatrix} + 6\begin{vmatrix} 1 & 4 \\ -1 & 2 \end{vmatrix}$

$= 2[4(-4) - 2(0)] + 4[1(-4) - (-1)(0)]$

$\qquad + 6[1(2) - (-1)4]$

$= 2(-16) + 4(-4) + 6(6)$

$= -32 - 16 + 36$

$= -12$

19. $\begin{cases} 4x + 3y = -23 \\ 3x - 5y = 19 \end{cases}$

The determinant D of the coefficients of the variables is

$D = \begin{vmatrix} 4 & 3 \\ 3 & -5 \end{vmatrix} = (4)(-5) - (3)(3) = -20 - 9 = -29$

Since $D \neq 0$, Cramer's Rule can be applied.

$D_x = \begin{vmatrix} -23 & 3 \\ 19 & -5 \end{vmatrix} = (-23)(-5) - (3)(19) = 58$

$D_y = \begin{vmatrix} 4 & -23 \\ 3 & 19 \end{vmatrix} = (4)(19) - (-23)(3) = 145$

$x = \dfrac{D_x}{D} = \dfrac{58}{-29} = -2$

$y = \dfrac{D_y}{D} = \dfrac{145}{-29} = -5$

The solution of the system is $x = -2$, $y = -5$ or $(-2, -5)$.

20. $\begin{cases} 4x - 3y + 2z = 15 \\ -2x + y - 3z = -15 \\ 5x - 5y + 2z = 18 \end{cases}$

The determinant D of the coefficients of the variables is

$D = \begin{vmatrix} 4 & -3 & 2 \\ -2 & 1 & -3 \\ 5 & -5 & 2 \end{vmatrix}$

$= 4 \begin{vmatrix} 1 & -3 \\ -5 & 2 \end{vmatrix} - (-3) \begin{vmatrix} -2 & -3 \\ 5 & 2 \end{vmatrix} + 2 \begin{vmatrix} -2 & 1 \\ 5 & -5 \end{vmatrix}$

$= 4(2 - 15) + 3(-4 + 15) + 2(10 - 5)$

$= 4(-13) + 3(11) + 2(5)$

$= -52 + 33 + 10$

$= -9$

Since $D \neq 0$, Cramer's Rule can be applied.

$D_x = \begin{vmatrix} 15 & -3 & 2 \\ -15 & 1 & -3 \\ 18 & -5 & 2 \end{vmatrix}$

$= 15 \begin{vmatrix} 1 & -3 \\ -5 & 2 \end{vmatrix} - (-3) \begin{vmatrix} -15 & -3 \\ 18 & 2 \end{vmatrix} + 2 \begin{vmatrix} -15 & 1 \\ 18 & -5 \end{vmatrix}$

$= 15(2 - 15) + 3(-30 + 54) + 2(75 - 18)$

$= 15(-13) + 3(24) + 2(57)$

$= -9$

$D_y = \begin{vmatrix} 4 & 15 & 2 \\ -2 & -15 & -3 \\ 5 & 18 & 2 \end{vmatrix}$

$= 4 \begin{vmatrix} -15 & -3 \\ 18 & 2 \end{vmatrix} - 15 \begin{vmatrix} -2 & -3 \\ 5 & 2 \end{vmatrix} + 2 \begin{vmatrix} -2 & -15 \\ 5 & 18 \end{vmatrix}$

$= 4(-30 + 54) - 15(-4 + 15) + 2(-36 + 75)$

$= 4(24) - 15(11) + 2(39)$

$= -9$

$D_z = \begin{vmatrix} 4 & -3 & 15 \\ -2 & 1 & -15 \\ 5 & -5 & 18 \end{vmatrix}$

$= 4 \begin{vmatrix} 1 & -15 \\ -5 & 18 \end{vmatrix} - (-3) \begin{vmatrix} -2 & -15 \\ 5 & 18 \end{vmatrix} + 15 \begin{vmatrix} -2 & 1 \\ 5 & -5 \end{vmatrix}$

$= 4(18 - 75) + 3(-36 + 75) + 15(10 - 5)$

$= 4(-57) + 3(39) + 15(5)$

$= -36$

$x = \dfrac{D_x}{D} = \dfrac{-9}{-9} = 1$, $y = \dfrac{D_y}{D} = \dfrac{9}{-9} = -1$,

$z = \dfrac{D_z}{D} = \dfrac{-36}{-9} = 4$

The solution of the system is $x = 1$, $y = -1$, $z = 4$ or $(1, -1, 4)$.

21. $\begin{cases} 3x^2 + y^2 = 12 \\ y^2 = 9x \end{cases}$

Substitute $9x$ for y^2 into the first equation and solve for x:

$3x^2 + (9x) = 12$

$3x^2 + 9x - 12 = 0$

$x^2 + 3x - 4 = 0$

$(x - 1)(x + 4) = 0$

$x = 1$ or $x = -4$

Back substitute these values into the second equation to determine y:

$x = 1$: $y^2 = 9(1) = 9$

$\qquad y = \pm 3$

$x = -4$: $y^2 = 9(-4) = -36$

$\qquad y = \pm\sqrt{-36}$ (not real)

The solutions of the system are $(1, -3)$ and $(1, 3)$.

22. $\begin{cases} 2y^2 - 3x^2 = 5 \\ \quad y - x = 1 \quad \Rightarrow \quad y = x + 1 \end{cases}$

Substitute $x + 1$ for y into the first equation and solve for x:

$$2(x+1)^2 - 3x^2 = 5$$
$$2(x^2 + 2x + 1) - 3x^2 = 5$$
$$2x^2 + 4x + 2 - 3x^2 = 5$$
$$-x^2 + 4x - 3 = 0$$
$$x^2 - 4x + 3 = 0$$
$$(x-1)(x-3) = 0$$
$$x = 1 \quad \text{or} \quad x = 3$$

Back substitute these values into the second equation to determine y:

$x = 1 : \ y = 1 + 1 = 2$

$x = 3 : \ y = 3 + 1 = 4$

The solutions of the system are $(1, \ 2)$ and $(3, \ 4)$.

23. $\begin{cases} x^2 + y^2 \le 100 \\ 4x - 3y \ge 0 \end{cases}$

Graph the circle $x^2 + y^2 = 100$. Use a solid curve since the inequality uses $\le$. Choose a test point not on the circle, such as (0, 0). Since $0^2 + 0^2 \le 100$ is true, shade the same side of the circle as (0, 0); that is, inside the circle.

Graph the line $4x - 3y = 0$. Use a solid line since the inequality uses $\ge$. Choose a test point not on the line, such as (0, 1). Since $4(0) - 3(1) \ge 0$ is false, shade the opposite side of the line from (0, 1). The overlapping region is the solution.

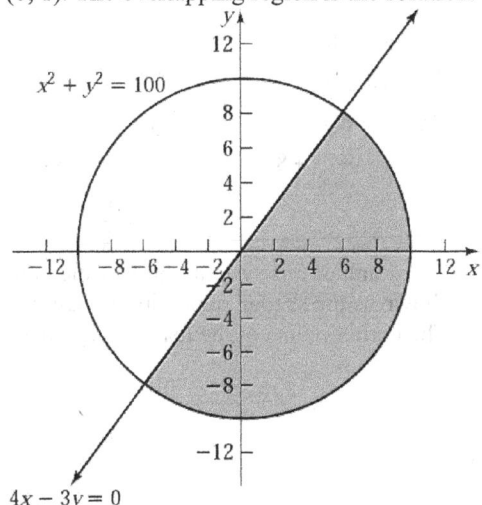

24. $\dfrac{3x+7}{(x+3)^2}$

The denominator contains the repeated linear factor $x + 3$. Thus, the partial fraction decomposition takes on the form

$$\frac{3x+7}{(x+3)^2} = \frac{A}{x+3} + \frac{B}{(x+3)^2}$$

Clear the fractions by multiplying both sides by $(x+3)^2$. The result is the identity

$$3x + 7 = A(x+3) + B$$

or

$$3x + 7 = Ax + (3A + B)$$

We equate coefficients of like powers of x to obtain the system

$$\begin{cases} 3 = A \\ 7 = 3A + B \end{cases}$$

Therefore, we have $A = 3$. Substituting this result into the second equation gives

$$7 = 3A + B$$
$$7 = 3(3) + B$$
$$-2 = B$$

Thus, the partial fraction decomposition is

$$\frac{3x+7}{(x+3)^2} = \frac{3}{x+3} + \frac{-2}{(x+3)^2}.$$

25. $\dfrac{4x^2 - 3}{x(x^2+3)^2}$

The denominator contains the linear factor x and the repeated irreducible quadratic factor $x^2 + 3$. The partial fraction decomposition takes on the form

$$\frac{4x^2-3}{x(x^2+3)^2} = \frac{A}{x} + \frac{Bx+C}{x^2+3} + \frac{Dx+E}{(x^2+3)^2}$$

We clear the fractions by multiplying both sides by $x(x^2+3)^2$ to obtain the identity

$$4x^2 - 3 = A(x^2+3)^2 + x(x^2+3)(Bx+C) + x(Dx+E)$$

Collecting like terms yields

$$4x^2 - 3 = (A+B)x^4 + Cx^3 + (6A + 3B + D)x^2$$
$$+ (3C + E)x + (9A)$$

Equating coefficients, we obtain the system

$$\begin{cases} A+B=0 \\ C=0 \\ 6A+3B+D=4 \\ 3C+E=0 \\ 9A=-3 \end{cases}$$

From the last equation we get $A=-\dfrac{1}{3}$.

Substituting this value into the first equation

gives $B=\dfrac{1}{3}$. From the second equation, we

know $C=0$. Substituting this value into the fourth equation yields $E=0$.

Substituting $A=-\dfrac{1}{3}$ and $B=\dfrac{1}{3}$ into the third

equation gives us

$$6\left(-\tfrac{1}{3}\right)+3\left(\tfrac{1}{3}\right)+D=4$$
$$-2+1+D=4$$
$$D=5$$

Therefore, the partial fraction decomposition is

$$\frac{4x^2-3}{x\left(x^2+3\right)^2}=\frac{-\dfrac{1}{3}}{x}+\frac{\dfrac{1}{3}x}{\left(x^2+3\right)}+\frac{5x}{\left(x^2+3\right)^2}$$

26. $\begin{cases} x\ge 0 \\ y\ge 0 \\ x+2y\ge 8 \\ 2x-3y\ge 2 \end{cases}$

The inequalities $x\ge 0$ and $y\ge 0$ require that the graph be in Quadrant I.
$$x+2y\ge 8$$
$$y\ge -\frac{1}{2}x+4$$

Test the point $(0,0)$.
$$x+2y\ge 8$$
$$0+2(0)\ge 8 \quad ?$$
$$0\ge 8 \text{ false}$$

The point $(0,0)$ is not a solution. Thus, the graph of the inequality $x+2y\ge 8$ includes the

half-plane above the line $y=-\dfrac{1}{2}x+4$. Because

the inequality is non-strict, the line is also part of the graph of the solution.

$$2x-3y\ge 2$$
$$y\le \frac{2}{3}x-\frac{2}{3}$$

Test the point $(0,0)$.
$$2x-3y\ge 2$$
$$2(0)-3(0)\ge 2 \quad ?$$
$$0\ge 2 \text{ false}$$

The point $(0,0)$ is not a solution. Thus, the graph of the inequality $2x-3y\ge 2$ includes the

half-plane below the line $y=\dfrac{2}{3}x-\dfrac{2}{3}$.

Because the inequality is non-strict, the line is also part of the graph of the solution. The overlapping shaded region (that is, the shaded region in the graph below) is the solution to the system of linear inequalities.

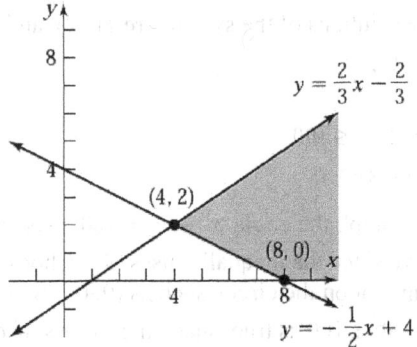

The graph is unbounded. The corner points are $(4,2)$ and $(8,0)$.

27. The objective function is $z=5x+8y$. We seek the largest value of z that can occur if x and y are solutions of the system of linear inequalities

$$\begin{cases} x\ge 0 \\ 2x+y\le 8 \\ x-3y\le -3 \end{cases}$$

$$2x+y=8 \qquad\qquad x-3y=-3$$
$$y=-2x+8 \qquad\qquad -3y=-x-3$$
$$\qquad\qquad\qquad\qquad y=\frac{1}{3}x+1$$

The graph of this system (the feasible points) is shown as the shaded region in the figure below. The corner points of the feasible region are

$(0,1)$, $(3,2)$, and $(0,8)$.

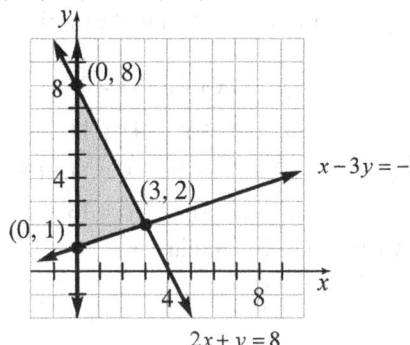

Corner point, (x,y)	Value of obj. function, z
$(0,1)$	$z = 5(0)+8(1) = 8$
$(3,2)$	$z = 5(3)+8(2) = 31$
$(0,8)$	$z = 5(0)+8(8) = 64$

From the table, we can see that the maximum value of z is 64, and it occurs at the point $(0,8)$.

28. Let j = unit price for flare jeans, c = unit price for camisoles, and t = unit price for t-shirts. The given information yields a system of equations with each of the three women yielding an equation.

$$\begin{cases} 2j+2c+4t = 90 & \text{(Megan)} \\ j \quad\quad +3t = 42.5 & \text{(Paige)} \\ j+3c+2t = 62 & \text{(Kara)} \end{cases}$$

We can solve this system by using matrices.

$$\begin{bmatrix} 2 & 2 & 4 & | & 90 \\ 1 & 0 & 3 & | & 42.5 \\ 1 & 3 & 2 & | & 62 \end{bmatrix} = \begin{bmatrix} 1 & 1 & 2 & | & 45 \\ 1 & 0 & 3 & | & 42.5 \\ 1 & 3 & 2 & | & 62 \end{bmatrix} \quad \left(R_1 = \tfrac{1}{2}r_1\right)$$

$$= \begin{bmatrix} 1 & 1 & 2 & | & 45 \\ 0 & -1 & 1 & | & -2.5 \\ 0 & 2 & 0 & | & 17 \end{bmatrix} \quad \left(\begin{matrix} R_2 = -r_1 + r_2 \\ R_3 = -r_1 + r_3 \end{matrix}\right)$$

$$= \begin{bmatrix} 1 & 1 & 2 & | & 45 \\ 0 & 1 & -1 & | & 2.5 \\ 0 & 2 & 0 & | & 17 \end{bmatrix} \quad \left(R_2 = -r_2\right)$$

$$= \begin{bmatrix} 1 & 0 & 3 & | & 42.5 \\ 0 & 1 & -1 & | & 2.5 \\ 0 & 0 & 2 & | & 12 \end{bmatrix} \quad \left(\begin{matrix} R_1 = -r_2 + r_1 \\ R_3 = -2r_2 + r_3 \end{matrix}\right)$$

$$= \begin{bmatrix} 1 & 0 & 3 & | & 42.5 \\ 0 & 1 & -1 & | & 2.5 \\ 0 & 0 & 1 & | & 6 \end{bmatrix} \quad \left(R_3 = \tfrac{1}{2}r_3\right)$$

The last row represents the equation $z = 6$.

Substituting this result into $y - z = 2.5$ (from the second row) gives
$$y - z = 2.5$$
$$y - 6 = 2.5$$
$$y = 8.5$$

Substituting $z = 6$ into $x + 3z = 42.5$ (from the first row) gives
$$x + 3z = 42.5$$
$$x + 3(6) = 42.5$$
$$x = 24.5$$

Thus, flare jeans cost $24.50, camisoles cost $8.50, and t-shirts cost $6.00.

Chapter 8 Cumulative Review

1. $2x^2 - x = 0$
$$x(2x-1) = 0$$
$$x = 0 \text{ or } 2x - 1 = 0$$
$$2x = 1$$
$$x = \frac{1}{2}$$
The solution set is $\left\{0, \frac{1}{2}\right\}$.

2. $\sqrt{3x+1} = 4$
$$\left(\sqrt{3x+1}\right)^2 = 4^2$$
$$3x + 1 = 16$$
$$3x = 15$$
$$x = 5$$
Check:
$$\sqrt{3\cdot 5 + 1} = 4$$
$$\sqrt{15+1} = 4$$
$$\sqrt{16} = 4$$
$$4 = 4$$
The solution set is $\{5\}$.

3. $2x^3 - 3x^2 - 8x - 3 = 0$
The graph of $Y_1 = 2x^3 - 3x^2 - 8x - 3$ appears to have an x-intercept at $x = 3$.

Using synthetic division:

$$3\overline{)2 \quad -3 \quad -8 \quad -3}$$
$$\underline{ \quad 6 \quad 9 \quad 3}$$
$$2 \quad 3 \quad 1 \quad 0$$

Therefore, $2x^3 - 3x^2 - 8x - 3 = 0$

$$(x-3)(2x^2 + 3x + 1) = 0$$
$$(x-3)(2x+1)(x+1) = 0$$
$$x = 3 \text{ or } x = -\frac{1}{2} \text{ or } x = -1$$

The solution set is $\left\{ -1, -\frac{1}{2}, 3 \right\}$.

4. $3^x = 9^{x+1}$

$$3^x = \left(3^2\right)^{x+1}$$
$$3^x = 3^{2x+2}$$
$$x = 2x + 2$$
$$x = -2$$

The solution set is $\{-2\}$.

5. $\log_3(x-1) + \log_3(2x+1) = 2$

$$\log_3\big((x-1)(2x+1)\big) = 2$$
$$(x-1)(2x+1) = 3^2$$
$$2x^2 - x - 1 = 9$$
$$2x^2 - x - 10 = 0$$
$$(2x-5)(x+2) = 0$$
$$x = \frac{5}{2} \text{ or } x = -2$$

Since $x = -2$ makes the original logarithms undefined, the solution set is $\left\{ \frac{5}{2} \right\}$.

6. $3^x = e$

$$\ln\left(3^x\right) = \ln e$$
$$x \ln 3 = 1$$
$$x = \frac{1}{\ln 3} \approx 0.910$$

The solution set is $\left\{ \frac{1}{\ln 3} \approx 0.910 \right\}$.

7. $g(x) = \frac{2x^3}{x^4 + 1}$

$$g(-x) = \frac{2(-x)^3}{(-x)^4 + 1} = \frac{-2x^3}{x^4 + 1} = -g(x)$$

Thus, g is an odd function and its graph is symmetric with respect to the origin.

8. $x^2 + y^2 - 2x + 4y - 11 = 0$

$$x^2 - 2x + y^2 + 4y = 11$$
$$(x^2 - 2x + 1) + (y^2 + 4y + 4) = 11 + 1 + 4$$
$$(x-1)^2 + (y+2)^2 = 16$$

Center: $(1, -2)$; Radius: 4

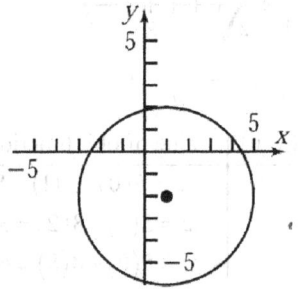

9. $f(x) = 3^{x-2} + 1$

Using the graph of $y = 3^x$, shift the graph horizontally 2 units to the right, then shift the graph vertically upward 1 unit.

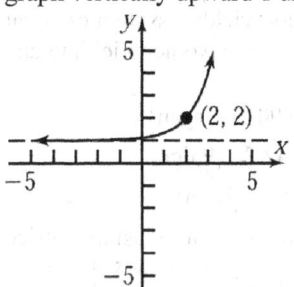

Domain: $(-\infty, \infty)$ Range: $(1, \infty)$

Horizontal Asymptote: $y = 1$

10. $f(x) = \frac{5}{x+2}$

$$y = \frac{5}{x+2}$$
$$x = \frac{5}{y+2} \qquad \text{Inverse}$$
$$x(y+2) = 5$$
$$xy + 2x = 5$$
$$xy = 5 - 2x$$
$$y = \frac{5-2x}{x} = \frac{5}{x} - 2$$

Thus, $f^{-1}(x) = \frac{5}{x} - 2$

Domain of $f = \{x \mid x \neq -2\}$

Range of $f = \{y \mid y \neq 0\}$

Domain of $f^{-1} = \{x \mid x \neq 0\}$

Range of $f^{-1} = \{y \mid y \neq -2\}$.

11. a. $y = 3x + 6$

The graph is a line.

x-intercept:	y-intercept:
$0 = 3x + 6$	$y = 3(0) + 6$
$3x = -6$	$= 6$
$x = -2$	

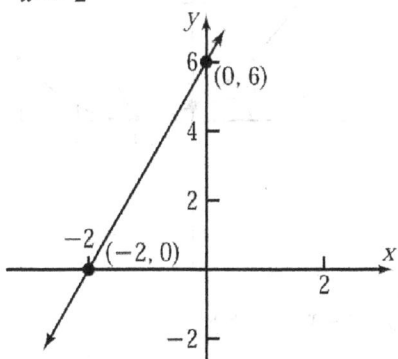

b. $x^2 + y^2 = 4$

The graph is a circle with center (0, 0) and radius 2.

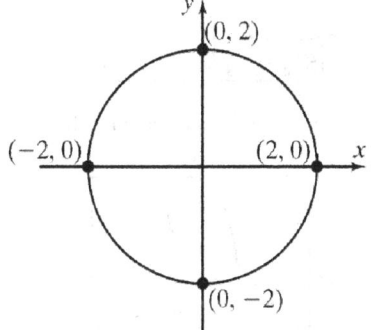

c. $y = x^3$

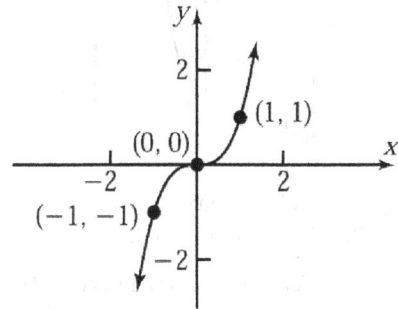

d. $y = \dfrac{1}{x}$

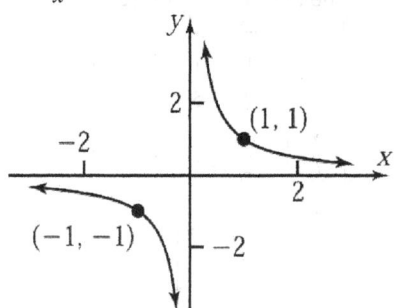

e. $y = \sqrt{x}$

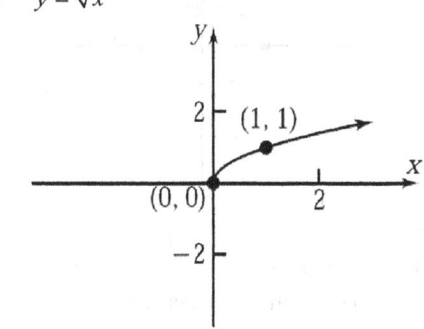

f. $y = e^x$

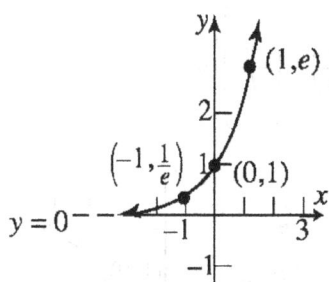

g. $y = \ln x$

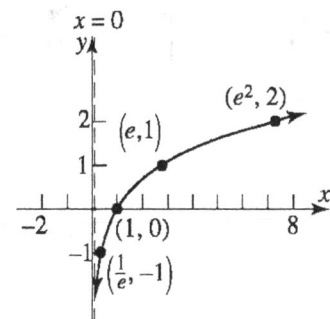

h. $2x^2 + 5y^2 = 1$

The graph is an ellipse.

$$\frac{x^2}{\frac{1}{2}} + \frac{y^2}{\frac{1}{5}} = 1$$

$$\left(\frac{x}{\frac{\sqrt{2}}{2}}\right)^2 + \left(\frac{y}{\frac{\sqrt{5}}{5}}\right)^2 = 1$$

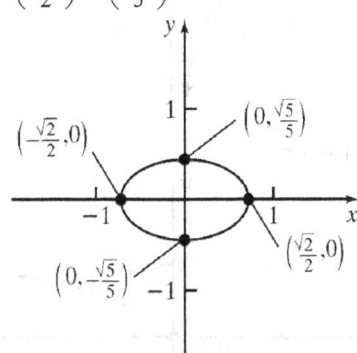

i. $x^2 - 3y^2 = 1$

The graph is a hyperbola

$$\frac{x^2}{1} - \frac{y^2}{\frac{1}{3}} = 1$$

$$\left(\frac{x}{1}\right)^2 - \left(\frac{y}{\frac{\sqrt{3}}{3}}\right)^2 = 1$$

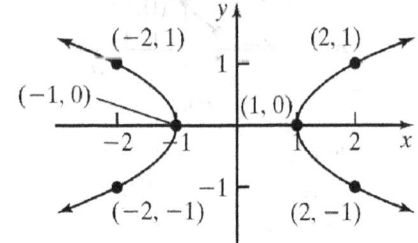

j. $x^2 - 2x - 4y + 1 = 0$

$$x^2 - 2x + 1 = 4y$$

$$4y = (x-1)^2$$

$$y = \frac{1}{4}(x-1)^2$$

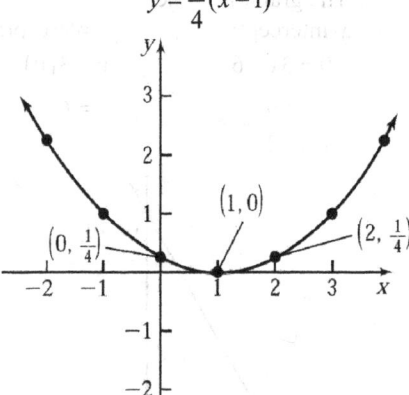

12. $f(x) = x^3 - 3x + 5$

a. Let $Y_1 = x^3 - 3x + 5$.

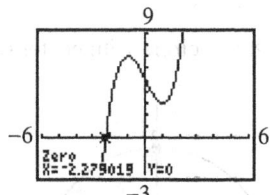

The zero of f is approximately -2.28.

b.

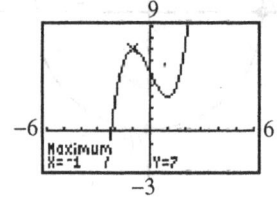

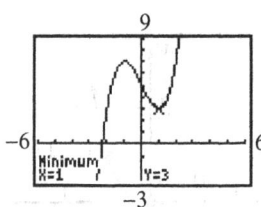

f has a local maximum of 7 at $x = -1$ and a local minimum of 3 at $x = 1$.

c. f is increasing on the intervals $(-\infty, -1)$ and $(1, \infty)$.

Chapter 8 Projects

Project I – Internet-based Project

1. $80\% = 0.80$ $18\% = 0.18$ $2\% = 0.02$
   $40\% = 0.40$ $50\% = 0.50$ $10\% = 0.10$
   $20\% = 0.20$ $60\% = 0.60$ $20\% = 0.20$

2. $\begin{bmatrix} 0.80 & 0.18 & 0.02 \\ 0.40 & 0.50 & 0.10 \\ 0.20 & 0.60 & 0.20 \end{bmatrix}$

3. $0.80 + 0.18 + 0.02 = 1.00$

   $0.40 + 0.50 + 0.10 = 1.00$

   $0.20 + 0.60 + 0.20 = 1.00$

   The sum of each row is 1 (or 100%). These represent the three possibilities of educational achievement for a parent of a child, unless someone does not attend school at all. Since these are rounded percents, chances are the other possibilities are negligible.

4. $P^2 = \begin{bmatrix} 0.8 & 0.18 & 0.02 \\ 0.4 & 0.5 & 0.1 \\ 0.2 & 0.6 & 0.2 \end{bmatrix}^2$

   $= \begin{bmatrix} 0.716 & 0.246 & 0.038 \\ 0.54 & 0.382 & 0.078 \\ 0.44 & 0.456 & 0.104 \end{bmatrix}$

   Grandchild of a college graduate is a college graduate: entry (1, 1): 0.716. The probability is 71.6%

5. Grandchild of a high school graduate finishes college: entry (2,1): 0.54. The probability is 54%.

6. grandchildren $\rightarrow$ k = 2.
   $v^{(2)} = v^{(0)} P^2$

   $= [0.319 \quad 0.564 \quad 0.117] \begin{bmatrix} 0.716 & 0.246 & 0.038 \\ 0.54 & 0.382 & 0.078 \\ 0.44 & 0.456 & 0.104 \end{bmatrix}$

   $= [0.584444 \quad 0.347274 \quad 0.068282]$

   College: $\approx 58.4\%$
   High School: $\approx 34.7\%$
   Elementary: $\approx 6.8\%$

7. The matrix totally stops changing at
   $$P^{30} \approx \begin{bmatrix} 0.64885496 & 0.29770992 & 0.05343511 \\ 0.64885496 & 0.29770992 & 0.05343511 \\ 0.64885496 & 0.29770992 & 0.05343511 \end{bmatrix}$$

Project II

a. $2 \times 2 \times 2 \times 2 = 16$ codewords.

b. $v = uG$
   u will be the matrix representing all of the 4-digit information bit sequences.

$$u = \begin{bmatrix} 0 & 0 & 0 & 0 \\ 0 & 0 & 0 & 1 \\ 0 & 0 & 1 & 0 \\ 0 & 0 & 1 & 1 \\ 0 & 1 & 0 & 0 \\ 0 & 1 & 0 & 1 \\ 0 & 1 & 1 & 0 \\ 0 & 1 & 1 & 1 \\ 1 & 0 & 0 & 0 \\ 1 & 0 & 0 & 1 \\ 1 & 0 & 1 & 0 \\ 1 & 0 & 1 & 1 \\ 1 & 1 & 0 & 0 \\ 1 & 1 & 0 & 1 \\ 1 & 1 & 1 & 0 \\ 1 & 1 & 1 & 1 \end{bmatrix}$$

(Remember, this is mod two. That means that you only write down the remainder when dividing by 2.)

$v = uG$

$$v = \begin{bmatrix} 0 & 0 & 0 & 0 & 0 & 0 & 0 \\ 0 & 0 & 0 & 1 & 1 & 1 & 0 \\ 0 & 0 & 1 & 0 & 1 & 0 & 1 \\ 0 & 0 & 1 & 1 & 0 & 1 & 1 \\ 0 & 1 & 0 & 0 & 0 & 1 & 1 \\ 0 & 1 & 0 & 1 & 1 & 0 & 1 \\ 0 & 1 & 1 & 0 & 1 & 1 & 0 \\ 0 & 1 & 1 & 1 & 0 & 0 & 0 \\ 1 & 0 & 0 & 0 & 1 & 1 & 1 \\ 1 & 0 & 0 & 1 & 0 & 0 & 1 \\ 1 & 0 & 1 & 0 & 0 & 1 & 0 \\ 1 & 0 & 1 & 1 & 1 & 0 & 0 \\ 1 & 1 & 0 & 0 & 1 & 0 & 0 \\ 1 & 1 & 0 & 1 & 0 & 1 & 0 \\ 1 & 1 & 1 & 0 & 0 & 0 & 1 \\ 1 & 1 & 1 & 1 & 1 & 1 & 1 \end{bmatrix}$$

c. Answers will vary, but if we choose the 6th row and the 10th row:

0101101
1001001
1102102 $\rightarrow$ 1100100 (13th row)

d. $v = uG$

$VH = uGH$

$$GH = \begin{bmatrix} 0 & 0 & 0 \\ 0 & 0 & 0 \\ 0 & 0 & 0 \\ 0 & 0 & 0 \end{bmatrix}$$

e. $rH = \begin{bmatrix} 0 & 1 & 0 & 1 & 0 & 0 & 0 \end{bmatrix} \begin{bmatrix} 1 & 1 & 1 \\ 0 & 1 & 1 \\ 1 & 0 & 1 \\ 1 & 1 & 0 \\ 1 & 0 & 0 \\ 0 & 1 & 0 \\ 0 & 0 & 1 \end{bmatrix}$

$= \begin{bmatrix} 1 & 0 & 1 \end{bmatrix}$

error code: 0010 000
r : 0101 000
0111 000

This is in the codeword list.

Project III

a. $A^T = \begin{bmatrix} 1 & 1 & 1 & 1 & 1 & 1 & 1 \\ 3 & 4 & 5 & 6 & 7 & 8 & 9 \end{bmatrix}$

b. $B = (A^T A)^{-1} A^T Y$

$B = \begin{bmatrix} -2.357 \\ 2.0357 \end{bmatrix}$

c. $y = 2.0357x - 2.357$

d. $y = 2.0357x - 2.357$

Project IV

Answers will vary.

Chapter 9
Sequences; Induction; the Binomial Theorem

Section 9.1

1. $f(2) = \frac{2-1}{2} = \frac{1}{2}$; $f(3) = \frac{3-1}{3} = \frac{2}{3}$

3. $A = P\left(1+\frac{r}{n}\right)^{n \cdot t}$

   $= 1000\left(1+\frac{0.04}{2}\right)^{2 \cdot 2}$

   $= 1000(1.02)^4$

   $= 1082.43$

   After two years, the account will contain $1082.43.

5. sequence

7. $n(n-1)\cdots 3 \cdot 2 \cdot 1$

9. summation

11. $10! = 10 \cdot 9 \cdot 8 \cdot 7 \cdot 6 \cdot 5 \cdot 4 \cdot 3 \cdot 2 \cdot 1 = 3,628,800$

13. $\frac{9!}{6!} = \frac{9 \cdot 8 \cdot 7 \cdot 6!}{6!} = 9 \cdot 8 \cdot 7 = 504$

15. $\frac{3! \cdot 7!}{4!} = \frac{3 \cdot 2 \cdot 1 \cdot 7 \cdot 6 \cdot 5 \cdot 4!}{4!}$

    $= 3 \cdot 2 \cdot 1 \cdot 7 \cdot 6 \cdot 5 = 1,260$

17. $s_1 = 1,\ s_2 = 2,\ s_3 = 3,\ s_4 = 4,\ s_5 = 5$

19. $a_1 = \frac{1}{1+2} = \frac{1}{3},\ a_2 = \frac{2}{2+2} = \frac{2}{4} = \frac{1}{2},$

    $a_3 = \frac{3}{3+2} = \frac{3}{5},\ a_4 = \frac{4}{4+2} = \frac{4}{6} = \frac{2}{3},$

    $a_5 = \frac{5}{5+2} = \frac{5}{7}$

21. $c_1 = (-1)^{1+1}(1^2) = 1,\ c_2 = (-1)^{2+1}(2^2) = -4,$

    $c_3 = (-1)^{3+1}(3^2) = 9,\ c_4 = (-1)^{4+1}(4^2) = -16,$

    $c_5 = (-1)^{5+1}(5^2) = 25$

23. $s_1 = \frac{2^1}{3^1+1} = \frac{2}{4} = \frac{1}{2},\ s_2 = \frac{2^2}{3^2+1} = \frac{4}{10} = \frac{2}{5},$

    $s_3 = \frac{2^3}{3^3+1} = \frac{8}{28} = \frac{2}{7},\ s_4 = \frac{2^4}{3^4+1} = \frac{16}{82} = \frac{8}{41},$

    $s_5 = \frac{2^5}{3^5+1} = \frac{32}{244} = \frac{8}{61}$

25. $t_1 = \frac{(-1)^1}{(1+1)(1+2)} = \frac{-1}{2 \cdot 3} = -\frac{1}{6},$

    $t_2 = \frac{(-1)^2}{(2+1)(2+2)} = \frac{1}{3 \cdot 4} = \frac{1}{12},$

    $t_3 = \frac{(-1)^3}{(3+1)(3+2)} = \frac{-1}{4 \cdot 5} = -\frac{1}{20},$

    $t_4 = \frac{(-1)^4}{(4+1)(4+2)} = \frac{1}{5 \cdot 6} = \frac{1}{30},$

    $t_5 = \frac{(-1)^5}{(5+1)(5+2)} = \frac{-1}{6 \cdot 7} = -\frac{1}{42}$

27. $b_1 = \frac{1}{e^1} = \frac{1}{e},\ b_2 = \frac{2}{e^2},\ b_3 = \frac{3}{e^3},$

    $b_4 = \frac{4}{e^4},\ b_5 = \frac{5}{e^5}$

29. Each term is a fraction with the numerator equal to the term number and the denominator equal to one more than the term number.

    $a_n = \frac{n}{n+1}$

31. Each term is a fraction with the numerator equal to 1 and the denominator equal to a power of 2. The power is equal to one less than the term number.

    $a_n = \frac{1}{2^{n-1}}$

33. The terms form an alternating sequence. Ignoring the sign, each term always contains a 1. The sign alternates by raising -1 to a power. Since the first term is positive, we use $n+1$ as the power.

    $a_n = (-1)^{n+1}$

35. The terms (ignoring the sign) are equal to the term number. The alternating sign is obtained by using $(-1)^{n+1}$.

$$a_n = (-1)^{n+1} \cdot n$$

37. $a_1 = 2, \ a_2 = 3+2 = 5, \ a_3 = 3+5 = 8,$
$a_4 = 3+8 = 11, \ a_5 = 3+11 = 14$

39. $a_1 = -2, \ a_2 = 2+(-2) = 0, \ a_3 = 3+0 = 3,$
$a_4 = 4+3 = 7, \ a_5 = 5+7 = 12$

41. $a_1 = 5, \ a_2 = 2 \cdot 5 = 10, \ a_3 = 2 \cdot 10 = 20,$
$a_4 = 2 \cdot 20 = 40, \ a_5 = 2 \cdot 40 = 80$

43. $a_1 = 3, \ a_2 = \dfrac{3}{2}, \ a_3 = \dfrac{\frac{3}{2}}{3} = \dfrac{1}{2},$

$a_4 = \dfrac{\frac{1}{2}}{4} = \dfrac{1}{8}, \ a_5 = \dfrac{\frac{1}{8}}{5} = \dfrac{1}{40}$

45. $a_1 = 1, \ a_2 = 2, \ a_3 = 2 \cdot 1 = 2, \ a_4 = 2 \cdot 2 = 4,$
$a_5 = 4 \cdot 2 = 8$

47. $a_1 = A, \ a_2 = A+d, \ a_3 = (A+d)+d = A+2d,$
$a_4 = (A+2d)+d = A+3d,$
$a_5 = (A+3d)+d = A+4d$

49. $a_1 = \sqrt{2}, \ a_2 = \sqrt{2+\sqrt{2}}, \ a_3 = \sqrt{2+\sqrt{2+\sqrt{2}}},$

$a_4 = \sqrt{2+\sqrt{2+\sqrt{2+\sqrt{2}}}},$

$a_5 = \sqrt{2+\sqrt{2+\sqrt{2+\sqrt{2+\sqrt{2}}}}}$

51. $\displaystyle\sum_{k=1}^{n} (k+2) = 3+4+5+6+7+\cdots+(n+2)$

53. $\displaystyle\sum_{k=1}^{n} \dfrac{k^2}{2} = \dfrac{1}{2}+2+\dfrac{9}{2}+8+\dfrac{25}{2}+18+\dfrac{49}{2}+32+\cdots+\dfrac{n^2}{2}$

55. $\displaystyle\sum_{k=0}^{n} \dfrac{1}{3^k} = 1+\dfrac{1}{3}+\dfrac{1}{9}+\dfrac{1}{27}+\cdots+\dfrac{1}{3^n}$

57. $\displaystyle\sum_{k=0}^{n-1} \dfrac{1}{3^{k+1}} = \dfrac{1}{3}+\dfrac{1}{9}+\dfrac{1}{27}+\cdots+\dfrac{1}{3^n}$

59. $\displaystyle\sum_{k=2}^{n} (-1)^k \ln k = \ln 2 - \ln 3 + \ln 4 - \cdots + (-1)^n \ln n$

61. $1+2+3+\cdots+20 = \displaystyle\sum_{k=1}^{20} k$

63. $\dfrac{1}{2}+\dfrac{2}{3}+\dfrac{3}{4}+\cdots+\dfrac{13}{13+1} = \displaystyle\sum_{k=1}^{13} \dfrac{k}{k+1}$

65. $1-\dfrac{1}{3}+\dfrac{1}{9}-\dfrac{1}{27}+\cdots+(-1)^6\left(\dfrac{1}{3^6}\right) = \displaystyle\sum_{k=0}^{6} (-1)^k \left(\dfrac{1}{3^k}\right)$

67. $3+\dfrac{3^2}{2}+\dfrac{3^3}{3}+\cdots+\dfrac{3^n}{n} = \displaystyle\sum_{k=1}^{n} \dfrac{3^k}{k}$

69. $a+(a+d)+(a+2d)+\cdots+(a+nd) = \displaystyle\sum_{k=0}^{n} (a+kd)$

or $= \displaystyle\sum_{k=1}^{n} (a+(k-1)d)$

71. $\displaystyle\sum_{k=1}^{40} 5 = \underbrace{5+5+5+\cdots+5}_{40 \text{ times}} = 40(5) = 200$

73. $\displaystyle\sum_{k=1}^{40} k = \dfrac{40(40+1)}{2} = 20(41) = 820$

75. $\displaystyle\sum_{k=1}^{20} (5k+3) = \sum_{k=1}^{20} (5k) + \sum_{k=1}^{20} 3 = 5\sum_{k=1}^{20} k + \sum_{k=1}^{20} 3$

$= 5\left(\dfrac{20(20+1)}{2}\right) + 3(20)$

$= 1050 + 60 = 1110$

77. $\displaystyle\sum_{k=1}^{16} (k^2+4) = \sum_{k=1}^{16} k^2 + \sum_{k=1}^{16} 4$

$= \dfrac{16(16+1)(2 \cdot 16+1)}{6} + 4(16)$

$= 1496 + 64 = 1560$

79. $\displaystyle\sum_{k=10}^{60} 2k = 2\sum_{k=10}^{60} 2k = 2\left[\sum_{k=1}^{60} k - \sum_{k=1}^{9} k\right]$

$= 2\left[\dfrac{60(60+1)}{2} - \dfrac{9(9+1)}{2}\right]$

$= 2[1830 - 45] = 3570$

81. $\displaystyle\sum_{k=5}^{20} k^3 = \sum_{k=1}^{20} k^3 - \sum_{k=1}^{4} k^3$

$$= \left[\frac{20(20+1)}{2}\right]^2 - \left[\frac{4(4+1)}{2}\right]^2$$

$$= 210^2 - 10^2 = 44,000$$

83. a. $p_1 = 1.03(2000) + 20 = 2080;$

$p_2 = 1.03(2080) + 20 = 2162.4$

There are approximately 2162 trout in the pond at the end of the second month.

b. Scrolling down the table, we find the trout population exceeds 5000 at the end of the 26th month when the population is 5084.

```
Plot1  Plot2  Plot3
nMin=0
·u(n)⊟1.03(u(n-1
))+20
u(nMin)⊟{2000}
·v(n)=
v(nMin)=
·w(n)=
```

```
 n    u(n)
22   4442.9
23   4596.2
24   4754.1
25   4916.7
26   5084.2
27   5256.8
28:
n=28
```

85. a. $r = \dfrac{0.05}{4} = 0.0125$

$A_0 = 1500, \; A_n = (1.0125)A_{n-1} + 750$

b. Scrolling down the table, we find the account exceeds $150,000 at the end of 99 quarters.

```
NORMAL FLOAT AUTO REAL DEGREE MP

Plot1   Plot2   Plot3
nMin=0
\u(n)⊟.0125*(u(n-1))+750
u(nMin)⊟{1500}
\v(n)=
v(nMin)=
\w(n)=
w(nMin)=
```

```
NORMAL FLOAT AUTO REAL DEGREE MP
PRESS + FOR △Tbl
 n      u(n)
95    140172
96    142674
97    145207
98    147773
99    150370
100   152999
101   155662
102   158358
103   161087
104   163851
105   166649
n=95
```

c. Scrolling down the table, we find the account value in 30 years (120 months) is $213,073.

```
NORMAL FLOAT AUTO REAL DEGREE MP
PRESS + FOR △Tbl
 n      u(n)
115   196628
116   199836
117   203084
118   206372
119   209702
120   213073
121   216487
122   219943
123   223442
124   226985
125   230572
n=115
```

87. a. $B_1 = 1.01(3000) - 100 = \$2930$

b. Put the graphing utility in SEQuence mode. Enter Y= as follows, then examine the TABLE:

```
Plot1  Plot2  Plot3
nMin=0
·u(n)⊟1.01(u(n-1
))-100
u(nMin)⊟{3000}
·v(n)=
v(nMin)=
·w(n)=
```

```
 n      u(n)
9     2344.2
10    2267.6
11    2190.3
12    2112.2
13    2033.3
14    1953.7
15    1873.2
n=14
```

From the table we see that the balance is below $2000 after 14 payments have been made. The balance then is $1953.70.

c. Scrolling down the table, we find that balance is paid off in the 36th month. The last payment is $83.78. There are 35 payments of $100 and the last payment of $83.78. The total amount paid is: 35(100) + 83.78(1.01) = $3584.62. (we have to add the interest for the last month).

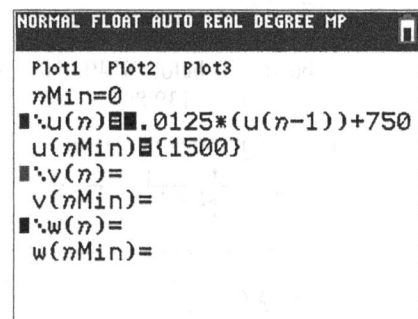

567

n	$u(n)$
31	470.71
32	375.42
33	279.17
34	181.96
35	83.781
36	-15.38
37	-115.5

$n=36$

d. The interest expense is:
3584.62 − 3000.00 = $584.62

89. a. Since the interest rate is 6% per annum compounded monthly, this is equivalent to a rate of 0.5% each month. Defining a recursive sequence, we have:
$a_0 = 150,000, \quad a_n = 1.005a_{n-1} - 899.33$

b. $1.005(150,000) - 899.33 = \$149,850.67$

c. Enter the recursive formula in Y= and create the table:

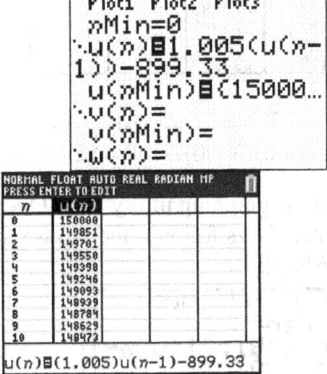

d. Scroll through the table:

n	$u(n)$
55	140573
56	140377
57	140180
58	139981
59	139782
60	139581
61	139380

$n=58$

After 58 payments have been made, the balance is below $140,000. The balance is about $139,981.

e. Scroll through the table:

n	$u(n)$
355	4425.9
356	3548.7
357	2667.1
358	1781.1
359	890.65
360	-4.231
361	-903.6

$n=360$

The loan will be paid off at the end of 360 months or 30 years.

Total amount paid = (359)($899.33) + $890.65(1.005) = $323,754.57.

f. The total interest expense is the difference of the total of the payments and the original loan: $323,754.57 - 150,000 = \$173,754.57$

g. (a) Since the interest rate is 6% per annum compounded monthly, this is equivalent to a rate of 0.5% each month. Defining a recursive sequence, we have:
$a_0 = 150,000, \quad a_n = 1.005a_{n-1} - 999.33$

(b) $1.005(150,000) - 999.33 = \$149,750.67$

(c) Enter the recursive formula in Y= and create the table:

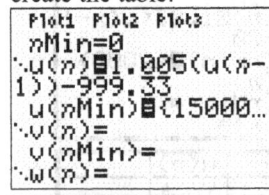

(d) Scroll through the table:

n	$u(n)$
34	140785
35	140489
36	140192
37	139894
38	139594
39	139293
40	138990

$n=37$

After 37 payments have been made, the balance is below $140,000. The balance is $139,894.

(e) Scroll through the table:

n	$u(n)$
273	5267.5
274	4294.6
275	3316.7
276	2333.9
277	1346.3
278	353.69
279	-643.9

$n=279$

The loan will be paid off at the end of 279 months or 23 years and 3 months.
Total amount paid = (278)($999.33) + 353.69(1.005) = $278,169.20

(f) The total interest expense is the difference of the total of the payments

and the original loan:
$$278,169.20 - 150,000 = \$128,169.20$$

h. Yes, if they can afford the additional monthly payment. They would save $44,586.07 in interest payments by paying the loan off sooner.

91. $a_1 = 1$, $a_2 = 1$, $a_3 = 2$, $a_4 = 3$, $a_5 = 5$,
$a_6 = 8$, $a_7 = 13$, $a_8 = 21$, $a_n = a_{n-1} + a_{n-2}$
$a_8 = a_7 + a_6 = 13 + 8 = 21$
After 7 months there are 21 mature pairs of rabbits.

93. 1, 1, 2, 3, 5, 8, 13
This is the Fibonacci sequence.

95. a. $f(1.3) = e^{1.3} \approx \sum\limits_{k=0}^{4} \frac{1.3^k}{k!}$

$$= \frac{1.3^0}{0!} + \frac{1.3^1}{1!} + ... + \frac{1.3^4}{4!}$$
$$\approx 3.630170833$$

b. $f(1.3) = e^{1.3} \approx \sum\limits_{k=0}^{7} \frac{1.3^k}{k!}$

$$= \frac{1.3^0}{0!} + \frac{1.3^1}{1!} + ... + \frac{1.3^7}{7!}$$
$$\approx 3.669060828$$

c. $f(1.3) = e^{1.3} \approx 3.669296668$

```
e^(1.3)
          3.669296668
```

d. It will take $n = 12$ to approximate
$f(1.3) = e^{1.3}$ correct to 8 decimal places.

```
          3.669296614
sum(seq(1.3^n/n!
,n,0,13))
          3.669296667
sum(seq(1.3^n/n!
,n,0,12))
          3.669296662
```

97. a. $a_1 = 0.4$,

$$a_2 = 0.4 + 0.3 \cdot 2^{2-2} = 0.4 + 0.3 = 0.7$$
$$a_3 = 0.4 + 0.3 \cdot 2^{3-2} = 0.4 + 0.3(2) = 1.0$$
$$a_4 = 0.4 + 0.3 \cdot 2^{4-2} = 0.4 + 0.3(4) = 1.6$$
$$a_5 = 0.4 + 0.3 \cdot 2^{5-2} = 0.4 + 0.3(8) = 2.8$$

$$a_6 = 0.4 + 0.3 \cdot 2^{6-2} = 0.4 + 0.3(16) = 5.2$$
$$a_7 = 0.4 + 0.3 \cdot 2^{7-2} = 0.4 + 0.3(32) = 10.0$$
$$a_8 = 0.4 + 0.3 \cdot 2^{8-2} = 0.4 + 0.3(64) = 19.6$$

The first eight terms of the sequence are 0.4, 0.7, 1.0, 1.6, 2.8, 5.2, 10.0, and 19.6.

b. Except for term 5, which has no match, Bode's formula provides excellent approximations for the mean distances of the planets from the Sun.

c. The mean distance of Ceres from the Sun is approximated by $a_5 = 2.8$, and that of Uranus is $a_8 = 19.6$.

d. $a_9 = 0.4 + 0.3 \cdot 2^{9-2} = 0.4 + 0.3(128) = 38.8$
$a_{10} = 0.4 + 0.3 \cdot 2^{10-2} = 0.4 + 0.3(256) = 77.2$

e. Pluto's distance is approximated by a_9, but no term approximates Neptune's mean distance from the Sun.

f. $a_{11} = 0.4 + 0.3 \cdot 2^{11-2} = 0.4 + 0.3(512) = 154$

According to Bode's Law, the mean orbital distance of 2003 UB 313 will be 154 AU from the Sun.

99. $\sqrt{5}$
We begin with an initial guess of $a_0 = 2$.

$$a_1 = \frac{1}{2}\left(2 + \frac{5}{2}\right) = 2.25$$

$$a_2 = \frac{1}{2}\left(2.25 + \frac{5}{2.25}\right) \approx 2.236111111$$

$$a_3 = \frac{1}{2}\left(2.236111111 + \frac{5}{2.236111111}\right)$$
$$\approx 2.236067978$$

$$a_4 = \frac{1}{2}\left(2.236067978 + \frac{5}{2.236067978}\right)$$
$$\approx 2.236067977$$

$$a_5 = \frac{1}{2}\left(2.236067977 + \frac{5}{2.236067977}\right)$$
$$\approx 2.236067977$$

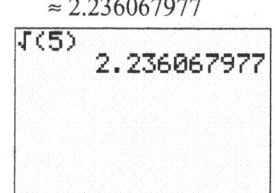

```
√(5)
          2.236067977
```

For both a_5 and the calculator approximation, we obtain $\sqrt{5} \approx 2.236067977$.

101. $\sqrt{21}$

We begin with an initial guess of $a_0 = 5$.

$a_1 = \frac{1}{2}\left(5 + \frac{21}{5}\right) = 4.625$

$a_2 = \frac{1}{2}\left(4.625 + \frac{21}{4.625}\right) \approx 4.58277027$

$a_3 = \frac{1}{2}\left(4.58277027 + \frac{21}{4.58277027}\right)$
≈ 4.582575699

$a_4 = \frac{1}{2}\left(4.582575699 + \frac{21}{4.582575699}\right)$
≈ 4.582575695

$a_5 = \frac{1}{2}\left(4.582575695 + \frac{21}{4.582575695}\right)$
≈ 4.582575695

```
√(21)
        4.582575695
```

For both a_5 and the calculator approximation, we obtain $\sqrt{21} \approx 4.582575695$.

103. $u_1 = 1$ and $u_{n+1} = u_n + (n+1)$: So

$u_1 = 1$

$u_2 = u_1 + (1+1) = 1 + 2 = 3$

$u_3 = u_2 + (2+1) = 3 + 3 = 6$

$u_4 = u_3 + (3+1) = 6 + 4 = 10$

$u_5 = u_4 + (4+1) = 10 + 5 = 15$

$u_6 = u_5 + (5+1) = 15 + 6 = 21$

$u_7 = u_6 + (6+1) = 21 + 7 = 28$

105. We know from number 97 and 98 that:

$u_{n+1} = \frac{(n+1)(n+2)}{2}$ and $u_n = \frac{(n)(n+1)}{2}$

Adding these together we get

$u_{n+1} + u_n = \frac{(n+1)(n+2)}{2} + \frac{(n)(n+1)}{2}$

$= \frac{(n+1)(n+2) + (n)(n+1)}{2}$

$= \frac{n^2 + 3n + 2 + n^2 + n}{2}$

$= \frac{2n^2 + 4n + 2}{2}$

$= \frac{2(n^2 + 2n + 1)}{2} = n^2 + 2n + 1 = (n+1)^2$

107. Answers will vary.

109. $x^3 + x^2 - 16x - 16 \geq 0$

$(x^3 + x^2) - (16x + 16) \geq 0$

$x^2(x+1) - 16(x+1) \geq 0$

$(x-4)(x+4)(x+1) \geq 0$

$x = -4, x = -1, x = 4$ are the zeros.

Interval	$(-\infty, -4)$	$(-4, -1)$	$(-1, 4)$	$(4, \infty)$
Number Chosen	-5	-3	0	5
Value of f	-36	14	-16	140
Conclusion	Neg	Pos	Neg	Pos

he solution set is $\left\{ x \mid -4 \leq x \leq -1 \text{ or } x \geq 4 \right\}$ or, using interval notation, $[-4, -1] \cup [4, \infty)$.

111. The vertex is $(-3, 4)$ and the focus is $(1, 4)$. Both lie on the horizontal line $y = 4$. $a = |-3 - 1| = 4$ and since $(1, 4)$ is to the right of $(-3, 4)$, the parabola opens to the right. The equation of the parabola is:

$(y-k)^2 = 4a(x-h)$

$(y-4)^2 = 4(4)(x-(-3))$

$(y-4)^2 = 16(x+3)$

Section 9.2

1. arithmetic

3. $12 = a_1 + (5-1)5 \Rightarrow a_1 = -8$

So

$a_6 = -8 + (6-1)5 = 17$

5. d

7. $d = s_n - s_{n-1}$

$= (n+4) - (n-1+4) = (n+4) - (n+3)$

$= n + 4 - n - 3 = 1$

The difference between consecutive terms is constant, therefore the sequence is arithmetic.

$s_1 = 1 + 4 = 5, \quad s_2 = 2 + 4 = 6, \quad s_3 = 3 + 4 = 7,$

$s_4 = 4 + 4 = 8$

9. $d = a_n - a_{n-1}$

$= (2n-5) - (2(n-1)-5)$

$= (2n-5) - (2n-2-5)$

$= 2n - 5 - 2n + 7 = 2$

The difference between consecutive terms is constant, therefore the sequence is arithmetic.

$a_1 = 2 \cdot 1 - 5 = -3, \quad a_2 = 2 \cdot 2 - 5 = -1,$

$a_3 = 2 \cdot 3 - 5 = 1, \quad a_4 = 2 \cdot 4 - 5 = 3$

11. $d = c_n - c_{n-1}$

$= (6 - 2n) - (6 - 2(n-1))$

$= (6 - 2n) - (6 - 2n + 2)$

$= 6 - 2n - 6 + 2n - 2 = -2$

The difference between consecutive terms is constant, therefore the sequence is arithmetic.

$c_1 = 6 - 2 \cdot 1 = 4, \quad c_2 = 6 - 2 \cdot 2 = 2,$

$c_3 = 6 - 2 \cdot 3 = 0, \quad c_4 = 6 - 2 \cdot 4 = -2$

13. $d = t_n - t_{n-1}$

$= \left(\frac{1}{2} - \frac{1}{3} n \right) - \left(\frac{1}{2} - \frac{1}{3}(n-1) \right)$

$= \left(\frac{1}{2} - \frac{1}{3} n \right) - \left(\frac{1}{2} - \frac{1}{3} n + \frac{1}{3} \right)$

$= \frac{1}{2} - \frac{1}{3} n - \frac{1}{2} + \frac{1}{3} n - \frac{1}{3} = -\frac{1}{3}$

The difference between consecutive terms is constant, therefore the sequence is arithmetic.

$t_1 = \frac{1}{2} - \frac{1}{3} \cdot 1 = \frac{1}{6}, \quad t_2 = \frac{1}{2} - \frac{1}{3} \cdot 2 = -\frac{1}{6},$

$t_3 = \frac{1}{2} - \frac{1}{3} \cdot 3 = -\frac{1}{2}, \quad t_4 = \frac{1}{2} - \frac{1}{3} \cdot 4 = -\frac{5}{6}$

15. $d = s_n - s_{n-1}$

$= \ln\left(3^n\right) - \ln\left(3^{n-1}\right)$

$= n \ln(3) - (n-1) \ln(3)$

$= (\ln 3)(n - (n-1)) = (\ln 3)(n - n + 1)$

$= \ln 3$

The difference between consecutive terms is constant, therefore the sequence is arithmetic.

$s_1 = \ln\left(3^1\right) = \ln(3), \quad s_2 = \ln\left(3^2\right) = 2 \ln(3),$

$s_3 = \ln\left(3^3\right) = 3 \ln(3), \quad s_4 = \ln\left(3^4\right) = 4 \ln(3)$

17. $a_n = a_1 + (n-1)d$

$= 2 + (n-1)3$

$= 2 + 3n - 3$

$= 3n - 1$

$a_{51} = 3 \cdot 51 - 1 = 152$

19. $a_n = a_1 + (n-1)d$

$= 5 + (n-1)(-3)$

$= 5 - 3n + 3$

$= 8 - 3n$

$a_{51} = 8 - 3 \cdot 51 = -145$

21. $a_n = a_1 + (n-1)d$

$= 0 + (n-1)\frac{1}{2}$

$= \frac{1}{2} n - \frac{1}{2}$

$= \frac{1}{2}(n-1)$

$a_{51} = \frac{1}{2}(51-1) = 25$

23. $a_n = a_1 + (n-1)d$

$= \sqrt{2} + (n-1)\sqrt{2}$

$= \sqrt{2} + \sqrt{2}n - \sqrt{2} = \sqrt{2}n$

$a_{51} = 51\sqrt{2}$

25. $a_1 = 2, \quad d = 2, \quad a_n = a_1 + (n-1)d$

$a_{100} = 2 + (100-1)2 = 2 + 99(2) = 2 + 198 = 200$

27. $a_1 = 1, \quad d = -2 - 1 = -3, \quad a_n = a_1 + (n-1)d$

$a_{90} = 1 + (90-1)(-3) = 1 + 89(-3)$

$= 1 - 267 = -266$

29. $a_1 = 2$, $d = \frac{5}{2} - 2 = \frac{1}{2}$, $a_n = a_1 + (n-1)d$

$a_{80} = 2 + (80-1)\frac{1}{2} = \frac{83}{2}$

31. $a_8 = a_1 + 7d = 8$ $\quad a_{20} = a_1 + 19d = 44$

Solve the system of equations by subtracting the first equation from the second:
$12d = 36 \Rightarrow d = 3$

$a_1 = 8 - 7(3) = 8 - 21 = -13$

Recursive formula: $a_1 = -13$ $\quad a_n = a_{n-1} + 3$

nth term: $a_n = a_1 + (n-1)d$

$\qquad = -13 + (n-1)(3)$

$\qquad = -13 + 3n - 3$

$\qquad = 3n - 16$

33. $a_9 = a_1 + 8d = -5$ $\quad a_{15} = a_1 + 14d = 31$

Solve the system of equations by subtracting the first equation from the second:
$6d = 36 \Rightarrow d = 6$

$a_1 = -5 - 8(6) = -5 - 48 = -53$

Recursive formula: $a_1 = -53$ $\quad a_n = a_{n-1} + 6$

nth term: $a_n = a_1 + (n-1)d$

$\qquad = -53 + (n-1)(6)$

$\qquad = -53 + 6n - 6$

$\qquad = 6n - 59$

35. $a_{15} = a_1 + 14d = 0$ $\quad a_{40} = a_1 + 39d = -50$

Solve the system of equations by subtracting the first equation from the second:
$25d = -50 \Rightarrow d = -2$

$a_1 = -14(-2) = 28$

Recursive formula: $a_1 = 28$ $\quad a_n = a_{n-1} - 2$

nth term: $a_n = a_1 + (n-1)d$

$\qquad = 28 + (n-1)(-2)$

$\qquad = 28 - 2n + 2$

$\qquad = 30 - 2n$

37. $a_{14} = a_1 + 13d = -1$ $\quad a_{18} = a_1 + 17d = -9$

Solve the system of equations by subtracting the first equation from the second:
$4d = -8 \Rightarrow d = -2$

$a_1 = -1 - 13(-2) = -1 + 26 = 25$

Recursive formula: $a_1 = 25$ $\quad a_n = a_{n-1} - 2$

nth term: $a_n = a_1 + (n-1)d$

$\qquad = 25 + (n-1)(-2)$

$\qquad = 25 - 2n + 2$

$\qquad = 27 - 2n$

39. $S_n = \frac{n}{2}(a_1 + a_n) = \frac{n}{2}(1 + (2n-1)) = \frac{n}{2}(2n) = n^2$

41. $S_n = \frac{n}{2}(a_1 + a_n) = \frac{n}{2}(7 + (2 + 5n)) = \frac{n}{2}(9 + 5n)$

43. $a_1 = 2$, $d = 4 - 2 = 2$, $a_n = a_1 + (n-1)d$

$70 = 2 + (n-1)2$

$70 = 2 + 2n - 2$

$70 = 2n$

$n = 35$

$S_n = \frac{n}{2}(a_1 + a_n) = \frac{35}{2}(2 + 70)$

$\qquad = \frac{35}{2}(72) = 35(36)$

$\qquad = 1260$

45. $a_1 = 5$, $d = 9 - 5 = 4$, $a_n = a_1 + (n-1)d$

$49 = 5 + (n-1)4$

$49 = 5 + 4n - 4$

$48 = 4n$

$n = 12$

$S_n = \frac{n}{2}(a_1 + a_n) = \frac{12}{2}(5 + 49) = 6(54) = 324$

47. $a_1 = 73$, $d = 78 - 73 = 5$, $a_n = a_1 + (n-1)d$

$558 = 73 + (n-1)(5)$

$485 = 5(n-1)$

$97 = n - 1$

$98 = n$

$S_n = \frac{n}{2}(a_1 + a_n) = \frac{98}{2}(73 + 558)$

$\qquad = 49(631) = 30,919$

49. $a_1 = 4$, $d = 4.5 - 4 = 0.5$, $a_n = a_1 + (n-1)d$

$$100 = 4 + (n-1)(0.5)$$
$$96 = 0.5(n-1)$$
$$192 = n - 1$$
$$193 = n$$
$$S_n = \frac{n}{2}(a_1 + a_n) = \frac{193}{2}(4 + 100)$$
$$= \frac{193}{2}(104) = 10,036$$

51. $a_1 = 2(1) - 5 = -3$, $a_{80} = 2(80) - 5 = 155$

$$S_{80} = \frac{80}{2}(-3 + 155) = 40(152) = 6080$$

53. $a_1 = 6 - \frac{1}{2}(1) = \frac{11}{2}$, $a_{100} = 6 - \frac{1}{2}(100) = -44$

$$S_{100} = \frac{100}{2}\left(\frac{11}{2} + (-44)\right)$$
$$= 50\left(-\frac{77}{2}\right) = -1925$$

55. $a_1 = 14$, $d = 16 - 14 = 2$, $a_n = a_1 + (n-1)d$

$$a_{120} = 14 + (120 - 1)(2) = 14 + 119(2) = 252$$
$$S_{120} = \frac{120}{2}(14 + 252) = 60(266) = 15,960$$

57. Find the common difference of the terms and solve the system of equations:

$$(2x+1) - (x+3) = d \Rightarrow x - 2 = d$$
$$(5x+2) - (2x+1) = d \Rightarrow 3x + 1 = d$$
$$3x + 1 = x - 2$$
$$2x = -3$$
$$x = -\frac{3}{2}$$

59. $d = 3$, $a_1 = 11$, and $S = 1092$

$$1092 = \frac{n}{2}[2(11) + (n-1)(3)]$$
$$1092 = \frac{n}{2}[22 + 3n - 3]$$
$$2194 = n[19 + 3n]$$
$$2194 = 19n + 3n^2$$
$$3n^2 + 19n - 2184 = 0$$
$$(3n + 91)(n - 24) = 0$$

So $n = 24$.

61. The total number of seats is:

$$S = 25 + 26 + 27 + \cdots + (25 + 29(1))$$

This is the sum of an arithmetic sequence with $d = 1$, $a_1 = 25$, and $n = 30$.
Find the sum of the sequence:

$$S_{30} = \frac{30}{2}[2(25) + (30 - 1)(1)]$$
$$= 15(50 + 29) = 15(79)$$
$$= 1185$$

There are 1185 seats in the theater.

63. The lighter colored tiles have 20 tiles in the bottom row and 1 tile in the top row. The number decreases by 1 as we move up the triangle. This is an arithmetic sequence with $a_1 = 20$, $d = -1$, and $n = 20$. Find the sum:

$$S = \frac{20}{2}[2(20) + (20 - 1)(-1)]$$
$$= 10(40 - 19) = 10(21)$$
$$= 210$$

There are 210 lighter tiles.

The darker colored tiles have 19 tiles in the bottom row and 1 tile in the top row. The number decreases by 1 as we move up the triangle. This is an arithmetic sequence with $a_1 = 19$, $d = -1$, and $n = 19$. Find the sum:

$$S = \frac{19}{2}[2(19) + (19 - 1)(-1)]$$
$$= \frac{19}{2}(38 - 18) = \frac{19}{2}(20) = 190$$

There are 190 darker tiles.

65. The air cools at the rate of $5.5°$ F per 1000 feet. Since n represents thousands of feet, we have $d = -5.5$. The ground temperature is $67°F$ so we have $T_1 - 67 - 5.5 = 61.5$. Therefore,

$$\{T_n\} = \{61.5 + (n-1)(-5.5)\}$$
$$= \{-5.5n + 67\} \text{ or } \{67 - 5.5n\}$$

After the parcel of air has risen 5000 feet, we have $T_5 = 61.5 + (5-1)(-5.5) = 39.5$.

The parcel of air will be $39.5°F$ after it has risen 5000 feet.

67. $a_1 = 35$, $d = 37 - 35 = 2$, $a_n = a_1 + (n-1)d$

$$a_{27} = 35 + (27-1)(2) = 35 + 26(2) = 87$$

$$S_{27} = \frac{27}{2}(35 + 87) = \frac{27}{2}(122) = 1647$$

The amphitheater has 1647 seats.

69. The yearly salaries form an arithmetic sequence with $a_1 = 35,000$, $d = 1400$, $S_n = 280,000$.

Find the number of years for the aggregate salary to equal \$280,000.

$$S_n = \frac{n}{2}[2a_1 + (n-1)d]$$

$$280,000 = \frac{n}{2}[2(35,000) + (n-1)1400]$$

$$280,000 = n[35,000 + 700n - 700]$$

$$280,000 - n(700n + 34,300)$$

$$280,000 = 700n^2 + 34,300n$$

$$400 = n^2 + 49n$$

$$n^2 + 49n - 400 = 0$$

$$n = \frac{-49 \pm \sqrt{49^2 - 4(1)(-400)}}{2(1)}$$

$$= \frac{-49 \pm \sqrt{4001}}{2} \approx \frac{-49 \pm 63.25}{2}$$

$n \approx 7.13$ or $n \approx -56.13$

It takes about 8 years to have an aggregate salary of at least \$280,000. The aggregate salary after 8 years will be \$319,200.

71. Answers will vary. Both increase (or decrease) at a constant rate, but the domain of an arithmetic sequence is the set of natural numbers while the domain of a linear function is the set of all real numbers.

73. $f(x) = x^4 + 5x^3 - 19x^2 - 29x + 42$; $c = -7$

$$f(-7) = (-7)^4 + 5(-7)^3 - 19(-7)^2 - 29(-7) + 42$$
$$= 2401 - 1715 - 931 + 203 + 42 = 0$$

Thus, -7 is a zero of f and $x + 7$ is a factor of f.

75. $A = \begin{bmatrix} 2 & 0 \\ 3 & -1 \end{bmatrix}$

Augment the matrix with the identity and use row operations to find the inverse:

$$\begin{bmatrix} 2 & 0 & | & 1 & 0 \\ 3 & -1 & | & 0 & 1 \end{bmatrix}$$

$$\rightarrow \begin{bmatrix} 1 & 0 & | & \frac{1}{2} & 0 \\ 3 & -1 & | & 0 & 1 \end{bmatrix} \quad \left(R_1 = \frac{1}{2}r_1\right)$$

$$\rightarrow \begin{bmatrix} 1 & 0 & | & \frac{1}{2} & 0 \\ 0 & -1 & | & -\frac{3}{2} & 1 \end{bmatrix} \quad \left(R_2 = -3r_1 + r_2\right)$$

$$\rightarrow \begin{bmatrix} 1 & 0 & | & \frac{1}{2} & 0 \\ 0 & 1 & | & \frac{3}{2} & -1 \end{bmatrix} \quad \left(R_2 = -1r_2\right)$$

Thus, $A^{-1} = \begin{bmatrix} \frac{1}{2} & 0 \\ \frac{3}{2} & -1 \end{bmatrix}$.

Section 9.3

1. $a_n = a_1 \cdot r^{n-1}$

3. c

5. b

7. False; the common ratio can be positive or negative (or 0, but this results in a sequence of only 0s).

9. $r = \dfrac{3^{n+1}}{3^n} = 3^{n+1-n} = 3$

The ratio of consecutive terms is constant, therefore the sequence is geometric.

$s_1 = 3^1 = 3$, $s_2 = 3^2 = 9$,

$s_3 = 3^3 = 27$, $s_4 = 3^4 = 81$

11. $r = \dfrac{-3\left(\dfrac{1}{2}\right)^{n+1}}{-3\left(\dfrac{1}{2}\right)^{n}} = \left(\dfrac{1}{2}\right)^{n+1-n} = \dfrac{1}{2}$

The ratio of consecutive terms is constant, therefore the sequence is geometric.

$a_1 = -3\left(\dfrac{1}{2}\right)^1 = -\dfrac{3}{2}, \quad a_2 = -3\left(\dfrac{1}{2}\right)^2 = -\dfrac{3}{4},$

$a_3 = -3\left(\dfrac{1}{2}\right)^3 = -\dfrac{3}{8}, \quad a_4 = -3\left(\dfrac{1}{2}\right)^4 = -\dfrac{3}{16}$

13. $r = \dfrac{\left(\dfrac{2^{n+1-1}}{4}\right)}{\left(\dfrac{2^{n-1}}{4}\right)} = \dfrac{2^n}{2^{n-1}} = 2^{n-(n-1)} = 2$

The ratio of consecutive terms is constant, therefore the sequence is geometric.

$c_1 = \dfrac{2^{1-1}}{4} = \dfrac{2^0}{2^2} = 2^{-2} = \dfrac{1}{4},$

$c_2 = \dfrac{2^{2-1}}{4} = \dfrac{2^1}{2^2} = 2^{-1} = \dfrac{1}{2},$

$c_3 = \dfrac{2^{3-1}}{4} = \dfrac{2^2}{2^2} = 1,$

$c_4 = \dfrac{2^{4-1}}{4} = \dfrac{2^3}{2^2} = 2$

15. $r = \dfrac{2^{\left(\frac{n+1}{3}\right)}}{2^{\left(\frac{n}{3}\right)}} = 2^{\left(\frac{n+1}{3} - \frac{n}{3}\right)} = 2^{1/3}$

The ratio of consecutive terms is constant, therefore the sequence is geometric.

$e_1 = 2^{1/3}, \quad e_2 = 2^{2/3}, \quad e_3 = 2^{3/3} = 2, \quad e_4 = 2^{4/3}$

17. $r = \dfrac{\left(\dfrac{3^{n+1-1}}{2^{n+1}}\right)}{\left(\dfrac{3^{n-1}}{2^n}\right)} = \dfrac{3^n}{3^{n-1}} \cdot \dfrac{2^n}{2^{n+1}}$

$= 3^{n-(n-1)} \cdot 2^{n-(n+1)} = 3 \cdot 2^{-1} = \dfrac{3}{2}$

The ratio of consecutive terms is constant,

therefore the sequence is geometric.

$t_1 = \dfrac{3^{1-1}}{2^1} = \dfrac{3^0}{2} = \dfrac{1}{2}, \quad t_2 = \dfrac{3^{2-1}}{2^2} = \dfrac{3^1}{2^2} = \dfrac{3}{4},$

$t_3 = \dfrac{3^{3-1}}{2^3} = \dfrac{3^2}{2^3} = \dfrac{9}{8}, \quad t_4 = \dfrac{3^{4-1}}{2^4} = \dfrac{3^3}{2^4} = \dfrac{27}{16}$

19. $a_5 = 2 \cdot 3^{5-1} = 2 \cdot 3^4 = 2 \cdot 81 = 162$

$a_n = 2 \cdot 3^{n-1}$

21. $a_5 = 5(-1)^{5-1} = 5(-1)^4 = 5 \cdot 1 = 5$

$a_n = 5 \cdot (-1)^{n-1}$

23. $a_5 = 0 \cdot \left(\dfrac{1}{2}\right)^{5-1} = 0 \cdot \left(\dfrac{1}{2}\right)^4 = 0$

$a_n = 0 \cdot \left(\dfrac{1}{2}\right)^{n-1} = 0$

25. $a_5 = \sqrt{2} \cdot \left(\sqrt{2}\right)^{5-1} = \sqrt{2} \cdot \left(\sqrt{2}\right)^4 = \sqrt{2} \cdot 4 = 4\sqrt{2}$

$a_n = \sqrt{2} \cdot \left(\sqrt{2}\right)^{n-1} = \left(\sqrt{2}\right)^n$

27. $a_1 = 1, \ r = \dfrac{1}{2}, \ n = 7$

$a_7 = 1 \cdot \left(\dfrac{1}{2}\right)^{7-1} = \left(\dfrac{1}{2}\right)^6 = \dfrac{1}{64}$

29. $a_1 = 1, \ r = -1, \ n = 9$

$a_9 = 1 \cdot (-1)^{9-1} = (-1)^8 = 1$

31. $a_1 = 0.4, \ r = 0.1, \ n = 8$

$a_8 = 0.4 \cdot (0.1)^{8-1} = 0.4(0.1)^7 = 0.00000004$

33. $a_1 = 7, \ r = \dfrac{14}{7} = 2, \ a_n = a_1 \cdot r^{n-1}$

$a_n = 7 \cdot 2^{n-1}$

35. $a_1 = -3, \ r = \dfrac{1}{-3} = -\dfrac{1}{3}, \ a_n = a_1 \cdot r^{n-1}$

$a_n = -3\left(-\dfrac{1}{3}\right)^{n-1} = \left(-\dfrac{1}{3}\right)^{n-2}$

37. $a_n = a_1 \cdot r^{n-1}$

$243 = a_1 \cdot (-3)^{6-1}$

$243 = a_1 (-3)^5$

$243 = -243 a_1$

$-1 = a_1$

Therefore, $a_n = -(-3)^{n-1}$

39. $\dfrac{a_4}{a_2} = \dfrac{a_1 \cdot r^{4-1}}{a_1 \cdot r^{2-1}} = \dfrac{r^3}{r} = r^2$

$r^2 = \dfrac{1575}{7} = 225$

$r = \sqrt{225} = 15$

$a_n = a_1 \cdot r^{n-1}$

$7 = a_1 \cdot 15^{2-1}$

$7 = 15 a_1$

$a_1 = \dfrac{7}{15}$

Therefore, $a_n = \dfrac{7}{15} \cdot 15^{n-1} = 7 \cdot 15^{n-2}$.

41. $a_1 = \dfrac{1}{4}, \ r = 2$

$S_n = a_1 \left(\dfrac{1-r^n}{1-r} \right) = \dfrac{1}{4} \left(\dfrac{1-2^n}{1-2} \right) = -\dfrac{1}{4} \left(1 - 2^n \right)$

$= \dfrac{1}{4} \left(2^n - 1 \right)$

43. $a_1 = \dfrac{2}{3}, \ r = \dfrac{2}{3}$

$S_n = a_1 \left(\dfrac{1-r^n}{1-r} \right) = \dfrac{2}{3} \left[\dfrac{1 - \left(\dfrac{2}{3} \right)^n}{1 - \dfrac{2}{3}} \right]$

$= \dfrac{2}{3} \left[\dfrac{1 - \left(\dfrac{2}{3} \right)^n}{\dfrac{1}{3}} \right] = 2 \left[1 - \left(\dfrac{2}{3} \right)^n \right]$

45. $a_1 = -1, \ r = 2$

$S_n = a_1 \left(\dfrac{1-r^n}{1-r} \right) = -1 \left(\dfrac{1-2^n}{1-2} \right) = 1 - 2^n$

47. Using the sum of the sequence feature:

```
(1/4)sum(seq(2^n
,n,0,14,1))
          8191.75
```

49. Using the sum of the sequence feature:

```
sum(seq((2/3)^n,
n,1,15,1))
        1.995432683
```

51. Using the sum of the sequence feature:

```
-1sum(seq(2^n,n,
0,14,1))
            -32767
```

53. $a_1 = 1, \ r = \dfrac{1}{3}$

Since $|r| < 1$, the series converges.

$$S_\infty = \frac{a_1}{1-r} = \frac{1}{\left(1-\dfrac{1}{3}\right)} = \frac{1}{\left(\dfrac{2}{3}\right)} = \frac{3}{2}$$

55. $a_1 = 8, \ r = \dfrac{1}{2}$

Since $|r| < 1$, the series converges.

$$S_\infty = \frac{a_1}{1-r} = \frac{8}{\left(1-\dfrac{1}{2}\right)} = \frac{8}{\left(\dfrac{1}{2}\right)} = 16$$

57. $a_1 = 2, \ r = -\dfrac{1}{4}$

Since $|r| < 1$, the series converges.

$$S_\infty = \frac{a_1}{1-r} = \frac{2}{\left(1-\left(-\dfrac{1}{4}\right)\right)} = \frac{2}{\left(\dfrac{5}{4}\right)} = \frac{8}{5}$$

59. $a_1 = 8, \ r = \dfrac{3}{2}$

Since $|r| > 1$, the series diverges.

61. $a_1 = 5, \ r = \dfrac{1}{4}$

Since $|r| < 1$, the series converges.

$$S_\infty = \frac{a_1}{1-r} = \frac{5}{\left(1-\dfrac{1}{4}\right)} = \frac{5}{\left(\dfrac{3}{4}\right)} = \frac{20}{3}$$

63. $a_1 = \dfrac{1}{2}, \ r = 3$

Since $|r| > 1$, the series diverges.

65. $a_1 = 6, \ r = -\dfrac{2}{3}$

Since $|r| < 1$, the series converges.

$$S_\infty = \frac{a_1}{1-r} = \frac{6}{\left(1-\left(-\dfrac{2}{3}\right)\right)} = \frac{6}{\left(\dfrac{5}{3}\right)} = \frac{18}{5}$$

67. $\displaystyle\sum_{k=1}^{\infty} 3\left(\frac{2}{3}\right)^k = \sum_{k=1}^{\infty} 3 \cdot \frac{2}{3}\left(\frac{2}{3}\right)^{k-1} = \sum_{k=1}^{\infty} 2\left(\frac{2}{3}\right)^{k-1}$

$a_1 = 2, \ r = \dfrac{2}{3}$

Since $|r| < 1$, the series converges.

$$S_\infty = \frac{a_1}{1-r} = \frac{2}{1-\dfrac{2}{3}} = \frac{2}{\dfrac{1}{3}} = 6$$

69. $\{n+2\}$

$d = (n+1+2)-(n+2) = n+3-n-2 = 1$

The difference between consecutive terms is constant. Therefore the sequence is arithmetic.

$$S_{50} = \sum_{k=1}^{50}(k+2) = \sum_{k=1}^{50} k + \sum_{k=1}^{50} 2$$
$$= \frac{50(50+1)}{2} + 2(50) = 1275 + 100 = 1375$$

71. $\{4n^2\}$ Examine the terms of the sequence: 4, 16, 36, 64, 100, ...

There is no common difference and there is no common ratio. Therefore the sequence is neither arithmetic nor geometric.

73. $\left\{3-\dfrac{2}{3}n\right\}$

$$d = \left(3-\frac{2}{3}(n+1)\right) - \left(3-\frac{2}{3}n\right)$$
$$= 3-\frac{2}{3}n-\frac{2}{3}-3+\frac{2}{3}n = -\frac{2}{3}$$

The difference between consecutive terms is constant. Therefore the sequence is arithmetic.

$$S_{50} = \sum_{k=1}^{50}\left(3-\frac{2}{3}k\right) = \sum_{k=1}^{50} 3 - \frac{2}{3}\sum_{k=1}^{50} k$$
$$= 3(50) - \frac{2}{3}\left(\frac{50(50+1)}{2}\right) = 150 - 850 = -700$$

75. 1, 3, 6, 10, ...

There is no common difference and there is no common ratio. Therefore the sequence is neither arithmetic nor geometric.

77. $\left\{ \left(\dfrac{2}{3}\right)^n \right\}$

$$r = \frac{\left(\dfrac{2}{3}\right)^{n+1}}{\left(\dfrac{2}{3}\right)^n} = \left(\frac{2}{3}\right)^{n+1-n} = \frac{2}{3}$$

The ratio of consecutive terms is constant.
Therefore the sequence is geometric.

$$S_{50} = \sum_{k=1}^{50} \left(\frac{2}{3}\right)^k = \frac{2}{3} \cdot \frac{1-\left(\dfrac{2}{3}\right)^{50}}{1-\dfrac{2}{3}} = 1.999999997$$

79. $-1, 2, -4, 8, \ldots$

$$r = \frac{2}{-1} = \frac{-4}{2} = \frac{8}{-4} = -2$$

The ratio of consecutive terms is constant.
Therefore the sequence is geometric.

$$S_{50} = \sum_{k=1}^{50} -1 \cdot (-2)^{k-1} = -1 \cdot \frac{1-(-2)^{50}}{1-(-2)}$$

$$\approx 3.752999689 \times 10^{14}$$

81. $\left\{ 3^{n/2} \right\}$

$$r = \frac{3^{\left(\frac{n+1}{2}\right)}}{3^{\left(\frac{n}{2}\right)}} = 3^{\left(\frac{n+1}{2}-\frac{n}{2}\right)} = 3^{1/2}$$

The ratio of consecutive terms is constant.
Therefore the sequence is geometric.

$$S_{50} = \sum_{k=1}^{50} 3^{k/2} = 3^{1/2} \cdot \frac{1-\left(3^{1/2}\right)^{50}}{1-3^{1/2}}$$

$$\approx 2.004706374 \times 10^{12}$$

83. Find the common ratio of the terms and solve the system of equations:

$$\frac{x+2}{x} = r; \quad \frac{x+3}{x+2} = r$$

$$\frac{x+2}{x} = \frac{x+3}{x+2} \rightarrow x^2 + 4x + 4 = x^2 + 3x \rightarrow x = -4$$

85. This is a geometric series with
$a_1 = \$42,000, \quad r = 1.03, \quad n = 5$. Find the 5th term:

$$a_5 = 42000(1.03)^{5-1} = 42000(1.03)^4 = \$47,271.37$$

87. a. Find the 10th term of the geometric sequence:

$$a_1 = 2, \ r = 0.9, \ n = 10$$

$$a_{10} = 2(0.9)^{10-1} = 2(0.9)^9 = 0.775 \text{ feet}$$

b. Find n when $a_n < 1$:

$$2(0.9)^{n-1} < 1$$

$$(0.9)^{n-1} < 0.5$$

$$(n-1)\log(0.9) < \log(0.5)$$

$$n-1 > \frac{\log(0.5)}{\log(0.9)}$$

$$n > \frac{\log(0.5)}{\log(0.9)} + 1 \approx 7.58$$

On the 8th swing the arc is less than 1 foot.

c. Find the sum of the first 15 swings:

$$S_{15} = 2\left(\frac{1-(0.9)^{15}}{1-0.9}\right) = 2\left(\frac{1-(0.9)^{15}}{0.1}\right)$$

$$= 20\left(1-(0.9)^{15}\right) = 15.88 \text{ feet}$$

d. Find the infinite sum of the geometric series:

$$S_\infty = \frac{2}{1-0.9} = \frac{2}{0.1} = 20 \text{ feet}$$

89. This is an ordinary annuity with $P = \$100$ and $n = (12)(30) = 360$ payment periods. The interest rate per period is $\dfrac{0.12}{12} = 0.01$. Thus,

$$A = 100\left[\frac{[1+0.01]^{360}-1}{0.01}\right] \approx \$349,496.41$$

91. This is an ordinary annuity with $P = \$500$ and $n = (4)(20) = 80$ payment periods. The interest rate per period is $\dfrac{0.08}{4} = 0.02$. Thus,

$$A = 500\left[\frac{[1+0.02]^{80}-1}{0.02}\right] \approx \$96,885.98$$

93. This is an ordinary annuity with $A = \$50,000$ and $n = (12)(10) = 120$ payment periods. The interest rate per period is $\dfrac{0.06}{12} = 0.005$. Thus,

$$50,000 = P\left[\dfrac{\left[1+0.005\right]^{120}-1}{0.005}\right]$$

$$P = 50,000\left[\dfrac{0.005}{\left[1+0.005\right]^{120}-1}\right] \approx \$305.10$$

95. This is a geometric sequence with $a_1 = 1$, $r = 2$, $n = 64$.
Find the sum of the geometric series:

$$S_{64} = 1\left(\dfrac{1-2^{64}}{1-2}\right) = \dfrac{1-2^{64}}{-1} = 2^{64}-1$$

$$= 1.845 \times 10^{19} \text{ grains}$$

97. The common ratio, $r = 0.90 < 1$. The sum is:

$$S = \dfrac{1}{1-0.9} = \dfrac{1}{0.10} = 10.$$

The multiplier is 10.

99. This is an infinite geometric series with $a = 4$, and $r = \dfrac{1.03}{1.09}$.

Find the sum: Price $= \dfrac{4}{\left(1-\dfrac{1.03}{1.09}\right)} \approx \$72.67$.

101. Given: $a_1 = 1000$, $r = 0.9$
Find n when $a_n < 0.01$:

$$1000(0.9)^{n-1} < 0.01$$

$$(0.9)^{n-1} < 0.00001$$

$$(n-1)\log(0.9) < \log(0.00001)$$

$$n-1 > \dfrac{\log(0.00001)}{\log(0.9)}$$

$$n > \dfrac{\log(0.00001)}{\log(0.9)} + 1 \approx 110.27$$

On the 111th day or December 20, 2015, the amount will be less than \$0.01.

Find the sum of the geometric series:

$$S_{111} = a_1\left(\dfrac{1-r^n}{1-r}\right) = 1000\left(\dfrac{1-(0.9)^{111}}{1-0.9}\right)$$

$$= 1000\left(\dfrac{1-(0.9)^{111}}{0.1}\right) = \$9999.92$$

103. First, determine the number of seats in the section:

$$S_n = \dfrac{n}{2}\left[2a_1 + (n-1)d\right]$$

$$= \dfrac{14}{2}\left[2\cdot 2 + 13\cdot 2\right] = 210 \text{ seats}$$

Now, find the sum of a geometric sequence with $a_1 = 0.01$ and $r = 1.05$ and $n = 210$.

$$S_{210} = 0.01\,\dfrac{1-(1.05)^{210}}{1-1.05} = \$5633.36$$

105. Find the sum of each sequence:
A: Arithmetic series with:
$$a_1 = \$1000, \quad d = -1, \quad n = 1000$$

Find the sum of the arithmetic series:

$$S_{1000} = \dfrac{1000}{2}(1000+1) = 500(1001) = \$500,500$$

B: This is a geometric sequence with
$$a\,1 = 1, \quad r = 2, \quad n = 19.$$
Find the sum of the geometric series:

$$S_{19} = 1\left(\dfrac{1-2^{19}}{1-2}\right) = \dfrac{1-2^{19}}{-1} = 2^{19}-1 = \$524,287$$

B results in more money.

107. The amount paid each day forms a geometric sequence with $a_1 = 0.01$ and $r = 2$.

$$S_{22} = a_1 \cdot \dfrac{1-r^{22}}{1-r} = 0.01 \cdot \dfrac{1-2^{22}}{1-2} = 41,943.03$$

The total payment would be \$41,943.03 if you worked all 22 days.

$$a_{22} = a_1 \cdot r^{22-1} = 0.01(2)^{21} = 20,971.52$$

The payment on the 22nd day is \$20,971.52.
Answers will vary. With this payment plan, the bulk of the payment is at the end so missing even one day can dramatically reduce the overall payment. Notice that with one sick day you would lose the amount paid on the 22nd day which is about half the total payment for the 22 days.

109. Answers will vary.

111. Answers will vary. Both increase (or decrease) exponentially, but the domain of a geometric sequence is the set of natural numbers while the domain of an exponential function is the set of all real numbers.

113. $P(x) = 8x^4 - 2x^3 + x - 8;$ $c = -2$

$f(-2) = 8(-2)^4 - 2(-2)^3 + (-2) - 8$
$= 128 + 16 - 2 - 8 = 134$

Thus, the remainder when $P(x)$ is divided by $x + 2$ is 134.

115.

$$\begin{vmatrix} 3 & 1 & 0 \\ 0 & -2 & 6 \\ 4 & -1 & -2 \end{vmatrix} = 3 \begin{vmatrix} -2 & 6 \\ -1 & -2 \end{vmatrix} - 1 \begin{vmatrix} 0 & 6 \\ 4 & -2 \end{vmatrix} + 0 \begin{vmatrix} 0 & -2 \\ 4 & -1 \end{vmatrix}$$
$$= 3(4 + 6) - 1(0 - 24) + 0$$
$$= 3(10) - (-24) = 30 + 24$$
$$= 54$$

Section 9.4

1. I: $n = 1:$ $2 \cdot 1 = 2$ and $1(1+1) = 2$

II: If $2 + 4 + 6 + \cdots + 2k = k(k+1)$, then
$2 + 4 + 6 + \cdots + 2k + 2(k+1)$
$= [2 + 4 + 6 + \cdots + 2k] + 2(k+1)$
$= k(k+1) + 2(k+1)$
$= (k+1)(k+2)$
$= (k+1)((k+1)+1)$

Conditions I and II are satisfied; the statement is true.

3. I: $n = 1:$ $1 + 2 = 3$ and $\dfrac{1}{2} \cdot 1(1+5) = 3$

II: If $3 + 4 + 5 + \cdots + (k+2) = \dfrac{1}{2} \cdot k(k+5)$, then
$3 + 4 + 5 + \cdots + (k+2) + [(k+1)+2]$
$= [3 + 4 + 5 + \cdots + (k+2)] + (k+3)$
$= \dfrac{1}{2} \cdot k(k+5) + (k+3)$
$= \dfrac{1}{2} k^2 + \dfrac{5}{2} k + k + 3$
$= \dfrac{1}{2} k^2 + \dfrac{7}{2} k + 3$
$= \dfrac{1}{2} \cdot (k^2 + 7k + 6)$
$= \dfrac{1}{2} \cdot (k+1)(k+6)$
$= \dfrac{1}{2} \cdot (k+1)((k+1)+5)$

Conditions I and II are satisfied; the statement is true.

5. I: $n = 1:$ $3 \cdot 1 - 1 = 2$ and $\dfrac{1}{2} \cdot 1(3 \cdot 1 + 1) = 2$

II: If $2 + 5 + 8 + \cdots + (3k - 1) = \dfrac{1}{2} \cdot k(3k+1)$, then
$2 + 5 + 8 + \cdots + (3k-1) + [3(k+1)-1]$
$= [2 + 5 + 8 + \cdots + (3k-1)] + (3k+2)$
$= \dfrac{1}{2} \cdot k(3k+1) + (3k+2) = \dfrac{3}{2} k^2 + \dfrac{1}{2} k + 3k + 2$
$= \dfrac{3}{2} k^2 + \dfrac{7}{2} k + 2 = \dfrac{1}{2} \cdot (3k^2 + 7k + 4)$
$= \dfrac{1}{2} \cdot (k+1)(3k+4)$
$= \dfrac{1}{2} \cdot (k+1)(3(k+1)+1)$

Conditions I and II are satisfied; the statement is true.

7. I: $n=1$: $2^{1-1}=1$ and $2^1-1=1$

II: If $1+2+2^2+\cdots+2^{k-1}=2^k-1$, then

$$1+2+2^2+\cdots+2^{k-1}+2^{k+1-1}$$

$$=\left[1+2+2^2+\cdots+2^{k-1}\right]+2^k$$

$$=2^k-1+2^k=2\cdot2^k-1$$

$$=2^{k+1}-1$$

Conditions I and II are satisfied; the statement is true.

9. I: $n=1$: $4^{1-1}=1$ and $\dfrac{1}{3}\cdot\left(4^1-1\right)=1$

II: If $1+4+4^2+\cdots+4^{k-1}=\dfrac{1}{3}\cdot\left(4^k-1\right)$, then

$$1+4+4^2+\cdots+4^{k-1}+4^{k+1-1}$$

$$=\left[1+4+4^2+\cdots+4^{k-1}\right]+4^k$$

$$=\frac{1}{3}\cdot\left(4^k-1\right)+4^k=\frac{1}{3}\cdot4^k-\frac{1}{3}+4^k$$

$$=\frac{4}{3}\cdot4^k-\frac{1}{3}=\frac{1}{3}\left(4\cdot4^k-1\right)$$

$$=\frac{1}{3}\cdot\left(4^{k+1}-1\right)$$

Conditions I and II are satisfied; the statement is true.

11. I: $n=1$: $\dfrac{1}{1(1+1)}=\dfrac{1}{2}$ and $\dfrac{1}{1+1}=\dfrac{1}{2}$

II: If $\dfrac{1}{1\cdot2}+\dfrac{1}{2\cdot3}+\dfrac{1}{3\cdot4}+\cdots+\dfrac{1}{k(k+1)}=\dfrac{k}{k+1}$, then

$$\frac{1}{1\cdot2}+\frac{1}{2\cdot3}+\frac{1}{3\cdot4}+\cdots+\frac{1}{k(k+1)}+\frac{1}{(k+1)(k+1+1)}=\left[\frac{1}{1\cdot2}+\frac{1}{2\cdot3}+\frac{1}{3\cdot4}+\cdots+\frac{1}{k(k+1)}\right]+\frac{1}{(k+1)(k+2)}$$

$$=\frac{k}{k+1}+\frac{1}{(k+1)(k+2)}=\frac{k}{k+1}\cdot\frac{k+2}{k+2}+\frac{1}{(k+1)(k+2)}$$

$$=\frac{k^2+2k+1}{(k+1)(k+2)}=\frac{(k+1)(k+1)}{(k+1)(k+2)}=\frac{k+1}{k+2}=\frac{k+1}{(k+1)+1}$$

Conditions I and II are satisfied; the statement is true.

13. I: $n=1$: $1^2=1$ and $\dfrac{1}{6}\cdot1(1+1)(2\cdot1+1)=1$

II: If $1^2+2^2+3^2+\cdots+k^2=\dfrac{1}{6}\cdot k(k+1)(2k+1)$, then

$$1^2+2^2+3^2+\cdots+k^2+(k+1)^2=\left[1^2+2^2+3^2+\cdots+k^2\right]+(k+1)^2=\frac{1}{6}k(k+1)(2k+1)+(k+1)^2$$

$$=(k+1)\left[\frac{1}{6}k(2k+1)+k+1\right]=(k+1)\left[\frac{1}{3}k^2+\frac{1}{6}k+k+1\right]=(k+1)\left[\frac{1}{3}k^2+\frac{7}{6}k+1\right]=\frac{1}{6}(k+1)\left[2k^2+7k+6\right]$$

$$=\frac{1}{6}\cdot(k+1)(k+2)(2k+3)$$

$$=\frac{1}{6}\cdot(k+1)\big((k+1)+1\big)\big(2(k+1)+1\big)$$

Conditions I and II are satisfied; the statement is true.

15. I: $n=1$: $5-1=4$ and $\frac{1}{2}\cdot 1(9-1)=4$

  II: If $4+3+2+\cdots+(5-k)=\frac{1}{2}\cdot k(9-k)$, then

$$4+3+2+\cdots+(5-k)+(5-(k+1))=\left[4+3+2+\cdots+(5-k)\right]+(4-k)=\frac{1}{2}k(9-k)+(4-k)$$

$$=\frac{9}{2}k-\frac{1}{2}k^2+4-k=-\frac{1}{2}k^2+\frac{7}{2}k+4=-\frac{1}{2}\cdot\left[k^2-7k-8\right]$$

$$=-\frac{1}{2}\cdot(k+1)(k-8)=\frac{1}{2}\cdot(k+1)(8-k)=\frac{1}{2}\cdot(k+1)\left[9-(k+1)\right]$$

Conditions I and II are satisfied; the statement is true.

17. I: $n=1$: $1(1+1)=2$ and $\frac{1}{3}\cdot 1(1+1)(1+2)=2$

  II: If $1\cdot 2+2\cdot 3+3\cdot 4+\cdots+k(k+1)=\frac{1}{3}\cdot k(k+1)(k+2)$, then

$$1\cdot 2+2\cdot 3+3\cdot 4+\cdots+k(k+1)+(k+1)(k+1+1)=\left[1\cdot 2+2\cdot 3+3\cdot 4+\cdots+k(k+1)\right]+(k+1)(k+2)$$

$$=\frac{1}{3}\cdot k(k+1)(k+2)+(k+1)(k+2)=(k+1)(k+2)\left[\frac{1}{3}k+1\right]$$

$$=\frac{1}{3}\cdot(k+1)(k+2)(k+3)$$

$$=\frac{1}{3}\cdot(k+1)((k+1)+1)((k+1)+2)$$

Conditions I and II are satisfied; the statement is true.

19. I: $n=1$: $1^2+1=2$ is divisible by 2

  II: If k^2+k is divisible by 2, then
$$(k+1)^2+(k+1)=k^2+2k+1+k+1$$
$$=(k^2+k)+(2k+2)$$
Since k^2+k is divisible by 2 and $2k+2$ is divisible by 2, then $(k+1)^2+(k+1)$ is divisible by 2.

Conditions I and II are satisfied; the statement is true.

21. I: $n=1$: $1^2-1+2=2$ is divisible by 2

  II: If k^2-k+2 is divisible by 2, then
$$(k+1)^2-(k+1)+2=k^2+2k+1-k-1+2$$
$$=(k^2-k+2)+(2k)$$
Since k^2-k+2 is divisible by 2 and $2k$ is divisible by 2, then $(k+1)^2-(k+1)+2$ is divisible by 2.

Conditions I and II are satisfied; the statement is true.

23. I: $n=1$: If $x>1$ then $x^1=x>1$.

  II: Assume, for some natural number k, that if $x>1$, then $x^k>1$.
Then $x^{k+1}>1$, for $x>1$,
$$x^{k+1}=x^k\cdot x>1\cdot x=x>1$$
$$\uparrow$$
$$(x^k>1)$$

Conditions I and II are satisfied; the statement is true.

25. I: $n=1$: $a-b$ is a factor of $a^1-b^1=a-b$.

II: If $a-b$ is a factor of a^k-b^k, show that $a-b$ is a factor of $a^{k+1}-b^{k+1}$.

$$a^{k+1}-b^{k+1}=a\cdot a^k-b\cdot b^k$$
$$=a\cdot a^k-a\cdot b^k+a\cdot b^k-b\cdot b^k$$
$$=a\left(a^k-b^k\right)+b^k(a-b)$$

Since $a-b$ is a factor of a^k-b^k and $a-b$ is a factor of $a-b$, then $a-b$ is a factor of $a^{k+1}-b^{k+1}$.

Conditions I and II are satisfied; the statement is true.

27. I: $n=1$: $(1+a)^1=1+a\ge 1+1\cdot a$

II: Assume that there is an integer k for which the inequality holds. We need to show that if $(1+a)^k\ge 1+ka$ then

$$(1+a)^{k+1}\ge 1+(k+1)a.$$

$$(1+a)^{k+1}=(1+a)^k(1+a)$$
$$\ge (1+ka)(1+a)$$
$$=1+ka^2+a+ka$$
$$=1+(k+1)a+ka^2$$
$$\ge 1+(k+1)a$$

Conditions I and II are satisfied, the statement is true.

29. II: If $2+4+6+\cdots+2k=k^2+k+2$, then

$$2+4+6+\cdots+2k+2(k+1)$$
$$=\left[2+4+6+\cdots+2k\right]+2k+2$$
$$=k^2+k+2+2k+2$$
$$=(k^2+2k+1)+(k+1)+2$$
$$=(k+1)^2+(k+1)+2$$

I: $n=1$: $2\cdot 1=2$ and $1^2+1+2=4\ne 2$

31. I: $n=1$:

$$a+(1-1)d=a \text{ and } 1\cdot a+d\frac{1(1-1)}{2}=a$$

II: If $a+(a+d)+(a+2d)+\cdots+[a+(k-1)d]$
$$=ka+d\frac{k(k-1)}{2}$$

then
$$a+(a+d)+(a+2d)+\cdots+[a+(k-1)d]+(a+kd)$$
$$=[a+(a+d)+(a+2d)+\cdots+[a+(k-1)d]]+(a+kd)$$
$$=ka+d\frac{k(k-1)}{2}+(a+kd)$$
$$=(k+1)a+d\left[\frac{k(k-1)}{2}+k\right]$$
$$=(k+1)a+d\left[\frac{k^2-k+2k}{2}\right]$$
$$=(k+1)a+d\left[\frac{k^2+k}{2}\right]$$
$$=(k+1)a+d\left[\frac{(k+1)k}{2}\right]$$
$$=(k+1)a+d\left[\frac{(k+1)((k+1)-1)}{2}\right]$$

Conditions I and II are satisfied; the statement is true.

33. I: $n=3$: $(3-2)\cdot 180°=180°$ which is the sum of the angles of a triangle.

II: Assume that for any integer k, the sum of the angles of a convex polygon with k sides is $(k-2)\cdot 180°$. A convex polygon with $k+1$ sides consists of a convex polygon with k sides plus a triangle. Thus, the sum of the angles is
$$(k-2)\cdot 180°+180°=((k+1)-2)\cdot 180°.$$

Conditions I and II are satisfied; the statement is true.

35. $\log_2\sqrt{x+5}=4$
$$2^4=\sqrt{x+5}$$
$$16=\sqrt{x+5}$$
$$256=x+5\rightarrow x=251$$

37.
$$4x + 3y = -7$$
$$2x - 5y = 16$$

Multiply each side of the second equation by -2 and add to eliminate x:

$$4x + 3y = -7$$
$$\underline{-4x + 10y = -32}$$
$$13y = -39 \rightarrow y = -3$$

Substitute and solve for x:
$$4x + 3(-3) = -7$$
$$4x - 9 = -7$$
$$4x = 2 \rightarrow x = \frac{1}{2}$$

The solution is $x = \frac{1}{2}$, $y = -3$ or $\left\{\frac{1}{2}, -3\right\}$.

Section 9.5

1. Pascal Triangle

3. False; $\displaystyle \binom{n}{j} = \frac{n!}{j!(n-j)!}$

5. $\displaystyle \binom{5}{3} = \frac{5!}{3!\,2!} = \frac{5 \cdot 4 \cdot 3 \cdot 2 \cdot 1}{3 \cdot 2 \cdot 1 \cdot 2 \cdot 1} = \frac{5 \cdot 4}{2 \cdot 1} = 10$

7. $\displaystyle \binom{7}{5} = \frac{7!}{5!\,2!} = \frac{7 \cdot 6 \cdot 5 \cdot 4 \cdot 3 \cdot 2 \cdot 1}{5 \cdot 4 \cdot 3 \cdot 2 \cdot 1 \cdot 2 \cdot 1} = \frac{7 \cdot 6}{2 \cdot 1} = 21$

9. $\displaystyle \binom{50}{49} = \frac{50!}{49!\,1!} = \frac{50 \cdot 49!}{49! \cdot 1} = \frac{50}{1} = 50$

11. $\displaystyle \binom{1000}{1000} = \frac{1000!}{1000!\,0!} = \frac{1}{1} = 1$

13. $\displaystyle \binom{55}{23} = \frac{55!}{23!\,32!} \approx 1.8664 \times 10^{15}$

15. $\displaystyle \binom{47}{25} = \frac{47!}{25!\,22!} \approx 1.4834 \times 10^{13}$

17. $(x+1)^5 = \binom{5}{0}x^5 + \binom{5}{1}x^4 + \binom{5}{2}x^3 + \binom{5}{3}x^2 + \binom{5}{4}x^1 + \binom{5}{5}x^0 = x^5 + 5x^4 + 10x^3 + 10x^2 + 5x + 1$

19. $(x-2)^6 = \binom{6}{0}x^6 + \binom{6}{1}x^5(-2) + \binom{6}{2}x^4(-2)^2 + \binom{6}{3}x^3(-2)^3 + \binom{6}{4}x^2(-2)^4 + \binom{6}{5}x(-2)^5 + \binom{6}{6}x^0(-2)^6$

$\qquad = x^6 + 6x^5(-2) + 15x^4 \cdot 4 + 20x^3(-8) + 15x^2 \cdot 16 + 6x \cdot (-32) + 64$

$\qquad = x^6 - 12x^5 + 60x^4 - 160x^3 + 240x^2 - 192x + 64$

26.
$\left(\sqrt{x} - \sqrt{3}\right)^4 = \binom{4}{0}\left(\sqrt{x}\right)^4 + \binom{4}{1}\left(\sqrt{x}\right)^3\left(-\sqrt{3}\right)^1 + \binom{4}{2}\left(\sqrt{x}\right)^2\left(-\sqrt{3}\right)^2 + \binom{4}{3}\left(\sqrt{x}\right)\left(-\sqrt{3}\right)^3 + \binom{4}{4}\left(-\sqrt{3}\right)^4$

$\qquad = x^2 - 4\sqrt{3}x^{3/2} + 6 \cdot 3x - 4 \cdot 3\sqrt{3}x^{1/2} + 9$

$\qquad = x^2 - 4\sqrt{3}x^{3/2} + 18x - 12\sqrt{3}x^{1/2} + 9$

21. $(3x+1)^4 = \binom{4}{0}(3x)^4 + \binom{4}{1}(3x)^3 + \binom{4}{2}(3x)^2 + \binom{4}{3}(3x) + \binom{4}{4}$

$\qquad = 81x^4 + 4 \cdot 27x^3 + 6 \cdot 9x^2 + 4 \cdot 3x + 1 = 81x^4 + 108x^3 + 54x^2 + 12x + 1$

23. $\left(x^2 + y^2\right)^5 = \binom{5}{0}\left(x^2\right)^5 + \binom{5}{1}\left(x^2\right)^4 y^2 + \binom{5}{2}\left(x^2\right)^3\left(y^2\right)^2 + \binom{5}{3}\left(x^2\right)^2\left(y^2\right)^3 + \binom{5}{4}x^2\left(y^2\right)^4 + \binom{5}{5}\left(y^2\right)^5$

$\qquad = x^{10} + 5x^8 y^2 + 10x^6 y^4 + 10x^4 y^6 + 5x^2 y^8 + y^{10}$

25. $\left(\sqrt{x}+\sqrt{2}\right)^6 = \binom{6}{0}\left(\sqrt{x}\right)^6 + \binom{6}{1}\left(\sqrt{x}\right)^5\left(\sqrt{2}\right)^1 + \binom{6}{2}\left(\sqrt{x}\right)^4\left(\sqrt{2}\right)^2 + \binom{6}{3}\left(\sqrt{x}\right)^3\left(\sqrt{2}\right)^3$

$\qquad + \binom{6}{4}\left(\sqrt{x}\right)^2\left(\sqrt{2}\right)^4 + \binom{6}{5}\left(\sqrt{x}\right)\left(\sqrt{2}\right)^5 + \binom{6}{6}\left(\sqrt{2}\right)^6$

$\qquad = x^3 + 6\sqrt{2}x^{5/2} + 15\cdot 2x^2 + 20\cdot 2\sqrt{2}x^{3/2} + 15\cdot 4x + 6\cdot 4\sqrt{2}x^{1/2} + 8$

$\qquad = x^3 + 6\sqrt{2}x^{5/2} + 30x^2 + 40\sqrt{2}x^{3/2} + 60x + 24\sqrt{2}x^{1/2} + 8$

27. $(ax+by)^5 = \binom{5}{0}(ax)^5 + \binom{5}{1}(ax)^4\cdot by + \binom{5}{2}(ax)^3(by)^2 + \binom{5}{3}(ax)^2(by)^3 + \binom{5}{4}ax(by)^4 + \binom{5}{5}(by)^5$

$\qquad = a^5 x^5 + 5a^4 x^4 by + 10a^3 x^3 b^2 y^2 + 10a^2 x^2 b^3 y^3 + 5axb^4 y^4 + b^5 y^5$

29. $n=10,\ j=4,\ x=x,\ a=3$

$\binom{10}{4}x^6\cdot 3^4 = \dfrac{10!}{4!\,6!}\cdot 81x^6 = \dfrac{10\cdot 9\cdot 8\cdot 7}{4\cdot 3\cdot 2\cdot 1}\cdot 81x^6$

$\qquad\qquad = 17{,}010x^6$

The coefficient of x^6 is 17,010.

31. $n=12,\ j=5,\ x=2x,\ a=-1$

$\binom{12}{5}(2x)^7\cdot(-1)^5 = \dfrac{12!}{5!\,7!}\cdot 128x^7(-1)$

$\qquad\qquad = \dfrac{12\cdot 11\cdot 10\cdot 9\cdot 8}{5\cdot 4\cdot 3\cdot 2\cdot 1}\cdot(-128)x^7$

$\qquad\qquad = -101{,}376x^7$

The coefficient of x^7 is $-101{,}376$.

33. $n=9,\ j=2,\ x=2x,\ a=3$

$\binom{9}{2}(2x)^7\cdot 3^2 = \dfrac{9!}{2!\,7!}\cdot 128x^7(9)$

$\qquad\qquad = \dfrac{9\cdot 8}{2\cdot 1}\cdot 128x^7\cdot 9$

$\qquad\qquad = 41{,}472x^7$

The coefficient of x^7 is 41,472.

35. $n=7,\ j=4,\ x=x,\ a=3$

$\binom{7}{4}x^3\cdot 3^4 = \dfrac{7!}{4!\,3!}\cdot 81x^3 = \dfrac{7\cdot 6\cdot 5}{3\cdot 2\cdot 1}\cdot 81x^3 = 2835x^3$

37. $n=9,\ j=2,\ x=3x,\ a=-2$

$\binom{9}{2}(3x)^7\cdot(-2)^2 = \dfrac{9!}{2!\,7!}\cdot 2187x^7\cdot 4$

$\qquad\qquad = \dfrac{9\cdot 8}{2\cdot 1}\cdot 8748x^7 = 314{,}928x^7$

39. The x^0 term in

$\sum_{j=0}^{12}\binom{12}{j}\left(x^2\right)^{12-j}\left(\dfrac{1}{x}\right)^j = \sum_{j=0}^{12}\binom{12}{j}x^{24-3j}$

occurs when:

$24-3j=0$

$\qquad 24=3j$

$\qquad\ j=8$

The coefficient is

$\binom{12}{8} = \dfrac{12!}{8!\,4!} = \dfrac{12\cdot 11\cdot 10\cdot 9}{4\cdot 3\cdot 2\cdot 1} = 495$

41. The x^4 term in

$\sum_{j=0}^{10}\binom{10}{j}(x)^{10-j}\left(\dfrac{-2}{\sqrt{x}}\right)^j = \sum_{j=0}^{10}\binom{10}{j}(-2)^j\,x^{10-\frac{3}{2}j}$

occurs when:

$10-\dfrac{3}{2}j=4$

$\qquad -\dfrac{3}{2}j=-6$

$\qquad\qquad j=4$

The coefficient is

$\binom{10}{4}(-2)^4 = \dfrac{10!}{6!\,4!}\cdot 16 = \dfrac{10\cdot 9\cdot 8\cdot 7}{4\cdot 3\cdot 2\cdot 1}\cdot 16 = 3360$

43. $(1.001)^5 = (1+10^{-3})^5 = \binom{5}{0}1^5 + \binom{5}{1}1^4 \cdot 10^{-3} + \binom{5}{2}1^3 \cdot (10^{-3})^2 + \binom{5}{3}1^2 \cdot (10^{-3})^3 + \cdots$

$= 1 + 5(0.001) + 10(0.000001) + 10(0.000000001) + \cdots$

$= 1 + 0.005 + 0.000010 + 0.000000010 + \cdots$

$= 1.00501 \quad$ (correct to 5 decimal places)

45. $\binom{n}{n-1} = \frac{n!}{(n-1)!(n-(n-1))!} = \frac{n!}{(n-1)!(1)!} = \frac{n(n-1)!}{(n-1)!} = n$

$\binom{n}{n} = \frac{n!}{n!(n-n)!} = \frac{n!}{n!\,0!} = \frac{n!}{n!\cdot 1} = \frac{n!}{n!} = 1$

47. Show that $\binom{n}{0} + \binom{n}{1} + \cdots + \binom{n}{n} = 2^n$

$2^n = (1+1)^n$

$= \binom{n}{0} \cdot 1^n + \binom{n}{1} \cdot 1^{n-1} \cdot 1 + \binom{n}{2} \cdot 1^{n-2} \cdot 1^2 + \cdots + \binom{n}{n} \cdot 1^{n-n} \cdot 1^n$

$= \binom{n}{0} + \binom{n}{1} + \cdots + \binom{n}{n}$

49. $\binom{5}{0}\left(\frac{1}{4}\right)^5 + \binom{5}{1}\left(\frac{1}{4}\right)^4\left(\frac{3}{4}\right) + \binom{5}{2}\left(\frac{1}{4}\right)^3\left(\frac{3}{4}\right)^2 + \binom{5}{3}\left(\frac{1}{4}\right)^2\left(\frac{3}{4}\right)^3 + \binom{5}{4}\left(\frac{1}{4}\right)\left(\frac{3}{4}\right)^4 + \binom{5}{5}\left(\frac{3}{4}\right)^5 = \left(\frac{1}{4}+\frac{3}{4}\right)^5 = (1)^5 = 1$

51.
$6^x = 5^{x+1}$

$\ln 6^x = \ln 5^{x+1}$

$x \ln 6 = (x+1)\ln 5$

$x \ln 6 = x \ln 5 + \ln 5$

$x \ln 6 - x \ln 5 = \ln 5$

$x(\ln 6 - \ln 5) = \ln 5$

$x = \frac{\ln 5}{\ln 6 - \ln 5} \approx 8.827$

53.
$x - y - z = 0$
$2x + y + 3z = -1$
$4x + 2y - z = 12$

Add the first equation and the second equation to eliminate y:

$x - y - z = 0$
$\underline{2x + y + 3z = -1}$
$3x + 2z \quad = -1$

Multiply each side of the first equation by 2 and add to the third equation to eliminate y:

$2x - 2y - 2z = 0$
$\underline{4x + 2y - z = 12}$
$6x - 3z \quad = 12$
$2x - z = 4$

Multiple the second derived equation by 2 and add the two results to eliminate z:

$3x + 2z = -1$
$\underline{4x - 2z = 8}$
$7x = 7$
$x = 1$

Substituting and solving for the other variables:

$2(1) - z = 4$
$-z = 2$
$z = -2$

$1 - y - (-2) = 0$
$1 - y + 2 = 0$
$-y = -3$
$y = 3$

The solution is $x = 1$, $y = 3$, $z = -2$ or $(1,3,-2)$.

Chapter 9 Review Exercises

1. $a_1 = (-1)^1 \dfrac{1+3}{1+2} = -\dfrac{4}{3}, \ a_2 = (-1)^2 \dfrac{2+3}{2+2} = \dfrac{5}{4}, \ a_3 = (-1)^3 \dfrac{3+3}{3+2} = -\dfrac{6}{5}, \ a_4 = (-1)^4 \dfrac{4+3}{4+2} = \dfrac{7}{6}, \ a_5 = (-1)^5 \dfrac{5+3}{5+2} = -\dfrac{8}{7}$

2. $c_1 = \dfrac{2^1}{1^2} = \dfrac{2}{1} = 2, \ c_2 = \dfrac{2^2}{2^2} = \dfrac{4}{4} = 1, \ c_3 = \dfrac{2^3}{3^2} = \dfrac{8}{9}, \ c_4 = \dfrac{2^4}{4^2} = \dfrac{16}{16} = 1, \ c_5 = \dfrac{2^5}{5^2} = \dfrac{32}{25}$

3. $a_1 = 3, \ a_2 = \dfrac{2}{3} \cdot 3 = 2, \ a_3 = \dfrac{2}{3} \cdot 2 = \dfrac{4}{3}, \ a_4 = \dfrac{2}{3} \cdot \dfrac{4}{3} = \dfrac{8}{9}, \ a_5 = \dfrac{2}{3} \cdot \dfrac{8}{9} = \dfrac{16}{27}$

4. $a_1 = 2, \ a_2 = 2 - 2 = 0, \ a_3 = 2 - 0 = 2, \ a_4 = 2 - 2 = 0, \ a_5 = 2 - 0 = 2$

5. $\displaystyle\sum_{k=1}^{4} (4k+2) = (4 \cdot 1 + 2) + (4 \cdot 2 + 2) + (4 \cdot 3 + 2) + (4 \cdot 4 + 2) = (6) + (10) + (14) + (18) = 48$

6. $1 - \dfrac{1}{2} + \dfrac{1}{3} - \dfrac{1}{4} + \cdots + \dfrac{1}{13} = \displaystyle\sum_{k=1}^{13} (-1)^{k+1}\left(\dfrac{1}{k}\right)$

7. $\{a_n\} = \{n+5\}$ Arithmetic
   $d = (n+1+5) - (n+5) = n + 6 - n - 5 = 1$
   $S_n = \dfrac{n}{2}[6 + n + 5] = \dfrac{n}{2}(n+11)$

8. $\{c_n\} = \{2n^3\}$ Examine the terms of the sequence: 2, 16, 54, 128, 250, ...
   There is no common difference; there is no common ratio; neither.

9. $\{s_n\} = \{2^{3n}\}$ Geometric
   $r = \dfrac{2^{3(n+1)}}{2^{3n}} = \dfrac{2^{3n+3}}{2^{3n}} = 2^{3n+3-3n} = 2^3 = 8$
   $S_n = 8\left(\dfrac{1-8^n}{1-8}\right) = 8\left(\dfrac{1-8^n}{-7}\right) = \dfrac{8}{7}\left(8^n - 1\right)$

10. 0, 4, 8, 12, ... Arithmetic $\quad d = 4 - 0 = 4$
    $S_n = \dfrac{n}{2}(2(0) + (n-1)4) = \dfrac{n}{2}(4(n-1)) = 2n(n-1)$

11. $3, \dfrac{3}{2}, \dfrac{3}{4}, \dfrac{3}{8}, \dfrac{3}{16}, \ldots$ Geometric
    $r = \dfrac{\left(\dfrac{3}{2}\right)}{3} = \dfrac{3}{2} \cdot \dfrac{1}{3} = \dfrac{1}{2}$
    $S_n = 3\left(\dfrac{1-\left(\dfrac{1}{2}\right)^n}{1-\dfrac{1}{2}}\right) = 3\left(\dfrac{1-\left(\dfrac{1}{2}\right)^n}{\left(\dfrac{1}{2}\right)}\right) = 6\left(1 - \left(\dfrac{1}{2}\right)^n\right)$

12. Neither. There is no common difference or common ratio.

13. $\displaystyle\sum_{k=1}^{30} (k^2 + 2) = \displaystyle\sum_{k=1}^{30} k^2 + \displaystyle\sum_{k=1}^{30} 2$
    $= \dfrac{30(30+1)(2 \cdot 30 + 1)}{6} + 2(30) = 9515$

14. $\displaystyle\sum_{k=1}^{40} (-2k+8) = \displaystyle\sum_{k=1}^{40} -2k + \displaystyle\sum_{k=1}^{40} 8$
    $= -2\displaystyle\sum_{k=1}^{40} k + \displaystyle\sum_{k=1}^{40} 8$
    $= -2\left(\dfrac{40(1+40)}{2}\right) + 40(8)$
    $= -1640 + 320 = -1320$

587

15. $\displaystyle\sum_{k=1}^{7}\left(\frac{1}{3}\right)^{k} = \frac{1}{3}\left[\dfrac{1-\left(\dfrac{1}{3}\right)^{7}}{1-\dfrac{1}{3}}\right] = \frac{1}{3}\left[\dfrac{1-\left(\dfrac{1}{3}\right)^{7}}{\left(\dfrac{2}{3}\right)}\right]$

$\quad = \dfrac{1}{2}\left(1-\dfrac{1}{2187}\right)$

$\quad = \dfrac{1}{2}\cdot\dfrac{2186}{2187} = \dfrac{1093}{2187} \approx 0.49977$

16. $\displaystyle\sum_{k=1}^{10}(-2)^{k} = -2\left(\dfrac{1-(-2)^{10}}{1-(-2)}\right)$

$\quad = -2\left(\dfrac{1-1024}{3}\right) = -\dfrac{2}{3}(-1023)$

$\quad = 682$

17. Arithmetic

$a_1 = 3, \ d = 4, \ a_n = a_1 + (n-1)d$

$a_9 = 3 + (9-1)4 = 3 + 8(4) = 3 + 32 = 35$

18. Geometric

$a_1 = 1, \ r = \dfrac{1}{10}, \ n = 11; \ a_n = a_1 r^{n-1}$

$a_{11} = 1\cdot\left(\dfrac{1}{10}\right)^{11-1} = \left(\dfrac{1}{10}\right)^{10}$

$\quad = \dfrac{1}{10,000,000,000}$

19. Arithmetic

$a_1 = \sqrt{2}, \ d = \sqrt{2}, \ n = 9, \ a_n = a_1 + (n-1)d$

$a_9 = \sqrt{2} + (9-1)\sqrt{2} = \sqrt{2} + 8\sqrt{2}$

$\quad = 9\sqrt{2} \approx 12.7279$

20. $a_7 = a_1 + 6d = 31 \qquad a_{20} = a_1 + 19d = 96$;

Solve the system of equations:

$a_1 + 6d = 31$

$a_1 + 19d = 96$

Subtract the second equation from the first equation and solve for d.

$-13d = -65$

$\quad d = 5$

$a_1 = 31 - 6(5) = 31 - 30 = 1$

$a_n = a_1 + (n-1)d$

$\quad = 1 + (n-1)(5)$

$\quad = 1 + 5n - 5$

$\quad = 5n - 4$

General formula: $\{a_n\} = \{5n-4\}$

21. $a_{10} = a_1 + 9d = 0 \qquad a_{18} = a_1 + 17d = 8$;

Solve the system of equations:

$a_1 + 9d = 0$

$a_1 + 17d = 8$

Subtract the second equation from the first equation and solve for d.

$-8d = -8$

$\quad d = 1$

$a_1 = -9(1) = -9$

$a_n = a_1 + (n-1)d$

$\quad = -9 + (n-1)(1)$

$\quad = -9 + n - 1$

$\quad = n - 10$

General formula: $\{a_n\} = \{n-10\}$

22. $a_1 = 3, \ r = \dfrac{1}{3}$

Since $|r| < 1$, the series converges.

$S_n = \dfrac{a_1}{1-r} = \dfrac{3}{\left(1-\dfrac{1}{3}\right)} = \dfrac{3}{\left(\dfrac{2}{3}\right)} = \dfrac{9}{2}$

23. $a_1 = 2, \ r = -\dfrac{1}{2}$

Since $|r| < 1$, the series converges.

$S_n = \dfrac{a_1}{1-r} = \dfrac{2}{\left(1-\left(-\dfrac{1}{2}\right)\right)} = \dfrac{2}{\left(\dfrac{3}{2}\right)} = \dfrac{4}{3}$

24. $a_1 = \dfrac{1}{2}, \ r = \dfrac{3}{2}$

Since $|r| > 1$, the series diverges.

25. $a_1 = 4, \; r = \dfrac{1}{2}$

Since $|r| < 1$, the series converges.

$$S_n = \frac{a_1}{1-r} = \frac{4}{\left(1-\dfrac{1}{2}\right)} = \frac{4}{\left(\dfrac{1}{2}\right)} = 8$$

26. I: $\quad n=1: \; 3 \cdot 1 = 3$ and $\dfrac{3 \cdot 1}{2}(1+1) = 3$

II: If $3+6+9+\cdots+3k = \dfrac{3k}{2}(k+1)$, then

$$3+6+9+\cdots+3k+3(k+1)$$
$$=\left[3+6+9+\cdots+3k\right]+3(k+1)$$
$$=\frac{3k}{2}(k+1)+3(k+1)$$
$$=(k+1)\left(\frac{3k}{2}+3\right)=\frac{3(k+1)}{2}\big((k+1)+1\big)$$

Conditions I and II are satisfied; the statement is true.

27. I: $\quad n=1: \; 2 \cdot 3^{1-1} = 2$ and $3^1 - 1 = 2$

II: If $2+6+18+\cdots+2 \cdot 3^{k-1} = 3^k - 1$, then

$$2+6+18+\cdots+2 \cdot 3^{k-1}+2 \cdot 3^{k+1-1}$$
$$=\left[2+6+18+\cdots+2 \cdot 3^{k-1}\right]+2 \cdot 3^k$$
$$=3^k - 1 + 2 \cdot 3^k = 3 \cdot 3^k - 1 = 3^{k+1} - 1$$

Conditions I and II are satisfied; the statement is true.

28. I: $\quad n=1:$

$$(3 \cdot 1 - 2)^2 = 1 \text{ and } \frac{1}{2} \cdot 1(6 \cdot 1^2 - 3 \cdot 1 - 1) = 1$$

II: If

$$1^2 + 4^2 + \cdots + (3k-2)^2 = \frac{1}{2} \cdot k\left(6k^2 - 3k - 1\right),$$

then

$$1^2 + 4^2 + 7^2 + \cdots + (3k-2)^2 + \big(3(k+1)-2\big)^2$$
$$=\left[1^2 + 4^2 + 7^2 + \cdots + (3k-2)^2\right]+(3k+1)^2$$
$$=\frac{1}{2} \cdot k\left(6k^2 - 3k - 1\right)+(3k+1)^2$$
$$=\frac{1}{2} \cdot \left[6k^3 - 3k^2 - k + 18k^2 + 12k + 2\right]$$
$$=\frac{1}{2} \cdot \left[6k^3 + 15k^2 + 11k + 2\right]$$
$$=\frac{1}{2} \cdot (k+1)\left[6k^2 + 9k + 2\right]$$
$$=\frac{1}{2} \cdot \left[6k^3 + 6k^2 + 9k^2 + 9k + 2k + 2\right]$$
$$=\frac{1}{2} \cdot \left[6k^2(k+1)+9k(k+1)+2(k+1)\right]$$
$$=\frac{1}{2} \cdot (k+1)\left[6k^2 + 12k + 6 - 3k - 3 - 1\right]$$
$$=\frac{1}{2} \cdot (k+1)\left[6(k^2+2k+1)-3(k+1)-1\right]$$
$$=\frac{1}{2} \cdot (k+1)\left[6(k+1)^2 - 3(k+1) - 1\right]$$

Conditions I and II are satisfied; the statement is true.

29. $\binom{5}{2} = \dfrac{5!}{2!\,3!} = \dfrac{5\cdot 4\cdot 3\cdot 2\cdot 1}{2\cdot 1\cdot 3\cdot 2\cdot 1} = \dfrac{5\cdot 4}{2\cdot 1} = 10$

30. $(x+2)^5 = \binom{5}{0}x^5 + \binom{5}{1}x^4\cdot 2 + \binom{5}{2}x^3\cdot 2^2 + \binom{5}{3}x^2\cdot 2^3 + \binom{5}{4}x^1\cdot 2^4 + \binom{5}{5}\cdot 2^5$

$= x^5 + 5\cdot 2x^4 + 10\cdot 4x^3 + 10\cdot 8x^2 + 5\cdot 16x + 1\cdot 32$

$= x^5 + 10x^4 + 40x^3 + 80x^2 + 80x + 32$

31. $(3x-4)^4 = \binom{4}{0}(3x)^4 + \binom{4}{1}(3x)^3(-4) + \binom{4}{2}(3x)^2(-4)^2 + \binom{4}{3}(3x)(-4)^3 + \binom{4}{4}(-4)^4$

$= 81x^4 + 4\cdot 27x^3(-4) + 6\cdot 9x^2\cdot 16 + 4\cdot 3x(-64) + 1\cdot 256$

$= 81x^4 - 432x^3 + 864x^2 - 768x + 256$

32. $n=9,\ j=2,\ x=x,\ a=2$

$\binom{9}{2}x^7\cdot 2^2 = \dfrac{9!}{2!\,7!}\cdot 4x^7 = \dfrac{9\cdot 8}{2\cdot 1}\cdot 4x^7 = 144x^7$

The coefficient of x^7 is 144.

33. $n=7,\ j=5,\ x=2x,\ a=1$

$\binom{7}{5}(2x)^2\cdot 1^5 = \dfrac{7!}{5!\,2!}\cdot 4x^2(1) = \dfrac{7\cdot 6}{2\cdot 1}\cdot 4x^2 = 84x^2$

The coefficient of x^2 is 84.

34. This is an arithmetic sequence with
$a_1 = 80,\ d=-3,\ n=25$

  a. $a_{25} = 80 + (25-1)(-3) = 80 - 72 = 8$ bricks

  b. $S_{25} = \dfrac{25}{2}(80+8) = 25(44) = 1100$ bricks
  1100 bricks are needed to build the steps.

35. This is an arithmetic sequence with
$a_1 = 30,\ d=-1,\ a_n = 15$

$15 = 30 + (n-1)(-1)$

$-15 = -n+1$

$-16 = -n$

$n = 16$

$S_{16} = \dfrac{16}{2}(30+15) = 8(45) = 360$ tiles

360 tiles are required to make the trapezoid.

36. This is a geometric sequence with

$a_1 = 20,\ r = \dfrac{3}{4}.$

  a. After striking the ground the third time, the
  height is $20\left(\dfrac{3}{4}\right)^3 = \dfrac{135}{16} \approx 8.44$ feet .

  b. After striking the ground the nth time, the
  height is $20\left(\dfrac{3}{4}\right)^n$ feet .

  c. If the height is less than 6 inches or 0.5 feet,
  then:

  $0.5 \ge 20\left(\dfrac{3}{4}\right)^n$

  $0.025 \ge \left(\dfrac{3}{4}\right)^n$

  $\log(0.025) \ge n\log\left(\dfrac{3}{4}\right)$

  $n \ge \dfrac{\log(0.025)}{\log\left(\dfrac{3}{4}\right)} \approx 12.82$

  The height is less than 6 inches after the
  13th strike.

  d. Since this is a geometric sequence with
  $|r| < 1$, the distance is the sum of the two
  infinite geometric series - the distances
  going down plus the distances going up.
  Distance going down:

  $S_{down} = \dfrac{20}{\left(1-\dfrac{3}{4}\right)} = \dfrac{20}{\left(\dfrac{1}{4}\right)} = 80$ feet.

  Distance going up:

$$S_{up} = \frac{15}{\left(1-\frac{3}{4}\right)} = \frac{15}{\left(\frac{1}{4}\right)} = 60 \text{ feet.}$$

The total distance traveled is 140 feet.

37. This is an ordinary annuity with $P = \$200$ and $n = (12)(20) = 240$ payment periods. The

interest rate per period is $\frac{0.10}{12}$. Thus,

$$A = 200\left[\frac{\left[1+\frac{0.10}{12}\right]^{240}-1}{\frac{0.10}{12}}\right] \approx \$151,873.77$$

38. This is a geometric sequence with $a_1 = 50,000$, $r = 1.04$, $n = 5$. Find the fifth term of the sequence:
$$a_5 = 50,000(1.04)^{5-1} = 50,000(1.04)^4$$
$$= 58,492.93$$
Her salary in the fifth year will be \$58,492.93.

Chapter 9 Test

1. $a_n = \dfrac{n^2-1}{n+8}$

$a_1 = \dfrac{1^2-1}{1+8} = \dfrac{0}{9} = 0$

$a_2 = \dfrac{2^2-1}{2+8} = \dfrac{3}{10}$

$a_3 = \dfrac{3^2-1}{3+8} = \dfrac{8}{11}$

$a_4 = \dfrac{4^2-1}{4+8} = \dfrac{15}{12} = \dfrac{5}{4}$

$a_5 = \dfrac{5^2-1}{5+8} = \dfrac{24}{13}$

The first five terms of the sequence are 0, $\dfrac{3}{10}$,

$\dfrac{8}{11}$, $\dfrac{5}{4}$, and $\dfrac{24}{13}$.

2. $a_1 = 4$; $a_n = 3a_{n-1} + 2$

$a_2 = 3a_1 + 2 = 3(4) + 2 = 14$

$a_3 = 3a_2 + 2 = 3(14) + 2 = 44$

$a_4 = 3a_3 + 2 = 3(44) + 2 = 134$

$a_5 = 3a_4 + 2 = 3(134) + 2 = 404$

The first five terms of the sequence are 4, 14, 44, 134, and 404.

3. $\displaystyle\sum_{k=1}^{3}(-1)^{k+1}\left(\frac{k+1}{k^2}\right)$

$= (-1)^{1+1}\left(\dfrac{1+1}{1^2}\right) + (-1)^{2+1}\left(\dfrac{2+1}{2^2}\right) + (-1)^{3+1}\left(\dfrac{3+1}{3^2}\right)$

$= (-1)^2\left(\dfrac{2}{1}\right) + (-1)^3\left(\dfrac{3}{4}\right) + (-1)^4\left(\dfrac{4}{9}\right)$

$= 2 - \dfrac{3}{4} + \dfrac{4}{9} = \dfrac{61}{36}$

4. $\displaystyle\sum_{k=1}^{4}\left[\left(\frac{2}{3}\right)^k - k\right]$

$= \left[\left(\dfrac{2}{3}\right)^1 - 1\right] + \left[\left(\dfrac{2}{3}\right)^2 - 2\right] + \left[\left(\dfrac{2}{3}\right)^3 - 3\right] + \left[\left(\dfrac{2}{3}\right)^4 - 4\right]$

$= \dfrac{2}{3} - 1 + \dfrac{4}{9} - 2 + \dfrac{8}{27} - 3 + \dfrac{16}{81} - 4$

$= \dfrac{130}{81} - 10 = -\dfrac{680}{81}$

5. $-\dfrac{2}{5} + \dfrac{3}{6} - \dfrac{4}{7} + \ldots + \dfrac{11}{14}$

Notice that the signs of each term alternate, with the first term being negative. This implies that the general term will include a power of -1. Also note that the numerator is always 1 more than the term number and the denominator is 4 more than the term number. Thus, each term is in the form $(-1)^k\left(\dfrac{k+1}{k+4}\right)$. The last numerator is 11 which indicates that there are 10 terms.

$-\dfrac{2}{5} + \dfrac{3}{6} - \dfrac{4}{7} + \ldots + \dfrac{11}{14} = \displaystyle\sum_{k=1}^{10}(-1)^k\left(\frac{k+1}{k+4}\right)$

6. $6, 12, 36, 144, \ldots$

$12 - 6 = 6$ and $36 - 12 = 24$

The difference between consecutive terms is not constant. Therefore, the sequence is not arithmetic.

$\dfrac{12}{6} = 2$ and $\dfrac{36}{12} = 3$

591

The ratio of consecutive terms is not constant.
Therefore, the sequence is not geometric.

7. $a_n = -\dfrac{1}{2} \cdot 4^n$

$$\dfrac{a_n}{a_{n-1}} = \dfrac{-\frac{1}{2} \cdot 4^n}{-\frac{1}{2} \cdot 4^{n-1}} = \dfrac{-\frac{1}{2} \cdot 4^{n-1} \cdot 4}{-\frac{1}{2} \cdot 4^{n-1}} = 4$$

Since the ratio of consecutive terms is constant, the sequence is geometric with common ratio

$r = 4$ and first term $a_1 = -\dfrac{1}{2} \cdot 4^1 = -2$.

The sum of the first n terms of the sequence is given by

$$S_n = a_1 \cdot \dfrac{1-r^n}{1-r}$$

$$= -2 \cdot \dfrac{1-4^n}{1-4}$$

$$= \dfrac{2}{3}\left(1-4^n\right)$$

8. $-2, -10, -18, -26, \ldots$

$-10 - (-2) = -8$, $-18 - (-10) = -8$,

$-26 - (-18) = -8$

The difference between consecutive terms is constant. Therefore, the sequence is arithmetic with common difference $d = -8$ and first term $a_1 = -2$.

$$a_n = a_1 + (n-1)d$$

$$= -2 + (n-1)(-8)$$

$$= -2 - 8n + 8$$

$$= 6 - 8n$$

The sum of the first n terms of the sequence is given by

$$S_n = \dfrac{n}{2}(a + a_n)$$

$$= \dfrac{n}{2}(-2 + 6 - 8n)$$

$$= \dfrac{n}{2}(4 - 8n)$$

$$= n(2 - 4n)$$

9. $a_n = -\dfrac{n}{2} + 7$

$$a_n - a_{n-1} = \left[-\dfrac{n}{2}+7\right] - \left[-\dfrac{(n-1)}{2}+7\right]$$

$$= -\dfrac{n}{2} + 7 + \dfrac{n-1}{2} - 7$$

$$= -\dfrac{1}{2}$$

The difference between consecutive terms is constant. Therefore, the sequence is arithmetic with common difference $d = -\dfrac{1}{2}$ and first term

$$a_1 = -\dfrac{1}{2} + 7 = \dfrac{13}{2}.$$

The sum of the first n terms of the sequence is given by

$$S_n = \dfrac{n}{2}(a_1 + a_n)$$

$$= \dfrac{n}{2}\left(\dfrac{13}{2} + \left(-\dfrac{n}{2}+7\right)\right)$$

$$= \dfrac{n}{2}\left(\dfrac{27}{2} - \dfrac{n}{2}\right)$$

$$= \dfrac{n}{4}(27 - n)$$

10. $25, 10, 4, \dfrac{8}{5}, \ldots$

$$\dfrac{10}{25} = \dfrac{2}{5}, \quad \dfrac{4}{10} = \dfrac{2}{5}, \quad \dfrac{\frac{8}{5}}{4} = \dfrac{8}{5} \cdot \dfrac{1}{4} = \dfrac{2}{5}$$

The ratio of consecutive terms is constant. Therefore, the sequence is geometric with common ratio $r = \dfrac{2}{5}$ and first term $a_1 = 25$.

The sum of the first n terms of the sequence is given by

$$S_n = a_1 \cdot \dfrac{1-r^n}{1-r} = 25 \cdot \dfrac{1-\left(\frac{2}{5}\right)^n}{1-\frac{2}{5}} = 25 \cdot \dfrac{\left(1-\left(\frac{2}{5}\right)^n\right)}{\frac{3}{5}}$$

$$= 25 \cdot \dfrac{5}{3}\left(1-\left(\dfrac{2}{5}\right)^n\right) = \dfrac{125}{3}\left(1-\left(\dfrac{2}{5}\right)^n\right)$$

11. $a_n = \dfrac{2n-3}{2n+1}$

$a_n - a_{n-1} = \dfrac{2n-3}{2n+1} - \dfrac{2(n-1)-3}{2(n-1)+1} = \dfrac{2n-3}{2n+1} - \dfrac{2n-5}{2n-1}$

$\qquad = \dfrac{(2n-3)(2n-1)-(2n-5)(2n+1)}{(2n+1)(2n-1)}$

$\qquad = \dfrac{(4n^2-8n+3)-(4n^2-8n-5)}{4n^2-1}$

$\qquad = \dfrac{8}{4n^2-1}$

The difference of consecutive terms is not constant. Therefore, the sequence is not arithmetic.

$\dfrac{a_n}{a_{n-1}} = \dfrac{\dfrac{2n-3}{2n+1}}{\dfrac{2(n-1)-3}{2(n-1)+1}} = \dfrac{2n-3}{2n+1} \cdot \dfrac{2n-1}{2n-5} = \dfrac{(2n-3)(2n-1)}{(2n+1)(2n-5)}$

The ratio of consecutive terms is not constant. Therefore, the sequence is not geometric.

12. For this geometric series we have $r = \dfrac{-64}{256} = -\dfrac{1}{4}$

and $a_1 = 256$. Since $|r| = \left|-\dfrac{1}{4}\right| = \dfrac{1}{4} < 1$, the series converges and we get

$S_\infty = \dfrac{a_1}{1-r} = \dfrac{256}{1-\left(-\frac{1}{4}\right)} = \dfrac{256}{\frac{5}{4}} = \dfrac{1024}{5}$

13. $(3m+2)^5 = \dbinom{5}{0}(3m)^5 + \dbinom{5}{1}(3m)^4(2) + \dbinom{5}{2}(3m)^3(2)^2 + \dbinom{5}{3}(3m)^2(2)^3 + \dbinom{5}{4}(3m)(2)^4 + \dbinom{5}{5}(2)^5$

$\qquad = 243m^5 + 5 \cdot 81m^4 \cdot 2 + 10 \cdot 27m^3 \cdot 4 + 10 \cdot 9m^2 \cdot 8 + 5 \cdot 3m \cdot 16 + 32$

$\qquad = 243m^5 + 810m^4 + 1080m^3 + 720m^2 + 240m + 32$

14. First we show that the statement holds for $n=1$.

$\left(1+\dfrac{1}{1}\right) = 1+1 = 2$

The equality is true for $n=1$ so Condition I holds. Next we assume that $\left(1+\dfrac{1}{1}\right)\left(1+\dfrac{1}{2}\right)\left(1+\dfrac{1}{3}\right)\cdots\left(1+\dfrac{1}{n}\right) = n+1$ is

true for some k, and we determine whether the formula then holds for $k+1$. We assume that

$\left(1+\dfrac{1}{1}\right)\left(1+\dfrac{1}{2}\right)\left(1+\dfrac{1}{3}\right)\cdots\left(1+\dfrac{1}{k}\right) = k+1$.

Now we need to show that $\left(1+\dfrac{1}{1}\right)\left(1+\dfrac{1}{2}\right)\left(1+\dfrac{1}{3}\right)\cdots\left(1+\dfrac{1}{k}\right)\left(1+\dfrac{1}{k+1}\right) = (k+1)+1 = k+2$.

We do this as follows:

$\left(1+\dfrac{1}{1}\right)\left(1+\dfrac{1}{2}\right)\left(1+\dfrac{1}{3}\right)\cdots\left(1+\dfrac{1}{k}\right)\left(1+\dfrac{1}{k+1}\right) = \left[\left(1+\dfrac{1}{1}\right)\left(1+\dfrac{1}{2}\right)\left(1+\dfrac{1}{3}\right)\cdots\left(1+\dfrac{1}{k}\right)\right]\left(1+\dfrac{1}{k+1}\right)$

$\qquad\qquad = (k+1)\left(1+\dfrac{1}{k+1}\right) \qquad$ (using the induction assumption)

$\qquad\qquad = (k+1)\cdot 1 + (k+1)\cdot\dfrac{1}{k+1} = k+1+1$

$\qquad\qquad = k+2$

Condition II also holds. Thus, formula holds true for all natural numbers.

15. The yearly values of the car form a geometric sequence with first term $a_1 = 31,000$ and common ratio $r = 0.85$ (which represents a 15% loss in value).

$a_n = 31,000 \cdot (0.85)^{n-1}$

The nth term of the sequence represents the value of the car at the beginning of the nth year. Since we want to know the value *after* 10 years, we are looking for the 11th term of the sequence. That is, the value of the car at the beginning of the 11th year.

$a_{11} = a_1 \cdot r^{11-1} = 31,000 \cdot (0.85)^{10} = 6,103.11$

After 10 years, the car will be worth $6,103.11.

16. The weights for each set form an arithmetic sequence with first term $a_1 = 100$ and common difference $d = 30$. If we imagine the weightlifter only performed one repetition per set, the total weight lifted in 5 sets would be the sum of the first five terms of the sequence.

$a_n = a_1 + (n-1)d$

$a_5 = 100 + (5-1)(30) = 100 + 4(30) = 220$

$S_n = \dfrac{n}{2}(a + a_n)$

$S_5 = \dfrac{5}{2}(100 + 220) = \dfrac{5}{2}(320) = 800$

Since he performs 10 repetitions in each set, we multiply the sum by 10 to obtain the total weight lifted.

$10(800) = 8000$

The weightlifter will have lifted a total of 8000 pounds after 5 sets.

Chapter 9 Cumulative Review

1. $\left| x^2 \right| = 9$

$x^2 = 9$ or $x^2 = -9$

$x = \pm 3$ or $x = \pm 3i$

2. a. graphing $x^2 + y^2 = 100$ and $y = 3x^2$.

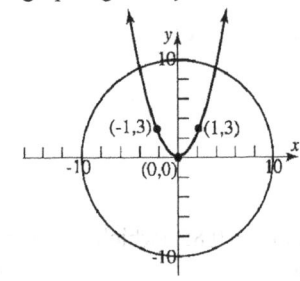

b. Solving

$\begin{cases} x^2 + y^2 = 100 \xrightarrow{\;3\;} 3x^2 + 3y^2 = 300 \\ y = 3x^2 \qquad\longrightarrow\; -3x^2 + y = 0 \end{cases}$

$3y^2 + y = 300$

$3y^2 + y = 300 \Rightarrow 3y^2 + y - 300 = 0$

$y = \dfrac{-1 \pm \sqrt{1^2 - 4(3)(-300)}}{2(3)} = \dfrac{-1 \pm \sqrt{3601}}{6}$

Substitute and solve for x:

$y = \dfrac{-1 + \sqrt{3601}}{6} \Rightarrow \dfrac{-1 + \sqrt{3601}}{6} = 3x^2$

$\dfrac{-1 + \sqrt{3601}}{18} = x^2 \Rightarrow x = \pm\sqrt{\dfrac{-1 + \sqrt{3601}}{18}}$

or

$y = \dfrac{-1 - \sqrt{3601}}{6} \Rightarrow \dfrac{-1 - \sqrt{3601}}{6} = 3x^2$

$\dfrac{-1 - \sqrt{3601}}{18} = x^2 \Rightarrow x = \pm\sqrt{\dfrac{-1 - \sqrt{3601}}{18}}$

undefined since $\dfrac{-1 - \sqrt{3601}}{18} < 0$

Therefore, the system has solutions

$x = \sqrt{\dfrac{-1 + \sqrt{3601}}{18}}, y = \dfrac{-1 + \sqrt{3601}}{6}$ and

$x = -\sqrt{\dfrac{-1 + \sqrt{3601}}{18}}, y = \dfrac{-1 + \sqrt{3601}}{6}$.

$\left\{ \left(\sqrt{\dfrac{-1 + \sqrt{3601}}{18}}, \dfrac{-1 + \sqrt{3601}}{6} \right), \right.$

$\left. \left(-\sqrt{\dfrac{-1 + \sqrt{3601}}{18}}, \dfrac{-1 + \sqrt{3601}}{6} \right) \right\}$

c. The graphs intersect at the points

$\left(\sqrt{\dfrac{-1 + \sqrt{3601}}{18}}, \dfrac{-1 + \sqrt{3601}}{6} \right) \approx (1.81, 9.84)$

$\left(-\sqrt{\dfrac{-1 + \sqrt{3601}}{18}}, \dfrac{-1 + \sqrt{3601}}{6} \right) \approx (-1.81, 9.84)$

3. $2e^x = 5$

$$e^x = \frac{5}{2}$$

$$\ln(e^x) = \ln\left(\frac{5}{2}\right)$$

$$x = \ln\left(\frac{5}{2}\right) \approx 0.916$$

The solution set is $\left\{\ln\left(\frac{5}{2}\right)\right\}$.

4. slope $= m = 5$; Since the x-intercept is 2, we know the point $(2,0)$ is on the graph of the line and is a solution to the equation $y = 5x + b$.

$$y = 5x + b$$
$$0 = 5(2) + b$$
$$0 = 10 + b$$
$$-10 = b$$

Therefore, the equation of the line with slope 5 and x-intercept 2 is $y = 5x - 10$.

5. Given a circle with center $(-1, 2)$ and containing the point $(3, 5)$, we first use the distance formula to determine the radius.

$$r = \sqrt{(3-(-1))^2 + (5-2)^2}$$
$$= \sqrt{4^2 + 3^2} = \sqrt{16 + 9}$$
$$= \sqrt{25}$$
$$= 5$$

Therefore, the equation of the circle is given by

$$(x-(-1))^2 + (y-2)^2 = 5^2$$
$$(x+1)^2 + (y-2)^2 = 5^2$$
$$x^2 + 2x + 1 + y^2 - 4y + 4 = 25$$
$$x^2 + y^2 + 2x - 4y - 20 = 0$$

6. $f(x) = \dfrac{3x}{x-2}$, $g(x) = 2x + 1$

a. $g(2) = 2(2) + 1 = 5$

$$f(5) = \frac{3(5)}{5-2} = \frac{15}{3} = 5$$

$$(f \circ g)(2) = f(g(2)) = f(5) = 5$$

b. $f(4) = \dfrac{3(4)}{4-2} = \dfrac{12}{2} = 6$

$$g(6) = 2(6) + 1 = 13$$

$$(g \circ f)(4) = g(f(4)) = g(6) = 13$$

c. $(f \circ g)(x) = f(g(x))$

$$= \frac{3(2x+1)}{(2x+1) - 2}$$

$$= \frac{6x + 3}{2x - 1}$$

d. To determine the domain of the composition $(f \circ g)(x)$, we start with the domain of g and exclude any values in the domain of g that make the composition undefined. $g(x)$ is defined for all real numbers and $(f \circ g)(x)$ is defined for all real numbers except $x = \dfrac{1}{2}$. Therefore, the domain of the composite $(f \circ g)(x)$ is $\left\{x \mid x \neq \dfrac{1}{2}\right\}$.

e. $(g \circ f)(x) = 2\left(\dfrac{3x}{x-2}\right) + 1$

$$= \frac{6x}{x-2} + 1$$

$$= \frac{6x + x - 2}{x - 2}$$

$$= \frac{7x - 2}{x - 2}$$

f. To determine the domain of the composition $(g \circ f)(x)$, we start with the domain of f and exclude any values in the domain of f that make the composition undefined. $f(x)$ is defined for all real numbers except $x = 2$ and $(g \circ f)(x)$ is defined for all real numbers except $x = 2$. Therefore, the domain of the composite $(g \circ f)(x)$ is $\{x \mid x \neq 2\}$.

g. $g(x) = 2x+1$

$$y = 2x+1$$
$$x = 2y+1$$
$$x-1 = 2y$$
$$\frac{x-1}{2} = y$$
$$g^{-1}(x) = \frac{x-1}{2}$$

The domain of $g^{-1}(x)$ is the set of all real numbers.

h. $f(x) = \frac{3x}{x-2}$

$$y = \frac{3x}{x-2}$$
$$x = \frac{3y}{y-2}$$
$$x(y-2) = 3y$$
$$xy - 2x = 3y$$
$$xy - 3y = 2x$$
$$y(x-3) = 2x$$
$$y = \frac{2x}{x-3}$$
$$f^{-1}(x) = \frac{2x}{x-3}$$

The domain of $f^{-1}(x)$ is $\{x \mid x \neq 3\}$.

7. Center: $(0, 0)$; Focus: $(0, 3)$; Vertex: $(0, 4)$; Major axis is the y-axis; $a = 4$; $c = 3$.

Find b: $b^2 = a^2 - c^2 = 16 - 9 = 7 \Rightarrow b = \sqrt{7}$

Write the equation using rectangular coordinates:

$$\frac{x^2}{7} + \frac{y^2}{16} = 1$$

8. The focus is $(-1, 3)$ and the vertex is $(-1, 2)$. Both lie on the vertical line $x = -1$. We have $a = 1$ since the distance from the vertex to the focus is 1 unit, and since $(-1, 3)$ is above $(-1, 2)$, the parabola opens up. The equation of the parabola is:

$$(x-h)^2 = 4a(y-k)$$
$$(x-(-1))^2 = 4 \cdot 1 \cdot (y-2)$$
$$(x+1)^2 = 4(y-2)$$

Chapter 9 Projects

Project I – Internet-based Project

Answers will vary based on the year that is used. Data used in these solutions will be from 2008.

1. I = net immigration = 887,168
   Population for 2008 = 303,824,640

2. r = 0.01416 − 0.00826 = 0.0059

3. $p_n = (1 + 0.0059)p_{n-1} + 887168$
   $p_n = (1.0059)p_{n-1} + 887168$
   $p_0 = 303824640$

4. $p_1 = (1.0059)p_0 + 887168$
   $p_1 = (1.0059)(303824640) + 887168$
   $p_1 = 306,504,373$
   The population is predicted to be 306,504,373 in 2009.

5. Actual population in 2009: 307,212,123. The formula's prediction was lower but fairly close.

6. Birth rate: 48.12 per 1000 population (0.04812)
   Death rate: 12.64 per 1000 population (0.01264)
   Population for 2008: 31,367,972
   I = net immigration = −6587
   $r = 0.04812 - 0.01264 = 0.03548$
   $p_n = (1 + 0.03548)p_{n-1} - 6587$
   $p_n = (1.03548)p_{n-1} - 6587$
   $p_0 = 31,367,972$

   $p_1 = (1.03548)p_0 - 6587$
   $p_1 = (1.03548)(31367972) - 6587$
   $p_1 = 32,474,321$
   The population is predicted to be 32,474,321 in 2009.

   Actual population in 2009: 32,369,558.
   The formula's prediction was higher but fairly close.

7. Answers will vary. This appears to support the article. The growth rate for the U.S. is much smaller than the growth rate for Uganda.

8. It could be but one must consider trends in each of the pieces of data to find if the growth rate is increasing or decreasing over time. The same thing must be examined with respect to the net immigration.

Project II

1. 2, 4, 8, <u>16</u>, <u>32</u>, <u>64</u>

2. length n ➜ 2^n levels
   This is a geometric sequence: $a_n = 2^n$
   Recursive expression: $a_n = 2a_{n-1}$, $a_0 = 1$

3. $256 = 2^n$
   $2^8 = 2^n$
   $n = 8$

Project III

1. $Q_{st} = -3 + 2P_{t-1}$, $Q_{dt} = 18 - 3P_t$
   $P_0 = 2$, $b = 2$, $d = 3$, $c = 18$
   $-a = -3 \rightarrow a = 3$

   $P_t = \dfrac{3 + 18 - 2P_{t-1}}{3} = \dfrac{21 - 2P_{t-1}}{3}$

   $P_t = 7 - \dfrac{2}{3}P_{t-1}$, $P_0 = 2$

2.

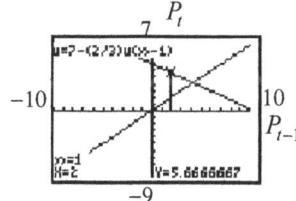

3. $P_1 = \dfrac{17}{3}$

   $Q_{s1} = -3 + 2(2)$ $\qquad Q_{d1} = 18 - 3\left(\dfrac{17}{3}\right)$

   $Q_{s1} = 1$ $\qquad\qquad Q_{d1} = 1$

$P_2 = \dfrac{29}{9}$

$Q_{s2} = -3 + 2\left(\dfrac{17}{3}\right)$ $\qquad Q_{d2} = 18 - 3\left(\dfrac{29}{9}\right)$

$Q_{s2} = \dfrac{25}{3}$ $\qquad\qquad Q_{d2} = \dfrac{25}{3}$

The market (supply and demand) are getting closer to being the same.

4. The equilibrium price is 4.20.

5. It takes 17 time periods.

6. $Q_{d17} = 18 - 3(4.20) = 5.40$
   $Q_{s17} = -3 + 2(4.20) = 5.40$
   The equilibrium quantity is 5.4.

Project IV

1. 1, 2, 4, 7, 11, <u>16</u>, <u>22</u>, <u>29</u>

2. It is not arithmetic because there is no common difference. It is not geometric because there is no common ration.

3. Scatter diagram

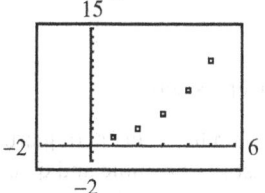

4. $y = 2.5x - 2.5$
   The graph does not pass through any of the points.
   $y_6 = 12.5$
   $y_7 = 15$
   $y_8 = 17.5$

$y_1 = 0$

$y_2 = 2.5$

$y_3 = 5$

$y_4 = 7.5$

$y_5 = 10$

$\sum_{i=1}^{5}(y_{r_i} - y_i)$

$= (0-1)+(2.5-2)+(5-4)+(7.5-7)+(10-11)$

$= -1$

This is the sum of the errors.

5. $y = 0.5x^2 - 0.5x + 1$

The graph passes through all of the points.

$y_6 = 16$

$y_7 = 22$

$y_8 = 29$

$y_1 = 1$

$y_2 = 2$

$y_3 = 4$

$y_4 = 7$

$y_5 = 11$

$\sum_{i=1}^{5}(y_{r_i} - y_i) = 0$

The sum of the errors is zero.

6. When trying to obtain the cubic and quartic polynomials of best fit, the cubic and quartic terms have coefficient zero and the polynomial of best fit is given as the quadratic in part e. For the exponential function of best fit,

$y = (0.59)(1.83)^x$.

$y_6 = 22.2 \qquad y_7 = 40.6 \qquad y_8 = 74.2$

The sum of these errors becomes quite large. This error shows that the function does not fit the data very well as x gets larger.

7. The quadratic function is best.

8. The data does not appear to be either logarithmic or sinusoidal in shape, so it does not make sense to try to fit one of those functions to the data.

Chapter 10
Counting and Probability

Section 10.1

1. union

3. True; the union of two sets includes those elements that are in one *or* both of the sets. The intersection consists of the elements that are in *both* sets. Thus, the intersection is a subset of the union.

5. subset; $\subseteq$

7. $n(A) + n(B) - n(A \cap B)$

9. $\varnothing, \{a\}, \{b\}, \{c\}, \{d\}, \{a,b\}, \{a,c\}, \{a,d\},$
   $\{b,c\}, \{b,d\}, \{c,d\}, \{a,b,c\}, \{a,b,d\},$
   $\{a,c,d\}, \{b,c,d\}, \{a,b,c,d\}$

11. $n(A) = 15, n(B) = 20, n(A \cap B) = 10$
    $n(A \cup B) = n(A) + n(B) - n(A \cap B)$
    $\qquad = 15 + 20 - 10 = 25$

13. $n(A \cup B) = 50, n(A \cap B) = 10, n(B) = 20$
    $n(A \cup B) = n(A) + n(B) - n(A \cap B)$
    $\qquad 50 = n(A) + 20 - 10$
    $\qquad 40 = n(A)$

15. From the figure: $n(A) = 15 + 3 + 5 + 2 = 25$

17. From the figure:
    $n(A \text{ or } B) = n(A \cup B)$
    $\qquad = 15 + 2 + 5 + 3 + 10 + 2 = 37$

19. From the figure:
    $n(A \text{ but not } C) = n(A) - n(A \cap C) = 25 - 7 = 18$

21. From the figure:
    $n(A \text{ and } B \text{ and } C) = n(A \cap B \cap C) = 5$

23. There are 5 choices of shirts and 3 choices of ties; there are (5)(3) = 15 different arrangements.

25. There are 9 choices for the first digit, and 10 choices for each of the other three digits. Thus, there are (9)(10)(10)(10) = 9000 possible four-digit numbers.

27. Let $A = \{\text{those who will purchase a major appliance}\}$ and
    $B = \{\text{those who will buy a car}\}$
    $n(U) = 500, \ n(A) = 200,$
    $n(B) = 150, \ n(A \cap B) = 25$
    $n(A \cup B) = n(A) + n(B) - n(A \cap B)$
    $\qquad = 200 + 150 - 25 = 325$
    $n(\text{purchase neither}) = n(U) - n(A \cup B)$
    $\qquad\qquad = 500 - 325 = 175$
    $n(\text{purchase only a car}) = n(B) - n(A \cup B)$
    $\qquad\qquad = 150 - 25 = 125$

29. Construct a Venn diagram:

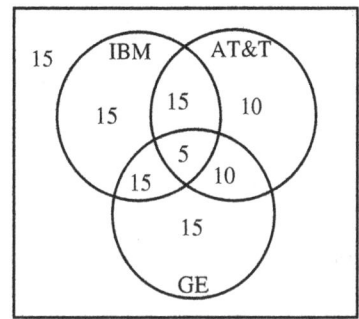

    (a) 15   (b) 15
    (c) 15   (d) 25
    (e) 40

31. **a.** $n(\text{widowed or divorced}) = n(\text{widowed}) + n(\text{divorced})$
    $\qquad\qquad = 3.1 + 10.7$
    $\qquad\qquad = 13.8$
    There were 13.8 million males 18 years old and older who were widowed or divorced.

    b. $n(\text{married, widowed or divorced})$
    $\quad = n(\text{married}) + n(\text{widowed}) + n(\text{divorced})$
    $\quad = 65.7 + 3.1 + 10.7 = 79.5$
    There were 79.5 million males 18 years old and older who were married, widowed, or divorced.

33. There are 8 choices for the DOW stocks, 15 choices for the NASDAQ stocks, and 4 choices for the global stocks. Thus, there are (8)(15)(4) = 480 different portfolios.

35. Answers will vary.

37. We move the graph horizontally left 3 units so the x value becomes $x = 3 - 3 = 0$. We stretch the graph by a factor of 2 and flip on the x axis so the y value becomes $y = -2(8) = -16$. Then we shift the graph vertically 5 units so the y value becomes $y = -16 + 5 = -11$. Thus the final point is $(0, -11)$.

39. $\log_3 x + \log_3 2 = -2$

$$\log_3 (2x) = -2$$

$$3^{-2} = 2x$$

$$\frac{1}{9} = 2x$$

$$x = \frac{1}{18}$$

The solution set is $\left\{ \frac{1}{18} \right\}$.

Section 10.2

1. $1; 1$

3. permutation

5. $P(n, r) = \dfrac{n!}{(n-r)!}$

7. $P(6, 2) = \dfrac{6!}{(6-2)!} = \dfrac{6!}{4!} = \dfrac{6 \cdot 5 \cdot 4!}{4!} = 30$

9. $P(4, 4) = \dfrac{4!}{(4-4)!} = \dfrac{4!}{0!} = \dfrac{4 \cdot 3 \cdot 2 \cdot 1}{1} = 24$

11. $P(7, 0) = \dfrac{7!}{(7-0)!} = \dfrac{7!}{7!} = 1$

13. $P(8, 4) = \dfrac{8!}{(8-4)!} = \dfrac{8!}{4!} = \dfrac{8 \cdot 7 \cdot 6 \cdot 5 \cdot 4!}{4!} = 1680$

15. $C(8, 2) = \dfrac{8!}{(8-2)!\, 2!} = \dfrac{8!}{6!\, 2!} = \dfrac{8 \cdot 7 \cdot 6!}{6! \cdot 2 \cdot 1} = 28$

17. $C(7, 4) = \dfrac{7!}{(7-4)!\, 4!} = \dfrac{7!}{3!\, 4!} = \dfrac{7 \cdot 6 \cdot 5 \cdot 4!}{4! \cdot 3 \cdot 2 \cdot 1} = 35$

19. $C(15, 15) = \dfrac{15!}{(15-15)!\, 15!} = \dfrac{15!}{0!\, 15!} = \dfrac{15!}{15! \cdot 1} = 1$

21. $C(26, 13) = \dfrac{26!}{(26-13)!\, 13!} = \dfrac{26!}{13!\, 13!} = 10,400,600$

23. {abc, abd, abe, acb, acd, ace, adb, adc, ade, aeb, aec, aed, bac, bad, bae, bca, bcd, bce, bda, bdc, bde, bea, bec, bed, cab, cad, cae, cba, cbd, cbe, cda, cdb, cde, cea, ceb, ced, dab, dac, dae, dba, dbc, dbe, dca, dcb, dce, dea, deb, dec, eab, eac, ead, eba, ebc, ebd, eca, ecb, ecd, eda, edb, edc}

$$P(5, 3) = \frac{5!}{(5-3)!} = \frac{5!}{2!} = \frac{5 \cdot 4 \cdot 3 \cdot 2!}{2!} = 60$$

25. {123, 124, 132, 134, 142, 143, 213, 214, 231, 234, 241, 243, 312, 314, 321, 324, 341, 342, 412, 413, 421, 423, 431, 432}

27. {abc, abd, abe, acd, ace, ade, bcd, bce, bde, cde}

$$C(5, 3) = \frac{5!}{(5-3)!\, 3!} = \frac{5 \cdot 4 \cdot 3!}{2 \cdot 1 \cdot 3!} = 10$$

29. {123, 124, 134, 234}

$$C(4, 3) = \frac{4!}{(4-3)!\, 3!} = \frac{4 \cdot 3!}{1!\, 3!} = 4$$

31. There are 4 choices for the first letter in the code and 4 choices for the second letter in the code; there are (4)(4) = 16 possible two-letter codes.

33. There are two choices for each of three positions; there are (2)(2)(2) = 8 possible three-digit numbers.

35. To line up the four people, there are 4 choices for the first position, 3 choices for the second position, 2 choices for the third position, and 1 choice for the fourth position. Thus there are (4)(3)(2)(1) = 24 possible ways four people can be lined up.

37. Since no letter can be repeated, there are 5 choices for the first letter, 4 choices for the second letter, and 3 choices for the third letter. Thus, there are $(5)(4)(3) = 60$ possible three-letter codes.

39. There are 26 possible one-letter names. There are $(26)(26) = 676$ possible two-letter names. There are $(26)(26)(26) = 17,576$ possible three-letter names. Thus, there are $26 + 676 + 17,576 = 18,278$ possible companies that can be listed on the New York Stock Exchange.

41. A committee of 4 from a total of 7 students is given by:
$$C(7,4) = \frac{7!}{(7-4)!\,4!} = \frac{7!}{3!\,4!} = \frac{7\cdot6\cdot5\cdot4!}{3\cdot2\cdot1\cdot4!} = 35$$
35 committees are possible.

43. There are 2 possible answers for each question. Therefore, there are $2^{10} = 1024$ different possible arrangements of the answers.

45. There are 5 choices for the first position, 4 choices for the second position, 3 choices for the third position, 2 choices for the fourth position, and 1 choice for the fifth position. Thus, there are $(5)(4)(3)(2)(1) = 120$ possible arrangements of the books.

47. The 1st person can have any of 365 days, the 2nd person can have any of the remaining 364 days. Thus, there are $(365)(364) = 132,860$ possible ways two people can have different birthdays.

49. Choosing 2 boys from the 4 boys can be done $C(4,2)$ ways. Choosing 3 girls from the 8 girls can be done in $C(8,3)$ ways. Thus, there are a total of:
$$C(4,2)\cdot C(8,3) = \frac{4!}{(4-2)!\,2!}\cdot\frac{8!}{(8-3)!\,3!}$$
$$= \frac{4!}{2!\,2!}\cdot\frac{8!}{5!\,3!}$$
$$= \frac{4\cdot3\cdot2!}{2\cdot1\cdot2!}\cdot\frac{8\cdot7\cdot6\cdot5!}{5!\,3!}$$
$$= 6\cdot56$$
$$= 336$$

51. This is a permutation with repetition. There are
$$\frac{9!}{2!\,2!} = 90,720 \text{ different words.}$$

53. a. $C(7,2)\cdot C(3,1) = 21\cdot3 = 63$

   b. $C(7,3)\cdot C(3,0) = 35\cdot1 = 35$

   c. $C(3,3)\cdot C(7,0) = 1\cdot1 = 1$

55. There are $C(100, 22)$ ways to form the first committee. There are 78 senators left, so there are $C(78, 13)$ ways to form the second committee. There are $C(65, 10)$ ways to form the third committee. There are $C(55, 5)$ ways to form the fourth committee. There are $C(50, 16)$ ways to form the fifth committee. There are $C(34, 17)$ ways to form the sixth committee. There are $C(17, 17)$ ways to form the seventh committee.
The total number of committees
$$= C(100,22)\cdot C(78,13)\cdot C(65,10)\cdot C(55,5)$$
$$\cdot C(50,16)\cdot C(34,17)\cdot C(17,17)$$
$$\approx 1.157\times10^{76}$$

57. There are 9 choices for the first position, 8 choices for the second position, 7 for the third position, etc. There are
$9\cdot8\cdot7\cdot6\cdot5\cdot4\cdot3\cdot2\cdot1 = 9! = 362,880$ possible batting orders.

59. The team must have 1 pitcher and 8 position players (non-pitchers). For pitcher, choose 1 player from a group of 4 players, i.e., $C(4, 1)$. For position players, choose 8 players from a group of 11 players, i.e., $C(11, 8)$. Thus, the number different teams possible is $C(4,1)\cdot C(11,8) = 4\cdot165 = 660$.

61. Choose 2 players from a group of 6 players. Thus, there are $C(6,2) = 15$ different teams possible.

63. a. If numbers can be repeated, there are $(50)(50)(50) = 125,000$ different lock combinations. If no number can be repeated, then there are $50\cdot49\cdot48 = 117,600$ different lock combinations.

b. Answers will vary. Typical combination locks require two full clockwise rotations to the first number, followed by a full counter-clockwise rotation past the first number to the second number, followed by a clockwise rotation to the third number (not past the second). This is not clear from the given directions. Perhaps a better name for a *combination* lock would be a *permutation* lock since the order in which the numbers are entered matters.

65. Answers will vary.

67. $R(x) = \dfrac{x+3}{x^2 - x - 12} = \dfrac{x+3}{(x+3)(x-4)}$ is in lowest terms. The denominator has zeros at -3 and 4. The degree of the numerator is $n = 1$ and the degree of the denominator is $m = 2$. Since $n < m$, the line $y = 0$ is a horizontal asymptote. Since the denominator is zero at 4, $x = 4$ is a vertical asymptotes. Since the factor $(x+3)$ cancels, $x = -3$ is not an asymptote.

69. $\dfrac{5}{x-3} \geq 1$

$\dfrac{5}{x-3} - 1 \geq 0 \qquad \dfrac{5 - 1(x-3)}{x-3} \quad 0 \qquad \dfrac{-x+8}{x-3} \quad 0$

$f(x) = \dfrac{-(x-8)}{x-3}$

The zeros and values where the expression is undefined are $x = 8$ and $x = 3$.

Interval	$(-\infty, 3)$	$(3, 8)$	$(8, \infty)$
Number Chosen	0	5	10
Value of f	$-\dfrac{8}{3}$	$\dfrac{3}{2}$	$-\dfrac{2}{7}$
Conclusion	Negative	Positive	Negative

The solution set is $\{ x \mid 3 < x \leq 8 \}$, or, using interval notation, $(3, 8]$.

Section 10.3

1. equally likely

3. False; probability may equal 0. In such cases, the corresponding event will never happen.

5. Probabilities must be between 0 and 1, inclusive. Thus, 0, 0.01, 0.35, and 1 could be probabilities.

7. All the probabilities are between 0 and 1.
The sum of the probabilities is
$0.2 + 0.3 + 0.1 + 0.4 = 1$.
This is a probability model.

9. All the probabilities are between 0 and 1.
The sum of the probabilities is
$0.3 + 0.2 + 0.1 + 0.3 = 0.9$.
This is not a probability model.

11. The sample space is: $S = \{HH, HT, TH, TT\}$.
Each outcome is equally likely to occur; so
$P(E) = \dfrac{n(E)}{n(S)}$. The probabilities are:

$P(HH) = \dfrac{1}{4}, \ P(HT) = \dfrac{1}{4}, \ P(TH) = \dfrac{1}{4}, \ P(TT) = \dfrac{1}{4}$.

13. The sample space of tossing two fair coins and a fair die is:
$S = \{HH1, HH2, HH3, HH4, HH5, HH6,$
$HT1, HT2, HT3, HT4, HT5, HT6, TH1,$
$TH2, TH3, TH4, TH5, TH6, TT1, TT2,$
$TT3, TT4, TT5, TT6\}$
There are 24 equally likely outcomes and the probability of each is $\dfrac{1}{24}$.

15. The sample space for tossing three fair coins is:
$S = \{HHH, HHT, HTH, HTT, THH, THT,$
$TTH, TTT\}$
There are 8 equally likely outcomes and the probability of each is $\dfrac{1}{8}$.

17. The sample space is: $S = \{$1 Yellow, 1 Red, 1 Green, 2 Yellow, 2 Red, 2 Green, 3 Yellow, 3 Red, 3 Green, 4 Yellow, 4 Red, 4 Green$\}$
There are 12 equally likely events and the probability of each is $\dfrac{1}{12}$. The probability of getting a 2 or 4 followed by a Red is

$P(2 \text{ Red}) + P(4 \text{ Red}) = \dfrac{1}{12} + \dfrac{1}{12} = \dfrac{1}{6}$.

19. The sample space is:
$S = \{$1 Yellow Forward, 1 Yellow Backward, 1 Red Forward, 1 Red Backward, 1 Green Forward, 1 Green Backward, 2 Yellow Forward, 2 Yellow Backward, 2 Red Forward, 2 Red Backward, 2 Green Forward, 2 Green Backward, 3 Yellow Forward, 3 Yellow Backward, 3 Red Forward, 3 Red Backward, 3 Green Forward, 3 Green Backward, 4 Yellow Forward, 4 Yellow Backward, 4 Red Forward, 4 Red Backward, 4 Green Forward, 4 Green Backward$\}$

There are 24 equally likely events and the probability of each is $\frac{1}{24}$. The probability of getting a 1, followed by a Red or Green, followed by a Backward is:

$P(1 \text{ Red Backward}) + P(1 \text{ Green Backward})$

$= \frac{1}{24} + \frac{1}{24} = \frac{1}{12}$

21. The sample space is:
$S = \{$1 1 Yellow, 1 1 Red, 1 1 Green, 1 2 Yellow, 1 2 Red, 1 2 Green, 1 3 Yellow, 1 3 Red, 1 3 Green, 1 4 Yellow, 1 4 Red, 1 4 Green, 2 1 Yellow, 2 1 Red, 2 1 Green, 2 2 Yellow, 2 2 Red, 2 2 Green, 2 3 Yellow, 2 3 Red, 2 3 Green, 2 4 Yellow, 2 4 Red, 2 4 Green, 3 1 Yellow, 3 1 Red, 3 1 Green, 3 2 Yellow, 3 2 Red, 3 2 Green, 3 3 Yellow, 3 3 Red, 3 3 Green, 3 4 Yellow, 3 4 Red, 3 4 Green, 4 1 Yellow, 4 1 Red, 4 1 Green, 4 2 Yellow, 4 2 Red, 4 2 Green, 4 3 Yellow, 4 3 Red, 4 3 Green, 4 4 Yellow, 4 4 Red, 4 4 Green$\}$

There are 48 equally likely events and the probability of each is $\frac{1}{48}$. The probability of getting a 2, followed by a 2 or 4, followed by a Red or Green is

$P(2\ 2 \text{ Red}) + P(2\ 4 \text{ Red}) + P(2\ 2 \text{ Green}) + P(2\ 4 \text{ Green})$

$= \frac{1}{48} + \frac{1}{48} + \frac{1}{48} + \frac{1}{48} = \frac{1}{12}$

23. A, B, C, F

25. B

27. Let $P(\text{tails}) = x$, then $P(\text{heads}) = 4x$

$x + 4x = 1$

$5x = 1$

$x = \frac{1}{5}$

$P(\text{tails}) = \frac{1}{5}, \quad P(\text{heads}) = \frac{4}{5}$

29. $P(2) = P(4) = P(6) = x$

$P(1) = P(3) = P(5) = 2x$

$P(1) + P(2) + P(3) + P(4) + P(5) + P(6) = 1$

$2x + x + 2x + x + 2x + x = 1$

$9x = 1$

$x = \frac{1}{9}$

$P(2) = P(4) = P(6) = \frac{1}{9}$

$P(1) = P(3) = P(5) = \frac{2}{9}$

31. $P(E) = \frac{n(E)}{n(S)} = \frac{n\{1,2,3\}}{10} = \frac{3}{10}$

33. $P(E) = \frac{n(E)}{n(S)} = \frac{n\{2,4,6,8,10\}}{10} = \frac{5}{10} = \frac{1}{2}$

35. $P(\text{white}) = \frac{n(\text{white})}{n(S)} = \frac{5}{5+10+8+7} = \frac{5}{30} = \frac{1}{6}$

37. The sample space is: $S = \{$BBB, BBG, BGB, GBB, BGG, GBG, GGB, GGG$\}$

$P(3 \text{ boys}) = \frac{n(3 \text{ boys})}{n(S)} = \frac{1}{8}$

39. The sample space is: $S = \{$BBBB, BBBG, BBGB, BGBB, GBBB, BBGG, BGBG, GBBG, BGGB, GBGB, GGBB, BGGG, GBGG, GGBG, GGGB, GGGG$\}$

$P(1 \text{ girl}, 3 \text{ boys}) = \frac{n(1 \text{ girl}, 3 \text{ boys})}{n(S)} = \frac{4}{16} = \frac{1}{4}$

41. $P(\text{sum of two dice is } 7) = \frac{n(\text{sum of two dice is } 7)}{n(S)}$

$= \frac{n\{1,6 \text{ or } 2,5 \text{ or } 3,4 \text{ or } 4,3 \text{ or } 5,2 \text{ or } 6,1\}}{n(S)} = \frac{6}{36} = \frac{1}{6}$

43. $P(\text{sum of two dice is 3}) = \dfrac{n(\text{sum of two dice is 3})}{n(S)}$

$= \dfrac{n\{1,2 \text{ or } 2,1\}}{n(S)} = \dfrac{2}{36} = \dfrac{1}{18}$

45. $P(A \cup B) = P(A) + P(B) - P(A \cap B)$

$= 0.25 + 0.45 - 0.15 = 0.55$

47. $P(A \cup B) = P(A) + P(B) = 0.25 + 0.45 = 0.70$

49. $P(A \cup B) = P(A) + P(B) - P(A \cap B)$

$0.85 = 0.60 + P(B) - 0.05$

$P(B) = 0.85 - 0.60 + 0.05 = 0.30$

51. $P(\text{theft not cleared}) = 1 - P(\text{theft cleared})$

$= 1 - 0.142$

$= 0.858$

53. $P(\text{does not own dog}) = 1 - P(\text{owns dog})$

$= 1 - 0.44$

$= 0.56$

55. $P(\text{never gambled online}) = 1 - P(\text{gambled online})$

$= 1 - 0.064$

$= 0.936$

57. $P(\text{white or green}) = P(\text{white}) + P(\text{green})$

$= \dfrac{n(\text{white}) + n(\text{green})}{n(S)}$

$= \dfrac{9+8}{9+8+3} = \dfrac{17}{20}$

59. $P(\text{not white}) = 1 - P(\text{white})$

$= 1 - \dfrac{n(\text{white})}{n(S)}$

$= 1 - \dfrac{9}{20} = \dfrac{11}{20}$

61. $P(\text{strike or one}) = P(\text{strike}) + P(\text{one})$

$= \dfrac{n(\text{strike}) + n(\text{one})}{n(S)}$

$= \dfrac{3+1}{8} = \dfrac{4}{8} = \dfrac{1}{2}$

63. There are 34 households out of 100 with an income of \$75,000 or more.

$P(E) = \dfrac{n(E)}{n(S)} = \dfrac{n(75,000 \text{ or more})}{n(\text{total households})} = \dfrac{34}{100} = \dfrac{17}{50}$

65. There are 48 households out of 100 with an income of less than \$50,000.

$P(E) = \dfrac{n(E)}{n(S)} = \dfrac{n(\text{less than } 50,000)}{n(\text{total households})} = \dfrac{48}{100} = \dfrac{12}{25}$

67. a. $P(1 \text{ or } 2) = P(1) + P(2) = 0.34 + 0.37 = 0.71$

b. $P(1 \text{ or more}) = 1 - P(\text{none}) = 1 - 0.09 = 0.91$

c. $P(3 \text{ or fewer}) = 1 - P(4 \text{ or more})$

$= 1 - 0.06 = 0.94$

d. $P(3 \text{ or more}) = P(3) + P(4 \text{ or more})$

$= 0.14 + 0.06 = 0.20$

e. $P(\text{fewer than } 2) = P(0) + P(1)$

$= 0.09 + 0.34 = 0.43$

f. $P(\text{fewer than } 1) = P(0) = 0.09$

g. $P(1, 2, \text{ or } 3) = P(1) + P(2) + P(3)$

$= 0.34 + 0.37 + 0.14 = 0.85$

h. $P(2 \text{ or more}) = P(2) + P(3) + P(4 \text{ or more})$

$= 0.37 + 0.14 + 0.06 = 0.57$

69. a. $P(\text{freshman or female})$

$= P(\text{freshman}) + P(\text{female}) - P(\text{freshman and female})$

$= \dfrac{n(\text{freshman}) + n(\text{female}) - n(\text{freshman and female})}{n(S)}$

$= \dfrac{18 + 15 - 8}{33} = \dfrac{25}{33}$

b. $P(\text{sophomore or male})$

$= P(\text{sophomore}) + P(\text{male}) - P(\text{sophomore and male})$

$= \dfrac{n(\text{sophomore}) + n(\text{male}) - n(\text{sophomore and male})}{n(S)}$

$= \dfrac{15 + 18 - 8}{33} = \dfrac{25}{33}$

71. $P(\text{at least 2 with same birthday})$

$= 1 - P(\text{none with same birthday})$

$= 1 - \dfrac{n(\text{different birthdays})}{n(S)}$

$= 1 - \dfrac{365 \cdot 364 \cdot 363 \cdot 362 \cdot 361 \cdot 360 \cdots 354}{365^{12}}$

$\approx 1 - 0.833$

$= 0.167$

73. The number of different selections of 6 numbers is the number of ways we can choose 5 white balls and 1 red ball, where the order of the white balls is not important. This requires the use of the Multiplication Principle and the combination formula. Thus, the total number of distinct ways to pick the 6 numbers is given by

$n(\text{white balls}) \cdot n(\text{red ball})$

$= C(59,5) \cdot C(35,1)$

$= \dfrac{59!}{5! (59-5)!} \cdot \dfrac{35!}{1! (35-1)!}$

$= \dfrac{59!}{5! \, 54!} \cdot \dfrac{35!}{1! \, 34!}$

$= \dfrac{59 \cdot 58 \cdot 57 \cdot 56 \cdot 55}{5 \cdot 4 \cdot 3 \cdot 2 \cdot 1} \cdot \dfrac{35 \cdot 34!}{34!}$

$= \dfrac{59 \cdot 58 \cdot 57 \cdot 56 \cdot 55 \cdot 35}{5 \cdot 4 \cdot 3 \cdot 2}$

$= 175,223,510$

Since each possible combination is equally likely, the probability of winning on a \$2 play is

$P(\text{win on \$2 play}) = \dfrac{1}{175,223,510}$

≈ 0.00000000571

75. $\sqrt[3]{24x^2 y^5} = \sqrt[3]{8 \cdot 3 \cdot x^2 \cdot y^3 \cdot y^2}$

$= 2y\sqrt[3]{3x^2 y^2}$

77. $\begin{cases} 3x + y + 2z = 1 \\ 2x - 2y + 5z = 5 \\ x + 3y + 2z = -9 \end{cases}$

Write the augmented matrix:

$\begin{bmatrix} 3 & 1 & 2 & | & 1 \\ 2 & -2 & 5 & | & 5 \\ 1 & 3 & 2 & | & -9 \end{bmatrix}$

$\rightarrow \begin{bmatrix} 1 & 3 & 2 & | & -9 \\ 2 & -2 & 5 & | & 5 \\ 3 & 1 & 2 & | & 1 \end{bmatrix} \quad (R_1 \leftrightarrow R_3)$

$\rightarrow \begin{bmatrix} 1 & 3 & 2 & | & -9 \\ 0 & 1 & -\frac{1}{8} & | & -\frac{23}{8} \\ 0 & -8 & -4 & | & 28 \end{bmatrix} \quad \left(R_2 = -\frac{1}{8}r_1 \right)$

$\rightarrow \begin{bmatrix} 1 & 0 & \frac{19}{8} & | & -\frac{3}{8} \\ 0 & 1 & -\frac{1}{8} & | & -\frac{23}{8} \\ 0 & 0 & -5 & | & 5 \end{bmatrix} \quad \left(\begin{matrix} R_3 = 8r_2 + r_3 \\ R_1 = -3r_2 + r_1 \end{matrix} \right)$

$\rightarrow \begin{bmatrix} 1 & 0 & \frac{19}{8} & | & -\frac{3}{8} \\ 0 & 1 & -\frac{1}{8} & | & -\frac{23}{8} \\ 0 & 0 & 1 & | & -1 \end{bmatrix} \quad \left(R_3 = -\frac{1}{5}r_3 \right)$

$\rightarrow \begin{bmatrix} 1 & 0 & 0 & | & 2 \\ 0 & 1 & 0 & | & -3 \\ 0 & 0 & 1 & | & -1 \end{bmatrix} \quad \left(\begin{matrix} R_2 = \frac{1}{8}r_3 + r_2 \\ R_1 = -\frac{19}{8}r_3 + r_2 \end{matrix} \right)$

The solution is $x = 2$, $y = -3$, $z = -1$ or $(2, -3, -1)$.

Chapter 10 Review Exercises

1. $\emptyset, \{Dave\}, \{Joanne\}, \{Erica\}, \{Dave, Joanne\},$
   $\{Dave, Erica\}, \{Joanne, Erica\},$
   $\{Dave, Joanne, Erica\}$

2. $n(A) = 8, \ n(B) = 12, \ n(A \cap B) = 3$
   $n(A \cup B) = n(A) + n(B) - n(A \cap B)$
   $\qquad\qquad = 8 + 12 - 3 = 17$

3. $n(A) = 12, \ n(A \cup B) = 30, \ n(A \cap B) = 6$
   $n(A \cup B) = n(A) + n(B) - n(A \cap B)$
   $\qquad 30 = 12 + n(B) - 6$
   $\qquad n(B) = 30 - 12 + 6 = 24$

4. From the figure: $n(A) = 20 + 2 + 6 + 1 = 29$

5. From the figure:
   $n(A \text{ or } B) = 20 + 2 + 6 + 1 + 5 + 0 = 34$

6. From the figure:
   $n(A \text{ and } C) = n(A \cap C) = 1 + 6 = 7$

7. From the figure:
   $n(\text{not in } B) = 20 + 1 + 4 + 20 = 45$

8. From the figure:
   $n(\text{neither in } A \text{ nor in } C) = n(\overline{A \cup C}) = 20 + 5 = 25$

9. From the figure: $n(\text{in } B \text{ but not in } C) = 2 + 5 = 7$

10. $P(8,3) = \dfrac{8!}{(8-3)!} = \dfrac{8!}{5!} = \dfrac{8 \cdot 7 \cdot 6 \cdot 5!}{5!} = 336$

11. $C(8,3) = \dfrac{8!}{(8-3)!\,3!} = \dfrac{8!}{5!\,3!} = \dfrac{8 \cdot 7 \cdot 6 \cdot 5!}{5! \cdot 3 \cdot 2 \cdot 1} = 56$

12. There are 2 choices of material, 3 choices of color, and 10 choices of size. The complete assortment would have: $2 \cdot 3 \cdot 10 = 60$ suits.

13. There are two possible outcomes for each game or $2 \cdot 2 \cdot 2 \cdot 2 \cdot 2 \cdot 2 \cdot 2 = 2^7 = 128$ outcomes for 7 games.

14. Since order is significant, this is a permutation.
    $P(9,4) = \dfrac{9!}{(9-4)!} = \dfrac{9!}{5!} = \dfrac{9 \cdot 8 \cdot 7 \cdot 6 \cdot 5!}{5!} = 3024$
    ways to seat 4 people in 9 seats.

15. Choose 4 runners –order is significant:
    $P(8,4) = \dfrac{8!}{(8-4)!} = \dfrac{8!}{4!} = \dfrac{8 \cdot 7 \cdot 6 \cdot 5 \cdot 4!}{4!} = 1680$
    ways a squad can be chosen.

16. Choose 2 teams from 14–order is not significant:
    $C(14,2) = \dfrac{14!}{(14-2)!\,2!} = \dfrac{14!}{12!\,2!} = \dfrac{14 \cdot 13 \cdot 12!}{12! \cdot 2 \cdot 1} = 91$
    ways to choose 2 teams.

17. There are $8 \cdot 10 \cdot 10 \cdot 10 \cdot 10 \cdot 10 \cdot 2 = 1,600,000$ possible phone numbers.

18. There are $24 \cdot 9 \cdot 10 \cdot 10 \cdot 10 = 216,000$ possible license plates.

19. There are two choices for each digit, so there are $2^8 = 256$ different numbers. (Note this allows numbers with initial zeros, such as 011.)

20. Since there are repeated colors:
    $\dfrac{10!}{4! \cdot 3! \cdot 2! \cdot 1!} = \dfrac{10 \cdot 9 \cdot 8 \cdot 7 \cdot 6 \cdot 5 \cdot 4 \cdot 3 \cdot 2 \cdot 1}{4 \cdot 3 \cdot 2 \cdot 1 \cdot 3 \cdot 2 \cdot 1 \cdot 2 \cdot 1 \cdot 1} = 12,600$
    different vertical arrangements.

21. **a.** $C(9,4) \cdot C(9,3) \cdot C(9,2) = 126 \cdot 84 \cdot 36$
    $\qquad\qquad\qquad\qquad\qquad\qquad = 381,024$
    committees can be formed.

    b. $C(9,4) \cdot C(5,3) \cdot C(2,2) = 126 \cdot 10 \cdot 1 = 1260$
    committees can be formed.

22. **a.** $365 \cdot 364 \cdot 363 \cdots 348 = 8.634628387 \times 10^{45}$

    b. $P(\text{no one has same birthday})$
    $= \dfrac{365 \cdot 364 \cdot 363 \cdots 348}{365^{18}} \approx 0.6531$

    c. $P(\text{at least 2 have same birthday})$
    $= 1 - P(\text{no one has same birthday})$
    $= 1 - 0.6531 = 0.3469$

23. **a.** $P(\text{unemployed}) = 0.062$

    b. $P(\text{not unemployed}) = 1 - P(\text{unemployed})$
    $\qquad\qquad\qquad\qquad = 1 - 0.062 = 0.938$

24. $P(\$1 \text{ bill}) = \dfrac{n(\$1 \text{ bill})}{n(S)} = \dfrac{4}{9}$

25. Let S be all possible selections, so $n(S) = 100$. Let D be a card that is divisible by 5, so $n(D) = 20$. Let PN be a card that is 1 or a prime number, so $n(PN) = 26$.

$$P(D) = \frac{n(D)}{n(S)} = \frac{20}{100} = \frac{1}{5} = 0.2$$

$$P(PN) = \frac{n(PN)}{n(S)} = \frac{26}{100} = \frac{13}{50} = 0.26$$

26. Let S be all possible selections, let T be a car that needs a tune-up, and let B be a car that needs a brake job.

   a. $P(\text{Tune-up or Brake job})$

$$= P(T \cup B)$$
$$= P(T) + P(B) - P(T \cap B)$$
$$= 0.6 + 0.1 - 0.02$$
$$= 0.68$$

   b. $P(\text{Tune-up but not Brake job})$

$$= P(\text{Tune-up}) - P(\text{Tune-up and Brake job})$$
$$= P(T) - P(T \cap B)$$
$$= 0.6 - 0.02$$
$$= 0.58$$

   c. $P(\text{Neither Tune-up nor Brake job})$

$$= 1 - P(\text{Tune-up or Brake job})$$
$$= 1 - \left(P(T) + P(B) - P(T \cap B) \right)$$
$$= 1 - (0.6 + 0.1 - 0.02)$$
$$= 0.32$$

Chapter 10 Test

1. From the figure:
$$n(\text{physics}) = 4 + 2 + 7 + 9 = 22$$

2. From the figure:
$$n(\text{biology or chemistry or physics})$$
$$= 22 + 8 + 2 + 4 + 9 + 7 + 15$$
$$= 67$$

Therefore,
$$n(\text{none of the three}) = 70 - 67 = 3$$

3. From the figure:
$$n(\text{only biology and chemistry})$$
$$= n(\text{biol. and chem.}) - n(\text{biol. and chem. and phys.})$$
$$= (8 + 2) - 2$$
$$= 8$$

4. From the figure:
$$n(\text{physics or chemistry}) = 4 + 2 + 7 + 9 + 15 + 8$$
$$= 45$$

5. $7! = 7 \cdot 6 \cdot 5 \cdot 4 \cdot 3 \cdot 2 \cdot 1 = 5040$

6. $P(10,6) = \dfrac{10!}{(10-6)!} = \dfrac{10!}{4!}$

$$= \frac{10 \cdot 9 \cdot 8 \cdot 7 \cdot 6 \cdot 5 \cdot 4!}{4!}$$
$$= 10 \cdot 9 \cdot 8 \cdot 7 \cdot 6 \cdot 5$$
$$= 151,200$$

7. $C(11,5) = \dfrac{11!}{5!(11-5)!} = \dfrac{11!}{5!6!}$

$$= \frac{11 \cdot 10 \cdot 9 \cdot 8 \cdot 7 \cdot 6!}{5 \cdot 4 \cdot 3 \cdot 2 \cdot 1 \cdot 6!}$$
$$= \frac{11 \cdot 10 \cdot 9 \cdot 8 \cdot 7}{5 \cdot 4 \cdot 3 \cdot 2 \cdot 1}$$
$$= 462$$

8. Since the order in which the colors are selected doesn't matter, this is a combination problem. We have $n = 21$ colors and we wish to select $r = 6$ of them.

$$C(21,6) = \frac{21!}{6!(21-6)!} = \frac{21!}{6!15!}$$
$$= \frac{21 \cdot 20 \cdot 19 \cdot 18 \cdot 17 \cdot 16 \cdot 15!}{6!15!}$$
$$= \frac{21 \cdot 20 \cdot 19 \cdot 18 \cdot 17 \cdot 16}{6 \cdot 5 \cdot 4 \cdot 3 \cdot 2 \cdot 1}$$
$$= 54,264$$

There are 54,264 ways to choose 6 colors from the 21 available colors.

9. Because the letters are not distinct and order matters, we use the permutation formula for non-distinct objects. We have four different letters, two of which are repeated (E four times and D two times).

$$\frac{n!}{n_1!n_2!n_3!n_4!} = \frac{8!}{4!2!1!1!}$$

$$= \frac{8 \cdot 7 \cdot 6 \cdot 5 \cdot 4!}{4! \cdot 2 \cdot 1}$$

$$= \frac{8 \cdot 7 \cdot 6 \cdot 5}{2}$$

$$= 4 \cdot 7 \cdot 6 \cdot 5$$

$$= 840$$

There are 840 distinct arrangements of the letters in the word REDEEMED.

10. Since the order of the horses matters and all the horses are distinct, we use the permutation formula for distinct objects.

$$P(8,2) = \frac{8!}{(8-2)!} = \frac{8!}{6!} = \frac{8 \cdot 7 \cdot 6!}{6!} = 8 \cdot 7 = 56$$

There are 56 different exacta bets for an 8-horse race.

11. We are choosing 3 letters from 26 distinct letters and 4 digits from 10 distinct digits. The letters and numbers are placed in order following the format LLL DDDD with repetitions being allowed. Using the Multiplication Principle, we get $26 \cdot 26 \cdot 23 \cdot 10 \cdot 10 \cdot 10 \cdot 10 = 155,480,000$
Note that there are only 23 possibilities for the third letter. There are 155,480,000 possible license plates using the new format.

12. Let A = Kiersten accepted at USC, and B = Kiersten accepted at FSU. Then, we get $P(A) = 0.60$, $P(B) = 0.70$, and
$P(A \cap B) = 0.35$.

a. Here we need to use the Addition Rule.
$$P(A \cup B) = P(A) + P(B) - P(A \cap B)$$
$$= 0.60 + 0.70 - 0.35$$
$$= 0.95$$
Kiersten has a 95% chance of being admitted to at least one of the universities.

b. Here we need the Complement of an event.
$$P(\overline{B}) = 1 - P(B) = 1 - 0.70 = 0.30$$
Kiersten has a 30% chance of not being admitted to FSU.

13. a. Since the bottle is chosen at random, all bottles are equally likely to be selected. Thus,
$$P(\text{Coke}) = \frac{5}{8+5+4+3} = \frac{5}{20} = \frac{1}{4} = 0.25$$
There is a 25% chance that the selected bottle contains Coke.

b. $P(\text{Pepsi} \cup \text{IBC}) = \frac{8+3}{8+5+4+3} = \frac{11}{20} = 0.55$
There is a 55% chance that the selected bottle contains either Pepsi or IBC.

14. Since the ages cover all possibilities and the age groups are mutually exclusive, the sum of all the probabilities must equal 1.
$0.03 + 0.23 + 0.29 + 0.25 + 0.01 = 0.81$
$1 - 0.81 = 0.19$
The given probabilities sum to 0.81. This means the missing probability (for 18-20) must be 0.19.

15. The sample space for picking 5 out of 10 numbers in a particular order contains
$$P(10,5) = \frac{10!}{(10-5)!} = \frac{10!}{5!} = 30,240 \text{ possible}$$
outcomes. One of these is the desired outcome. Thus, the probability of winning is:
$$P(E) = \frac{n(E)}{n(S)} = \frac{n(\text{winning})}{n(\text{total possible outcomes})}$$
$$= \frac{1}{30,240} \approx 0.000033069$$

16. The number of elements in the sample space can be obtained by using the Multiplication Principle:
$6 \cdot 6 \cdot 6 \cdot 6 \cdot 6 = 7,776$
Consider the rolls as a sequence of 5 slots. The number of ways to position 2 fours in 5 slots is $C(5,2)$. The remaining three slots can be filled with any of the five remaining numbers from the die. Repetitions are allowed so this can be done in $5 \cdot 5 \cdot 5 = 125$ different ways.
Therefore, the total number of ways to get exactly 2 fours is
$$C(5,2) \cdot 125 = \frac{5!}{2! \cdot 3!} \cdot 125 = \frac{5 \cdot 4 \cdot 125}{2} = 1250$$
The probability of getting exactly 2 fours on 5 rolls of a die is given by
$$P(\text{exactly 2 fours}) = \frac{1250}{7776} = \frac{625}{3888} \approx 0.1608.$$

Chapter 10 Cumulative Review

1. $3x^2 - 2x = -1$

$$3x^2 - 2x + 1 = 0$$

$$x = \frac{-b \pm \sqrt{b^2 - 4ac}}{2a}$$

$$= \frac{-(-2) \pm \sqrt{(-2)^2 - 4(3)(1)}}{2(3)}$$

$$= \frac{2 \pm \sqrt{4 - 12}}{6} = \frac{2 \pm \sqrt{-8}}{6}$$

$$= \frac{2 \pm 2\sqrt{2}i}{6} = \frac{1 \pm \sqrt{2}i}{3}$$

The solution set is $\left\{ \dfrac{1-\sqrt{2}i}{3}, \dfrac{1+\sqrt{2}i}{3} \right\}$.

2. $f(x) = x^2 + 4x - 5$

$a = 1, b = 4, c = -5$. Since $a = 1 > 0$, the graph opens up. The x-coordinate of the vertex is

$$x = -\frac{b}{2a} = -\frac{4}{2(1)} = -2.$$

The y-coordinate of the vertex is

$$f\left(-\frac{b}{2a}\right) = f(-2) = (-2)^2 + 4(-2) - 5.$$

$$= 4 - 8 - 5 = -9$$

Thus, the vertex is $(-2, -9)$. The axis of symmetry is the line $x = -2$. The discriminant is:

$$b^2 - 4ac = (4)^2 - 4(1)(-5) = 16 + 20 = 36 > 0.$$

So the graph has two x-intercepts.
The x-intercepts are found by solving:

$$x^2 + 4x - 5 = 0$$

$$(x+5)(x-1) = 0$$

$$x = -5 \text{ or } x = 1$$

The x-intercepts are -5 and 1.
The y-intercept is $f(0) = (0)^2 + 4 \cdot (0) - 5 = -5$.

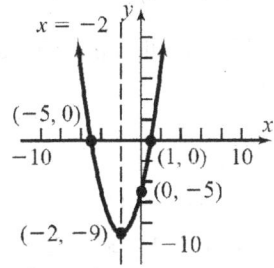

3. $y = 2(x+1)^2 - 4$

Using the graph of $y = x^2$, horizontally shift to the left 1 unit, vertically stretch by a factor of 2, and vertically shift down 4 units.

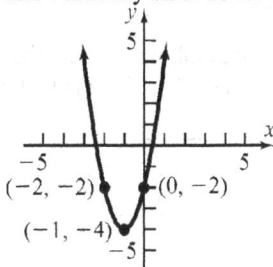

4. $|x - 4| \le 0.01$

$$-0.01 \le x - 4 \le 0.01$$

$$-0.01 + 4 \le x \le 0.01 + 4$$

$$3.99 \le x \le 4.01$$

The solution set is

$$\{x \mid 3.99 \le x \le 4.01\} \quad \text{or} \quad [3.99, 4.01]$$

5. $f(x) = 5x^4 - 9x^3 - 7x^2 - 31x - 6$

Step 1: $f(x)$ has at most 4 real zeros.

Step 2: Possible rational zeros:

$$p = \pm 1, \pm 2, \pm 3, \pm 6; \quad q = \pm 1, \pm 5;$$

$$\frac{p}{q} = \pm 1, \pm \frac{1}{5}, \pm 2, \pm \frac{2}{5}, \pm 3, \pm \frac{3}{5}, \pm 6, \pm \frac{6}{5}$$

Step 3: Using the Bounds on Zeros Theorem:

$$f(x) = 5\left(x^4 - 1.8x^3 - 1.4x^2 - 6.2x - 1.2\right)$$

$$a_3 = -1.8, \ a_2 = -1.4, \ a_1 = -6.2, \ a_0 = -1.2$$

$$\text{Max }\{1, |-1.2| + |-6.2| + |-1.4| + |-1.8|\}$$

$$= \text{Max }\{1, 10.6\} = 10.6$$

$$1 + \text{Max }\{|-1.2|, |-6.2|, |-1.4|, |-1.8|\}$$

$$= 1 + 6.2 = 7.2$$

The smaller of the two numbers is 7.2. Thus, every zero of f lies between -7.2 and 7.2.

Graphing using the bounds: (Second graph has a better window.)

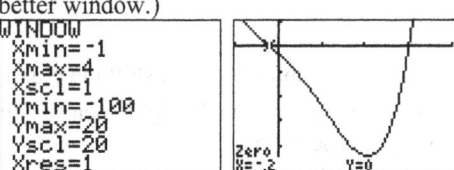

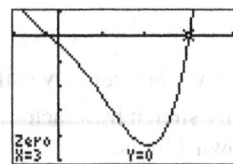

<u>Step 4:</u>   From the graph we see that there are x-intercepts at -0.2 and 3. Using synthetic division with 3:

$$3\overline{)\begin{array}{rrrr} 5 & -9 & -7 & -31 & -6 \\ & 15 & 18 & 33 & 6 \\ \hline 5 & 6 & 11 & 2 & 0 \end{array}}$$

Since the remainder is 0, $x-3$ is a factor. The other factor is the quotient: $5x^3+6x^2+11x+2$. Using synthetic division with 2 on the quotient:

$$-0.2\overline{)\begin{array}{rrrr} 5 & 6 & 11 & 2 \\ & -1 & -1 & -2 \\ \hline 5 & 5 & 10 & 0 \end{array}}$$

Since the remainder is 0, $x-(-0.2)=x+0.2$ is a factor. The other factor is the quotient:

$5x^2+5x+10=5(x^2+x+2)$.

Factoring, $f(x)=5(x^2+x+2)(x-3)(x+0.2)$

The real zeros are 3 and -0.2.
The complex zeros come from solving
$x^2+x+2=0$.

$$x=\frac{-b\pm\sqrt{b^2-4ac}}{2a}=\frac{-1\pm\sqrt{1^2-4(1)(2)}}{2(1)}$$

$$=\frac{-1\pm\sqrt{1-8}}{2}=\frac{-1\pm\sqrt{-7}}{2}$$

$$=\frac{-1\pm\sqrt{7}i}{2}$$

Therefore, over the set of complex numbers, $f(x)=5x^4-9x^3-7x^2-31x-6$ has zeros

$$\left\{-\frac{1}{2}+\frac{\sqrt{7}}{2}i,-\frac{1}{2}-\frac{\sqrt{7}}{2}i,-\frac{1}{5},3\right\}.$$

6. $g(x)=3^{x-1}+5$

    Using the graph of $y=3^x$, shift the graph horizontally 1 unit to the right, then shift the graph vertically 5 units upward.
    Domain: All real numbers or $(-\infty,\infty)$
    Range: $\{y\,|\,y>5\}$ or $(5,\infty)$
    Horizontal Asymptote: $y=5$

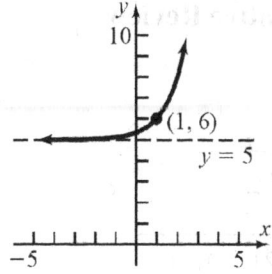

7. $\log_3(9)=\log_3(3^2)=2$

8. $\log_2(3x-2)+\log_2 x=4$

    $\log_2\left(x(3x-2)\right)=4$

    $x(3x-2)=2^4$

    $3x^2-2x=16$

    $3x^2-2x-16=0$

    $(3x-8)(x+2)=0$

    $x=\dfrac{8}{3}$ or $x=-2$

    Since $x=-2$ makes the original logarithms undefined, the solution set is $\left\{\dfrac{8}{3}\right\}$.

9. Multiply each side of the first equation by -3 and add to the second equation to eliminate x; multiply each side of the first equation by 2 and add to the third equation to eliminate x:

    $$\begin{cases} x-\ \ 2y+z=\ \ 15 \\ 3x+\ \ y-\ 3z=\ -8 \\ -2x+4y-\ \ z=-27 \end{cases}$$

    $$\begin{array}{r} -3x+6y-3z=-45 \\ 3x+\ y-3z=\ \ -8 \\ \hline 7y-6z=-53 \end{array}$$

    $$\begin{array}{l} x-2y+\ z=15 \xrightarrow{\ \ 2\ \ } 2x-4y+2z=\ \ 30 \\ -2x+4y-\ z=-27 \longrightarrow \underline{-2x+4y-z=-27} \\ \hspace{6cm} z=3 \end{array}$$

    Substituting and solving for the other variables:
    $z=3\Rightarrow 7y-6(3)=-53$

    $$7y=-35$$
    $$y=-5$$
    $$z=3, y=-5\Rightarrow x-2(-5)+3=15$$
    $$x+10+3=15\Rightarrow x=2$$

    The solution is $x=2$, $y=-5$, $z=3$ or $(2,-5,3)$.

10. $-3, 1, 5, 9, \ldots$ is an arithmetic sequence with $a = -3, \ d = 4$.

Using $a_n = a + (n-1)d$,

$$a_{33} = -3 + (33 - 1) \cdot 4$$
$$= -3 + 32 \cdot 4$$
$$= -3 + 128$$
$$= 125$$

To compute the sum of the first 20 terms, we use

$$S_{20} = \frac{20}{2}(a + a_{20}).$$

$$a_{20} = -3 + (20 - 1) \cdot 4$$
$$= -3 + 19 \cdot 4$$
$$= -3 + 76$$
$$= 73$$

Therefore,

$$S_{20} = \frac{20}{2}(a + a_{20})$$
$$= \frac{20}{2}(-3 + 73)$$
$$= 10 \cdot 70$$
$$= 700.$$

Chapter 10 Projects

Project I

1. Table 3 is a probability model since the total of the probabilities is 1.

Cash Prize	Probability
Jackpot	0.00000000386
$1,000,000	0.00000005408
$5000	0.00000135192
$500	0.00001892689
$50	0.00009328256
$5	0.00342036036
$2	0.01770813839
$1	0.04674948535
$0	0.93200839659
Total	1.00000000000

2. There are 75 numbers to choose for the first 5 'white' numbers and 15 numbers to choose for

the last 'gold' number. Thus there are $_{75}C_5$ possibilities for the 'white' numbers and $_{15}C_1$ possibilities for the 'gold' number. Multiply these together to find the total possibilities.

$$_{75}C_5 \cdot {}_{15}C_1 = (17259390)(15)$$
$$= 258,890,850$$

So the probability of winning is:

$$P(\text{win}) = \frac{1}{258,890,850}$$
$$\approx 0.00000000386$$

3. To calculate the expected value, multiply each numeric outcome by its corresponding probability and then add these products.

Cash Prize	Probability	prize $\cdot$ prob
$20,000,000	0.00000000386	0.07720000000
$1,000,000	0.00000005408	0.05408000000
$5000	0.00000135192	0.00675960000
$500	0.00001892689	0.00946344500
$50	0.00009328256	0.00466412800
$5	0.00342036036	0.01710180180
$2	0.01770813839	0.03541627678
$1	0.04674948535	0.04674948535
$0	0.93200839659	0.00000000000
expect value		0.25143473693

So the expected cash prize is 0.25.

4. The expected financial result from purchasing one ticket is $\$0.25 - \$1.00 = -\$0.75$. Therefore, your expected profit from one ticket is $-\$0.75$.

5. Use the same procedure replacing the Jackpot with $100,000,000.

Cash Prize	Probability	prize $\cdot$ prob
$100,000,000	0.00000000386	0.38600000000
$1,000,000	0.00000005408	0.05408000000
$5000	0.00000135192	0.00675960000
$500	0.00001892689	0.00946344500
$50	0.00009328256	0.00466412800
$5	0.00342036036	0.01710180180
$2	0.01770813839	0.03541627678
$1	0.04674948535	0.04674948535
$0	0.93200839659	0.00000000000
expect value		0.56023473693

The expected financial result from purchasing one ticket is $\$0.56 - \$1.00 = -\$0.44$. Therefore, your expected profit from one ticket is $-\$0.44$.

6. We need to solve the expected value equation for the Jackpot amount that would make the expected value equal to $\$1.00$. Thus:

Cash Prize	Probability	prize · prob
JP	0.00000000386	$J(0.00000000386)$
$1,000,000	0.00000005408	0.05408000000
$5000	0.00000135192	0.00675960000
$500	0.00001892689	0.00946344500
$50	0.00009328256	0.00466412800
$5	0.00342036036	0.01710180180
$2	0.01770813839	0.03541627678
$1	0.04674948535	0.04674948535
$0	0.93200839659	0.00000000000
expect value		1.00000000000

$J(0.00000000386) + 0.05408 + 0.0067596$
$\quad + 0.009463445 + 0.004664128$
$\quad + 0.0171018018 + 0.03541627678$
$\quad + 0.04674948535 = 1$
$J(0.00000000386) + 0.17423473693 = 1$
$J(0.00000000386) = 0.82576526307$
$J \approx \$213,928,825$

7. Answers will vary.

Project II

1. 0 bit errors: 1011

   1 bit errors: 0011
   1111
   1001
   1010

   2 bit errors: 0111
   0001
   0010
   1101
   1110
   1000

3 bit errors: 0110
0101
0000
1100

4 bit errors: 0100

2. $P(\text{symbol received correctly}) = \left(\dfrac{2}{3}\right)^4 = \dfrac{16}{81}$

3. # of received symbols with 2 bit errors:
   $C(8, 2) = 28$

   $P(\text{received correctly}) = \left(\dfrac{2}{3}\right)^8 = \dfrac{256}{6561}$

   $P(\text{received incorrectly}) = 1 - P(\text{received correctly})$
   $\qquad\qquad = \dfrac{6305}{6561}$

4. Let k = # of errors, n = 8 = length of symbol. Probability of k errors :

$$P(n, k) = \binom{n}{k} (p)^k (1 - p)^{n-k}$$

$$P(8, k) = \binom{8}{k} \left(\frac{1}{3}\right)^k \left(\frac{2}{3}\right)^{8-k}$$

Since this parity code only detects odd numbers of errors,

$P(\text{error detected})$
$= P(8, 1) + P(8, 3) + P(8, 5) + P(8, 7)$

$$-\binom{8}{1}\left(\frac{1}{3}\right)^1\left(\frac{2}{3}\right)^7 + \binom{8}{3}\left(\frac{1}{3}\right)^3\left(\frac{2}{3}\right)^5$$

$$+ \binom{8}{5}\left(\frac{1}{3}\right)^5\left(\frac{2}{3}\right)^3 + \binom{8}{7}\left(\frac{1}{3}\right)^7\left(\frac{2}{3}\right)^1$$

$= 0.156464 + 0.272992 + 0.068044 + 0.002423$
$= 0.499923$

To find the probability that an error occurred but is not detected, we need to assume that an even number of errors occurred:

$P(\text{error occured, but not detected})$
$= P(8, 2) + P(8, 4) + P(8, 6) + P(8, 8)$

$$= \binom{8}{2}\left(\frac{1}{3}\right)^2\left(\frac{2}{3}\right)^6 + \binom{8}{4}\left(\frac{1}{3}\right)^4\left(\frac{2}{3}\right)^4$$

$$+ \binom{8}{6}\left(\frac{1}{3}\right)^6\left(\frac{2}{3}\right)^2 + \binom{8}{8}\left(\frac{1}{3}\right)^8\left(\frac{2}{3}\right)^0$$

$= 0.273402 + 0.170364 + 0.016985 + 0.000151$
$= 0.460951$

Project III

Answers will vary.

Project IV

   e. Answers will vary, depending on the L_2 generated by the calculator.

   f. The data accumulates around $y = 0.5$.

Project V

One simulation might be:

Woman has	Woman told you about	Probability
Boy-Boy	Older boy	$\frac{1}{4}$
Boy-Boy	Younger boy	$\frac{1}{4}$
Boy-Girl	Younger boy	$\frac{1}{4}$
Girl-Boy	Older boy	$\frac{1}{4}$

We leave out the combinations where she would have to tell you about a girl. Thus, the probability that she has 2 boys is $\frac{1}{4}+\frac{1}{4}=\frac{1}{2}$.

Man has	Man told you about	Probability
Boy-Boy	Older boy	$\frac{1}{2}$
Girl-Boy	Older boy	$\frac{1}{2}$

Thus the probability he has two boys is $\frac{1}{2}$. The probabilities are the same.